Post-Harvest Management of Mango

(*Mangifera indica L.*)

Post-Harvest Management of Mango

(*Mangifera indica L.*)

Edited by

R. Srihari Babu

Ph.D (Hort.) (BHU)

Professor of Horticulture (Retd.),
Acharya N.G. Ranga Agricultural University,
Rajendranagar, Hyderabad – 500 030,
Andhra Pradesh, India.

A unit of **BSP Books Pvt. Ltd.**

4-4-309/316, Giriraj Lane, Sultan Bazar,
Hyderabad - 500 095
Phone : 040 - 23445605, 23445688

Published by

BS Publications
A unit of **BSP Books Pvt. Ltd.**
4-4-309, Giriraj Lane, Sultan Bazar,
Hyderabad - 500 095
Phone : 040 - 23445605, 23445688
e-mail : info@bspbooks.net

ISBN: 978-93-85433-83-2 (HB)

Dr. M.V. Rao
Member of A.P. Legislative Council
President. Agri Biotech Foundation
Former Special Director-Genral, ICAR
Former Special Secretary, DARE, GOI
Former Vice Chancellor, ANGRAU

Off: Agri Biotech Foundation
ANGRAU Campus.
Rajendranagar,
Hyderabd - 500 030.
phone: (O) : 91-040-64562146
(R) : 91-040-27954748
Telefax: 040-24002147
Mobile: 9849477088
Mangina2008@gmail.com
agribiotechfoundation@gmail.com

Foreword

Mango (*Mangifera indica.L*) which is considered as the "king of fruits" is indigenous to India and is widely cultivated since 4000 years B.C. Most of the states in the country cultivate a variety of mangoes with varying taste, color, shape and size. From India the crop has migrated to more than 100 countries. Because of its taste, flavour and aroma it received royal patronage. Inida accounts for 60% of the mango production in the world. It is consumed not only in India Council of Agriculture Research and the State Agricultural Universities made efforts to collect and preserve the germ plasm of different mango varieties from all over the country.

Mango during different stages of harvesting, packing, handling and marketing suffers serious post-harvest loses. The losses are estimated to the order of 25% to 30%. This is a colossal loss to the grower as well as the country. Therefore, in recent years increasing emphasis has been placed on post-harvest management of mango.

A few books have been published on mango since 1960 which, however, have dealt the production technology for the most part. In recent years a lot of technological information has been generated all over the world on postharvest management and product development of mango. This widely dispersed information needs compilation for ready reference. I am happy that this lacuna is fulfilled by the publication of this book "Post-harvest management of mango" by Dr. Srihari Babu and his associates.

This book has been divided into 14 well defined chapters each dealling with a particular aspect of post-harvest management of mango fruit. All the chapters are written in simple and lucid manner. Mango can be good income generator for farmers in the villages if they start small industries to process mango in to various products and supply to the bigger factories. In this way it will improve the economy of the villagers and generate employment.

I congratulate Dr.R. Srihari Babu former Professor and Head of the Department of Horticulture, S.V. Agriculture College ANGRAU Tirupathi and also the other contributes of the book. Dr. Srihari Babu also published a monograph "Citricculture" and received Life Time Acheivement award in citrus.

I warmly congratulate not only Dr. Srihari Babu but all the scholars who contributed different chapters. I am sure tha this book will be of great practical value and interest not only to the growers and processors but also to the horticulture students, teachers, and researchers quality monitoring authorities, exporters and other concerned with mango cultivation and post-harvest management and product development. I wish the publication all success.

Hyderabad **Padmasri Dr. M. V. Rao.**

Preface

Mango (*Mangifera indica* L.) is an important and commercial fruit of the world particularly in Indian sub continent. It is rightly claimed as 'King of fruits' because of its delicacy in taste, flavour and nutrition. The cultivation of mango is now taken up in more than 100 countries in the tropical and sub tropical regions of the world. Its cultivation in the Indian subcontinent dates back to 4000 years B.C. However, its cultivation was lagely confined to elite – Maharajas, Rajas, Emperors, Navabs, Princes, Jameendars and Jagirdars, for Millinia. Mango enjoyed royal patronage during Moghal era, though it remained confined in their gardenes only. Moghul period is considered to be the 'Golden period' for mango. The legacy of patronage continued during the colonial rigime also. During this period, the mango came out of the royal gardens to the gardens of common man. The commercial cultivation started barely 200 years back. However, systematic cultivation started only during the beginning of 20th Century. Since then more and more common men took cultivation of mango.

Research on mango also started during the start of 20th Century after the establishment of Fruit Research Stations by British Government under Imperial Council of Agricultural Research. S.R. Gandhi published a farm bulletin 'Mango in India' in 1955 for ICAR, New Delhi. The first monograph on 'Mango' was published in 1960 by Dr. L.B. Singh of Sahranpur, which includes his vast research work and experiences on mango, and which is considered as the Bible of mango. Later in 1967, Indian Council of Agricultural Research, New Delhi published 'A Hand Book on Mango' and in 1968 Dr. R.N. Singh published a book 'Mango' through ICAR. New Delhi. After that till 1995 no book on mango was broughtout, except as chapter in several pomology books. In 1997, R.E. Litz edited a book on 'Mango' and published through CAB, International, Willingford, U.K. In 1998, R.P. Srivastava from Lucknow brought out a book on mango entitled "Mango Cultivation' through International Book Distributing Company, Lucknow. Thereafter a few fancy books were published on mango. In all these books much emphasis was given on botany, taxanomy and cultivation of mango. Very little attention was paid to post-harvest aspects of mangoes. Since the establishment of Fruit Research Stations, National Institutions like CFTRI, IIHR, CISH etc. and SAUs lot of information on different post harvest aspects of mangoes has been generated. All these information is available in the journals, in the form of research papers, reviews etc. Much information is also brought into light through National and International Symposia on mango-published by International Society for Horticulture through Acta Horticulturae. So to say the information on post-harvest aspects of mango is scattered. Many workers and enthusiastic entrepreneurs desired that all such scattered information should be brought at one place and published. Realising the importance and gravity of their desire I have attempted to fulfil their long-cherished desire. The present book entitled "Post-harvest management of mango" is the result of such effort. For compiling the information on different post-harvest aspects of mango, information was freely drawn from Research Journals, all over the world, reviews published in different journals, Annual Reviews, Acta Horticulturae, Proceedings of National and International Symposia on mango etc.

The book contains 14 chapters. Each chapter was contributed by Teachers and Research workers of A.N.G. Ranga Agricultural University of Andhra Pradesh and D.Y.S.R Horticultural University Hyderabad. The first chapter, Introduction – Introduces the crop and briefs the contents of different chapters. The second chapter, Post-harvest handling, deals with various post-harvest handling aspects of mango, grading and transportation. The third chapter deals with storage of fresh fruits, in which various methods of storage – traditional, low temperature storage, controlled atmosphere, modified atmosphere storage etc. are discussed. The fourth chapter – "Fruit ripening" presents the need for ripening of fruits, different methods of ripening, changes during ripening, regulation of ripening etc. The fifth chapter –Preservation and Processing – furnishes the various ways of preservation and processing of both green, unripe and ripe fruits, the methodology of processing of various bye products from fruits. The sixth chapter deals the mango varieties suitable for processing of different products. The seventh chapter – Microbiology of processed products, explains the microbiological spoilage of processed products and ways to avoid the spoilage. This is followed by different methods of storage of processed products till their sale in eighth chapter – storage of processed products. Production of quality products is very essential and is demanded by the present day consumer. Quality control is done through quality standards and several Acts by the Government. Mango fruits not only in field but also after harvest are invaded by different pathogens. They cause different rots – post-harvest diseases, which are elaborately dealt in the 10th chapter with suitable management measures. In the chapter 11th – various post-harvest physiological disorders of mango fruits are presented with causes and management. The post-harvest losses of mango fruit at different levels and their economic importance are discussed in chapter 12. Mango industy generates lot of waste material in the kitchen as well as at processing units etc., whose efficient utilization helps in minimizing the pollution hazards; provides very valuable and useful bye-products, brings down the cost of processing, with the possibility of providing additional employment and income. All these are discussed in chapter 13. Marketing of fresh mango fruits, market channels, role of intermediars, evils of present marketing system, betterment of marketing system etc. are discussed in the last chapter 14.

Efforts were made to present information on different aspects of post-harvest management of mango in whole and systematic manner. This book is mainly intended for P.G. students, Research workers and mango processors. The learned readers are requested to make suggestions for the betterment of this book in future editions. The editor welcomes such suggestions whole heartedly with sincer thanks.

- Editor

Acknowledgement

Suitable words are not available to express my heartfelt thanks and gratitude to all the contributors of this book "Post-harvest management of mango" for contributing chapters, and their untired efforts in gathering and compiling the gathered information neatly and systematically. Numerous references on each topic were collected painstakingly by the contributors from libraries etc. I am highly thankful to them for their cooperation, dedication and sincere efforts in bringing my long felt desire and dreams into a book form. I am immensely thankful to Dr. (Smt.) Veena Joshi, Scientist (Hort.), Vegetables Research Institute, YSR Horticulture University, Rajendranagar, Hyderabad, one of the contributors, who took great pains in making corrections and typing the final copies of the chapters and also in bringing uniformity in the book. I am also indebted to all those computer assistants who have typed neatly with utmost attention and care. I am thankful to Sri. Veeranjaneyulu, University Librarian, Central Library, A.N.G. Ranga Agricultural University, Hyderabad and all the librarians at Central Institute of Sub-Tropical Horticulture, Lucknow, Indian Agricultural Research Institute, New Delhi, Indian Institute of Horticultural Research, Bangalore and of SAUs for providing me opportunity to consult their librarires and xeroxing the references from Journals, Annual Reviews, Acta Horticulturae etc. I am also grateful to my colleagues in the University and friends in the above Institutiions who encouraged to compile this book. I am thankful to Dr. M. Rama Rao, Professor of Horticulture (Rtd.) for his help in shaping this book. Further, I am thankful to Sri. K. Balakrishna Reddy, Proprietor, Sawera Processed Foods Privated Limited, Chittoor for allowing me to visit his mango pulp factory and to take photos of pulp making in the factory.

Last, but not least, I am very much thankful to Dr. M.V. Rao, former Vice-chanceller. A.N.G. Ranga Agriculatural University, Hyderabad, A.P. to kindly write an excellent "Foreward" for the book.

I will be failing in my duty, if I do not express my sincere thanks to my better-half (wife) Smt. R. Nalini Devi, who encouraged and cooperated whole heartedly and to a full extent in bringing out this book. I am highly thankful to her.

I also whole heartedely express my thanks to the Publishers BS Publication, Hyderabad who has taken pains in bringing this book in print in time.

R. Srihari babu
Editor

Contents

List of Tables

List of Flow Charts

List of Photos

List of Contributors

S.No	Name and Address	Topic
1.	DR.R.SRI HARI BABU Professor of Horticulture (Rtd). Apt.No.23084 Janapriya- Utopia Apts, Hyderguda, Attapur (P.O.) Rajendranagar (Mandal), Hyderabad- 500048 Andhra Pradesh (A.P.)	(i) Introduction (ii) Preservation and Processing (iii) Post-Harvest Diseases
2.	DR.V. KRISHNA RAO Professor & Head, Dept. of Plant Pathology, College of Agriculture, A.N.G.R Agricultural University, Rajendranagar (P.O.) Hyderabad- 500030. (A.P.)	(i) Post-Harvest Diseases
3.	DR.K.VANAJALATHA Professor (Hort), College of Horticulture, Dr.Y.S.R Horticultural University, Rajendranagar, Hyderabad – 500 030 (A.P.)	(i) Varietal Suitability for Processing of Mango
4.	DR.M. PADMA Principal Scientist (Hort.), Vegetable Research Station, Rajendranagar Dr.Y.S.R Horticultural University, Hyderabad-500030	(i) Post-Harvest Handling
5.	DR.L. NARAM NAIDU Principal Scientist (Hort.), Horticultural Research Station, Dr.Y.S.R Horticultural University, Lam, Guntur (A.P.)	(i) Fruit Ripening
6.	DR.M. MADHAVI Professor (Hort.), College of Horticulture, Dr.Y.S.R Horticultural University, Mojerla, Mahaboobnagar (A.P.)	(i) Storage of Fresh Fruits (ii) Storage of Processed Products

S.No	Name and Address	Topic
7.	DR.B. NEERAJA PRABHAKAR Professor (Hort.), Dept. of Horticulture, College of Agriculture, A.N.G.R Agricultural University, Rajendranagar, Hyderabad – 500 030 (A.P.)	(i) Microbiology of Processed Products (ii) Marketing of Mango Fruits and its Products
8.	DR.K.M. YUVARAJ Scientist (Horticulture), Citrus Improvement Project, Dr.Y.S.R Horticultural University, Tirupati – 517 502 (A.P.)	(i) Post-Harvest Physiological Disorders of Mango Fruits
9.	SRI. M. HANUMAN NAIK Scientist (Horticulture) & Head, J.V.R. Horticultural Research Station, Dr.Y.S.R Horticultural University, Malyal, Warangal – 506 101 (A.P.)	(i) Utilization of Mango Waste Material
10.	SMT.G. JYOTHI Assistant Professor (Hort.), College of Horticulture, Dr.Y.S.R Horticultural University, Rajendranagar, Hyderabad – 500 030 (A.P.)	(i) Post-Harvest Losses
11.	DR.VEENA JOSHI Scientist (Horticulture) , Grape Research Station, Dr.Y.S.R Horticultural University, Rajendranagar, Hyderabad – 500 030 (A.P.)	(i) Quality Control of Processed Products

CHAPTER 1

Introduction

R. Srihari babu

Mango (*Mangifera indica* L.) is the most popular and ancient fruit of old world. It is one of the most widely known fruit in tropical and subtropical regions of the globe. Mango is one of the six major fruits of the world. It occupies relatively the same position in the tropics as is enjoyed by the apple in the temperate zones. It is the choiest of all the fruits in the world and is loved by one and all for its delicious taste, nutritive and medicinal values. No other fruit of the world excels mango fruit in deliciousness, flavour, taste and nutrition, hence it is rightly called "King of fruits". Throughout the ages it has been aknowledged as an excellent fruit by one and all. It is the national fruit of India.

Mango (*Mangifera indica* L.) belongs to the family, Anacardiaceae, the most important member of the family, followed by Cashewnut (*Anacardium accidentale* L.), another important member of the family.

Mango has originated in the Indo-Mynmar region of South East Asia (De Condalle, 1904., Poponoe, 1920., Mukherjee, 1951., Gangolly *et al* 1957). From the place of its origin, mango travelled and spread far and wide in all directions, through travelers, missionaries, traders etc. Now it is grown in more than 100 countries and in all the five continents.

Mango, thus enjoys world wide distribution. It is grown on either side of the equator, in both dry and wet tropical and subtropical low land areas up to 23° North and South of the equator. Although, at present it is being cultivated in more than 100 countries, no where it achieved the same premier position as in the Indian sub-conteinent, which is the largest producer of mango in the world. Heritage and culture of the country (India) are closely associated with mango and interwoven.

Nutritive value

The mango fruit is a highly nutritious one, rich in sugars, vitamins and minerals (Table 1.1).

Table 1.1: Chemical composition of ripe mango (per 100 g pulp)

S.No.	Constituent	Measure	Quantity
1.	Protein	gram	0.6
2.	Fat	gram	0.4
3.	Fibre	gram	0.7
4.	Carbohydrates	gram	16.9
5.	Energy	Kilo calories	74.0
6.	Vitamin-A	IU	3890
7.	Vitamin-B	mg	40
8.	Riboflavin	mg	50
9.	Vitamin-C	mg	16
10.	Total carotene	mg	2210
11.	Minerals	gram	0.4
12.	Potassium	mg	205.0
13.	Sodium	mg	26.0
14.	Calcium	mg	14.0
15.	Iron	mg	1.3
16.	Phosphorus	mg	16.0

Ripe mango is the most inexpensive source providing the human body with energy and nutrition. Mango fruit provides two and half-a-time more calories per ha, than wheat. In nutritive value no other fruit can match mango. It is an outstanding sources of vitamin A and C. Besides, the mango fruits are rich source of carotenoids, specialy beta-carotene (pre-cursor of Vitamin A). Sugars constitute bulk of carbohydrates and most of the soluble solids in the ripe mango fruits. Good mango varieties contain 20% total sugars, in which non-reducing sugars are more than reducing sugars. The principal sugars are glucose and fructose. The acid content of the ripe fruit varies from 0.2 to 0.5 %. Mangoes are high in fibre content, but low in calories, fat and sodium and form a good staple food in human daily diet (Chauhan *et al* 1998). The fruit derives its characteristic taste from the blend of acid, sugar and tannins (Bhatnagar and Subramanyam, 1973).

Consumption of a medium sized mango fruit provides daily requirement of vitamin A and C and 40% of fibre.

Uses

Mango tree and its different parts such as wood, leaves, inflorescences, flowers, fruits, seeds and kernlas have several uses and have been used in several ways for the last several millinia in India and other South East Asian countries. Fruits, in particular are put to multiferous uses right from first stage of growth and development to maturity and ripening stage. Wood of the tree is used for making doors, windows, beams, furniture etc. The leaves, inflorescences and twigs are used in various religious rituals and functions in India and other South East Asian countries.

Mango fruits are processed and preserved in many ways. Mango fruit is utilized at all stages of its growth and development both in its immature, mature and ripe stages. Raw and green immature fruits are used for making chutney, pickle, amchoor and juices like Panna. The ripe fruits besides being used as dessert are also utilized for preparing several products like squashes, syrups, nectars, jam, jellies, leather etc.

The mango seed kernel contains 8-10% good quality fat, which is used for making soaps and also as a substitute for Cola in confectionary. Seeds including kernels are used as cattle and poultry feed. Tribals make starch out of kernels at times of food scarcity and after mixing with wheat/maize flour use the mixture for making Chapaties.

The medicinal properties of mango fruits and other parts of the plant are well known and have been put to use since long throughout the world. The different medicinal uses of mango tree and its various parts were reviewed (Nair, 1995., Budhar, 2002). Almost all parts of the plant viz., roots, bark, twigs, leaves, inflorescences, fruits, seeds and kernels are used as medicines in one way or the other. Both unripe and ripe mangoes are used in treating stomach problems and to stimulate bile formation and in blood related diseases and disorders. The powder made from the skin (peel) of unripe fruit is used to cure bleeding, dysentery, diarrhoea, sore throat, cholera and in many gynecological troubles. Consumption of ripe fruit is useful to overcome nightblindness and to protect the health of the human skin. It helps to increase digestive capacity of the human body. The ripe fruit is very good for treating nervous dyspepsia and constipation. The fruit pulp is useful against haemorrhiage problem. The seed kernel is useful to control diarrhoea, dysentery, haematomesis and piles. Mango seeds, leaves and bark are used in the treatment of diarrhoea and disorders in reproduction system of women. It is also claimed that mango takes care of the sexual debility, defects of kidney, anaemia, brain weaknesses, chronic diarrhoea, dysentery and inflammation of spleen. Mango also contains an enzyme with stomach soothing properties similar to papain of papaya.

Cultivation

According to De Condalle (1884), cultivation of mango originated in India and dates back atleast 4000 years ago. However, Tjiptano *et al* (1984) stated that ASEAN countries were among the earliest of other countries to commence cultivation of mango in the world. Cultivation of mango on orchard scale in India started only during the period of Moghuls in the 15th Century AD. But till the patronage of colonial Europeans, the cultivation of mango was taken up by effluent people like Princes, Nawabs, Zameendars, Jagirdars etc. Commercial cultivation of mango started during colonial regime in India, from which time it passed from the above elite people to the common man. Now mango growing is in the hands of common man, marginal and small farmers also.

Mango is cultivated under diverse agroclimatic conditions of the world. Different varieties of mango are cultivated in different regions. In many regions, the crop is facing with the problems of biotic and abiotic stresses. The crop faces many challenges in production, post-harvest management and supply chain management (marketing). The post-harvest behaviour of mango fruit is strongly influenced by cultivar, harvest maturity, post-harvest treatments, diseases, disorders and storage conditions. One of the impediments of mango industry is the lack of infrastructure for post-harvest management in terms of pack houses, storage, processing, waste utilization, marketing etc.

Need for post-harvest management

Efforts to enhance production of crops, including mango, by increasing the area under the crop and adoption of improved agro-techniques and plant protection measures will be futile, unless the post-harvest losses are reduced, extended the availability of fruits, through storage and value addition is made. Minimising the post-harvest losses has, therefore, been recognized as an effective means of increasing the availability of mangoes without additional inputs. Since the cost involved in preventing the losses will be less than the production, post-harvest management attains greater significance. In order to reduce the post-harvest losses there is a need to adopt proper strategies, involving new technologies in the post-harvest management. Reduction of post-harvest losses both quantitatively and qualitatively could be achieved by adoption of good pre-harvest production practices including pre-harvest fungicide spray schedules for preventing post-harvest diseases, harvesting at proper maturity, proper harvesting methods, adoption of good packhouse practices like sorting, washing, grading, packing, pre-cooling etc. and adoption of refrigerated transportation, post-harvest treatments, storage at optimum conditions and sound marketing practices. Losses can also be reduced by preservation and processing of mango fruits, which add value to the fruits and fetch higher returns. The issue of supply chain (marketing) management also needs due focus for their improved supply and profitability to the growers/traders.

Brief Account of Post-harvest Management Aspects

Post-harvest losses in mango have been estimated, as ranging from 25 to 40% from harvesting to consumption stage (FAO, 2003), due to improper handling after harvest and lack of infrastructure for carrying out various post-harvest handling operations. The main aim of post-harvest handling is to protect the fruit from both biotic and abiotic hazards, extend their storage life, and maintain quality and freshness till it reaches the consumer. Post-harvest operations include operations at harvest, packhouses, sorting, grading, pre-cooling, storage, ripening, transportation etc.

The available season of fresh fruits of mango is very short in any producing region (2-4 months), but mangoes are produced throughout the year, in one region or the other of the world. However, there is deamand for the fruit alround the year in all regions. Further, because of rhythum of alternate bearing, in certain years, there wil be glut in the market in which case the surplus supply may go waste for want of adequate processing. Mango is highly perishable and does not keep for more than 10 days at room temperature at mature green stage. The supply throughout the year is possible by storage of fresh fruits for extended periods. Extension of storage life can be achieved by different approaches like delaying ripening and checking biochemical activity in the fruits by exposing them to low temperature, altered gaseous atmosphere around the fruit, by irradiation etc. However, each method has its own merits and demerits. For example, the chilling sensitivity of mango fruits limits its long duration storage and transportation at temperatures below 12-13 °C.

Mangoes for table as well as processing purposes are generally harvested at hard-green and unripe stage and are ripened artificially by different methods. Ripening of fruit is a dramatic event in the life of the fruit, which enables the fruit edible and also helps in the dispersal of the seed for perpetuation. Mango is classified as a climacteric fruit. Ripening is a complex phenomenon triggered by endogenous ethylene after harvest. Under tropical conditions fruit ripening is faster as compared to subtropical conditions. However, regulation of ripening (accelerating or delaying) becomes necessary at times. The fruit should remain unripe until it reaches the destination market/consumer. This needs delaying of ripening. Once it reaches, the destination market, the ripening fruit undergoes changes in its biochemical constituents and these changes render the fruit edible. Checking these changes is one way of delayng ripening of fruits.

During glut in the market, the prices decline to the disadvantage of the farmer as well as the trader and much of the produce is likely to lost due to one or other reason. Further as mentioned above, the mangoes are available for a shorter period of 2-4 months in a season. Fresh fruits are not available always and do not give much income to the grower, traders etc. To get higher income, to avoid glut in the market, and to extend the availability of mango (in a different form) value addition to them is necessary and this is possible through preservation and processing. Mangoes can be consumed raw as dessert fruit or processed into various products. However, consumption as a dessert fruit up to over 90% is very common and predominant and only 5-6% of the production is processed in India (Naidu and Hari Babu, 2009). Mango is a versatile fruit that can be utilized at different stages of its growth and development. Green and unripe mangoes are used for making different products such as amchoor, pickles, chutney, panna etc., while ripe mangoes are used for making different types of drinks, jams, jellies, bars etc. From the start, different methods like sun-drying etc. and till now several methods of preservation and processing of both raw, green, unripe and ripe mangoes were developed. Processing, however, faces many problems in procurement of raw material, suitable variety(s) for making different products, disposal of finished products, and commercilisation of the industry and market expansion. Processing provides employment, results in increase in value addition and is quite remunerative. Though there are more than 1000 named varieties in mango, all however, are not suitable for preservation and processing into different products. A few varieties were identified for processing into different products. Despite this, Bangalora (Totapuri) is mostly used for processing.

Preserved and processed products are often spoiled by microbes, because of high sugar content, higher acidity and low pH in the products. Contamination of fruits from the field and their non-removal before processing, inadequate quantities of chemical preservatives etc. are some of the causes of spoilage of processed products. Unhyegenic conditions in the industry and storage places, lack of proper packaging also contribute to microbial spoilage of processed products. Methods of avoiding such microbial spoilage of processed products have been worked out.

Processed products are very much liable for spoilage due to microorganisms, biochemical changes and enzymatic activity. They would be on the shelf for a longer time till they are disposed off by sale. Hence, their proper storage is very much essential and is needed for profits. Besides extending the shelf life, the processed products need to retain/preserve their nutritive value and acceptability. Their storage can be achieved by packaging, thermal preservation, sulphiting and addition of preservatives etc.

Quality is a measure of the degree of excellence or a degree of consumer acceptability. The production of high quality food products has been the important goal since the start of the processing. Additionally quality control is essential. Quality standards have been laid out by different governmental organizations in each country, through enacting several Acts for the quality control of processed products. Quality standards have been laid out for each processed product at each step of its processing.

Post-harvest diseases of fruits due to microorganisms (fungi and bacteria) are responsible to a large extent for spoilage of fruits after harvest. Some of them are carried from the field as latent infections. Post-harvest diseases cause huge post-harvest losses to the great disadvantage of sellers particularly. They are serious constraints in trading, particularly in exporting mango to global markets. Diseased fruits become unfit for either table or processing purpose. Hence, their management/control, both in the field and after harvest is inevitable. The management strategies include pre-harvest as well as post-harvet ones, like cultural, physical and chemical and of late biological methods.

Mangoes also suffer from certain serious physiological disorders during their post-harvest life. Physiological disorders are mostly due to imbalances in the metabolism induced by some abiotic factor(s) in the pre-and post-harvest environments, that render the fruit unfit for consumption as well as for processing. Some of them limit the chanes of exporting mangoes to the international markets. Efforts were made to understand and elucidate the causal factors and to ward off of such disorders in the initial stages by avoiding various causal conditions.

Post-harvest losses in case of mango in India account for 25-40%. The losses may occur at any stage right from the field at harvest, till the fruit is delivered at market, in transportation, storage etc. The post-harvest losses may be due to physical, physiological and pathological reasons. Both quantitative and qualitative losses occur at all stages in the post-harvest system. The losses will be severe in the tropical and sub-tropical production centres, where high temperatures and relative humidity prevail during the harvesting and post-harvesting periods. The post-harvest losses reduce the availability of produce to the consumer, financial loss to all the concerned in the trade. Besides these quantum losses, the quality loss before consumption is serious and enormous and can not be estimated. Therefore, the post-harvest losses have to be prevented. Preventive methods have been devised to a satisfactory level.

After consumption as dessert fruit and/or industrial processing of the fruit, large quantities of waste in the form of peels, pulp, seeds etc. get generated. Such waste accounts up to 30-35% of the fruit depending upon the variety, size of the fruit and efficiency of the processing industry. Mango waste is not used for any useful purpose even today and is largely wasted. However, this waste can be converted into food ingredients, biofuels etc. Edible products from mango waste such as jelly grade pectin, edible fibre, vinegar, citric acid, oil, butter, kernel powder, alcohol (wine) etc. can be synthesized. Kernel oil can be used in costmotics and soap industries. The kernel flour after mixing with wheat/maize flour is used for making chapaties. About 10% alcohol could be obtained from kernel by co-culture fermentation. Enzymes such as cellulase and pectinase from mango kernel could be produced by microbial fermentation. Mango waste material also provides raw material for the gas reactors, organic manures and extraction of several useful products. Left unattended, these wastes may become the source of atmospheric pollution and contamination of soil and ground water etc. Efficient disposal and/or recycling of these wastes may help in minimizing the pollution hazards, supply of vital nutritional components in our food and feeds and bringing down the cost of processing with the possibility of providing additional income.

Marketing is as important as production. The marketing system of mango is very unorganized, because of the presence of a large number of intermediaries, mainly private individuals (pre-harvest contractors, commission agents, wholesalers and retailers) and lack of efficient supervision of Government officials. At present marketing of fruits including mango is under private entrepreneurs or collectors, who buy the fruit gardens from the farmers, arrange for harvesting and transportation to suitable collection centres, where sorting and grading to their advantage are carried out, after which the fruit is consigned to domestic or global markets. About 80% of the mango growers in the country sell their orchards to pre-harvest contracters/lease them for varying periods (1-5 years). Marketing plays a key role in the post-harvest handling operations of fruits. A perfect and efficient marketing system covers all aspects of handling from the stage of harvest till the commodity reaches the consumer. However, marketing of perishable products like fruits presents more problems as compared to other non-perishable agriculture commodities like grains, pulses etc. Generally the interests of the producer and consumer are ignored/poorly served in the present marketing system. The growers get less returns for their produce and consumer has to pay more than what is necessary and justifiable. Due to the presence of middle men, the price of the fruit is 50-100% higher at terminal markets than at farm gate. The middle men manipulate the situation by artificially creating glut/scarcity and offer low prices to growers. They often syndicate for their advantage. They also falsely reject the produce (fruits) as substandard and indulge in malpractices in weighments etc. and delay payments to the growers. The cooperative systems play a very important role in the marketing of agriculture produce, particularly fresh perishable produce.

AREA AND PRODUCTION

World

The mango is cultivated throughout the tropical and sub-tropical regions of all the 5 continents of the world. According to FAO data (FAO 2003) currently more than 100 countries grow mango in the world and the number of countries, likely to take up cultivation of mango may increase further in the future. However, the number of countries which produce more than one million metric tones of mangoes are only few during 2004 (Table 1.2).

Table 1.2 : Area and Production in major mango producing countries during 2004 (Galan Sauco, 2004)

S.No.	Country	Area (m. ha)	Production (mill. mt)
1.	India	1.600	10,800
2.	China	0.419	3,582
3.	Thailand	0.270	1,700
4.	Mexico	0.174	1,503
5.	Pakistan	0.105	1,089
6.	Indonesia	0.170	1,006
7.	Philippines	0.159	0.968
8.	Brazil	0.068	0.850
9.	Egypt	0.045	0.327

It is clear from the above data, that India is the largest producer of mango with largest area under the fruit, followed by China at a distance. India is not only the largest producer of mango, but also the largest consumer of the fruit, however, is a poor exporter. Mexico, on the other hand, stands at the 4th place in area and production, but is the largest exporter of mango to the International markets, especially to USA and continues to be so. The area under mango in the world is 4.72 million ha, with a production of 34.89 million metric tones during 2009-10 year. India was the largest producer of the fruit accounting for 49% of the area and 49.1% of the world production in the year 2009-10 (FAO). However, its share in the world area and production has been dwindling over the years.

The mango industry gained a great impetus in the past decade and continues to boom further in future, mainly due to expansion of its cultivation to newer areas of the world and the International Trade Liberalization.

India

India is the traditional grower of mango since millennia. Mango is successfully cultivated throughout the length and breadth of the country under diverse agro-climatic conditions. Each state in the country grows mango. The data on state-wise area and production of mango in the country during 2008-09 are furnished in the Table 1.3.

Table 1.3: State-wise area and production of mango in India during 2008-09

S.No.	State	Area (m. ha)	Production (m. T)
1.	Andhra Pradesh	0.498	2.522
2.	Maharashtra	0.457	0.713
3.	Uttar Pradesh	0.271	3.466
4.	Orissa	0.164	0.450
5.	Bihar	0.144	1.330
6.	Tamilnadu	0.149	0.821
7.	Karnataka	0.141	1.284
8.	Gujarat	0.166	0.300
9.	West Bengal	0.066	0.549
10.	Kerala	0.077	0.445
11.	Others	0.206	0.870
	Total	**2.309**	**12.750**

A perusal of the above state-wise area and production indicates that Andhra Pradesh stands first in area, whereas Uttar Pradesh stands first in production. Though the area under mango is highest in Andhra Pradesh, its production is less because of lower productivity (5.1 t/ha). On the other hand, Uttar Pradesh stands 3rd in area, but first in production due to highest productivity (12.8 t/ha).

REFERENCES

Bhatnagar, H.C. and Subramanyam, H. 1973. Some aspects of preservation, processing and export of mango and its products. Indian Food Packer, 27(4) : 33-52.

Budhar, K 2002. Romance of the Mango. Penguin Books, India (P) Ltd., New Delhi, India.

Chauhan, S.K., Lal, B.B. and Joshi, V.K. 1998. Development of a Protein Rich Mango Beverage. Journal of Food Science and Technology, 35 (6) : 521-523.

De Condalle, A.P. 1884. Origin of cultivated plants. London.

De Condalle, A.P. 1904. Origin of cultivated plants. Kegaul Part, London.

FAO 2003. CCP : BA/TFO3/13. Evolucion reciente del mercado Chino de las frutes tropicals. Grupo FAO Interguberamental sobre elbanao ylas frutes tropicals. Puerto dela Cruz, Espan, 11-15/12/03 (Celebrated on 22-26, March, 2004).

Galan Sauco, V. 2004. Mango production and world market : Current situation and future prospects. Acta Horticulturae, 645 : 107-116.

Gangolly, S. R., Singh, R., Katyal, S. L. and Singh, D. 1957. The Mango. Indian Council of Agricultural Research, New Delhi.

Mukherjee, S.K. 1951. Origin of mango. Indian Journal of Genetics and Plant Breeding, 11 : 56.

Naidu, L.N. and Hari Babu K. 2009. Development and Availability of Mango varieties for processing. Paper presented at National Seminar on Horticulture Based Food Processing Industry Trends and Prospects. 19th December, 2009, Bangalore, India.

Nair, P.T. 1995. The Mango in Indian life and culture. Part I. Bisheri Singh Mahendrapal Singh, Dehradun, 617-667

Poponoe, W. 1920. Manual of Tropical and Sub-tropical fruits. Newyork.

Tjiptono, P., Lam, P.F., Kosiyachinda, S., Mendoza Jr. D. B. and Leong, P.C. 1984. Status of the Mango Industry in ASEAN. In : Mango-Fruit Development, Post-harvest physiology and Marketing in ASEAN (Ed). D.B. Mendoza Jr. and Wills, R.B.H. ASEAN Food Handling Bureau, Kuala Lumpur, Malaysia.

CHAPTER 2

Post-Harvest Handling

M. Padma

Mango is a perishable produce and gets spoiled fast unless properly handled and cared after harvest. The post-harvest losses in mango have been estimated within a range of 25 to 40 percent from harvesting to consumption stage (FAO, 2003). Hence proper care and attention are needed in its post-harvest period to ensure that the quality of the fruit at harvest is preserved till the point of consumption. In between the harvest and consumption, mangoes are subjected to several treatments/practices or handled, which are collectively termed as **"POST-HARVEST HANDLING"** practices/operations. The main aim of these post-harvest handling operations is to protect the fruit from both biotic and abiotic hazards, extend its shelf life, maintain its quality and freshness as best as possible till it reaches the consumer.

The two major goals of post-harvest handling of perishables like mango are:

1. loss prevention and
2. value addition to the produce through preservation and processing

Fruits are cleaned, graded and then conditioned for storage. Value is added through integration of cleaning, grading, packaging, transport and storage operations which enables the grower earn more income for his produce. Thus, post-harvest handling systems include operations like transport, desapping, fungicide treatment/hot water treatment, grading, packing, labelling, pre-cooling, palletization, storage, ripening, marketing etc., elaborately. Such elaborate steps, however, are not usually followed for domestic markets, but followed for International markets. For domestic markets, harvested fruits are simply transported to markets, after bleeding in a rude way and ripened at the destination markets. This results in huge losses up to 30 percent (Sawant and Shinde, 2000). There is scope for following appropriate post-harvest handling operations. The vast scope of exporting mangoes to the International markets can be gainfully tapped by improvement and carrying of post-harvest operations on scientific lines. Some of the improvements in post-harvest handling practices are: the CFB boxes, pre-treatments, pre-cooling and development of cool chains and cold storage facilities etc. The post-harvest handling practices for mangoes vary from production area to area.

However, the post-harvest handling practices for mango follow a more or less common pattern in the country with little variation. The number and the sequence vary depending on the cultivar, prevailing conditions, market (domestic/global), local practices and requirements of importing country. The typical post-harvest handling practices (flow-charts) as per different sources/agencies in India are as follows (Flow chart 2.1):

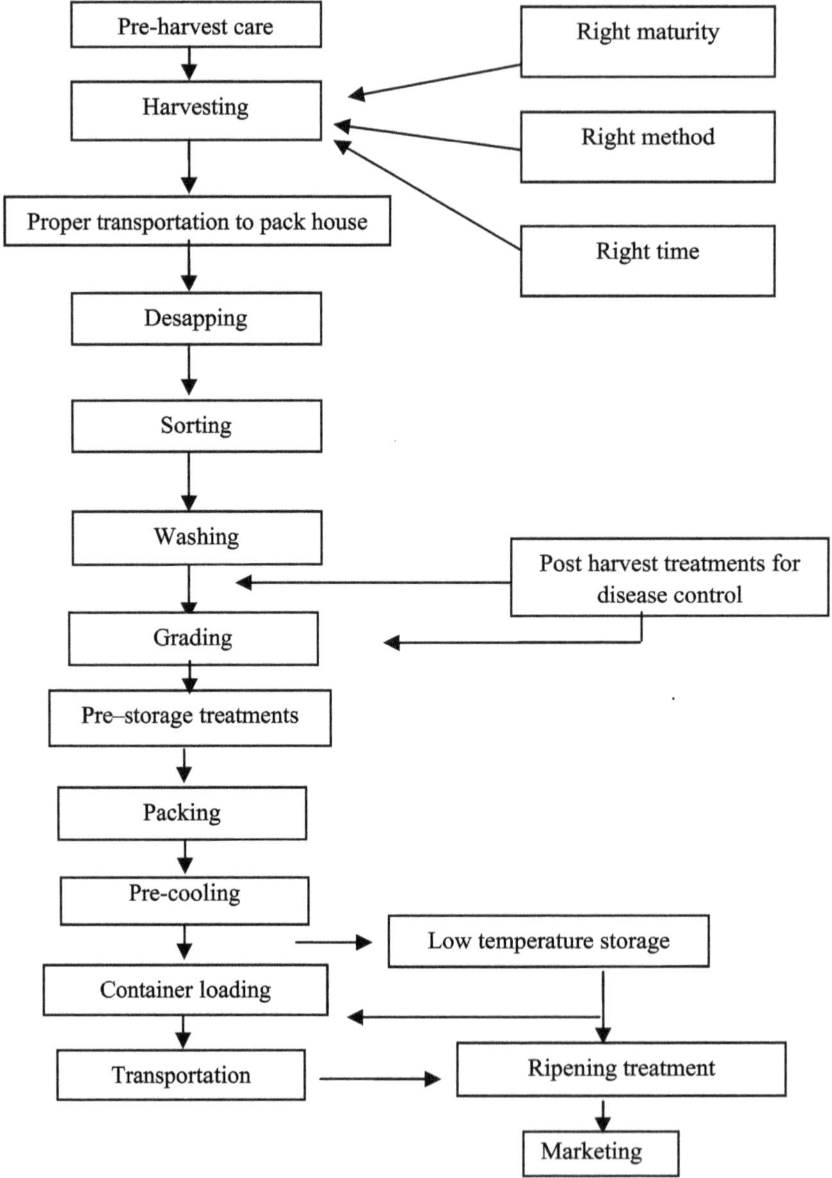

Flow chart: 2.1. Integrated Post-Harvest management operations (Anon,2005)

Transportation to pack houses

The harvested fruits should not be loaded loose in the vehicles, but loaded into plastic crates and transported carefully in clean vehicles as rapidly as possible and without exposure to sun, preferably in closed vans, rather than in open lorries, trucks etc. If the pack house is located far away from the orchards, the fruits should be transported in the refrigerated/insulated vehicles, with care to avoid/minimize damage to fruits from rough handling. Roads should be smooth, vehicle tyres correctly inflated, if necessary suspension may be provided to avoid vibrations (Johnson *et al* 1997). Driving should be smooth to prevent bruising of the fruits in the crates. No other product should be loaded on to the produce and on to the vehicle other than mango and no human and/or animal is allowed over the mango fruits. The fruits may be covered with suitable material with proper ventilation to avoid exposure to sun etc; during transit, in case open vehicles are used. The produce at the pack house should be unloaded carefully in a clean and cool place and the fruit temperature should be noted. Then both tare weights and net weights of crates are to be noted.

Operations at the Pack houses

Many of the post-harvesting handling operations are carried at the pack house. Pack house facilities and operations can range from basic operations with just wash tanks and grading tables to highly sophisticated systems with auto-mated washing, hot water treatment and vapour treatments and size/weight grades etc. Pack house for export purpose normally uses wash tanks, roller conveyers, grading tables or size graders and a vapour heat treatment unit.

The objective of the pack house operations is to select and pack freshly harvested fruit in the consumer acceptable form. This is usually a case of cleaning, grading and packaging of the fruit. Post-harvest operations in the pack house are to be carried under clean hygienic conditions and completed within a maximum period of 6 hours. The temperature and humidity conditions in the pack house should be congenial and maintained at 25 °C and 65-85% relative humidity till the operations are completed.

Delivery inspection

Samples of fruit should be evaluated immediately after arrival at the pack house for maturity, blemishes, diseases and pest incidences. Acceptance inspections are the most-critical part of quality control in the food industry. Entry-controls regulate product appearance and can be used to motivate producers, pickers and traders to deliver high quality produce. Processing plants usually require pre-harvest field inspection and entrance acceptance as in the case of export trade.

Desapping (Bleeding)

Spilling of sap (latex) on the fruit surface(skin) during harvesting/subsequent handling operations causes stains or sap injury (sapburn) and also increases the chance of disease incidence, thereby reduces the appearance of the fruit, fruit quality and its shelf life. To avoid all the above, desapping (bleeding of sap) should be done. It should be done in a manner which minimizes the incidence of sapburn. Several systems of desapping were assessed (Brown *et al* 1986., O'Hare and Prasad, 1993., Holmes *et al* 1993., Lim and Kuppeweiser, 1993., O'Hare 1994., Shorter and Joyce, 1994) like:

(i) desapping in CaOH (1%) solution,

(ii) washing fruit in AlK_2SO_4 (1%) solution,

(iii) applying surface coatings to fruit prior to desapping,

(iv) trimming pedicel and desapping at the pack house followed by inversion on a stationary rack or a roller-conveyer running below water (water/detergent) sprays for 20 mts and

(v) inversion in the soil beneath a tree in the shade, immediately after harvest for 30 mts.

Options (i) and (ii) are not used commercially. Trimming and desapping (option iv) at the pack house have been widely adopted commercially in India and Australia. Johnson *et al* (1994) showed that soil bleeding caused a significant increase in the incidence and earlier appearance of stem-end-rot resulting from the soil borne inoculum and therefore not advisable.

Desapping may be done either at the field immediately after harvest or at pack house, the later being mostly preferable as the pack house has facilities. Fruits harvested with long stems (8-10 cm) are transferred to pack house. At the pack house the stalks (stems) are clipped to about 1 cm length with sharp scissors holding the fruits inverted with stem-end down and placed on desapping benches, specially made with bamboo poles, leaving gaps between poles to allow desapping (Photo 2.1). Desapping requires 2-3 hours depending upon the quantity of flow of the sap. The level of latex flow depends upon the maturity of the fruit; (younger fruits exude more latex), time of the day of harvest (more latex flows in early morning harvests) and level of rainfall (more latex after rainfall). The stalk should never be pulled with hand or otherwise, the practice may cause skin injury, ultimately leading to fungal infection. A fine water spray over the fruit helps to reduce the chances of sapburn. Dipping the fruit in water containing detergents reduces the risk of sapburn even further. After every desapping operation the bamboo racks should be cleaned to ensure that they are free from spilled latex and fungal inoculum on the racks.

Sorting

Not all the fruit harvested or reaching the pack house or the processing unit will be of good/export quality or suitable for the purpose for which it is intended. It is therefore, essential to sort out unwanted/poor quality fruit/unsuitable fruit in the lot. In the field or on arrival at the pack house, the fruits should be sorted out to remove (by visual appearance) immature, under-sized, mis-shaped, damaged, bruised, scared, sap stained, diseased or ripe fruits. Mature, unripe fruits are separated from immature and ripe fruits.

Such sorted out fruits are either discarded or diverted to alternate purposes (like processing). Much quality sorting is done by visual assessment on tables or in large mechanized pack houses on conveyers (Photo 2.2).

Photo 2.1 : Desapping of mango fruits

GRADING

Mangoes at this stage are re-checked to ensure that quality specifications have been met before being separated into groups for packaging by count. Fruits are graded according to size by weight either manually or mechanically as per the laid norms for domestic markets and of importing countries as the case may be. Each importing country has developed their own grade standards. Fruits are graded according to their size, weight, colour and maturity. The following counts with fruit weight (± 30 g) can be used as a guideline for 4 kg packaging (Table 2.1).

Table 2.1 : Size grades (weight) in mango and counts in a pack

Fruit wt (g)	Fruit count in a pack
500	8 pieces
400	10 pieces
330	12 pieces
290	14 pieces
250	16 pieces

Photo 2.2 : Sorting of mangoes for packing

The minimum quality requirements and classifications are defined by the CODEX standards (CAC 2005) for mangoes. Based on the tolerance to minor visual defects, the mango fruits are grouped as

(i) Extra class, (ii) Class I and (iii) Class II.

Based on size the fruits are grouped as Grade A (200-300 g), Grade B (351-550 g) and Grade C (551-800 g) (CAC-2005) **(2003-2008)**, for international markets (see grading). In case of national markets in the country different markets follow different grades (see grading).

Sizing

The purpose of sizing fruits is to categorize them into size or weight groups for packing. It is done manually (by eye) often as the fruits are being packed and prior to disinfection/disinfections with hot-water/vapour heat. Mechanical sizers (for both weight and dimension) are developed (see grading). Automatic graders separate fruit by weight, shape or colour into groups which correspond to pre-determined categories (Schoorl and Holt, 1982).

Washing (Cleaning)

Fruits need to be washed (cleaned) to remove latex stains, spray residues, sooty moulds, dust, dirt and any other extraneous matter on the fruit surface. Cleaning is done by washing the fruits in detergent/fungicidal water which is free from heavy metals and microbes. The detergents and sanitizers should be chosen carefully, as some of them damage the fruits or promote diseases (Korsten *et al* 1993). Sodium hypochlorite (100 ppm) is added to water to wash and/or rinse in pack houses.

The fruits are first washed in clean water to remove dust, dirt etc; and then transferred to Tank-I containing solution of neutral detergent like Teepol, Sandovit or Indtron (0.1%) and permitted fungicide (benomyl, prochlorez or thiobendazole at 250 ppm) depending upon the importing country. The temperature of the washing solution should be 27 °C. The mangoes are rubbed smoothly in the solution and cleaned with soft muslin cloth. Soft rubber gloves need to be used during washing of fruits for rubbing. Washing should be completed within 3-5 minutes and the solution in the tank must be changed for every 3 tonnes of fruit. After this cleaning, the fruits are transferred to Tank-II, containing clean water at 27 °C and rinsed for 5 minutes to remove extra detergent adhering the fruits using muslin cloth and air-driers to remove the surface moisture.

In some countries fruits are brushed after hot water and fungicide treatments (Johnson *et al* 1997). Brushing removes superficial deposits of fungicides on the fruit surface **sprayed in** the field, and in pack houses (Wells and Littlemore, 1989., Lonsdale,1993). Soft and non-damaging brushes should be used and replaced frequently.

Post-harvest-disease control

The post-harvest diseases of mangoes include:

(i) anthracnose (*Colletotrichum gloeosporioides Penz and C. aculatum Sim*),

(ii) stem-end-rot (*Lasiodiplodia theobromae Pat* etc),

(iii) black mould rot or Aspergillus rot (*Aspergillus niger*),

(iv) phomopsis rot (*Phomopsis mangiferae* Ahmad) and

(v) alternaria rot (*Alternaria alternata*) which cause huge losses during ripening, storage and transport and hinder both internal and external trade.

Their prevention/control is necessary. Most of these diseases come from the field as latent infections. Hence, the better approach would be to integrate both pre-and-post-harvest strategies (See Post-harvest diseases).

Pre-harvest strategies include keeping the trees healthy by providing required nutrition in time and canopy management, keeping the orchard clean through good sanitation, pre-harvest bagging of fruits, pre-harvest fungicidal sprays and harvesting and transportation of fruits with utmost care to avoid injury and subsequent infections.

Post-harvest strategies include heat treatments (hot-water, vapour-heat or forced air) for eliminating both fungal pathogens and insect pests and fungicidal treatments alone for eliminating fungal pathogens.

Hot-water treatment

Hot-water with or without fungicides is used to eradicate quiescent infections of fungi that have become established on or beneath the cuticle and within the pedicel in the field prior to picking (Coates and Gowanlock,1994).

Mango fruits after washing are subjected to hot-water treatment (HWT). The treatment involves immersion of fruit in hot-water (52-53 °C) for 5-10 minutes depending on the cultivar and fruit size. It requires the use of hot-water-bath with temperature control for effectiveness of the treatment and for prevention of heat damage to the fruit. For this, hot-water treatment plants are utilized having thermo-stat control to maintain a constant desired temperature. HWT helps clean fruit but can contribute to increased heat injury and shortens the post-harvest life if not properly managed (Cooke and Johnson, 1994) and accelerates ripening with more yellow colour of the skin (Suhardjo and Yuniarti, 1993). Fruits are moved through hot- water bath for 5 minutes (48°-50 °C) in less mature fruit and hot-water susceptible cultivars (Zill, Irwin) and 50-55 °C for mature fruit and less susceptible cultivars (Anon, 1994b). HWT at 49 °C for 10 minutes or at 51 °C or higher for 5 minutes gave the best suppression of anthracnose of mangoes (Suhardjo and Yuniarti,1993). Anthracnose is controlled readily as compared to stem-end-rot (Johnson and Coates, 1993., Johnson 1994,) and phomopsis rot (Johnson *et al* 1997) by HWT. Treatment for 3 minutes is adequate for the control of anthracnose, while immersion up to 7 minutes is required for control of stem-end-rot (Muirhead and Grattidge, 1986).

Fungicidal treatments

Post-harvest fungicide treatment either alone or included in HWT can also reduce the incidence of disease infection of fruits. The choice of the fungicide, however, depends on the exporting and importing country's regulations. Domestically neither HWT nor fungicides are used for the control of post-harvest diseases in India. Permitted fungicides are: benomyl, prochloraz, and thiobendazole. The fruit dipping in benomyl solution (9500 ppm) or thiobendazole (1000 ppm) before storage reduces anthracnose from 29 to 5 percent (Sohi *et al* 1973). Pre-harvest spraying with Bavistin (0.1%) or Topsin –M (0.1%) or Prochloraz (0.1%) followed by dipping in the same after harvesting was also effective (Rawal, 1997). Permitted fungicide (500 ppm) in acetone is added to the hot water and the pH of the water bath is adjusted to 4.5 by using phosphoric acid. In this water bath, fruits are immersed for 2-3 minutes in perforated plastic crates and dried.

Where acceptable, benomyl is added to the hot water at the recommended rate (500-1000 ppm) to control stem-end-rot and anthracnose at a temperature of 52 °C (Johnson *et al* 1990., Wells and Littlemore, 1989., Sanghote, 1991). Effective control of anthracnose was obtained at 51.5 °C with benomyl (500 ppm) and 48.5 °C with benomyl (1000 ppm) (Muirhead, 1976). Better control of anthracnose is possible within 24-48 hours of harvest only with hot benomyl and prochloraz (Dodds *et al* 1991).

Benomyl is unstable in very dilute aqueous solutions or in fruit waxes (Eckert, 1983). The alkalinity of mango latex may also interfere with the efficacy of benomyl in controlling stem-end-rot. Addition of gaur gum with benomyl improved control of stem-end-rot of mangoes, as it helps to keep benomyl in suspension (Johnson *et al* 1990). Dipping of fruit in hot, dirty latex contaminated water can increase the incidence of phytotoxicity and lenticel damage during ripening. Benomyl should be replenished often, since the level declines rapidly because of fruit stripping and interaction between mango latex and benomyl (Wells and Littlemore, 1989).

Hot-Thiobendazole (TBZ) was as effective as hot-benomyl in controlling stem-end-rot, but not effective against anthracnose (Coates *et al* 1993). Imazalil, in combination with hot water provided control of anthracnose and stem-end-rot (Spalding and Reeder, 1986., McGuire and Campbell, 1993) but was inferior to benomyl (Sanghoti, 1991)

Prochloraz also provides good control of anthracnose and alternaria rot, but does not give adequate control of stem-end-rot (Johnson *et al* 1990., Johnson and Coates 1993). A hot water and benomyl combination treatment followed by a prochloraz spray provides effective control of anthracnose, stem-end-rot and alternaria rot during long storage (Johnson *et al* 1989, 1990).

Pandey and Tandon (2007) suggested dipping of mango fruits in hot water (52 ± 1 °C) with carbendazim (0.025%) for 10 min against post-harvest diseases.

When fungicides are used in the pack house, spent dip-suspensions and fungicide containers should be disposed off using appropriate methods.

Post-harvest disinfestations of Insect Pests

Some of the mango insect pests such as internal pulp and seed feeders (fruit fly larva, seed pests, seed weevils) and external pests (scales, mealy bugs, thrips etc) are very important and concerned particularly in regard to export of fresh fruits to global markets. External pests pose less problems because of their presence on the fruit surface and can easily be seen and removed during sorting before the fruits are subjected to other post-harvest handling operations. It is not, however, the case with internal pulp and seed feeders. The larvae of seed/stone weevils cause severe damage to the pulp and are difficult to kill without damaging the market quality of the treated fruits. Entry of mangoes from countries having the seed weevils and fruit flies is restricted/prohibited into the countries free of these insects; unless the fruits are disinfested by some means before entry/export. So post-harvest disinfestation of fruits constitutes a practice in post-harvest handling operations of mango as these are considered to be quarantine pests. Post-harvest disinfestation of insect pests in mango is accomplished through operations viz., heat treatments (hot-air, hot-water and vapour-heat), irradiation, quick freezing, fumigation etc.

The major constraint in disinfestation has been the susceptibility of mangoes to heat, cold and irradiation damage and oxygen depletion. Treatment's efficiency is influenced by cultivar, fruit size and environment and their combination (Quimio and Quimio, 1973., Jacobi *et al* 1995). Small fruits are damaged by heat more readily than larger ones, which need to be treated longer to satisfy disinfestation requirements. So while size-grading, fruits of even size should be selected. Pre-conditioning treatments for fruit, delaying treatments for 24 hours after harvest and selection of more mature fruit can reduce the fruit damage to a large extent (Jacobi and Wong, 1992., Jacobi *et al* 1995). Some treatments such as methyl bromide fumigation (Spalding *et al* 1977) and cold temperature storage (Kane and Marcellin, 1978) cannot be used because they lower the quality at dosages that are lethal to insect pests.

Hot water treatment

Hot-water immersion is an environmentally safe and efficient approach to kill mango internal pulp and seed feeders. Nascimento *et al* (1992) developed a hot-water treatment against fruit flies in mangoes in Brazil. Smith (1992) showed that immersing mangoes in hot water (48 °C) for 3 minutes killed eggs and larvae of B. *aquilonis* in Australia. However, weevils in 'Alphonso' mangoes in India were not killed when infested fruits were immersed in hot water (48 °C - 52 °C) for even up to 90 min. and at 54-70 °C for up to 50 min. (Shukla and Tandon, 1985).

Vapour heat treatment

Countries that are free from quarantine pests (fruit flies, seed weevils) do not accept mango fruits that are not disinfested and made free from such pests. Vapour heat treatment (VHT) is the only eco-friendly treatment available at present for this purpose. Various mango exporting countries now demand the use of this technique (VHT) to treat mangoes for import, particularly Japan. VHT has been accepted as the "International Quarantine Treatment" against different mango fruit flies (Gaffeney *et al* 1990).

VHT is generally done at 46-49 °C (fruit temperature) for 20-30 minutes depending on the variety. Heated air at 95% relative humidity is circulated through the crates of fresh mangoes through condensation of water vapour. Total treatment time for mangoes is 195 minutes. This consists of 125 minutes approach period, 10 minutes holding time at 50 °C, 4 minutes air-cooling and 20 minutes treatment with a cooling shower. In case of Indian mangoes meant for export to Japan, the temperature of VHT chamber is escalated to 50 °C, at which, the temperature of the innermost fruit pulp will be raised to 47.5 °C and thereafter maintained at 47 °C for 20 minutes (Anon, 2004).

All fruit fly immature stages were killed in mangoes from Puerto-Rico after exposure to vapour at 43 °C for 4 hours (Sein, 1935). Koidsumi (1937) used VHT to disinfest mangoes from Taiwan. VHT was approved for disinfestation of mango fruits from Mexican fruit fly for exporting mangoes to USA (Balock and Starr, 1945). VHT at 46.5 °C disinfested mangoes of oriental fruit fly (*Bactrocera dorsalis* Hendel) in Thailand (Unahawutti *et al* 1986). Japan requires mangoes from the Philippines and Thailand to be treated with vapour heat until seed surfaces are 46 °C and 46.5 °C respectively and held at the respective temperatures for 10 minutes (Anon, 1987). A VHT schedule was approved against Queensland fruit fly (*Bactrocera tryoni*) in "Kensington Pride" mangoes from

Australia for Japanese markets (Heather and Jacobi, 1994). USA approved VHT as a quarantine treatment for Mexican fruit fly and other *Anastrepho* sp. in 'Manila' cultivar and mangoes from Taiwan infested with oriental fruit fly for import into USA (Anon, 1994a).

VHT enhances the ripening process and results in better marketability of fruits because of uniform peel colour (Roy and Pal, 2000). However, one should be cautious with Indian mango Cv. 'Alphonso' since it is prone to development of spongy tissue, unless it is cooled to 15 °C immediately after VHT (Roy and Pal, 2000).

Hot air

Conditioned (hot) air is forced over the surface of mangoes and allowed to heat the pulp slowly. Temperature in the unit is maintained a few degrees below the dew point temperature, not to allow condensation formation on fruit surface or inside the unit. Relative humidity should be high above 50%, but never allowed to go high to produce condensation and resultant desiccation of mango tissue (Johnson *et al* 1997). A mean centre pulp temperature of over 47 °C killed all stages of West Indian fruit fly in Mexico grown mangoes treated with hot air (Mangan and Ingle, 1992). Similarly a mean centre pulp temperature of over 48 °C killed all stages of Caribbean fruit fly in Florida grown mangoes treated with hot air (Sharp, 1992). Forcing of air at 50 °C over the mangoes killed all stages of Mexican fruit fly, West Indian fruit fly and *A. serpentina* as the seed surface temperature reached 48°C (Anon, 1994a).

Irradiation

While Japan prefers VHT for disinfestation of mango fruits from pulp and seed feeders for importing into their country, USA insists on irradiation of mango fruits for import into the country. Irradiation is a high-tech method gainfully employed to reduce post-harvest losses, extend shelf-life and for disinfestations of mango fruits from insect pests. Irradiation could be a viable alternative to HWT quarantine treatment for mangoes. Irradiation appears to be the only feasible quarantine treatment for mango infested with seed weevil.

Irradiation involves the exposure of mango fruits to ionizing energy such as gamma rays, X-rays, electron and microwaves to disinfest mangoes to ensure quarantine security. Radiation treatments were developed against fruit flies in mangoes from Florida (USA), Mexico and Australia and later from India.

In Hawaiian mangoes, 206 and 329 Gy killed mango stone weevil (Seo *et al* 1974). In South Africa a dose of 500 Gy killed all immature stages of mango seed weevil, while 750 Gy prevented the emergence of adults from infested mangos (Thomas, 1975). A dose slightly above 500 Gy disinfested mangoes of seed weevil. (Milne *et al* 1977), however, the same dose did not disinfest 'Alphonso' mangoes of seed weevil in India (Shukla and Tandon, 1985).

Exposing mangoes to 76 Gy disinfested them of Caribbean fruit fly eggs and larvae (Von Windeguth, 1996). Third instar of Mediterranean fruit fly larvae in Mexican mangoes irradiated with 250 Gy did not emerge from pupae and 60 Gy applied to third instar of *A.serpentina*. Irradiation of Mexican mangoes prevented the adult emergence of Mexican fruit fly and west-Indian fruit fly (Bustos *et al* 1992). Queensland fruit fly eggs, larvae and pupae in the laboratory were killed when exposed to 800 Gy and 50 Gy prevented adult emergence (Macfarlane, 1966). "Kensington" mangoes infested with eggs and larvae of Queensland fruit fly and *Bactrocera jarvist* (Tryon) were disinfested with 74-101 Gy (Heather *et al* 1991).

Irradiation by gamma rays also exhibited some promise in respect of disease control in mangoes. However, only marginal control was obtained at the highest non-deleterious doses from mature green fruits (300 Gy). Effect of irradiation is additive only with regard to disease control (Johnson *et al* 1990).

Irradiation of fruits enhances the shelf-life of fresh fruits besides delaying ripening and senescence. Inspite of its benefits, its high expensive installation, technical difficulties and consumer apprehensions and objections are coming in the way of its utilization in developing countries.

Quick freezing

In processing factories, mango fruits are disinfested using quick freezing at 17 °C. The process, however, is not approved for importing mangoes with seeds into USA from Oceania, Hawaii, South East Asia, Philippines and South Africa, because of fear of probable presence of mango stone weevil in the seeds (Anon, 1994a).

Fumigation

Fumigation to disinfest insect pests in the importing produce (particularly in case of grains and dry products), is an old practice, which is banned later in USA, New Zealand etc. However, fumigation is no longer in vogue.

Chemical dip

Post-harvest treatment of 'Kensington' mangoes in dimethoate was effective against Queensland fruit fly (Swaine *et al* 1984). This treatment is required for fruit entering Australia from Queensland and New South Wales. Countries like USA do not allow the use of chemicals to disinfest mangoes due to chemical residues in the fruits.

Atmosphere Modifications

Atmosphere with low Oxygen (O_2) and/or very high Carbon-di-oxide (CO_2) is insecticidal in nature. Potentiality of controlled and modified atmosphere regimes for disinfesting mangoes has been recognized well (Yahia, 1993., Ke and Kader, 1992). It is advantageous to use insecticidal atmospheres as they do not leave any toxic chemical on the fruit and environmentally safe and are comparable in cost with chemical fumigants. But require longer time to kill insects than fumigants (Gould and Sharp, 1990). They reported that the time needed to disinfest Florida grown mangoes infested with Caribbean fruit fly eggs and larvae far exceeded the shelf life of shrink-wrapped mangoes.

Atmospheric treatments are limited to regions which do not adversely affect ripe fruit quality. These extreme atmospheres may cause anaerobiosis and fermentation in fruits. Thus the use of these atmospheres is only feasible if they do not harm the fruit. Mangoes stored in insecticidal low O_2 and/or high CO_2 up to 5 days at 20 °C did not develop off-flavour or tissue injury and ripened normally. Therefore, it is possible that insecticidal atmosphere could be used to control insect pests in mangoes. However, there is still no information available on the control of fruit flies with insecticidal atmosphere (Yahia, 1993).

Packaging

Mangoes must be packed to protect them from the hazards of transportation, storage and handling. Besides, packaging of fruits puts them into unitized volumes for easy handling. Several types of packages are in vogue all over the world. However, bags, sacs, baskets etc. should be avoided for packing mangoes as these generally result in mechanical damage and bruising. Wooden boxes are commonly used for packaging and transportation. Cartons/boxes made of card board, solid fiber board or corrugated fiber board (CFB) are the latest developments in packing of fruits and vegetables. These meet the quality, hygiene, ventilation and resistance characteristics to ensure suitable handling, shipping and preserving of the fruit (Photos 2.3 and 2.4).

Photo 2.3: Packages of Kesar mangoes in Gujarat

CFB boxes are particularly used for exports, but also necessary even for the domestic markets due to over-felling of trees for wood and greater concern of the country over environmental preservation. CFB boxes of 5 kg and 10 kg capacity for packing and shipping of mango fruits were designed and developed in India and put to use presently. However, net weight requirements depend on the variety and the destination markets and mode of transport (Anon, 2004).

Photo 2.4 : C.F.B. Box for packing mangoes prepared by Central Institute of Sub-Tropical Horticulture, Lucknow

Ex:	Alphonso	:	320 × 230 × 90 mm
	Banganapalli	:	390 × 260 × 115 mm
	Chausa	:	390 × 280 × 105 mm.

The typical carton external dimensions will be (L × W × H): 400 × 300 × 100 mm for holding 4 kg of fruits.

The carton should have ventilation and handle holes to provide adequate ventilation and ease of handling. For sea shipment and pre-cooling, the carton should have ventilation in the sides, top and bottom for allowing horizontal air movement, being used during forced air, pre-cooling and vertical air-movement being used in most modern refrigerated containers. The size and placement of holes in the cartons should be suitable for proper air-circulation. Carton must have 4 holes each on top and bottom, 2 holes each side. The holes should be covered with nylon screens to check the entry of insects etc., in the pack houses, transportation, storage etc. The pack outer wall should have a water-proof coating to prevent damage due to high humidity in cold rooms. The material used inside the pack should be of permitted food contact material, meeting the recommended standards (Anon, 2004). The cushioning material varies. Either cartons or individual fruits are cushioned. Cushioning material also varies with the mode of transport. For air-shipment, a layer of shredded paper at the base of the carton is used as cushioning material (Photos 2.5 to 7). This is not recommended in sea-shipment as it can restrict ventilation and may harbour fungal spores. Instead a bubble sheet at the bottom or enclosing individual fruits with polystyrene foam sleeves are used as cushioning material to avoid vibration damage. Alternately, each alternate mango in the carton be wrapped in a tissue paper to reduce fruit to fruit contact and rubbing. Individual fruit wrapping with tissue paper imparts uniform surface colour, reduces shrinkage and spread of diseases. Organic materials such as paper, leaves or shredded wood(wood wool) are also sometimes used to cushion individual fruits in the carton. But not advisable, since organic material such as paper, leaves or wood wool often harbour pathogens. Further, wood-created micro-wounds in the fruit skin provide access to the pathogen. Losses were more when fruits were removed from cold storage and condensation had developed on the fruit as well as on wood wool (Muirhead and Grattidge, 1986).

In some countries plastic or cardboard liners that had depressions designed to accommodate fruits of particular size are in use. The depressions provide some support for individual fruit during packing while cardboard lines also provide some additional buffering against impact damage. The pattern of depressions facilitates most efficient packing and utilization of the space within the carton. Low density polythene (LDPE) has also been found beneficial as it maintains humidity and results in lesser shrinkage of fruits during storage. CFB boxes with partition are best for packing and transporting of "Alphonso" mangoes (Roy and Pal, 1991). Cartons of interlocking types are ordinarily used for packing mangoes.

The packages should have adequate mechanical strength to survive against pre-cooling/cold storage procedures and be able to withstand the weight of the top boxes without getting distorted during the normal stacking under the storage conditions and palletization. The paper/material used for making boxes should not become soggy after exposure to high relative humidity (90-95%) during storage.

Some countries use 'Modified Atmosphere Packaging' wherein fruits are packed in plastic bags or films, trays covered with polythene sheet. This reduces the rate of respiration and ripening in storage. Their permeability will produce varying results depending on the variety of fruits, cooling systems used and storage temperature (Johnson *et al* 1997).

Photo 2.5 : Arrangement of mango fruits in a package cushioned with news papers

Photo 2.6 : Individual wrapping of mango fruits with colour tissue paper in a package

Photo 2.7 : Cushioning of mango fruits with paper strips in a package

The packages must be free of all foreign materials and odour. The carton must be clean and of a quality such as to avoid any external or internal damage to the fruits. The carton should have sufficient mechanical strength (bursting strength of 250-275 lb/inch). The materials used inside the package must be fresh, clean and of good quality so that it can avoid any external and internal damage to the fruit. The mangoes should be packed in a single layer with stem-end facing upwards and slightly bent on one side, rather than directly on the base. Mangoes normally soften from the base and in long-term storage, this method helps to prevent bruising when ripening begins during shipment. Only sound and uniform grade fruit should be packed in a carton. They should be of the same origin, variety, quality and size (CAC, 2005). Mixing of different grades, varieties, sizes and bruised/cut fruits, fruits of different maturities and ripening stages should not be done. Cartons should not be **under or over-filled**. The filled cartons must be kept under shade.

Labelling

Labelling of cartons is necessary now-a-days for providing information about the contents, attraction and tracing the origin of the product. The general guidelines to be observed for labelling cartons are outlined in the Codex Alimentarius Document (AC/RCP 44-1995., AMD -1-2004.).

Carton labelling/printing may be done with a non-toxic ink or glue. Each package shall bear the particulars in letters grouped on the same side legibly and indelibly marked and visible from the outside. The labelling should indicate the following details (Anon, 2004).

A. Identification

(i) Paker name and address or officially issued or accepted code mark.

(ii) And/or Dispatcher

B. Nature of produce

(i) Fresh mangoes, if contents are not visible from outside

(ii) Name of the variety

C. Origin of produce

(i) Country of origin and optionally district where grown or national, regional or local place name.

D. Commercial classification of produce

(i) Class

E. Official control mark (optional)

F. Net weight

G. Date code.

Note:

1. All information shall be on the same side of the package. The marking may be done by using an ink stamp, by printing onto a package, by means of a label firmly fixed to the package or by combination of these methods.
2. Information to be marked on both ends of the package
3. In addition to the details of exporter, the details of importer shall also be given on the label.

Individual labels can be pasted/affixed on the fruit for identification and market appeal.

Palletization (Photo 2.8)

Stacking of filled cartons in vertical column on a pallet (stand) with corner boards and strapping is"palletization". It is essential to minimize fruit damage from multiple handling and has several advantages. Handling of mangoes on **pallet** allows convenient movement of large volumes of fruit in pack houses and/or during temporary storage. There will be savings in handling costs (up to 40-45%); reduced handling times, more utilization of storage space and reduction in post-harvest losses. However, the disadvantages are; high cost, lower number of cartons per a sea-container and loss (Johnson *et al* 1997).

Photo 2.8 : Palletization

For the export of mangoes by sea pallatization is necessary. Pallet should be strong enough to ensure transit till the final destination. Precision stacking with each box/carton fitted exactly on top of the one below minimizes the risk of damage. Collapsed or lopsided pallet stacks have usually been due to careless stacking. Pallet slots should not block ventilation holes in the cartons. They should be stacked in such a manner that ventilation is continuous. Link sheets which bind the cartons together at intervals should also be designed so as to ensure continuous ventilation through the pallet. In the cold room pallets should not be stacked against a wall or placed directly against each other (Boelema,1987). Cartons should be stacked in vertical columns of 18-20 levels (depending on carton depth) with corner boards and strapping to prevent shifting during transport. Stability of pallets load is an important factor and essential and all earlier efforts would be futile if this process is overlooked. **Pallets** should be secured by plastic or fibre corner boards. Strappings and clips used should preferably be of plastic materials. The commonly used strapping methods include:

(i) interlocking strapping,

(ii) tension netting,

(iii) horizontal strapping and

(iv) full length corner post protectors and strapping. Use of stretch wrapping is not recommended as it prevents ventilation.

The pallets are fork-lifted on to the refrigerated sea containers (reefer containers) and set at 12.5 °C or controlled atmosphere containers for shipment by sea.

The carton specifications for palletisation should be comparable with the International Standards Organization (ISO). Pallet dimensions, 1000 × 1200 mm and 800 × 1200 mm (only to export to Europe) should be used. However, relevant markets and transporters should be consulted concerning the acceptability of pallet dimensions and appropriate access for fork-lift system. The correctly sized pallet. e.g. the ISO one, which fits snugly into a standard sea-container and gets accommodated the standard interlocking box (ILB) should be used for the local markets as well (Johnson *et al* 1997).

Non-introduction of pallets in trading of fruits and vegetables is one problem that needs immediate attention in India. So India should consider seriously the introduction of pallets at least for export purposes. It is also essential to introduce mechanical loading and unloading particularly with the use of fork-lift trucks while handling pallets.

PRE-COOLING

Field-heat accumulated by the mango fruit during its pre-harvest life on the tree, will cause rapid ripening and deterioration of the fruit after it is harvested, particularly in tropical regions. Therefore, it is highly essential to remove the field heat from the harvested fruit as rapidly as possible to avoid above problems. Prompt removal of field heat from harvested fruits is the best way of retaining the freshness of fruits for a longer time. Pre-cooling reduces the cooling demand on any in-transit cooling system. The choice of cooling depends upon the mode of shipment to over-seas markets. Mangoes meant for sea-shipment require pre-cooling prior to loading into containers as the journey time is longer, compared to air-shipment, which requires shorter journey time and pre-cooling is optional.

Mangoes are generally pre-cooled to 13 °C by placing the fruits in pre-coolers, where temperature is set at 13 °C with 90-95% relative humidity. However, studies in India indicated that pre-cooling of mangoes to 13 °C could be carried out 2 to 3 times faster by setting the pre-cooler temperatures at 5 °C without any low temperature injury to the fruits (Anon, 2005). This would help to pre-cool more number of batches in a day. Pre-cooling of mangoes is generally done 1-3 days following harvesting. Pre-cooling of mangoes 6-12 hours after harvest was better than pre-cooling after 24 hours for extending storage life (20 days) and maintaining firmness (Anon, 2005).

Methods of pre-cooling

Several methods of pre-cooling were developed.

Forced-air-cooling system is more efficient wherein cooling can be carried in 4 or 6 hours. With the palletized fruit this method requires specially designed unit and compatible packing. High humidity systems are preferred in some cases as they reduce fruit water loss. Hydro-cooling is also effective, but can increase the risk of infection by wound pathogens. Roy and Pal (1991) described a low cost cooling system for use at field level. Hydro-cooled or forced-air-cooled mangoes to 13 °C could be stored for 3 weeks in unripe conditions at 13 °C followed by one week at room temperature (Anon, 2005).

Pre-cooling of hot-water treated mangoes (cv. Alphonso) either by hydro-cooling or forced-air-cooling results in reduced water loss compared to non-cooled hot-watered fruits during storage at 13 °C. Though hydro-cooling is faster than forced-air-cooling in bringing down the fruit temperature of hot-watered fruits, spoilage is more in hydro-cooling method than in forced-air-cooling (Anon, 2005).

Pre-cooling is sometimes coupled with disease management. Pre-cooling of mangoes (12-15 °C) with Bavistin (500 ppm) increases the shelf life and also reduces the incidence of spongy tissue in 'Alphonso' mangoes. Significant reduction in respiration rate and ripening rate and good surface colour of mangoes can be achieved by hydro-cooling of fruits with Bavistin (Roy and Pal,2000).

After pre-cooling, the cooled fruits should be shifted to cold storage immediately once the fruit temperature reaches 13 °C to avoid further reduction in fruit temperature and resultant chilling injury. Packaging of pre-cooled fruits considerably enhances the shelf life of fruits by reducing the fruit metabolism and the accompanied physiological and biochemical changes in the cooled fruits.

For successful pre-cooling of fruits certain precautions should be observed. Temperature of harvested fruits should never be allowed to rise before it is cooled and stored. Therefore, the temperature in the pack house should be maintained at 25 °C and the mango fruits should be transported to pre-cooling centers within 6 hours of harvest. The temperature of the harvested fruits should be brought down to 13 °C within 8 hours of harvest in the pre-cooling chambers. The delay in bringing down the temperature to this level will reduce the keeping quality of mangoes proportionately (Anon, 2004). Use of refrigerated containers for pre-cooling of the fruit should be avoided, because these are designed to maintain the fruit temperature at a desired particular level and not to remove the field heat from the fruits.

STORAGE

Being a living entity and perishable, mango fruit is extremely susceptible (after harvest) to irreversible damage in quality in response to its environment. Hence, its storage is necessary and essential. Storage of mangoes may be necessary when picked fruits are held prior to transportation or when the entry into markets is delayed and also to spread the duration of availability. The duration of storage required may be shorter or longer depending very much on the logistics of the marketing chain. In general mangoes are exported as rapidly as possible after harvest since they have short post-harvest lives and highly perishable. These may be relatively short period of storage at either the pack house or the air or sea-port while awaiting for transportation. Thereafter the storage period depends on the method of transport. In air-shipment, fruits are stored for a shorter period in contrast to sea-transport. However, air-freighted fruits should not be stored for more than 3-5 days. On the other hand, sea- shipment being of longer duration, requires longer periods of storage. Storage of mangoes is required to tide over the glut, to prolong the availability of fruits etc (see Storage chapter). The primary objective of an optimal storage protocol is to arrest the damage in quality and extend the shelf-life for as long period as possible.

Several methods of extending the shelf-life of mangoes are available (see Storage chapter). However, the choice of the method depends on cultivar, its response to the method, duration of storage, mode of transport, the purpose (fresh consumption /processing and domestic/export) etc. Low-temperature storage, controlled atmosphere and modified atmosphere storage are often employed for storage in pack houses prior to their disposal to markets/transport.

Low temperature (cool/cold) storage

After pre-cooling, the packages shall be moved to the cold storage and held till the pallets are loaded into the refrigerator containers. Cool storage is essential particularly for sea-shipment. The optimum conditions (temperature and relative humidity) at which the mangoes are held in cold storage vary from cultivar to cultivar, length of storage period, mode of transport etc. Trials conducted at Bangalore (India) indicated an optimum temperature for cold storage of mangoes as 10-15 °C and 85-95% relative humidity depending on the cultivar (e,g. Banganapalli-12 °C, Chausa-10°C, Dashehari-12 °C, Langra-15 °C) which will maintain the fruit in an acceptable condition and allow satisfactory ripening on transfer to high temperature (20 °C). However, "Banganpalli" mangoes for export purpose could be stored at 12-13 °C and 85-90% RH in an acceptible condition which ripen satisfactorily on transfer to high temperature (20 °C) at destination. The temperature and humidity in the cold storage should strictly be maintained throughout storage and transit. Cold-stored fruits may be transferred gradually to room temperature to minimize sweating. Otherwise, stored fruits exhibit higher level of spoilage and poor quality. Hence, strict control of temperature, with humidity and quality control for cold storage of fruits is essential. For this cold chain must be maintained until the fruits reach destination/consumer.

Chilling injury sensitiveness of mango fruits does not allow cold storage below 10 °C (see Physiological disorders). However, mangoes can be successfully stored below 10 °C for longer periods with Modified Atmosphere Packing (MAP) as well as under Controlled Atmosphere of elevated CO_2 and lowered O_2 levels.

Individual shrink-wrapping

A method for extension of storage life of fruits at lower temperature (8 °C) by overcoming the problem of chilling has been standardized using "Individual shrink wrapping" technique. In this method each fruit individually is loosely sealed in a flexible film using a sealer and then the packs are passed through a heat shrink tunnel where they are exposed to hot blown air for a period of few seconds. The loosely sealed film shrinks tightly over the surface of the fruits in the tunnel and comes out individually tight wrapped. Selective film wraps like PVC and LDPE have increased the shelf life of mango besides alleviating chilling injury (Castro *et al* 2005).

Studies carried out at Bangalore (India) indicated the possibility of applying this technique in case of mangoes. Individually shrink-wrapped mangoes could be stored for 30-35 days at 8 °C in harvest fresh, unripe hard conditions without chilling injury. The fruits of 'Banganapalli' and 'Alphonso', thus stored ripened normally after removing the film and shifted to ambient temperature within a week of removal and had a good surface yellow colour, firmness and acceptable organoleptic quality. The suggested flow chart for individually shrink wrapping is as follows (Flow chart 2.2) (Anon, 2005).

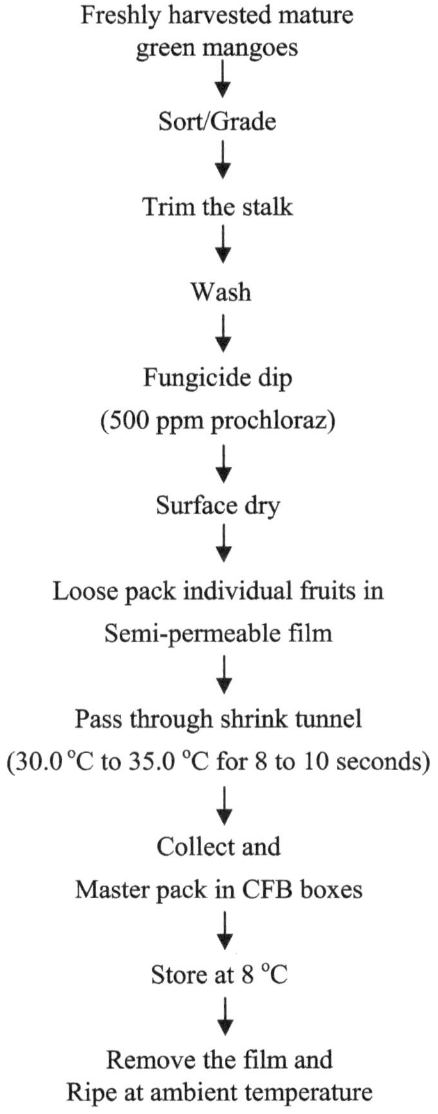

Flow chart 2.2: For individual shrink wrapping of mango fruits

Controlled Atmosphere Storage (CAS)

Some countries employ controlled atmosphere (CA) refrigerator containers for long-term storage (up to 4 weeks) and sea-shipment of mangoes. This method involves maintaining the fruit in a controlled atmosphere having a high level of CO_2 and a low level of O_2 around the fruits. The optimum gas composition for different cultivars of mango in different growing areas was worked out (see Storage chapter). The optimum gas composition recommended for mangoes under South Indian conditions is 5 percent each of CO_2 and O_2 (Anon, 2005). Exposure below 2 percent O_2 and/or above 8 percent CO_2 may induce skin discolouration, grayish flesh colour and off-flavours.

CAS delays ripening and reduces respiration and ethylene production rates, thereby extends the storage life of fruits, which could be preserved for 4-5 weeks in CA at 13 °C depending on cultivar and maturity of fruit as compared to 2-3 weeks at room temperature. 'Banganapalli' and 'Alphonso' cultivars could be stored in unripe green condition for 30-35 days at 13 °C and for 45 days at 8 °C without chilling injury under CA condition of 5 percent O_2 and 3-5 percent CO_2. The stored fruits ripened normally within a week of moving to ambient condition, but with chilling injury. The suggested flow chart for CA storage of mangoes is as follows (Flow chart 2.3) (Anon, 2005).

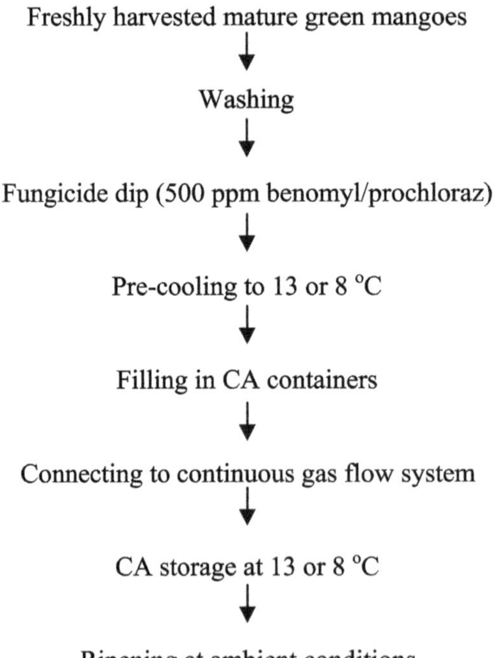

Flow chart 2.3 : For Controlled Atmosphere Storage of mango fruit

Modified Atmosphere Package (MAP)

Packing the mature green fruit in semi-permeable plastic bags/flexible films is referred to as "Modified Atmosphere Package". MAP reduces the respiration rate, weight loss and delays ripening in storage/transport (see Storage chapter). Studies at Bangalore showed that individual MAP of mature green 'Banganapalli' mangoes in polyethylene (PE) extended the shelf life to 25 days at 13 °C followed by one week for ripening under ambient conditions. At 8 °C the storage life of 'Alphonso' and 'Banganapalli' mangoes could be further extended to 5 weeks without chilling injury by packing individually in micro-perforated PE or polypropylene (PP) films. The stored fruits ripened normally within a week when shifted to ambient condition, whereas the mangoes packed in bulk in CFB boxes lined with PP films (one side open) could be stored for one month at 8 °C followed by 5 days for ripening at room temperature. The suggested flow chart for MAP of mango fruits is as follows (Flow chart 2.4) (Anon, 2005).

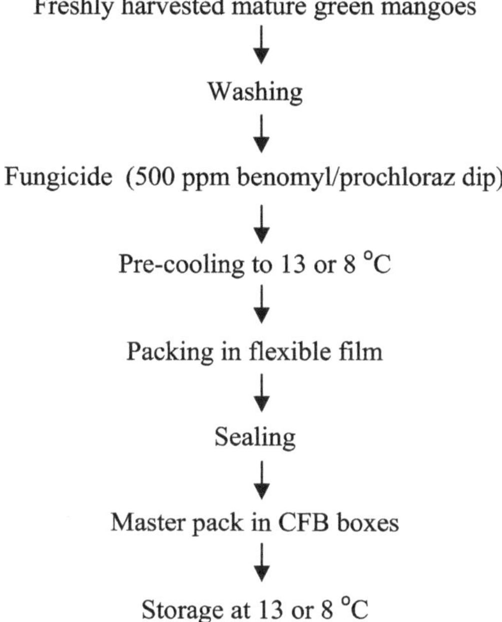

Flow chart 2.4 : For Modified Atmosphere Storage of mango fruit

Ripening

Fruits in pack houses or at the destination/markets are ripened prior to sale. The optimum temperature recommended for ripening mangoes is 20-25 °C, with 90-95% relative humidity. Temperatures higher above 25-30 °C result in ripe fruits with off-flavour and mottling of the peel. Ripening could also be achieved with ethylene gas and hot water treatment. Initiation and synchronization of ripening and shortening of the ripening period can be achieved with exposure to ethylene gas. For hastening of ripening, the fruits are exposed to ethylene gas, either directly from cylinder or as liberated gas from ethephon (ethrel) when utilized with a catalytic generator in ripening chambers. Ethylene required during exposure for ripening is 10 to 100 ppm of ethylene with an exposure time of 12-24 hours (Anon, 2005). Pandey and Tandon (2007) suggested the use of ethylene gas (100 ppm) in air-tight rooms with exposure for 24-48 hours under controlled conditions of temperature and humidity to ripen mangoes (cv. Dashehari) at Lucknow (India). Alternately fruits may be ripened by dipping the fruits in ethrel/ethephon (250-750 ppm) solution in hot water (52 °C ± 2 °C) for 5 minutes. Mere HWT at 52 °C – 55 °C for 5 minutes alone also results in uniform and accelerated ripening, reduces spoilage (up to 50%) and weight loss (Pandey and Tandon,2007).

Concentrations of ethylene gas are controlled by adjusting flow deliveries or the volume of ethylene delivery liquids such as ethephon. Air-tight room should have adequate air-circulation to enable uniform distribution of ethylene gas throughout the chamber and fruits. Build up of CO_2 has to be avoided as this reduces the efficacy of ethylene and will have a detrimental effect on fruit ripening. Under optimum conditions, the air should be changed regularly (once in every hour) and ethephon be replaced.

Transportation to markets

After storage/ripening/from ports (air or sea) the fruits are transferred to local/distant markets for dispersal, in various types of conveyances depending on the distance/journey time and condition of the fruits. For local markets (less than 3 hours journey) fruits may be transported in unrefrigerated carriers provided the fruit had been pre-cooled and the carriers (trucks or trains) travel at night. However, fruit must be sheltered from direct sun and rain as the situation requires when open trucks/lorries are used. The journey must be completed in as minimum time as possible with minimum stops. In case of export to global markets, the fruit is transported in refrigerated containers by road/rail up to the sea/air-port and from there by sea or air, when the sea/air is more than 1 hour from pack houses. The following road transport recommendations may be followed (Johnson *et al* 1997).

1. The refrigerated truck should be clean and in a good mechanical condition. Insulation and floor should be in a sound condition, the door seals must be intact and the doors must close very tightly.
2. The refrigerator equipment must be correctly set on air delivery and must be calibrated for each journey.
3. Close monitoring of temperatures on the vessel is essential. A device (recorder) may be arranged along with each container to record the temperature and humidity at regular intervals for monitoring the cold chain conditions.
4. The boxes may be arranged in the vehicle in such a way to allow proper air circulation.
5. The fruits may be covered properly leaving access for ventilation.
6. Large containers for packaging and transport may be avoided.

GRADING

Considerable variation in the morphological characters (shape, size, colour etc) and quality of harvested fruits and vegetables and their varieties exists due to genetical, environmental and agronomical factors. Further, the produce of the same tree sometimes show variation in morphology and quality. Plant age also influences the variation. It is generally observed that graded produce fetches higher prices, about 20-25% than ungraded one. Therefore, grading of produce is necessary and required to get uniformity in the produce and remunerative returns through higher prices. Further, produce of even size, shape and colour are more appealing to the customer and such produce allows ease in packing.

Need for grading

Grading of fruits and vegetables etc., is needed for the following reasons:

(i) for better packing,

(ii) for realizing higher prices,

(iii) for ease in marketing,

(iv) for ease in price fixing,

(v) for compliance of national and international standards,

(vi) for ease in quantifying,

(vii) for sales appeal

(viii) for separation of diseased and damaged fruits to arrest the multiplication of microbial load and spread of diseases,

(ix) for efficient peeling and cutting in processing and

(x) for uniform hydro-thermal treatment while thermal processing is adopted.

The need for grading in mango has become important now-a-days in the context of increased awareness in the consumer and of an expanding exports to international markets. When the fruit has to travel long distances, the need for grading of mango fruits etc; assumes greater importance as ungraded/low graded fruits are uneconomical to transport.

GRADING

Grading generally implies the sorting of healthy fruits/vegetables from defective, blemished or diseased ones, thereafter, they are graded (separated) according to size or weight, external factors like colour, freedom from any mechanical injuries (blemishes etc) or phytological defects. Thus sorting of the fruits into different quality grades is referred to as 'grading'. Appearance and quality of produce are best judged according to grades. Grading involves balanced assessment of all the properties of the produce, which affect its acceptance as food and as working substance for the processor (Brennan *et al* 1976).

Grading has two functions: (i) to eliminate completely all unsatisfactory ones and (ii) consolidation. The purposes of grading are also two fold: (i) to sort out fruit in well defined categories of uniformity (usually shape, size and absence of physical defects) and (ii) to divert-out off-grade fruit from the pack line to a second grade line or processing line (Johnson, *et al* 1997).

Advantages of Grades

Grading or grade standards constitute one of the most important factors governing modern marketing, since grading has the following advantages (Pruthi, 1992).

1. Grading provides a sound basis for a system of trading on both national and international level in which inspection of individual packages becomes unnecessary thereby affecting a considerable economy in labour, time and at all stages of marketing.
2. Fruit escapes damage which is caused by unnecessary handling at different stages of distribution.
3. Fruits keep well for a comparatively longer period.
4. Grading also provides a sound means for intelligent comparison of prices according to quality.
5. Speeds up distribution which is particularly advantageous for perishables like mango.
6. Beneficial for transportation, marketing and to fetch high premium prices.
7. Brings out uniformity in size, shape and absence of physical and pathological defects.

In nutshell, grading is of considerable importance in successful marketing as properly graded fruits infuse confidence in the buyers and thus create an excellent good will in the markets and directly helps in standardization of the quality throughout the markets. Grading adds value to the produce/product and gives better economic gains. Grading is done on the basis of some norms and is important for marketing availability of high quality produce to the consumer (Varshney *et al* 2002).

Basis for grading

Grading in mangoes is based on specific gravity, stage of maturity, weight (size), shape, colour, external appearance etc. Specific gravity grading by water and brine floatation though appears to be useful and more reliable, than the other methods in vogue, is seldom adopted on commercial scale. Grading for appearance is based on the absence of defects, on the surface and the amount and intensity of colour. The most commonly adopted base (criterion) both for national and international markets is the size (weight) of the produce within the cultivars also. This is the most obvious gradable factor. Fresh fruit quality standards are generally based on physical characteristics such as size or weight. The term 'quality' indicates that the fruit shall have reached a stage of maturity which will permit subsequent completion of ripening in the ordinary course of transportation and marketing and that fruit shall be reasonably uniform in colour and shall have size and shape typical to the cultivar in question, shall be free from disease or insect pest, mechanical injuries or blemished scars or cell-sap burns etc. (Prathi, 1992)

Grading based on Specific Gravity of the fruit

Several workers (Subramanyam *et al* 1976., Cancel and Gorcia De Perez, 1979., Prasad Rao and Chadha,1982., Joshi and Roy,1985., Tandon *et al* 1988., Singh *et al* 1989) have successfully tried the segregation gravity or grading of fruits based on the 'floatation technique' or specific fruits which shrink in water followed by further dipping in 3 percent brine (salt) solution. The floaters in water were presumed to have specific gravity of less than 1.0 (Grade I), floaters in 3 percent brine had a specific gravity of 1.00 to 1.02 (Grade II) and sinkers, having more than 1.02 specific gravity (Grade III). The fruits of the Grade I were the biggest, while those of Grade III were the smallest, which ripened earliest (5 days) with less shelf life as compared to Grade I and Grade II fruits. Grade I mangoes ripened with better quality and better shelf life and retained marketability even after 11 days of storage. However, the findings of Tandon *et al* (1988) are somewhat contradictory to their later findings (Singh *et al* 1989). That 'Dashehari' mangoes graded with specific gravity more than 1.0 did not have any advantage and were prone to higher PLW, senescence and other losses. However, grading based on specific gravity is not employed on commercial scale.

Grading based on fruit diameter

Some growers, contractors and merchants use standard sieves having circular holes of 8.0, 7.5 and 7.0 cm diameter to segregate the fruits into different grades. This is mechanized by using vibration sieves of above specifications. Mechanical graders developed at Bangalore, grades mangoes into four grade viz., (i) 5.0-5.7 cm, (ii) 5.7-6.4 cm, (iii) 6.4-7.0 cm and (iv) > 7.0 cm.

Grading based on weight (size)

Since the mass in the fruit does not match with the diameter of the fruit, weight is commonly considered as a criterion for grading mango fruits and such grading is called 'size grading'. 'Size' here does not mean a product with physical dimensions, such as length, width and breadth (thickness) but means "weight" (g or kg) of the product. 'Size grading' of mangoes is most common and easiest being followed universally both for national and international markets (Pruthi 1992)., (Baez-Sanndo *et al* 1999., CAC 2005). Size grades vary with the producing country and also variety and between national and international markets.

Grading factors

Grading involves combinations of physical parameters as follows:

1. Size and shape-as functional and acceptability factors,
2. Maturity- ripeness of fruits,
3. Purpose-suitability of the product for the end-use e.g. canning and freezing properties of the fruits,
4. Colour affecting process and consumer acceptance,
5. Freedom from blemishes-bruises etc.,
6. Freedom from contaminants-soil and pesticide residues,
7. Conformity with legal standards,
8. Freedom from undesired plant materials like, stalks etc.

Methods of grading

A. Manual Grading

Grading on small scale, particularly for local markets is usually done manually by visual inspection of individual fruit, according to size or weight, colour and absence of blemishes, spots etc. In ASEAN countries, mangoes are graded manually with emphasis on latex stains, softness or over-ripeness, cuts and injuries due to insect pests and diseases. Manual grading is also done at retailer level. However, manual grading is laborious, labour intensive, time consuming and uneconomical. Moreover it is not standardized. Quality perceptions vary from person to person. Hence, the uniformity of graded material becomes doubtful.

B. Mechanical Grading

In view of the drawbacks of manual grading (as above), mechanical grading was developed and is found to be an alternative and useful option in which the uniformity of the product is achieved. Mechanical grading is done based on size, weight, shape and diameter of the produce without reference to internal quality. However, in mechanical grading, bruising of the produce is possible and sometimes comes in the way of its adoption. The differences between colour, blemishes etc is not possible in mechanical grading in which the principle of optics is blacking at present and is needed. Both simple and composite machines have been developed in different countries for mechanical grading of mangoes.

However, mechanical grading is not used extensively in India for grading mangoes. Some attempts have been made in this direction and some mechanical graders were developed and are available in the country (see below). The feed back regarding the mechanical graders from growers/traders/packers/exporters is not encouraging. Most of them are of opinion that mechanical grading of mango is not feasible especially when fruits are consumed as ripened ones. Upon ripening, the mechanically graded mango fruits do not get golden yellow colour or full colour of the typical variety in question on the surface of the fruit, which reduces the price of the fruit considerably (Mangaraj and Varshney, 2006). Other reasons observed are:

(i) the consumers are less aware of the importance of such grading, hence less interested,

(ii) the consumer is satisfied with whatever rough grading of fruits done manually in the domestic markets,

(iii) exports to global markets are negligible as of now and

(iv) high expensive nature of machines and mechanical grading.

Grade Quality Requirements

According to the "Organization of Economic Cooperation and Development", the minimum requirements of quality of fruits intended for International trade are:

(i) the fruit should be intact, firm, fresh in appearance, sound, clean, free from black stains and bruising, from low temperature damage, pest and disease damage, carefully picked at the right stage of physiological development which will allow transport and handling and

(ii) continuation of ripening process so as to arrive in a satisfactory condition at the destination (Anon, 1993).

Quality classification of Indian mangoes

Depending on the fruit quality (physical) "The Organization for Economic Cooperation and Development" for export purpose, classified Indian mangoes into three standards (categories) as follows:

A. **Extra Class (551-800 g):** Mangoes in this class must be of superior quality and must be of characteristic of the cultivar, with respect to shape, colour etc. They must be virtually free from visible defects with the general appearance of the produce, quality, keeping quality and the presentation in the package. The maximal permissible differences allowed is 75 g.

B. **Class I (351-550 g):** Mangoes in this class must be of good quality and must be of characteristic of the cultivar. However, the following slight defects may be allowed, provided these do not affect the general appearance of the produce quality, the keeping quality and presentation in the package.

1. Slight defect of shape, 2. Slight defect of the skin due to rubbing or sun-burn, 3. Suberized stains due to resin exudation, 4. Healed bruises not exceeding 3, 4, 5 cm^2 for size groups A,B,C respectively. The maximal permissible difference is 100 g.

C. **Class II (200-350 g):** This class includes mangoes which do not qualify for inclusion in the higher (above) classes but satisfy the minimum requirements specified. The following defects may be allowed provided that the mangoes retain their essential characteristics as regards the quality, the keeping quality and presentation in the package.

1. Defects in shape, 2. defects of skin due to rubbing or sun burn, not exceeding 40% of the fruit surface, 3. healed bruises, not exceeding 5, 6, 7 cm^2 in size groups A,B,C respectively. 4. free from any sign of necrosis. The maximal permissible difference is 125 g.

In class I and II scattered suberized rusty lenticels as well as yellowing of green varieties due to exposure to direct sunlight, not exceeding 40 percent of the surface and not showing any necrosis are allowed (Table 2.2)

Table 2.2 : Quality grades and size tolerances in Indian mangoes for International markets (CAC, 2005)

S.No	Code size	Weight/each fruit (g)	Permissible size (g)
1	Extra class	551-800	426-925
2	Class I	351-550	251-650
3	Class II	200-350	180-425

$\leq$ 10% of the fruit/package exceeding the normal fruit size range.

The Table 2.3 indicates the grade designations and quality of mangoes with label colour specifications to be used for different grades of Indian 'Alphonso' mangoes for export purpose.

Table 2.3: Grade designations and label colour specifications of 'Alphonso' mangoes for export

<table>
<tr><th rowspan="2">Grade Designation</th><th rowspan="2">Colour of label</th><th colspan="2">Weight of each fruit (g)</th><th rowspan="2">Definitions of quality</th></tr>
<tr><th>Maximum</th><th>Minimum</th></tr>
<tr><td>I</td><td>White</td><td>338</td><td>280</td><td>1. The fruits shall be firm and entirely free from damage blemishes or malformation</td></tr>
<tr><td>II</td><td>Red</td><td>280</td><td>222</td><td rowspan="2">2. Each fruit shall be packed in colour without traces of yellow colour at the time of packing.</td></tr>
<tr><td>III</td><td>Black</td><td>222</td><td>163</td></tr>
</table>

A 5 percent tolerance level is allowed in all the 3 grades.

Under Agmark quality, 26 mango cultivars growing in India were included. The grades for some of them formed by Government of India are presented in annexure.

Mango Grading Scenario in India

Grading of mangoes is done on scientific lines in some mango producing and exporting countries while in India it is not practised. Though Government of India has framed grade specifications for different varieties of mango, unfortunately by and large grading is not practised, commercially. This is attributed to the absence of systematic effort on the part of Government to enforce the Act and put it in practice. Lack of enthusiasm on the part of the growers/traders is also attributed to non-adoption of the grades in mango. Similarly lack of awareness and indifference attitude of the consumer with regard to grading are also considered for lack of grading in mango. However, of late marketing agencies and buyers, particularly the educated consumers, are welcoming these innovations in the fruit trade. The conciousness and increased purchasing power of the buyers are said to be the cause of enthusiasm for graded as well as quality fruit.

Not withstanding the above, some amount of grading is done in the country. Fruits in India are graded on the compulsory basis for export and on voluntary basis for the internal trade. However, the process of grading of fruits, including mango in the country has not progressed much due to poor legislation, negligible implementation, poor financial support and lack of extension services for dissemination of grading and its significance.

In India, the producer seldom grades his produce, but pre-harvest contractors, whole salers and retailers particularly grade the fruit according to size, appearance and variety before sending to the markets and sale. However, traders expressed (in a survey) that they grade mango fruits according to area from which the produce has come, besides based on variety and size of the fruit (Sawant and Shinde, 2000).

Grading is also prompted by the value of the particular cultivar and export. The mangoes are generally sorted into 3 or 4 grades according to size and stage of maturity and those of approximately the same size are packed in one container. This practice, however is confined to the fruits of superior quality and resultant high value, which are intended for export to overseas markets and sometimes to distant markets across the country. Such practice is common in Maharashtra and Gujarat states in particular for 'Alphonso'; in Chitoor district (A.P) and Vellore district (Tamil Nadu) for 'Bangalora', 'Pairi' and 'Alphonso' and in Uttar Pradesh for 'Dashehari' (Pruthi, 1992). In Lucknow region (Uttar Pradesh) at packers level manual grading is carried out. The grades followed at Lucknow, Bihar and Maharashtra are presented in the Table 2.4, 2.5.

Table 2.4: Grades of different mango cultivars at different markets in India (Singh, 2004)

S. No	Lucknow- Dashehari		Bihar- Gulabkhas		Maharashtra- Alphonso	
	Grade	**Weight / each fruit (g)**	**Grade**	**Weight / each fruit (g)**	**Grade**	**Weight /each fruit (g)**
1	Grade-I	>175	A Grade (large)	>210	Grade-I	300 and above
2	Grade-II	161	B Grade (medium)	170±20	Grade-II	250-300
3	Grade-III	112	C Grade (small)	130±20	Grade-III	<250

Table 2.5: Grades of different varieties of mango and their specifications in India (Mangaraj and Varshney, 2006)

S. No	Grade	Alphonso		Langra		Totapari		Dashehari	
		E.D* (mm)	Weight * (g)	E.D (mm)	Weight (g)	E.D (mm)	Weight (g)	E.D (mm)	Weight (g)
1	Grade-I	65	176	68	237	73	292	61	200
2	Grade-II	63	155	65	208	69	272	55	180
3	Grade-III	59	142	63	183	62	227	49	135
4	Grade-IV	56	112	56	133	55	169	44	120

ED *: Average Equitorial Diameter; Weight *: Average fruit weight.

TRANSPORTATION

Mangoes are grown in several countries of the world, but consumed all over the world. Production centres are generally in the rural/interior areas, away from the markets/consumer centres/pack houses; cold storage units, processing factories etc. Further during the past half a century mango has become an export oriented crop in many countries like Australia, Brazil, Israel, Mexico, South Africa etc. Importing countries impose quarantine treatments which are carried at places away from the orchards. All these need movement of mangoes (transport) from production centres. Transport plays a vital role in post-harvest handling of mangoes. Success of the marketing of perishable produce depends on efficient and fast transportation system. Quick transport of produce from the field to the packing shed is very essential as the time taken between harvesting and storage at optimum cold storage conditions should be minimum for long term storage. Otherwise, the said delay will reduce the storage life at low temperatures. Thus, it is the key link in marketing the produce and connecting production centres with consuming centres.

Mode of transport

Perishables are transported by different ways. The mode (type) of transport is based on destination, value of the produce, degree of produce perishability, amount of produce to be transported, recommended storage temperature and relative humidity outside, temperature conditions at origin and destination points, time taken to transport to reach destination, transport freight rates and type and quality of transportation services. Basically perishables like mangoes are transported by land (road), water ways (including sea) and of late by air.

Transport by land (road)

Depending upon the distance to be covered, mangoes are carried as head loads or on the back of animals, bullock carts, tractor trollies, tempos, trucks etc. (Photos 2.9 and 10). Road transport particularly by trucks is faster and is preferred over train transport as it has not proved beneficial, because of lack of ventilated wagons or air-cooled wagons in India. Mangoes from hilly areas are generally transported as head loads packed in baskets/gunny bags etc. Mangoes are transported in certain parts of the world on pack animals (camels, donkeys, mules, bullocks, horses etc) over a distance of 15-25 Km. Transport by carts (bullock, horse etc) is very common in rural areas and is generally adopted up to 20-50 km. Tractor trollies and trucks (lorries) (open or closed) tempos, are used for transporting mangoes. Tractor trollies and tempos are used to transport mangoes from godowns to assembling points within a locality. Whereas trucks are employed for local as well as distant markets, to pack houses, irradiation and vapour heat treatment plants, processing units, to rail roads, sea and air ports. In many cases, fruits are carried in trucks over distance of 300 to more than 1000 km, connecting the producing centres of one state with markets of the other states. The advantage of road transport by trucks is that they generally travel faster and during nights, thus avoiding the excessive heat of the day, which is hazardous to mango in transit as fruits are loaded loose without any coverage (in India) and transport losses by truck are very small (about 1%) . It allows the merchant or his agent to travel with the produce and ensure the produce reaches the market safely and without damage or loss. Road transport by trucks is also preferred because of ready availability of trucks unlike railway wagons, their arrival to destination is in time, absence of pilferage enroute and facilities of road at the door step of the traders and direct delivery to the markets.

Though road transport is comparatively faster than other modes of road transport, delays may occur in truck transport. The roads connecting the orchards to the main roads being rough/metal roads, their movement is slow often and hence the vehicular movement is often delayed. In Bangladesh, vehicular movement is delayed as they have to wait for ferries, poor transport facilities, bad (rough) roads (Quroshi and Meah, 1991). Similar may be the conditions in many developing countries.

Inspite of certain constraints in road transport by trucks, it is still preferred over rail transport because of the reasons mentioned above. Hence, there is a need to improve road transport infrastructure to provide maintenance of quality of produce till it reaches the consumer. A practicable alternative would be to introduce double roofed and well ventilated vans as used in Australia, where the temperature conditions are similar to those prevailing in India. In these vans, the temperature of the produce is controllable and does not rise above ambient temperatures and there is sufficient cooling at night (Sethi and Maini, 1989). Road services can be considerably improved by introducing wider roads with smooth surface having the facility of one-way traffic, by avoiding long-waits at level crossings and at inter sections, through introduction of fly-overs at these places. All these increase the speed and movement of produce by road. At intervals of few hundred kilometers, facilities for halting with provision of parking trucks, gas (petrol/diesel) stations, food plazas and cool sheds go long way in reducing the time of transportation and minimizing the post-harvest losses of perishable produce.

Photo 2.9 : Transport of mangoes in a open truck without packing

Photo 2.10 : Transporting mango fruits loosely in a tempo

Transport by rail

Within country, mangoes are transported often by rail, for example: mangoes from southern states of India are transported to northern markets more often by rail, since for long distance transport within the country, rail is a cheaper mode as compared to road. Fruits reach the destination in 3-4 days depending on the distance and cultivar, by which time, they are ripe and nearly ready to eat. However, when the fruit is transported by rail, particularly over long distances, there is a loss of 10-20 percent in the consignment, mainly an account of frequent delays in the movement of the wagons. Hence, inspite of concessional frieght rates offered by railways, for fresh mango, road transport is continued. Despite the disadvantages, rail transport still prevails when the commodities have to be transported in bulk or when the distances involved are too long.

For rail transport the fruits are generally packed in bamboo baskets, lined with paddy straw/cut paper. The wagon is insulated by placing dried palmyra leaves on the top and sides. A total of 700 baskets (12.5%) are loaded per wagon in 5-6 layers high. The wagon is partitioned by erecting scaffoldings for avoidance of pressure on the bottom layers.

For rail transport, steel wagons are not suitable as they get heated (40 °C - 50 °C) during the journey and lack ventilation and cause huge losses. Long distance transport of fruits by rail involves much time during which deterioration of fresh fruits under normal conditions occurs due to (a) excessive heat developed by respiration of the fruit and by radiation (b) mould infection (c) wilting of the fruits (d) faster ripening and softening and (e) excessive weight loss.

To keep the fruit fresh during the transit, the wagons have to be ventilated and also other measures have to be taken to neutralize the heat developed by the produce and/or heat accumulated due to radiation inside the wagon. This could be achieved by employing ventilated wagons, coated with aluminum paint outside. The deterioration detailed above could be minimized by suitable pre-treatment of the produce, adequate package, and transporting in well-ventilated wagons/refrigerated vans (reefer containers). Best approach would be to adopt refrigerated transport to avoid spoilage over long distances and to tide over seasonal glut. Pre-cooling of the fruit or stage-cooling of the carrier system with ice is also beneficial. Use of air-conditioned coaches is increasing now-a-days, as they reduce spoilage and extend the keeping quality of fruits. These coaches maintain a temperature of 12.8 to 15.6 °C with relative humidity of 60%.

Railways can play a significant role in the transport of perishable produce by improving the movement of trains, by making provision of cooling and ventilation, providing better handling facilities at stations and providing special storage space till the produce is lifted/delivered. The development of transport wagons or containers, ideal for maintaining the quality of produce is indeed an important and urgent need in the country. For quick transport by rail, wagons may be attached to the passenger trains, to avoid delays, and damages during transit. Evaporative cooling systems may be designed and attached to wagons to retain quality of produce during transit (Sethi and Mani, 1989).

Transport by waterways

In some countries, where rivers, canals, streams etc., exist, fruits, vegetables etc., are transported by these water ways. For instance, in Bengal, North Eastern states and Kerala in India transport of mangoes and other commodities by boats is a common practice. There is also considerable coastal sea movement of mangoes along the West Coast of India, by country crafts, boats, steamers etc. This mode of transport is the cheapest. The fruits or other produce is generally loaded loose on scaffoldings in the boat etc. Mumbai receives considerable quantities of mangoes by coastal steamers from Konkan tract of Maharashtra state. Transport by water ways also includes sea-transport.

Transport by sea

Transport of mangoes by sea to world-markets is done by some countries. Shipment by sea is much slower and requires considerable period of time to reach the destination. However, untreated fruits get spoiled during this transit, under normal conditions, leading to huge losses. Mangoes have very short storage life, hence cannot be transported by sea without certain conditions. Firstly, the fruits must be stored in unripe condition. For this it is essential to harvest the fruit at the optimal physiological age and kept under optimal environmental conditions (temperature, humidity and atmosphere) to slow down the post-harvest changes (ripening, water loss, pathogenic growth and senescence). Secondly in extending the storage life, refrigerated transport using controlled atmospheric storage (CAS) technique should be employed. Refrigerated transport by containers with fresh produce to any part of the world is now feasible by sea. The refrigerated sea-transport using CAS technique is considered to be the answer for high perishability of mango fruits and high air-freight charges and for export purposes in many countries. In India, trials carried by Agricultural and Processed Products Export Development Authority (APEDA) to export mangoes (Alphonso, Banganapalli and Kesar) from Andhra Pradesh and Maharashtra states by sea transport using CAS with refrigeration to the United Kingdom however, met with partial success. The major problems encountered in these trials were: the incidence of diseases (Anthracnose and stem-end-rot), CO_2 injuries, exudation of the sap (latex) and mechanical damage. Although a shelf-life of 30-40 days has been achieved, the results of the trails indicated that more work is needed regarding the pre- and post-harvest practices for the disease control, sorting, grading and packaging. The requirements of continuous and optimum gas levels throughout the transport period in CA containers is very essential to have good quality fruits at the end of the transport and storage.

In refrigerated transportation by sea, equipment availability is one of the limiting factor at the market place. However, for over-seas export, ship is the most economical way of transport. The cost of sea transport can be further economized by two way trips with full-container utilization, using the new innovations (pre-cooling processes, CA containers, air-delivery systems, air-exchange systems, humidity and temperature conditions as well as scrubbers). For example, apples are shipped from the Pacific North West to Asia and the container is returned to the Candian West with mangoes. This two-way utilization certainly improves the economics of refrigerated sea-transport (Haines,1991). Fresh fruits are largely transported in temperature controlled cargo space either in break bulk or in containers, "reefer containers". Besides the above, modified atmosphere containers and controlled atmosphere containers are used.

Between sea and air transport, export by ship is the best alternative provided there are cooling facilities, shipping containers and storage atmosphere in the ship. Measures to avoid situations which may accelerate respiration should be taken in shipping mangoes. These include: the adherence of transport temperature specifications, avoidance of delays in the placement of fruit in cold storage after harvest and the avoidance of situations during transit which might effect a rise in fruit temperature (Oosthuyse,1997).

Transport by Air

Because of the longer period required by sea-transport to reach the over-seas destination, short shelf-life, higher chances of ripening and post-harvest losses in sea-transport, air-transport is preferred. Air-transport is rapid, often taking only a few hours, although storage conditions are less controllable and space is limited. This is further complicated if the produce has to be transported between places at an intermediary air-ports. Air-transport is costly and is charged based on gross-weight of the consignment.

The fruits are usually packed in cartons, which inturn are packed into either closed containers or net-covered pallets, which are specially designed to fit into specific type of air-craft. The pallets or containers are loaded with produce and then taken to the air-craft for loading. Containers come in various shapes and sizes which are designed to maximize the use of the interior of the air-craft. Air-crafts have pressurized cargo compartments which are maintained at low humidity (20-40% RH) and the same temperature as the passenger compartment. Containers which are used in air-craft have facilities to control temperature and gasses.

Hazards during transportation

Major hazard experienced during transportation of perishable is the mechanical damage (bruising, cutting, scratching). While transporting perishables like fruits they are dumped into lorries (trucks)/bullock carts etc without packing, which increases the risk of physical damage. This can be avoided by harvesting fruit by hand and not to drop on the ground, taking care to transport them after packaging and in vehicles having tyres correctly installed and having special suspensions. Sea-transport involves hazards like higher relative humidity, and consequent rots, and these can be avoided by using refrigeration transport with CAS facility. New *et al* (1978) listed the different transport hazards as follows (Table 2.6).

Table 2.6: Transport hazards in perishables

S.No	Hazard	Examples
1	Impact vertical	Packages dropped, while unloading trucks
2	Impact, horizontal vibration	Shunting rail, trucks
3	Compression static	Stack in warehouse or vehicle
4	Compression, transient	Packages slaked in rolling pitching ship
5	Racking or deformation	Uneven support due to poor floor or uneven lifting
6	Puncturing, sagging	Projections on vehicles
7	Temperature-high	Sun, ships boiler
8	Temperature-low	Cold storage
9	Water, liquid	Rains, sprays, condensation
10	Water-vapour	Humidity of atmosphere, natural or artificial
11	Biological	Pilferage, inspection

Precautions: Because of the above hazards, it is important that certain precautions are followed in transporting perishables like mango fruits to minimize the hazards and resultant losses to the extent possible and to deliver the produce in as fresh and sound conditions as possible.

For local markets (<3 hours access) transportation of fruit in unrefrigerated carriers may be feasible, particularly if fruit has been pre-cooled and trucks/trains travel at night. However, fruit must be sheltered from direct sun and rain and the journey must be completed with minimum stops. For sea export, it is essential that fruit pulp be cooled down to the required level of temperature (or within 2 °C thereof) of vessel carrying the fruit and that the cold chain be maintained without interruption until the fruit is delivered and disposed. When the export dispatch facility (sea or air) is more than one hour away from the pack-house, the following road transport recommendations apply.

1. The refrigerated truck should be clean and in a good mechanical condition. The insulation and floor should be in a sound condition, the door seals must be intact and the door must close very lightly. The vehicles must be pre-cooled prior to loading.
2. The refrigeration equipment must be correctly set on air-delivery and must be calibrated for each journey. Air should be delivered at the set point. There should be minimum temperature rise during transit. Refrigerated vehicles should be filled with a thermocouple in the delivery air and the temperature monitored from outside by connecting an electronic thermometer to the thermo-couple. Refrigerated vehicles are not usually designed for or capable of lowering fruit temperature. The mangoes must be at the relevant shipping temperature, when loaded. The commodity should be pre-cooled before packing and transport to remove the field heat.

3. Because of the shorter time involved, air-transported fruit may have less stringent temperature requirement than sea-export fruit. Air-lines carrying cargo may need to be consulted concerning the normal hold temperatures in the air-craft.
4. Sea-export fruit should be held under refrigeration until loading of the vessel. Sea-transport can be in refrigerated vessels with entire refrigerated deck filled with pallets or in a sea-container, each of which is linked to a central refrigeration system. Alternately, integral containers with their own individual cooling systems or in integral controlled atmosphere container may be used.
5. Close monitoring of temperature on the vessel is essential by monitoring delivery air-temperature (DAT) and return air-temperature (RAT), it is possible to assess whether fruiting is heating up due to respiration or inadequate pre-cooling and to take the necessary steps to adjust DAT (Eksteen,1990). While most refrigerated container vessels monitor individual container air-temperature, including DAT and RAT, it is sometimes advisable to include additional monitoring devices such as a 'Grant Squirrel', multi-data logger, which can be used to measure air-and fruit pulp temperatures throughout the journey period.
6. Mixed consignments i.e. transporting mangoes with other perishable in the same truck, cart etc should be avoided since such mixed consignments are hazardous to mango fruits as they cause faster ripening or diseases etc depending on the other perishables mixed with mango (O'Hare *et al* 1994).

REFERENCES

POST-HARVEST HANDLING

Anonymous, 1987. Standards established by The Ministry of Agriculture, Forestry and Fisheries on Fresh Nangklavngwun Variety of Mangoes from Thailand. Notification No:187. In: Collection of Ministry Notification of Plant Protection Laws and Regulations. Plant Protection Division, Japan Plant Quarantine Association. Tokyo pp.177-178.

Anonymous, 1993. International Standardization of Fruit and Vegetables: Mangoes. Organization for Economic Co-operative and Development (OECD) Brochure.

Anonymous, 1994a. Plant Pest and Quarantine Treatment Manual. United States Department of Agriculture, Animal and Plant Healh Inspection Service, Hay Hsville.

Anonymous, 1994b. Mango Farmer/Packhouse/Exporter Guidelines. South American Mango Growers Association.

Anonymous, 2004. Post-Harvest Manual for Export of Mangoes. Agricultural and Processed Food Products Export Development Authority. New Delhi, India.

Anonymous, 2005. Post-Harvest Practices for Mango, Banana, Tomato and Capsicum. Indian Institute of Horticultural Research. Bangalore, India. Technical Bulletin 27: 1-13.

Baez-Sanudo, R., Blingas, T.E., Ojeda. Co., and Lavi, U. 1997. Mexican fresh mango quality standard grades and application methodology. Proceedings of the 5th International Mango Symposium. Tel Aviv, Israel, 1-6 September 1996, Acta Horticulturae, 455: 726-731.

Balack, J.W. and Starr, D.F. 1945. Mortality of the Mexican Fruit Fly in Mangoes treated by The Vapour Heat Process. Journal of Economic Entomology, 38: 646-651.

Boelema.T. 1987. Long Distance Transport of Avocadoes, South African Avocado Growers Association Year Book, 10:153-156.

Brennan, J.G., Butters, J.R., Gowett, N.D. and Lilly, A.E.V. 1976. Food Engineering Operations. Second Edition. Applied Science Publishers. Ltd, London: 34-57.

Brown, B.I., Wells, I.A., and Murray, C.F. 1986. Factors affecting the incidence and severity of mango sapburn and its control. ASEAN Food Journal, 2: 127-132.

Bustos, M.E., Enkerlin, H.W., Toledo, A.J., Reyes, F.J and Casimivo. G.A. 1992. Irradiation of Mangoes as a Quarantine Treatment. In: Use of Irradiation as a Quarantine Treatment of Food and Agricultural Commodities. Proceedings of Final Research Co-Ordination Meeting, IAEA, Vienna.

CAC. 2005. Codex Standard for Mangoes. Codex Alimentarius Commission. Codex Standards 184-1993, AMD 1-2005.

Cancel, H.L. and Garcia De Perez, 1979. Effect of fruit size on the ripening of Edward Mangoes. Journal of Agricultural University, Puerto Rico, pp. 63.

Castro, J.V., Praffen Ioach, L.B., Carvalho, C.R.L., and Rossetto, C.J. 2005. Effects of film packaging and cold storage on post-harvest quality of 'Tommy Atkins' Mangoes. Acta Horticulturae, 682: 1683-1688.

Coates, I.M and Gowanlock, D.1994. Infection process of *Colletotrichum* species in sub-tropical fruits. In: B.C. Champ,. E. Highley and G.I. Johnson (REds) Post-harvest Handling of Tropical Fruits. ACIAR proceedings.50. Camberra, pp 162-168.

Coates, I.M., Johnson, G.I and Cooke, A.W.1993. Post-harvest disease control in mangoes using high humidity hot air and fungicide treatments. Annals of Applied Biology, 123: 441-448.

Cooke, A.W. and Johnson, G.F 1994. Mango post-harvest disease control: Effect of rain at harvest, fungicides treatment and fruit brushing on fruit appearance. In: B.C. Champ., E. Highley and G.I. Johnson (Eds) Post-harvest Handling of Tropical Fruits. ACIAR Proceedings, 50, Canberra, pp. 448-449.

Dodds, J.C., Bugante, R., Koomen, I., Jefries.P and Jeger, M.J. 1991. Pre- and Post-harvest control of mango anthracnose in the Philippines. Plant Pathology, 40: 576-583.

Eckert, J.W. 1983, Control of post-harvest diseases with anti-microbial agents. In: Diseases of Fruits and Vegetables. NATO Advisory Study Institute Series A- Life Sciences, Newyork, 46, pp. 265-285.

Eksteen, G.I. 1990. Temperature changes during sea-shipment of mangoes inducted containers. South African Mango Growers Association Year Book,10:26-27.

FAO 2003. Compendium on post-harvest operations. Chapter XX Mango: Post-harvest operations. Food and Agricultural Organization http://www. fao.org/inpho/content/compend/text/ch 20 Sec2. htm, P.93-103.

Gafferney, J.J., Hallman, G.I. and Sharp, J.L. 1990. Vapour heat research Unit for insect Quarantine treatments. Journal of Economic Entomology, 83: 1965-1971.

Gould, W.P. and Sharp, J.L. 1990. Caribbean fruit fly (Diptera: Tephritidae) mortality induced by shrink-wrapping infested mangoes. Journal of Economic Entomology, 83: 2324-2326.

Haines, W.1997. The world Mango Situation- A market perspective. Acta Horticulturae, 291:1-11.

Heather, N.W and Jacobi, K. 1994. Who will be first to Japan? Mango Care, 11:4-5.

Heather, N.W., Corcoran, R.J. and Banos, C. 1991, Disinfestation of mangoes with gamma irradiation against two Australian fruit flies (Diptera: Tephritidae). Journal of Economic Entomology, 84: 1304-1307.

Holmes, R.J., Ladger, S.N and Macleod, W.N.B. 1993. Handling systems to reduce sapburn. Acta Horticulturae, 341: 528-532.

Jacobi, K.K and Wong, I.S 1992. Quality of 'Kensington' mango (*Mangifera indica Linn*) following hot water and vapour heat treatments. Post Harvest Biology and Technology, 1: 349-359.

Jacobi, K.K, Wong, I.S. and Giles, J.E 1995. Effect of fruit maturity on quality and physiology of high humidity hot air-treated 'Kensington' mango (*Mangifera indica Linn*) Post Harvest Biology and Technology, 5: 149-159.

Johnson, G.I 1994. Mango, Stem-End-Rot, Part III In: R.C Ploetz *et al* (ed) Compendium of Tropical Fruit Diseases. APS Press, St. Paul, pp 39-41.

Johnson, G.I., Boag, T.S., Cooke, A.W., Izard, M., Pantz, r. and Sangchote, S. 1990. Interaction of post-harvest disease control treatments and gamma radiation an mangoes. Annals of applied Biology. 116: 245-257.

Johnson, G.I. and Coates, I.M. 1993. Post-harvest diseases of mango. Post Harvest News and Information, 4: 27N-34N.

Johnson, G.I., Muirhead, I., Mayers. P. and Cooke. A.W 1989. Diseases. In: R. Rideway, (ed) Mango Pests and disorders. Department of Primary Industries, Brisbane, pp. 1-9.

Johnson, G.I., Sanghote. S. and Cooke, A.W. 1990, Control of Stem-end-rot (*Dothiorella dominicana*) and other post-harvest diseases of mangoes (cv. Kansignton Pride) during short and long-term storage. Tropical Agriculture, 67: 183-187.

Johnson, G.J., Sharp. J.L., Milne, D.L and Oosthuyse, S.A. 1997. Chapter 14. Post-Harvest Technology and Quarantine Treatments. In: R.E. Litz (Ed) The Mango-Botany, Production and Uses. CAB International. Willnig ford, 447-507.

Joshi, G.D. and Roy, S.K. 1985. Effect of integrated post-harvest handling on biochemical changes in Alphonso mango fruits. Progressive Horticulture, 17: 56-63.

Kane, O. and Marcellin.P 1978. Incidence of ripening and chilling on the mitochondria from mango fruits. Plant Physiology, 61: 634-638.

Ke. D. and Kader, A.A. 1992. Potential of Controlled atmosphere for post-harvest insect disinfestation of fruits and vegetables. Post-harvest News and Information,3:31N-37N.

Koid Sumi.K. 1937. Heat sterilization of Formosan fruits for fruit flies (111). Japanese Society of Tropical Agriculture, 9: 275-286.

Korsten, I., Devilliers, E.E. and Lonsdale, J.H. 1993. Biological control of mango post-harvest disease in the pack house. South African Mango Growers Association Year Book, 13: 117-121.

Lim, T.K. and Kuppelweiser.W. 1993. Mango Sapburn amelioration in the Northern Territory. Acta Horticulturae, 341:518-527.

Lonsdale. J.H, 1993. Strategies for the Control of post-harvest diseases of mango. Year Book of South African Mango Growers Association, 13: 109-116.

Macfarlane, J.J. 1996. Control of the Queensland fruit fly by gamma irradiation. Journal of Economic Entomology, 59: 884-889.

Mangan, R.L. and Ingle, S.J. 1992. Forced hot air quarantine treatment for mangoes infested with West Indian Fruit fly (Diptera: Tephritedae). Journal of Economic Entomology, 85: 1859-1864.

Mangaraj, S. and Varshney, A.C. 2006. Fruit Grading In India-A review: Indian Food Industry, 25(3): 46-52.

Mc Guire, R.G and Campbell. C.A. 1993. Imazalil for post-harvest control of anthracnose on mango fruits. Acta Horticulturae, 341:371-376.

Milne, D.I., KoK, I.B., Thomas, A.C. and Swarts, D.H. 1977. Inactivation of mango seed weevil, *Sternochatus mangiferae* by gamma radiation. Citrus and Subtropical Fruit Journal, 518: 11:13-15.

Muirhead, I.F. 1976. Post-harvest control of mango anthracnose with benomyl and hot water. Australian Journal of Experimental Agriculture and Animal Husbandry, 16.600-603.

Muirhead I.F. and Grattidge. R. 1986. Post-harvest diseases of Mango-The Queensland Experience. Proceedings of the First Australian Mango Workshop. CSIRO, Melbourne. pp 248-252.

Nascimento, A.S., Malavasi.A., Morgante, J.S. and Duarte, A.L.A. 1992, Hot water immersion treatment for mangoes infested with *Anastrepha fraterculus*, *A.oblique* and *Ceratitis capitata* (Diptera: Tephritidae) in Brazil. Journal of Economic Entomology, 85: 456-460.

New, J.H., Proctor, J.E. and Hewit, V.S. 1978. Packaging of Horticulture Produce for Export. Tropical Science, 20(1):213.

O'Hare, T.J. 1994. The susceptibility of Thai and Australian mango cultivars to sap injury and possible means of control. In: G.I. Johnson and Highley, E.(eds). Development of Post-Harvest Handling Technology for Tropical Tree Fruits. ACIAR Proceedings 58, Caniberra, pp.21-24.

O'Hare, T.J., Mc Laughlan, R.I. and Turner, C.D. 1994. Effect of co-shipment of exotic tropical fruit on existing fruit transported in bulk-shipment. Proceedings of the Australian Post-harvest Conference, University of Queensland, Gatton, pp. 245-250.

O'Hare, T.J. and Prasad.A. 1993, The alleviation of sap-induced skin injury by Calcium hydroxide. Acta Horticulturae, 321:372-381.

Oosthuyse, S.A. 1997. In transit temperature relationships of two shipping containers loaded with mangoes, South African Mango Growers Association Year Book.17:90-94.

Pandey, B.K. and Tandon, D.K. 2007. Protocol for post-harvest management of mangoes. Indian Horticulture, 52 (3):18-19.

Prasad Rao, G.S. and Chadha, K.L. 1982. Studies on grading of Dashehari mango for size. Indian Journal of Horticulture, 39(1/2):36-39.

Pruthi, J.S.1992. The Mango Book-I. Production, Post-Harvest Technology and Export Marketing: Agricultural and Processed Food Products Export Development Authority (APEDA) New Delhi, 105-112.

Quimio, A.J. and Quimio, T.H. 1973. Post-harvest control of Philippine mango anthracnose by hot water treatment, Philippine Agriculturist, 58: 138-146.

Quroshi, S.V. and Meah,M.B.1991. Post-harvest loss in mango owing to stem-end-rot. International Journal of Tropical Agriculture, 9(2):98-105.

Rawal, R.D. 1997. Management of fungal diseases in Tropical Fruits. In: Tropical Fruits in Asia. Proceedings of IPGRI-ICAR-UTFANET Regional Training Course, IIHR, Bangalore, India, p. 217-218.

Roy, S.K. and Pal, R.K. 1991. A Low-cost zero energy cool chamber for short term storage of mango. Acta Horticulturae, 291: 519-524.

Roy, S.K. and Pal, R.K. 2000. Post-harvest handling of fresh Horticultural Produce: The Hi-Techway. Indian Horticulturae, 45 (1): 13-17.

Sanghote, S. 1991. Botrydiplodia stem-end-rot of mango and its control. Acta Horticulturae, 291: 296-303.

Sawant, C.K. and Shinde, S.B. 2000. Post-harvest handling of Alphonso mangoes in Maharashtra: India, Acta Horticulturae, 509: 459-466.

Schoorl.D. and Holt, J.E. 1982. Planning for change in horticulture- A case study in packing shed design. Agricultural System, 8: 77-86.

Sein, F.J. 1935. Heat utilization of mangoes and guavas for fruit flies. Journal of the Agriculture of the University of Puerto Rico, 19: 105-115.

Seo, S.T., Kobayashi, R.M., Chambers, D.L., Steiner, L.F., Lee, C.Y.L and Komura.M. 1974. Mango Weevil, Cobalt 60 gamma irradiation of packaged mangoes. Journal of Economic Entomology, 67: 504-505.

Sethi, V. and Maini, S.B.1989. Appropriate Technology for reducing Post-harvest losses in Fruits and Vegetables. Indian Food Packer,43 (2):42-56.

Sharp, J.L. 1992. Hot-Air Quarantine Treatment for mango infested with carribean fruit fly (Diptera: Tephritidae). Journal of Economic Entomology, 85: 2302-2304.

Shorter. A. and Joyce, D.C. 1994. Effect of surface coatings on sapburn of 'Kensington Pride' mango Fruit . Tropical Agriculture, 71: 243-246.

Shukla, R.P. and Tandon, P.L 1985. Bio-ecology and Management of the mango weevil. *Sternochetus mangiferae* (Fabricius) (Coleoptera: Curculionidae). International Journal of Tropical Agriculture, 3: 293-303.

Singh, B.P., Kalra, S.K. and Tandon, D.K. 1989. Effect of size grading on storage of mangoes. Indian Journal of Horticulture, 46 (2): 54-60.

Singh, S. 2004. Effect of size grading on quality and storage of mango cv.Gulabkhas. Scientific Horticulture, 9:15-20.

Smith, E.S.C., 1992. Fruit fly disinfestations in mangoes by hot water dipping. In: P.A.C. Ooi., G.S. Lim and P.S. Teng. (eds). Proceedings of the 3rd International Conference on Plant Protection in the Tropics. Malaysian Plant Protection Society No. 6. Kaula Lumpur, PP. 188-195.

Sohi, H.S., Sokhi, S.S. and Tewari, RP. 1973. Studies on storage rot of mango caused by *Colletotrichum gloeosporioides* and its control. Phytopathologia Mediterranea, 1291: 114-116.

Spalding, D.H., Benschoter, C.A., von Windeguth, D.L., King. J.R., Reeder, W.F and Burdtt, A.K.Jr.1997. Methyl bromide and phosphine fumigation injury to avocado and mangoes. Proceedings of the Florida State Horticultural Society, 90: 268-270.

Spalding, D.H., and Reeder, W.F. 1986. Decay and acceptability of mangoes treated with combinations of hot water, imazalil, and radiation. Plant Disease, 70: 1149-1156.

Subramanyam, H., Gauri, S. and Krishnamurthy, S. 1976. Ripening behavior of mango fruits graded on gravity basis. Journal of Food Science and Technology, 13: 84-86.

Suhardjo.C. and Yuniarti. 1993. Hot water control of anthracnose on mango varieties, Arumanis, Golek and Mandagi. Proceedings of International Conference on 'Post Harvest Handling of Tropical Fruits' held at Chiang Mai, Thailand, 19-23 July 1993. pp. 453-454.

Swaine, G., Melksham, K.J. and Corcoran. R.J. 1984. Dimethoate dipping of 'Kensington' mango against Queensland fruit fly. Australian Journal of Experimental Agriculture and Animal Husbandary, 14:620-623.

Tandon, D.K., Kalra, S.K. and Singh, B.P. 1988. Ripening pattern of specific gravity graded 'Dashehari' mangoes. Indian Journal of Horticulture, 45 (3/4):219-223.

Thomas. A.C. 1975. The application of Ionizing Radiation to the shelf life extension of mangoes in South Africa. Republic of South Africa Energy Board, Pelinsdaba, Pretoria PEL: 244.

Unahawutti. U., Chettanas Chitara., Poomthong. M *et al* 1986. Vapour heat treatment for Nang Klangwam mango, *Mangifera indica* Linn infested with eggs and larvae of the oriental fruit fly *Dacus dorsalis* Hendel and the melon fly; *D. cucurbitae Coquillett* (Diptera: Tephritidae). Agriculture Regulation Division, Department of Agriculture, Ministry of Agriculture and Co-Operatives, Bangkok.

Varshney, A.C., Mangaraj.S. and Birla.S. 2002. Fruit Grading practices-Status and future assessment. Technical Bulletin No. CIAE/2002/90. Central Institute of Agricultural Engineering, Bhopal, India.

von Windeguth, D.L. 1986. Gamma irradiation as a quarantine treatment for Caribbean fruit fly infested mangoes. Proceedings of Florida State Horticultural Society, 99: 131-134.

Wells.I. and Littlemore, J. 1989. Dipping mangoes, Watch the Benlete concentration. Queensland Fruit and Vegetables News, 60:18.

Yahia, E.M 1994. The Potential use of insecticidal Atmosphere for Mango, Avocado and Papaya Fruits. In: B.C Champ., E. Highley and G.T Johnson (eds) Post-Harvest Handling of Tropical Fruits. ACIAR Proceedings 50, Caniberra, pp.373-374.

ANNEXURE

Schedule II
(see Rules 2, 3 and 4)
Grade Designations and Definition of Quality of the ALPHONSO variety of Indian Mangoes

Grade Designations	Definition of Quality	
	Special Characteristics	**General Characteristics**
	Minimum weight of each fruit*	
Choice	280 g	1. Each Mango shall
Special Good Standard	230 g 160 g 120 g	(a) Have reached a stage of maturity which will permit the subsequent completion of ripening in the ordinary course of transport and marketing.
		(b) Have developed the characteristic colour of the variety,viz yellow/red skin when ripe, and all the fruits shall be reasonably uniform in colour throughout the pack.
		Have the shape normal to the variety and be free from malformation.
		(c) Have good keeping quality and be firm reasonably developed and in good condition
		(d) Be free from defects due to diseases or insects or injury affecting the keeping quality of the fruits.
		2. Windfall and Shrivelled fruits shall not be packed
		3. Blemishes not affecting the keeping quality shall be permitted. Provided no single mark is larger than 1-6 sq-cm. in area.
		4. The stalk shall be removed close to the fruit but not pulled out.
A tolerance of 10% shall be allowed for accidental errors in grading in respect of fruits which are below the specified limit to the extent of 30 g. A tolerance of 5% shall be allowed for accidental errors in grading. Blemishes include marks due to fungus diseases, insect pests, hailstorms spray etc.		

Schedule III

(see Rules 2,3 and 4)

Grade Designations and Definition of Quality of the DASHEHARI variety of Indian Mangoes

Grade Designations	Definition of Quality	
	Special Characteristics	**General Characteristics**
	Minimum weight of each fruit*	
Choice	250 g	1. Each Mango shall
Special Good	150 g 100 g	(a) Have reached a stage of maturity which will permit the subsequent completion of ripening in the ordinary course of transport and marketing.
		(b) Have developed the characteristic colour of the variety,viz yellow/red skin when ripe, and shape normal for the variety.
		(c) Have good keeping quality and be firm reasonably developed and in good condition
		(d) Be free from malformation and the defects due to diseases or insects or injury affecting the keeping quality.
		2. Windfall and Shrivelled fruits shall not be packed
		3. The stalk shall be removed close to the fruit but not pulled out.
		4. Blemishes not affecting the keeping quality shall be permitted. Provided no single mark is larger than 1-6 sq-cm. in area.

A tolerance of 10% shall be allowed for accidental errors in grading in respect of fruits which are below the specified limit to the extent of 30 g.

A tolerance of 5% shall be allowed for accidental errors in grading.

Blemishes include marks due to fungus diseases, insect pests, hailstorms spray etc.

Schedule V

(see Rules 2 and 3)

Grade Designations and Definition of Quality of the BANGALORA and TOTAPURI variety of Indian Mangoes

Grade Designations	Definition of Quality	
	Special Characteristics	**General Characteristics**
	Minimum weight of each fruit*	
Choice	410 g	1. Each Mango shall
Special Good	350 g 290 g	(a) Have reached a stage of maturity which will permit the subsequent completion of ripening in the ordinary course of transport and marketing.
		(b) Have developed the characteristic colour of the variety,viz yellow/red skin when ripe, and shape normal for the variety.
		(c) Have good keeping quality and be firm reasonably developed and in good condition
		(d) Be free from malformation and the defects due to diseases or insects or injury affecting the keeping quality.
		2. Windfall and Shrivelled fruits shall not be packed
		3. The stalk shall be removed close to the fruit but not pulled out.
		4. Blemishes not affecting the keeping quality shall be permitted. Provided no single mark is larger than 1-6 sq-cm. in area.
A tolerance of 10% shall be allowed for accidental errors in grading in respect of fruits which are below the specified limit to the extent of 30 g. A tolerance of 5% shall be allowed for accidental errors in grading. Blemishes include marks due to fungus diseases, insect pests, hailstorms spray etc.		

Schedule VI

(see Rules 2 and 3)

Grade Designations and Definition of Quality of the BANGANAPALLI and SAFEDA variety of Indian Mangoes

Grade Designations	**Definition of Quality**	
	Special Characteristics	**General Characteristics**
	Minimum weight of each fruit*	
Choice	165 g	1. Each Mango shall
Special Good	130 g 95 g	(a) Have reached a stage of maturity which will permit the subsequent completion of ripening in the ordinary course of transport and marketing.
		(b) Have developed the characteristic colour of the variety, viz yellow/red skin when ripe, and shape normal for the variety.
		(c) Have good keeping quality and be firm reasonably developed and in good condition
		(d) Be free from malformation and the defects due to diseases or insects or injury affecting the keeping quality.
		2. Windfall and Shrivelled fruits shall not be packed
		3. The stalk shall be removed close to the fruit but not pulled out.
		4. Blemishes not affecting the keeping quality shall be permitted. Provided no single mark is larger than 1-6 sq-cm. in area.
A tolerance of 10% shall be allowed for accidental errors in grading in respect of fruits which are below the specified limit to the extent of 30 g. A tolerance of 5% shall be allowed for accidental errors in grading. Blemishes include marks due to fungus diseases, insect pests, hailstorms spray etc.		

Schedule VII

(see Rules 2 and 3)

Grade Designations and Definition of Quality of the FAZLI variety of Indian Mangoes

Grade Designations	Definition of Quality	
	Special Characteristics	**General Characteristics**
	Minimum weight of each fruit*	
Choice	410 g	1. Each Mango shall
Special Good	350 g 290 g	(a) Have reached a stage of maturity which will permit the subsequent completion of ripening in the ordinary course of transport and marketing.
		(b) Have developed the characteristic colour of the variety,viz yellow/red skin when ripe, and shape normal for the variety.
		(c) Have good keeping quality and be firm reasonably developed and in good condition
		(d) Be free from malformation and the defects due to diseases or insects or injury affecting the keeping quality.
		2. Windfall and Shrivelled fruits shall not be packed
		3. The stalk shall be removed close to the fruit but not pulled out.
		4. Blemishes not affecting the keeping quality shall be permitted. Provided no single mark is larger than 1-6 sq-cm. in area.
A tolerance of 10% shall be allowed for accidental errors in grading in respect of fruits which are below the specified limit to the extent of 30 g. A tolerance of 5% shall be allowed for accidental errors in grading. Blemishes include marks due to fungus diseases, insect pests, hailstorms spray etc.		

Schedule VIII

(see Rules 2 and 3)

Grade Designations and Definition of Quality of the BANGANAPALLI, KHAJRI and MIGGO variety of Indian Mangoes

Grade Designations	Definition of Quality	
	Special Characteristics	**General Characteristics**
	Minimum weight of each fruit*	
Choice	410 g	1. Each Mango shall
Special Good Standard	350 g 290 g 235 g	(a) Have reached a stage of maturity which will permit the subsequent completion of ripening in the ordinary course of transport and marketing.
		(b) Have developed the characteristic colour of the variety, viz yellow/red skin when ripe, and shape normal for the variety.
		(c) Have good keeping quality and be firm reasonably developed and in good condition
		(d) Be free from malformation and the defects due to diseases or insects or injury affecting the keeping quality.
		2. Windfall and Shrivelled fruits shall not be packed
		3. The stalk shall be removed close to the fruit but not pulled out.
		4. Blemishes not affecting the keeping quality shall be permitted. Provided no single mark is larger than 1-6 sq-cm. in area.
A tolerance of 10% shall be allowed for accidental errors in grading in respect of fruits which are below the specified limit to the extent of 30 g. A tolerance of 5% shall be allowed for accidental errors in grading. Blemishes include marks due to fungus diseases, insect pests, hailstorms spray etc.		

Schedule IX
(see Rules 2 and 3)
Grade Designations and Definition of Quality of the GOPAL BHOG, AGRA MOHAN BHOG, MURSHAPET, LAKSHMAN BHIG, KUSPAHARI and BHARTI variety of Indian Mangoes

Grade Designations	Definition of Quality	
	Special Characteristics	General Characteristics
	Minimum weight of each fruit*	
Choice	350 g	1. Each Mango shall
Special Good Standard	290 g 235 g 175 g	(a) Have reached a stage of maturity which will permit the subsequent completion of ripening in the ordinary course of transport and marketing.
		(b) Have developed the characteristic colour of the variety,viz yellow/red skin when ripe, and shape normal for the variety.
		(c) Have good keeping quality and be firm reasonably developed and in good condition
		(d) Be free from malformation and the defects due to diseases or insects or injury affecting the keeping quality.
		2. Windfall and Shrivelled fruits shall not be packed
		3. The stalk shall be removed close to the fruit but not pulled out.
		4. Blemishes not affecting the keeping quality shall be permitted. Provided no single mark is larger than 1-6 sq-cm. in area.
A tolerance of 10% shall be allowed for accidental errors in grading in respect of fruits which are below the specified limit to the extent of 30 g. A tolerance of 5% shall be allowed for accidental errors in grading. Blemishes include marks due to fungus diseases, insect pests, hailstorms spray etc.		

Schedule X

(see Rules 2 and 3)

Grade Designations and Definition of Quality of the SWARNA REKHA and MUNDAPPA variety of Indian Mangoes

Grade Designations	Definition of Quality	
	Special Characteristics	**General Characteristics**
	Minimum weight of each fruit*	
Choice	290 g	1. Each Mango shall
Special Good	235 g 175 g	(a) Have reached a stage of maturity which will permit the subsequent completion of ripening in the ordinary course of transport and marketing.
		(b) Have developed the characteristic colour of the variety,viz yellow/red skin when ripe, and shape normal for the variety.
		(c) Have good keeping quality and be firm reasonably developed and in good condition
		(d) Be free from malformation and the defects due to diseases or insects or injury affecting the keeping quality.
		2. Windfall and Shrivelled fruits shall not be packed
		3. The stalk shall be removed close to the fruit but not pulled out.
		4. Blemishes not affecting the keeping quality shall be permitted. Provided no single mark is larger than 1-6 sq-cm. in area.

A tolerance of 10% shall be allowed for accidental errors in grading in respect of fruits which are below the specified limit to the extent of 30 g.

A tolerance of 5% shall be allowed for accidental errors in grading.

Blemishes include marks due to fungus diseases, insect pests, hailstorms spray etc.

Schedule XI
(see Rules 2 and 3)
Grade Designations and Definition of Quality of the MALDA variety of Indian Mangoes

Grade Designations	Definition of Quality	
	Special Characteristics	**General Characteristics**
	Minimum weight of each fruit*	
Choice	350 g	1. Each Mango shall
Special Good	290 g 235 g	(a) Have reached a stage of maturity which will permit the subsequent completion of ripening in the ordinary course of transport and marketing.
		(b) Have developed the characteristic colour of the variety,viz yellow/red skin when ripe, and shape normal for the variety.
		(c) Have good keeping quality and be firm reasonably developed and in good condition
		(d) Be free from malformation and the defects due to diseases or insects or injury affecting the keeping quality.
		2. Windfall and Shrivelled fruits shall not be packed
		3. The stalk shall be removed close to the fruit but not pulled out.
		4. Blemishes not affecting the keeping quality shall be permitted. Provided no single mark is larger than 1-6 sq-cm. in area.
A tolerance of 10% shall be allowed for accidental errors in grading in respect of fruits which are below the specified limit to the extent of 30 g. A tolerance of 5% shall be allowed for accidental errors in grading. Blemishes include marks due to fungus diseases, insect pests, hailstorms spray etc.		

Schedule XII

(see Rules 2 and 3)

Grade Designations and Definition of Quality of the MALGOA variety of Indian Mangoes

Grade Designations	Definition of Quality	
	Special Characteristics	General Characteristics
	Minimum weight of each fruit*	
Choice	250 g	1. Each Mango shall
Special	200 g	(a) Have reached a stage of maturity which will permit the subsequent completion of ripening in the ordinary course of transport and marketing.
		(b) Have developed the characteristic colour of the variety, viz yellow/red skin when ripe, and shape normal for the variety.
		(c) Have good keeping quality and be firm reasonably developed and in good condition
		(d) Be free from malformation and the defects due to diseases or insects or injury affecting the keeping quality.
		2. Windfall and Shrivelled fruits shall not be packed
		3. The stalk shall be removed close to the fruit but not pulled out.
		4. Blemishes not affecting the keeping quality shall be permitted. Provided no single mark is larger than 1-6 sq-cm. in area.

A tolerance of 10% shall be allowed for accidental errors in grading in respect of fruits which are below the specified limit to the extent of 30 g.

A tolerance of 5% shall be allowed for accidental errors in grading.

Blemishes include marks due to fungus diseases, insect pests, hailstorms spray etc.

CHAPTER 3

Storage of Fresh Fruits

M. Madhavi

Introduction

Concerted research efforts have aided to increase the production of mango during the last five decades, inspite of the inherent alternate bearing tendency in certain commercial cultivars, mango malformation etc., but the purpose of obtaining maximum production will not serve good unless the increased production is supplemented with similar efforts to minimize their post-harvest losses and extend their storage life, particularly in the years of higher production, in order to extend the supply of fruits to the consumer in a better quality condition. High temperature and low humidity prevailing at harvesting result in fast ripening of fruits after harvest. Under these conditions, it becomes difficult to hold the fruit long in fresh condition, without deterioration in appearance and quality.

Extension of storage life of fruits is achieved by delaying their ripening and checking their biochemical activity, so as to keep them in fresh condition after they have fully matured, for as long period as possible without impairing their quality. The storage of fruits involves reduction of temperature, maintaining the right humidity, reduction of respiration by reducing the oxygen levels and removal of ethylene around the fruit. Storage is generally done for an orderly marketing to extend the duration of their availability and tide over the glut. The post-harvest losses, which range between 30 and 40 % in mango (Salunkhe and Desai, 1984) can be minimised by storage. The spoilage during storage can be checked or reduced by controlling the post-harvest environmental conditions around the fruits by skin coatings, wrappings, fungicidal treatments and storage at low temperature.

The principal aim of storage of fruits and vegetables is to control the rate of respiration, transpiration, ripening, any other biochemical changes and disease infection and to preserve the commodity in its most acceptable and usable form for consumption by consumers. The object of fruit storage is not only to prolong the life of the fruit, but also to maintain the characteristics like flavour, taste and appearance.

Thus the storage of fruits is required :

(i) to avoid glut in the market during peak production years,

(ii) to extend the availability of fruit beyond the main season,

(iii) to avoid post-harvest losses, as they are highly perishable and prone to incidence of diseases,

(iv) to supply clean and healthy fruits of better quality,

(v) for long distance transport and

(vi) to regulate the supply to the market.

Since ancient days, a number of techniques of preservation, and storage have been developed (ambient storage, refrigerated storage, air-cooled storage, zero-energy storage, controlled atmosphere storage, modified atmospheric storage, hypobaric storage etc) and are being used for fruits and vegetables depending upon the nature of commodity and the storage period intended. The technologies capable of extending the post-harvest life of climactric fruits (banana, mango etc.) at ambient temperature are urgently needed, particularly in developing countries, where the availability of cold storage facilities and controlled/modified storage facilities are either lacking/limited.

METHODS OF STORAGE

The several methods of storage of perishables (fruits and vegetables) can be classified as (i) physical and (ii) chemical methods and also,

(i) low cost and

(ii) high cost storage methods.

Physical methods include waxing, wrapping, skin coatings, storage in polyethylene bags, hypobaric storage and irradiation, while chemical methods include treatments with calcium, growth regulators and use of ethylene absorbants.

A. Physical Methods

Waxing

In nature, a waxy-coating develops on the fruit surface, while still it is on the tree, as a protective measure against inclement weather and microbial incidence. However, this waxy layer slowly disappears after harvest during various handling operations and the fruit becomes increasingly vulnerable to spoilage. Hence, it was conceived that by providing an artificial waxy layer to the fruit surface, it can be protected from spoilage and the ripening can be delayed and thereby storage life of the fruit can be extended.

A cheaper system of delaying ripening and extending the shelf life of mango fruits etc. is the use of waxes and other skin coating materials. Waxing is especially important in case of fruits and vegetables having tiny/minute injuries and scratches on their surface. Waxing also helps in extending the transit period and availability of the commodity in the market.

Purposes of waxing

1. Reduction of transpirational loss of moisture (30-35%) from the harvested produce by waxing the stem-end-scar and fruit surface.
2. Reduction in the rate of respiration, moisture and weight loss and modifying internal atmosphere (Brown, 1986) and thereby extension of shelf life.
3. Maintenance of quality by inhibition of dehydration and impart shining.
4. Prevention of microbial spoilage, as waxing seals all the entries (stomata, lenticels, injuries etc.) on the surface of the commodity and avoids access to the microbes.
5. Enhancement of marketability of the perishable produce by giving glossy appearance.
6. Overcoming the glut in the markets by extending shelf life.

Advantages

1. Improved and attractive appearance of the fruits and vegetables.
2. Reduction in moisture losses and retardation of wilting and shrivelling during transport and storage.
3. Reduction in spoilage especially due to chilling injury and browning in low temperature storage.
4. Reduction in the rate of respiration, through lowering the availability of oxygen as result of creation of a diffusion barrier on the surfaces.
5. Protection of commodities from microbial infections.
6. Absence of interference with the natural appearance and the quality of the commodity. Waxing, however, does not improve the quality of any inferior product.
7. Cost effective substitute in the reduction of spoilage in the absence of refrigerated storage.
8. Can be employed both under normal and refrigerated condition.
9. Waxing of the commodity overcomes the problem of market glut and long distance transportation of the perishables.

Disadvantages

1. Maintenance of chlorophyll levels for longer time, which detracts the consumer to purchase (Ilangantileke *et al* 1989).
2. Development of off-flavour due to inhibition of O_2 and CO_2 exchange, resulting in anaerobic respiration and elevated ethanol and acetaldehyde contents (Sharma and Singh, 2000).
3. Disrupts respiration and ethylene production.

Types of waxes

Waxes are esters of higher fatty acids with monohydric alcohols and hydrocarbons and some free fatty acids. There are several types of waxes from which wax emulsions (WE) are made viz., microcrystalline, paraffin, carnauba, sisal, sugarcane, bee-wax, shellac, wood resins and unemulsifiable polyethylene. Emulsion waxes are composed of a natural wax such as carnauba or paraffin or synthetic wax like polyethylene emulsion. A suitable emulsion can be made out of refined hard sugarcane wax which is a constituent of press mud or lime sludge.

Waxes are mainly classified as follows :

Solvent wax: A wax based on one or more resins dissolved in a blend of petroleum solvents with plasticisers and film forming agents added to improve shine and film flexibility. The solvent readily evaporates and a thin film is left on the surface of the produce.

Water wax: There are two kinds of water waxes i.e., resin solution and emulsion based one. Resin solution waxes are composed of one or several soluble resins or resin-like materials, such as shellac or natural gums. They can be dissolved in water with or without the aid of other water-miscible solvents such as alcohols and glycerine. Emulsion waxes are composed of a natural wax such as carnauba or paraffin or a synthetic wax like oxidised polyethylene emulsifier in a soap or detergent (Ben-Yehoshua, 1991).

Water wax emulsions are safer to use than solvent waxes, which are highly flammable. Water wax emulsion can be used without drying the fruit or vegetable prior to the treatment, therefore, washing and wax treatments can be combined.

Carnauba wax, sugarcane wax, thermoplastic resins, chitosan based coating and shellac are used commercially; however, waxes of plant origin are preferable (Ketsa and Prabhasavat, 1992., Bower *et al* 2003).

Commercial wax emulsions are usually proprietary formulations, based on paraffin, bees wax or carnauba wax, emulsified in water, with or without shellac, gum, resin or mineral oil. Food grade waxes are used to replace some of the natural waxes removed during harvesting and post-harvest handling operations.

Preparation of wax emulsion (WE): Emulsion is a homogenous suspension of tiny droplets of oil and water or of water in oil. Wax is emulsified by different emulsifiers like ammonium linoniate etc.

The WE is prepared by melting micro-crystalline paraffin and carnauba wax along with emulsifiers. The WE is made from 500g wax blocks. The block is dropped in 500 ml boiling water and dissolved. When the solution is completely cool, about 11 litres soft cold water is added to it. The emulsion is now ready for use. The emulsion also comes in tin containers with 12% solids, which is to be diluted with 3 parts of cold soft water before use. Where soft water is not available, river water may be used (Pruthi, 1992).

Concentrated WE containing 12% solids, is diluted with cold soft water to make 4% emulsion, suitable for mangoes.

Wax emulsion with fungicides and growth regulators: In order to reduce the incidence of diseases/or to minimize spoilage in bruised fruits and vegetables, fungicides are at times mixed with WE before application. Sodium orthophenyl phenate (SOPP) (1%) or Flit 406 (50% n-trichloromethyl thio tetra phthelamide) (0.2%) could be added to WE. It checks the microbial spoilage. However, if fruit carries incipient infection fungicidal WE may only delay the development of such infection but fails to control the decay caused by seed infection during ripening.

The addition of growth-regulators like maleic hydrazide (1000-1500 ppm) further reduces the ripening rate, thus extending the storage life of the produce (Srivastava, 1967). Bleached shellac or a non-ionic resin can be mixed with the WE to impart extra gloss to the produce (Pruthi, 1992).

Treatment of mango fruits with wax emulsion: Waxing may be applied by foaming, spraying, dipping or brushing of which dipping is the most simple, cheap and effective one and is most commonly employed. Foaming is a satisfactory means of application, because it leaves a very thin coating of wax on the fruits after drying. Waxy layer can be imparted to the fruits by dipping them in an aqueous WE for a specified duration and then taking out and drying the surface thoroughly to be free from residual moisture.

Dipping method: Fruits intended to be waxed after harvest are sorted and dipped in the diluted WE of stipulated strength for a minute, then taken out, drained to get rid of extra emulsion, arranged in rows to dry under shade either on floor plastered with cow dung or on waste news papers and allowed to dry for 10-15 minutes. The fruits should be given 2-4 turnings while drying. Drying can also be done quickly with the help of an electric air-blower or a fan.

Each fruit after treatment picks up enough WE. The treated fruit can be distinguished only by its glossy and better appearance. The amount of wax left on the fruits being extremely small, there is no need to wash it off.

Mechanical wax-emulsion applicator

A simple device for application of WE in orchards where power is not available, has been developed at the Central Food Technological Research Institute, Mysore (India). It consists of a tank with movable perforated carrier, which after loading is immersed in the WE. After treatment, the carrier moves up in an inclined position when a handle is moved and the treated fruits roll out on a gravity conveyor. At the end of the conveyer, the fruits are arranged in rows for drying before packing. This unit will help in treating large number of fruits in a short time and can be operated in remote places, being portable (Pruthi, 1992).

Concentration of Wax Emulsion for mango

The concentration of WE for mangoes recommended/suggested varied from 3% (Sharma and Kapoor, 1967) to 12% (Sadasivam *et al* 1970). With WE (8%) fruits stored better (Sadasivam *et al* 1970). However, when the concentration of WE is 5% or higher, treated fruits developed off flavour (Passam, 1982).

However, the effective concentration of WE varied with the cultivar. WE containing 8, 10 and 12% solids was satisfactory in case of 'Bangalora', in delaying ripening and prolonging shelf life, while in case of 'Neelum', the results are in favour of 6 and 8% WE (Sadasivam *et al* 1970). The fresh 'Alphonso', fruits treated with 6% WE showed the lowest spoilage. 'Alphonso' fruits dipped in 1.7 to 2.7% aqueous emulsion of fungicidal wax containing 0 – phenyl phenol showed less PLW and extended storage life. The differential response exhibited by different cultivars may be due to the differences in the characteristics of the fruit surface. Hence, it is possible that different mango cultivars may need different concentrations of WE (Sadasivam *et al* 1970). Late cultivars do not respond as satisfactorily as early-or mid season cultivars to WE treatment (Ram and Date, 1971). Hence tolerance levels of each cultivar to the concentration of WE should be worked out.

Precautions

1. WE should be prepared and diluted with soft cold water.
2. Fruits should be dry and healthy, otherwise foaming of water emulsion waxes may occur.
3. Diseased fruits should not be used.
4. An uniform flow of fruit through the wax emulsion applicator needs to be maintained. Fluctuations result in some fruit receiving too much/too little wax.
5. Complete coverage of the entire fruit surface is essential. Patchy application results in less effectiveness of wax.
6. The brushes in mechanical applicator should be kept soft by regular washing with hot water.

Mode of action of wax emulsion

Application of WE to fruits protects them against excessive loss of moisture, high rate of respiration, build up of heat or thermal decomposition as a result, texture and quality of fresh produce are maintained as in the fresh condition as long as possible. When the rate of respiration is reduced, all the biochemical processes/changes taking place in the fruit are retarded/checked, there by the ripening is delayed and the shelf life of the treated fruit is prolonged (Agnihotri *et al* 1963., Garg and Ram, 1973). WE imparts an appealing glossyness to the fruit surface and this was retained even after three weeks by which time the unwaxed fruits were dull and completely wilted (Sadasivam *et al* 1970). Increased appearance of waxed fruit was mainly due to maintenance of turgidity in the fruits (Fonseca *et al* 2000).

Use of waxes with or without others for storage

Wax emulsion (WE) with or without fungicides, growth regulators etc. is used for retarding/delaying ripening and to extend the storage life of mango fruits (see below).

Several workers have demonstrated extension of storage life of treated mango fruits with wax mainly by delaying the ripening of fruits. Dipping of `Fazli' mangos of `B' maturity stage for 10 seconds in molten paraffin wax at 80 °F, extended the storage life of the fruits to 44 days when stored for 42 days at 55 °F (12.5 °C) and 90% relative humidity and thereafter 2 days at room temperature. Treated fruits showed the same flavour, texture and colour as that of fresh mangoes (Bose and Basu, 1953). Coating mangoes with paraffin WE (7% solids) resulted in increased shelf-life (Mathur and Srivastava, 1955). Shelf-life of mangoes treated with 2.7% WE was 19 days compared with 12.5 days in untreated fruits (Mathur and Subramanyam, 1956). There has been an increase in shelf life by application of sugarcane wax (4%) emulsion at room temperature in cvs. Chausa and Langra (Srivastava, 1967). Shelf-life of waxed mangoes in Pakistan increased by more than 50% with less mould development, higher degreening and delayed ripening at ambient temperature (Sheikh *et al* 1977). WE treatment was the most effective treatment in increasing the storage life of `Langra' and `Himsagar' fruits to 9-10 days against 6 days in control (Roy *et al* 1980).

Least wastage was observed in waxed fruits (12% solids) at room temperature for 6 days. The unwaxed fruits stored for the same period at room temperature, showed a wastage of 24%. Fruits stored at low temperature after waxing with 9 and 12% WE showed no spoilage and least PLW as compared to the unwaxed ones (Sastry and Habibunnisa, 1979).

Mangoes pre-cooled to 15 °C and treated with WE (0.5%) could be kept up to 30 days in hypobaric storage without loss in quality after ripening (Ilangantileke *et al* 1989). 'Banganapalli' mango fruits dipped in 6% WE and stored under zero energy cool chamber (ZECC) condition recorded greater shelf life (Gautam *et al* 2003). Storage of waxol (8%) treated mango fruits under ZECC storage conditions showed maximum profit/benefit) (1.17 : 1 for Dashehari and 1.09 : 1 for Langra) compared with calcium chloride and maleic hydrazide treatments (Singh and Pandit, 2004). Fruits treated with waxol (6%) plus calcium chloride (2%) plus carbendazim (0.1%) could be kept up to 22 days in cool chamber compared to 16 days at room temperature followed by those treated with waxol (6%), plus calcium chloride (2%) plus Captan (0.2%) (Masalkar *et al* 2006).

Wax coating of `Kesar' mango fruits and packing in 100 gauge polyethylene bags and pre-cooling to 13 °C followed by storage at 18 °C enhanced the storage life up to 28 days without much change in quality and were highly marketable even at the end of the storage period of 28 days, whereas fruits packed in 350 and 225 gauge polyethylene bags decayed early. Wax coated un-packed fruits stored at 13 °C and 18 °C also showed promising results, but the fruits showed shrivelling and ripening during storage (Antala *et al* 2008).

The most acceptable storage life for uncoated fruits was 10 days while that of fruits coated with lac-based emulsion formulations was up to 18-20 days (Giri *et al* 2003). Cultivars showed differential response to different types of waxes. In case of 'Keitt' shellac and TAG were most effective in extending shelf life. Shellac decreased ground discolouration in `Keitt' and `Kent' cultivars. Disease incidence was enhanced and TSS content decreased in `Sensation' fruits. In `Keitt' TSS decreased with TAG. In `Sensation' taste was poor with shellac when all the quality parameters were collectively considered. Treatment with citrishine orange or TAG was associated with the highest percentage of good quality fruits (Oosthuyse, 1997c). The polysaccharide based coatings retarded the development of colour, lowered the quality of `Alphonso' fruits, but showed greater firmness values compared to waxol and control. Evolution of CO_2 and loss in weight were also reduced significantly. Chitoson based coatings were much superior in prolonging the shelf life and maintaining the quality of treated mango (Alphonso) fruits (Kittur *et al* 2001).

Coating of fruits with refined mineral oil, on the other hand, resulted in fruit injury (Mathur and Srivastava, 1955). The oil coating appeared to decrease respiration more than did waxing, resulting in anaerobic condition, severe enough to injure the fruit (Mathur and Subramanyam, 1956).

In order to reduce the incidence of diseases and/or to minimize spoilage of fruits, fungicides are often added to the WE before or after application. SOPP (1%) or Flit 406 (0.2%) could be effectively added to check the microbial spoilage in many commodities including mango. A combination of wax and fungicidal coatings will be more effective in enhancing the shelf life of fruits as compared to those fruits receiving either of them separately (Dalal *et al* 1970). Dipping of immature green mangoes (cv. Alphonso) in 1.7 and 2.7% aqueous solution of a fungicidal wax, containing 5% 0-phenyl phenol, micro-crystalline, petroleum-wax, terpene resin, Oleic acid and trietha nolamine, extended the shelf life (12-19 days) besides increasing the retention of ascorbic acid and moisture, delaying ripening, reduction of respiration and transpiration rates and delaying climacteric peak by 4 days. The treated fruits were acceptable till the end of storage life, although there was delay in colour development (Mathur and Subramanyam, 1956). Dipping mango fruits in Benomyl (0.5 g/litre) or in Peerless citrus wax (5%), Mobillor-C and Stay fresh and storage at 20 °C, extended the storage life, but disrupted respiration and ethylene production. The application of Peerless citrus wax to the pre-ripened fruits extended the shelf life from 11 days in control to 14 days with comparable quality (Brown, 1986). 'Kesar' mango fruits treated with wax coupled with carbendazim (0.6%) could be kept up to 26 days in cool chamber as against 20 days at room temperature (Dhemre and Wasker, 2003). 'Kesar' mango fruits treated as above and pre-cooled to 10 °C for 12 hours and then stored in cool storage recorded maximum shelf life of 50 days as compared to the un-pre-cooled ones which recorded a shelf life of 46 days, whereas cool chamber and room temperature-stored fruits recorded minimum storage life of 26 and 20 days respectively (Dhemre and Waskar, 2003).

The fruits of 'Langra' and 'Dusheri' stored well up to 18 days and 30 days respectively at ambient temperature following the dipping in an aqueous solution of Maleic hydrazide (1000 ppm) or gibberellic acid and coated with fruit wax (2.5%) and then packed in perforated cartons (Manzoor *et al* 1977). 'Beneshan' mango fruits dipped in WE (6%) containing GA_3 (200 ppm) and Bavistin (0.1%) and kept in perforated polyethylene bags remained longer in storage, with reduction in weight loss, acidity and respiration rate and increase in sugar content (Rangavalli *et al* 1993).

Hot water dip (52° $\pm$ 1 °C) for 5 minutes and wax coating with Frutox (8%) emulsion and dried in air, significantly increased the keeping quality of 'Neelum' mango fruits by over 50%. Hot-water treatment (HWT) showed pronounced effect on the total rot and anthracnose rot with respective reductions from 61.6 to 26.9% and from 37.5 to 16.0%. The combined treatment alone could record 59.3% of sound fruits, while the percent of sound fruits were as low as 2.0 and 4.0 respectively with HWT and control. HWT followed by skin coatings delayed the ripening by 8 and 12 days respectively (Narayana *et al* 1989).

Effect of waxing with or without adjuncts on quality parameters

Waxing of fruits and vegetables aims not only at extending the shelf life of treated fruits and vegetables but also at retaining their quality. Waxing coupled with others (fungicides, growth regulators etc.) showed influence on the different quality parameters, but in an inconsistent manner. The majority of quality contributing factors affected by waxing with or without adjuncts include : reduction in PLW, respiration rate, post-harvest spoilage and maintenance of quality in extended storage, firmness, colour, texture, acidity etc. and ripening processes are influenced to different extents by wax coatings.

The reduction of PLW in wax-treated mango fruits was due to controlling the transpiration and respiration processes of mango fruit during storage. The minute particles of WE clog the pores of fruit skin and thus reduce the water loss through transpiration from the fruit, in addition to formation of a continuous layer of wax on the fruit surface. This reduction in the moisture loss reduces dehydration, shrinking, wrinkling of treated fruits (Mathur and Srivastava, 1955., Mathur and Subramanyam, 1956; Garg and Ram, 1973., Bhullar *et al* 1984).

Waxing checked the loss of ascorbic acid, titratable acidity, while enhancing TSS and sugars in treated fruits (Garg and Ram, 1973). By retarding the rate of respiration and ripening, waxing showed minimum decline in acidity.

One of the disadvantages of using wax for delaying ripening as well as for prolonging storage was the retention of green colour of the fruit skin. Wax treatments maintained chlorophyll levels for upto 20 days, but high concentration injured the fruit skin after 20 days of hypobaric storage (Ilangantileke *et al* 1989). Breakdown of chlorophyll requires O_2 and the wax provided a physical barrier between the fruit and atmosphere, resulting in less availability of O_2, thereby reduced breakdown of chlorophyll and delayed degreening. Waxing reduces the synthesis of carotenoid pigments in the fruit pulp (Fonseca *et al.*, 2004a) again for want of O_2 (Spurgeon and Porter, 1980). However, waxing did not affect the carotenoid content of the peel (Fonseca *et al* 2004a).

Skin coatings

Waxes and other surface coatings have been used in handling fruits and vegetables. However, because of ever growing public resentment of toxic residues and adverse effects of certain coatings (off-flavour) (Cohen *et al* 1990) and ineffectiveness of the surface coatings in completely preventing weight loss (Ben-Yehoshua, 1985) surface coatings were discontinued and instead skin coatings were introduced. The emergence of so called "edible coatings" in the 1980's and natural origin coatings such as Chitosan (Stossel and Leuba, 1984; El-Ghaouth *et al* 1991, 1992) has renewed interest in skin coatings. Unlike surface coatings, 'skin coatings' with edible materials were considered as another possible packaging method for extending the post-harvest storage life of perishable. These coatings are generally made from lipids, resins, proteins and polysaccharides.

Skin coatings extend the shelf life of fruits and vegetables, retard water loss and reduce fungal decay (Baldwin, 1994). Skin coatings increase the shelf life under non – refrigerated conditions also. "Cling films" (PVC and LDPE films) are very thin polyethylene material, which adhere on the surface of the fruits and vegetables. Coatings with "Cling films" lead to reduction in PLW and increase in shelf life. By forming an extra coating on the surface of the fruits and vegetables, they check the respiration and transpiration losses. They also reduce decay losses by checking the entry and the growth of various microflora due to complete isolation of commodities from atmospheric conditions. Applying a semi-permeable membrane like edible coatings, can reduce the high respiration, ethylene production and moisture loss and they also serve as carriers of antimicrobial agents.

However, skin coatings cause physiological damage to the commodities, due to suspension of respiratory enzymes. This can be overcome by the spurt of respiration activity resulting from radiation. Skin coatings help to keep the respiration activity of irradiated fruits (mangoes) to the optimum, reducing physiological losses and increasing storage life of the fruits (Dharkar *et al* 1966b).

Properties of skin coatings

The skin coatings must possess the following functional properties:

1. Barrier properties in terms of moisture and gases (O_2 and CO_2).
2. Palatability and non-toxicity
3. Good mechanical properties.
4. Adequate stability in a moderate range of temperature and relative humidity.
5. Organoleptic properties similar to the food stuff with which it is to be used.

Types of skin coatings

Edible packagings (coating) are of two types depending on the technique used to pack the material viz., coatings and packaging in films. Coatings are applied and formed directly on the food stuff, whereas films are independent structures added to the food at a later stage (Kester and Fennema, 1986).

Coating/Filming Technology

Coating is applied by either brushing, dipping in the coating solution or by spraying the solution on to the produce. The setting of the coating material is achieved by drying or cooling. Cooling temperature/drying conditions should be regulated so that the film gets attached firmly, but not too rapidly, other wise, it results in defects such as formation of cracks, pinholes etc. Water based gloss-coating composition dries slowly.

Recently, Indian Institute of Technology, Delhi, India has developed a non-refrigerator coating technology that can retain freshness and can improve the shelf life of perishable up to 14 days without refrigeration and 30 days with refrigeration or irradiation. This coating technology, christined as 'Fru Wash" is made out of local raw materials (names not given, but said to be available only in India and Thailand) that can be converted into a water soluble, biodegradable emulsion. "Fru wash" is stable (2 years), cost effective (1% of the fruit cost), ecofriendly, edible and non-toxic. It forms a very thin (0.001-0.002 mm) layer on the surface of fruits and vegetables. It can be applied by dip or spray coating techniques with or without additives. The coated produce retains all its natural characteristics of quality (colour, texture, TSS, juice content, chewiness and softness etc.) and shows marked reduction in water loss. The thin film of the "Fru wash" coating protects the produce surface from diseases and pests and stops post-harvest changes due to oxidation. Thus, it gives better results than currently available wax based coatings which give an artificial appearance to the produce. The 'Fru wash' was very effective for non refrigerate shelf life extension of several fruits and vegetables at the Institute.

Mode of action of skin coatings

During coating (filming) operations, the liquid coating material flows in to stomatal, lenticel and other openings and plugs them effectively restricting the transport/permeability of gases and ethylene.

Skin coatings and mango

Use of edible coatings (Tal-Prolong, Semper fresh etc.) has made it possible to extend the shelf life of mangoes in various production centres of mango. Polysaccharide based coatings have shown other benefits, besides prolonging the shelf life.

'Tal-Prolong', a sugar based chemical (mixture of sucrose, esters of fatty acids and the sodium salt) is used for extension of shelf life of mangoes. A permeable membrane of coating can be produced on fruit surface by dipping the fruits in aqueous dispersion of 'Tal-Prolong'. Mangoes (cv. Julie) were treated with 0.75 and 1.0% (w/v) aqueous suspension of 'Tal-Prolong' and stored at 25 °C and 85-95% relative humidity and the following results were noted :

(i) treatment with 0.75% did not adversely affect the sensory quality of the coated mango fruits,

(ii) reduction in weight loss without adversely affecting the sensory quality of the coated mangoes,

(iii) treatment with 1.0% level increased ethanol formation in the fruit pulp, and

(iv) retardation of ripening and resultant extension of storage life (Dhalla and Hanson, 1988).

Mango fruits (cv. Dashehari and Mallika) treated with 1.5% 'Tal-Prolong' exhibited least PLW, during storage followed by 1% concentration. Both concentrations reduced the rate of increase in TSS throughout the storage period. Though effective concentration differed with the cultivar and the concentration differed in their influence on quality parameters of both cultivars, fruits subjected to 1% treatment retained better overall acceptability as compared to 1.5% in respect of colour, flavour, texture and taste (Hudder *et al* 1989). Treatment with 'Prolong' and ethylene application of 'Haden' mango fruit resulted in good colour and uniform ripening after 20 days storage at 12° or 18 °C, but had no effect on TSS (Mangano *et al* 1997), but delayed ripening of 'Alphonso' fruits by 4 to 5 days at 1% concentration. This delayed ripening was correlated with reduced softening, development of yellow colour of the peel, acidity and increase in TSS in treated fruits. However, normal ripening of fruits was affected with 1.5 and 2.0% concentration of 'Prolong'. The treated fruits remained green and soft with higher percentage of spoilage. Delayed ripening of 'Alphonso' mangoes with 1% 'Prolong' was primarily due to inhibition of yellow colour development and delaying of some of the ripening changes (Santha, 1989).

Mango fruits treated with 2 and 3% 'Semper fresh' caused off flavour within 6 days of treatment in storage, while 1% treated fruits showed normal ripening and were acceptable (Ketsa and Prabhasauat, 1992). 'Samper fresh' coated fruits showed minimum PLW and it was least with 3% concentration which was due to reduction of transpiration rate, as a result of blockage of lenticles etc. and stem-end-scar (Ketsa and Prabhasauat, 1992).

'Haden' mangoes treated with various fatty acid solutions, took longer time to reach climacteric peak, but did not show differences in weight loss, pulp colour development, TSS accumulation and acidity loss. However, firmness loss after 15 days storage was less in treated fruits, but their appearance was better. Fruits coated with 2% fatty acids plus 1% carbohydrates plus calcium cascinate retained a fresh appearance for the longest time (Baez sanuado *et al* 1997a).

Mangoes coated with chitosan (0.5%) and carboxymethyl cellulose (1%) maintained best quality, while butter emulsion produced off flavour. All these three coatings with propylene glycerol (2%) and butter emulsion (butter : water = 4 : 1) reduced water loss and respiration rate, delayed softening, and colour development of peel and pulp. TSS, reducing and total sugars and titratable acidity were lower in coated fruits. The respiratory pattern of fruits coated with chitosan (0.5%) and CMC (1%) showed climacteric rise. Fermentation was induced in fruits coated with butter emulsion (Phaiphan and Raltanapanone, 2008).

Although skin coatings have been widely used since long, there exist several problems, which impede their application. First, current consumer trends towards additive- free foods; second, coatings some times inadequately reduce transpiration, but are effective in restricting gases diffusion, causing off flavours ; third, coatings faster some physiological disorders such as scald in some fruits like in pears. However, the emergence of edible and natural-origin coatings have renewed interest in skin coatings and greater commercial application in future is anticipated. Additionally, as a relatively low cost technology, skin coatings may have more potential application in developing countries, where refrigerated storage is either lacking or not affordable if available.

WRAPPINGS

Wrapping of fruits is a type of packaging intended to protect them from spoilage, shrivelling, moisture and weight loss, to prevent or delay ripening and finally to extend the storage life. Wrapping is done with materials like plastic films, polyethylene films, PVC film, heat shrinkable film etc. Commodities are wrapped individually or as a lot. Wrapping of fruits and vegetables provides a modified atmosphere around the fruit in which concentration of oxygen is lowered and that of carbondioxide is increased. Wrapping also maintains attractive appearance of the commodities (Hall *et al* 1975). The feasibility of wrapping with various plastic film etc. as a technique for extending the storage life of mango fruits was investigated by many workers individually or with combination. The use of selective films for prolonging the shelf life of mango fruits has increased greater in recent years, in view of International trade. Selective film wraps like PVC and LDPE have increased the storage life of mango and also alleviated chilling injury in the fruits (Castro *et al* 2005). Storage of mangoes in wrapping paper or films regulates humidity and the concentration of O_2, CO_2 and ethylene inside the package; reduces weight loss and colour development and eventually delays ripening (Macnish *et al* 1997).

Wrapping of mangoes in polyethylene film will increase their storage life (Mukherjee, 1961), but wrapped fruits had lower soluble solids, higher acidity, less peel and pulp colour and less disease development (Thompson, 1977). However, in another trial, wrapping with polyethylene film showed a decrease in weight loss without increase in storage life (Miller *et al* 1983), which was attributed to poor sealing of the wrapper and the presence of pin holes in the film, both of which would have prevented major changes in O_2 and CO_2 levels within the wrap. Polyethylene wrapping of fruits reduces the rate of respiration, by creating a modified atmosphere around the fruits and thereby retards the senescence and ripening, leading to extension of storage life (Narayana *et al* 1991., Roy and Pal, 1993). 'Nam Dok Mai' mango fruit wrapped in PVC (Polyvinyl chloride) films could be stored for up to 28 days (Ketsa and Raksritong, 1992).

Mango is sensitive to chilling injury when stored below 8 °C-13 °C, which limits the adoption of low-temperature storage for long-distance transport, particularly marine transport for export. In view of this, of late, another type of packaging technique viz., 'Shrink-wrapping' for extending the shelf life, for alleviation of chilling injury as well as for consumer appeal has been attempted. The shrink-wrapping essentially consists of wrapping a heat shrinkable film around the commodity and subsequently heating the film to a desired shrinkage so as to make it tight-fit. Miller *et al.* (1983) held 'Tommy Atkins' fruits for 14 days at 12.8 °C in different heat shrinkable films and found that wrapped fruits were firmer and greener than unwrapped fruit after initial storage, but when ripened, the wrapped fruits showed serious off flavour. The shrink-wrapped mangoes could be stored for 30 days at 8 °C in harvest fresh unripe hard green condition without any chilling injury symptoms (Rao and Shankara, 2004).

Wrapping of the mango fruits (cv. Banganapalli) with HM-film alone did not affect the storage life, but the fruits had better quality in terms of appearance and gloss, without shrivelling. A combination of Bavistin dip, stem-end waxing and film wrapping was the best for enhancement of shelf life at low temperature up to 23 days. HM-film wrapping cuts down the rate of respiration (Narayana *et al* 1996).

Polyethylene wrapper showed maximum delaying effect on ripening of fully matured and firm fruits of mango, followed by Cellophane paper (Deol, 1985). Weight loss in 'S.B. Chausa' fruits was least with fungicidal wax and polyethylene wrappers. Losses due to disease and physiological disorders were least with fungicidal wax and tissue paper wrappers. Wrapped fruits were acceptable for up to 12-14 days (Deol and Bhullar, 1972).

Wrapping mango fruits (cv. Alphonso) in food grade 'Klin' wrap film resulted in more number of fruits remaining in semi-ripe condition after 21 days of storage. The wrapped fruits retained about 40% chlorophyll, while unwrapped ones turned completely yellow. The PLW was reduced in wrapped fruits. More than 70-80% fruits were marketable at the end. However, fungal spoilage was slightly enhanced in "Klin" film wrapped fruits. Fungicidal treatment of wrapped fruits with Contef (1%) for 5 minutes could not completely eliminate the spoilage (Janave and Sharma, 2005).

Storage in polyethylene bags

Packaging of mangoes in polyethylene bags extends the storage life (Mukherjee, 1961., Reddy and Thimmaraju, 1989., Singh and Narayana, 1995., Ahmed and Singh, 2000) by delaying ripening (Narayana *et al.*, 1991). This is brought about through creation of modified atmosphere inside the bags. High level of relative humidity in the polyethylene bags (poly bags) maintains the gloss and prevents shrivelling of fruits during storage. However, excess buildup of CO_2 inside the poly bags leads to anaerobiosis and results in fermentation under extended storage. Fruits in unventilated poly bags often get spoiled due to latent infections as a result of build up of moisture in such bags.

Mature fruits of 'Kensington' mangoes stored in sealed poly bags at 20 °C did not develop normal skin colour and flesh colour on ripening and were unacceptable due to off-flavours. The gaseous atmosphere inside the bags varied between the treatments, but often CO_2 exceeded 20% (v/v) and O_2 was lower than 5% ((v/v). The above observed ripening abnormalities in poly bag stored fruit were ascribed to the modified atmosphere that developed in the sealed poly bags. Further, the storage life was not extended in poly bags as compared to control fruits (Chaplin *et al* 1982). The above demerits of storage in unventilated bags can be avoided by providing ventilation to the bags, which allows gaseous exchange and escape of moisture and ethylene (Bhullar *et al* 1984., Reddy and Thimmaraju, 1989., Ahmed and Singh, 2000).

Storage of fruits in perforated poly bags increased delayed rotting of fruits and decreased the percentage of rotted fruits, compared to the unventilated bags, due to the accumulation of CO_2 within the bag and preservation effect of CO_2 (Hardenburg, 1956). Packaging of 'Dashehari' and 'Langra' fruits in perforated poly bags of 100 gauge extended the storage life up to 12 days and the general appearance of the fruit was best in perforated bags even at the end of 12 days storage (Bhullar *et al* 1984). Storage in perforated poly bags also resulted in the reduction of PLW, by minimizing dehydration of fruits, but fruits showed highest acidity (Bhullar *et al* 1984). The reduced weight loss in fruits stored in ventilated poly bags was attributed to the build up of humidity inside the bags (Singh and Narayana, 1995), to the decline in the rate of moisture loss and rate of respiration (Reddy and Thimmaraju, 1989). Fruits packed in ventilated poly bags showed a slow rate of ripening (Garg *et al* 1971., Chaplin *et al* 1982., Singh and Narayana, 1995). However, the rate of ripening and senescence were faster as the level of ventilation was increased. To avoid the above and fermentation of packed fruits, a minimum of 0.5 % ventilation is required. The delay in ripening may be due to decreased production of ethylene under higher concentration of CO_2 due to respiration in poly bags (Garg *et al* 1976). Storage of 'Banganapalli' mangoes in ventilated (1%) poly bags of 250 gauge decreased quality in terms of TSS, TSS/acid ratio, non-reducisng and total sugars and increase in acidity and ascorbic acid content, which may be due to retardation of ripening processes and associated biochemical changes. Fruits stored in thicker (350 gauge) bags failed to ripen properly and were unacceptable due to poor quality and strong aldehydic flavour (Gautham and Neeraja, 2005).

With the extension of storage period beyond 10 days, in perforated poly bags, some of the fruit became anaerobic and developed off-flavours (Passam, 1982). Removal from the bags and transfer to higher temperature prevented off-flavour. Therefore, removal of fruits to ambient temperature is essential to prevent build up of CO_2 and associated off-flavours (Thompson, 1971). The incidence of decay that appears in unventilated poly bags can be avoided by treating the fruits with fungicides before packaging. 'Amrapali' fruits treated with Bavistin (500 ppm) and stored in perforated poly bags (200 gauge) with 20% ventilation could be stored well up to 11 days at ambient temperature (33 °C - 37 °C) without decay (Ahmed and Singh, 2000).

'Amrapali' fruits treated with GA_3 (50 ppm) and stored in perforated poly bags (200 gauge) having 20% vents could be stored well up to 11 days, with minimum PLW and delayed ripening (Ahmed and Singh, 2000). Mango fruits (cv. Himsagar and Langra) could be successfully stored for comparatively longer period by post-harvest dip in calcium chloride (8%) solution in combination with perforated (20%) poly bags at room temperature (Munshi *et al* 2004).

Granthy *et al* (1982) observed serious off-flavours and lack of normal peel colour on ripening of fruits, that were sealed in air-tight poly bags for various durations at 20 °C. However, Bhullar *et al* (1984) did not report any off-flavours in the ripened fruits following 12 days storage in perforated poly bags at ambient temperature (26-34 °C). This was due to the use of perforated poly bags, which permitted exchange of gases, unlike in air-tight poly bags.

Storage of mango fruits (cv. Mallika) in ventilated LDPE bags (0.25 and 0.5% ventilation) at low temperature showed lower microbial load on the fruit surface, compared to fruits stored in open baskets. Lesser microbial counts in fruits in ventilated LDPE bags (0.25%) was due to less ventilation and resultant accumulation of CO_2 which inhibited microbial growth as compared to fruits packed in 0.5% ventilated poly bags and fruits kept in open (Garg *et al* 2000).

Hypobaric or low pressure storage

Low pressure (LP) storage refers to holding the commodity under a reduced atmospheric pressure in a vacuum-tight refrigerated container generally less than 200 mm Hg. It is a form of controlled atmosphere storage. In this system, oxygen concentration is reduced proportionately with atmospheric pressure. It works on the principle that chamber pressure is linearly related to the oxygen composition of the chamber. In this the produce is stored in air-tight containers at a constant low pressure and temperature (Lougheed *et al* 1978). Containers are continuously evacuated with a vacuum pump, till the pressure is reduced to the desired level. Consequently, the levels of oxygen, ethylene etc. are reduced. The onward diffusion of ethylene, within the commodity tissue, is also enhanced (Jamieson, 1980). Low pressure storage consists of placing a commodity in a flowing stream of air, subsequently saturated with water (at RH 80-100%) at a reduced pressure (4-400 mm Hg absolute) and controlled temperature (Burg and Burg, 1966).

Sub-atmospheric pressure (low pressure, hypobaric) storage methods were introduced by Burg and Burg (1966) and have since been used to prolong the storage life of several perishable commodities including mango. Hypobaric (Hypo = less than, baric = barometric pressure) has been studied for the long term storage of mango fruits, which revealed the possibilities of prolonging the shelf life of mango fruits at sub-atmospheric pressure (Burg and Burg, 1966). The positive results with mango suggested that LP system was useful for extending the marketability of mangoes. It economically holds the fruits for shippers and packers at the point of production for distribution in areas 3-4 days away from the growing centres. Such storage would be especially helpful during years of excessive production or supply (glut) when the prices are depressed. LP storage is also needed to retard ripening and decay development when mangoes are in marketing channels and transit for 2-3 weeks (Spalding and Reeder, 1977).

Some advantages have been demonstrated in storing mangoes under sub-atmospheric pressure. LP storage extends storage life by delaying ripening, which is inversely related to the pressure (Apelbaum *et al* 1977). Weight loss is minimum in LP stored mangoes due to high relative humidity and none of the fruits showed shrivelling and softening after storage. Stored fruits remained greener than those stored at normal pressure. LP stored fruits were firmer and required longer time to soften after storage. Decay was less during LP storage than at normal pressure. LP storage is also helpful in retarding the development of insects, which cause spoilage under normal storage. LP storage lowers the sensitivity of mango fruits to ethylene. The percentage of acceptable fruits after hypobaric storage was also high (93-100%) (Burg and Burg, 1966).

Desiccation of the commodity in LP storage due to the water vapour pressure deficit between the commodity and the chamber atmosphere as a result of the reduced water vapour pressure within the chamber, is a minus point in LP storage, which, however, has been controlled by ventilating the chamber with water saturated air (Dilley, 1982).

'Haden' mangoes ripened four times slower at 150 mm Hg and 15°C than at normal pressure of 760 mm Hg. The delay in ripening was attributed to the reduction in oxygen partial pressure in the storage chamber, which resulted in a reduction of ethylene production as well as its activity (Burg and Burg, 1966). Another consequence of reduced oxygen partial pressure is the retardation of fruit aerobic respiration which lowers the rate of fruit substrate catabolism and generation of energy required to power bio-chemical reactions associated with the ripening processes (Apelbaum *et al* 1977). Similar results were reported from Florida (Spalding and Reeder, 1977) from Israel (Apelbaum *et al.*, 1977) and India (Rameshwar, 1988). Marked delay in ripening of mangoes at reduced atmospheric pressure (< 100 mm Hg) in the storage chamber was observed with resultant extension of shelf life (Apelbaum *et al* 1977). Control fruits stored at 760 mm Hg started to ripen after 10 days in storage at 13 °C, while fruit stored at 100 and 75 mm Hg pressure remained unripe for 35 days. 'Banganapalli' mangoes stored at normal atmospheric pressure and 30 °C lasted for only 11 days while hypobaric stored ones lasted for 46 days (Rameshwar, 1988).

The storage behaviour of mango fruit varied with changes in atmospheric pressure, indicating an inverse relationship between storage period and the pressure. When the pressure was above 200 mm Hg, there was no effect on ripening, while storage at pressure below 50 mm Hg resulted in substantial desiccation and shrivelling of the fruit. Fruit softening was delayed at sub-atmospheric pressure, resulting in retention of firmness for longer time (Apelbaum *et al* 1977).

Hypobaric storage also influences the fruit quality. Florida mangoes were the most green, least red and least yellow at 760 mm Hg storage (Spalding and Reeder, 1977). However, green fruits of the coloured cultivars (cv. Haden and Maya) stored at sub-atmospheric pressure for a prolonged period did not develop their typical red or orange colour during storage, but turned pale instead. Ethylene exposure, upon removal from LP storage slightly improved colour development in these fruits (Apelbaum *et al* 1977). The characteristic increase in TSS of the fruit pulp is markedly delayed in LP storage. Similarly the reduction in titratable acidity was slower in LP stored fruits. Further, fruits stored at sub-atmospheric pressure showed no skin damage during prolonged storage, but exhibited excellent quality and taste (Apelbaum *et al* 1977).

Hypobaric conditions of 600 and 500 mm Hg pressure were found to be better for storing "Chausa" mangoes as they delayed ripening, maintained pulp firmness and slowed down the development of TSS and higher level of acidity and extended shelf life of harvested fruits under ambient conditions in India. On the other hand, fruits stored at 300 and 100 mm Hg pressure remained green, unripe with blemished brown skin, which turned later dark brown and emitted off-flavour on prolonged storage (Anon, 2002-03b).

LP systems were used commercially for a short period in the 1970's for the transport of some food products, flowers and fruits (Burg, 1975). However, currently this system is not used commercially for storage of fruits (like mango) and vegetables in tropical countries including India, because this system is more expensive than modified and controlled atmosphere storage. Further development of colour and aroma was poor in fruits stored under low pressure compared to fruits stored at normal atmospheric pressure (Spalding and Reeder, 1977).

IRRADIATION

Exposure of fruits and vegetables etc. to ionizing radiation from different sources is 'Irradiation'. It is the process which involved exposing the perishables/food either packed or in bulk to carefully controlled doses of ionizing radiation for a specified period to achieve certain desirable objectives like delaying ripening, extending shelf life, conserve the flavour, disinfestation of insect pests and disinfection of fungal diseases. The potential of irradiation in the post-harvest handling of perishables lies in two directions viz.,

(i) the extension of shelf life of the commodity and

(ii) disinfestation of insects pests and disinfection of diseases (as quarantine treatment). The first objective will be dealt in this chapter on storage of mango fruits.

Advantages

1. It is a physical, non-additive process causing minimal changes in perishables.

2. It is highly effective compared to chemicals and fumigants.
3. It does not leave harmful residues in food.
4. It can be applied to bulk as well as pre-packed food.
5. It is a cold process and preserves food in natural form.
6. It does not destroy heat-labile aroma constituents of food.
7. The process is safe to workers and friendly to environment.

Fruit irradiation, thus offers effective means of reducing the post-harvest losses at low cost, thereby increasing their availability, stimulating global exports and improving nutritional and public health standards. There is an added bonus of irradiation, particularly in case of mango. Elimination of seed weevil by the highly penetrating radiation is the only satisfactory solution for a quarantine problem which has come in the way of export of mango fruits to USA, Japan etc. However, irradiation is not without disadvantages like causing skin damage.

Sources of radiation: The sources of ionizing radiations can be classified into two broad categories viz., radio isotopes and machines.

Radioisotope sources: Most common source of ionizing radiation is Cobalt-60 (CO-60), but Cesium-137 is another effective source, which emits gama rays for a longer period. However, limited availability of Cesium-137 restricts its large scale use. While, easy availability and high penetration activity of CO-60 makes it of wide spread commercial use.

Process: Radiation consists of a flow of atomic and sub-atomic particles of waves. Radiation process involves the controlled exposure of the perishables and food to short wave energy. Goods to be irradiated are conveyed to the irradiation chamber through a labyrinth and exposed to a specifically worked out dose and time and removed after the exposure from labyrinth.

Machine sources

Machine sources used in food irradiation include various types of electron accelerators. The electron beams, emerging from the accelerator can be either used directly or converted into X-rays. In both types, electrons are accelerated close to the speed of light in an evacuated tube. Electrons emitted from an electron source are pushed from the negative end of the tube and are attracted by the positive end. A scanning magnet at the end of the accelerator tube deflects the mono-magnetic electron beam on to the material being irradiated.

Dose: Threshold doses of irradiation vary with cultivar, purpose and region of production of mangoes. It is, therefore, necessary to establish the dose requirement for each cultivar for the local conditions.

Lower doses of radiation (25-50 KR) are effective in delaying the ripening in fruits, with attendant extension of storage life. In India, mangoes are generally irradiated at

0.25 (minimum) to 0.75 (maximum) kGy for extension of shelf life and as quarantine treatment for export trade. Higher doses cause deterioration in fruit quality. A dose of 25 Krads is optimum since it causes least spoilage during storage of 20 days. There was progressive deterioration with increase in irradiation dose as evidenced by more spotting and with 200 krads, there was complete blackening of fruits (Dharkar *et al* 1966a). Therefore, irradiation of mangoes at doses of 300 Gy or more is not recommended. Doses of 75 Gy can be used without adverse effect on marketability (Mc Launchlah *et al* 1990). However, gamma irradiation of pre-climacteric mango (cv. Alphonso) fruits at 100 Gy extended the shelf life by 5-6 days at ambient temperature (28 °C - 32 °C) indicating that the extension of shelf life is dose dependent (Janave and Sharma, 2005). The optimum dose for delaying ripening of 'Alphonso' mangoes is 250 Gy and doses higher than this resulted in physiological damage and earlier ripening (Srinivasan *et al* 1971).

Spalding and von Windeguth (1988) recommended that 'Tommy Atkins' and 'Keitt' mangoes can be irradiated at doses less than 250 Gy in USA. A dose of 750 Gy resulted in peel injury of the above cultivars (Spalding and Reeder, 1986), manifested as tissue blackening, which was attributed to the activation of polyphenol oxidase (Thomas and Janave, 1973).

Fruits irradiated under a nitrogen atmosphere could withstand doses as high as 2 kGy, without injury, while fruits exposed to these doses in air turned completely dark.

Of several rates tested, 60 Krad ionizing radiation gave the minimum decay in 'Carabao' mangoes when stored at 30 °C for 3 weeks after irradiation. Control fruits spoiled to an extent of 67%, compared with 5.5% of the irradiated fruits (Pablo *et al* 1971). Doses above 250 Gy impaired palatability of Irwin mangoes (Hatton *et al* 1961) and caused skin browning while dose above 1000 Gy caused pulp damage in green 'Haden' mangoes (Akamine and Goo, 1979., Thomas, 1986). However, in case of 'Tommy Atkins' mangoes, doses of 150-1500 Gy resulted in fruit injury consisting of breakdown with or without spongy tissue and flesh breakdown (Spalding and von Windeguth, 1988).

Storage

Storage of mangoes by means of irradiation in India was attempted by Mathur and Lewis (1961). They studied the effect of Y-rays (12×10^3 rads) on 'Alphonso' mangoes with cobalt 60 at a dose of 12 Krads/minute and found that the fruits could be stored for 24 days in sound condition as against 16 days in unirradiated control at 23 °C - 39 °C. Irradiated fruits on ripening remained shelf stable for longer periods as compared to the unirradiated control fruits. The total shelf life extension that could be accomplished was 6-8 days under ambient conditions (25 °C - 30 °C) (Thomas and Desai, 1976). The storage life of mangoes could be further enhanced by 6 days with a combination of skin coating (with a 9% emulsion of an acetylated monoglyceride) and irradiation (Dharkar *et al* 1966b). Treatment of mangoes by 30 Krad gamma rays increased storage life from 10 to 15 days in Egypt (Hassabulla *et al* 1984).

Extension of storage life of mango fruits on irradiation not only depends on the dose of irradiation, but also on the stage of maturity of fruit. Irradiation of fruits in pre-climacteric stage produces a greater response than fruits beyond the onset of climacteric. Delay in ripening of less mature fruits is characterised by inhibition of skin degreening and slow reduction of titratable acidity, while the fruits at climacteric stage were unaffected (Boag *et al* 1990).

Irradiation when coupled with other techniques of storage has further extended the storage life of mangoes. Skin coatings alone cause delayed ripening with loss of flavour and taste, but the combined use of irradiation resulted in additional delay in ripening and extension of storage period, with retention of organoleptic properties in the fruit (Dharkar and Srinivasan, 1972). 'Tommy Atkins' mango fruits showed extended post-harvest conservation when fruits were irradiated with gamma rays and coated with Carnauba wax prior to putting them in cold storage. This treatment increased the fruit resistance to cold storage and improved its conservation on transfer to ambient temperature (Durigam *et al* 2000). Gamma irradiation (100-300 Gy) in combination with 'Klin' film wrapping further extended the shelf life of mangoes (Alphonso, Dasheri) at room temperature by 10-15 days as compared to that in unwrapped irradiated mangoes resulting in total shelf life of 25-30 days by delaying ripening as indicated by higher retention of chlorophyll and reduction of PLW (Janave and Sharma, 2005). Irradiation (25 Krad) in combination with antitranspirant vapour gard (Hassabulla *et al* 1984) and maleic hydrazide (Rashid and Farooqui, 1984) showed better influence, but with refrigeration did not show any added benefit in extending the storage life (Ahmad *et al* 1972). A hot benomyl dip in combination with irradiation (75 and 600 Gy) preserved fruit quality attributes during storage in Australia (Jessup *et al* 1988). Mango fruits that received HWT with irradiation retained firmer texture than those irradiated only (Cucras – Ruiz *et al* 1968).

Detrimental effects

Doses of irradiation, especially higher than threshold levels show detrimental effects on mango fruit skin. Scald like peel injury increased with dosage, especially at 500 Gy or higher. Initial breakdown of pulp was slight, but increased at 250 Gy or higher. Hollow packets in the pulp and flesh darkness were increased by irradiation at 1500 Gy (Spalding and von Windeguth, 1988). Damage to 'Alphonso' mangoes (skin spotting or blackening) occurs at 0.5 to 2.0 KGy doses of irradiation (Dharkar *et al* 1966a). Irradiation of 'Kensington Pride' mangoes at 0.6-1.2 KGy showed unacceptable levels of lenticel spotting and skin damage (Mc Lauchlan *et al* 1990).

Conditions for irradiation

For irradiation, fruit should be harvested at the hard mature green state and must have entered the climacteric state. Irradiation, done after the start of climacteric phase, does not delay ripening and may even accelerate it. On the other hand, irradiation of too early stage may hamper normal ripening of the fruit.

Zero energy cool chamber

Although refrigerated storage is ideal for long term storage of perishables, on farm storage is economical for short term storage. Long term storage is expensive and requires a high level of technical knowledge of the process. Being a non-refrigerated storage method on farm storage does not need energy. The cost of storage is much less compared to refrigerated storage etc. Therefore, it can be very useful for short term storage of mangoes. On farm storage can be used effectively by the farmer to avoid distress sale, especially during glut and to get better returns for his produce. If the quantity of the produce harvested at a time is small, on farm storage for a few days is required to accumulate sufficient quantity to send to market. On farm storage is also necessary for a few days when the prices are low and are likely to increase in a few days time. On farm storage of perishables etc is a traditional one. It may be less effective as compared to modern methods, but their efficiency can be improved by modern modifications.

One of the short term on farm storage methods is cool chamber, utilizing the principle of evaporative cooling. Roy and Kudriya (1983) developed a cool chamber (which they called "Zero energy cool chamber' since it does not require any energy, either electrical or mechanical, as in cold storage) at New Delhi for on farm storage of perishables. It is an alternative to refrigerated storage. It is a low cost cooling technology for storing the perishables immediately after harvest for a shorter period. It maintains low temperature and high humidity, checks spoilage and increases storage life of stored perishables. Zero energy cool chamber (ZECC) can also be constructed in assembling centres and market yards for temporary storage of perishables till they are disposed off or sent for storage in other systems of storage/to pack houses.

Advantages

The major advantage of ZECC is the maintenance of firmness of fruits by lowering the PLW and other metabolic processes. The other advantages include: uniform and slow ripening process by slowing down of physico-chemical activities, absence of shrivelling of fruits and negligible disease incidence. The added advantages are: easy construction, maintenance and low cost. It is an ideal on farm storage facility for maintaining quality and market acceptability of different fruits including mango (Dhemre and Waskar, 2003).

Construction of Zero Energy Cool Chamber

The ZECC can be constructed using locally available raw materials like bricks, sand, bamboo, dry grass, jute cloth etc. The chamber is an above ground double-walled structure made up of bricks with a gap of 7.5 cm between the two walls. The vacant space (cavity) is filled with river-bed-sand. The floor of the chamber is made with a single layer of bricks. The lid is made by using dry grass/straw on a bamboo frame. Instead, jute cloth can also be used, but needs frequent wetting. The size of the chamber depends upon the quantum of produce. The chamber walls are wetted by sprinkling water at intervals for evaporative cooling.

The cool chamber maintains considerable and relatively low temperature compared to field and room temperature respectively. The rise in relative humidity (90% or more) and fall in temperature (10 °C - 15 °C) from the ambient temperature are achieved by wetting the chamber twice a day (Roy and Pal, 1991b). Cool chamber can reduce the temperature by 17 °C to 18 °C even during peak summer, maintain a very high humidity (> 90%) throughout the year even when atmospheric relative humidity falls below 20%. This is because of evaporative cooling facility, which consists of porous sand through which air is drawn and humidified by evaporation of water (Roy and Khudriya, 1986).

Storage of Mango in ZECC

The harvesting season of mango in India generally commences with the onset of summer season in different parts of the country, due to which the ripening of fruit is faster and the post-harvest losses sometimes are as high as 40%. Cold storage of mango is limited owing to its susceptibility to chilling injury, besides the general constraints of refrigerated storage. The entire produce of mango in tropical countries like India, is mainly handled at a high temperature often under open conditions after harvest resulting in fast ripening and spoilage. Hence, proper management of this commercially important fruit is essential not only for delaying ripening and extending the storage life, but also to reduce the post-harvest losses in tropics.

Reducing the temperature in cool chamber, slows down the rate of respiration, ethylene production and its action, biochemical changes and the biological heat production. It also slows down transpiration thereby reducing the desiccation of the fruit and also checks the development of pathogen. The cumulative effect of all the above is extension of shelf life in ZECC. The cool chamber provides more marketable fruit than at the ambient temperature by checking desiccation and diseases (Dhaka *et al* 2001).

Cool chamber storage delayed ripening in ‘Alphonso’ ‘Pairi’ and seedling mangoes and extended their shelf life to 15, 10 and 16 days respectively as compared to storage at ambient temperature 12, 9 and 14 days respectively) (Gole, 1986). The storage life of `Kesar’, `Totapuri’ and `Vanraj’ mango fruits could be extended up to 25, 36 and 31 days in cool chamber as against 17, 21 and 19 days respectively at room temperature (Waskar and Masalkar, 1997). Under ZECC condition, the mango fruits could be stored up to 10 days with acceptable minimum rotting and quality as against 5 days shelf life and higher incidence of decay at room temperature (Jain and Mukherjee, 2001).

However, the extension of storage life of mango fruits stored in ZECC varies with stage of maturity/ripeness of the fruit at the time of storage. The shelf life of firm ripe (cv. Amrapali) mango fruits in ZECC could be extended up to 8 days with only 4.65% PLW as compared to 4 days with 9.40% PLW at ambient temperature (Anon, 1985). The mature green mangoes (cv. Baneshan) stored well up to 12 days in ZECC as against 6 days under ambient and open condition. Ripe mangoes when stored in ZECC had 9 days shelf life against 6 days under ambient conditions. Therefore, ripening and storage life of fruits in ZECC are inversely related (Roy and Pal, 1991b).

Storage life of mango fruits in cool chamber can further be extended by combining the cool chamber with fungicides, pre-packaging, waxing etc. Bavistin (500 ppm) dip treated fruits in cool chamber stored well up to 9 days (Roy and Pal, 1991b). The storage life of ripe mangoes cv. Beneshan could be increased upto 12 days in ZECC (23 °C $\pm$ 3 °C and 85-90% RH) against 8 days at room temperature (29.4 °C-33.81 °C and 56-66% RH) following the treatments with Bavistin (500 ppm) and sealing the stem end with paraffin wax (Narayana *et al* 1996). Similarly, the storage life of 'Kesar' mangoes was extended up to 26 days when treated with a combination of wax (6%) plus carbendazim (0.1%) and stored in cool chamber (21.36 to 23.55 °C with 87.0 to 91.5% RH), while control fruits showed a shelf life of 20 days at room temperature (24.18-30.57 °C and 47.5 to 74.0% RH) (Dhemre, 2001). The shelf life of mango fruits treated with calcium chloride (2%) plus waxol (6%) plus carbendazim (0.1%) was extendable up to 22 days in ZECC (Masalkar *et al* 2006). Pre-packed 'Neelum' mangoes showed colour development in 8-9 days and were ripe in another day in ZECC against 4-7 days in control (Thangaraj and Irulappan, 1993).

Mango fruits kept in cool chambers showed delayed ripening, longer shelf life, less shrivelling due to less PLW, less decay and better retention of original fruit quality. The fruits stored in cool chamber were fresher, glossy and attractive. Ripening changes were slower in fruits stored in ZECC (Badar, 1990). However, mango fruits stored in ZECC lost their taste and flavour (Dhemre and Waskar, 2003).

Freezing storage

Investigations made at Mysore (India) have shown that mango fruits could be frozen, whole or slices in sugar syrup, dry slices coated with sugar or as pulp (Mathur *et al* 1952., Singh, 1970., 1974). Even raw green mangoes could be frozen preserved for more than a year for chutney or pickle manufacture during off-season and could also be frozen in grated form, peeled or unpeeled (Srivastava, 1967). Mango fruits of `Neelum' and `Mulgoa' cultivars were successfully stored for 9 months when packed in poly bags and frozen at –20 °F (Mathur *et al* 1952).

Preservation by freezing is based on the principle that extreme cold retards the growth of microorganisms and slows down enzyme activity and oxidation. However, freezing does not sterilize the fruit.

The storage life being short, freezing preservation will prove helpful to make mangoes available without a break. Further, the cost of frozen mangoes, would not be high as they do not require any processing or elaborate packing and would be definitely within the reach of the people in towns and cities where the sale of frozen fruits and products is possible through hotels, super markets, malls etc.

Method of freezing

Mango fruits meant for freezing storage should be of a good edible maturity, fully mature and evenly ripe. It is advisable to ripen the freeze-intended fruits between 19.5° and 21 °C. Ripe mangoes with firm flesh having an attractive colour and flavour should be used for freezing storage after washing. Freezing should be done as fast as possible. Otherwise in slow freezing, big pointed crystals appear in the cells of the fruit, puncture the cells and cause oozing of cell sap. Such a product upon thawing crumbles and loses all the juice and flavour, along with the biting properties.

Among the various methods available for quick freezing, the blast tunnel freezing seems to be suitable for mangoes under Indian conditions. The intended, commodity is moved through the tunnel where a temperature of –40 °C to –29 °C is maintained. The commodity moves on a trolley and by the time it reaches the other end of the turnel, it gets completely frozen in less than one hour. The frozen fruits are packed in cans, glass jars with screw lids, waxed-paper cartons or poly bags and stored at –18 °C.

For marketing, frozen mangoes have to be transported under refrigeration at –18 °C and displayed in a freezing display cabinet maintained at the same temperature. It is important that during storage, transport and/or marketing, the temperature is properly maintained always at –18 °C. The frozen fruits after removal from storage can be thawed at room temperature, which takes a considerable time. For quick thawing, the frozen fruit may be kept in running water or luke warm water. Care should be taken that outside water does not enter the package. After thawing the fruits can be utilized as usual.

The preservation of mangoes by freezing was found suitable in USA, Trinidad and India. Frozen mangoes are very popular in USA. There is, therefore, a great potential for export of frozen mangoes from India to USA etc.

LOW TEMPERATURE STORAGE

As mentioned earlier, in tropical production regions like India, the prevailing high temperature and low relative humidity during harvesting season, cause faster ripening and deterioration of mango fruits resulting in short shelf life and huge post-harvest losses. Such a situation also occurs in sub-tropical production areas, but to lesser magnitude. This calls for an efficient storage technique to check/slow the ripening and deterioration and prevent post-harvest losses and consequently extend the storage life. The common method employed for increasing the storage life of mangoes is to reduce the storage temperature to slow down the biochemical and physiological processes and the rate of respiration and finally extend the post-harvest life. Refrigeration is the only known economical method for long term storage of fresh fruits and vegetables including mango. Low temperature storage (cold storage, refrigerated storage) is essential to delay ripening and quality deterioration during transport, to regulate the entry of the fruits and vegetables into the market, to ensure regulated distribution and extending the duration of their availability and to minimise the post-harvest losses and to ensure better returns to the grower (Johnson *et al* 1997).

The main principle underlying low temperature storage is to slow down the physiological and biochemical activities of the fruits and vegetables leading to senescence and increase their storage life. Prolonging storage life at low temperature is mainly due to slow activity of ripening associated enzymes. Ethylene activity declines several fold at low temperature (Reddy and Srivastava, 1999). The functions of cold storage are

(i) to retard the natural physiological and biochemical processes and deterioration occurring in the living system,

(ii) to retard fungal activities,

(iii) to reduce water loss and

(iv) to maintain acceptable quality.

Disadvantages with cold storage of mangoes are

(i) decay,

(ii) chilling injury and

(iii) improper ripening, resulting in fruits with poor colour and flavour (Hatton *et al* 1965).

This is because mangoes produce high quantities of CO_2 during storage and ripening. In closed system, the accumulated level of CO_2 is beyond tolerance levels leading to toxic metabolism and ripening disorders.

However, the possibilities of using low temperature storage are limited by the high capital cost of installation and the susceptibility of mangoes to chilling injury. Not withstanding the above, the importance of low temperature lies in the fact that most other methods of storage of perishables are adjuncts to cold storage and are not exclusive techniques.

Storage temperature

Temperature is one of the most important factors that regulates the natural processes of maturation and ripening of fruits such as colour changes, softening, carbohydrate metabolism and storage life. For this reason optimum/best storage temperature needs to be defined so that mango fruits can be stored for as long as possible, without any type of losses and injuries. The ideal temperature for storage is the lowest one, that does not permit chilling injury. The optimum temperature is the one that delays senescence and maintains quality without chilling, freezing or other injuries. Thus the ideal optimum/ best storage temperature will be above the critical temperature of chilling injury, but at the same time it should not cause undue losses through respiration and transpiration and extend the shelf life to the maximum extent. Optimum storage temperature (for mangoes) depends upon the cultivar, fruit maturity, susceptibility of cultivar to chilling injury as well as to post-harvest diseases and disorders etc.

According to Cheema *et al* (1950), 48 °F was the optimum temperature for cold storage of mangoes in India. The optimum conditions for the cold storage of 'Totapuri' mangoes are a temperature of 42-45 °F and a RH of 85-90% at which they could be stored for 7 weeks (Singh *et al* 1954). Krishnamurthy *et al* (1963) recommended a temperature range of 7.2-9.2 °C as optimum range for cold storage of all mango cultivars in India. The optimum refrigerated storage temperature requirement for mango was worked as 8.3-10.0 °C for 'Alphonso' and 5.5-7.2 °C for 'Raspuri' with 4 and 6 days storage life respectively at CFTRI, Mysore (India) (Dalal and Subramanyam, 1970). However, cold storage studies on mango in India recommended 7.9 °C for 'Alphonso' mangoes with anticipated storage life of 7 weeks (Singh, 1974) and the best storage temperature to be in the range of 5.5-9.0 °C (Mukherjee, 1958). Temperature below 12° and 13 °C, however, was considered as the optimum for cold storage of mango in India (Kalra and Tandon, 1983). Of late, an optimum storage temperature of 13 °C with 85-95% RH has been suggested for mango, which would maintain the fruits in an acceptable condition and satisfactory ripening on transfer to higher temperature (Anon, 2005). However, mangoes in India are generally stored at 13 °C to prevent chilling injury and retard decay development. Temperatures below 13 °C are not generally recommended for the storage or shipping of mangoes because of their susceptibility to chilling injury. However, mangoes can be successfully stored below 13 °C for longer periods with modified atmosphere package and controlled atmosphere storage of elevated CO_2 and O_2 levels (Anon, 2005).

According to Wardlaw and Leonard (1936) the optimum temperature for mango was 45 °F in Florida. For Floridan cultivars, the optimum temperatures were worked out as 12 °C for 'Tommy Atkins', 'Zill' and 'Haden' (Monchu *et al* 1975), 'Momey' (Samayoa and de Arricola, 1973); 'Kent' (Zauberman *et al* 1979). Florida mango cultivars store best at 12.8 °C, while others successfully store at 10 °C, but for specific periods (Hatton *et al* 1965). Therefore, temperature around 12 °C was recommended for storage of mangoes in Florida (Medlicott and Jegar, 1987).

There is considerable variation in the minimum temperature at which fruits of different cultivars can be stored. The minimum temperature for storage of West Indian mango cultivars has been stated to be 4-7 °C (Shill, 1934) for Indian cultivars 10 °C (Benerjee *et al* 1934) or 6° - 7 °C (Mathur *et al* 1953), for Florida cultivars as 8.6 °C (Wardlaw and Leonard, 1936) or 10 °C (Popenoe, 1956). The lowest storage temperature for major mango cultivars in Israel ranged between 12° and 14 °C (Apelbaum *et al* 1977). Below these temperatures in cold storage or on subsequent removal to higher temperature, the fruits fail to ripen normally. Wrapped and coated fruits withstand lower temperature better without adverse effects (Wardlaw and Leonard, 1936). Storage temperatures below their minimum tolerance temperature, render the fruits more susceptible to fungal attack. Fruits at 13 °C kept for longer periods than those at 20 °C, but their ripening qualities were inferior, particularly their skin failed to turn yellow. Storage of fruits at 6 °C hindered ripening, especially colour development, showed chilling injury and rotting and slower breakdown of starch (Chatpar *et al* 1971).

Storage at lower temperatures for longer periods is more beneficial than storage for shorter periods at higher temperatures. Mango fruits (cv. Harumanies) which were stored at 15 °C for 1 or 2 weeks were of better quality than fruits stored at 10 °C, but fruits stored at 10 °C for 3 or 4 weeks were better than those stored at 15 °C (Lam and Wong, 1988).

Factors influencing the storage temperature

The storage temperature required to achieve best results in case of mango, varies greatly with cultivar, fruit maturity, method of harvesting, time delay (gap) in placement of fruits in cold storage etc. and they are some what interrelated.

Cultivar

Of the many cultivars screened 'Alphonso' was the best for keeping in cold storage. 'Bangalora' fruits could be stored upto 8 weeks at 5° – 7 °C and 'Alphonso' fruits for 4-7 weeks at 7°-10 °C (Srivastava, 1967). However, at this temperature 'Alphonso' fruits showed chilling injury symptoms in Sudan (Musa and Mirghanii, 1974). Some mango cultivars like 'Dashehari' and 'Langra' can be safely stored at 7°-8 °C up to 25 days (Mann and Singh, 1976); others, however, require temperature above 10 °C (Farooqui *et al* 1985., Fornaris Rullan *et al* 1989) or even 13 °C (Sancedevolez *et al* 1977). Therefore, temperatures around 12 °C were recommended to reduce the risk of chilling injury in Florida cultivars (Peacock, 1984., Medlicott and Jegar, 1987). Mango fruits of 'Momey', 'Tommy Atkins' 'Zill' and 'Haden' were successfully stored at 12 °C for 21, 33, 28 and 21 days respectively. At 7 °C 'Momey' were successfully stored for 15 days, but at 5 °C the fruits showed skin damage (Monchu *et al* 1975). Fruits of 'Dashehari', 'Mallika'; 'Amrapali' and 'Kesar' can be stored at 12 °C, 'Baneshan' at 13 °C, 'Langra' at 14 °C and 'Chausa' at 8 °C with 80-90% RH for 3-4 weeks in India (Negi, 2000).

Differences in the requirement of optimum storage temperature between different cultivars of mango could be related to ethylene sensitivity of the cultivars or to levels of endogenous ethylene production in storage (Seymour *et al* 1990), besides inherent differences among the cultivars.

Fruit maturity

In commercial practice, mangoes are harvested either at pre-mature stage or full mature stage depending upon the purpose, the distance and time taken for transport to (distant) markets. Therefore, the stage of maturity at which the fruits are picked is of considerable importance for successful cold storage. Best results were obtained when fruits of 'C' (picked just prior to softening) were stored at 45 °F with minimum wastage and higher flavour (Wardlaw and Leonard, 1936). Fruits of 'A' stage maturity (green and just mature) and of 'D' stage maturity (tree-ripe) are unsuitable while 'B' stage of fruit (green and mature) are suitable for cold storage, which is considered as the optimum stage of maturity for cold storage over a period of 3-4 weeks. However, Sellar and Tomeres, (1999) did not find any relation between stage of fruit maturity and their storage at low temperature, as they obtained an extended storage life of 35 days, with fruits of 'Tommy Atkins' harvested at different stages of development and stored at 10 $\pm$ 1 °C and 85% RH, after pre-cooling.

Excessive wrinkling of the skin and inferior flavour are the characteristics of younger fruits, while shorter shelf life is characteristic of more matured fruits (Thompson, 1971). In case of 'Amelie' mangoes from Senegal and 'Tommy Atkins' from Brazil, immature fruits stored better than mature fruit, but they did not ripen on return to ambient temperature. In view of the above, it was concluded that harvesting at the half maturity stage would be the best compromise between storage potential and the quality characteristics of the ripe fruit (Medlicott *et al* 1990). The 'Tommy Atkins' mango fruits picked at maturity stage-3, stored for 18 days at 8 °C and 65 $\pm$ 5% RH developed best quality during ripening at room temperature without showing any symptoms of chilling injury. Fruits picked at maturity stage-2 were highly cold sensitive. Thus, immature fruits stored at low temperature despite having a longer storage life than fruits picked at more advanced stage of maturity do not develop the same organoleptic characters during ripening (Morais and Assis, 2004). However, fruits of advanced stage ('D' stage) can be retained in cold storage for shorter periods only (Akamine, 1963., Thompson, 1971., Mukherjee, 1972., Mann and Singh, 1975a,b). Mature green and partially ripe fruits of 'Haden' did not ripen normally when stored at 35° and 45 °F, but ripe fruits were in good condition for 4 weeks at these temperatures. Mature green, ¼ ripe, ½ ripe and ¾ ripe fruits stored at 55 °F ripened well and remained in good condition for 2 weeks, whereas ripe fruits remained in good condition for only one week at 55 °F (Akamine, 1963). Low temperature storage (5 °C) of physiologically matured mango fruits did not inhibit fully the development of climacteric peak and the fruit ripened more rapidly, when returned to ambient temperature, showing less developed colour and no loss in firmness (Laborem *et al* 1992). Pre-climacteric, hard green 'Alphonso' mangoes stored at 10 °C for 30 days developed chilling injury (skin pitting) in the ripened fruit (Thomas and Oke, 1983). Ripe and mature green fruits of `Carabao' and `Pico' were stored well at 10 °C for 18-21 and 23-26 days respectively (Valmayor, 1972). Semi-ripe fruits of `Dashehari' were stored for 2 weeks where as full ripe fruits for one week only without any sign of deterioration, but beyond these periods, fruits of either maturity exhibited over-ripe taste, shrivelling and deterioration in fruit quality when stored at 12 °C with 85-90% RH (Mann and Singh, 1975b). Just ripened fruits (cv. Sensation) may be stored at 4 °C for 3 weeks without deterioration in quality, than at 8 °C, however, deterioration in quality may be expected on exposure of ripe mangoes to room temperature after their storage at low temperature in South Africa (Straten and Oosthuyse, 1994). Fully ripe mangoes (Beneshan, Arka Puneet) could be stored for one week at 6°-8 °C followed by one day at ambient temperature without chilling injury and quality deterioration (Anon, 1999-2000).

Green mangoes for chutney making can be stored for 40 days at 40°-43 °F and 85-90% RH. Temperatures of 32-35 °F caused physiological breakdown of the fruit making it unfit for chutney making. But, the percentage yield of slices from fruit increased with storage temperature from 32-35 °F (Anand and Johar, 1953). Ripe juicy varieties stored only for a short period in cold storage as compared with table varieties (Srivastava, 1967).

Although storage life may be further extended by harvesting slightly mature fruit than 'B' stage and by lowering the temperature to 10 °C, care should be taken to avoid chilling injury or impairment of flavour due to immaturity and irregular ripening (Wardlaw and Leonard, 1936., Hulme, 1971., Lakshminarayana, 1973).

The earlier picked fruits (lettuce green in colour) showed greater loss in weight, while later picked (light yellow in colour) fruits showed higher spoilage (Mann and Singh, 1976). Oosthuyse (1997a) recommended the harvesting of decay sensitive cultivars at a more advanced stage of maturity (2-3 weeks later than at present) and then holding the fruit at 8 °C, to prevent further ripening during storage.

However, the physiological and biochemical reasons for differences in cold storage behaviour of mangoes at different stages of maturity are not clearly understood.

Method of harvesting

Shelf life of fruits in cold storage is also influenced by the method of harvesting. Generally, fruits are harvested either by hand or beating the tree with sticks, which invariably cause mechanical damage, which accelerates ripening and decaying and thereby reduce their storage life. The improved method of harvesting with recently developed harvesters avoids such mechanical injury to a large extent, thereby reduce decaying, post-harvest losses and extend the shelf life. Abu-Gouk and Mohamad (2004) reported such improvement in storage life of 'Totapuri' mangoes in Sudan. This is probably due to lack of bruising etc. which stimulate the respiration rate and decaying (Kader *et al* 1985). The fruits harvested by improved methods were firmer at harvest and this was retained during storage (Abu-Gouke and Mohamad, 2004).

The usual practice of harvesting mango fruits is to harvest without stalk. However, harvesting the fruits with stalk has several benefits. 'Dashehari' mango fruits harvested with 8-10 mm stalks, stored at 12 °C with 85-90% RH had the storage life of 27 days with minimum spoilage (Negi, 2000).

Time delay in placement of fruits in cold storage

In general, there will be a time gap between picking and placing of harvested fruits in cold storage, particularly when production areas are at a distance from cold storage units. This gap, however, is inevitable as the harvested fruits, during this period undergo pack line treatment, sorting, grading etc. A delay in placement of fruits in cold storage affects the ripening, skin colouration, their shelf life etc. Oosthuyse (2000) identified differences in the time taken from harvest until mangoes are placed in cold storage as a possible cause for different outcomes in terms of fruit ripening and quality after cold storage.

Hard green fruits (cv. Dashehari) cold stored immediately after harvesting and those put in cold storage 2 days after keeping at room temperature storage were in good condition after 35 days of cold storage. While those placed in cold storage after 4 days retention at room temperature were in good condition only up to 25 days. However, weight loss and spoilage were higher after 20 days in directly stored fruit and those stored after 2 days holding at room temperature prior to cold storage, which is normally required between harvest and cold storage (Mann and Singh, 1975b).

Storage life decreased with delay in time of fruit placement in cold storage after harvest, due to the influence on the rate of ripening during storage. Ripening rate during the first week of cold storage increased with delay prior to cold storage. Undue delays in placing mango fruits in cold storage after harvesting can result in softening of fruit on arrival at destination market and distribution areas, particularly under prolonged cold storage. In view of the above, it was suggested that mangoes be harvested, treated, sorted and packed in as short a period as possible prior to their placement in cold storage to maximize storage life and quality (Oosthuyse, 2000).

Problems in cold storage

Major problems encountered in cold storage of fruits are:

(i) chilling injury

(ii) abnormal ripening,

(iii) quality deterioration.

Chilling injury

When commodities of tropical and sub-tropical origin (banana, mango, papaya, citrus etc.) are held at temperatures above their freezing point but below 5-10 °C, chilling injury occurs depending on the type of commodity. Mangoes are sensitive to chilling temperature regimes ranging from 3° to 12 °C (Medlicott *et al* 1990). Storage of 'Totapuri' mangoes in Sudan at 6°C resulted in chilling injury (Musa, 1974). Efforts were made to avoid chilling injury in mangoes in cold storage. Storage of mango fruits of 'Dashehari' and 'Langra' in polyethylene bags at 45 °F reduced chilling injury and physiological losses. The fruits remained fresh during 6 week storage and on removal to room temperature ripened normally (Mukherjee, 1961). It has been demonstrated that by rising TSS in fruits, it is possible to alleviate chilling injury in cold stored fruits (Mathur *et al* 1953., Mukherjee and Prasad, 1972). The TSS can be raised by gradual cooling of the fruits (Mukherjee and Srivastava, 1978). Therefore, by gradual cooling of fruits, chilling injury is avoided for storage below the conventional critical limit of 7 °C besides extending the storage life of the fruits (See chapter- Physiological Disorders).

Pre-storage treatments such as pre-cooling, irradiation, storage at higher temperatures for a short period, heat conditioning were reported to overcome chilling injury in fruits. Pre-cooled 'Langra' and 'Dashehari' mangoes could be stored at 7°–9 °C for one month with proper ripening and without chilling injury after removal from storage (Mann and Singh, 1975a).

Pre-storage heat conditioning of fruits alleviates cold storage disorders such as chilling injury in mango fruits (McCullum *et al* 1993). Temperature conditioning or cold adaption of mango fruits also alleviates chilling injury (Thomas and Oke, 1983).

Abnormal ripening

Prolonged storage of mango fruits at low temperature and high humidity in cold storage impaired the ripening process. Improper ripening with poor colour and flavour are other problems encountered in cold storage of mango fruits (Hatton *et al* 1965). This is because mangoes produce high quantity of CO_2 during storage and ripening. In a closed system, the accumulated level of CO_2, is beyond tolerance levels of mangoes, leading to toxic metabolism resulting in ripening disorders. At 10 and 15% CO_2, there were traces of CO_2 injury as evidenced by grayish patches and softening. As storage period increased brownish exudation from the stalk ends was observed. With increase in CO_2 concentration in the storage atmosphere there was reduction in TSS, sugar contents and carotenoid synthesis. Spoilage of fruits increased with increase in concentration of CO_2 (Lakshminarayana and Subramanyam, 1970).

Abnormal ripening was observed in 'Momey' fruits, when stored at 8-12 °C, with poor colour and flavour (Rolz *et al* 1971).

Low temperature damages

Low temperature has deleterious effects on skin, flesh and ripening, which limit the range of temperature and time combination for mango cold storage (Oosthuyse, 1994). It was concluded that refrigerated transport and low temperature storage are not recommendable for 'Alphonso' and `Pairi' mangoes grown in India, because of their susceptability to the above damages (Subramanyam *et al* 1975). Browning of the pulp is another problem in cold storage of mangoes (Krugar *et al* 1996). Prolonging the hot- water treatment (15-20 minutes instead of 5 minutes) and lowering of the temperature inhibit browning.

Precautions

In a cold storage, the temperature and humidity control and their maintenance at required level are very important and vital. Any deviation from the desired temperature or humidity is detrimental. The maintenance of high RH is important to maintain the freshness of the fruit and preserve the fruit quality as RH has considerable bearing on the keeping quality of the fruit. However, it is very difficult to control the moisture in the air. A small temperature difference between the cooling coil and the stored fruit is required to maintain an adequate control of RH in storage chambers. In order to increase the RH, water may be sprayed into the cold storage room. Storage of fruits at high RH may however, pre-disposes the fruits to microbial pathogenetic infection. Thus proper temperature and humidity management is very crucial and is an effective tool in ensuring the good condition of the fruits throughout period of storage. Hence, utmost care should be bestowed.

GAS STORAGE

The tropical regions are characterised by high temperatures and high humidity, which cause faster ripening of fruits, higher post-harvest losses etc., that reduce their post-harvest (storage) life and quality. Under these conditions, low temperature storage is adopted for storage of tropical fruits. However, many tropical fruits are chilling sensitive limiting the use of low temperature storage. Nevertheless tropical fruits must be shipped to distant markets, usually by air/sea. Air transport is expensive. Sea transport requires longer time to reach the distant markets (Table 3.1).

Table 3.1 : Time required for sea transport of mangoes

S.No.	From	To	Days required	Reference
1.	Eastern Australia	South East Asia	21	McGlosser, 1989
2.	Eastern Australia	Japan, North America	28	McGlosser, 1989
3.	Eastern Australia	Europe	42	McGlosser, 1989
4.	Mexico	Europe	18	Yahia, 1995
5.	Mexico	Japan	21	Yahia, 1995

Therefore, it is essential to assure a sufficiently long post-harvest life for the tropical fruits to reach the over-seas markets in sound condition. In this context, controlled atmosphere/modified atmosphere storage have shown promise in prolonging the post-harvest life of fruits and vegetables, besides preserving the quality of these perishables and ameliorating chilling sensitivity of tropical commodities during transit-storage (Wang, 1990). These storage methods suppress the rate of respiration and ethylene synthesis of fruits and vegetables, as the gases viz., oxygen (O_2) and carbondioxide (CO_2) are altered in their composition around the produce, which permits the extension of storage life of stored produce, through reduction of respiration rate, delaying ripening and restricting other metabolic processes. Thus, these techniques made it possible to meet the demands of long distance transportation, distribution as well as maintenance of product quality.

Benefits of CA / MA storage

Gas storage includes controlled atmosphere (CA) and modified atmosphere storage (MAS).

1. Delay in fruit ripening during long sea transportation.

2. Reduction in some physiological disorders (Hatton and Spalding, 1990).
3. Reduction in the growth and survival of pathogens and some insect pests directly or indirectly (El-Gourani and Sommer, 1981., Shetty *et al* 1989., Jang, 1990., Ke and Kader, 1992).
4. Reduction of maturation, ripening and senescence (Hatton and Spalding, 1990).
5. Alleviation and/or controlling chilling injury (Wang, 1990).
6. Reduction in quality and post-harvest losses.
7. Extension of storage life (Yahia, 1995). Storage under CA/MAS is known to extend the shelf life of produce by 1.5-4.0 fold than under refrigeration depending on the commodity (Zagory and Kader, 1988).

Problems

1. CA/MAS stored fruits can develop various physiological disorders (Yahia, 1998).
2. Reduction in the biosynthesis of aromatic volatiles (Lalel *et al* 2001b).
3. Irregular ripening.
4. Development of off-flavours and odours (Burdan, 1997).
5. Inadequate atmosphere can increase microbial growth (Kader, 1986).

Usually modification of atmosphere during storage involves reduction of oxygen and/ or elevation of carbondioxide level. Addition or removal of gasses resulting in an alteration in atmospheric composition different to that of normal air, such as in CA, MAS or hypobaric storage has been employed for storage of perishables. CA/MAS either alone or in combination with refrigerated storage have been recommended for storage of perishables. Thus the use of CA/MAS is considered as a supplementary practice to cold storage and been reported to improve storage life of fresh fruits while minimizing quality deterioration (Kader, 1986). The combination of CA/MAS and low temperature storage markedly retards respiratory activity and delays softening, senescence and changes in quality and colour of the stored fruits.

Mode of action

In case of fruits, both CA/MAS delayed ripening and increased the activities of ATP, phosphofructo kinase, alcohol dehydrogenase and pyruvate decarboxylase (Yahia and Vazquez-Moreno, 1993). Atmospheric modifications through reduction of O_2 and elevation of CO_2 help to retard senescence and ripening by reducing respiration, ethylene production and sensitivity. Elevated CO_2 atmosphere inhibits activity of aminocyclopropane-l-carboxylic acid (ACC) synthase (Key regulatory site for ethylene synthesis), while ACC oxidase activity is stimulated at low CO_2 and inhibited at elevated CO_2 and/or low O_2 levels. Optimum atmospheric composition retards biosynthesis of carotenoids and anthocyanin and biosynthesis and oxidation of phenolic compounds. CA also slows down the activity of cell wall degrading enzymes influencing softening of fruit tissue. Low O_2 and/or elevated CO_2 levels influence flavour by reducing loss of acidity, starch to sugar conversion, inter-conversion of sugars and biosynthesis of flavour volatiles. Thus, CA/MAS help in extending the storage life of stored commodities.

Differences between CA/MA storage

Terms like 'CA storage', 'MA storage' and 'gas storage' are frequently used. These terms imply the addition or deletion of gases resulting in atmospheric composition different from that of normal air. Thus the levels of CO_2, O_2, N and ethylene and carbon monoxide in the storage atmosphere are manipulated.

'Controlled atmosphere storage' (CAS) generally refers to decreased O_2 and increased CO_2 concentrations and implies precise control of these gases, whereas the term "modified atmosphere storage' is used when the composition of storage atmosphere is altered, but not closely controlled. Thus CA/MA storages differ only in the degree of control. CAS is more exact. Further, MA refers to atmosphere that differs from ambient air composition, whereas CA refers strictly controlled atmosphere (Smock, 1979., Kader, 1986).

The proportion of each gas in CA is maintained (controlled) at the original level throughout the storage regardless of the temperature or other environmental variation within gas tight containers. CA technique is used primarily for the bulk storage and transport of produce and requires constant monitoring and control of the gas composition within the storage chamber. In the case of MA storage, the commodity is packed in side a selectively permeable packaging film and the atmosphere inside the film is altered by active (gas flushing) or passive (produce-generated) means.

CONTROLLED ATMOSPHERE STORAGE

Controlled atmospheric storage (CAS) is a method for extending the shelf life of perishables and semi-perishables by altering relative proportion of atmospheric gases that surround the commodity to a desired level and controlled at this level throughout the transport and storage distribution cycle, regardless of temperature or other environmental variations. CAS creates water saturated or near saturated atmosphere around the stored product, which reduces water loss and shrinkage. CA technology has been well known for many years for prolonging the ripening period for extension of storage for prolonged transport and control of pests and diseases. This method involves maintaining the fruit in a controlled atmosphere with a high level of CO_2 and a low level of O_2. Apart from maintenance of temperature and humidity, reduction of ethylene production by reducing respiration rate and also removal of ethylene from storage atmosphere are of vital importance in storage of perishables. CAS takes care of all such needs, as the levels of O_2 and CO_2 are maintained at pre-determined optimum levels. In case of tropical fruits, CAS reduces respiration, inhibits ethylene production and its action to retard ripening and improves and maintains nutritive quality of stored commodity during long term storage (Yahia, 1998).

CAS is adopted singularly or together with other technology such as low temperature for long term storage or other treatments for short term storage periods.

Essential features of CAS

1. Mechanical refrigeration maintains required temperature.
2. The CAS room is gas tight.
3. Nitrogen gas is introduced into the storage chamber to hasten the reduction of O_2 after the chamber is filled and sealed.
4. Then CO_2 is added into storage chamber from CO_2 gas cylinder.
5. Excess CO_2 in the storage chamber is removed by dry-hydrated lime.
6. O_2 level is held at required level.
7. Atmospheric samples are taken daily for monitoring the levels of CO_2 and O_2 at required levels.

Benefits of CAS

1. CAS prevents or delays ripening, softening, degreening and sugar loss.
2. Avoids the risk of chilling injury and enables the storing of climacteric fruits at higher temperature.
3. Reduces respiration rate, ethylene production and sensitivity to ethylene.
4. Reduces the incidence of certain physiological disorders such as scald, chilling injury and ethylene induced disorders.
5. Delays fruit ripening and senescence.
6. Retains firm texture of the fruit, thereby helps to resist pathogenic invasion.

All above benefits of CAS are attributed to reduction of O_2 concentration in the storage atmosphere (Thompson, 1998).

Process

In CAS, the availability of O_2 is reduced either by reducing the percentage of O_2 or by increasing the content of CO_2 in the atmosphere surrounding the fruit. In commercial practice, the fruits are stored in a sealed gas-tight rooms, into which a known amount of CO_2 is introduced. Gas composition in the storage chambers is monitored through drawing and analysing the samples frequently. The excess accumulated CO_2 (as a result of respiration of stored produce) is removed by scrubbing the air, from outside the storage room (Srivastava, 1967).

Controlled atmosphere system should have an effective humidification system to reduce shrinkage and weight loss. The ethylene produced, should be removed promptly using ethylene scrubbers like potassium permanganate. The design and construction of CA stores require precise control of the system. The CA storage should be gas-tight and fitted with a reliable refrigeration system, with a means of measuring and controlling the concentration of both O_2 and CO_2.

Gas composition

Composition of gases (O_2 and CO_2) in the storage atmosphere is very important for successful storage of perishables in a CAS system. Optimum gas composition for different perishables is very variable and depends on many factors like type of produce, their physiological age, temperature, duration of treatment, tolerability of the produce to altered gas composition etc. (Isenberg, 1979., Smock, 1979). In general, the concentration of O_2 is reduced while that of CO_2 is elevated. Concentration of oxygen between 2 to 5% and that of CO_2 between 1 and 20% are used for CAS of fruits. However, beneficial and harmless ranges of O_2 and CO_2 concentration differ with the commodity and even amongst cultivars of the same commodity (Harvey, 1978). Exposure of horticultural products to O_2 and/or CO_2 levels above their optimum tolerable range can cause initiation and/or aggregation of certain physiological disorders like irregular ripening, increased susceptibility to decays, development of off-flavours etc. (Kader, 1986). Off-flavours and skin discolouration occur at 1% O_2 or high (15%) CO_2 in the storage atmosphere. The usual lower limit tolerated by most cultivars of mango is 2% O_2. Below this level, anaerobic respiration occurs resulting in the development of off-flavour and failure of fruit ripening after removal from the CAS.

The vast majority of CAS studies with mango have been conducted with the goal of determining gas composition and their combination with temperature that allows the longest possible storage life to be attained without loss of quality.

A gas composition having higher O_2 and lower CO_2 was tried with CAS of mangoes with variables, often with contradictory results. The best combination for storage of 'Rad' mango fruits in Thailand was 6% O_2 plus 4% CO_2, followed by ripening at 13 °C (Noomhorm and Taisuwan, 1988, 1995). Low O_2 levels had little effect on ethylene production, firmness and sugar levels in mature green fruits. Off-flavours and skin discolouration were observed in fruits stored in CAS containing 1% O_2 and such effect is called 'low oxygen injury' (Hatton and Reeder, 1966). 'Low oxygen injury' also caused irreversible inhibition of ripening, especially chlorophyll degradation with resultant abnormal colour development on ripening. The most sensitive indicator of O_2 stress in mango was elevated ethanol production (anaerobiosis). At low levels of O_2, amylase and PG activities were depressed while alcohol dehydrogenase and pectin methyl esterase activities increased due to switch over to anaerobic metabolism. Ethanol and acetaldehyde, which increase by 5 and 15 fold respectively might be responsible for death of tissue at low O_2 levels (Lizada, 1991).

Fruits stored in 2% oxygen exhibited greatest suppression in CO_2 production after transfer to ambient air, compared to higher levels of O_2. The respiration mechanism and chemical composition of ripe fruit were not affected as a result of storage in 2% and 5% O_2 atmosphere for up to 4 weeks at 15 °C (Shukor *et al* 2000). Low O_2 level atmosphere showed a potential to maintain quality and slow ripening of mature green mangoes during marine transport (Bender *et al* 2000b).

Similar to lower levels of O_2, higher levels of CO_2 in CAS are detrimental to stored produce. Accumulation of toxic products (aldehyde and alcohol) was observed at high levels (5, 10 and 15%) of CO_2 in case of 'Alphonso' mangoes (Lakshminarayana and Subramanyam, 1970). CO_2 levels of 10 to 15% in a closed system of CAS (where the original concentration was modified by fruit respiration) caused 'CO_2 injury' in the form of greyish patches and softening and poor flavour. Symptoms of 'CO_2 injury' appeared in mangoes when exposed to 50 and 70% CO_2 during storage for sea transport of 2-3 weeks period (Bender *et al* 1994) – like irreversible inhibition of ripening upon transfer to air at the ripening-conducive temperature of 20 °C which was characterized by abnormal greyish epidermal colour, inhibition of normal aroma development and appearance of off-flavour (Bender *et al* 2000a). Fermentative decarboxylation with the formation of aldehyde etc., due to higher concentration of CO_2 impairs the consumers acceptability of CA stored mango fruits. CO_2 at 10% level alleviated chilling symptoms in 'Kensington Pride' fruit, but higher concentration (> 10%) was ineffective in alleviating chilling injury and tended to cause tissue injury. High levels of ethanol in the pulp could be either due to compounding injury (chilling + CO_2) or due to greater sensitivity of fruits to CO_2 (O'Hare and Prasad, 1993). The effects of elevated CO_2 concentration on delayed ripening of fruits have been attributed to its inhibitory action on ethylene biosynthesis (Bender *et al* 1995).

CA stored fruits particularly at lower concentration of O_2 and CO_2 showed abnormal ripening and fruit injury. In a CAS system, reduction of O_2 levels down to 3% had little effect on ethylene and ethanol production, while CO_2 levels of 45% and above caused severe damaged abnormal colour development and fruit injury and stored fruits in such atmosphere failed to recover after transfer to air (Bender *et al* 1995) and showed internal browning and off-flavours in atmospheres containing 6 or 8% CO_2 (Noomhorm and Taisuvan, 1995). CAS comprising CO_2 (> 6%) and O_2 (< 2%) induced anaerobic respiration and led to the development of off flavours and increased the concentration of acetaldehyde, ethanol and some esters (Singh *et al* 2004). The effects of reduced O_2 and elevated CO_2 on fruits are additive and can be greater than the effects of either one alone (Zagory and Kader, 1988).

CA storage at 6% CO_2 plus 2% O_2 was promising for extending the shelf life of 'Kensington' mangoes up to 35 days in Australia (Lalel *et al* 2001a). Fruit quality was maintained at this combination of gases, but CA comprising O_2 and CO_2 each at 2% seemed to be better for maintaining the aroma compounds in the ripe fruits (Lalel *et al* 2003). However, in a later study, CA storage at 2% O_2 and 3% CO_2 or 3% O_2 plus 6% CO_2 at 13 °C seemed to be promising for extending the storage life of 'Kensington Pride' and 'Delta R2E2' mangoes while still allowing the fruit to ripen normally with yellow colour, good taste, higher TSS and TSS : acid ratio and a high level of sugar content (Lalel *et al* 2005).

Mango fruits (Haden and Keitt) stored well in an atmosphere of increased CO_2 and reduced O_2 at 13 °C for 20 days, but required longer time to soften when removed to 21 °C than those stored in a normal atmosphere (Hatton and Reeder, 1966). Optimum atmosphere appears to be around 4% CO_2 plus 2-4% O_2 for 'Kensington Pride' mangoes in Australia (Mc Lauchlan *et al* 1992).

The optimum CAS gas composition for mango is recommended as 5% O_2 plus 3-5% CO_2 at 13 °C under Indian conditions (Santha and Rao, 2001). However, recently, the optimum gas concentrations recommended for mangoes (Alphonso and Banganapalli) under South Indian conditions are 5% each of CO_2 and O_2 (Anon, 2005). Under American conditions the optimum gas composition appears to be 3.5% each of CO_2 and O_2 at 12°-15 °C (Kader, 1990). The best gas composition for CAS of 'Heidi' mangoes in South Africa was 4% CO_2 plus 6% O_2 for extending the storage (Kruger *et al* 1997). According to Kader (1997) the typical recommended CA condition for mango fruits was 10-15 °C, 3-7% O_2 and 5-8% CO_2. However, a 5% O_2 and 5% CO_2 appears a more appropriate condition under which mangoes can be stored and ripened without injury and O_2 needs of the ripening fruit are met.

It is not advisable to store mango fruits in CAS below 5% O_2 at 5 and 10 °C, below 10% O_2 at 15 and 20 °C and below 15% O_2 at 25 °C, because of anaerobic respiration, which sometimes induces skin discolouration and off-flavours (Yamashita, 1991).

Factors influencing the success in CAS

The success in CAS of mango fruits depends on gas composition, storage temperature, cultivar, maturity of fruit, susceptibility to gases etc. of which maturity of fruit with references to its tolerance to altered gas atmosphere in the storage chambers is most important.

For successful refrigerated CA storage, the fruits at correct maturity should be harvested and used. 'Badami' and 'Raspuri mangoes at 'B' stage of maturity are suitable for CAS (Srivastava, 1967). 'Heidi' fruits in South Africa harvested at an advanced maturation, developed pulp browning in CAS. In fruits harvested at the conventional stage of maturation, CAS reduced surface scald, pulp browning and internal breaking and enhanced taste and shelf life (Oosthuyse, 1997b). Mature green fruits benefit from CA storage, but tree-ripe fruits are not suitable both in America (Bender *et al* 2000a) and in Japan (Nakamura *et al* 2003).

Ideally the mango fruits for CAS should be harvested and handled with the strictest possible attention to maturity, sanitation, avoidance of physical damage, proper temperature, management and packaging. Improper maturity, presence of decay organisms, bruising, fruits that are not cooled at the shipping temperatures, inadequate package venting or storage pattern that do not allow adequate air exchange or packages that do not adequately protect fruits from physical damage during handling or shipping are all factors that play role in successful CA storage (Bender *et al* 2000a).

CAS does not seem to be a promising one for many tropical fruits including mango, as compared to temperate fruits. This is due to several reasons including those related to crop availability, and quantity, pre-and post-harvest handling and availability of appropriate CA technology (Yahia, 1998). However, the studies on CAS of mango indicated the possibility of using this technique for mango, particularly for long distance transport by sea. The general conclusion of the studies has been that only moderate or no benefits are possible from CAS of mangoes. This is due mainly to limited disease control by moderate O_2 (5%) and CO_2 (5%) levels tolerated by mango fruit over extended storage period at 12-15 °C temperature, required to avoid chilling injury. Trinidad *et al* (1997) concluded that CAS did not have any benefit over refrigeration in extending the storage period of mangoes, since the fruits in CAS develop off-flavours and off-odours. However, Oosthuyse and Dodd (1998) from South Africa concluded that CAS is beneficial for mangoes. Commercial viability of CAS for mangoes has not been demonstrated at least in India. Trials conducted by APEDA for sea export of mangoes from India have met partial success only.

However, in view of short storage life of mangoes in tropical region, the application of CAS as an alternative to or in conjunction with cold storage assumes greater significance and appears as a boon in these regions in marketing mangoes throughout the year.

MODIFIED ATMOSPHERE STORAGE

'Modified atmosphere' means any deviation from the normal atmosphere gas composition around the produce/product. Modified atmosphere (MA) may designate any synthetic atmosphere but often means there is little or no possibility of adjusting gas composition within storage atmosphere during storage or transit. The storage atmosphere in which the commodities (fruits and vegetables) are held can be modified directly by artificially generating a specific atmosphere and flushing a storage container or by restricting the movement of oxygen to the commodity and carbondioxide from the commodity and then allow the commodity to generate its own atmosphere through normal respiration.

Sealing the commodity (fruits, vegetables etc.) in polyethylene film, wax, wrappers etc. is one of the means to achieve MA situation. Such technique is often referred to as "Modified atmosphere packaging" (MAP). It is a technique which creates and maintains a modified atmosphere around the commodity through the use of flexible film packages (Ben-yehoshua, 1985). It is the replacement of air in a pack with single or a mixture of gases, the proportion of each component is fixed when the mixture is introduced into the pack. No control is extended over the initial gas composition, which is likely to change with time owing to the diffusion of gases in and out of the packages and also due to the effects of product and microbial metabolism (Rai *et al* 2002).

Unlike controlled atmosphere storage, where the complete control on the gas composition is exercised, MA storage implies no or little control of gas concentration in the storage atmosphere surrounding the commodity. The main feature distinguishing

MAP from CA is that in case of MAP, active human involvement stops at the moment of sealing. MAP is an inexpensive method compared to CA storage and transport (Mc Glosser, 1989).

The ability of MAP to extend the storage life of foods has been recognised for over 100 years, but emerged in the 1970's as a new technology that may extend post-harvest life of perishables and reduce spoilage and decay. Since then there has been a substantial expansion in its use. Concerted efforts in MAP technology resulted in its practical application during export, temporary storage or even long term storage for fresh market or processing. MAP can also be utilized for transporting the produce from production centres to markets/distribution centres far and near. Long term shipping of mango in MA has been commercially used on limited extent by some countries like Mexico (Yahia, 1993).

Advantages/Benefits

1. Delay of ripening of fruits due to reduction in ethylene production and its action.
2. Reduction of weight loss, desiccation and shrivelling due to maintenance of proper humidity in the surroundings of the produce.
3. Alleviation of chilling injury.
4. Quality advantages such as colour, moisture, flavour and maturity retention.
5. Reduction of decay of produce due to diseases and pests.
6. Finally extension of storage life of the produce.
7. Reduction of labour and waste at the retail level.
8. Reduction of handling and distribution of unwanted or low grade produce.
9. Excellent branding options.
10. Use of MAS where refrigeration is not available or not economical.

Limitations

However, there are certain limitations and problems in the adoption of MA storage (in case of mangoes) if not properly carried out (Lizada *et al* 1984., Miller *et al* 1986).

1. Irregular ripening of fruits.
2. Development of off-flavours and off-odours (fermented)
3. Physiological breakdown of tissue probably due to CO_2 toxicity after prolonged storage (Thompson, 1971).
4. Failure of development of characteristic internal and external colour, aroma and sweetness.
5. Failure of softening.
6. Development of peel and pulp discolouration.

7. Risk of spoiled produce (decays/rots) due to improper packaging or abuse of temperature.
8. Possible occurrence of new risks of microbiological safety due to possible development of anaerobic pathogenic flora.
9. Requirement of additional investment in machinery and labour in the packing line.

Methods of creating MA conditions

1. Commodity-generated or passive modified atmosphere

In this method, the stored commodity consumes the available oxygen inside the package and liberates carbondioxide. After sometime there will be only CO_2 inside the package. This type of atmosphere must be established rapidly and without creating anoxic condition or injurious levels of carbondioxide to prevent any physico-chemical damage to the stored produce. When the rate of respiration of the stored commodity and permeability of film package are suitably matched it can create appropriate modified atmosphere in the package. The film package must allow O_2 to enter inside the package at a rate to off-set by the consumption of O_2 by the packaged produce and CO_2 must be vented out from the package to off-set production of CO_2.

2. Active modified atmosphere

The passively established modified atmosphere has a drawback that, it can not be regulated effectively and is largely dependent upon film permeability characteristics and stored-produce respiration rates. Therefore, it necessitates that MAP should be actively established and adjusted. This can be done by putting a slight vacuum and replacing the package atmosphere with desired gas mixture. This gas mixture can be adjusted through the use of absorbers in the package to scavenge O_2, CO_2 and ethylene. This method ensures the rapid establishment of desired atmosphere in the package (Kader, 1992).

(i) ***Oxygen absorbers*:** Most commonly available oxygen absorbers are : Ferrous oxide. Iron powder is the main active ingredient in powdered form which becomes Fe_2O_3 and Fe_2O_4 and their hydroxides, after absorption of O_2.

(ii) ***Carbondioxide absorbers*:** Hydrated lime, activated charcoal and magnesium oxide are some of the CO_2 absorbers which can prevent the build up of CO_2 to injurious levels in the packages.

(iii) ***Ethylene absorbers*:** Perishables produce ethylene which triggers ripening. Its prompt removal checks the ripening and extends the storage life. Potassium permanganate ($KMnO_4$) is a well known ethylene absorber. First it is adsorbed on celite, vermiculite, silica gel or alumina pellets. Then they are placed in side the package. They oxidise ethylene (C_2H_4) to CO_2 and H_2O. These scavengers are held in small sachets within the packages or impregnated in the wrappers or into porous materials like vermiculite. They help to ensure the delay of the climacteric rise in respiration.

Plastic films for MAP

Plastic films play a great role in creation of MA condition around the commodities when they are packed in them. Many different types of polymers are used for the purpose. But the most common ones are different types of polyethylene film, made from basic polymers like polyvinyl chloride (PVC), polyethylene terephoshealete (PET), polyphenyl (PP) and polyethylene (PE). Since the maximum CO_2 and minimum O_2 concentrations tolerated by different fruits and vegetables vary, a film should be selected which can maintain the desired gas concentration for a specific commodity for a desired length of storage.

The selected packaging material must be suitable in terms of optimal gaseous exchange which would help in creation of an appropriate atmosphere around the produce, thus increasing the shelf life of the packaged produce. The development of edible films with selective gas permeabilities would be very promising for controlling respiratory exchange and improving the conservation of fresh or minimally processed fruits and vegetables e.g. hydrocolloid based films such as gluten films, appear to have suitable barrier properties while remaining relatively permeable to CO_2.

The films used for tropical fruits like mango should have a relatively higher permeability to gases and water vapour. As O_2 content in a package is reduced from 21% to 2.5%, there is danger that CO_2 will increase from 0.03% to 16-19% which would be injurious to fruits. Therefore, an ideal film is one which lets more exit of CO_2 than O_2 entry. The CO_2 permeability should be somewhat 3-5 times greater than O_2 permeability depending upon the desired atmosphere (Zagory and Kader, 1988). The type of package further depends on the type of disposal, retail catering or for sea transport (Brody, 1993).

The package should allow enough O_2 to pass through and reach the product in order to eliminate the potential for anaerobic conditions. This should be done at such a rate that O_2 concentration in the head space is maintained at a much lower level than atmospheric O_2, for the maintenance of desired quality of the produce. Similarly, CO_2 is vented out of the package while the optimum level of CO_2 is maintained inside (Lioutas, 1988).

Modified atmosphere storage of mangoes

The technique of MA storage most commonly involves storing mangoes in sealed poly bags of suitable thickness (0.05-0.08 mm). The desired atmosphere is attained and maintained by an equilibrium between rates of respiration and diffusion of CO_2 and O_2 through the polyethylene film with or without scavengers. This MA condition is generally characterised by increased level of CO_2 and decreased level of O_2 created by respiration of fresh produce, reduced rate of respiration, delayed production of ethylene and reduction in ethylene activity (Kader, 1986). Reduction in the rate of respiration results in slow metabolism and potentially longer storage life. Similarly, a reduction in ethylene production associated with the MAS can delay the onset of climacteric and prolong the storage life of the fruits. Ethylene production is reduced by either low O_2, high CO_2 or both (Zagory and Kader, 1988). Further, reduced O_2 or elevated CO_2 can overcome the impact of the low temperature injury on the ripening process.

The feasibility of using MA technology for prolonging the storage life and marketing of mangoes has been demonstrated in Mexico (Gonzalez *et al* 1994). Packaging of mango fruits in LDPE film was more effective in reducing weight loss than wax emulsion coating (Diazsobac *et al* 1996). This was due to the development of a saturated atmosphere in the LDPE bags which prevented the weight loss. The MA created in LDPE bags was effective in delaying ripening of mangoes up to 16 days. Ripeing was further delayed up to 21 days in the presence of ethylene scavenger ($KMnO_4$) (Illeperuma and Jayasurya, 2002).

Wrapping of mangoes in plastic films for extending the shelf life did not provide promising results. Miller *et al* (1983) concluded that there were no practical benefits associated with wrapping of mangoes in polyethylene film and storing them at 21 °C or even at low temperature, though there were no detrimental effects on carbohydrate metabolism. Chaplin *et al* (1982) and Miller *et al* (1986) did not favour the wrapping of mango fruits in polyethylene film due to development of off-flavours. However, the studies of Mc Collum *et al* (1992) indicated that 'Tommy Atkins' mango fruits could be stored under MAP conditions in cold storage beneficially for retarding ripening process and achieve the longest possible storage period with retention of quality and improvement in marketability. 'Alphonso' and 'Baneshan' mangoes packed in PEBA `X' film after pre-cooling were kept in unripe condition for 25 days at 13 °C (Anon, 2002-03a).

'Tommy Atkins' fruits individually sealed in heat shrinkable films and stored for 2 weeks at 12.8 °C and then ripened at 21 °C had less weight loss, did not show loss in firmness, skin colour development, decay, but showed more off-flavours than unwrapped fruits (Miller *et al* 1983). In contrast, fruits of 'Keitt' individually packed in LDPE and HDPE and kept at 20 °C and 27 °C for 4 weeks, showed delay in ripening, reduction in weight loss, but not off-flavour in Mexico (Gonzalez *et al* 1990). The storage life of 'Alphonso' mangoes was extended to 30 days at 8 °C, when the fruits were packed individually, in micro-perforated PE or PP film without chilling injury (Anon, 2003-04).

In non-perforated polyethylene bags, there will be build up of CO_2 and moisture to toxic levels both of which should be vented out for success of MA storage, particularly when no scavengers are employed. Water saturated atmosphere around the fruit in non-perforated polyethylene bags favours the development of decays. In addition, there will be an accumulation of ethylene inside the packages which accelerates ripening. Accumulation of CO_2 coupled with low O_2 in non-perforated bags leads to 'CO_2 injury' to the fruits. All these can be avoided by perforation of the polyethylene film bags etc. Perforation retains many of the benefits of sealing like reduction in water loss and alleviation of water stress, without deleterious effects of anaerobiosis (off-flavours, fermentation, CO_2 damage). Singh *et al* (1998) found that perforated polyethylene film wrapping was the most effective treatment for improving the storage life of 'Amrapali' mango fruits in India. There is a need for effective disease control treatments and effective methods for absorbing of water and ethylene in the packages with saturated atmosphere (Kader *et al* 1989). The growth and activity of microorganism can be retarded by elevated CO_2 and lowered O_2 concentration in the packages. Elevated CO_2 levels (10-15%) can be used to provide fungistatic effects on commodities that tolerate such CO_2 levels (Kader *et al* 1989). Combination of hot benomyl (1000 ppm) treatment at 55 °C for 5 minutes and seal packaging in 0.01 mm thick PVC film extended the storage life of 'Nan-Dok Mai' mango fruits stored at 13 °C without decay incidence (Sornsri vichai *et al* 1992). Minimum spoilage was observed in fruits treated with Bavistin (500 ppm) stored under MA condition (Lal *et al* 2003). The carbendazim (0.05%) treated fruits of 'Dashehari' and 'Amrapali' (with 8-10 cm stalks) packed in 2% ventilated polyethylene bags of 200 gauge thickness showed complete control of post-harvest diseases (stem-end-rot and anthracnose) up to 10 and 12 days of storage respectively (Anon, 2001-02).

Less peel and pulp colour development of MA stored fruits was observed, which are dependent on the type of film, perforation, O_2 level and pore area. Fruits packed in all types of films showed unsatisfactory peel and flesh colour upon prolonged storage at 13 °C (Sornsrivichai *et al* 1992., Ben-yehoshua, 1995). Perforation is needed for the development of colour of both peel and flesh (Sornsrivichai *et al* 2000). Colour development of both peel and flesh is controlled by O_2 level in the package (Yantarsri *et al* 1994). A concentration of 16% O_2 is essential to develop the peel colour. To develop the marketable colour of the peel at 28 °C a minimum total pore area (TPA) of 0.64 cm^2 is required, while for flesh (pulp) a TPA of 0.08 cm^2 is needed to bring O_2 to about 17% and 9% for proper colour development respectively (Sornsrivichai *et al* 2000).

Factors affecting MAS/MAP

1. ***Temperature and relative humidity*:** Ambient temperature of the surrounding atmosphere affects the temperature of the commodity. Temperature changes also affect the permeability of the film, which increases with increase in temperature. CO_2 permeability responds more than O_2 permeability. Storage temperature also affects the influence of MA, whose inhibitory effect on pathogen growth decreases as the temperature rises and follows the solubility of CO_2 (Dixon and Kell, 1989).

   Relative humidity has less effect on permeability of film package. Most common films are good barriers to moisture and vapour, because they maintain high internal humidity even in dry ambient condition (Kader *et al* 1989). Condensation inside the package adversely effects the gas permeability of the film. Therefore, proper temperature should be maintained throughout the post-harvest handling for preventing condensation within the packages.

2. ***Sanitation*:** The high humidity maintained within MAP may enhance the growth of decay pathogens. So care must be taken to ensure proper sanitation and to avoid

conditions favourable to growth and reproduction of such microorganisms. Fungicidal treatment of packaged fruits is thus very important in MA storage.

***Others*:** Various other factors determine the viability of a product under MAP. Quality of raw produce coupled with proper temperature control throughout the food chain are the two important factors that account for 50-60% of the total success accomplished in MAP system. The packing materials, packing equipment and gas mixture are the other factors, which are more product specific and account for the remaining 40% of the total success (Rai *et al* 2002).

Therefore, many factors must be considered while trying to develop a MAS/MAP systems including type, thickness and method of fabrication of the film, package size, temperature, humidity, length of storage, type, quantity and physiological state of fruit and tolerance of each type of fruit to different gases in the MAP.

STORAGE BY USING CHEMICALS

The possibilities of using low temperature storage for mango are limited by the high capital investment and susceptibility of mangoes to chilling injury at temperatures below 12-15 °C (Santha and Joshi, 1989). CA/MAP of mangoes are also limited due to high incidence of rots, fermented odours, off-flavours and internal breakdown of fruits (Gautam and Lizada, 1984). Therefore, alternate techniques for extension of storage life of fruits, including mango were sought. In these attempts chemicals like calcium and plant growth regulators were tested and found as possible means of extending the post-harvest life of treated fruits, besides avoiding physiological disorders of fruits. Chemical storage treatments involve the use of dip in chemicals like calcium salts, fungicides, ethylene inhibitors or plant growth regulators with or without the combination of more than one treatment. The treatment efficiency depends on the cultivar, maturity, storage regime and infection set in the fruit at the time of harvest/treatment.

Calcium

In the process of seeking alternate methods for extending the storage life of perishables, the role and importance of calcium (Ca) in extending the storage life as well as ameliorating the physiological disorders of fruits have come to the light, particularly in temperate fruits like apple. And the same had been extended successfully to tropical fruits like mango.

Role and importance of calcium

Calcium (Ca) is an essential secondary plant nutrient, involved in a number of physiological processes in the plants. It is believed to be also involved in anti-ripening and anti-senescence processes in fruits (Lester and Grusak, 1999). It is known to play an important role in quality retention of fruits and vegetables by maintaining the firmness, reducing the respiration rate and ethylene evolution (Poovaiah, 1986) and decreasing rots and disorders (Conway and Sams, 1984). Ca protects cell membranes, being a part of the cell wall, from disintegration and maintains cell integrity and protein synthesis of cells (Faust, 1975). Ca is therefore, essential for the maintenance of cell wall structure, especially in fruits and vegetables, that are stored for prolonged period (Poovaiah, 1988).

Ca is perhaps the most important mineral element determining the storability of fruits, which depends on the levels of Ca in the fruit. Deficiency of Ca leads to reduction in storage life and higher incidence of physiological disorders (Wills *et al* 1998). Fruits with enhanced levels of Ca have shown greater resistance to the decay development (Yuen, 1993), maintained firmness, delayed ripening, reduced respiration and extended storage life (Tirmazi and Wills, 1981., Wills *et al* 1998., Yuen *et al* 1993., Singh *et al* 1993). Low-Ca fruit has very high respiratory rate and the respiration is inversely proportional to Ca concentration (Faust and Shear, 1972). The high respiration rate decreases the storability of fruits. Ethylene production is also inversely related to Ca concentrations in the fruit (Conway and Sams, 1987).

Role of calcium in fruit firmness

Ca (application) retains firmness of fruit, which is an important quality parameter during storage. Ca components make the middle lamella of cell wall more thick by increasing the deposition of calcium pectate and thus retains the firmness of fruits. Ca serves as in intermolecular binding agent that stabilizes pectin-protein complexes of the middle lamella (Dey and Brinson, 1984). Ca delays softening (in apples) by virtue of its ability to delay degradation of cell wall polymers (Sams and Conway, 1984). Thus, fruits treated with Ca are generally firmer than untreated ones. The fruits treated with 8% calcium chloride ($CaCl_2$) solution were 25% firmer than untreated ones (Conway and Sams, 1987).

Calcium and ethylene evolution

Ca is known to inhibit ethylene bio-synthesis (ripening hormone), which is considered as one factor contributing to delayed ripening and senescence. Tingwa and Young (1974) found that Ca reduces respiration at both pre- and post-climacteric stages of ripening and delays the onset of ethylene production. Ca can also stabilize ethylene production in senescent fruits (Ferguson, 1983). It also reduces CO_2 evolution (Poovaiah, 1986).

Sources of calcium

Two different sources of calcium viz., calcium chloride ($CaCl_2$) and calcium nitrate ($Ca(NO_3)_2$) were tested as pre-and post-harvest treatments to extend the storage life of mangoes. Pre-harvest sprays of $Ca(NO_3)_2$ and $CaCl_2$ extended the storage life of 'Amrapali' fruits more than 10 days, by delaying the onset of senescence during storage, while fruits dipped in aqueous solution (post-harvest) of above chemicals maintained their quality and marketability up to 8 days as against 4 days in control (Singh *et al* 1987).

Post-harvest dips in either 4% (Tirmazi and Wills, 1981) or 8% (Mootoo, 1991) $CaCl_2$ solution have delayed ripening of green mature mangoes and thereby extended their post-harvest life. Post-harvest application of Ca $(NO_3)_2$ reduced the loss in weight and rotting of fruits of cv. 'Dashehari' at 2% concentration (Kaushik and Kumar, 1991).

Of the two sources of calcium, $CaCl_2$ was superior to Ca $(NO_3)_2$ as its impact on various storage parameters was pronounced. $CaCl_2$ (6%) was most effective in reducing the weight loss and in improving colour and quality of the fruit and the treated fruits were marketable even after 12 days of storage at room temperature (30 $\pm$ 2 °C, 60-65% RH) as compared to 7 days in control fruits (Anon, 1990-91).

Time of application

Time of application of Ca to fruits is an important consideration, since the response differs. Basically fruits for extension of storage life are treated either before harvest (pre-harvest) or after harvest (post-harvest). Both showed promising results. Pre-harvest applications are in the form of sprays, whereas post-harvest application is through dips or infiltration.

Pre-harvest application

Pre-harvest sprays of Ca have distinct advantage over post-harvest application, in that it would not need special structures for treating the fruit, but requires larger quantities of chemicals, since the whole trees are to be sprayed, whereas in post-harvest application, smaller quantities are needed, since only fruits are treated, and the same solution can be repeatedly used. Another demerit of pre-harvest application of Ca is loss of Ca from fruits to another parts of the tree due to mobilisation. Further, pre-harvest application of $CaCl_2$ (1-2%) causes scorching of the leaves, which was attributed to chloride ions in the salt. This results in reduced absorption of Ca into the fruit pulp (Singh *et al* 1993). Pre-harvest sprays of Ca $(NO_3)_2$ (1 or 2% Ca) or $CaCl_2$ (0.6 – 1.2% Ca) 20 and 30 days before harvest delayed ripening, extended storage life, lowered weight loss, reduced respiration rates and showed a favourable effect on fruit quality during storage (Singh *et al* 1993).

Post-harvest application

Post-harvest application of Ca includes two methods :

(i) dipping or immersing of fruits in Ca solutions and

(ii) pressure or vacuum infiltration of Ca salts.

In the former, harvested fruits are dipped/immersed in aqueous solution at a particular concentration for a specified period and then stored.

Use of post-harvest immersion (dip) in $CaCl_2$ solution to delay ripening and to extend storage life has a potential commercial application in transportation of fruits to distant markets as well as greater flexibility in marketing (Santos *et al* 2004).

Pressure/vacuum infiltration of Ca into the fruits improves the Ca uptake, effective in reducing decay, physiological disorders and retaining fruit firmness (Yuen, 1993). The benefits are more evident in case of fruits harvested in the early maturity stages (Santos *et al* 2004).

However, no marked differences in effect were observed between the two methods, but dipping is recommended being easier to adopt and apply (Zambreno and Manzono, 1995). However, fruits infiltered with $CaCl_2$ (4-6%) solution under vaccum gave better extension in shelf life compared to control (Hudder and Reddy, 1989).

Calcium Treatments and Mango

The primary aim of Ca application to mango fruits is to extend the storage life without impairment of quality. The reported results indicate the promising effects of calcium treatments in prolonging the storage life of treated fruits. Ca treatments extended the storage life of fruits through delaying ripening, senescence and colour development, reducing the rate of respiration, maintaining fruit firmness and delaying softening. Besides the above, Ca treatments also reduced weight loss during storage and minimized the disease incidence by strengthening the cell walls, making them more resistant to decays (Conway and Sams, 1984). However, Ca treatments caused surface injuries sometimes, limiting the use of Ca treatments for extending the storage life of fruits (Tirmazi and Wills, 1981., Hudder and Reddy, 1989., van Eeden, 1992., Shorter and Joyce, 1998). Post-harvest treatment of Ca at higher concentrations resulted in skin damage around the lenticels (lenticel spotting) manifested as lenticel blackening (van Eeden, 1992; Shorter and Joyce, 1998) due to accumulation of Ca ions (Joyce *et al* 2001). Fruits treated with $CaCl_2$ (4%) by infiltration, showed skin black spots, soaked areas and decay. Fruit injury was shown in the form of brown, circular succulent lesions on the fruit surface and decay. Internal disorder due to excess Ca in the fruit was manifested as internal breakdown and hollow packets in the flesh. The breakdown was localized in the area above the seed in the apex of the fruit (Santos *et al* 2004). The injury might have been due to slow movement of Ca from epicarp to mesocarp tissues on prolonged storage at low temperature (12 °C) (Singh and Narayana, 1999). However, the above fruit injuries were not observed in dip treated fruits (Mootoo, 1991). The peel injury due to infiltration of Ca can be reduced by increasing the temperature of the flesh or the $CaCl_2$ solution during infiltration or by packaging the fruits in sealed polyethylene bags or cling films or shrink wraps after Ca treatment (Yuen *et al* 1993).

Combination of Ca with other storage techniques

Often, other storage techniques are coupled with Ca treatments for additional benefits. The hypobaric infiltration of $CaCl_2$ (dehydrated) on the storage behaviour of 'Dashehari' mangoes under ambient conditions indicated that 4% $CaCl_2$ at 0.25% sub-atmospheric conditions increased storage life for 9-10 days (Anon, 1991). These treatments reduced the weight loss, controlled degreening and arrested carotenoid development. The Ca infiltration through the hypobaric route was thus superior to hypobaric storage alone for enhancing the storage life. Post-harvest treatment of Ca coupled with MAP prolonged the storage life of mangoes besides improving the quality of mango fruits (Singh *et al* 2000).

Use of Ethylene Absorbants

Many volatile compounds, evolved by the produce and from other sources in the storage atmosphere, accumulate during the storage of perishable produce. Ethylene is the most important of these volatile compounds. Its accumulation above critical levels accelerates ripening and reduces the storage life of stored fruits/vegetables. So its removal from the storage atmosphere or inhibition of its biosynthesis assume greater significance in extending the storage life of these perishables. One of the means of its removal is the use of some chemicals, which absorb the evolved ethylene. A few chemicals were identified for the purpose (e.g., Potassium permanganate, ethysord), which are called ethylene absorbants (removers).

Use of ethylene absorbants like potassium permanganate ($KMnO_4$) was helpful in enhancing the shelf life of bananas (Scott *et al* 1970). Ethylene inhibitors like amino-ethoxy Vinyl glycine (AVG) enhanced storage life of pears by slowing down ripening process (Romani *et al* 1982).

Raje *et al* (1997) attempted the use of preservatives, ethylene absorbers etc. to extend the shelf life of 'Alphonso' mangoes with two approaches :

(i) adsorption of ethylene and

(ii) chemical destruction of ethylene.

Fruits were stored at 32°-36 °C and 70-75% RH in CFB cartons with four test materials, viz., (a) Ethysord (ethylene absorber), (b) Stop rot (SO_2 releaser), (c) Ethylene oxidizer and/or (d) halogen release. Out of these, ethysord preserved the fruits up to 16 days against 8-days in untreated control, with minimum PLW, higher sensory score and lower microbial load. Combination of ethysord and stop rot (SO_2 releaser) achieved the best results. Ethylene oxidizer and halogen releaser slowed ripening after the above two and resulted in better overall fruit condition compared with control fruits.

Packing of the "Banganapalli" mango fruits in cardboard box containing permanganate silica gel (10 g) and stored in zero energy cool chamber was most effective in delaying the spoilage of the fruits and extending their storage life (Gautam *et al* 2003).

PLANT GROWTH REGULATORS

A number of synthetic plant growth regulators (auxins, gibberellins etc.) has been tested to prolong the storage of mangoes, as these retard ripening and senescence of fruits. Some proprietary products containing the plant growth regulators were also screened for storage of mango fruits.

Auxins

Mango fruits (cv. S.B. Chausa) were successfully stored for a period of 20 days at room temperature, following the treatment of fruits with 2, 4-D/2, 4, 5-T. The greatest delay was with 2, 4-D (400 ppm) followed by 2, 4, 5-T (60 ppm) (Deol, 1985). Kumar *et al* (1992) observed extension of storage life of mango fruits with post-harvest treatment of 2, 4, 5-T besides improvement in the quality, in terms of TSS, total sugars, reducing sugars during storage.

NAA (100, 200, 300 ppm) as a water dip-treatment was effective in prolonging the storage life of 'Kesar' mangoes (Wavhal and Athale, 1988). The shelf life of mango fruits could be effectively prolonged by 4-8 days by treating them with NAA (200 ppm) and packing in polyethylene bags or with $KMnO_4$ in polyethylene bags (Pruthi, 1992).

Gibberellins

Pre-harvest sprays of gibberellic acid (GA_3 at 100-300 ppm) significantly retarded ripening of mango fruits for up to 6 days of storage at ambient temperature, allowed retention of more ascorbic acid during storage (Kader, 1991). The storage life of 'Baneshan' mango fruits was considerably increased by pre-harvest sprays of GA_3 (15 ppm) coupled with carbendazim (0.5%) and fruits under this treatment were free from pathogenic infection even at advanced stage of ripening (Sudhavani *et al* 2000).

Several studies have shown that post-harvest treatments of fruits with GA_3 with or without other storage techniques extended the storage life. Mango fruits treated with GA_3 (50 ppm) and sealed in perforated polyethylene bags were stored longer than those in control (Joshi and Roy, 1985., Singh, 1988., Gautam and Chundawat, 1990., Ahmed and Singh, 2000). GA_3 (200 ppm) treated fruits of 'Bangalora' and 'Neelum' stored in ventilated polyethylene bags with ethylene absorbent showed significant increase in the extenstion of the storage life for 18 and 17 days respectively (Srinivasa Reddy and Haripriya, 2002).

Maleic hydrazide

'Dashehari' mango fruits treated with maleic hydrazide (1000 ppm) and stored in perforated polythene bags remained stored for 12 days and were acceptable at the end of the storage period. Dipping mango fruits in a solution of maleic hydrazide (1000 ppm) and wrapping in news paper recorded a storage life of 11 days (Singh and Gupta, 2004).

1-Methyl cyclo propane (1-MCP)

Ethylene enhances and accelerates ripening of climacteric fruits. The ripened fruits do not keep long. Recent studies have shown that 1-Methyl cyclopropane (1-MCP) could inhibit fruit ripening by binding irreversibly to ethylene-binding sites in the fruit. As a result ethylene is unable to bind and elicit subsequent signal transduction and translation (Serek *et al* 1994). Fruits treated with 1-MCP remain insensitive to ethylene (Sisler and Serek,

1997). The inhibiting effect of 1-MCP on fruit ripening has been widely observed in various fruits. Mangoes treated with 1-MCP showed extended storage life without undesirable effects on fruit quality (Wang *et al* 2006).

Polyamines

Post-harvest application of polyamines such as putrescine, spermine and spermidine with 'Tween 20' (0.01%) as surfactant on mango fruits has shown a positive influence on the shelf life of the fruit without adverse effect on fruit quality (Malik *et al* 2003).

INTEGRATED STORAGE

As seen from the foregoing account on storage of mango fruits, various efforts have been made to enhance the storage life of mangoes using different methods alone or in combination. Though the extension of storage life was achieved, each approach had one or other limitations for commercial adoptation. Therefore, an integrated approach was conceived and was tested to see the cumulative effect of combination of various treatments. The results indicated that the mango fruits harvested with stalks (8-10 cm) could be stored successfully up to 27 days (21 days at 12 °C with 85-90% RH and 6 days at ambient temperature) following the dipping of fruits in a fungicide solution in acceptable good condition without spoilage compared with control fruits which could be stored for 17 days only. The increase in storage life was due to synergestic effect of intact stalk, which slowed down the respiration, and low temperature which further reduced the rate of respiration, ethylene production and other metabolic activities associated with ripening process (Singh and Narayana, 1999).

A combination of Bavistin dip, stem-end-waxing and HM film wrapping proved best for enhancement of storage life of mango fruits (Garg *et al* 1971., Narayana *et al* 1996). When this treatment was coupled with low temperature, the fruits of `Baneshan' had a storage life of 23 days and were acceptable and marketable at the end of storage (Narayana *et al* 1996).

Low cool, low input storage method like evaporative cool chamber for the on farm storage and pre-cooling of 'Kesar' mango fruits at 10 °C for 12 hours and storage in cool store (Walk-in-cool) contributes towards the retention of post-harvest quality up to a substantial period of 26 and 50 days in cool chamber and cool store respectively after harvest when simple fungicidal wax treatment (Waxol 6%) plus carbendazim (0.1%) was given prior to storage (Waskar and Dhemre, 2003).

CHANGES DURING STORAGE

Stored fruits undergo physical, chemical, physiological and biochemical changes during the period of storage. The changes, however, depend on fruit maturity at harvest, methods of storage, duration of storage, storage temperature etc. In general, TSS, proteins, β-carotene, total sugars, TSS : acid ratio, amylase activity and PLW increase, while moisture content, firmness, starch, acidity, ascorbic acid, respiration and enzyme activity decrease during storage. Skin colour also shows change. The ultimate quality of fruit after storage depends on all the above changes. The duration of storage depends on these changes. When the changes occur rapidly, the storage life of the produce is less. On the other hand, slow rate of changes leads to enhancement of storage life of the produce.

Physical changes

At low temperature (10 °C) the respiration of mango fruit is markedly inhibited. The respiratory rates in early stages of storage were high compared to fruits before storage, which decline as storage period progressed (Ueda *et al* 2001).

The change in skin colour from green to yellow is delayed during low temperature storage (Li-anni *et al* 1993). Similarly white green colour of the pulp turns yellow-orange at the end of the storage period (Baez-Sanudo *et al* 1997b). The change of skin as well as pulp colour was faster at ordinary room temperature, due to high temperatures.

Moisture content of the fruit decreased gradually during storage due to transpiration, with concurrent physiological loss in weight till the end of storage period (Lakshminarayana *et al* 1974., Joshi and Roy, 1988., Li-Anni *et al* 1993., Baez-Sanuado *et al* 1997b). Higher the temperature, more is the loss in weight due to increased respiration and transpiration and these losses increase progressively during storage, irrespective of the storage method. Along with loss of weight, flesh firmness decreases with increase in storage duration.

Chemical Changes

Total soluble solids (TSS)

The content of TSS increases during storage. The increase is dependent on storage temperature. Increase in TSS of stored mangoes was more rapid at room temperature than at low temperature due to slow conversion of starch into sugar and slow degradation of pectic acids, fats and protopectin at low temperature. At 1.7°-3.3 °C the content of TSS in mangoes remained practically the same throughout the storage period; at 4.4°-7.2 °C, there was an increase in the content of TSS, followed by a decrease till the end of storage period (Singh *et al* 1954). Salama (1959) found rapid increase in TSS at higher temperature ranges and after attaining a peak, the TSS declined very slowly till the end of storage.

Acidity

Acidity of the stored fruit decreases over the storage period, rendering the fruit palatable. However, the rate of decrease depends on the storage temperature and stage of fruit maturity at harvest. Acidity decreases rapidly in fruits at room temperature, as a result of high temperature and low humidity, at which faster degradation of organic acids into sugars and utilisation of acids in respiration occurs in mango fruits (Sahani *et al* 1994). A positive correlation between storage temperature increase and acidity loss was observed during storage of 'Pairi' mangoes (Abou-Aziz *et al* 1975). Total titratable acidity showed decline during transport and storage (Joshi and Roy, 1988). Chandler (1950) reported that the acid content of green fruits declined rapidly when they were held at 7.2 °C or higher. There is a correlation between the acid contents of green fruit and the length of storage life, the latter being short in the case of fruits with higher acidity (Mathur *et al* 1953).

Starch and sugars

Starch content decreases while certain sugars show increase in fruits during storage. The decrease is faster at room and higher temperature storage. The initial content of starch in fruits determines the post-harvest life of fruits. The decline in starch content is mainly due to its hydrolysis into sugars (Pandey *et al* 1974).

Total sugar content increases with prolonged storage and the rate of increase is rapid at higher temperature and at room temperature. In storage, sucrose tends to disappear and is replaced by reducing sugars (Abou-Aziz *et al* 1975). Fructose content increased while polysaccharides decreased with storage at 7° to 25 °C (Ueda *et al* 1999). Reducing, non-reducing and total sugars increased up to a certain period and after attaining a peak, reducing and total sugars declined slowly, whereas non-reducing sugars remained constant till the end of storage life.

Ascorbic acid

Like acidity, ascorbic acid content decreases in the stored fruits till the end of storage life (Lakshminarayana, 1973., Thomas and Oke, 1980). However, Dhaka *et al* (2001) reported an increase in the content of ascorbic acid in 'Totapuri' mangoes during storage.

BIOCHEMICAL CHANGES

Alcohol-insoluble solids : The percentage of alcohol insoluble solids (AIS) decreased in the mango fruits of 'Hindi' and 'Pairi' cultivars in storage at various temperatures, the decrease being greater at the higher temperature range (Salama, 1959., Ueda *et al* 1999). Storage of mangoes at 0 °C slowed down the rate of loss of AIS (Abou-Aziz *et al* 1975). A gradual decrease in AIS together with a parallel decrease in starch content occurs. Starch appears to be the major alcohol insoluble matter in mango and as it gradually converts to sugars, the insoluble matter is diminished. Decrease in soluble pectin occurs at low temperature (Mukherjee, 1969-71).

Carotenoids: These pigments show increase in storage due to their biosynthesis and degradation of chlorophyll. The rate of increase of carotenoids depends on the storage temperature. The rate was faster in fruits held at room temperature due to their accelerated biosynthesis and faster degradation of chlorophyll as a result of higher temperature and lower humidity (Roy and Pandey, 1983., Dhemre, 2001., Masalkar *et al* 2006).

Enzymes and their activity

Several enzymes are associated with ripening of fruits (See chapter-Fruit Ripening) some of them disappear, while others appear in storage. The activities of PG and PE in peel and pulp increase, being higher in the peel than in the pulp. The activity of PPO, catechol oxidase during storage gradually decreased and the activity of POD increased. The activity of cellulase in the peel decreased, but increased in the pulp, reaching a peak and then declined (Chen *et al* 1998). Chlorophyllase in the peel increases the degradation of chlorophyll. There was a boost in amylase activity, but after sometime in the storage, its activity declined (Kalra and Tandon, 1983).

FACTORS AFFECTING THE STORAGE LIFE

The storage life of mango fruits is influenced by several factors : Pre-and post-harvest factors, method of storage, storage conditions (temperature, humidity, gases) etc. Effective preservation of fruits in the fresh condition, with no or least loss of quality is governed by factors such as, stage of maturity at harvest, time of harvest, handling and transport conditions, method and length of storage, storage temperature and relative humidity, cultivar, infections and injuries at the time of harvest, rate of cooling, rate of respiration, rate of accumulation of CO_2 and ethylene, pre-packing etc. Controlling or managing the above factors checks the ripening, senescence and spoilage of fruits and enables to extend the storage life of fruits.

Pre-harvest factors

Some of the pre-harvest factors, like balanced nutrition, timely and amount of irrigation and management of insect pest and diseases determine the post-harvest life and quality of fruits. Heavy doses of nitrogen applied for achieving higher yields, reduce the fruit quality as well as its storage life. Similarly, irrigation a few days prior to harvest also reduces the quality and storage of fruits. Incidence of pests and diseases also reduces the quality and storability of fruits. Bagging the fruits during the development was found beneficial in avoiding the pests and diseases and extending the shelf life (Joyce *et al* 1997).

Stage of maturity

The maturity of fruit at harvest determines its quality and storability. The fruits must be harvested at correct stage of maturity. The fruits harvested prior to the optimum stage of maturity show erratic ripening and store poorly. Fruits harvested at late maturity stages show reduced storage life with greater susceptibility to diseases and storage disorders during storage. The stage of maturity at harvest, however, depends on whether they are intended for domestic or export markets, method of storage, storage temperature, gaseous composition, duration and kind of transport and duration of storage.

'Alphonso' mangoes with shoulders out-growing the stem end and olive green in colour were the best for export (Cheema and Dani, 1934) and such fruit show long storage life and ripen uniformly and retain their colour, taste and flavour after cold storage.

Prolonged storage of immature fruit is not possible due to rapid rate of ripening even at reduced temperatures and such fruit does not attain optimal quality on ripening (Medlicott *et al* 1988). Such fruits often fail to soften to the same extent as fruit harvested at optimum maturity.

The physiological and biochemical activity of over mature fruits differs with respect to the rate of respiration, transpiration, conversion of starch into sugar and storage life. Two days delay in harvesting after the attainment of physiological maturity shortened storage life by two weeks (Mukherjee, 1969-71).

Harvesting time

For long term storage of mangoes, early season harvests are better. In subsequent harvests, fruits exhibit substantial ripening changes during storage, which effectively reduce the storage life. In mid-and late-season harvests, half-mature fruits are preferred, which can be stored at 10 °C for 21 days (Hatton *et al* 1965).

Method of harvesting

Method of harvesting has considerable bearing on the post-harvest life of fruits. In India, mangoes are generally hand picked, sometimes branches are vigorously shaken or beaten and dropped fruits are collected. When the branches are shaken or beaten, the dropped fruits get bruised or cracked, leading to huge losses even before storage/marketing. Hand picking of mangoes was considered superior in terms of quality and decay control than harvesting by shaking or beating the branches (Yagi *et al* 1978).

Harvesting devices have been developed by different research institutions in India, which should be utilized for harvesting mangoes to overcome the demerits of hand picking etc. Harvesting of mango fruits by right technique of using harvesting aids reduces mechanical injury, post-harvest losses and enhances post-harvest life of fruits. Mango fruits (cv. Kesar) harvested by different harvesters developed in the country recorded equal storage life of 15 days at ambient conditions (24.1° - 38.1 °C and 42.2 - 81.3% RH) compared with 11 days storage in case of manually harvested fruits (Wasker *et al* 1997). The improved method of harvesting using a pole equipped with a long cloth sleeve held open by a ring, greatly improved fruit quality, reduced post-harvest losses and extended the storage life of mango fruits in Sudan (Abou-Goukh and Mohamed, 2004).

In most cases fruits are harvested without stalk, since the growers believe that stalks do damage or come in the way in handling, packing etc., of fruits. However, during hand harvesting of mangoes, the latex (sap) trickles down on the fruit surface from the point of detachment of stalk and imparts shabby appearance to the fruits (Mukherjee, 1972). Besides, the latex attracts spoilage organisms during storage/transport. Hence, it is advised to harvest the fruits with stalk (8-10 cm long) irrespective of the method of harvesting (Photo 3.1). Pedicellate fruits showed less infection than non-pedicellate fruits upon ripening and stored for 2-3 days more at ambient conditions (35 $\pm$ 2 °C, 60-65% RH) with minimum spoilage (Pathak and Srinivasa, 1969., Roy and Pal, 1991a., Singh *et al* 1993). Further, the fruits harvested with stalk had good appearance and marketability up to 10 days, mainly due to slower respiration, glossy appearance and freedom from decay (Singh *et al* 1993).

Post-harvest factors

Post-harvest changes in fruits, method of storage and atmosphere conditions in storage etc. are some of the post-harvest factors which influence the storability of harvested fruits.

Post-harvest changes

After harvest of fruits, losses of respirable structures and moisture are not replaceable, as in the case of intact fruits, therefore, deterioration occurs, soon followed by senescence. The biological activities in the harvested fruits, cause decline in quality and limit the post-harvest life. Besides physiological and biochemical changes, other factors also influence the limited shelf life of perishables which include mechanical damage, pathological (diseases) causes etc. (Burton, 1982).

Loss of water : Transpirational losses of water is one of the main processes that affects the physiological and commercial deterioration of perishables after harvest, resulting in reduced storage life. The moisture loss adversely affects the appearance, texture, flavour and weight of the produce. Most noticeable effect of moisture loss is the softening of the tissues due to loss of turgidity (Ben-Yehoshiua, 1987).

***Respiration and metabolism*:** Perishables (fruits and vegetables) are living commodities and continue to respire and carry other metabolic activities even after harvest. Respiration uses the stored food leading to its depletion and consequently to the loss of quality and storability of the produce (Burton, 1974).

Microbial spoilage : Fruits and vegetables are susceptible to the infection of a variety of decay causing microorganisms, leading to substantial decay losses, which reduce the storage life and their marketability. (See chapter: Post-harvest diseases).

Methods of storage

Storability of mango fruits is influenced by different methods of storage. Storage life of ripe fruits (cv. Baneshan) was 9 days in cold storage, 6 days in ZECC and 4 days at room temperature. The storage life was further extended when the fruits were treated with Bavistin (500 ppm) prior to storage and the stem end was sealed with molten paraffin wax, followed by individual wrapping in HM film to 23 days in cold storage, 12 days in ZECC and 8 days at room temperature (Narayana *et al* 1996) 'Kesar' mango fruits treated with Waxol (6%) plus carbendazim (0.1%), pre-cooled at 10 °C for 12 hours and then stored in cool storage recorded maximum storage life of 50 days as compared to cool chamber and room temperature stored fruits which recorded the minimum shelf life of 26 and 20 days respectively. This was due to low temperature coupled with high humidity in cool chamber as compared to room temperature (Waskar and Masalkar, 1997., Waskar and Dhemre, 2003).

Photo 3.1 : Fruits harvested with pedicel (Stalk)

Storage conditions

Storage temperature : Temperature is the most important factor affecting the physiology and storage life of the perishables. Respiration rate increases exponently with increase in storage temperature. The rate of a chemical reaction approximately doubles for each 10 °C rise in temperature. Therefore, the storage life of perishables is highly variable and is related to wide range of respiration rates according to the storage temperature.

Product temperature at harvest is another and major determinant of respiratory rate of fruits. Good cooling and temperature management, are critical for slowing physiological deterioration, due to increase in temperature.

Temperature affects both the rate of ethylene production, its activity and the sensitivity of produce to ethylene. Rapid cooling and good temperature management are very vital in extending the storage life through delay in ripening and other deterioration process.

Relative humidity

Inadequate humidity in storage will have an adverse effect on the quality and storage of fruits. For perishables (fruits and vegetables) the relative humidity in storage is designed in the range of 90-95%. Below this level, there will be unacceptable moisture loss (desiccation) especially in the presence of high temperature, resulting in shorter storage life. On the other hand, there will be excessive growth of microorganisms and surface cracking, if relative humidity is maintained close to 100%. Therefore, control of humidity has become an important feature of some storage methods, as it is related to the keeping quality of fruits etc.

Gaseous composition

The composition of gases in the storage atmosphere affects the storage life of perishable produce. As mentioned earlier, even after separation from the parent tree fruits respire and thereby CO_2 accumulates in the storage atmosphere. Excess CO_2, beyond the tolerance leads to detrimental effects of "CO_2 injury'. Many volatile compounds, produced by stored commodities and/or from other sources may accumulate in the storage atmosphere, of which 'ethylene' is the most important one, which adversely affects the storage life of the produce. In the presence of higher amount of 'ethylene' fruits ripen at a faster rate and store poorly and for shorter period.

Cultivar

Cultivar is also considered as one of the factors influencing the storage life of the fruit. Out of the 28 cultivars of mango from different regions of India evaluated for cold storage 'Alphonso' proved to be the best keeper in cold storage (Cheema *et al* 1950). `Raspuri' fruits were stored for a maximum period of 42 days at 42-48 ^{o}F in green condition (Mathur *et al* 1953). 'Neelum' cultivar has shown a better keeping quality at lower temperature by maintaining a higher content of total sugars, ascorbic acid, TSS and lower acidity than 'Rumani' cultivar (Sachan and Srivastava, 1968).

'Dashehari' and 'Langra' had a shelf life of 10 days in storage at $34^{o} \pm 4$ °C with 50-65% RH. Maximum weight loss in 'Sipiashah Pasand' was attributed to higher density of lenticels on the fruit surface (Joshi and Roy, 1985). Similarly the variation in the extent of rot in different cultivars was attributed to the differential size of lenticels and status of stomata in the fruit skin. Higher content of Ca in 'S.B. Chausa' and 'Bombay green' was considered responsible for maintaining good firmness till 8 days of storage (Kaushik and Kumar, 1992).

The hybrids 'Arka Anmol' and 'Arka Neelkiran' had very good post-harvest life followed by 'Mallika' in comparison to 'Alphonso' and was considered to be due to longer post-ripening life and more number of days required for ripening (Gowda and Hudder, 2000). 'Mallika' was most efficient in remaining in good condition in storage under ambient conditions. Lower rates of respiration and less incidence of fungal spoilage were considered to be the main causes of longer storage life of 'Mallika' fruits. 'Amrapali' was stored up to 9 days showing a highest rate of respiration with highest fungal spoilage. 'Ratna' exhibited good storage life of 11 days at ambient temperature (Hoda *et al* 2000).

Precautionary measures to obtain better storage life and quality fruits

High quality fruits with better storage life from field stores and storage can be obtained by following certain precautionary measures. Primary factors in extending the storage life (defined as time interval from harvest to consumption, during which the produce remains safe and wholesome under specific production and storage condition) of perishables have been listed (Zagory and Kader, 1988).

1. Harvesting at optimum maturity without or minimum injures, with proper harvesters
2. Selection of good quality fruit by sorting, grading followed by pre-storage treatments (pre-cooling, pre-packaging), degreening, curing etc. for avoiding mechanical injuries.
3. Quick reduction of field heat of the produce to check the physiological changes and chilling injury (pre-cooling and evaporative cooling).
4. Usage of refrigerated transport from field to pack houses/store houses for checking various types of deterioration.
5. Usage of mechanical graders and modern techniques of post-harvest handling and storage to maintain good quality of the produce, along with proper sanitation.
6. Usage of appropriate chemicals/fungicides to check post-harvest fungal spoilage.
7. Providing optimum storage conditions-meticulus maintenance of the desired temperature and humidity in low temperature storage (refrigerated storage), meticulous maintenance of gas composition in storage atmosphere in case of CA and MA storage systems, all the time during storage and marketing channels.
8. Adoption of proper packaging, good hygiene and sanitation during post-harvest management to avoid microbial spoilage.
9. Waxing and pre-packaging in plastic film of fresh produce to protect skin coatings for enhancing the storage life and checking the spoilage.
10. Avoiding over storage i.e. beyond the optimum storage life, to check physiological deterioration and disorders and for satisfactory ripening and quality.

Conclusion

Each of the above storage techniques has its own inherent advantages, disadvantages and demerits and techno-economic constraints and limitations, which need to be taken into account before its adoption. Techniques like hypobaric and irradiation storage have shown promising in the laboratory studies and need to be pursued on a large scale before their commercial application. Cryogenic refrigerants like liquid nitrogen have ample potential for adoption, due to their large refrigeration capacity and inertness. Research efforts to tap these potentials for mango storage need to be actively pursued (Ravindra and Goswami, 2007).

REFERENCES

Abou-Aziz, A.R., Eloabawy, S.M., Abdel Kader, A.S. and Abdel-Waheb, F.K. 1975. Effect of chilling and non-chilling temperatures on some chemical compounds of mango fruits. Egyptian Journal of Horticulture, 2(2) : 173-185.

Abou-Gouk, A.A. and Mohamad, H.J. 2004. Effect of harvesting method on quality and shelf life of mango fruits. Tropical Science, 44 : 73-76.

Agnihotri, B.N., Kapur, K.L. and Srivastava, J.C. 1963. Physico-chemical changes in Dashehari mango during storage. Punjab Horticultural Journal, 3(2-4) : 286-291.

Ahmad, M., Naqvi, M.H. and Hussain, A. 1992. Induction of ripening delay in mango var. Dashehari by gamma irradiation and refrigeration. Pakistan Journal of Scientific and Industrial Research, 15(4/5) : 314.

Ahmed, M.S. and Singh, S. 2000. Studies on extension of storage life of Amrapali mango. Orissa Journal of Horticulture, 28(2) : 73-76.

Akamine, E.K. 1963. Haden Mango storage. Hawaii Farm Science, 12 : 6-7.

Akamine, E.K. and Goo, T. 1979. Effects of ionizing irradiation on 'Haden' mangoes. Hawaii Agricultural Experiment Station Research Report. 205.

Anand, J.C. and Johar, D.S. 1953. Cold storage of green mangoes for preparation of chutneys. Indian Food Packer, 7(2) : 9-11.

Anonymous. 1985. Zero energy cool chamber. The Indian Agricultural Research Institute, New Delhi, Bulletin No. 43 : 24.

Anonymous. 1990-91. Annual Report. Central Institute of Horticulture for Northern Plains, Lucknow, India, pp. 15.

Anonymous. 1991. Annual Report. All India Coordinated Research Project on Post-Harvest Technology of Fruits and Vegetables, IARI, New Delhi, India.

Anonymous. 1999-2000. Annual Report of Indian Institute of Horticultural Research, Bangalore, India, pp. 25.

Anonymous. 2001-02. Annual Report. Central Institute for Sub-tropical Horticulture, Lucknow, India, p. 31.

Anonymous. 2002-2003a. Annual Report of Indian Institute of Horticultural Research, Bangalore (India) for the year 2002-2003, pp. 17.

Anonymous. 2002-03b. Annual Report. Central Institute for Subtropical Horticultural, Lucknow, India, p. 39.

Anonymous. 2003-04. Annual Report of Indian Institute of Horticultural Research, Bangalore, India.

Anonymous. 2005. Mango. In : Post harvest practices for mango, banana, tomato and capsicum. Indian Institute of Horticultural Research, Bangalore. Technical Bulletin No. 27 : 7-8, 11-12.

Antala, D.K., Sangani, V.P., Cholera, S.P., Vyas, D.M., Bhatt, V.M., and Varshney, A.K. 2008. Effect of wax coating and packging at low temperature on the storage behaviour of Kesar mango. Indian Food Parker, 62(6): 121-127.

Apelbaum, A., Zauberman, G. and Fuchs, Y. 1977. Sub atmospheric pressure storage of mango fruits. Scientia Horticulturae, 7 : 153-160.

Badar, R.S. 1990. Studies on maturity indices, grading and storage of mango (*Mangifera indica* L.) fruits cv. Ratna and Kesar. M.Sc. (Ag.) Thesis, Konkan Krishi Vidya Peeth, Dapoli, India.

Baez-Sanuado, R., Bringas, T.E., Gonzalez, A.G., Ojeda, C.J., Mendozo, W.A. and Ramos, C.G. 1997a. Effects of edible films on the post harvest life of mangoes. Proceedings of the Inter-American Society for Tropical Horticulture, 41 : 172-178.

Baez-Sanuado, R, Bringas, T.E. and Ojeda C.J. 1997b. Determination of the maturity index of '*Ataulfo*' mango. Proceedings of the Inter-American Society for Tropical Horticulture, 41 : 179-183.

Baldwin, E.A. 1994. Edible coatings for fresh fruits and vegetables, past, present and future. In : Krocht, J., Baldwin, E.A. and Nisperos, M.O. (eds.). Edible Coating and Films to improve food quality. Carriedo. Technomic Publishing Co, Lancaster, pp. 25-64.

Banerjee, B.N., Karmarkar, D.V. and Row, G.R. 1934. Investigations on the storage of mangoes. Agriculture and Livestock in India, 4 : 36-53.

Bender, R.J., Brecht, J.K. and Campbell, C.A. 1994. Responses of 'Kent' and 'Tommy Atkins' mangoes to reduced O_2 and elevated CO_2. Proceedings of Florida State Horticultural Society, 107 : 274-277.

Bender, R.J., Brecht, J.K. and Sargent, S.A. 1995. Inhibition of ethylene production in mango fruit by elevated CO_2 and recovery during subsequent air-storage. Proceedings of the Florida State Horticultural Society No. 108 : 279-285.

Bender, R.J., Brecht, J.K., Sargent, S.A. and Huber, D.J. 2000a. Low temperature controlled atmosphere storage for tree-ripe mangoes (*Mangifera indica* L.). Acta Horticulturae, 509 : 447-458.

Bender, R.J., Brecht, J.K., Sargent, S.A. and Huber, D.J. 2000b. Mango tolerance to reduced oxygen levels in controlled atmosphere storage. Journal of American Society for Horticultural Science, 125(6) : 707-713.

Ben-Yehoshua, S. 1987. Transpiration, water strees and gas exchange. In: J.W. Zihman (ed) Post-harvest physiology of vegetables. Marcel Dekker Inc. Newyork, PP. 113-170.

Ben-Yehoshua, S. 1991. New developments in applying individual seal packaging for improved handling of fruits and vegetables. In : B.Y. Huang and Q. Yang (ed.). Proceedings of International Citrus Symposium. Guangzhou, China, November, 1990. Beijing International Academic Press, 757-771.

Ben-Yehoshua, S. 1985. Individual seal packaging of fruits and vegetables in plastic film – a new post harvest technique. Hort.Science, 20 : 32-37.

Ben-Yehoshua, S. (Ed) 1995. Project on extending storage life of tropical and sub-tropical fruits. Final Report submitted to the European Community by ARO. The Volcanic Centre, Institute of Technology and Storage of Agricultural Products, Israel.

Bhullar, J.S., Dhillon, B.S. and Randhawa, J.S. 1984. Ambient storage of Langra and Dashehari mangoes. Journal of Research, Punjab Agricultural University, Ludhiana, 21(1) : 33-38.

Boag, T.S., Thompson, J.J., Izard, M., Murray, C. and Fitz, K.C. 1990. Physiological responses of mangoes cv. Kensington Pride to gamma irradiation treatment as affected by fruit maturity and ripeness. Annals of Applied Biology, 116(1) : 177-187.

Bose, A.N. and Basu, G. 1953. Extenstion of storage life of Fazli mango by coating with pareffin. Science and Culture, 19: 263-264.

Bower, J. P., Denninson, M.T. and Fowler, K. 2003. Avocado and mango cold storage damage as related to water loss control. Acta Horticulturae, 628 : 401-406.

Brody, A.L. 1993. The Market. In : Principles and applications of modified atmosphere packaging of foods. R.T. Parry (ed), Blackie Academic and Professional, London. p. 114

Brown, B.I. 1986. Some effects of the post-harvest use of commercial wax emulsions on common mango. Proceedings of First Australian mango research workshop. Melbourne, Australia, CISRO : 290-294.

Burden, J.N. 1997. Post-harvest handling of tropical and subtropical fruit for export. In : Post-harvest physiology and storage of tropical and sub tropical fruits (Mitra, S.K. ed.). CAB International, London, U.K.1-20.

Burg, E.K. 1975. Hypobaric storage and transportation of fresh fruits and vegetables. In: Post-Harvest biology and handling of fruits and vegetables. Av west-post, conn.

Burg, S.S. and Burg, E.A. 1966. Fruit storage at sub-atmospheric pressure. Science, 153 : 314-360.

Burton, W.G. 1974. Some bio-physical principles under lying the controlled atmosphere storage of plant material. Annals of Applieed Biology, 78: 149-168.

Burton, W.T. 1982, Post-harvest physiology of food groups. Long man, London.

Castro, J.V., Pfafrenbach, L.B., Carralho, C.R.L. and Rossetto, C.J. 2005. Effects of film packaging and cold storage on post-harvest quality of Tommy Atkins mangoes. Acta Horticulturae, 682 : 1683-1688.

Chandler, W.H. 1950. Evergreen orchards. Henry Kimpton, London.

Chaplin, G.R., Scott, K.J. and Brown, B. I. 1982. Effects of storing mangoes in polyethylene bags at ambient temperature, Singapore Journal of Primary Industries, 10(2) : 84-88.

Chatpar, H.S., Mattoo, A.K. and Modi, V.V. 1971. Bio-chemical studies on chilling injuries in mango. Phyto-chemistry, 10 : 1007-1009.

Cheema, G.S. and Dani, P.G. 1934. Report on export of mango to Europe in 1932-33. Bulletin No. 170. Department of Agriculture, Bombay, India.

Cheema, G.S., Karmarkar, D.V. and Joshi, B.M. 1950. Investigation on the cold storage of mangoes. Indian Journal of Agricultural Science, 20 : 259-325.

Chen, Z.H., Zhang, R.L. and Chen, H.M. 1998. The physiological changes of mango fruit during post-harvest storage. Plant Physiology Communications, 34(3) : 178-180.

Cohen, E., Shaloam, Y. and Rosenberger, 1. 1990. Post-harvest-ethanol build up and off-flavour in 'Murchi' Tangerine fruits. Journal of American Society for Horticultural Science, 115: 776-778,

Conway, W.S. and Sams, S.E. 1984. Possible mechanisms by which post-harvest calcium treatments reduce decay in apples. Phytopathology, 24 : 208-210.

Conway, W.S. and Sams, S.E. 1987. The effect of post-harvest infiltration of calcium, magnesium or strontium on decay, firmness, respiration and ethylene production in apples. Journal of American Society for Horticultural Sciences, 112 : 3.

Cucras-Ruiz, J., Luse, R.A. and Graham, H.D. 1968. Preservation of tropical food stuffs by irradiation. Proceedings of Tropical Region. American Society for Horticultural Science, 12 : 246-256.

Dalal, V.B., Eiepsen, W.E. and Sigh, N.S. 1970. Wax emulsion for fresh fruits and vegetables to extend their storage life. Indian food Packer, 25(5): 9-15.

Dalal, V.B. and Subramanyam, H. 1970. Refrigerated storage of fresh fruits and vegetables. Climate Control, 3 : 37-50.

Deol, I.S. 1985. A study on chemical changes during the storage life of Samar Bahist Chausa mango (*Mangifera indica* L.). Punjab Horticultural Journal, 25(1-4): 5-11.

Deol, I .S. and Bhullar, S.S. 1972. Effect of wrappers and growth regulators on the storage life of mango fruit. Punjab Horticulture Journal, 12 (2/3) : 114-119.

Dey, R.M. and Brinson, K. 1984. Plant Cell. Advanes in Carbohydrate Chemistry and Biochemistry, pp. 382.

Dhaka, R.S., Verma, M.K. and Agrawal, M.K. 2001. Effect of post-harvest treatments on physico-chemical characters during storage of mango cv. Totapuri. Haryana Journal of Horticulture Science, 30(1&2) : 36-38.

Dhalla, R. and Hansor, S.W. 1988. Effect of permeable coatings on the storage life of fruits II pre-cooling of mangoes (*Mangifera indica* L. cv. Julie). Journal of Food Science and Technology, 23 : 107-112.

Dharkar, S.D., Saragaon, K.A., Srirangarajan, A. N. and Sreenivasan, A. 1966a. Irradiation of mangoes II. Radiation effects on skin-coated Alphonso mangoes. Journal of Food Science 31 : 870-877.

Dharkar, S.D., Saragaon, K.A., Srirangarajan, A.N. and Sreenivasan, A. 1966b. Irradiation of Mangoes, I. Radiation-induced delay in ripening of Alphonso mangoes. Journal of Food Science, 31 : 863-869.

Dharker, S.D. and Sreenivasan, A. 1972. Irradiation as a method for improved storage and transportation of mangoes. Acta Horticulturae, 24 : 259-269.

Dhemre, J.K. 2001. Studies on extending the shelf life of mango (*Mangifera indica* L.) cv. Kesar, Ph.D thesis, Mahatama Phule Krishi Vidyapeeth, Rahuri, India.

Dhemre, J.K. and Waskar, D.P. 2003. Effect of post-harvest treatments on shelf life and quality of mango in Evaporative Cool Chamber and Ambient conditions. Journal of Food Science and Technology, 40(3) : 316-318.

Diaz Sobac, R., Luna, A.V., Beristain C.L. and Grus, J.D.L. 1996. Emulsion Coating to extend Post-harvest life of mango (*Mangifera Indica* L) Cr. Monila. Journal of Food Proceesing and Preservation, 20: 191-202.

Dilley, D.R. 1982. Principles and effects of hypobaric storage of fruits and vegetables. American Society Heat Refrigeration Air Condition, 88 : 1461-1478.

Dixon, N.M. and Kell, D.B. 1989. The inhibition of CO_2 on the growth and metabolism of microorganisms. Journal of Applied Bacteriology, 67 : 109-136.

Durigam, J.F., Teixeire, G.H.A., Castanharo, N.M. and Domarco, R.E. 2000. Post-harvest conservation of 'Tommy Atkins' mango fruit-influenced by Gamma Radiation, Wax, Hot water and Refrigeration. Acta Horticulturae, 645 : 601-604.

El-Ghaouth, A., Arul, J., Grenier, J. and Asselin, A. 1992. Antifungal activity of Chitosan on the post-harvest pathogens of Strawberry fruits. Phytopathology, 82 : 398-402.

El-Ghaouth, A., Arul, J. and Ponnampadam, R. 1991. Use of Chitosan coating to reduce water loss and maintain quality of cucumber and bell pepper fruits. Journal of Food Processing and Preservation, 15 : 359-368.

El-Gourani, M.A. and Sommer, N.F. 1981. Effects of modified atmosphere on post-harvest pathogens of fruits and vegetables. Horticultural Reviews, 3 : 412-468.

Farooqui, Q.A., Sattar, A., Dand, K and Hussain, M. 1985. Studies on the post-harvest chilling sensitivity of mango fruits (*Mangifera indica* L.). Proceedings of the Florida State Horticultural Society, 98 : 220-221.

Faust, M. 1975. The role of calcium in the respiration mechanism and senescence of apples. Fact aura et regulation de maturation des fruits Colloguium International. C.N.R.S. No, 238. Pp. 87-92.

Faust, M. and Shear, A.B. 1972. The effect of calcium on respiration of apples. Journal of American Society for Horticultural Science, 97 : 437-439.

Fergusan, I.B. 1983. Calcium stimulation of ethylene production induced by 1-aminocyclopropane-1-carboxylic acid and indole-3-acetic acid. Journal of Plant Growth Regulation, 2 : 205-214.

Fonseca, M. J.de., Cecon, P.R., Salamao, L.C.C. and Pushmann, R. 2004a. Pulp and skin pigments in mango 'Haden' treated with fungicides and Wax. Acta Horticulturae. 645 : 633-638.

Fonseca, M.J.de., Salomao, L.C.C., Cecon, P.R. and Pushmann, R. 2004b. Fungicides and wax in post-harvest preservation of mango `Haden'. Acta Horticulturae, 645: 557-563.

Fornaris-Rullan, G.R., Guadalupe, L. and Chao de Baez, C. 1989. Shelf life and quality components of Parvin mango after 15 days storage. Journal of Agriculture University of Puerto Rico. 73 : 295-298.

Garg. N., Singh, B.P. and Narayana, C.K. 2000. Effect of modified atmosphere on aerobic mesophilic and pyychrophitic micro-flora of mango fruits. Indian Food Packer, 54(5): 75-76.

Garg, R.C. and Ram, H.B. 1973. A note on the effect of wax emulsion treatment on mango cv. Lucknow Safeda. Progressive Horticulture, 5 : 35-39.

Garg, R.C., Ram, H.B., Singh, S.K. and Singh, R.V. 1976. Effect of some plant growth regulators on the storage behaviour, rate of respiration and general quality of mango cv. Dashehari. Progressive Horticulture, 8 : 51-55.

Garg, R.C., Srivastava, R.K., Ram, H.B. and Verma, R.A. 1971. Role of pre-packaging and skin coating on the storage behaviour of mango variety Dashehari (*Mangifera indica* L.). Progressive Horticulture, 3(2) : 49-67.

Gautam, B. and Neeraja, G. 2005. Effect of polythene bag storage on shelf life and quality of Banganapalli mangoes. Orissa Journal of Horticulture, 33(2) : 89-91.

Gautam, B., Sarkar, S.K. and Reddy, Y.N. 2003. Effect of post-harvest treatments on shelf life and quality of Banganapalli mango. Indian Journal of Horticulture, 60(2): 135-139.

Gautam, D.M. and Lizada, M.C.C. 1984. Internal breakdown in Carabao mango subjected to modified atmosphere IV. Post-Harvest Research Notes, 1(3) : 90-92.

Gautam, S.K. and Chundawat, B.S. 1990. Post-harvest changes in Sapota cv. Kalipatti II. Effect of various post-harvest treatments on physico-chemical attributes. Indian Journal of Horticulture, 47 : 264-269.

Giri, S.K., Prasad, N. and Kumar, K.K. 2003. Lac-wax based emulsion coating to extend post-harvest life of mango. Indian Food Packer, 57(3) : 62-68.

Gole, R.S. 1986. Studies on fruit development and some aspects of post-harvest handling of mango (*Mangifera indica* L.) fruits. M.Sc. (Agric.) Thesis, Konkan Krishi Vidyapeeth, Dapoli, India.

Gonzalez, A.G., Vasquez, C., Felix, L., Baez, R. and Siller, J. 1994. Low O_2 treatment before storage in normal or modified atmosphere packaging of mango. Proceedings of International Symposium on recent advances in Post-harvest physiology, pathology and technologies for Horticultural Commodities, held at Agadir, Morocco, 16-21, January, 1994; 185-189.

Gonzalez, G., Yahia, E.M. and Higuera, 1990. Modified atmosphere packaging of mango and avocado fruit. Acta Horticulturae, 269 : 335-344.

Gowda, I.N.D and Hudder, A.G. 2000. Evaluation of mango hybrids for storage behaviour and sensory qualities. Journal of Food Science Technology, 37(6) : 620-623.

Granthy, C.R., Scott, K.G. and Brown, B.I. 1982. Effects of storing mangoes in polyethylene bags at ambient temperature. Singapore Journal of Primary Industries, 10 : 84-88.

Hall, C.W., Hardenburg, R.E. and Pantastico, Er. B. 1975. Consumer packaging with plastic, pp. 303-313. In : B. Pantestico (ed.), Post-harvest Physiology, Handling and Utilization of Tropical and Subtropical Fruits and Vegetables. The AVI Publishing Co., Inc., West Port, C.T.

Hardenburg, R.E. 1956. Polythene film box liner for reducing weight losses and shriveling of Golden Delicious apples in storage. Proceedings of the American Society for Horticultural Society, 67 : 82-90.

Harvey, J.M. 1978. Reduction of losses in fresh market fruits and vegetables. Annual Review of Phytopathology, 16 : 321-341.

Hassabulla, I.A., Ibrahim, M.M., Abdel-Aziz, A.Z., Sharaf, M.M. and El-Kady, S.M. 1984. Effect of gamma radiation, some growth regulators. Benlate and Vapor Gard on physical properties of 'HindiBe-Sinnara' mango fruits during storage. Annals of Agriculture Science, Moshtobar, 20(2) : 279.

Hatton, T.T. Jr., Beraha, L. and Wright, W.R. 1961. Preliminary trials of gamma radiation on mature green Irwin and Sensation mangoes. Florida Mango Forum, p. 15.

Hatton, T.T. Jr., and Reeder, W.F. 1966. Controlled atmosphere storage of Keitt mango. Proceedings of American Society for Horticultural Science, Caribbean Region, 10: 114-120.

Hatton, T.T. Jr., and Reeder, W.F. and Campbell, C.W. 1965. Ripening and storage of Florida mangoes. USDA Marketing Research Report, 725.

Hatton, T.T., Jr. and Spalding, D.H. 1990. Controlled atmosphere storage of some tropical fruits, p. 301-313. In : M.Calderon and R. B. Golen. (eds.). Food Preservation by modified atmospheres. CRC Press, Boca Raton, Florida, USA.

Hoda, M.N., Yadav, G.S., Singh, S., and Singh, J. 2000. Storage behaviour of mango (*Mangifera indica*) hybrids. Indian Journal of Agricultural Sciences, 71(7): 469-472.

Hudder, A.G. and Reddy, D.M. 1989. Post-harvest calcium chloride treatment of mangoes for extension of shelf life at ambient temperature. Souvenir, 3rd International Symposium on 'Mango' held at Darwin. Australia, September, 24-29, p. 92 (Abstract)

Hudder, A.G., Bharati, B.C. and Thimmaraju, K.R. 1989. Note on extension of storage life of mango fruits by Tal Prolong. Acta Horticulturae, No. 231 : 668-669.

Hulme, A.C. 1971. The mango, In : A.C. Hulme (Ed.). The Biochemistry of Fruits and their products. Vol. 2, Academic Press, London, New York : 233-254.

Ilangantileke, S.G., Turla, L. and Chen, R. 1989. Pre-treatment and hypobaric storage for increased storage life of mango. Canadian/American Society of Agricultural Engineering. Paper No. 89 : 6036, Quebec, Canada, June 25-28.

Illeperuma, C.K. and Jayasurya, P. 2002. Prolonged storage of 'Karutha Colomban' mango by modified atmosphere packaging at low temperature. Journal of Horticultural Science and Biotechnology, 77(2) : 153-157.

Isenberg, F.M.R. 1979. Controlled atmosphere storage of vegetables. Horticultural Reviews, 1 : 337-394.

Jain, S.K. and Mukherjee, S. 2001. Zero-Energy-Cool Chamber storage studies on Mango. Indian Food Packer, 55(4) : 55-56.

Jamieson, W. 1980, Hypobaric inter modal Containers. CSIRO Food Research Quarterly, 40: 129-137.

Janave, M.T. and Sharma, A. 2005. Extended storage of gamma-irradiated mango at tropical ambient temperature by film wrap packaging. Journal of Food Science and Technology, 42(3) : 230-233.

Jang, E.B. 1990. Fruit fly disinfestations of tropical fruits using semi-permeable shrink wrap film. Acta Horticulturae, 269 : 453-458.

Jessup, A.J., Rigney, C.J. and Wills, P.A. 1988. Effects of gamma irradiation combined with hot dipping on quality of 'Kenington Pride' mangoes. Journal of Food Science, 53 : 1488-1489.

Johnson, G.I., Sharp, J.L., Milne, D.L. and Oosthuyse, S.A. 1997. Post-harvest technology and quarantine treatments. In : The Mango- Botany, Production and uses. RE Litz (Ed). CAB International, Willingford : 447-507.

Joshi, G.D. and Roy, S.K. 1985. Effect of Integrated Post-harvest handling on biochemical changes in 'Alphonso' mango fruits. Progressive Horticulture, 17 : 56-63.

Joshi, G.D. and Roy, S.K. 1988. Influence of maturity, transport and cold storage on biochemical composition of Alphonso mango fruits. Journal of Maharashtra Agricultural Universities, 13(1) : 12-15.

Joyce, D.C., Beasley, D.R. and Shorter, A. J. 1997. Effect of pre-harvest bagging on fruit calcium levels and storage and ripening characteristics of 'Sensation' mangoes. Australian Journal of Experimental Agriculture, 37(3) : 383-389.

Joyce, D.C., Shorter, A.J. and Hockings, P.D. 2001. Mango fruit calcium levels and the effect of post-harvest calcium infiltration at different maturities. Scientia Horticulturae, 91 : 81-99.

Kader, A.A. 1986. Biochemical and physiological basis for effects of controlled and modified atmospheres on fruits and vegetables. Food Technology, 40 : 99-100, 102-104.

Kader, A.A. 1990. The effect of storage methods on the quality of horticultural crops. Tondazione Cesebra Agricultural Cesena, pp. 11-26.

Kader, A.A. 1992. Post-harvest biology and technology : An overview. In : Post-harvest technology of Horticultural Crops. Kader, A.A. (Ed). Division of Agriculture and Natural Resources, University of California, Publication – 3311, California, pp. 15-20.

Kader, A.A., Kasmire, R.E., Mitchell, F.G., Reid, M.Z., Sommer, N.F. and Thompson, J.F. 1985. Post-harvest technology of Horticultural Crops. Special Publication 3311. Cooperative Extension, University of California, Division of Agriculture and Natural Resources, USA.

Kader, A.A., Zagory, D. and Kerbel, E.D. 1989. Modified atmosphere packaging of fruits and vegetables. Critical Reviews in Food Science and Nutrition, 28 : 1-30.

Kader, S.E.S.A, 1991. Effect of pre-harvest application of GA_3 on Post-harvest behaviour of mango fruits. Scientia Horticulturae, 47 (3-4): 317-321.

Kalra, S.K. and Tandon, D.K. 1983. Ripening behaviour of 'Dashahari' mango in relation to harvest period. Scientia Horticulturae 19 : 263-269.

Kaushik, R.A. and Kumar, R. 1991. Effect of post-harvest application of 2, 4-5 trichloro phenoxy acetic acid, maleic hydrazide and calcium nitrate on the storage behaviour of Dashehari mango. Haryana Agricultural University Journal of Research, 21(4) : 287-291.

Kaushik, R.A. and Kumar, R. 1992. Physico-chemical changes in different mango cultivars during storage. Haryana Journal of Horticultural Science. 21(3-4): 166-173.

Ke, D. and Kader, A.A. 1992. Potential of controlled atmospheres for post-harvest insect disinfestation of fruits and vegetables. Post harvest News and Information, 3(2) : 31N-37N.

Kester, J.J. and Fennema, O.R, 1986. Edible Films and Coatings: A Review Food Technology No: 40.

Ketsa, S. and Prabhasavat, T. 1992. Effect of skin coating on shelf life and quality of 'Nang klangwan' mangoes. Acta Horticulturae, 322 : 764-770.

Ketsa, S. and Raksritong, T. 1992. Effect of PVC film wrapping and temperature on storage life and quality of Nam Dok Mai mango fruits on ripening. Acta Horticulturae, 321 : 756-763.

Kittur, F.S., Saroja, N., Habibunnisa and Taranathan, R.N. 2001. Polysaccharides based composite coating formulations for shelf life extension of fresh banana and mango. European Food Research and Technology, 213 : 4-5.

Krishnamurthy, S., Singh, S. and Katyal, S.L. 1963. Fruit Culture in India. Indian Council of Agricultural Research, New Delhi (India), pp. 361-362.

Kruga, F.o, Louw M and Good child, L. 1996. Aetiology and potential stragegy for preventing post-harvest physiolocal ailments in 'Heidi' mango fruits Year Book of South Africa Mango Groners' Association, 16: 12-16.

Kruger, F.J., Kritzinger, M. and Bezuidenhour, M 1997. Trials aimed at improving the storage potential of 'Heidi'. Year Book of South African Mango Growers' Association, 17: 100-104.

Laborem, G., Reyes, F.J. and Romzel, L. 1992. Maduracion del mango previo dmaconamiento a boja temperature. Fruits, 47: 419-423.

Lakshminarayana, S. 1973. Respiration and ripening pattern in the life cycle of the mango fruit. Journal of Horticultural Science, 48 : 227-233.

Lakshminarayana, S., Krishna Prasad, C.A. and Shetty, M.S. 1974. Hot water treatment to regulate ripening and reduce spoilage of Alphonso mangoes. Journal of Horticultural Science, 49(4) : 365-371.

Lakshminarayana, S. and Subramanyam, H. 1970. Carbondioxide injury and fermentative Decarboxylation in mango fruit at low temperature storage. Journal of Food Science and Technology, 7 : 148-152.

Lal, G., Singh, R.V., Saran, P.L. and Singh, Y.V. 2003. Shelf life of mango as influenced by various chemicals under different storage conditions. Paper presented at National Seminar on mango held at Junagadh, 14-15, June, 2003, p. 71 (Abstract).

Lalel, H.J.D., Singh, Z. and Tan, S.C. 2001a. Elevated levels of CO_2 in controlled atmosphere affects shelf life, carotenoid and fruit quality of mango. Acta Horticulturae, 553(2) : 605-606.

Lalel, H.J.D., Singh, Z. and Tan, S.C. 2001b. Elevated levels of CO_2 in controlled atmosphere storage affects shelf-life, fruit quality and aroma volatiles of mango. Acta Horticulturae, 628 : 407-413.

Lalel, H.J.D., Singh, Z and Tan, S.C. 2005. Controlled atmosphere storage affects fruit ripening and quality of 'Delta R_2E_2' mango. Journal of Horticulture Science and Biotechnology 80 : 551-556.

Lam, P.F. and Wong, L.S. 1988. Eating quality of ethylene – ripened Harumanis mangoes after cold storage. MARDI Research Journal, 16(1) : 85-90.

Laogheed, E.C., Murr, D.P and Berard, Z. 1978, Low pressure storage for horticultural crops. Hortscience. 13(1): 21-27.

Lester, G.E. and Grusak, M.A. 1999. Post-harvest application of Calcium and Magnesium to Honeydew and netted muskmelons : effects on tissue ion concentrations, quality and senescence. Journal of American Society for Horticultural Science, 124 : 545-552.

Li-Anni., Deng, Y.C., Zhu, H.Y., Ku, My and Li, M.Q. 1993. A study of physiological and biochemical changes of mangoes during cold storage. Guandong Agricultural Sciences, No. 6 : 16-17.

Lioutas, T.S. 1988. Challenges of controlled and modified atmosphere packaging : a food company's perspective. Food Technology, 49 : 78-86.

Lizada, M.C.C. 1991. Post-harvest physiology of the mango – A Review. Acta Horticulturae, 291 : 437-453.

Lizada, M.C.C., Kosiyachinda, S. and Mendoza, D.B.Jr. 1984. Physiological disorders of mango. In : Mango – fruit development, post-harvest physiology and marketing in ASEAN. ASEAN Food Handling Bureau, Kaula Lumpur, p. 67-74.

Macnish, A.T., Joyce, D.C. and Hetherington, S.E. 1997. Packaging to reduce water loss can delay ripening of mango (*Mangifera indica* L.) cv. 'Kensington Pride' fruit. Australian Journal of Experimental Agriculture, 37 : 463-467.

Malik, A.V., Singh, Z., and Dhaliwal, S.S. 2003. Exogenous application of putrescine, affects mango fruit quality and shelf life. Acta Horticulturae, 628 : 121-127.

Mangano, J.E., Perez, Y. and Rojas, E. 1997. Coating waxes on Haden mango fruits (*Mangifera indica* L.) for export. Acta Horticulturae, 455 : 738-746.

Mann, S.S. and Singh, R.N. 1975a. Studies on cold storage of mango fruits (*Mangifera indica* L. cv. Langra). Indian Journal of Horticulture, 32(1/2) : 7-14.

Mann, S.S. and Singh, R.N. 1975b. Cold storage of hardgreen Dashehari mangoes (*Mangifera indica* L.). Progressive Horticulture, 7(3) : 11-17.

Mann, S.S. and Singh, R.N. 1976. The cold storage life of 'Dashehari' mangoes. Scientia Horticulturae, 5(3) : 249-259.

Manzoor V., Haq, M.A and Bhatti, M.S. 1977, A new technique to boost export of grafted mango fruit by expanding its marketing season from June to November by the use of water soluble wax and bio-regulators at room temperature. Journal of Agricultural Research, 15(3): 311-314.

Masalkar, S.D., Gaikwad, R.S. and Dhemre, J.K. 2006. Effect of post-harvest treatments on shelf life and quality of Alphonso mango fruits in cool chamber and ambient conditions. Orissa Journal of Horticulture, 34(2): 69-73.

Mathur, P.B. and Lewis, W.F. 1961. Storage behaviour of gamma irradiated mangoes. International Journal of Applied Radiation Isotopes, 11 : 435.

Mathur, P.B., Singh, K.K. and Kapur, N.S. 1952. Refrigerated preservation of mangoes. Indian Food Packer, 6(9) : 15-19.

Mathur, P.B., Singh, K.K. and Kapur, N.S. 1953. Cold storage of mangoes. Indian Journal of Agricultural Science, 23 : 65-77.

Mathur, P.B. and Srivastava, H.C. 1955. Effect of skin coatings on storage behaviour of mangoes. Food Research, 20 : 559-566.

Mathur, P.B. and Subramanyam, H. 1956. Effect of fungicidal wax coatings on storage behavior of mangoes. Journal of Science of Food and Agriculture, 7 : 673-676.

McGlosser, W.B. 1989. MA packaging : A practical alternative to CA shipping containers. Proceedings of 5th CA Research Conference, vol. 2. other commodities and storage recommendations. Wenatahee, Washington, 14-16, June, 1989 : 235-240.

McLauchlan, R.L., Barker, L.R., Johnson, G.I. and Highley, E. 1992. Controlled atmosphere for Kensington mango storage : Chemical atmospheres. Proceedings of Workshop on Development of post-harvest handling technology for tropical tree fruits held in Bangkok, Thailand, 16-18 July, 1992 : 41-44.

McLauchlan, R.L., Mitchell, G.E., Johnson, G.I. and Wills, P.A. 1990. Irradiation of 'Kensington Pride' mangoes. Acta Horticulturae, 269 : 469-476.

McCullum, T.G., D'Aquino, S. and McDonald, R.E. 1993. Heat treatment inhibits mango chilling injury. Hort Science, 28 : 197-198.

McCullum, T.G., D'Aquino, S., Miller, W.R. and McDonaldo, R.E. 1992. Individual shrink film wrapping of mangoes. Proceedings of the Florida State Horticultural Society, 105 : 103-105.

Medlicott, A.P. and Jegar, M.J. 1987. The development and application of post-harvest handling treatments to manipulate ripening in mangoes. In : Mangoes - A review. (Ed. By RT. Primsley and G.A. Tucker). Common Wealth Science Council, London, 56-77.

Medlicott, A.P., Reynolds, S.B., New, S.W. and Thompson, A.W. 1988. Harvest maturity effects on mango fruit ripening. Tropical Agriculture, 65 : 153-157.

Medlicott, A.P., Sigrist, J.M.M. and Sy, O. 1990. Ripening of mangoes following low temperature storage. Journal of the American Society for Horticultural Science, 115 : 430-434.

Miller, W.R., Hale, P.W., Spalding, D.H. and Davis, P. 1983. Quality and decay of mango fruit wrapped in heat-shrinkable film. Hort.science, 18(6) : 957-958.

Miller, W.R., Spalding, D.H. and Hale, P.W. 1986. Film wrapping mangoes at advancing stages of post-harvest ripening. Tropical Science, 26 : 9-17.

Mootoo, A. 1991. Effect of post-harvest calcium chloride dips on ripening changes in Julie mangoes. Tropical Science, 31 : 243-248.

Morais, P.L.D. de and Assin, J.S. de 2004. Quality and conservation of mango Cv. 'Tommy Atkins' as affected by maturity stage and storage temperature. Acta Horticulturae, 645 : 639-643.

Mukherjee, P.K. 1958. Cold storage of mango. Horticulture Advance, 2 : 44-48.

Mukherjee, P.K. 1961. The behaviour of mango varieties Dashehari and Langra under low temperature. Annual Report Fruit Research Station, Saharanpur for 1954-56, pp. 60-64.

Mukherjee, P.K. 1969-71. Post-harvest biochemical and physiological changes in mango under low temperature. Horticulture Advance, 8 : 40-45.

Mukherjee, P.K. 1972. Harvesting, storage and transport of mango. Acta Horticulturae, 24 : 251-258.

Mukherjee, P.K. and Prasad, A. 1972. Post-harvest physiology of mango. Paper presented at 3rd International Symposium on Tropical and Sub-tropical Horticulture, Bangalore, India.

Mukherji P.K. and Srivastana, R.B. 1978. Increasing the storage life of Mango (*Mangifera indica* L) by lowering the critical temperature. Progressive Horticulture, 10(4): 63.-69.

Munshi, P.S., Maharana, S. and Bhuyan, N.K. 2004. Effect of different concentrations of $CaCl_2$ with wrapping materials on storage life of mango. Paper presented at First Indian Horticulture Congress, held at New Delhi, 6-9 November 2004, p. 364 (Abstract).

Musa, S.K. 1974. Preliminary investigations on the storage and ripening of 'Totapuri' mangoes in Sudan. Tropical Science, 16(2) : 67-73.

Musa, S.K. and Mirghani, M.A. 1974. Preliminary investigation of grading, storage and ripening of Alphonso mangoes. Sudan Journal of Food Science and Technology, 6 : 18-32.

Nakamura, N., Sudhakar Rao, D.V., Shiina, T. and Nawa, Y. 2003. Effects of temperature and gas composition on respiratory behaviour of Tree-Ripe 'Irwin' mango. Acta Horticulturae, 600(1) : 425-429.

Narayana, C.K., Pal, R.K. and Roy, S.K. 1996. Effect of pre-storage treatments and temperature regimes on shelf-life and respiratory behaviour of ripe Baneshan mango. Journal of Food Science and Technology, 33(1) : 79-82.

Narayana, C.K., Roy, S.K. and Pal, R.K. 1991. Use of high molecular weight – high density polyethylene films (HM – Films) in transit of mango (cv. Baneshan). Acta Horticulturae, 291 : 489-492.

Negi., J.P. 2000. Post-harvest management of horticultural produce in India. Proceedings of National Seminar on post-harvest management of crops and cold storage in the 21st Century, Hyderabad.

Noomhorm, A. and Taisuwan, N. 1988. Effect of controlled atmosphere storage for mango. American Society of Agricultural Engineers, 88 - 6589, pp.8.

Noomhorm, A. and Taisuwan, N. 1995. Controlled atmosphere storage of mango fruit, *Mangifera indica* L. cv. Rad. Journal of Food Processing and Preservation, 19(4) : 271-281.

O'Hare, T.J. and Prasad, A. 1993. The effect of temperature and carbondioxide on chilling symptoms in mango. Acta Horticulturae, 343 : 244-250.

Oosthuyse, S.A. 1994. Quality of mature Zill mangoes after long term refrigerated storage as determined by pre-storage ripeness and cold storage regime. Year Book of South African Mango Growers' Association, 14 : 37-42.

Oosthuyse, S.A. 1997a. Effect of cool storage delays after harvest on the extent of ripening during and fruit quality after cool storage. Year Book of South African Mango Growers' Association, 17 : 81-85.

Oosthuyse, S.A. 1997b. Effect of long term controlled atmosphere storage on fruit quality in Heidi mango. Year Book of South African Mango Growers' Association, 17 : 95-99.

Oosthuyse, S.A. 1997c. Effect of waxing and of a number of waxes on weight loss, shelf life and fruit quality of mangoes after four weeks of cool storage. Year Book of South African Mango Growers' Association, 17 : 105-110.

Oosthuyse, S.A. 2000. Effect of cold storage delays after harvest on the extent of ripening during and fruit quality after cool storage. Acta Horticulturae, 509 : 395-400.

Oosthuyse, S.A. and Dodd, M.C. 1998. Experiments on extended controlled atmosphere storage of Heidi mango. Year Book of South African Mango Growers' Association, 18 : 26-29.

Pablo, I.S., Manalo, J.A. and Cardeno, Y.A. 1971. Sensory, chemical and nutritional evaluation on the effect of ionizing radiation on mangoes (*Mangifera indica* L.) Cv. Carabao. Disinfestation of fruit by irradiation. Vienna Austria International Atomic Energy Agency No. STS/Pub./299:101-111.

Pandey, R.M., Rao, M.M. and Singh, R.N. 1974. Biochemical changes in the developing mango fruits Cv. Dashehari. Progressive Horticulture, 5(4) : 47-59.

Passam, H.C. 1982. Storage of some local and introduced mango cultivars grown in Trinidad. Scientia Horticulturae, 16 : 171-177.

Pathak, V.N. and Srinivasa, D.N. 1969. Epidemiology and prevention of Diplodia stem end rot of mango fruits. Phytopathology Zournal, 65 : 164-175.

Peacock, B.C. 1984. Post-harvest handling of mangoes. In : Proceedings of the First Australian Mango Research Workshop, Cairs, Queensland, Australia, 295 : 313.

Phaiphan, A. and Rattamapanone, N. 2008. Effects of Edible Coatings on Quality of Mango fruit (*Mangifera indica* L) 'Chok-Aman' during storage. Acta Horticulturae, 773:227. 231.

Poovaiah, B.W. 1986. Role of calcium in prolonging storage life of fruits and vegetables. Food Technology, 40 : 86-89.

Poovaiah, B.W. 1988. Molecular and cellular aspects of calcium action in plants. Hort Science, 23 : 267-271.

Popenoe, J. 1956. Storage of Keitt mango at different temperatures – 1955. Proceedings Florida Mango Forum, pp. 13-16.

Pruthi, J.S. 1992. The Mango Book, I. Production, Post-harvest – Technology and Export Marketing. Published by APEDA. Ministry of Commerce, Government of India, New Delhi.

Rai, D.R., Oberoi, H.S. and Baboo, B. 2002. Modified atmosphere packaging and its effect on quality and shelf life of fruits and vegetables – An overview. Journal of Food Science and Technology, 39(3) : 199-207.

Raje, L., Sherlekar, S., Ramakrishnan, K., Malshe, V.C. and Subbalakshmi, G. 1997. Post-harvest preservation of mangoes by controlled chemical release agents and absorbent. Acta Horticulturae, 455 : 622-628.

Ram, H.B. and Date, W.B. 1971. Studies on the storage behaviour of two late varieties of mangoes treated with fungicidal wax emulsion at room temperature. Indian Journal, of Horticulture, 28(3) : 206-209.

Rameshwar, A. 1988. 'Packaging of mango' In : L. Venkataratnam (Ed) ' Packaging of Fruits and Vegetables in India'. Agri-Horticultural Society (Regd.). Public Gardens, Hyderabad, Andhra Pradesh, India.

Rangavalli, K., Ravishankar, C. and Hari Prasad, P. 1993. Post harvest changes in mango fruit Cv. Baneshan under different storage conditions. Andhra Agricultural Journal, 4(1-4) : 23-26.

Rao, D.V.S. and Shankara, K.S.S. 2004. Individual shrink wrapping of mango fruits for alleviation of chilling injury and extension of storage life. First Indian Horticulture Congress, New Delhi, 6-9 November, 362 (Abstract).

Rashid, A. and Farooqui, W.A. 1984. Studies on the effect of gamma irradiation and maleic hydrazide on the shelf life of mangoes. Journal of Agricultural Research Pakistan, 22(2) : 151.

Ravindra, M.R. and Goswami, T.K. 2007. Post-harvest handling and storage of mangoes - An overview. Journal of Food Science and Technology, 44(5) : 449-458.

Reddy, L.S. and Thimmaraju, K.R. 1989. Effects of pre-packaging and post harvest treatments on the storage behaviour of mango fruits Cv. Alphonso. Acta Horticulturae, 231 : 670-674.

Reddy, V.V. and Srivastava, G.C. 1999. Regulation of cell wall softening enzymes during ripening of mango fruits. Indian Journal of Plant Physiology, 4(3) (NS) : 194-196.

Rolz, C, Flores, M.C., De Ariola, M.C., Mayorga, H. And Menches, H. 1971. Development and cold storage of mango – Momey – a Guatemalan local variety. Report of UNIDO expert group meeting, Bahia, Brazil, pp. 77-94.

Romani, R., Puschmann, R. and Finch, J. 1982. Effect of pre-harvest application of AVG on ripening of pears with and without cold storage. Hortscience, 17(2) : 214-215.

Roy, B.N., Saha, S.K., Pramanick, S. and Biswas, 1980. Effects of wax emulsion and M.H. on the storage behaviour of mangoes (*Mangifera indica*. L.) Cv. Langra and Himsagar. Progressive Horticulture, 11(4) : 35-38.

Roy, S.K. and Khudriya, D.S. 1986. Studies on evaporatively cooled zero energy input cool chambers for the Horticultural Produce. Indian Food Packer, 40(6) : 26.

Roy, S.K. and Kudriya, D.S. 1983. Zero-energy cool-chamber for storage of Horticultural produce. Science in Service of Agriculture. Indian Agricultural Research Institute, New Delhi.

Roy, S.K. and Pal, R.K. 1991a. Multilocational studies to reduce post-harvest losses during harvesting, handling, packaging, transportation and marketing of mango in India. Acta Horticulturae, 291 : 499-507.

Roy, S.K. and Pal, R.K. 1991b. A low cost zero energy cool chamber for short-term storage of mango. Acta Horticulturae, 291 : 519-524.

Roy, S.K. and Pal, R.K. 1993. Use of plastic in post-harvest technology of fruits and vegetables. Indian Food Packer, 47 : 27-48.

Roy, S.K. and Pandey, R.M. 1983. Effects of existing methods of harvesting on ripening and storage of mango (*Mangifera indica* L.). Paper presented at Silver Jubilee Celebration Seminar on Plant Physiology in Coming Years, New Delhi. Indian Society of Plant Physiology, 22.

Sachan, B.P. and Srivastava, D.C. 1968. Chemical composition and changes during low temperature storage of South Indian mango (*Mangifera indica* L.) varieties – Neelum and Rumani. Madras Agricultural Journal, 55 : 457-460.

Sadasivam, R., Muthuswamy, S., Sundararaj, J.S. and Daniel, S.D. 1970. Experiments with waxing of mango fruits in common storage. Madras Agricultural Journal, 57 : 694-699.

Sahari, C.K., Khurdiya, D.S. and Dalal, M.A. 1994. Effect of ripening aids on the quality of mango. Maharashtra journal of Horticulture, F(1) : 72,

Salama, S.B. 1959. Physiological studies on the changes of 'Hindi' and 'Pairi' mango fruits during growth, storage and ripening. Ph.D. Thesis, Cairo University, Egypt.

Salunkhe, D.K. and Desai, B.B. 1984. Post-harvest biotechnology of fruits. Vol.1 CRC Press, Bocabaton, Florida, USA.

Samayoa de Arricola, M.C. 1973. Storage of mango. Proceedings of Tropical Region American Society for Horticultural Science, 17 : 104-124.

Sams, C.E. and Conway, W.S. 1984. Effect of calcium infiltration on ethylene production, respiration rate, soluble polyuronide content and quality of 'Golden Delicious' apple fruit. Journal of American Society for Horticultural Science, 109: 53-57.

Santha, K. 1989. Effects of 'Tal prolong' on shelf life and quality attributes of mango. Acta Horticulturae, 231 : 678-688.

Santha, K. and Joshi, S.S. 1989. Studies on low temperature storage of Alphonso mangoes. Journal of Food Science and Technology, 26 : 177-180.

Santha, K. and Rao, D.v.s. 2001. Studies on Post-harvest Management of fruits. Indian Journal of Horticulture, 58 (1 & 2): 153-163.

Santos, A.F. dos., Silva, S. de. M., Mendonca, R.M.N., Alues, R.E. and Martins, L.P. 2004. Storage of mango fruits Cv. Rosa treated with calcium chloride after harvest at different maturity stages. Acta Horticulturae, 645 : 663-670.

Sastry, M.V. and Habibunnisa, 1979, Studies on Indian pickles-part V. Stroge 2 Raw mangoes for pickling. Indian food packer, 33(6): 10-18.

Saucedovelz, C., Torres, E.E. and Lakshminarayana, L. 1976 b. Effect of refrigerated temperatures on the incidence of chilling injury and ripening quality of mango fruit. Proceedings of the Florida State Horticultural Society, 90: 205-200.

Scott, K.J., Glarson, M.C. and Robats, E.A. 1970. Potassium permanganate as an ethylene absorbent in polythene bags to delay ripening of Banana during storage. Australian Journal of Experimental Agriculture, 10 : 237-239.

Sellas, Jr. De. J. and Tomeres, J.C. 1999. Post-harvest life of mango (*Mangifera indica* L. cv. Tommy Atkins) : the influence of temperature and state of maturity. Revista Brasileira de Fruiti cultura, 21(2) : 171-176.

Serek, M., Sisler, E.C and Reid, M.S. 1994. Novel gaseous ethylene binding inhibition prevents ethylene effects in potted flowering plants. Journal of the American Society for Horticultural Science 119 : 1230-1233.

Seymour, G.B., Diaye, M.N., Wainwright, H. and Tucker, G.A. 1990. Effects of cultivar and harvest maturity on ripening of mangoes during storage. Journal of Horticulture Science, 65(4) : 479-483.

Sharma, C. D. and Kapoor, N.S. 1967. Role of the skin coating on the transportation of perishables IV. Studies on the transportation of mangoes. Indian Food Packer, 21 : 31-36.

Sharma, R.M. and Singh, R.R. 2000. Harvesting, post-harvest handling and physiology of fruits and vegetables. Chapter 3. In : L.R. Verma and V.K. Joshi (ed.) Post-harvest Technology of Fruits and Vegetables. pp. 94-147, Indian Publishing Co., New Delhi, India.

Sheikh, I.A., Ali, S.S., Ehteshamuddin, A.T. M. and Haque, MY. I. 1977. Preservation of mangoes with fungicidal wax emulsion. Pakistan Journal of Scientific and Industrial Research, 20(3) : 198-200.

Shetty, K.K., Klowden, M.J., Jang, E.B. and Koshan, W. 1989. Individual shrink wrapping technique for fruit fly disinfestation in tropical fruits. HortScience, 24 : 317-319.

Shill, A.C. 1934. Report of the marketing officer, Trinidad, October 1932 – June 1934, pp. 21-24. Port of Spain, Government Printing Office.

Shorter, A. and Joyce, D.C. 1994. Effect of surface coatings on sapburn of Kensington Pride mango fruit. Tropical Agriculture, 71 : 243-246.

Shukor, A.R.A., Razali, M. and Omar, D. 2000. Post-storage respiratory suppression and changes in chemical composition of mango fruit after storage in low oxygen atmosphere. Acta Horticulturae, 509 : 467-470.

Singh, B.K. 1988. Post-harvest studies of mango (*Mangifera indica* L.). Ph.D. Thesis submitted to Rajendra Agricultural University, Patna, Bihar, India.

Singh, B.P. and Narayana, C.K. 1995. Storage behaviour of 'Dashehari' mango in ventilated poly bags. Indian Food Packer, 49 : 29-31.

Singh, B.P. and Narayana, C.K. 1999. An integrated approach for storage of mango. Indian Journal of Horticulture, 56(1) : 5-9.

Singh, B.P., Kalra, S.K. and Tandon, D.K. 1993. Effect of method of harvesting on storage behaviour of mango. Indian Journal of Horticulture, 50(1) : 5-9.

Singh, B.P., Tandon, D.K. and Kalra, S.K. 1993. Changes in post-harvest quality of mangoes as affected by pre-harvest application of calcium salt. Scientia Horticulturae, 54 : 211-219.

Singh, D.B. and Pandit, A.H. 2004. Economics of various post-harvest treatments in enhancing shelf life of mango. First Indian Horticulture Congress, New Delhi, November 6-9 : 354 (Abstract).

Singh, K.K. 1970. Studies in Frozen Storage of Some Fruits. Climate and Control, 3(3): 33-36.

Singh, K.K. 1974. Refrigerated storage of mango in India. Climate and Control, 7(4) : 33-39.

Singh, K.K., Kapoor, M.S. and Mathur, P.B. 1954. Further studies on the cold storage of mangoes. Indian Journal of Agriculture Science, 24 : 137-148.

Singh, P.V. and Gupta, R. 2004. Effect of carbendazim, GA. and MH. on quality and shelf life of mango cv. Dashehari. Paper presented at 1st Indian Horticulture Congress, held at New Delhi, 6-9, November, 2004, pp. 353 (Abstract).

Singh, R.N., Singh, G., Mishra, J.S. and Rao, O.P. 1987. Studies on the effect of pre- and post-harvest treatment of calcium nitrate and calcium chloride on the storage life of Amrapali mango. Progressive Horticulture, 19(1-2) : 1-9.

Singh, S., Brahamachari, V.S. and Jha, K.K. 1998. Effect of calcium and polythene wrapping on storage life of mango. Indian Journal of Horticulture, 55(3) : 218-222.

Singh, Z., Janes, J. and Tan, S.C. 2000. Effects of different surfactants on calcium uptake and its effects on fruit ripening, quality and post-harvest storage of mango under modified atmosphere packaging. Acta Horticulturae, 509 : 413-417.

Singh, Z., Lalel, H.J.D. and Nair, S. 2004. A review of mango fruit aroma volatile compounds – State of the Art Research. Acta Horticulturae, 645 : 519-527.

Sisler, E.C. and Serek, M. 1997. Inhibition of ethylene resources in plants at the receptor level : Recent developments. Physiologia Plantarum, 100 : 577-582.

Smock, R. 1979. Controlled atmosphere storage of fruits. Horticultural Reviews, 1 : 301-336.

Sornsrivichai, J., Gomolmanee, S., Boonyakiat, D., Uthaibutra, J., Boon – Long, P and Gemma, H. 1992. Seal packaging by plastic film as a technique for limiting fungal decay of mangoes. Acta Horticulturae, 296 : 23-32.

Sornsrivichai, J., Uthai butra, J. and Yantarasri, T. 2000. Controlling of peel and flush colour development of mango by perforation of Modified Atmosphere package at different temperatures. Acta Horticulturae, 509 : 387-394.

Spalding, D.H. and Reeder, W.F. 1977. Low pressure (Hypobaric) storage of mangoes. Journal of American Society for Horticultural Science, 102 : 367-369.

Spalding, D.H. and Reeder, W.F. 1986. Decay and acceptability of mangoes treated with combinations of hot water, Imazalil and gamma radiation. Plant Disease, 70 : 1149-1151.

Spalding, D.H. and von Windeguth, D.L. 1988. Quality and decay of irradiated mangoes. Hortscience 23 : 187-189.

Spurgeon, S.L. and Porter, J.W. 1980. Carotenoids, p. 419-483. In : P.K. Stumpf and E.E. Conn. The Biochemistry of Plants - A Comprehensive Treatise. Vol. 4 : Academic Press, Newyork.

Srinivasa Reddy, N. and Haripriya, K. 2002. Extension of storage life of mango (cvs). Bangalora and Neelum. South Indian Horticulture, 50(1-3) : 7-8.

Srinivasan, A., Thomes, P. and Dhankar, S.D. 1971. Physiological effects of gamma radiation on some tropical fruits. In : Disinfestation of fruit by irradiation. Proceedings of a Panal on the use of Irradiation to solve Quarantine Problems in the International Fruit Trade, Hawaii, December, 1970. International Atomic Energy Agency, Vienna, 113-124.

Srivastava, H.C. 1967. Grading, Storage and Marketing. In : The Mango : A Hand Book. Indian Council of Agricultural Research, New Delhi, p. 99-149.

Stossel, P. and Leuba, J.L. 1984. Effect of Chitosan, Chitin and some amino sugars on growth of various soil borne phytopathogenic fungi. Phytopathology Zhurnal, 171 : 82-90.

Straten, B. van, and Oosthuyse, S.A. 1994. The effect of cold storage at 4 or 8°C on the rate of quality deterioration of ripe Sensation mango fruit. Year Book of South African Mango Growers' Association, 14 : 34-36.

Subramanyam, H., Krishnamurthy, S. and Paria, H.A.B. 1975. Physiology and biochemistry of mango fruit. Advances in Food Research, 24 : 223.

Sudhavani, V., Ravishankar, C. and Rao, B.V. 2000. Effect of pre-harvest sprays on the storage behaviour of Baneshan mango. Andhra Agricultural Journal, 47 (1&2) : 108-112.

Thangaraj, T. and Irulappan, I. 1993. Studies on pre-packaging and cool chamber storage of mango (*Mangifera indica*). Paper presented at Golden Jubilee Sympsoium on Horticultural Research-Changing Scenario, held at Bangalore, May 24-28, p. 355 (Abstract)

Thomas, P. 1986. Radiation preservation of tropical fruits – bananas, mangoes and papayas. Critical Reviews in Food Science and Nutrition, 23 : 147-205.

Thomas, P. and Desai, S.R.P 1976. Improvement of shelf-life and quality of mangoes by gamma irradiation. Indian Food Packer, 30(5): 83-86.

Thomas, P. and Janave, M.T. 1973. Polyphenoloxidase activity and browning of mango fruits induced by gamma irradiation. Journal of Food Science, 38 : 1149-1152.

Thomas, P. and Oke, M.S. 1980. Technical note : Vitamin C content and distribution during ripening. Journal of Food Technology, 15(6) : 669-672.

Thomas, P. and Oke, M.S. 1983. Improvement in quality and storage of Alphonso mangoes by cold adaption. Scientia Horticulturae, 19(3/4) : 257-262.

Thompson, A.K. 1971. The storage of mango fruits. Tropical Agriculture, Trinidad, 48 : 63-70.

Thompson, A.K. 1977. Transport of West Indian Mango Fruits. Tropical Agriculture, 48 : 71-77.

Thompson, A.K. 1998. Controlled atmosphere storage of fruits and vegetables. CAB International, Willingford, U.K.

Tingwa, P.D. and Young, R.E. 1974. The effect of calcium on the ripening of avocado (*Persea americana* Mill.) fruits. Journal of American Society for Horticultural Science, 99 : 540-542.

Tirmazi, S.I.H. and Wills, R.B.H. 1981. Retardation of ripening of mangoes by post-harvest application of calcium. Tropical Agriculture, 58 : 137-141.

Trinidad, M., Bosquez, E., Escalona, H., Diaz de Leon, F., Perez, F.L., Kenberk, C. Ponce de Leon, L., Munos, C. and Perez, L. 1997. Controlled atmosphere (5 % CO_2 5 % O_2 and 10 % CO_2 – 5 % O_2) do not significantly increase the storage life of refrigerated Kent mangoes. Acta Horticulturae, 455 : 643-653.

Ueda, M., Sasaki, K., Utsumomiya, N., Inaba, K. and Shimbayashi, Y. 1999. Effects of storage on fruit colour, flesh firmness, respiratory rate and several principal compounds in fully matured mango fruit (*Mangifera indica* L. 'Irwin') cultured in plastic house. Journal of the Japanese Society for Food Science and Technology, 46(1) : 16-23.

Valmayor, R.Y. 1972. The Philippine mango industry. Acta Horticulturae, 24 : 19.

Van Eeden, S.J. 1992. Calcium infiltration as a possible post-harvest treatment to increase storage potential of mango fruit. Year Book of South African Mango Growers' Association, 12 : 26-27.

Wang, B.C., Jiang, W.B., Liu, H.X., Lin, L and wang, J.H 2006. Enhancing the post-harvest qualities of mango fruit by Vacuum infilteration treatment with 1-Methyl cyclo propene. Journal Horticultural Sicence and Biotechnology, 81(1): 163-167.

Wang, C.V. 1990. Chilling injury of Horticultural Crops. CRC Press, Boca Raton, Florida, USA.

Wardlaw, C.W. and Leonard, E.R. 1936. The storage of West Indian mangoes. Low Temperature Research Station, I.G.T.A., Trinidad, Memoir, 3.

Waskar, D.P. and Dhemre, J.K. 2003. Advanced techniques in post-harvest handling of Kesar mango fruits. Processed Food Industry, 6(12) : 15-17.

Waskar, D.P. and Masalkar, S.D. 1997. Effect of hydrocooling and Bavistin dip on the shelf life and quality of mango during storage under various environments. Acta Horticulturae, 455 : 687.

Waskar, D.P., Masalkar, S.D., Gaikwa., R.S. and Damme, S.V. 1997. Effect of method of harvesting on post-harvest loss reduction and extension of shelf life of mango. Acta Horticulturae, 455 : 662-668.

Wavhal, K.N. and Athele, P.W. 1988. Studies to prolong shelf life of mango fruits. Acta Horticulturae, 231 : 771-775.

Wills, R., Mc Glasson, B. Graham D., and Joyce, D. 1998. Post-harvest : An Introduction to the Physiology and Handling of Fruit, Vegetables and Ornamentals, UNSW Press, Sydney, 262 pp.

Yagi, M.I., Habish, H.A. and Agab, M.A. 1978. Effect of the method of harvest on the development of microorganism and quality of mango fruit. Sudan Journal of Food Science and Technology, 10 : 83-89.

Yahia, E.M. 1993. Modified / Controlled atmosphere storage in Mexico, P. 570-579. In : CA'93. Proceedings of the 6th International CA Research Conference, Ithaca, New York, June 14-16.

Yahia, E.M. 1995. Post-harvest handling of horticultural crops in Mexico. In : A. Ait. Qubahou and M.El. Otmani (eds). Post-harvest physiology, pathology and technologies in horticultural crops : recent advances. Institute of Agronomique and Veterinaire, Hasson II, Agadia, Morocco.

Yahia, E.M. 1998. Modified and Controlled atmospheres for tropical fruits. Horticultural Reviews, 22 : 123-185.

Yahia, E.M. and Vazquez-Moreno, L. 1993. Response of mango to insecticidal oxygen and carbon dioxide atmospheres. Food science and Technology (Labensm. Wiss. U-Technology), 26: 42-48.

Yamashita, I. 1991. Effects of gas concentration on the quality of horticultural commodities. In : Seikabutsu Kaki sendo hand book (Hand book. Fruit, Vegetable and Flowers) eds. Iwamoto, M. *et al.* Science Forum, Tokyo, 73-78 (In Japanese).

Yantarasri, T., Uthaibutra, J., Sornsrivichai, J., Kumpuran, W., Sardsud, V. and Kanathum, N. 1994. Modified atmosphere packaging by perforated polymeric film and its effect on physical properties of mango fruit p. 438-440. In : B.R. Chapman. E. Highley. and G.I. Johnson (eds.). Proceedings of International Conference in Post-harvest handling of Tropical Fruits, held at Chiangmai, Thailand, 19-23 July, AICAR, Proceedings, 50.

Yuen, C.M.C. 1993. Calcium and fruit storage potential, P. 218-227. In : Post-harvest handling of tropical fruits. B.R. Chapman, E. Highley, and G.I. Johnson (Eds), ACIAR, Brisbane, Australia.

Yuen, C.M.C., Tan, S.C., Jayce, D. and Chettri, P. 1993. Effect of post-harvest calcium and polymeric films on ripening and peel injury in Kensingston Pride mango. ASEAN Food Journal, 8 : 110-113.

Zagory, D. and Kader, A.A. 1988. Modified atmosphere packaging of fresh produce. Food Technology, 42 : 70-74, 76-77.

Zambreno, J. and Manzono, J. 1995. Effect of applying calcium salts on the ripening of mango fruits. Agronomia Tropical, Maracay, 45(3) : 407-415.

Zauberman, G., Fuchs, R., Prusky, D. and Yanko, U. 1979. Mango storage – 1978. Preliminary Report, Agriculture Research Organization. Bet Dagan No. 774 : 11 pp.

CHAPTER 4

Fruit Ripening

L. Naram Naidu

In most cases mangoes are harvested at hard green and unripe stage both for table and processing purposes, while immature fruits are harvested for making products like amchur, pickles, chutney etc. Hard green and unripe fruits are ripened by different methods for commercial purpose, to get a good quality edible ripe fruit. In nature ripening of fruits is needed for the dispersal of the seed through human beings, animals like pre-mates, monkeys, bats, birds etc for perpetuation. Ripening is the final stage of the fruit development followed by senescence. Ripening makes an astringent or acidic (sour) fruit, sweet and edible through changes in physico-chemical and bio-chemical parameters. The ripening of fruits may be defined as a "sequence of changes in colour, flavour, taste and texture" which lead to the acceptable edible stage. Based on the respiratory behaviour, mango is categorized as a "climacteric fruit" which undergoes rapid increase in respiration, concomitant with ethylene production followed by marked changes in their composition, texture, colour etc. These fruits show a progressive decline in respiration for the entire period of their lives once harvested. Ripening is an irreversible event.

HOW A FRUIT RIPENS

Fruit ripening is a complex process which includes the following:

(i) Increase in respiration

(ii) Synthesis or formation of new sets of enzymes.

(iii) Bringing out changes characteristic of ripening such as softening, pigmentation and quality changes.

The ripening process includes colour changes usually from green to yellow or red, loss of chlorophylls and production of carotenoids and/or anthocyanins; textural changes from hard firmness to softness due to solubilization and degradation of pectin and changes in flavour from astringent and tart to sweetness due to alteration of their chemical composition, like conversion of starch to sugar, increase in sugar content, and altered proportions of sucrose, fructose, and glucose, decrease in acidity and phenolic contents and synthesis of new volatiles giving characteristic aroma to the ripe fruit.

All the above changes are closely related to the action of endogenous hormones. Ethylene is a natural hormone involved in fruit ripening. The auto catalytic production of endogenous ethylene is considered to trigger off the ripening process in these climacteric fruits, which produce large quantities of ethylene during ripening. Auxins decline, at the same time the level of abscissic acid begins to increase. The enzymes viz., indole acetic acid oxidase and peroxidase which breakdown the native auxin, increase during ripening. By inactivating the inhibition of catalase and peroxidase in mango, ethylene initiates the ripening process (Mattoo and Modi,1 969a).

METHODS OF RIPENING

Mango fruits are not allowed to ripe on the tree, since tree-ripe fruits show uneven ripening, pale skin colour, poor edible quality, take longer period for ripening and have a shorter post-harvest life. Moreover, the retention of fruits on tree beyond the final maturity attracts the attack of bats, birds, insects etc and even human pilferage. Further, all the fruits do not come to ripening at one time, resulting in multiple pickings which increase harvesting cost. There will be uneven supply of fruits to the markets. Since the fruits are sent to distant places, tree ripe fruits get easily spoiled/over ripe by the time they reach the destination and huge losses may occur. In view of the above, fruits are harvested in unripe condition and ripened artificially using different methods.

(a) **Traditional Methods:** Traditionally, mango fruits are ripened using paddy straw (Photo 4.1) wild palmrosa grass, barbon grass, paper cuttings etc. These materials conserve heat and the contaminating moulds produce ethylene in sufficient amounts to stimulate ripening in the fruit. In South India, paddy straw and wild palmrosa grasses are traditionally and still used for local markets, while barbon grass is used in North India, which gives satisfactory results. Paper cuttings were found as the best media as judged by palatable ratings for ripening 'Dashehari' and 'Langra' fruits. (Mann and Singh, 1975), but Srivastava (1961) reported poor quality of 'Dashehari' mangoes after ripening with paper cuttings.

In Karnataka and other places of South India mango fruits are ripened by smoking in closed room or by heaping the fruits and covering with paddy straw in ventilated room, particularly by processing units. In these cases, fruit quality is poor due to uneven ripening and also there will be spoilage losses due to spread of latent infections of anthracnose, stem-end-rot etc., as conducive conditions prevail, especially when fruits were kept for ripening without pre-treatments against latent infections (Chikkasubbanna and Huddar, 1982)

Mann and Singh (1975) considered saw dust as inferior material, for ripening mango fruits of 'Dashehari' and 'Langra', while Teaotia *et al* (1968) reported good results with sawdust. Guha and Bhuiyan (1997) from Bangladesh also found that fruits ripened well, but slowly in sawdust and attributed to lower respiration and ethylene production as a result of generation of lower heat in this treatment. Fruits ripened faster in ria straw and was the best medium in terms of fruit colour, texture, sweetness and flavour (Ola, 1978., Guha and Bhuiyan, 1997). Fresh acacia leaves induced faster ripening (Ola, 1978). Fresh banana leaves reduced weight loss but enhanced the incidence of diseases due to high content of moisture in the leaves (Guha and Bhuiyan, 1997).

In many parts of Nigeria, dipping of harvested fruits in hot-wood ash (55 °C) is one of the pre-treatment practiced for ripening (Inyang and Agbo 1995). This treatment leads to higher content of sugar, ascorbic acid and carotenoids, higher p^H, decrease in titratable acidity, and moisture content (Morga *et al* 1979., Dhalla and Hanson 1988., Inyang and Agbo, 1995). The increase in carotenoid content was due to heat treatment (Inyang and Agbo 1995) as in the case of hot water treatment (Lakshminarayana *et al* 1974).

Hot water treatment (HWT) has been employed for artificial ripening of mangoes. It hastens the ripening but results in higher physiological loss of weight due to activation of enzymatic process at higher temperature (Kumar and Dhawan, 1995). However, HWT of mangoes at 52 °C - 55 °C for 5 min. has been recommended for an uniform ripening which also reduces the microbial spoilage by more than 50 percent (Krishnamurthy and Sudhakar Rao, 2001).

Photo 4.1 : Traditional method of ripening of mangoes under the layer of paddy straw at a processing unit

(b) Chemical Ripening: As harvested fruits vary in the degree of maturity, ripening (without any treatment) is not uniform and poses a major and serious problem in commercial marketing. Further they take longer time for ripening under ambient conditions. Therefore, mangoes harvested while still hard green or at colour break are ripened artificially with chemicals. Ethylene and related compounds have long been used to ripen the mango fruits and has become an established practice all over the world. In many countries, including India, both calcium carbide and ethylene are used for ripening mangoes. Indian traders generally ripen mature as well as immature mango fruits by using calcium carbide to fetch higher returns in the early markets.

(i) ***Calcium carbide:*** Calcium carbide (CaC_2) is obtained by heating pulverized lime stone or quicklime with carbon and is commonly used in gas welding and now used for ripening of fruits. It is imported from countries such as China, Taiwan and South Africa. The low price of calcium carbide (Rs.15/250 g) encourages its indiscriminate usage for ripening of climacteric fruits.

Calcium carbide pellets or nuggets in paper or cloths are kept in between layers of fruits in containers or heaps of mango fruits, after moistening for the release of ethylene gas which is supposed to induce ripening of the fruits. The chemical when hydrolyzed produces heat and acetylene gas along with traces of ethylene, which hasten the ripening process. In West Bengal, calcium carbide pellets are usually kept in boxes or baskets of mango fruits, covered with news paper and kept in closed space for 48-72 hours for ripening (Mitra, 2008). A simple method is to place the fruits in basket lined with banana leaves with calcium carbide, and this gives uniform colour within 2-3 days at ambient conditions in Malaysia (www.ikisan.com). In South Africa fruits are placed in a room maintained at 21 °C - 26.7 °C and 85-90 per cent relative humidity and calcium carbide is introduced at 10Z/72 cubic feet, for 1-8 days with ventilation at every 4 hours. A similar practice is commercially adopted in mangoes for air-fright in Brazil (www.ikisan.com).

Large quantities of calcium carbide are used for this purpose and quantity varies with place and cultivar. But 2-4 grams of calcium carbide was found to be optimum for ripening 1 kg of fruit. In a ripening trial in Costa Rica, calcium carbide at 5 g/kg fruit caused over ripening, while 3.8 g/kg or less delayed ripening and caused weight loss (Valverde *et al* 1986). Calcium carbide at 2 g/kg accelerated ripening, but resulted in higher percentage of weight loss during storage in paper wraps and perforated wooden boxes in Pakistan (Abbas *et al* 1989). Mature green 'Dashehari' fruits treated with calcium carbide at 2-4 g/kg had higher concentration of carotenoids than ethrel treatment (Kumar and Dhawan, 1995). Physiological loss in weight and decay losses were maximum in fruits treated with calcium carbide at

4 g/kg fruit (Joon *et al* 2001). At 2 g/kg fruit it ensured uniform and early ripening, the time being reduced to 7 days from 14-17 days in the untreated control (Nagaraj *et al* 1984). Acceleration of peel colour by calcium carbide treatment was enhanced by increasing the rate further from 5 to 15 g per 15 fruits in Philippines. However, increasing the rate further resulted in fruits, in which most ripening changes in the peel lagged behind pulp colour development (Lizada, 1991).

Calcium carbide is most effective in reducing fruit pressure (firmness) by causing quick softening (Kumar and Dhawan, 1995) and reducing the ripening period (Bhullar, 1982., Guha and Bhuiyan, 1997., Joon *et el* 2001). Calcium carbide treated fruits showed a sharp decline in acidity and increase in pH and TSS (Nagaraj *et al* 1984). Calcium carbide alters the aroma and flavour and the ripened fruits exhibit a sharp and abrupt change in the physico-chemical parameters during ripening. The mature mango fruits ripened by calcium carbide show rapid red colour development and become attractive and get higher price (Sahni and Khurdia, 1992). The rate of carotenoid synthesis in the pulp of treated fruits was higher in the initial stage of ripening, but the content of carotenoid was lower than the control at full ripening. Aroma was also less in the treated fruits (Nagaraj *et al* 1984).

Several studies indicated adverse effects of calcium carbide on fruit quality. The treated fruits are poor in taste (insipid) and lack real taste of the fruit. Ripening is not uniform, degreening of the peel is not uniform, but more at pedicel and less on distal end. Treatments tend to reduce the ascorbic acid and total sugar contents (Pandey and Singh, 1976). Although, softening was accelerated by calcium carbide treatment, it was observed that at a given peel colour, treated fruits were more firm than untreated fruits (Nagaraj *et al* 1984). Maximum decay was noticed in treated fruits, due to increased rate of acetylene and ethylene production, which spontaneously increase the various physiological processes leading to more rotting (Sivastava and Mukherjee 1961., Bhullar, 1982., Kumar and Dhawan, 1995., 1998). Calcium carbide treatment reduces the shelf life of fruits (Nagaraj *et al*). Calcium carbide treated fruits were unmarketable after 4 days of storage (Joon *et al* 2001). The adverse effects of calcium carbide on the taste of the fruit is more with the acidic cultivars (Alphonso, Pairi) (Krishnamurthy and Sudhakar Rao, 2001).

The use of calcium carbide is hazardous, as released acetylene gas burns brilliantly in air and is toxic when inhaled. Its contact severely irritates and burns the eyes and skin causing permanent damage to eyes and ulcers on the skin. Exposure to the gas can severely irritate the mouth, nose and throat causing sores, cough and wheezing. It is a flammable and reactive chemical with a dangerous explosion hazard. It causes cessation of vital functions or suspended aminosity owing to any cause, interfering with respiration and higher concentration may cause drowsiness or unconsciousness (Tandon and Kalra, 2001). Calcium carbide ripened fruits cause mouth ulcers and even food poisoning after consumption. Because of the above negative effects of calcium carbide on fruit quality and hazardness to human beings, particularly the use of the chemical for ripening of fruits is banned by Government of India during 1954.

(ii) ***Ethylene:*** 'Ethylene' is a gas at physiological temperature and is produced endogenously. It is a natural hormone involved in fruit ripening. However, ripening by endogenous ethylene is slow and is ununiform. Therefore, for hastening and uniform ripening exogenous application of ethylene becomes necessary and assumes significance with commercial point of view. Ethylene is applied either in the form of gas or in the form of liquid (ethephon) exogenously (Photos 4.2, 3 & 4). Ethephon (2-chloro ethane phosphoric acid) is strongly acidic in water solution. When in solution above a pH of 5, the ethephon molecules spontaneously hydrolyze, liberating ethylene. The amount of ethylene will increase as pH and/or relative humidity increases. Ethephon is commercially available (ethrel, florel, Cepa) and used for enhancing post-harvest ripening, Ethephon releases ethylene on exposure. Ethephon has to be applied to the fruit as a spray or dip treatment. Fruit response to ethylene was slightly faster than response to ethephon (Barmore, 1975). However, dipping of mangoes in ethephon may be a useful alternative in areas where the cost of constructing gas chambers is prohibitive. However, in several countries like, Australia, United States of America etc., mango fruits are commercially treated with ethylene gas (Sergent *et al* 1993). Exogenous application of ethylene not only hastens ripening, but also improves fruit quality (Chaplin, 1988., Lizada, 1991., Kalra *et al* 1995). On the contrary, ethrel (500 ppm) treated fruits ripened slowly with lower carotenoid content. Such fruits maintained proper texture and colour even up to 8th day of storage by reducing physiological loss in weight and decency losses (Kumar and Dhawan, 1995., Joon *et al* 2001). Reduction in decay loss in ethrel treated fruits has been assigned to the slow release of ethylene (Chundawat *et al* 1973., Chikkasubbanna and Huddar, 1982., Kumar and Dhawan, 1995). The acceptable colour in ethrel treated fruits was ascribed to complete conversion of chloroplast into chromoplasts during the climacteric peak by the action of released ethylene (Chikkasubbanna and Huddar, 1982). During storage, the rupture force in mango fruits decreased with ethylene treatments (Chundawat *et al* 1973., Adinarayana, 1975). Quality of ethrel treated fruits was best with enhanced TSS, but without a fall in the titratable acidity (Chikkasubbanna and Huddar, 1982). Ethrel treatment caused green mangoes to develop full colour in 7-10 days depending on the degree of maturity, where as untreated fruits took 10-15 days for the same.

Photo 4.2 : Ripening chambers

Photo 4.3 : Ethylene gas cylinders

Photo 4.4 : Mango fruits in crates in ripening chamber

The most obvious effects of ethylene are faster and more uniform ripening and changes in pigments (Chaplin, 1988) and reduction in spoilage (Chikkasubbanna and Hudder (1982). Barmore (1975) obtained similar responses with Florida-grown mangoes treated with ethylene and ethephon for the fresh market. Subsequently, Barmore and Mitchell (1976) made successful shipments of mangoes that were pretreated with ethylene to major North American countries from Florida. The advantages of post-harvest treatment of mangoes with ethylene destined for processing have been demonstrated by Chikkasubbanna and Huddar (1982), who found that 'Alphonso' mangoes treated with ethephon ripened sooner and more uniformly with less spoilage.

Doses of ethrel/ethylene vary with the cultivars and areas of production, besides the quantity of fruits. Recommended doses include 500 or 1000 ppm ethephon (Chundawat *et al* 1973., Lakshminarayana *et al* 1975) or 10-20 ppm ethylene (Baramore and Mitchell 1975) Among all the concentrations of ethephon (0-750 ppm) tested 500 ppm ethephon gave the best fruit quality in case of 'Dashehari' cultivar (Chundawat *et al* 1973). Baramore (1975) observed that mature fruit dipped in 500-1000 ppm of ethrel for one minute ripened earlier. Adinarayana (1975) concluded that ripening of 'Alphonso' mango fruits could be enhanced by treating them with 1000-2000 ppm ethrel. Apparently no adverse effects are caused by high ethylene (1000-1500 ppm) levels, but the ripening was reduced (Shanmugavelu *et al* 1976). Ethrel at 5000 ppm caused early ripening and secured highest organoleptic score in cv.Bangalora at Basti, Uttar Pradesh (India) (Singh, *et al* 1979). Among different levels ethrel 750 ppm was found to be optimum for obtaining the fruits with desirable colour, flavour etc., in 'Alphonso' mangoes at Bangalore, India (Chikkasubbanna and Huddar, 1982). Ethephon at 750 ppm ripened fruits faster (8-12 days) and improved fruit flavour and appearance in Costa Rica, but ethephon above 1500 ppm caused over ripening (Valverde *et al* 1986). The threshold concentration of ethylene for initiation of ripening in 'Tommy Atkins' fruits was between 1.25 and 10 ppm (Medlicott *et al* 1987). Ethephon treatment at 500 ppm resulted in a slow rate of ripening (Kumar and Dhawan, 1995). Ethrel at 200 or 400 ppm resulted in better ripening attributes (Pal, 1998). Ethrel (100-500 ppm) dip treatment in hot water (52° + 2 °C) for 5 minutes was effective in accelerated and uniform ripening, better surface colour, TSS and acid blend and carotenoid development in case of pre-mature and/or mature 'Dashehari' fruits at Lucknow, India (Tandon and Kalra, 2001).

The time taken for ripening varies with the cultivar, fruit maturity, temperature, concentration of ethrel/ethylene etc. Generally 24 hours exposure is sufficient for the fruits picked at the proper maturity, but for immature fruits, 48 hours exposure to ethylene is required. Irwin fruits harvested 90 days after anthesis (DAA) could be ripened to an acceptable quality after exposure to 100 ppm ethylene for 3 days, with enhancement in the development of both peel and pulp colour. This effect was observed over a wide range of maturity (76-90 DAA). The development of yellow colour in less matured fruit (68-76 DAA) was also greatly enhanced by ethylene treatment. However, ethylene treatment was not able to promote the development of pulp colour of less matured fruit to the same level as the fully matured fruit and the fruit quality was still inferior to that of higher maturity fruit. Only those fruits with maturity of 90 DAA or higher were able to reach or even surpass the commercially acceptable eating quality after ripening (Wang and Shiesh, 1990).

Pre-mature mango fruits meant to catch early market could be ripened uniformly by dipping the harvested fruit in 750 ppm ethrel in hot water (52 ± 2 °C) for 5 min. within 4-8 days under ambient condition. Mature fruits could similarly be ripened, but with lower doses of ethrel for faster ripening and uniform colour development in North India (Negi, 2000). In 'Alphonso' mango fruits ripening was very much hastened when the mature fruits were kept in a chamber saturated with ethylene gas released from 10,000 ppm ethrel solution containing NaOH pellets (www.ikisan.com).

'Banganapalli' mango fruits ripened in 48 hours after picking when treated with 10,000 ppm ethrel. 'Mulgoa' fruits ripened 60 hours after picking with 5000 and 10000 ppm under normal conditions. Skin and pulp colour were markedly improved with concurrent increase in carotenoid content in treated fruits (Shanmugavelu and Selvaraj, 1974). Bhuva *et al* (1978) reported variation in the response of fruits to ethephon treatment according to season of varieties. Fruits of early season ripened in 48-65 hours and mid-late varieties ripened in 70-92 hours.

Response to ethrel/ethylene treatments is temperature dependent. At low temperatures (15-18 °C) longer exposure to ethylene results in more green colour on the skin of the ripe fruit. Treatment with higher level (100 μL/l) of ethephon at 15 °C for 72 hours caused softening more quickly than other ripening changes (loss of green colour and flesh acidity) so that the fruits reach eating-soft stage before flesh acidity and skin colour were acceptable. Fruits ripened at 20 °C and ethephon (10-100 μL/l) had better skin colour than those ripened at 25 °C but had more acid flesh. Treatment with lower concentration of ethrel (10 μL/l) and above influences skin colour development during ripening at 20 °C and 25 °C. The treatment of 10 μL/l ethylene for 72 hours at 20 °C resulted in the best degreening treatment. However, care is required to ensure that temperatures are maintained above at least 15 °C (especially when high concentration and long duration are used) and to use low concentration for longer times rather than high temperatures for short duration (Nguyen *et al* 2002).

The ripening temperature recommended for mangoes is 29-31 °C with ethrel at 10 microns per litre for 24 hours with 85-90 percent relative humidity. In another work 21°-24 °C was recommended as the optimum ripening temperature with ethrel (www.ikisan.com).

For domestic/local markets, on farm treatment with ethylene is used to ensure that fruits have colour, when they are displayed at the market, 48-72 hours after dispatch from the farm. Ethylene treatment is a common practice in Israel for ripening of fruits for the local markets. For more distant markets, ethylene induction of ripening is undesirable because fruits arrive at the market, too ripe for handling and sale, with anthracnose and stem-end-rot development. Close-control of over exposure to ethylene is required to prevent pre-mature initiation of ripening on farm or in transit. However, shipping ethylene treated fruit for longer distances is not recommended since these fruits soften faster than do untreated fruit at comparable temperature (Mitra and Baldwin, 1997). Ethylene treatment can be given at either the packing house or at the retailers point of distribution.

Ethylene Vs Calcium carbide

As seen above, both ethylene and calcium carbide are used for ripening of immature as well as mature mangoes in several production regions. Few studies have been conducted to compare the above two chemical aids with reference to fruit ripening, quality and spoilage. However, the results are variable and contradictory making it difficult to draw a conclusion. Many workers concluded that ethylene was superior over calcium carbide (Singh *et al* 1979., Bhullar, 1982., Sahni *et al* 1994., Kumar and Dhawan, 1995., Pal, 1998 a,b) where as others favoured calcium carbide.

Ethylene treatments were superior over calcium carbide in terms of earliness in ripening and fruit quality (Singh *et al* 1979). The results of a comparative trial indicated that ethrel dips for 5 min. were relatively superior to calcium carbide (Sahni *et al* 1994). PLW was highest and lowest in calcium carbide and ethylene treated fruits respectively (Bhullar, 1982., Nagaraj *et al* 1984) but TSS and ascorbic acid were higher in ethylene treated fruits. The ethylene treated fruits showed a slow and steady synthesis of carotenoids. Sensory quality score of the ethylene-treated fruits was higher as compared to calcium carbide treated fruits (Sahni *et al* 1994). Changes in TSS and acidity were slow in ethylene treated fruits. Ripening was fast and ununiform in calcium carbide treatment but ripening was uniform in ethylene treated fruits (Joon *et al*, 2001). Highest respiration rate and maximum decrease in acidity were noted in calcium carbide treated fruits (Bhullar, 1982).

On the other hand, it was reported that the fruits ripened faster with calcium carbide treatment and were superior in quality to those ripened with ethrel, they were more palatable, had better colour and greater content of TSS, sugars and carotenoids (Mann and Dhillon, 1974). Sampaio (1981 a) obtained best ripening of mangoes with calcium carbide in Brazil. Ripening was faster, but the quality of fruits was inferior to that of ethrel treated fruits.

Not withstanding the above positive effects of calcium carbide on ripening and quality of fruit, the use of calcium carbide has been prohibited, because of the negative effects on fruit ripening and quality (See calcium carbide) and hazardousness to human beings. Further, ethylene treatment is costly as the chemical is not available in commercial packings. The cost of the treatment, however, could be reduced by repeated use of ethrel solution (Tandon and Kalra, 2001).

Ethylene Vs Acetylene

Acetylene, an ethylene analogue, has a lower biological activity than ethylene and is also used for ripening mangoes in some countries. However, a higher concentration of acetylene is required for the same period of exposure (Burg and Burg, 1967). Exogenous application of ethylene initiates ripening of mangoes, without detrimental effects on the overall fruit quality (Lakshminarayana *et al* 1975., Medlicott *et al* 1987). Acetylene treatments enhanced softening, peel colour development (Medlicott, *et al* 1987) but results in insipid taste and poor quality (Mann and Dhillon, 1974). Ethylene and acetylene treatments cause ripening of the treated fruits irrespective of their degree of maturity, however, ripening remains incomplete in immature fruits with acetylene. Acetylene treatment was more effective with early season harvests, notably for peel colour development than the late season fruits. Textural changes were most rapidly affected, while the rate of acidity loss was often reduced in ethylene treated fruits. Acetylene treated fruits showed delayed ripening of 'Tommy Atkins' fruits in storage. Thus both ethylene and acetylene have potential in the commercial ripening of half mature and mature fruits of mango ((Medlicott and Reynolds, 1988). However, in commercial use of acetylene, it should be ensured that the fruits are in contact with sufficient acetylene for an adequate period of time to initiate ripening (Medlicott *et al* 1990a).

CHANGES DURING RIPENING OF FRUITS

Ripening is a dramatic event in the life of a fruit and marks the completion of its development and the commencement of senescence. The fruit undergoes several definite physico-chemical and bio-chemical changes during ripening which make it edible. The nutritive value and quality of the fruits depend on these changes that occur during ripening. The decrease in firmness, alcohol-insoluble solids, tannins and increase in pulp colour are the general physico-chemicals features accompanying the ripening of mango fruit. (Selvaraj and Kumar, 1989 a). The main changes observed during ripening are reduction in fruit weight, volume, length, thickness, firmness, pulp content, starch and vitamins-C content and increase in peel colour, TSS, total sugars, sugar-acid ratio and p^H. During ripening fruit passes through a series of changes in texture, flavour and colour which are indicative of compositional changes. Textural changes during ripening result from changes in the structure and composition of cell walls (Gowda and Huddar, 2001). Similar as well as variable changes have been reported by several workers and the discrepancies were attributed to cultivar, stage of maturity, conditions under which the fruits are grown, stored and ripened etc. The various changes that occur in the ripening fruits are addressed below under (i) physical (ii) chemical and (iii) bio-chemical changes.

I. PHYSICAL CHANGES

The physical changes that occur during ripening of mango fruits include: changes in peel and pulp colour, physiological loss in weight, volume, specific gravity, respiration and firmness of pulp.

Peel Colour

Colour of the fruit peel (skin) plays an important role in consumer's acceptability of the product. Visible colour change in the peel colour is the commercially applied indicator of estimation of the degree of ripening of mango fruit. The hard green colour of the harvested fruits changes to olive-green, reddish, pinkish, orange-yellow or yellow during ripening according to the cultivar. The colour change is due to degradation of chlorophyll pigments in the peel (degreening). Chloroplasts in the peel are changed to chromplasts having red and yellow pigments. The loss of green colour and development of yellow colour is associated with many-fold increase in carotenoids (John *et al* 1970., Lakshminarayana, 1980., Lizada, 1993). Development of red blush on the shoulders of fruits of some cultivars was attributed to anthocyanin pigments (Medlicott *et al* 1986a., Lizada, 1993). In a study of peel colour changes in ripening 'Tommy Atkins' mangoes, Medlicott *et al* (1986a) observed progressive loss in chlorophyll and increase in carotenoids indicating that the transition from a green to yellow colour does not involve in mere unmasking of carotenoids, but synthesis of these pigments as well. They also observed a decrease in anthocyanins which led them to conclude that the development of red blush in "Tommy Atkins" can be attributed to the unmasking rather than the synthesis of these pigments. The decrease in chlorophyll content, was suggested due to degradation of chlorophyll followed by synthesis of new pigments in 'Dashehari' mangoes (Gupta *et al* 1979). Chlorophyllase enzyme is mainly responsible for degradation of chlorophyll (Looney and Patterson, 1967). Besides the above enzyme, chlorophyll oxidase and chlorophyll peroxidase also degrade the chlorophyll molecule (Neeraj *et al* 2002). Substantial losses in peel chlorophyll content of "Keitt" mangoes occurred after the fruit began to soften (Medlicott and Thompson, 1985), whereas the level of carotenoids in "Tommy Atkins" mangoes increased with a gradual decrease in anthocyanins (Medlicott *et al* 1986 a).

Pulp Colour

The visible changes in pulp colour occur from harvest maturity to half-ripe stage and maximum colour development coincides with edible ripe stage of fruits (Selveraj and Kumar, 1989a). The pulp colour changes from white to pale-yellow, deep-yellow, orange-yellow depending on the cultivar (Gowda and Huddar, 2001). The colour change in the pulp is due to carotenoids, which increase in direct proportion as the ripening progresses. (Medlicott, 1985., Verma *et al* 1986., Yoneya *et al* 1990). The pulp carotenoid level in the ripe fruit varies with the cultivar. In ripe 'Alphonso' mangoes 16 different carotenoid fractions were observed (Jungalwala and Cama, 1963). A significant increase in all carotenoids (Total carotene, alpha-carotene, beta-carotene and xanthophyll) was observed (Zhou *et al* 1994). Changes in carotenes coincides with the changes in flesh colour from light yellow to yellow and were accompanied by a decrease in pressuro-meter reading, acidity and ascorbic acid content and increase in sugars and TSS (Verma *et al* 1986). The total carotene content reached a maximum level on 6th day of ripening and dropped later (Rama Sarma *et al* 1946). A steady increase occurs in the beta-carotene and xanthophyll content during ripening (Sadana and Ahmad, 1949).

Respiration and ethylene production

Even after separation from the tree, fruits respire till their senescence. The respiration decreases as the fruit matures and rises with the commencement of ripening. Total respiration increases during ripening from mature green stage up to the ripe stage, there after it declines in full ripe stage and is not detectable in over-ripe fruits nearing senescence (Reddy and Srivastava, 1998). The marked increase/upsurge in post-harvest respiration, known as the 'Climacteric' is recognized as an essential event of the ripening process in a number of fruits, which are categorized as; 'Climacteric fruits' and include mango besides banana, papaya etc. Climacteric peak in respiration is often described as a turning point in the life of the fruit and is associated with enhanced glycolysis (Chalmers and Rowan, 1971). The climacteric fruits undergo rapid increase in respiration, concomitant with ethylene production followed by changes in the composition, texture, colour etc of fruits (Akamine and Goo, 1973., Krishnamurthy and Subramanyam 1970a., Salunkhe and Desai, 1984).

The time of occurrence of climacteric peak in harvested fruits varies with the cultivar. In 'Alphonso' the respiratory peak was observed on 5th day of harvest and fruits ripened with in 7/8 days thereafter (Karmarkar and Joshi, 1941), while in 'Kent' and 'Haden' cultivars, the peak was observed at the 9th and 11th day of harvest (Krishnamurthy and Subramanyam, 1970 a). The climacteric peak occurred 2 days after harvest (DAH) in 'Alphonso', 4 DAH in 'Banganpalli' 'Fazli' and 'Totapuri' and 5 DAH in 'Dashehari' 'Langra' and 'Suvarnarekha' (Selveraj and Kumar, 1989 b).

Respiration in climacteric fruits is classified as

(i) pre-climacteric rise extending up to 3 days with a slow release of CO_2,
(ii) a climacteric rise up to 6 days with a sudden spurt in CO_2 production,
(iii) a climacteric peak between 6 and 10 days with softening of the fruit and
(iv) a post-climacteric phase lasting from 10-14 days with a decrease in CO_2 production resulting in edible ripeness of fruit followed by senescence.

The pattern of respiration and ripening varies within the fruit. At the first climacteric peak, the mesocarp (pulp) adjacent to the endocarp (seed) begins to ripen and the second climacteric peak brings about the ripening of the remaining part of the mesocarp (Kosiyachinda and Panakasemsum, 1990). However, the pattern of respiration and ripening behaviour of fruits may vary among the cultivars, climatic conditions, and places of growing the fruit (Krishnamurthy and Subramanyam, 1970 a). 'Alphonso' showed high respiration rate; while 'Banganapalli', 'Dashehari', 'Fazli' 'Langra' and 'Suvarnarekha' showed intermediate rate and 'Totapuri' exhibited lowest rate of respiration (Selvaraj and Kumar, 1989 b).

With the commencement of climacteric peak in respiration, a host of changes in physical, chemical and biochemical characteristics take place, of which concomitant ethylene production is one (Krishnamurthy *et al* 1971). Although, Burg and Burg (1962) stated that ethylene production rises at or before the rise of CO_2 in ripening mangoes, Bialte and Young (1981) included mangoes among the fruits in which ethylene rises after the rise of the CO_2 production. Burg and Burg (1962) in 'Haden' and Lam *et al* (1982) in 'Golak' cultivars observed that the climacteric pattern of respiration is associated with the onset of ethylene production with the former preceding the later. In contrast, the onset of ethylene production in 'Carabao' coincides with the climacteric rise (Cua and Lizada, 1990). It appears that for the majority of fruits, ethylene biosynthesis commences before and is experimentally inseparable from respiratory changes (Bialte and Young, 1981) and go hand in hand together during ripening (Reddy and Srivastava, 1999a).

Ethylene production by mango fruit tissues as in many other climacteric fruits, is maximal at the onset of the climacteric phase of fruit ripening (Burg and Burg, 1962., Mattoo and Modi, 1969 a) and starts declining with the advancement of ripening. The levels of endogenous ethylene before the onset of the climacteric are low (0.01 ppm) (Burg and Burg 1962). In the 'Carabao' mango, the internal level of ethylene rises to only about 0.06 ppm during the climacteric (Cua, 1989). The internal ethylene level in some cultivars of mango is only about 3 ppm (Burg and Burg, 1962) while externally the detectable level is about 0.1 ppm (Peacock *et al* 1986). Nevertheless, these small amounts of ethylene are enough to trigger ripening in mango.

Normally, climacteric fruits have a single peak of ethylene production. However, cultivars like 'Carabao' in the Philippines (Cua and Lizada, 1990) 'Kneaw Sawoey' in Thailand (Kosiyachinda and Pankasemsum, 1990) and 'Amrapali' and 'Dashehari' in

India (Reddy and Srivastava, 1999a) exhibited three and two ethylene peaks respectively during ripening; one being at early stage (colour break) and the other at the last stage (when the peel colour is fully yellow) of ripening (Cua and Lizada, 1990). A third peak of ethylene production was observed in 'Carabao' mango fruit apparently from embryo, since *in vitro* studies showed neither the mesocarp nor the peel producing ethylene at the latter stages (Cua and Lizada, 1990). The second peak of ethylene production in 'Amrapali' and 'Dashehari' mango fruits was attributed to embryo (Reddy and Srivastava, 1999 a). In 'Amrapali' the first ethylene peak occurs preceding the respiratory burst and the second peak coincided with respiratory peak. In case of 'Dashehari' the two ethylene peaks occurred at 3 days and 7 days after harvest and coincided with respiratory peaks (Reddy and Srivastava, 1999 a). Ethylene production in 'Carabao' fruit occurred prior to full maturation/harvest (Mendoza, 1981) and declined as the fruit approaches full maturity (Cua and Lizada, 1990). Ethylene production ceases as the fruit becomes over-ripe. This pattern of ethylene production has been attributed to an impairment of the enzyme converting ACC to ethylene as membrane integrity is lost (Yang and Hoffman, 1984).

The production of ethylene is not uniform in all portions of the fruit. Differences in production pattern (in Carabao) among the inner and outer portion of the pulp and peel were observed (Cua and Lizada, 1990). Similar observations in 'Amrapali' and 'Dashehari' mango were reported (Reddy and Srivastava, 1999 a). The fruit peel showed highest level of ethylene and ethylene forming enzyms (EFE) and less ACC accumulations than outer and inner mesocarp at mature green stage. The inner mesocarp showed less EFE activity and higher ACC accumulation as compared to peel during ripening process. Outer mesocarp fell in between the peel and inner mesocarp. Thus the ripening process appears to start in the mesocarp and proceeds towards the peel (Reddy and Srivastava, 1999 a).

Respiration rate and activities of ethylene biosyntesis, however, differ among cultivars. Among the two cultivars studied 'Dashehari' showed higher level of ACC-synthase and ACC accumulation and low level of ACC oxidase and ethylene production compared to 'Amarpali' during ripening. Respiration rate, ethylene biosynthesis enzyme activities were more in 'Amrapali' than in Dashehari' mangoes. These metabolic activities are associated with early ripening in 'Amrapali' compared to 'Dashehari' mangoes. Therefore, approaches to regulate respiration and ethylene biosynthesis at the initiation of post-climacteric stage in mango may lead to slow down ripening process and enhance its shelf life (Reddy and Srivastava, 1999 a).

The evolved ethylene from the fruit tissues stimulates respiration enzymes and inactivates the enzyme inhibitors, leading to fruit ripening (Mattoo and Modi, 1969 a)

Softening of the Pulp

Ripening of the mango fruit is characterized by softening of the flesh (mesocarp). It is a key qualitative change, a fundamental guide and is commonly measured by finger pressure and scientifically by pressure-meters. The changes in the mesocarp (pulp) softness during ripening follows roughly sigmoid curve (Chaplin *et al* 1990). Among the four steps (optimum blend of colour, taste, aroma and texture) involved in the transformation of fruit from unripe stage to ripe edible stage, textural change is very important as it significantly affects the handling, processing and consumer acceptance (Bhagyalakshmi *et al* 2002). However, full knowledge about the mechanism of mango fruit softening is lacking, although it was postulated that textural changes in the mesocrap are resultant of enzyme activity, notably by ethylene.

An interesting feature of fruit softening in mango is the occurrence of differential softening and a differential distribution of cell wall softening and respiratory enzymes (Lazan *et al* 1993). Softening of inner mesocarp occurs first followed by the outer mesocarp near the peel. This pattern was considered as a characteristic feature of mango fruit ripening. Sometimes, their variation leads to uneven ripening of the fruit (Chaplin *et al* 1990., Lazan *et al* 1993., Mitcham and Mc Donald, 1992). A few workers attempted to elucidate the reasons for such differential pattern of ripening. It was suggested that the inner mesocarp (pulp) adjacent to the seed is more metabolically active than the outer mesocarp and probably this allows the initiation of ripening process earlier in the inner mesocarp (Lazan *et al* 1993). It may also be possible that the earlier synthesis of ethylene in the inner mesocarp is responsible for the initiation of ripening first in the inner mesocarp (Reddy and Srivastava, 1999b). Polygalacturonase (PG) is differentially distributed and its activity correlates positively with variation in tissue softness. Beta-galactosidase and phosphophenol pyruvate carboxylase (PC) too are differentially distributed and their activity increases more rapidly in the softer inner than in the firmer outer mesocarp. NADP-Malic enzyme activity also shows a faster increase in the inner tissue and this may partly account for a more rapid decline in malate level, in that tissue during ripening. The metabolic preponderance of the inner mesocarp suggests that ripening of mango is initiated in the inner tissues and progresses outwardly. The under laying mechanism that regulates this differential ripening phenomenon remains to be ascertained (Lazan *et al* 1993). In some cases the inner mesocarp of ripe fruit was softer than the mid-or the outer region. In the 'Kensington Pride' and 'Mulgoa' cultivars the mesocarp at the stylar end of fruit was softer than at the mid region or pedicel end, which may be attributed to cultivar differences (Chaplin *et al* 1990., Mitcham and Mc Donald, 1991).

Softening in fleshy fruits is primarily due to cell wall modifications during ripening. Changes in texture leading to softening is closely related with break down of proto-pectin into soluble pectins and conversion of starch to sugars. However, limited information is available on mango cell walls and the softening process during ripening. Softening of mango fruit is characterized by an increase in the solubility of cell wall pectins (Lazan *et al* 1986., Nasrijal, 1993). The general observation is that softening of flesh is accompanied by solubilization of pectins in cell walls, involving action of PG and PE enzymes (Muda *et al* 1995). This notion gets support from the reports of changes in cell wall pectins in ripening mangoes (Nour, 1978., Roe and Brummer, 1981) and in tomato (Besford and Hobson, 1972). As in other fruits, mango fruit softening has been primarily attributed to depolymerization and solubilization of polyuronide in the middle lamella of cell wall by cell wall hydrolases (Huber, 1983). While pectin solubilization in inner mesocarp and outer mesocarp tissues is comparable, pectin deploymerization appeared to begin earlier in the inner mesocarp than in the outer mesocarp tissues (Lazan and Ali, 1993). Roe and Brummer (1981) suggested that PG and cellulase are the degrading enzymes responsible for softening of pulp during ripening of mango fruits. However, Abu-Sarra and Abu-Goukha (1992) found no significant role of cellulose in tissue softening of mango fruit. Ketsa *et al* (1998) observed increased activity of PG and β-galactosidase, while PE activity decreased continuously during softening and reported that softening of ripening fruits was more closely related to changes in β-galactosidase activity than PG and PE activities. It appears that pectin polymers become less tightly bound in the cell walls during ripening and that this loosening involves hydrolysis of galactose-containing polysaccharides. Pectin polymer composition differed between fruits of cv.'Sensation' which soften quickly during storage and those of Cv. 'Tommy Atkins' which gave a long storage life (Seymour *et al* 1989).

The studies of Selveraj and Kumar (1989a) indicated that all the three enzymes viz., pectin esterage, polygalacturonase and cellulase are related to softening of the pulp, which were designated as 'key enzymes' of softening. These enzymes were found to increase from harvest maturity to colour break stage.

Softening of mangoes is accompanied by a rapid decline of alcohol insoluble solids (AIS) (Roe and Brummer, 1981., Subramanyam *et al* 1972) with a concomitant increase in sugar and TSS. However, decrease in AIS also reflects the conversion of pectin substances to alcohol soluble compounds. Both the water and alkali-soluble fractions decreased and the ammonium oxalate fraction increased slightly (Roe and Brummer, 1981). Since the alkali-soluble pectin is structurally bound to the cell wall these relationships suggest that flesh softening in the ripening process results from disintegration of the cell wall caused by degradation of the structurally bound pectin.

Contrary to the above, Lazan *et al* (1986), reported that softening of fruit was accompanied by an increase in water-soluble, ethanol insoluble pectins and concluded that PG might be one of the enzymes involved in mango fruit softening. In 'Keitt' mangoes, ripened at 21° and 26 °C, PG activity correlated with textural changes. However, the observation that vaporgard persistently affected PG activity but not decrease in firmness suggests that PG may not be the significant enzyme involved in the softening process.

Ali *et al* (1995) suggested that β-galactosidase might be playing an important role in cell wall pectin modification and softening of mango fruit during ripening as they observed a close correlation between changes in β-galactosidase activity, tissue softness and increased pectin solubility and degradation in ripening fruits of cv. 'Harumanies'

Enzyme activity increased parellely with increased softness during ripening; glyconases such as mannase, galactonase and arabinanase were recognized as the 'key enzymes' for textural degradation leading to softening of mesocarp during fruit ripening of mango (Bhagyalakshmi *et al* 2002).

Mitcham and McDonald (1992) indicated that enzymatic and/a non-enzymatic processes, in addition to PG activity, are involved in the extensive softening of mango fruits. They studied the cell wall modification during ripening of 'Keitt' and 'Tommy Atkins' mango fruits and observed that cell wall neutral sugars, particularly arabinosyl, rhamnosyl and galactosyl residues decreased with ripening of both cultivars. 'Keitt' had more loosely associated chelated soluble pectin, accumulated more soluble polyuronoids and retained more total pectin at the ripe stage than did 'Tommy Atkins'. Both cultivars had similar PG activity, which increased with ripening. In 'Sensation' mangoes galactose was the only cell wall neutral sugar to show a significant decrease during ripening (Seymour *et al* 1989).

Bhagyalakshmi *et al* (2002) found a gross increase in uronoids and a steady decease in starch, cellulase, hemicellulases and pectins during ripening of 'Alphonso' fruits. They also observed a major drop in molecular weights of carbohydrate polymers during ripening with complete disappearance of low molecular weight polysaccharides and considered that this information is useful for monitoring at the bio-chemical level, the textural loss during ripening without inhibiting ethylene associated natural ripening.

Different mango cultivars exhibit different degrees of softening during ripening. This may be simply due to their different ethylene synthesis rate or to differential response to ethylene. It could be due to the fact that mango cultivars have inherently different softening characteristics. Thus some cultivars are inherently softer than others when fully ripe. This may in turn be related to their cell wall structure and/or degrading enzymes. A preliminary study on a good storing cultivar 'Tommy Atkins' has shown some differences in cell wall changes from those reported for 'Sensation' cultivar, a poor storing one. If this is the case, then cell wall characteristics may be useful in predicting storage capacity and altering these characteristics may improve their storage capacity in some cases (Tucker and Seymor, 1991).

As regards the other physical changes, physiological loss in weight (PLW) increases with progress in ripening, due to transpiration of moisture from the fruit, while the specific gravity of the fruit decreases correspondingly. Fruit pressure decreases as the fruit ripened due to softening of the pulp.

II. CHEMICAL CHANGES

The chemical changes that occur in the ripening of fruit include increase in total soluble solids and reducing sugars and decrease in the contents of starch, titratable acidity and ascorbic acid.

Total Soluble Solids (TSS)

The content of total soluble solids (sugars, minerals and acids) increases during ripening and then declines till senescence. Increase in TSS during ripening is associated with conversion of starch and other insoluble sugars. Conversion of certain cell wall materials like hemicellulose and pectin substrates from water insoluble to water-soluble pectic fraction also contribute towards rise in TSS (Bhullar *et al* 1981).

The fall of TSS after peak ripening and towards senescence phase was attributed to the further utilization of soluble solids and sugar for respiration (Sahni and Khurdiya, 1989., Hoda *et al* 2001).

Organic acids (Acidity)

Non-volatile organic acids (citric, malic, tartaric etc) are among the major cellular constituents undergoing changes in fruits during ripening. They are considered as reserves of energy to the fruit as they are utilized as respiratory substitutes. Their loss/decrease in the fruit indicates the decrease in the respiration. There will be accumulation of organic acids with maturity which undergo active metabolism and decline during fruit ripening (Krishnamurthy *et al* 1960., Sarkar and Munshi, 1981., Medlicott and Thompson, 1985). Loss in acidity during ripening is indicated by decreased titratable acid contents. A 6-9 fold decrease in acidity from harvest maturity to ripe stage was observed in all seven cultivars studied (Selveraj *et al* 1989). Certain individual acids show greater changes than others (Ulrich, 1970) while certain others undergo rapid metabolic changes in the course of ripening of mangoes (Kumar *et al* 1993).

Certain differences exist with type of acid(s) and changes during ripening. There was a large decrease in citric acid and a small reduction in malic acid level in 'Keitt' mangoes (Medlicott and Thomson, 1985). An increase in malic acid was observed in 'Badami' during ripening (Shashirekha and Patwardhan, 1976). Tartaric and malic acids remained fairly stable in Floridan mangoes (Yoney *et al* 1990). The decline in acidity during ripening, however, was attributed mainly to a decrease in citric acid (Dubery *et al* 1984). A 2-6 fold decrease in citric acid was observed during ripening in the cultivars studied (Selveraj *et al* 1989).

During the course of ripening of mangoes, acid components are more or less actively metabolized, succinic acid being the first to disappear, closely followed by oxalic and citric acids. At full ripeness acidity is represented by malic acid alone (Kumar *et al* 1993). In general, levels of citrate and succinate gradually decrease during ripening, while malate shows variation with different cultivars (Lizada, 1993). The acidity, expressed as 'anhydrous' citric acid, was higher in mature fruits as compared to ripe fruits and decreased continuously thereafter (Upadhyay and Tripathi, 1985).

The acids are converted into sugars during ripening and their derivatives are used in respiration and this utilization is responsible for the decreasing trend of acidity during ripening (Bhullar *et al* 1981., Kumar and Dhawan, 1995., Hoda *et al* 2001).

Decline in acidity brings shifting of pH from 2.0 to 5.5 in mango (Modi and Reddy, 1967). The reduction in acidity and increase in soluble sugars results in a substantial increase in sugar : acid ratio, which serves as an index of degree of ripeness and basically determines the flavour of the fruit.

Ascorbic Acid

Ascorbic acid (Vitamin C) increases with the age of the fruits to maturity, but decreases during ripening due to oxidation and enzymatic degradation of ascorbic acid (Nandini and Oommen, 2002). The rate at which this change occurs is largely conditioned by fruit pH, as the pH increases, ascorbic acid content in ripe fruit than in mature unripe fruits.

Starch and Sugars

Carbohydrates constitute the major components of mango fruits. During maturation polysaccharides are actively synthesized and are accumulated as starch till full maturity as reserve food. Starch serves both as substratum for respiration and as the main source of sugars responsible for sweetness of the fruit. The main change that occurs during ripening is the break down of starch into sugars. In general, the starch level is reduced from harvest to eating-ripe stage due to its break down. Ripe fruits have minimum or no content of starch and was almost undetectable (Bernarday *et al* 2003). The hydrolysis of starch into sugars occurs in the chloroplasts. The sugars converted from starch are important factors in determining the palatability of fruits as the balance between sugars and acid imparts flavour to the fruit.

The information on sugar accumulation during ripening of mango fruits is very limited and is contradictory. Starch is hydrolyzed into three kinds of sugars viz., sucrose, fructose and glucose. All these, in general increase during ripening but show a shortfall in over-ripe fruits due to fall in sugar conversion and continued respiratory loss (Selveraj *et al* 1989., Zeng *et al* 1995).

Among the three kinds of sugars, sucrose has been the predominant one (Krishnamurthy and Subramanayam, 1970 a., Selveraj *et al* 1989., Kumar, *et al* 1994., Yoneya *et al* 1990). Sucrose constituted 57 percent of the total sugars in ripe 'Keitt' mangoes, with fructose and glucose making up 28 and 15 per cent respectively (Medlicott and Thompson, 1985). However, comparable concentrations of the three sugars were reported in Indian mangoes (Shashikala and Patwardhan, 1976., Tandon and Kelra, 1983) and Israli mangoes (Fuchs *et al* 1980). Among the three sugars greatest increase was shown by sucrose followed by glucose and fructose (Lakshminarayana and Subramanyam, 1966). The increase of sucrose coincides with the disappearance of starch suggesting sucrose synthesis and its progressive accumulation were directly related to the hydrolyzed by-products of starch such as glucose.

However, several workers (Krishnamurthy *et al* 1971., Lakshminarayana, 1973, 1975., Sashirekha and Patwardan, 1976) suggested simultaneous increase of all the three sugars during ripening, but Vazquez – Salinas and Lakshminarayana (1985) observed a gradual reduction in both glucose and fructose and a continuous increase of sucrose during ripening in 'Haden' 'Irwin' 'Kent' and 'Keitt' cultivars.

Although 'sucrose' has been reported to be the predominant sugar, the predominance of a particular type of sugar appears to differ with the stage of ripening. At unripe stage fructose was the major sugar. At the half-ripe stage and full ripe stages sucrose was the major sugar. At all the above three stages of ripening glucose was comparatively low. However, fructose content remained fairly constant at different stages of ripeness (Yoney *et al* 1990). However, Kumar *et al* (1994) reported that sucrose was by far more abundant than glucose and fructose at the two-third mature and over-ripe stages. Glucose was always lower than fructose during the whole period of ripening. However, all these three sugars were present at all stages of ripening (Selveraj *et al* 1989).

III. BIOCHEMICAL CHANGES

Most of the work in mango fruit for the last 60 years has largely been confined to post-harvest analysis and technology. The fruit has received little attention from the biochemist. The biochemical changes are initiated by autocatalytic production of ethylene and increase in respiration. Main biochemical changes during ripening are addressed below.

Pectins

Pectins are the major class of cell wall polysaccharides, which are degraded by both solubilization and depolymerization (Chaimanee *et al* 2000). Solubilization of pectic substances in the cell wall due to the activity of PE and PG results in decrease of pressure of fruit during ripening (Sahni and Khurdiya, 1989). Changes occur in pectin fraction and alcohol insoluble solids (AIS) during ripening of mango fruits (Mizuta Subramanayam, 1973., Roe and Brummer, 1981). AIS declined markedly during ripening (Roe and Brummer, 1981) along with alkali soluble (Proto pectin) fraction, while water-soluble and ammonium oxalate soluble pectin fraction increased during ripening (Subramanayam. *et al* 1972). Proto pectin is an important pectin fraction which increases before physiological maturity and then decreases with ripening and softening. It, upon hydrolysis, yields water soluble pectin and its content increases in ripe mango fruit (Mizuta and Subramanayam, 1973).

Phenolics

Astringency of the fruit and enzymatic browning are related to the presence of phenolics in the fruit (Goldstein and Swain, 1963). Phenolics also play an important role in disease resistance in plants (Cruikshank and Perrin,1964). The polyphenol content decreases during eating ripe stage (Lakshminarayana *et al* 1970). The peel of the fruit had higher phenolic content than the pulp at all stages of development (Lakshminarayana *et al* 1970). The higher content of polyphenol in unripe mango fruit declines during ripening (Selveraj and Kumar, 1989 a).

Lipids and fatty acids

The lipid content of the pulp was correlated with the flavour characteristics (Bandopadhyay and Gholap, 1973 a). Changes in pulp lipid composition was observed to be associated with a progressive change during fruit ripening, which is dependent on post-harvest storage temperature. The ripening of 'Alphonso' mangoes at ambient temperature was accompanied by a sharp increase in triglyceride content, together with the development of strong aroma and flavour. On the other hand, ripening at low temperature (10 ± 2 °C) did not bring about any such changes in lipids (Bandopadhyay and Gholep, 1973 b).

The increase in fatty acid during ripening was reported in certain mango varieties, which have been suggested to be involved in the development of the characteristic aroma in the ripe mango fruit (Bandopadhyay and Gholep, 1973 a,b). Ripening of 'Alphonso' mangoes in a well ventilated room at ambient temperature (25 °C - 30 °C) was associated with a change in fatty acid composition (Bandopadhyay and Gholep, 1973b). Total fatty acids increased, whereas free fatty acids and phospholipids decreased during ripening of mangoes. The content of fatty acids varies with the cultivar. 'Alphonso' had high total lipids content followed by 'Langra 'cultivar (Selveraj *et al* 1989).

Proteins and amino acids

Changes in protein content during fruit development and ripening had scarcely been investigated. Lopez and Gomez-Lim (1992) noted that certain proteins increased while others decreased during natural ripening of mango fruits. The quantity of extractable proteins decreased from pre-climacteric to ripening stage (Marin and Cano, 1992). On the other hand, Lopez and Gomez-Lim (1993) reported that proteins increased during natural ripening and thereafter decreased in 'Manila' fruits held at 25 °C.

A considerable variation in free amino acid content was observed in the pulp of different mango cultivars. Initially the free amino acid content increased suggesting the hydrolysis of proteins caused by high protease activity (Selveraj *et al* 1989) followed by a decline (Zeng *et al* 1995). Both total and individual amino acid levels increased during the early ripening period and then fell. Organoleptic maturity coincided with the stage of peak amino acid and sugar levels (Elahi and Khan, 1973). The ripening process is associated with a decrease in all the free amino acids, except alanine, ornithine and glycine, which increase while threomine remained unchanged (Passera and Spettol, 1981). Thus the concentration of individual amino acids changed markedly from harvest maturity to eating ripe stage (Selveraj *et al* 1989).

Aroma volatile compounds

Aroma, a component of flavour is one of the important quality attributes of ripe mango fruit, which attracted the consumer since antiquity. Aroma is contributed by changes in sugar-acid blend and is a result of production of volatile compounds such as organic acids, alcohols, esters, carbonyl compounds, hydrocarbons and terpenoides. Aroma volatile compounds are responsible for odour and contribute to the overall flavour of the fresh and processed fruit (Brackman *et al* 1993).

Production of most volatile compounds rises many fold during fruit ripening. During ripening of 'Kensington Pride' mango, production of monoterpenes and sesquite terpenes followed ethylene production pattern, whilst esters continuously increased (Lalel *et al* 2002).

Polyamines

These are the recently added group of phytohormones, influencing several growth and development phenomena in plants. They are implicated in fruit growth and development and ripening (Smith, 1985). Three principal polyamines viz., Putrascine (PUT), Spermidine (SPD) and Spermine (SPM) were detected during fruit ripening. Total polyamines increased in the skin and pulp tissue along with the climacteric rise of ethylene and reached maximum level on fourth day of ripening. Skin exhibited higher mean polyamines than the pulp. The increase in endogenous polyamines especially PUT is simultaneous with climacteric ethylene production (Malik and Singh, 2003). The simultaneous increase of ethylene and polyamines during fruit ripening suggests that their biosynthesis from a common precursor may not be competitive and free polyamines may have evolved as a response to increased biosynthesis of ethylene. The increase in SPD and SPM at later stages of fruit ripening may be explained on the basis of PUT being metabolized into SPD and SPM (Malik and Singh, 2004).

In 'Nam Dok Mai' mango fruit, the levels of polyamines differed among the three stages of ripening. For the stage I fruits, PUT level increased reaching maximum values during 15 days stage at 5 °C and then decreased. In contrast, stage II and III fruits (at 13 °C) had high PUT levels than stage I fruits. SPD levels in stage II and III fruits decreased with storage time while they showed increase in stage I fruits after a fortnight storage. When fruits were transferred to 25 °C for ripening PUT and SPD levels increased in fruits at all stages (Wangchai *et al* 2000).

Enzymes

Fruit ripening is a metabolically active process which results in the transformation of fruit into an edible product. The metabolic changes during ripening are highly enzyme-mediated phenomena (Grierson, 1985). Ripening of fruits is associated with increase in the activity of several hydrolytic and gluconogenic enzymes (Tucker and Grierson, 1987). Several enzymes associated with carbohydrate and cell wall metabolism such as invertase and pectin esterage (Hulme, 1971) and cellulase and polygalacturonase (Lazan *et al* 1986) have been detected in mango fruits during ripening.

(a) Enzymes related to softening of the fruit pulp

Softening of fruit is accompanied by solubilization of pectin in cell walls involving the action of the enzymes such as polygalacturonase, pectinesterase and cellulase (Mc Cready and Mc Comb, 1954). Cell wall degradation results in softening of the fruits and the responsible enzymes such as polygalacturonase and pectin methyl esterase have been reported to increase during ripening (Tuker and Grierson, 1987). Similarly, increase in cellulase and its activity were reported (Abu-Sarra and Abu-Goukha, 1992).

Polygalacturonase

During ripening polygalacturonase (PG) activity increased significantly and then declined (Zhou *et al* 1996). However, the activity of PG varied with the stage of ripening as well as with the variety. PG activity in 'Banganapalli' and 'Fazli' cultivars decreased from harvest to ripe and/or ¾ the ripe stage. In cultivars 'Alphonso' 'Dashehari' 'Langra' 'Suvarnarekha' and 'Totapuri', the activity increased from harvest to half-ripe stage. Thereafter the activity declined a little in ¾ the ripe stage and declined markedly in eating ripe stage fruits (Selveraj and Kumar, 1989 a). Thus it appears that the activity of PG follows the progress of softening/ripening of the fruit. Roe and Brummer (1981) observed a correlation between the increase in PG activity with increase in soluble pectin and resultant softening.

PG can be present as either endo-acting and exo-acting enzyme. In mango it seems that only exo-acting PG is present (Lazan *et al* 1986., Brinson *et al* 1988). However, Chaimanee *et al* (2000) detected both exo-PG and endo-PG during the process of ripening of mango fruits and suggested that the endo-PG was the key enzyme in ripening associated with pectin degradation. The increase in exo-and endo-PG activities correlated well with increasing ripeness of 'Nam Dok Mai' mango fruits, ripened at ambient temperature of 27 °C - 28 °C (Chaimanee and Subhadrabandhu, 1992).

Pectin esterase

Contradictory reports were made on the changes of pectin esterase (PE) activity during ripening of mango fruits. The activity of this enzyme has been shown to decrease (Abu-Sarra and Abu-Goukha, 1992., El-Zoghbi, 1994., Ketsa *et al* 1998), to remain constant (Ashraf *et al* 1981) or to increase (Abu-Sarra and Abu-Goukha, 1992). A number of studies have shown that PG activity increased during the ripening period but the activity pattern varied with the cultivar (Selveraj, *et al* 1989., Abu-Sarra and Abu-Goukha 1992., Gomez-Lim 1993, 1997). PE activity in ripening mango cultivars showed three distinct patterns. In cvs 'Alphonso' and 'Totapuri' the activities increased gradually from harvest maturity to ¾ the ripe stage and then declined. 'Dasheri' 'Fazli' 'Langra' and 'Suvarnarekha' showed increased activity levels from harvest maturity to half-ripe stage, little decline during ¾ the ripe stage and considerable decrease in ripe fruits. In 'Banganapalli' PE activity declined more or less steadily from harvest to ripe stage. However, the decline in activity was sharp from ¾ the stage to eating ripe stage (Selveraj and Kumar, 1989 a).

Cellulase

Cellulase is another type of enzyme, which has been related to fruit ripening. Contradictory reports were made in changes with levels of cellulase during ripening of mango fruits. Cellulase levels and their activities in unripe fruits are generally low, but increase dramatically during ripening (Roe and Brummer, 1981., Abu-Sarra and Abu-Goukha, 1992., El-Zoghbi, 1994). The increase in cellulase activity during ripening correlated well with decrease in firmness of the fruit (Roe and Brummer, 1981). However, Lazan *et al* (1986) observed a decrease in the activity of cellulase during ripening of 'Harumanies' cultivar. Recently Bhagyalakshmi *et al* (2002) also made similar observation in the activity of cellulase during ripening. On the other hand, Tahir and Malik (1977) noticed no change in cellulase activity during ripening of 'Pairi' mangoes.

The increase/decrease in cellulase activity varied with stage of ripening as well as the cultivar. Cellulase activity in 'Alphonso' 'Dasheri' 'Fazli' 'Suvarnarekha' and 'Totapuroi' increased from harvest maturity to half-ripe stage and decreased at ¾th and eating ripe stages, whereas in 'Banganapalli' and 'Langra' cellulase activities increased up to 3/4th ripe stage and then declined in ripe fruits (Selveraj and Kumar, 1989 b).

(b) Other enzymes related to cell wall degradation

Mattoo and Modi (1969 b) showed an increase in pectin methyl esterase (PME) activity during ripening of mango fruits. Its activity increases as the fruit reaches climacteric and shows a decrease on further ripening (Subramanayam *et al* 1976). The peak activity of PME occurred before that of respiration and gradually decreased as the fruit ripened (Zeng *et al* 1995., Zhou *et al* 1996).

Brinson *et al* (1988) detected significant quantities of galacturonic acid, arabinose and galactose among the water soluble polysaccharides released from the cell wall during ripening of monoenbyonic 'Ngowe' mango fruit. Mitcham and McDonald (1992 a) found that cell wall neutral sugars particularly arabinosyl, rhamnosyl and galactosyl residues as well as the amount and molecular hemicellulose decreased with ripening monoenbyonic 'Keitt' and 'Tommy Atkins'. The activity of beta-galactosidase increased dramatically during ripening of mango fruit (Seucedo-Arias, 1997). The amount and molecular weight of cell wall hemicellulose decreased during ripening (Mitcham and Mc Donald, 1992).

(c) Enzymes related to carbohydrate metabolism

Sucrose and starch metabolism in fruits is mediated by different enzymes during fruit ripening, one of them is α-amylase (Mattoo and Modi, 1969 a., Fuchs *et al* 1980). It increased four-fold in activity during ripening of mangoes (Mattoo *et al* 1975., Majumdar *et al* 1981). The increased activity of α-amylase is associated with starch hydrolysis (decrease in starch content) (Fuchs *et al* 1980., Tandon and Kalra, 1983).

Sucrose phosphotasynthase (SPS) increases in activity by ten-fold during ripening (Castrillo *et al* 1992., Hubbard *et al* 1991). This enzyme is responsible for sucrose synthesis from hexose phosphatase. In contrast, sucrose-synthase (SS) an enzyme responsible for sucrose breakdown, remained relatively constant throughout the ripening period (Castrillo *et al* 1992). The activities of neutral and acid invertase enzymes were low in comparison with those of enzymes of sucrose synthesis. However, acid invertase activity increased and later decreased during ripening (Castillo *et al* 1992) but it does not seem to contribute significantly to sucrose breakdown (Mattoo *et al* 1975). In contrast, neutral invertase did not show noticeable change during ripening (Castrillo *et al* 1992). The high activities of both sucrose synthatase and invertase in the mesocarp during ripening are indicative of sucrose metabolism (Kumar *et al* 1994).

Mango fruit ripening involves increases in the activity of some gluconcogenic enzymes such as glucose-6-phosphatase and fructose-1,6-diphosphatase (Kumar and Selveraj 1988., 1990., Parikh and Modi, 1990). The activity of glucose -6-phosphatase was reported to increase up to $3/4^{th}$ ripe stage, whereas fructose-1,6-diphosphatase activity increased thereafter to full-ripe stage (Kumar and Selveraj, 1990). The glycolytic enzyme, hexokinase activity was also detectable at the ripe stage. The phosphofructo kinase showed maximum activity in the ripe stage; while pyruvate kinase activity was found to increase until $3/4^{th}$ ripe stage and then declined at ripening (Selveraj and Kumar, 1994).

(d) Enzymes related to lipid metabolism

The occurrence of an increase in ATP: citrate oxalo acetic acid lyase (Citrate cleaving enzyme) activity during ripening led to the suggestion that acetyl-COA and oxalo acetic acid (OAA) formed in the action of the enzyme as citrate may contribute to the synthesis process of lipid, taking place during ripening and that natural lipid breakdown products might regulate the activity of the enzyme *in vivo* (Mattoo and Modi, 1970a). The activity of lipase, lipoxygenase, and alcohol dehydrogenase, which is high in unripe fruits decreases when the fruits were more than half-ripe (Selveraj, 1989). Other fatty acid related enzymes, whose activity change during ripening are thiolase and orcyl COA oxidase (Mitra and Baldwin, 1997).

(e) Other enzymes related to other ripening processes

Besides the above, several other enzymes have been detected during mango fruit ripening such as catalase, peroxidase, phosphofructokinase, aldolase, phosphogluconate dehydrogenase, maliate dehydrogenase, citrate cleavase, geranylkinase, pyruvate decorboxylase, acid phosphotase, ascorbic acid oxidase etc (Modi and Reddy, 1967., Mattoo and Modi, 1970 b., Mattoo *et al* 1975., Dubery, *et al* 1984., Kumar and Selveraj, 1988., Cardello and Cardello, 1998).

Peroxidase activity increased in pulp of mango fruits up to the climacteric stage of ripening in storage at 12 °C and 85-90 percent relative humidity. Pyruvate kinase (PK) activity decreased from harvest to half-ripe stage and increased at 3/4th ripe stage and declined in ripe stage in 'Banganapalli' 'Dasheri' and 'Fazli' cultivars. However, in cvs. 'Alphonso' 'Langra' and 'Totapuri' the PK activities increased from harvest maturity to half-ripe stage/or 3/4th ripe stage and then declined at ripe stage. 'Suvarnarekha' registered a decreased activity pattern from harvest maturity to half-ripe stage and increased activity thereafter until ripe stage. Though the PK activities declined at ripe stage in most mango cultivars studied, they remained fairly high in them. Higher relative PK activities during ripening were observed in cvs. 'Banganapalli' 'Langra' and 'Suvarnarekha' as compared to other cultivars. In general, the respiratory climacteric is accompanied by an increased level of PK activity in mango during fruit ripening (Selveraj and Kumar,1989 b).

Malic enzyme catalyses the decarboxylation of the malic acid, the major organic acid for respiration during ripening in climacteric fruits like mango (Goodenough *et al* 1985). Malic enzyme was detectable in 3/4th ripe stage and ripe stages in mango (Selveraj and Kumar, 1994). The activity of the enzyme increased gradually from the mature green to ripe stage and declined thereafter (Reddy and Srivastava, 1998). Kumar *et al* (1990) showed that the internal temperature of 'Alphonso' increased from 28 °C to 38.9 °C over the ripening period. It was hypothesized that the enhanced generation of heat during fruit ripening helps to volatilize aroma-giving compounds in fruits (Gomej-Lim, 1997).

***ATP*:** Phospho fructokinase (PFK) has been suggested as a regulatory enzyme in ripening mango fruits and that it plays a crucial role in the regulation of glycolysis at 3/4th ripe stage (Kumar and Selveraj, 1989). Activity of PFK was maximum at eating ripe stage and could not be detected at harvest maturity and half-ripe stage.

The absence of enzyme activity at earlier stages could be due to

(i) inactivation of enzyme *in vivo* by changes in its regulatory properties,

(ii) low level below the detection threshold and

(iii) the use of extracts of acetone powder and not the fresh fruit pulp.

The increased activity at the eating ripe stage could be due to increased synthesis of enzyme and/or activation of metabolites (Kumar and Selveraj, 1989).

During ripening ascorbate oxidase activity increased resulting in decreased content of ascorbic acid showing a significant inverse linear correlation in 'Haden' fruits harvested at hard green stage and stored up to 14 days at 20 °C (Cardello and Cardello, 1998).

Phosphotase activity increases with the progress of ripening due to stimulation of beta-carotene implying that carotene might regulate this enzyme during ripening. Since, a concomitant increase in phohshotase activity was observed along with the accumulation of mevalonic and geraniol before the rise of climacteric, Mattoo *et al* (1968) concluded that phosphotase activity was an important regulatory factor in mango cartenogenesis.

The polyphenol oxidase (PPO) is an enzyme related to phenolics. Its activity increased slightly from harvest maturity to the half ripe stage and then declined in 'Dasheri'

'Banganapalli' 'Fazli' and 'Langra' mangoes, whereas in 'Alphonso' 'Suvarnarekha' and 'Totapuri' mangoes, its activity decreased from harvest maturity to the half-ripe stage (Selveraj and Kumar, 1989 a). Increased PPO activity during ripening of 'Mulgoa' and 'Harumanies' paralleled the decrease in the ascorbic acid content of the fruit (Lazan *et al* 1986).

A very high activity of antioxidant enzymes Super oxide dismutase (SOD) and Peroxidase (POD) at mature green stage was noticed, which markedly declined as the ripening proceeded (Reddy and Srivastva, 2003). In 'Alphonso' mangoes the level of malic dehydrogenase and succinic dehydrogenase increased with the onset of ripening, whereas the level of citranate synthase decreased markedly (Baqui *et al* 1974). In 'Pairi' mangoes maximum concentration of α-oxoglutarate and pyruvate were reached before the climacteric peak and there was no accumulation as the fruit ripened. Asparlate and glutamate concentrations increased for about three days after harvest and then decreased as the peak of climacteric reached (Krishnamurthy *et al* 1971). Varietal differences have been observed with activity pattern of phosphoenol pyruvate carboxylase (PEPC) and pyruvate decorboxylase during ripening of mangoes. A relatively high level of PEPC in 'Alphonso' and 'Langra' and a low level in 'Dasheri' and 'Totapuri' during ripening were observed (Selveraj and Kumar, 1994).

FACTORS AFFECTING FRUIT RIPENING IN MANGO

Several factors influence the ripening behaviour and ripening changes, leading either to accelerating or retarding/delaying ripening. Understanding these factors and their influence on ripening is essential for effective regulation of ripening, storage and transportation.

Cultivar

More than 1,500 varieties exist in mango allover the world. They show great diversity not only in fruit but also in their post-harvest characteristics, ripening behaviour and storage behaviour. Each variety manifests different ripening behaviour not only in the time taken for ripening but also in the pattern of ripening, changes in physico-chemical and bio-chemical characteristics; temperature requirement for ripening etc.

'Pairi' mango fruits ripen earlier than those of 'Alphonso'. This was attributed to the presence of higher number and levels of amino acids in 'Pairi' as compared to 'Alphonso' fruits (Gupta *et al* 1979). 'Sensation' mangoes ripen rapidly in storage irrespective of fruit maturity at harvest as compared to 'Kent' and 'Amelie' cultivars (Seymour *et al* 1990). 'Tommy Atkins' mangoes developed more red and yellow pigmentation in peel and pulp tissues respectively than 'Keitt'. The outer mesocarp of 'Keitt' remained firm longer than 'Tommy Atkins' and the inner mesocarp was softer than the outer mesocarp. However, the decrease in cell wall neutral sugar and the amount and molecular weight of cell wall hemi-cellulose and increase in PG activity were similar in both cultivars during ripening (Mitcham and McDonald, 1992). Fruits of cv.'Kitchner' ripen faster than those of 'Abu-Samaka' and 'Dr.Knight' hence may not be suitable for long distance transportation. In contrast 'Abu-Samaka' retained a firm texture for longer period and hence could be able to withstand long distance transportation. These cultivars have also shown variation in ripening period and respiration rate. 'Abu-Samaka' had the lowest respiration rate and ripened in 15 days after harvest whereas 'Kitchner' with highest respiration rate ripened in 6 days and 'Dr.Knight' fell between the two, having intermediate respiration rate and ripened in 8 days (Abu-Sarra and Abu-Goukh, 1992). Studies on three mango cultivars (Dashehari, Amrapali and Ratual) indicated that during ripening of fruits, considerable variation exists in physiological phenomena like respiration, loss in fresh weight and sugar content and enzymes associated with ripening such as PME and PG. Further, the highest rate of respiration was noted in 'Amrapali' while the lowest was in 'Dashehari' (Zeng *et al* 1995). Loss in fresh weight, respiration rate, ethylene content and its biosynthetic enzyme activities were more in 'Amrapali' and were associated with early ripening compared to 'Dashehari' (Reddy and Srivastava, 2001). The slow ripening (10 days) of 'Ramkela' as compared to 'Amrapali' was due to its low rate of respiration for about 10 days and low activity of PG and ME enzymes during these 10 days. In contrast, 'Amrapali' showed a burst in the respiration within a couple of days after harvest, besides higher activity of PG and ME, because of which, it ripened faster and earlier than 'Ramkela' (Pandey *et al* 1998).

Cultivars also show variation in the kind of enzymes, their levels and activities. This could probably explain the fact that unlike most other fruits, each cultivar/variety of mango when ripe has distinguishing characterization in terms of texture, colour and flavour (Selveraj and Kumar, 1989 a). 'Dashehari' mangoes exhibited higher PG activity than 'Amarpali' and 'Rataul' (Zeng *et al* 1995). Between the two cultivars 'Amrapali' and 'Dashehari', the latter showed high level of ACC-synthase and ACC accumulation and low level of ACC-oxidase and ethylene production compared to the former during the process of ripening. Higher respiration rate accompanied by ethylene production was recorded in 'Amrapali' than in others (Reddy and Srivastava, 2001).

Stage of fruit maturity

The degree of maturity of fruit at the time of picking is one of the most important factors influencing the green life span, ripening, quality, storage (storability) suitability etc. The age of the fruit at the time of harvesting has a profound effect on ripening. Ripening changes during low temperature were also influenced by the stage of physiological maturity of fruits (Medlicott *et al* 1990 a). Stage of maturity also determines the market value of the fruit. Therefore, the fruit must be of optimum physiological maturity at picking to achieve the desirable sensory qualities upon ripening and to have good market value. However, optimum stage of maturity for harvest and ripening seems to vary with cultivar.

However, 'Alphonso' mangoes picked at any stage of maturity and stored at room temperature (RT) show typical respiratory changes characteristics of a fully matured fruit

(Lakshminaryana, 1973). It is possible that immature fruits may show high respiratory climacteric rates, as metabolism is high in fruits during the developmental stages. However, the studies of van Lely Veld and Smith (1979) in South Africa indicated that normal physico-chemical changes donot occur fully during ripening in immature fruit and confirm that mangoes have to reach a certain physiological stage of maturity before successful initiation of ripening can occur, particularly with exogenous application of ripening gases. Fruits plucked during the earlier and later stages of development show respiration hike within 6-10 days of harvest. During mid-stages of development pre-climacteric trough continues to extend and the climacteric occurs after 10 days. Mango fruits picked from 11th week onwards are more or less similar with ripe stage with regard to their chemical composition although the blend in sugar: acid ratio, fruit colour and aroma components will only reach very high acceptability standards from the 13th week onwards. It has also been noted that the susceptibility of fruits to diseases becomes greater when it reached 15th week age. Ripe fruits of 6-10 weeks development were highly acidic and those of 11th week age harvested were less acidic. Optimal acidity was found in ripe fruits at or after 13 weeks. Ascorbic acid content showed higher loss during the ripening of fruits of advanced maturity than in fruits from younger age. The total sugars of the ripe fruit increased as maturity advanced. Maximum sweetness was found in fruits harvested after 16 weeks of development. Carotenoid changes in the ripe fruit of all stages suggest that 'Alphonso' fruits harvested at any stage of development undergo normal process of ripening. However, fruits harvested at the later stage of maturity attain better colour development during ripening from 13th week onwards (Lakshminaryana, 1973).

Kalra and Tandon (1983 b) studied the ripening behaviour of 'Dashehari' mango fruits in relation to harvest period at ambient temperature. Fruits were harvested at 85, 90 or 95 days after fruit set (DAFS) and stored at ambient temperature (31.5 to 37.5 °C, 39-90% RH). Each successive picking stage had higher beta-carotene, TSS, reducing and total sugars with little variation in peel chlorophyll contents, than the earlier stage of picking. The fruits picked at 85 DAFS stored better with lower fruit weight loss, less fungal spoilage, with superior quality. Therefore, 'Dashehari' mangoes in North India (where temperatures go high up to 45 °C during the mangoes season) could be harvested after approximately 85 DAFS and stored for 6-9 days at 35-37 °C with RH above 70 percent.

Immature fruits show extended storage life, but do not ripen normally. Wardlaw and Leonard (1936) recognized three stages of maturity at which fruits ripen satisfactorily. They are:

(a) Fruits almost fully grown, green in colour with the shoulders of the fruit level with the stem insertion,

(b) Shoulders rised above the stem insertion due to further fruit growth and the skin a lighter green in colour,

(c) No further growth, but fruits on the point of becoming soft and coloured in the skin.

The rate of ripening is also influenced by the stage of fruit maturity at harvest. Mangoes harvested at various stages of physiological maturity showed a faster ripening with increasing harvest maturity. Thus a direct relation exists between the stage of fruit maturity and ripening and its rate (Medlicott *et al* 1987). Fruits harvested at an early stage of maturity ripened more slowly and therefore, had a longer storage life compared with fully mature fruit (Lskhsminarayana *et al* 1974). Although fruits harvested prior to climacteric maturity have a longer storage life, they do not ripen uniformly and are subjected to heavy post-harvest spoilage (Lakshminarayanan, 1973).

'Sensation' mango fruits ripened rapidly in storage irrespective of their maturity at harvest. When mature fruits, harvested at different dates during the season were compared, it became apparent that the extent of ripening in storage increased as the harvesting season progressed (Seymour *et al* 1990). Though immature fruits of 'Amelie' in Senegal and 'Tommy Atkins' in Brazil stored well better than mature fruit, they did not ripen rapidly on return to ambient temperature (Medlicott *et al* 1988). However, the results of studies by Seymour *et al* (1990) confirm the earlier finding of Medlicott *et al* (1988) that the mango fruits need to reach a certain stage of maturity before successful initiation of ripening can be achieved using exogenous application of ethylene.

Matured fruits of 'Tommy Atkins' ripened faster than both half-mature and immature fruits when treated with exogenous ethylene or acetylene. Acetylene treatments appeared more effective with early season harvests, notably the peel colour development. The overall effect was less apparent in late season fruits (Medlicott and Raynolds, 1988). Thus treatment with ethylene or acetylene initiated ripening at all three stages (immature, half-mature and mature) although ripening remained incomplete in immature fruits.

The maturity of fruit at harvest in 'Tommy Atkins' fruits was found to have an important effect on the development of fruit quality via ripening (Medlicott and Raynolds, 1988). Fruits harvested at the mature and half mature stages developed good quality characteristics, but immature fruits showed only limited changes during ripening, resulting in slow ripening and poor quality. In commercial situation where transport over longer distances is involved, mangoes are generally harvested before optimum/full maturity which may result in fruits of reduced quality (Medlicott *et al* 1987).

Specific gravity

Ripening of fruits showed positive correlation with specific gravity in 'Alphonso' (Lakshminaryana *et al* 1974., Roy and Joshi, 1988) and 'Dashehari' (Roy and Pandey, 1983). Fruits of lower maturity with a specific gravity less than 1.0 ripen in weeks and those with a specific gravity more than 1.02 take about a week to attain edible ripe stage. (Bhatnagar and Subramanayam, 1973). The fruits with specific gravity less than one did not ripen normally until the end of storage and those with specific gravity of 1.02 showed impaired ripening. At optimum maturity, fruits with specific gravity ranging from 1.00 to 1.02 showed better ripening and suitability for transport and storage (Kapse and Katrodia, 1997). Fruits having specific gravity of more than 1.02 ripen faster, with reduced storage life, but are suitable for consumption. Fruits with specific gravity of 1.00 to 1.02 require longer period for ripening, have longer storage life and also superior for dessert purposes as well as for processing. Fruits having specific gravity lower than 1.0 show increased susceptibility to disease infection and the quality is often poor in fresh or processed form (Anon 1968-70).

Packing materials

Different packing materials like paper cuttings, dry leaves, paddy straw, sawdust, green leaves of mango etc. are used for packaging mango fruits both for storage and transport. These packing materials have been found to influence the ripening of fruits and their quality.

Among the cushioning/packing materials leaves of banana, mango and *Albizzia lebbeck*, the last one gave best appearance on ripening, but mango leaves imparted best flavour and good appearance whereas packing in banana leaves caused damage to the fruits (Srivastava, 1961). In 'Dashehari' the most favourable fruit was that stored in paper and least palatable was that stored in saw dust (Mann and Singh, 1975). Packaging of 'Langra' fruits in wooden boxes containing calcium carbide (2 g) resulted in quick ripening (1.5 days). Paddy straw, paper wraps, dry sand, saw dust, paper layers and green mango leaves resulted in good ripening (Kumar *et al* 1996). Guha and Bhuiyan (1997) studied the effects of fresh banana leaves, dried banana leaves, rice straw, saw dust, rice hulls, fresh leaves of *Albizzia lebbeck* and calcium carbide on ripening and quality of mango fruits of cv.'Ashwina' in Bangladesh. Calcium carbide induced earlier ripening (3 days) closely followed by rice straw (5 days). Fresh banana leaves reduced weight loss, but increased disease incidence. The fruits ripened with calcium carbide had the highest content of TSS (19%) but rice straw gave better results in terms of most qualitative characters. Fruit packaging to reduce water loss could delay ripening and reduce colour development in mango. In a preliminary study, the results showed that when mango fruits were stored, wrapped in various types of wrapping papers and bags, physiological loss of weight and fruit colour were reduced, but TSS content and fruit firmness were improved (Singh *et al* 2001). Storage of fruits in wrapping of mango fruits in wrapping papers or bags can regulate relative humidity and the concentration of CO_2, O_2 and ethylene inside the package at various levels, thus affecting the mango fruit ripening process.

Skin coatings

Skin coatings with various materials are used for extending the shelf life as well as for protection of fruits. Incidentally these skin coatings influence the ripening and quality of coated fruits. The effect of different types of skin coatings on the rate of ripening of un-refrigerated mango fruits was investigated. For the most part, waxed fruits take slightly longer time to ripen than unwaxed fruits (Garg and Ram, 1973). Treatment with paraffin wax and Delvecoat reduced weight loss but the colour did not develop very well especially in case of the paraffin wax treatment (Musa, 1974). Dipping of mature fruit in

5 percent concentration of three commercial waxes (Peerless, Mobil cur-C and Stay fresh) and storing them at 20 °C extended the shelf life, but affected ripening by disrupting respiration and ethylene production (Brown, 1984). Wax emulsion (WE) treatments resulted in the lowest weight loss and also in the lowest TSS (Paramar and Chundawat, 1988). In 'Semper fresh' treated fruits, titratable acidity, firmness, and green colour were higher while, weight loss, Brix and pH were lower (Lopez *et al* 1995) but the ripening was not influenced (Carrillo *et al* 1995). The use of wax reduced the synthesis of carotenoid pigments in the pulp. Wax coatings also reduced chlorophyll breakdown of the skin, resulting in a lesser development of yellow colour (Fonseca *et al* 2004).

Lazan *et al* (1990) investigated the effect of surface coating with 'Vapor Gard' an antitranspirant, on ripening of mango cv.'Harumanis' with particular reference to the variation in response of tissue at different depths of the fruit pulp. Treatment with Vapor Gard (1.3% v/v) reduced water loss, retarded firmness decrease and inhibited the increase in PG enzyme activity during ripening in both tissues. Vapor Gard also inhibited malic enzyme activity as well as reduced the decrease in ascorbic acid. Inner mesocarp had higher PG activity than outer mesocarp. It also had more acid, but lower ascorbic acid, which resulted in the delay of decline in titrable acidity during ripening. Thus, inner mesocarp seems to be more responsive to Vapor Gard treatment than outer mesocarp. The marked inhibition of malic enzyme is detrimental to flavour development due to reduced de-carboxylating capacity and adverse effect on sugar acid ratio of the fruit.

Irradiation

The use of irradiation not only influences the ripening of fruits, but also influences quality parameters which may lead to delay in ripening. Irradiated fruits (0.75 KGY) showed high titratable acidity values and a smaller percentage of physiological mass loss per day than did control fruits. These aspects became more pronounced with the progress of ripening. The increase in NADP-malic enzyme activity was significantly diminished, but not delayed by gamma radiation (Dubery *et al* 1984). Differences in L-Phenyl alamine ammonialyase, polyphenol oxidase, peroxidase and catalse activities were observed in the gamma irradiated fruits of cultivar 'Kent' in the pre-climacteric phase. Thus, it appears that ripening is delayed/checked by irradiation by regulating the different enzymes of ripening (Frylink *et al* 1987).

The respiratory pattern of irradiated fruits appeared to result from an initial stress response and a reduced, but not delayed climacteric in 'Haden' fruits (Dubery *et al* 1984). Irradiation (0.5-0.9 KGY) of 'Keitt' mangoes significantly reduced the rate of respiration in fruits, but had no effect on TSS, or vitamin C or in the organoleptic quality. However, the loss of firmness during ripening was greater in irradiated fruits than un-irradiated control fruits (Lacroix *et al* 1990).

Temperature

Temperature, both during ripening and storage, is important in the ripening process. Temperature influences many events of fruit ripening, like development of colour, softening of the pulp, ripening time, ripening enzymes, production of aroma volatiles and carotenoids etc.

Optimum range of temperature for mangoes differs with cultivar as well as stage of maturity. For Indian mangoes the optimum range of temperature was reported to be 19 °C - 21° C (Singh and Mathur, 1952) whereas for Floridan cultivars it has been stated to be 21 °C - 24 °C (Hatton *et al* 1965) and temperature above 26 °C seriously impairs ripening (Campbell 1959., Hatton *et al* 1965). Colour development is better at 15 °C - 18 °C but decay is likely to be extensive with immature fruit (Hatton *et al* 1965). At 26.7 °C (80 °F) fruits ripen with a strong flavour and show some mottling of the skin (Campbell 1959., Hatton *et al* 1965). At 60°-65 °F Florida mangoes developed highest and most attractive skin colour, but required an additional 2-3 days at 70-75 °F to attain sweet flavour. 'Kent' and 'Keitt' mangoes ripened at 80 °F without mottling and strong flavour. In general, the average time for mangoes to soften decreased as the ripening temperature increased to 80 °F. Ripening is retarded at 90 °F. The higher the temperature during ripening, the grater the weight loss (Hatton *et al* 1965).

Mangoes ripen with good quality characteristics at 25 °C (Pantastico *et al* 1984). In 'Tommy Atkins', ripening at a temperature range of 22-32 °C, provided good quality fruits. Above and below this range, acidity and sugar contents are adversely affected. For completion of softening of the pulp of 'Tommy Atkins' fruits required 9 days at 22 °C (Medlicott *et al* 1986 b). However, a temperature of 25 °C inhibits skin colour and pulp colour development and lowers the flavour, texture, soluble solids of the ripe fruit and general acceptability (Brown *et al* 1986).

Preclimacteric fruits of 'Kensington Pride' ripened at 18°-22 °C after treatment with ethylene, achieved a high quality score with all quality parameters reaching a maximum in a couple of days. Fruits ripened at 13° or 30 °C had low skin colour and quality score, related to poor carotenoid development and chlorophyll retention respectively. The poor carotenoid development at 13 °C also resulted in lower pulp quality score. Eating quality was significantly lower at 13 °C and 30 °C than at intermediate temperatures, which are related to the slow decline of titratable acidity and poor flavour respectively (O'Hare, 1995). In Taiwan, mango fruits were unable to complete the ripening process in up to 4 days at 25 °C while ripening was accelerated at 38 °C. Fruits ripened at 28 °C had a good skin colour and flesh colour, bright orange yellow at day 6. Those ripened at 38 °C also showed orange yellow skin but green patches and some black spots appeared 8 days after treatment. Based on the above, it was concluded that ripening at 25 °C could attain a better post-harvest quality than ripening at 38 °C (Lee *et al* 1998). High rate of respiration and ethylene production were observed at 30 °C and 35 °C ripening temperature in 'Kensington Pride' mangoes (Lalel *et al* 2004).

Temperature plays a significant role in the activity of softening enzymes. At an ambient temperature (25 ± 1 °C), the activity of enzymes is enhanced, while at 10 °C PG activity was declined in control as well as in ethylene–treated fruits of 'Amrapali' and 'Dashehari' (Reddy and Srivastava ,1999).

Ripening temperature has direct influence on the synthesis of aroma volatiles and/or indirect influence through biosynthesis of their precursors. Ripening temperatures below 25 °C adversely affected the development of typical aroma, flavour and carotenoid syntehsis of 'Alphonso' mangoes (Thomas, 1975). The 'Tommy Atkins' mangoes when ripened at 22°, 23° and 32 °C exhibited good characteristics including full chlorophyll breakdown, higher pulp carotenoids, good texture and a balanced sugar: acid ratio. Similar characteristics were shown at 37 °C although the peel was mottled and sugar: acid ratio was lower. (Medlicott *et al* 1986b). The elevated level of carotenoids in fruit pulp with increased ripening temperature was attributed to the increased bio-synthesis of carotenoids (Vazquez-Salinas and Lakshminarayana, 1985., Medlicott *et al* 1986b).

An increase in the temperature markedly increased ethylene production in fruit and total fatty acids in the pulp. However, the pulp of the fruit ripened at 25 °C exhibited significantly higher concentration of individual fatty acids. The increase in ripening temperature significantly increased the rate of respiration and total caroteniods in pulp. A ripening temperature of 20 °C seems to be the best for ripening and biosynthesis of aroma volatile compounds. The concentration of aromatics was significantly higher in fruit ripened at 15 °C, which may be due to a higher level of polyphenols and related compounds (Lalel *et al* 2004). Increase in temperature during ripening of 'Kensington Pride' mango also increased the biosynthesis of sesquiterpenes, alcohols, esters and nariprenoids. However, an increase in temperature during ripening lowered the synthesis of monoterpenes (Lalel *et al* 2004).

Similar to air temperature, the storage temperature also influences the process of ripening, its rate and other quality parameters. Refrigerated storage prior to ripening reduced the rate of ripening followed by inhibited formation of sugars, carotenioids and flavour, with least PLW. Further, refrigerated fruits showed an extended pre-climacteric respiratory trough lasting throughout in storage period until they ripened (Thomas, 1975). Hatton *et al* (1965) reported two weeks storage at 12.8 °C (RH > 85%) as optimum and 21 °C as the best ripening temperature for most mango cultivars in United States of America. They found considerable variation in the number of days for fruit to reach soft-ripe in the same harvested lot in addition to differences between cultivars in ripening time. The time taken for fruits to ripen and the development of flavour and yellow colour of skin depends on the temperature at which fruits are held. The temperature of 20-23 °C is considered best for appearance, palatability and low decay of fruit. At this temperature, it takes approximately 13 days to achieve the maximal yellow characteristics of the cv. 'Kensington Pride' (Jobin-Décor, 1989).

Mango fruits held at 12 °C and transported by air to UK did not ripen to full eating quality even within 16 days. Sugar levels in these fruits were comparable with those fruits at other temperatures, but acid levels were high. The fruits retained some green colouration and showed incomplete ripening and lower pulp carotenioids. After 17 days at 17 °C fruits softened and degreened well, but had a poor pulp colour, high acidity and low sugar : acid ratio (Medlicott *et al* 1986 b). On the other hand, 'Dashehari' mango fruits stored at 12 °C ripened normally when transferred to ambient temperature, while those stored at 8° or 10 °C ripened slowly with poor colour development (Narayana and Singh, 1998). Fruits stored at 4 °C showed abnormal ripening (Balasubramanyam *et al* 1991a). Thomas (1975) found that the storage temperature below ambient (25 °C) would adversely affect the development of typical aroma, flavour and carotene formation of 'Alphonso' mango on ripening. However, the 'Alphonso' fruits cold stored at 11°-12 °C with 85-90 per cent RH failed to ripen normally when brought to RT (27°-32 °C). Further the low temperature storage caused loss of colour, texture and flavour and sour taste, with heavy spoilage (Lakshminarayana and Subramanayam, 1970).

Storage of mangoes (cv.Alphonso) at ambient temperature (26-32 °C) favoured the development of typical organoleptic qualities and flesh carotenoids on ripening. In contrast storage at low temperature (7-20 °C) adversely affected the above qualities, when fruits were subsequently ripened under ambient temperature (Thomas, 1975). However, the quality of 'Totapuri' fruits, ripened after storage at 20 °C for 20 days was as good as tree ripened fruits in Sudan (Musa, 1974).

Rapid increase in TSS at ambient temperature as compared to low temperature storage was observed in 'Totapuri' mangoes (Kapse *et al* 1977) which was attributed to

1. slow conversion of starch to sugar and slow degradation of pectic acids, fats and proteins (Singh *et al* 1954);
2. slow and less formation of beta-carotene in the fruits stored at low temperature was attributed to the slow hydrolysis of starch and synthesis of beta-carotene from the carotenes (Thomas, 1975);
3. slower loss of ascorbic acid at low temperature than at ambient temperature, due to the inactivation of enzymes responsible for oxidation and/or slow degradation of ascorbic acid;
4. rapid decrease in acidity at ambient temperature as compared to low-temperature stored fruits where the respiration rate was least;
5. very rapid increase at ambient temperature till 10th day of storage which later decreased gradually;
6. as the starch was converted into sugar and declined in quantity gradually reduction of sugars as the fruits reached senescence stage;
7. very slow increase in sugar content up to 37th day of storage and slow decrease thereafter, as there was suppression in the rate of respiration and in the enzyme activity during ripening in storage at ambient temperature and at low temperature (10 °C) storage (Kapse *et al* 1977).

Vazquez–Salinas and Lakshminarayana (1985) followed the chemical changes in four mango cultivars (Haden, Irwin, Kent and Keitt) in Mexico to determine the optimum storage and ripening conditions. Their study showed that the pattern of chemical changes were strikingly similar in all the cultivars. Breakdown in acidity during ripening was slower at 16 °C resulting in higher acidity in fruits and starch together with slightly lower p^H, sucrose and a subdued taste, flavour and aroma. At temperature between 22 °C and 28 °C, the reactions controlling these changes proceeded at a faster rate resulting in sweeter fruit, with bright yellow pulp colour and strong aroma. Rate of ripening was moderately slow at 16°-22 °C, which enhanced the shelf life to some extent, considered to be beneficial to retailers. Temperature above 25 °C is not desirable as it accelerated the rate of ripening, leading to increased weight loss and decreased storage life, though it may be ideal for obtaining excellent fruit quality.

Low temperature (8 °C) storage conditions inhibited the development of some of the ripening processes, particularly acidity loss in 'Tommy Atkins' fruits. The reduction in acid loss may be the result of disruption of acid metabolism during storage (Medlicott, *et al* 1990 b). In an earlier study, Medlicott *et al* (1986b) observed changes in acid as one of the slower processes that occur during ripening of mangoes. Pulp colour development and therefore, carotenoid content were impaired during ripening after storage at 8 °C, which inturn, was related to the production of volatiles.

Several workers noted a lower carotenoid content in fruits of different cultivars when ripened at room temperature subsequent to storage at low temperature (Thomas, 1975., Mann and Singh, 1976). Thomas (1975) observed as much as 20-70 percent reduction in the total carotenoid content of the flesh of 'Alphonso' mangoes during ripening after storage at low temperature.

Storage temperature influenced the content of ascorbic acid on ripening. 'Alphonso' mangoes stored at room temperature retained about 32 percent of the original ascorbic acid content at the eating-ripe stage, while those held for 16 days at low temperature and subsequently ripened at room temperature showed 67 to 97 percent retention (Thomas, 1975). Ascorbic acid in mature mango fruit kept at RT declined rapidly during ripening, whereas the decline was slow at 10 °C (Kapse *et al* 1977).

Fruit firmness was adversely affected by high ambient temperature (Balasubramanyam *et al* 1991 a). Ripening was delayed in mango fruits that were held at above 35 °C in storage due to adverse effect on normal respiratory pattern (Balasubramanyam *et al* 1991b).

Growth regulators

Ever since the discovery of plant growth regulators, they are put to regulate different plant processes. In the absence of threshold levels of phyto-hormones several plant processes are adversely affected. The deficiency of phyto-hormones can be made up by exogenous application of concerned growth regulator(s) to take advantage on the affected plant process like ripening of fruits. Exogenous application of growth hormones causes acceleration or retardation of ripening of fruits with concomitant changes in chemical and biochemical composition of the treated fruits. Following is a brief account of effects of growth regulators on mango fruit ripening and internal changes in chemical and biochemical composition.

Ethylene: Since its discovery, ethylene has been recognized as the principal and indispensive hormone that is directly involved in the process of ripening of fruits and vegetables. Though, in succeeding years, several other roles of ethylene have been discovered, it is often referred to as "ripening hormone". Until physiological maturity of climacteric fruits (like mango) the synthesis of ethylene is negligible, after which very small quantities of ethylene are traceable. However, even this minute quantities of ethylene increase the respiration, trigger and hasten the ripening of climacteric fruits (Baile *et al* 1954).

Exogenous application of ethrel (ethephon) either by spray or dip treatment or by release of ethylene gas in the ripening chambers causes ripening and influences several ripening processes. Tree-intact fruits of 'Dashehari' fruits dipped in ethrel (500 ppm) ripened two days later with much higher sugar : acid ratio (Chundawat *et al* 1973). On the other hand, harvested 'Neelum' fruits dip-treated with ethrel ripened 8 days after treatment with excellent sensory quality attributes, where as untreated control fruits ripened in 12 days. The respiratory peak in ethrel treated fruits reached characteristic peak on 4th day of treatment as against 6th day by untreated control fruits. Physico-chemical characteristics such as TSS, p^H, reducing and total sugars and total carotenoids showed increasing trend up to 8 days and a decreasing trend in fruit pressure (texture), specific gravity, total acidity and ascorbic acid during ripening. Sensory characteristics with respect to colour, taste and flavour of ethrel treated fruits were excellent (Kulkarni *et al* 2004). Ethylene treatment gives a deep colour to fruits and enhances the taste (Fuchs *et al* 1975). Ethephon affects the flavour of the ripened fruit (Sergent *et al* 1993). Development of skin colour and decrease in texture were faster in ethylene-treated (ethephon) fruits (Medlicott *et al* 1988., Kulkarni *et al* 2004). However, ethylene treated fruits showed higher PLW which could be due to increased transpiration losses and enhanced respiratory activities (Ram *et al* 1984., Kulkarni *et al* 2004). The loss of ascorbic acid was more in ethylene-treated fruits (Chikkasubbanna and Hudder, 1982). Ethylene treatments also result in degradation of starch and pectin (Peacock and Brown, 1984., Lazan *et al* 1986). Though the colour development of peel was uniform and characteristic of the cultivar, the texture of the pulp was more soft in ethrel treated fruits compared to the fruits ripened by conventional method (Gupta *et al* 1979).

Singh and James (2001) studied the effect of ethephon treatment of mango fruits in sealed poly bags in Australia. They found that the ripening time, fruit firmness and acid content were significantly reduced whereas colour index and TSS were markedly increased as compared to untreated controls. Eating quality was improved. During the storage period, the concentration of O_2 increased and that of CO_2 decreased in the sealed bags due to ethephon treatments. The concentration of ethylene in sealed bags increased as the concentration of exogenous ethephon increased. These results showed that ethephon induced fruit ripening combined with MAP improved fruit quality and prolonged shelf life of treated mango fruits.

Abscisic acid (ABA) **:** Enzyme activity is influenced by exogenous application of growth regulators. Post-harvest treatment with abscisic acid (ABA) increased the activity of PG, but did not show consistent effect on PME activity during ripening (Zhou *et al* 1996). Dip treatment of ABA (10^{-4}) increased the activity of hydrolytic enzymes (amylase, cellulase and invertase) during ripening in 'Alphonso' mangoes, besides increasing the level of total and reducing sugars and peel carotone (Majumdar *et al* 1981).

Gibberellic Acid

Application of gibberellic acid (GA_3) delays ripening of mango fruits, by retarding PLW, loss of chlorophyll and decline in ascorbic acid content and reducing the activities of amylase and peroxidase enzymes during ripening of 'Alphonso' mango fruits (Krishnamurthy and Rao, 1982). GA_3 delayed the activity of PG enzyme, but showed no consistent effect on the changes of PME activity during ripening (Zhou *et al* 1996).

1-Methyl cyclo propene (1-MCP)

This is a new chemical which found a place among the growth regulators and is used for regulating ripening of fruits. It is mainly employed for delaying ripening of fruits (See later).

The effects of 1-MCP include: inhibition of firmness losses, delay in colour changes and decreases in respiration and ethylene production rates, reduced fresh weight loss and higher total titratable acidity (Harris *et al* 2000., Mir *et al* 2001., Jacomino *et al* 2002., Kluge and Jacomino, 2002., Alves *et al* 2004). Mathur and Srivastava (2005) studied the effect of 1-MCP on malic enzyme during ripening of 'Amrapali' mangoes and observed complete inhibition of the activity of malic enzyme for 8 days when 1-MCP lost its control to inhibit endogenous ethylene production, whose autocatalytic evolution started, triggering host of bio-chemical changes leading to ripening of fruits.

The response to 1-MCP treatments depends on the maturity stage of fruits (Harris *et al* 2000., Mir *et al* 2001., Jacomine *et al* 2002., Kluge and Jacomino, 2002). Effects of 1-MCP were noticed on ripening of mangoes harvested at S_2 stage, as delayed climacteric peak, reduced respiration rate, reduced fruit weight loss, firmer pulp and high titratable acidity compared to untreated fruits of same maturity stage. Generally, fruits harvested in more advanced maturity stage (S_3) are less responsive to 1-MCP application (Alves *et al* 2004).

1-MCP was tried in combination with MAP using polyethylene bags. Exposure of prochloraz treated mango fruit to increasing concentration of 1-MCP for 12 hours in polybags resulted in the reduced softening of pulp, when subsequently held in air for 7 days following ethylene treatment. In the absence of ethylene stimulation, the natural ripening of mangoes held in polybags, was delayed by prior exposure to 1-MCP for 12 hrs. Treatment with 1-MCP allowed storage of mango fruit in polyethylene bags at 20 °C for 30 days. (Jiang and Joyce, 2000). Lowest rate of respiration was observed in mango fruits treated with 1-MCP alone and also with MAP. 1-MCP treatment kept the fruits more acidic with higher vitamin C content, but had no effect on TSS (Cocozzo *et al* 2004b).

Others: Post-harvest dip-application of (1 minute dip) of kinetin or cycloheximide to fully mature Zardalu fruits 80 days after anthesis increased sucrose synthase activity compared with control fruits dipped in distilled water. At the 1/3 and 2/3 ripe stages, invertase activity was slightly affected by kinetin and cycloheximide, but at later stages it was reduced in fruits treated with kinetin (1μgml^{-1}) and eventually increased in those treated with 100 μgml^{-1} (Kumar *et al* 1994).

Acetylene, an analogue of ethylene is also frequently used for ripening of climacteric fruits including mangoes. Treatment with acetylene also influences ripening and composition of fruits. A sharp decline in acidity and increase in p^{H} and TSS were observed in acetylene (through calcium carbide) treated fruits. The rate of carotenoid synthesis in the pulp of acetylene-treated fruit was higher in the initial stages, but the total carotenoid content was lower than in the control after the fruits were fully ripe. Aroma was less in the treated fruits (Nagaraj *et al* 1984).

REGULATION OF RIPENING

One of the greatest problems in post-harvest handling of climacteric fruits such as mango in tropical and sub-tropical areas, particularly during transit is their rapid ripening, which does not allow large quantity of fruit to transport and to reach the distance markets without ripening and spoilage. Regulation of ripening of mangoes is therefore, necessary because

(i) Mangoes are highly perishable and can hardly be stored for 7 to 10 days without any treatment at ambient conditions after harvest without ripening and spoilage.

(ii) Hard green mangoes ripen rapidly during transit and reach in half-ripe or ripe condition at the destiny depending on the maturity of the fruit, storage condition (temperature), distance from the production centre to consuming centre and the time taken to reach the destination, besides how the commodity is harvested and handled after harvest.

(iii) Mangoes are generally handled at ambient temperatures which are high in the season, particularly in tropical and sub-tropical regions, leading to a considerable quantity of fruits being spoiled due to physiological degradation and diseases resulting in huge economic losses.

(iv) Mangoes for the most part are not amenable for cold storage as in case of temperate fruits such as apple, pear etc., since they are highly susceptible to chilling injury.

In view of the above, regulation of ripening of mangoes particularly, delaying is very much desirable for increasing the shelf life and for export and would be worthwhile in regulating the marketability leading to subsequent higher commercial return and for commercialization of mango production. Any means to prevent or delay the ripening during transit would go a long way in extending the availability of mango.

A. Delaying ripening

Based on physiological approach, delay of ripening and extension of shelf-life of harvested fruit can be achieved by lowering its respiratory processes through various methods like irradiation, modified/controlled atmosphere, low temperature and skin coatings individually or a combination of these methods, which are essentially storage methods. Others include removal of ethylene or inhibition of production or action of ethylene through chemicals. Delaying of ripening was also attempted by the application of different growth regulators, chemicals such as calcium, silver thiosulphate (STS), amino oxyacetic acid (AOA), dichlo hexyla carbodimide (DCCD) which potentially inhibit/delay ripening and on set of wilting (Singh and Mishra, 2003) (Chart 4.1).

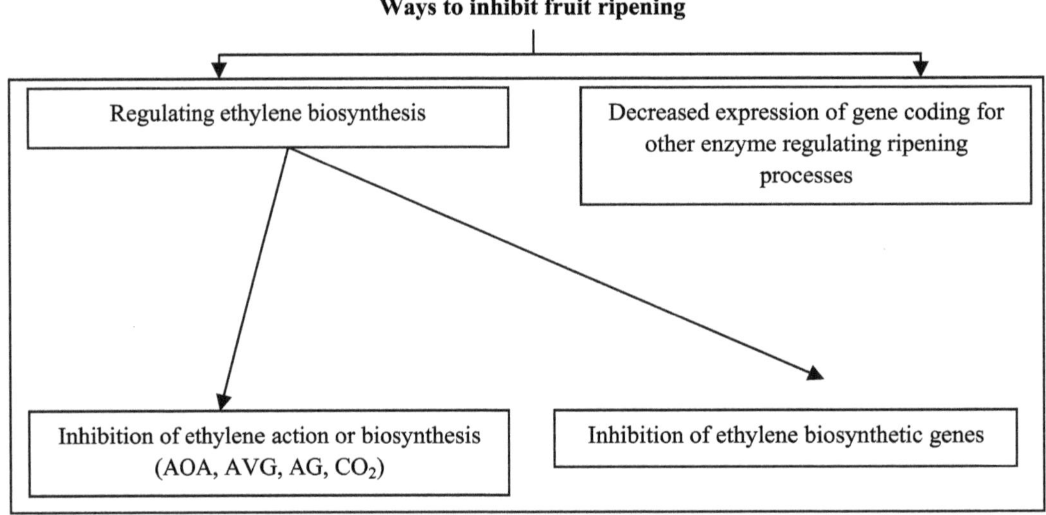

Chart 4.1 : Ways to delay the ripening process in Fruits

I. Irradiation

This is a physical method of delaying ripening in fruits. Exposure of fruits to radiation is one way of delaying/retarding/checking ripening of fruits. By delaying ripening the shelf life of fruit is extended. An extension of shelf life through delaying of ripening would be beneficial for export trade and for extending the availability of produce. However, the response of treated fruits to irradiation depends on the stage of fruit maturity at the time of treatment, as well as its dose, besides cultivar and source of radiation. Gamma irradiation delays the process of ripening and senescence in climacteric fruits like banana, mango etc.

The climacteric stage of the fruit at the time of irradiation determines to a large extent its response to the irradiation treatment. Both banana and mango show ripening delay only if irradiated in the pre-climacteric stage. However, at this stage fruits are less tolerant to irradiation treatment and exhibit phytotoxicity if the dose exceeds threshold levels. On the other hand, if irradiated after the onset of climacteric these fruits can withstand relatively higher dose of radiation, but donot show delay in ripening (Thomas, 1993). Irradiation at 0.25 KGY delayed the ripening of pre-climacteric mature green 'Alphonso' mangoes stored at 5, 10, 15 or 20 °C with a delay of 10 days. At 5 °C covering with paper shreds or dry paddy straw further extended the green pre-climacteric life of both control and irradiated mangoes due to accumulation of respired CO_2 creating a modified atmosphere around fruits (Sreenivasan *et al* 1971., Farkas *et al* 1972). They further found that 100 percent of 'Alphonso' fruits, skin-coated with 6 percent wax emulsion and then irradiated, packed in bamboo baskets in paddy straw and transported over 1000 Km in 36 hours at up to 42 °C were marketable as against 60 percent of only irradiated fruits and none of the control fruits. In further experiments, similarly treated 'Alphonso' mangoes wrapped individually in tissue paper and packed in plywood boxes, were air-freighted from Bombay to Budapest in 4 days. An assessment of treated fruits on arrival and during subsequent storage for 4 days at 20-22 °C showed striking delay in ripening and senescence with coating and irradiation. Such fruits remained firmer with substantial delay in disappearance of chlorophyll and formation of carotenoids. Weight loss in storage was lower and reduction of TTA and formation of sugar took place later (Farkas *et al* 1972). Exposure of the fruits in the pre-climacteric stage to an optimal dose of 25 Krad delayed the onset of ripening as judged by chlorophyll disappearance, softening, changes in organic acids and formation of sugars. When fully mature 'Alphonso' mangoes individually packed in perforated poly ethylene bags of 200 gauge and exposed on the day of harvest to 0.12 KGY of gamma rays, delayed ripening was observed during storage for 24 days, as compared to 10 days in control. Non-irradiated fruits ripened faster as judged by change in skin colour and sugar levels (Mathur and Lewis, 1961). Dharkar *et al* (1966) treated mature green 'Alphonso' mangoes one day after harvest with 0.12 to 2.0 KGY gamma rays and observed that the dose of 0.12 to 0.75 KGY delayed ripening. Treated fruits were fully ripe 16 days after storage at 25-30 °C as against after 10 days in control. Though the changes in sugar and starch transformation in the pulp progressed at a similar rate, in terms of fruit texture and skin colouration, irradiated fruits at 0.25 KGY in nitrogen atmosphere were less ripe than those not irradiated. Irradiation of pre-mature fruit caused an initial increase in respiration rate and delay in the climacteric peak by about 4 days. Fruits which were partially ripe in their climacteric rise were unaffected by irradiation.

Irradiation of 'Irwin' mangoes at 100 GY and 250 GY delayed ripening (Hatton *et al* 1961). Ripening of hard green mangoes could best be delayed by irradiation from a cobalt source at a dose of 30 Krad at room temperature (25 ^{o} ± 2 ^{o}C) and were still acceptable after 2 weeks of storage (Ali *et al* 1968). Ahmed *et al* (1972) achieved a 7 days delay in ripening of mangoes at 35-38 ^{o}C storage with a radiation dose of 30 Krad. Irradiated fruits placed in either wooden boxes lined with tissue paper or poly ethylene or in perforated wooden shelves and held in air at 20-25 ^{o}C or 30-38 ^{o}C and 80-87 percent RH remained in good condition for 15 days whereas untreated fruits ripened within 5 days. Holding the fruits in poly ethylene lined boxes also, delayed ripening (Farooki *et al* 1974). 'Kensington Pride' mangoes exhibited delayed ripening when irradiated at 300 or 600 GY but not at 750 GY (Mc Lanchlan *et al* 1990). A further delay in ripening of 3-5 days at 20 ^{o}C from the above doses could probably be increased by reducing the storage temperature to around 15 ^{o}C. Maximum delay in ripening was observed in mango fruits exposed to a dose of 25 Krads. At higher doses fruits ripened earlier, which may be due to activation or distortion of respiratory enzymatic system or to differences in degree of ethylene production (Dharkar *et al* 1966).

Mangoes (Tommy Atkins and Keitt) irradiated with X-rays from 60 cobalt at doses of 150-1500 Gy, then held 3 days at 13 ^{o}C followed by ripening at 24 ^{o}C developed less peel colour than un-irradiated mangoes. Ripening of 'Tommy Atkins' was delayed for 2-3 days by 150 and 250 Gy, unaffected by 500 and 750 Gy and accentuated by 1500 Gy (Spalding and Von Windeguth,1988). Inhibition of ripe peel colour development in 'Tommy Atkins' mangoes irradiated at 750 Gy was reported (Spalding and Reeder, 1986). Ripening of 'Alphonso' (Dharkar *et al* 1966) and Keitt (Dennison and Ahmad 1967) mangoes was delayed by irradiation dose up to 250 Gy especially in terms of degreening of the peel. Spalding and Reeder (1986) confirmed the above results, who reported a delay in ripening of mango fruits with an irradiation dose of 150, 250 and 750 Gy. A later study also showed delayed and reduced respiratory climacteric and degreening of the peel with a dose of 200 Gy for 'Kensington Pride' mangoes.

Sreenivasan *et al* (1971) reported that the optimum dose of irradiation for delaying ripening of 'Alphonso' mangoes was 250 Gy and doses higher than this resulted in earlier ripening and physiological damage. 'Kensington Pride' mangoes exhibited delayed ripening with increased external injury (lenticel damage) following irradiation at 300 or 600 Gy, but not at 75 Gy. Hence, irradiation of pre-climacteric mangoes at doses of 300 or more Gy is not recommended. A dose of 75 Gy can be used without adversely affecting marketability (Mc Lanchlan *et al* 1990).

Irradiation suppresses CO_2 production rate (CPR) (Mc Lanchlan *et al* 1990) which does not agree with the published literature. On the other hand, it was reported that there would be immediate increase in CPR from irradiation (Boag *et al* 1989., Urbain, 1986). In pre-climacteric fruits (firm, dark green with negligible ethylene production) irradiation suppressed the respiratory climacteric. However, no explanation/hypothesis for the depressed CPR in irradiated fruits was advanced (Mc Lanchlan *et al* 1990). Another theory is that ripening is not really delayed, but that certain bio-chemical processes are distorted. For example fruits irradiated with 750 Gy showed higher titratable acidity and lower p^H values coupled with a diminished increase in the activity of malic enzyme and respiratory climacteric (Dubery *et al* 1984). Irradiation of pre-climacteric fruit caused an initial increase in respiration rate and a delay in climacteric peak of about 4 days (Mc Lanchlan *et al* 1990).

II. Low temperature

Post-harvest refrigeration slows down the rate of ripening and is employed for short-term storage of fruits. The usual temperature range employed for storage of mango fruits is 10-15 °C. To inhibit ripening for long term storage, temperatures around 5 °C are necessary, but such low temperatures attract chilling injury (Thompson, 1971., Veloz *et al* 1977). Thomas (1975) reported that cultivars ripened at lower than 18 °C had significantly reduced skin colour development, while at temperatures greater than 24 °C many cultivars develop mottled skin (Hatton *et al* 1965). Fruits ripened at 15 °C had more green colour on the skin when ripe than those ripened at 20° and 25 °C (Medlicott *et al* 1986b). The ripening process in 'Mallika' was delayed at 10° and 12 °C storage. At low temperature, softening process continued at slow pace and the development of yellow colour was hampered. Similarly storage of 'Langra' and 'Dashehari' fruits at low temperature (9.0 - 10.5 °C) delayed ripening by 8 weeks as compared to control fruits, which took 4 weeks for ripening (Mukherjee and Srivastava, 1979). However, refrigeration did not delay ripening of 'Dusseri' mangoes in Pakistan (Ahmed *et al* 1972).

Ripening of fruits at controlled temperature (22-25 °C with 85-90% RH) extended the shelf life and reduced PLW and spoilage due to rots, slowed catabolic activities, postponement of respiratory peak and ethylene evolution in 'Kesar' and 'Amrapali' mangoes. This led to higher fruit quality with high level of sugars, TSS and Vitamin C. It was concluded that for international trade mangoes can be handled at 22-25 °C (Singh and Chundawat, 1991). At low temperatures (15-18 °C) longer exposure to ethylene resulted in more green colour on the skin of ripe fruit (Nguyen *et al* 2002). They have also shown that temperature is an important factor influencing the ripening process of mangoes. Ripening of fruits can be retarded more effectively in immature fruits than in mature fruits by low temperature (Seymour *et al* 1990).

III. Modified atmospheric package and storage

The post-harvest life of several fruits is influenced by changes in the atmosphere of packages and storage. Atmosphere generated by modified atmosphere packages delays the ripening of mango. Many workers reported that fruit ripening was delayed by modified atmospheric package and storage alone or in combination with other treatments. However, the response of mangoes to such packages and storage has been variable.

Valmeyor (1972) obtained a significant extension of ripening of mangoes when stored at 5 percent CO_2 level at 10 °C. The most obvious response of mango (cv. Kensington

Pride) was that the fruit stored in sealed poly ethylene bags (containing CO_2 more than 20% and O_2 less than 2%), apparently lost much of their capacity to degrade chlorophyll in the peel and to synthesize carotenoids in the pulp during ripening (Chaplin, 1988). Sornasrivichai *et al* (1989) reported that mango ripening was inhibited in MAP, but fruit quality was badly affected. Rosa *et al* (2001) also observed delayed ripening in MAP storage as expressed by less colour development in Israel. In a study, Kane and Marcellin (1979) showed that, chlorophyll breakdown and fruit yellowing processes in 'Amelie' and 'Julie' mangoes were retarded leading to delay in ripening in a storage atmosphere comprising of oxygen (5%) and carbon dioxide (5 or 10%). Mango fruits of 'Tommy Atkins' and 'Keitt' stored in MAP with polyethylene or X-tend TM film delayed ripening as expressed by reduced softening and colour retardation in mangoes in Israel.

Stead and Chithambo (1980) showed that fruit of mango, stored with potassium permanganate ($KMnO_4$) in closed polyethylene bags at ambient temperature (20-30 °C) took 10 days to ripen, whereas the control fruit in the open took only 5 days to ripen. Prabha *et al* (1987) studied the ripening behaviour of mango fruits, stored in MA using labeled compounds and observed delayed ripening through disturbed metabolism. Addition of ethylene absorbent sachets in MAP showed a further effect of delaying – softening (ripening) as compared to the delay due to the MAP alone (Rosa *et al* 2001). Adding ethylene absorbent (conserver, 21) reduced the ethylene and CO_2 levels in the storage atmosphere. Ethylene absorption inhibited chlorophyll breakdown on the green cheek, but there was no effect on anthocyanin development on the red cheek. MAP showed dramatic reduction in the weight loss (Rosa *et al* 2001). MAP plus ethylene extended the shelf life of 'Kensington Pride' mangoes in addition to improving fruit quality in Australia (Singh and James, 2001). MAP alone or in combination with refrigerated storage has been shown to reduce post-harvest losses and maintain superb quality of treated fruits of tropical region (Kader, 1993). The main factors that maintain mango fruit quality in MAP are the increased CO_2 levels and decreased O_2 levels, which reduce respiration rate and prevent water loss, thereby retard ripening of fruits.

IV. Skin coatings

Cold storage, modified atmosphere and controlled atmosphere storage for checking fruit ripening are relatively expensive particularly in developing countries like India. A relatively inexpensive method to modify the internal atmosphere of the fruit and to slow down its metabolism is more desirable in such countries. Surface or skin coatings with different materials such as wax, answer the need. Many commercial products are available for the purpose. Skin coatings alone or in combination with others are employed for retarding ripening as well as for storage. Ripening of fruits can be retarded by skin coatings of wax or, along with proper packing or use of growth retardants or by storing at a low temperature.

Waxed fruits ripened by 12th day, whereas un-waxed fruits ripened in 6 days. Ripening of 'Totapuri' fruits stored at 30° and 32 °C was delayed by the use of paraffin wax and Delvocoat. The delay in ripening caused by waxing and use of Delvocoat was the result of modification of the atmosphere around the fruits, where the CO_2 concentration increased and that of O_2 decreased (Sadasivam *et al* 1970). Dalal *et al* (1971) also achieved delayed ripening of mango fruits along with improvement in gloss and storability with wax emulsion coatings. Dipping of 'Lucknow Safeda' fruits in 6 percent wax emulsion (WE) showed retarded ripening, as shown by slow increase in TSS and sugar contents, better retention of ascorbic acid and a reduction in respiration rate (Garg and Ram, 1973). From Sudan, Musa (1974) also reported that ripening of 'Totapuri' mangoes was delayed with paraffin wax and fungicidal wax at 20° and 32 °C. However, though dipping of fruits in 6 percent waxol was highly effective in retarding the rate of ripening, marketability and quality of these waxed fruits were lowered due to higher levels of disease infections in waxed fruits (anthracnose, stem-end-rot and soft rot) and the quality was poor due to higher acidity as a consequence of accumulation of CO_2 inside the fruit tissues in the absence of gaseous exchange. Therefore, usefulness of waxol alone is questionable and doubtful in spite of delaying ripening. However, this bottle neck can be overcome by using fungicidal wax (by using Bavistin etc) or hot water treatment prior to placing them for ripening (Paramar and Chundawat, 1988) 'Fruitex' (fungicidal wax) retarded ripening in 'Alphonso' and 'Pairi' mango fruits, but they turned soft at ripening. Combination of waxol and hot water treatment resulted in fruits with uniform ripening and extended shelf life (Sethi and Maini, 1989).

'Tal-prolong' is a product of Tal chemicals and mainly consists of a mixture of sucrose, esters, glycerols and cellulose and is often used to retard ripening of fruits. The ripening of both 'Dashehari' and 'Mallika' fruits treated with 'Tal-prolong' (1.5%) was delayed as evidenced by retention of yellowish green colour and greater firmness of pulp (Huddar *et al* 1988). Krishnamurthy (1988) also reported delayed ripening of 'Alphonso' mangoes by 4 or 5 days with 'Tal-prolong' (1%). This delayed ripening was correlated with reduced softening, development of yellow colour of the peel, acidity and increase in TSS and carotenoids in the pulp. Prolong treated fruits showed less loss in weight. Normal ripening was affected at higher concentration of prolong (1.5 and above) and the fruits remained green and soft with high percentage of spoilage. They also showed improper ripening and poor development of carotenoids and higher acidity. Quality attributes of 1 percent prolong treated fruits were comparable with those of control fruits except the dull colour of the peel in treated fruits. Delay of ripening of 'Alphonso' mango fruit was primarily due to inhibition of development of yellow peel colour and delay in some of the ripening changes. Prolong treated (1%) fruits were ripe after 18 days of storage. The quality of treated fruits though acceptable, was slightly inferior to control fruits due to less red colour. Delayed ripening may be due to differential changes in the permeability of the peel to O_2 and CO_2 in treated fruits (Banks, 1984). It was therefore, concluded that prolong delays ripening at lower concentrations and use of higher concentrations should be avoided for delaying ripening of mangoes (Krishnamurthy, 1988).

'Vapor Gard' an anti-transpirant, is also used as a skin coating for extending the storage life of perishables through affecting their ripening. But Vapor Gard treated fruits are unacceptable due to reduced flavour (Sugar: acid ratio) and much loss of ascorbic acid during ripening (Lazan *et al* 1990).

'Semper fresh' is an edible film, often used as a skin-coating of fruits for retardation of ripening. Skin coat treatment with 'Semper fresh' delayed ripening of mango fruits by retarding softening, development of yellow colour and the chemical changes associated with ripening (Dhalla and Hanson, 1988., Kalra *et al* 1988., Ketsa and Prabhsavat, 1992). However, the colour development of these fruits was dull yellow and their taste was inferior compared to untreated control fruits (Ketsa and Prabhsavat, 1992). Mango fruits coated with 3 percent 'Semper fresh' remained green with off-odour. This indicated that 'Semper fresh' concentration above 1 percent leads to anaerobiosis, forming ethanol in the pulp (Dhalla and Hanson, 1988). Low oxygen has been shown to delay ripening of mango fruits, probably by inhibiting ethylene bio-synthesis (Hatton and Reeder, 1965., Spalding and Reeder, 1977). By forming a thin film around the fruits 'Semper fresh' presumably reduces exchange of gases, thus creating a modified atmosphere with lowered O_2 tension around the fruit (Mathur and Srivastava, 1955., Banks, 1984., Smith and Stow, 1984). 'Tommy Atkins' and 'Haden' mangoes treated with 'Semper fresh' (0.8 to 2.4%) were green and firm, for extended period, but 'Haden' fruits showed higher anthracnose development. However, 'Semper fresh' did not retard ripening of 'Keitt' mango fruits (Lopez *et al* 1995). 'Semper fresh' reduced water loss, but did not decrease vitamin C content. Treated fruits had higher titratable acidity, as a result of accumulation of CO_2 inside the fruit, due to acidosis, after dissolving and forming carbonic acid (Lopez *et al* 1995).

'Seal-packaging' of commodities individually is a new means of storage of fruits, particularly in citrus to extend the storage life and retarding ripening (Benyehoshua *et al* 1993). It was also tried in case of mangoes (Yantarasri *et al* 1995). 'Seal-packaging' of fruits in plastic film reduced shrivelling, provide better appearance and extended storage life of 'Tommy Atkins' 'Keitt' and 'Nam Dok Mai' mangoes stored at 13^0C (Ketsa and Raksritong 1992., Sornsrivichai *et al* 1992., Rodov *et al* 1993).

Perforated films having modified atmosphere also offer similar benefits, without possible anaerobiosis and avoid poor ripening found in un-perforated fruits held at ambient temperature (Rodov *et al* 1993., Yantarasri *et al* 1993). This method alleviates the need to remove the film to allow fruits to ripen properly. Keeping the above in view, Yantarasri *et al* (1995) attempted the delay of ripening of 'Nam Dok Mai' mangoes, using perforated PVC film. 'Seal-packaging' of mango fruits in PVC films delayed ripening as evidenced by the delay in the decrease of titrable acidity and in the colour development of both peel and flesh, particularly in case of un-perforated packages and the packages with small degree of perforation. This delay was attributed to the limited O_2 level to complete the proper ripening processes. Miller *et al* (1986), Sornsrivichai *et al* (1989, 1992) and

Ketsa and Raksritong (1992) showed that film-sealed packaging of 'Tommy Atkins' 'Keaw Sawoey' and 'Nom Dok Mai' delayed some processes of ripening. Their study indicated that higher level of O_2 is needed for the colour development of the peel than that of flesh.

The fact that most of the plastic films have lower permeability to gases than PVC film, suggests that most of the available plastic films cannot supply the O_2 needed for ripening by O_2 influx through film alone, hence, ventilation of plastic film is required. Perforation of the film enhanced the inhibition of ripening in 'Tommy Atkins' as compared to non-perforated packages (Ben-Yehoshua *et al* 1996).

V. Plant growth regulators

There is mounting evidence that plant hormones other than ethylene may also be involved in the regulation of fruit ripening. Exogenous application of auxnis, gibberellins and cytokinin was shown to retard ripening of many fruits including mango (Roy *et al* 1972., Krishnamurty and Rao, 1982). The treatment of fruits with plant growth regulators controls the fruit ripening processes without detrimental effects as those associated with CA storage and cold storage (Charles and Tung, 1973). However, none of the plant growth regulators tested so far has been commercially used for regulating (delaying) ripening in mango.

***Auxins*:** Treatment with 2,4,5-trichloro propionic acid (2,4,5-TP) delayed ripening of 'Pairi' mango fruits, without improvement in skin colour (Krishnamurty and Subramanyam, 1970b). Dipping 'Alphonso' mangoes in aqueous solution of indole acetic acid (IAA) at 10^{-4}M delayed ripening (Majumdar *et al* 1981). Post-harvest treatment with 2,4-dichlorophenoxy acetic acid (2,4-D) at 40 ppm was most effective in delaying the ripening processes and reducing PLW, maintaining the marketability of the fruits and extending the shelf life of 'Kesar' mangoes in Gujarat (Yuvaraj and Ughreja 1998). WE (6%) with 2,4-D (200 ppm) delayed ripening of mango fruits (Subramanyam *et al* 1962).

***Gibberellins*:** Out of the several gibberellins detected in plants, only gibberellic acid 3 (GA_3) and 7 (GA_7), notably the former one is extensively used in agriculture in general and in horticulture in particular. Several studies have demonstrated that post-harvest application of GA_3 retarded/delayed ripening of mango fruits though the concentration varied with cultivar (Dilley 1969., Majumdar *et al* 1981., Krishnamurty and Rao 1982., Khader *et al* 1988., Paramar and Chundawat, 1988., Khader, 1989, 1991., Kumar 1998., Singh *et al* 2000a). However, different workers explained different reasons for delaying of ripening by GA_3.

Post-harvest dips of 'Alphonso' mango fruits in aqueous solution of GA_3 delayed ripening as indicated by lower levels of reducing sugars, peel carotene and hydrolytic enzyme, which, according to Majumdar *et al* (1981) explains the senescence retarding effect of GA_3. However, GA_3 treatment did not allow ripening of 'Alphonso' fruits at 28 °C under controlled conditions as seen by the low Brix: acid ratio and firmer pulp

(Krishnamurty and Rao, 1982). Post-harvest dip-treatment of 'Mallika' fruits with GA_3 (100 or 200 ppm) significantly delayed ripening in storage at 35° ± 2 °C and 29° ± 2 °C (Khader *et al* 1988).

Delayed ripening due to GA_3 treatment (200 ppm) was assigned to consequent restriction of ethylene accumulation in the treated fruits (Dilley, 1969). GA_3 could have retarded the biochemical changes involved in the process of ripening. Alternately GA_3 would have inhibited the synthesis of additional enzymes, which are synthesized *denova* to accelerate ripening following the on set of climacteric peak (Dilley 1969). According to Paramar and Chundawat (1988) the delayed ripening of 'Kesar' mangoes treated with GA_3 (200 ppm) after harvest was through controlled respiration and ripening processes, whereas Khader (1991) was of the opinion that the delay in ripening of GA_3 treated fruits might be owing to reduced degradation of chlorophyll, ascorbic acid, less carotenoid synthesis and decrease in enzyme activities. Slow catabolic activities, postponement of the peeks of respiratory climacteric and ethylene evolution might be the causes for delayed ripening of 'Kesar' and 'Amrapali' fruits consequent to the post-harvest dips in GA_3 solutions (150 ppm). However, fruits were greener at the ripe stage due to slower degradation of chlorophyll and synthesis of carotenoids in GA_3 treated fruits. GA_3 dip-treatments were suggested for local movement of mangoes (Singh and Chundawat, 1991).

Pre-harvest foliar spray of GA_3 (100-300 ppm) retarded ripening of mango fruits for up to 6 days of storage at ambient temperature. Treated fruits also displayed lower levels of soluble solids, a lower sugar: acid ratio, lower total carotenoids and lower peroxidase and amylase activities at harvest. In contrast, total acidity, ascorbic acid and chlorophyll levels in the peel were higher in treated fruits (Khader, 1991). Hence, pre-harvest sprays of GA_3 are not generally advised.

Growth inhibitors and retardants

Delayed ripening of mango fruits following the post-harvest application of Malic hydrazide (MH) was reported (Date and Mathur, 1959., Srivastava, 1967., Krishnamurty and Subramanyam, 1970b., Musa, 1974). However, MH-treated fruits showed undesirable off-flavours (Date and Mathur, 1959). The effect of MH appears to be temperature dependent. MH extended the storage life of 'Totapuri' fruits under ambient conditions, but the effect diminished at 20^0C (Musa, 1974).

Growth retardants, such as Alar and Cycocel were also tried successfully for delaying ripening of mango fruits. Their influence differed with stages of ripening and also with the development of carotenoids in the fruit pulp. Pre-harvest dip treatment with cycocel delayed ripening in the early stages of ripening, but not later. A reduction in carotenoid content in the cycocel treated fruits and enhanced carotenoid content in Alar-treated fruits were observed (Subramanyam and Sebastian, 1970., Krishnamurty and Rao, 1982).

Benzylamino purine (BA) delayed ripening by 2 days as compared to the untreated 'Bjoribo' mango fruit and altered the composition of fruits on ripening. However, total and reducing sugar contents were increased and organic acids (citric, pyruvic and oxalic acids) and free amino acid contents were decreased after ripening in both treated and untreated fruits (Passera and Spettoli, 1981).

Polyamines: Recently, polyamines have been reported to extend the shelf life of mango fruits (Krammer *et al* 1991) by delaying ripening and senescence of fruits. (Galston and Kaur-Sawhney, 1995). Pressure infiltration of polyamines delayed mango fruit softening (Purwoka *et al* 1998). In a preliminary study, polyamine (Putrescine) retarded the ripening of 'Kensington Pride' mangoes and significantly reduced colour development, total and non-reducing sugars and total carotenoids compared to the untreated control treatment (Malik and Singh, 2003).

VI. Ethylene absorbents and ethylene action inhibitors

Ethylene is the main chemical (Phytohormone) which induces ripening. Ethylene is synthesized in the fruit tissues particularly after detachment from the tree. Its activity increases as the fruit ripens. In other words, fruit ripening progresses at a rapid pase due to increased synthesis and activity of ethylene in the tissues of the fruit and/or in the surrounding area of the fruits. So it is logically envisaged that by reducing the biosynthesis of ethylene and/or lowering its activity, it is possible to retard ripening. It is expected that any chemical which could have an inhibitory effect on ethylene production and its activity will delay ripening. Similarly any chemical which absorbs the ethylene on its evolution from the fruit, resulting in its removal in the surrounding area of the fruit, also retards/check the ripening.

Potassium permanganate ($KMnO_4$) is an ethylene absorbent. It as well as its commercial formulations are known to delay fruit ripening by removal of ethylene as it is released from the fruit. Esquerra *et al* (1978) showed that perlite- $KMnO_4$ inserts reduced ethylene concentration around the fruits and retarded the ripening of 'Carabao' mangoes enclosed in polyethylene bags at 10 °C. Exposed fruits had a lower TSS and higher acidity and showed a delay in the softening of the fruit. One sachet of 'Green keeper' (a commercial ethylene absorbant) per 2 kg of fruits was effective to delay ripening of 'Langra' mangoes having more than 1.0 specific gravity, characterized by slow development of skin colour, lower TSS and yellowness index value, with reduced loss in weight and reduction of shrinkage. However, firmness, pH and organoleptic quality remained unaffected (Singh and Narayana, 1998).

1-MCP has been recognized as an ethylene inhibitor, that has been found to block ethylene production by its irreversible binding to the ethylene receptive protein (Sisler *et al* 1996., Sisler and Serek, 1997) and preventing adverse ethylene responses in plant tissues for extended periods (Fan *et al* 1999., Selveraja *et al* 2001) and severely reduces ripening related changes (Faubion, 2000) in fruits and vegetables. 1-MCP totally blocks ethylene-mediated ripening processes in climacteric fruits (Ollveira Neto 2002., Fan *et al* 1999., Feng *et al* 2000). This compound is non-toxic, odourless and effective in very low concentrations as low as ppb levels (Sisler and Sereck, 1997). This chemical, therefore, provides a valuable tool and has tremendous commercial potential to extend keeping quality of climacteric fruits, in which ethylene is the key player that promotes ripening and to control ethylene-induced post-harvest disorders (Selveraja *et al* 2001). As an effective inhibitor of ethylene biosynthesis, in a wide range of fruits and vegetables 1-MCP attracted the Indian scientists for the past few years.

1-MCP mediated inhibition of endogenous ethylene directly causes inhibition of respiratory climacteric by two-folds in mango fruits. It was effective for 8 days in delaying fruit ripening in mango (Mathur and Srivastava, 2005). According to Hofman *et al* (2001) 1-MCP has the potential to reduce the risk of pre-mature ripening of mango due to accidental exposure to ethylene. However, additional precautions would be necessary to reduce the disease severity associated with 1-MCP treatment.

Treating 'Rosa' mango fruit with 1-MCP slowed down the senescence processes caused by the impact resulting from dropping of fruits from the tree at harvest (Santos *et al* 2004). Post-harvest application of 1-MCP delayed colour development and fruit softening of 'Zihna' mangoes stored at 20 °C (Jiang and Joyce, 2000).

Response to 1-MCP varies with the cultivar and physiological maturity in mango. In a study in Brazil, 1-MCP was capable of delaying ethylene induced ripening processes in all mango cultivars tested. Mature green 'Rosa' and 'Espada' mangoes treated with 1-MCP showed low weight losses, while maintaining TSS, titratable acidity and better external appearance. By the end of evaluation period, in case of 'Jasmine' cultivar, better external appearance was more evident in case of pre-climacteric fruits. In general, titratable acid was higher in mature-green, 1-MCP treated fruits. In treated fruits, colour progressed during storage, unlike in un-treated mangoes in which colour declined during storage. However, no disease incidence was observed in treated mature-green or pre-mature mangoes during the entire storage period (Silva *et al* 2004).

Variable results were reported when 1-MCP treatment was coupled with modified atmosphere storage in mango. While Jiang and Joyce (2000) obtained good results, Cocozza *et al* (2004b) did not get encouraging results. Neither MAS nor 1-MCP treatment delayed the climacteric peak during storage (Garcia-Estrada *et al* 2001). 1-MCP treatment resulted in more acidic fruits with high vitamin C content (Garcia-Estrada *et al* 2001., Cocozza *et al* 2004a). However, 1-MCP has showed no effect on TSS.

However, in spite of certain above negative effects particularly disease incidence and higher acidity with 1-MCP treatment, the use of 1-MCP alone is welcome, as it would result in lower post-harvest costs, and less post-harvest handling steps.

VII. Calcium

Regulation of cell membrane permeability and strengthening of cell walls are the main structural functions of calcium in plants. Calcium is important for structural integrity of plant cells and cell membranes. Elevated levels of calcium delay ripening and can inhibit fungal decay (Conway and Sams, 1984). Similarly higher concentration of calcium sometimes prevent fruit ripening resulting in firmer fruits (Fergusan, 1984). Pre-or post-harvest applications of calcium delay fruit ripening. In case of pre-harvest application, generally the entire trees are sprayed with calcium compounds like calcium chloride. On the other hand, post-harvest applications are made through either dip-treatment or vacuum/pressure infiltrations. Application either way delays ripening and senescence of fruits.

Increasing Ca^{+}_{2} content in fruits, either through pre-harvest spraying at fruit developmental stage or post-harvest dipping in calcium chloride leads to prevention of ripening (Wills *et al* 1977) besides an increase in the firmness of the fruit (Cooper and Bangreth, 1976). All pre-harvest sprays of calcium delayed ripening and improved fruit quality during storage at ambient temperature (35 ± 3 °C, 65 ± 5% RH) (Singh *et al* 1993). Gautam (1984), however, observed that 'Carabao' mangoes treated with pre-harvest calcium sprays showed only delayed peel colour development.

Under normal conditions, the transpirational losses from the fruit surface occur resulting in shrinkage of fruit and also probably loss of calcium. Since fruit bags are likely to offer a resistance to water vapour movement, fruit bagging might be anticipated to reduce transpirational flux of water and therefore, of calcium to developing fruit. Reduction in shelf life and early colour development of 'Sensation' mango fruit bagged during development with plastic bags were associated with rapid loss of water from the fruit (Joyce *et al* 1997). Plastic bags used during fruit development, modified the micro-environment around the mango fruit surface, thereby interfering with cuticle and/or lenticel development. The net result was the acceleration of post-harvest weight (water) loss. On the other hand, paper bags had no influence on weight loss or ripening characteristics of 'Sensation' mangoes. Calcium concentrations in flesh and skin tissues of 'Sensation' mango fruit were un-affected by bagging with either plastic or paper bags (Joyce *et al* 1997).

Dipping of unripe mangoes of 'Kensingston Pride' in a solution of 4 percent (w/v) calcium chloride under reduced pressure (250 mm Hg) and subsequent storage at 25 °C resulted in delayed ripening by one week compared to control. The degree of retardation showed relation to the amount of calcium applied. The use of calcium chloride dips of 8 percent (w/v) or higher delayed ripening, with skin injury and secondary microbial growth development on the fruits. A concentration of 4 percent, calcium chloride as dip appears to be the safe limit for 'Kensignton Pride' mangoes for delaying ripening. Calcium treated fruits on ripening had good eating quality, due to maintenance of titrable acidity, soluble solids and p^H and higher level of ascorbic acid. Thus, the use of calcium chloride, as dip-treatment to delay ripening of mangoes seems to have potential for commercial application.

Delaying of ripening means extension of shelf life. Calcium-treated mango fruits showed an extended shelf-life of about 3 weeks compared to 2 weeks of normal fruit. This time extension of one week would be sufficient to allow produce to be transported to distant markets and allows greater flexibility in local marketing. An added advantage in the use of calcium chloride treatment is that the fruit would be treated shortly after harvest, and would not require any other special storage or packaging during transport and marketing. The technique fits in the current handling system also (Tirmazi and Wills, 1981).

Pressure and vacuum infiltration of mature green 'Kensignton Pride' mango fruits with calcium chloride at different concentrations (2-8%) and storing them at 20 °C in boxes lined and covered with poly film delayed ripening by 12 and 8 days respectively, compared with fruits infiltered with water (Yuen *et al* 1993).

Calcium treatments result in delaying of ripening besides maintaining the firmness of fruits. The delay of ripening due to vacuum infiltration of calcium chloride was due to inhibition of activity of polygalacturonase (Reddy and Srivastiva, 1999). The maintenance of firmness of calcium treated fruits was due to reduced activity of softening enzymes viz., polygalacturonase and pectin methyl esterase (Brady *et al* 1985., Zambrano and Manzome, 1995). Tissues deficient in calcium showed increased activity of polygalacturonase activity (Konno *et al* 1984).

However, calcium chloride treatments caused some peel injury. Vacuum infiltration caused more injury than pressure infiltration. Peel injury was not reduced by reducing the p^H of calcium chloride solution during pressure infiltration from 9 to 4. However, the peel injury could be reduced by increasing the temperature of the fruit flesh or the calcium chloride solution during pressure infiltration (from 13° to 25 °C) or by packaging the fruits in sealed polyethylene bags or cling or shrink wraps after calcium chloride treatment (Yuen *et al* 1993). Due to poor uptake of calcium and the risk of peel damage, the commercial application of calcium by reduced pressure appeared to be doubtful (Joyce and Shorter, 1995). However, the use of surfactants to increase the uptake of calcium by fruit skin and flesh was suggested to overcome the constraint (Singh *et al* 2000b).

The combination of calcium treatment with modified atmosphere packaging (Singh *et al* 2000b), hot water treatment (52° ± 2 °C) (Abdul-Gofar *et al* 1997) and wax emulsion (8%) in cool chamber, not only delayed the ripening of mango fruits, but also improved the appearance and quality of treated fruits, besides reducing spoilage (Eeden, 1992., Dhaka *et al* 2001).

VIII. Wrappings

Fruits are wrapped for protection from external factors, which incidentally influences the ripening and storage of wrapped fruits. Wrapping of individual fruits with tissue paper significantly delayed ripening, caused less shrinkage and minimized damage in transit (Pruthi, 1968., Paramar and Chundawat, 1988). Fruits wrapped with high molecular weight high density polyethylene films (popularly known as HM films) took a relatively longer time for ripening in transit. The development of carotenoids and reduction in starch and tannins were also slow with HM film wrapped fruits in transit, compared to un-wrapped control, as there was considerable reduction in the physiological activity of wrapped 'Baneshan' fruits (Naryana *et al* 1991). Polyethylene-long-life individual bags or low density polyethylene individual bags were more effective in reducing the rate of softening and delaying ripening during storage, besides reducing the percentage of PLW and improving TSS content (Singh *et al* 2001).

B. Accelerating ripening

Accelerating ripening of mangoes becomes necessary when the grower intends to catch early market; to avoid erratic ripening in some cultivars, and also to hasten ripening in certain slow ripening cultivars. Ripening can be accelerated by the use of calcium carbide, ethylene, ABA etc., and hot-water treatment.

I. Ethylene

Ethylene and related compounds have long been used to accelerate or to enhance fruit ripening in mangoes in all the production/marketing areas. Incidentally there was reduction in/no spoilage of fruits with ethylene treatments.

Kalyanasundaram and Parthasarathy (1979) showed that 'Padiri' mangoes ripened within 5 days with a combined treatment of ethephon (7500 ppm) and Thiabendazole (1%) and had better control of stem-end rot than control fruits. Ethylene treatment at a concentration of 1000 ppm shortened the period of ripening from picking time (Fuchs *et al* 1975). Ripening of 'Alphonso' and 'Pairi' mango fruits was very much hastened, when they were exposed to ethylene (10,000 ppm). Dip method of treatment of ethephon reduced the ripening period by 50 percent that of controls (Gupta *et al* 1979). Dipping of fruits in solutions of ethephon (250-2000 ppm) hastened ripening by 2-3 days (Sampaio, 1981b). Erratic ripening in 'Mallika' fruits was overcome with ethrel and uniform and accelerated ripening were obtained with ethrel (Kalra and Tandon, 1983a).

Mango fruits are prone to microbial spoilage during ripening in storage. Therefore, it becomes necessary to combine both accelerating of ripening and controlling of spoilage. Ethylene treatment in hot water accelerated ripening and reduced spoilage in mango (Barmore, 1975., Lakshminarayana *et al* 1975., Sergent *et al* 1993). The treatment also greatly advanced the period of maximum production of CO_2 during respiration. Fruits treated with ethylene in hot water presented excellent surface and pulp colour

(Lakshminarayana *et al* 1975., Kalra and Tandon, 1983b). In contrast, fruits treated with ethylene in cold water showed no marked changes in external colour, but exhibited similar changes in ripening and respiration. Higher concentration of ethephon in hot water induced discolouration (Lakshminarayana *et al* 1975). However, hot water treatment alone produced fruits with best appearance (Kalra and Tandon, 1983b). Ethrel (500 ppm) treatment of mango fruits, harvested 2 weeks prior to optimum maturity, for 5 minutes in hot water (52° + 2 °C) enhanced colour development, increased softening and improved fruit quality as indicated by better TSS: acid blend (Tandon and Kalra, 1997). These authors earlier (1995) reported that it was possible to ripen the early harvested mangoes to an acceptable quality with post-harvest ethrel treatment. Although post-harvest ehtrel application improved the quality of early season fruits, the optimum quality parameters like flavour associated with mature fruits were not attained (Tandon and Kalra, 1997). Early harvested 'Dashehari' mangoes could be ripened to an acceptable quality superior to control when treated with ethrel and packed in wooden boxes. However, these fruits showed higher PLW, with concomitant shrinkage as compared to those ripened with ethrel and packed in polyethylene bags (Tandon and Kalra, 1997).

The simple treatment with ethrel can be adopted by the mango processing industry, which requires uniformly ripened fruits within a short period of time for processing purpose. Also the early and uniform ripening reduces the number of sortings for ripe fruits for processing, which is labour intensive and costly (Kulkarni *et al* 2004).

II. Abscisic acid

Abscisic acid (ABA) has been shown to promote ripening in tomatoes (Kader *et al* 1973) and grapes (Combe, 1976), which prompted to test ABA with mangoes also for accelerating ripening. A postharvest dip of mature unripe mangoes of 'Alphonso' cultivar in aqueous solution of ABA (10^{-6}M) hastened the ripening as judged by high level of total reducing sugars, carotene content of the peel and hydrolytic enzymes (amylase, cellulase, and invertase) during storage at 30° ± 2 °C (Majumdar *et al* 1981). The higher levels of total reducing sugars and hydrolytic enzymes in ABA-treated fruits might be the result of an enhanced rate of ripening, probably due to the increased production of ethylene (Libermann *et al* 1978). Promotion of ripening in ABA-treated fruits was also associated with decreasing levels of acids and elevation of sugars and sugar: acid ratio. The action of ABA is linked with protein synthesis. It was also suggested that an interaction between ABA and other hormones may have a significant role to play in regulating the process of ripening in mango. Blocking the effect of ABA in hastening ripening in mango by vacuum infiltration of cycloheximide (a protein synthesis inhibitor) also supports the role of ABA in fruit ripening (Palejwala *et al* 1989).

III. Paclobutrazol

Foliar application of paclobutrazol (PBZ) (2000-3000 ppm) before the commencement of flowering advanced ripening of 'Dashehari'mangoes in India. This indicates that ripening processes were initiated in the tree itself before harvesting. PBZ, as a potent specific inhibitor of gibberellin synthesis could have restricted the synthesis of gibberellins in leaves and fruits facilitating the rise of ethylene evolution and concentration to a threshold level needed for triggering the ripening processes. In such fruits, significantly greater TSS values, less total acidity, better carotenoid development, a higher activity of amylase and peroxidase were registered at harvest (Khader, 1990). However, the grower should have a pre-plan for getting such an advancement in fruit ripening with an eye to take advantage of early market.

IV. Hot water treatment

At the start, hot water treatment was designed as a quarantine measure to free the fruits from insect pests and for reduction/control of post-harvest diseases on the fruit surface. However, it was soon learnt, that hot water treatment (HWT) could regulate/hasten ripening of fruits. HWT provides dual benefits: hastening of fruit ripening and freedom from insect pests and diseases in treated fruits (Singh and Chundawat, 1991).

Hot water dip treatment ($54^{o} \pm 1\ ^{o}C$) for 5 minutes quickned the processes of ripening with reduced spoilage and PLW in 'Kesar' and 'Amrapali' cultivars (Singh and Chundawat, 1991). Treated fruits showed lower levels of total sugars and TSS and vitamin C compared to the control fruits. The reduction of ascorbic acid by HWT was attributed to highly sensitive nature of ascorbic acid to the heat and destruction by HWT (Mann and Singh, 1975). However, HWT reduced the shelf life of treated fruits by quickening the process of ripening. Due to quick turnover of metabolites the quality of treated ripe fruits was poor but colour development was better due to rapid and complete degradation of chlorophyll and higher synthesis of carotenoids (Lakshminarayana *et al* 1974., Paramar and Chundawat, 1988., Singh and Chundawat, 1991).

The response of mango to HWT, however, depends upon the stage of maturity of fruits at harvest, besides the cultivar. In fruits at an early stage, HWT hastened the onset of respiratory climacteric peak by 3-6 days in advance, but had no effect on more mature fruit. Subramanyam *et al* (1969) reported the hastening effect of HWT on 'Alphonso' mangoes at commercial maturity, whereas Lakshminarayana *et al* (1974) reported the effect of HWT on 'Alphonso' fruit harvested at various stages of maturity, from 11 weeks after fruit set to commercial maturity. The HWT of 12 weeks-old mature fruits of 'Dashehari' enhanced the ripening after 6 days of storage (Tandon and Kalra, 1998). These results indicate that HWT is effective even on half-mature fruits. However, there was no advantage in hastening ripening as early as 11 or 12 weeks after fruit set, because the resultant ripe fruits were more acidic, have a lower sugar content and a tendency to shrivel earlier. Further, as mentioned above, the shelf life of fruits ripened by HWT was shorter (Subramanyam *et al*, 1969., Subramanyam and Sebastian, 1970).

Shortened shelf life of HW treated fruits can be extended by incorporating maleic hydrazide in the dip water. HWT with or without MH and 2,4,5-TP increased the carotene content of flesh without altering the chemical composition or fruit quality

(Krishnamurthy and Subramanayam , 1970b). The combination of HWT plus ethrel (500 ppm) was effective in minimizing the spoilage of fruits during storage and could be employed for inducing ripening without impairing taste and flavour. HWT was, however, the better of the two techniques for reduction of microbial spoilage. Further, HW treated-mangoes were organoleptically superior over all others (Ram *et al* 1984).

In contrast to the above, Roy *et al* (1980) reported that HWT (52-58 °C), with Captan (0.5%) for 5 minutes retarded ripening in 'Langra' and 'Safdar Pasand' cultivars and thereby extended the shelf life by 10 days.

REFERENCES

Abbas, T., Iqbal, J. and Musahib-ud-Din, K.1989. Effect of calcium carbide on physical changes in mango fruits (cvs.Desi and Dusehri) during ripening. Journal of Agricultural Research – Lahore, 27(2): 121-126.

Abdul Gofar., Shafique, M.Z., Heleli, M.O.H., Ibrahim, M., Rahman, M. M and Alam, MS. 1997. Studies on extension of post-harvest storage life of mango (*Mangifera indica*). Bangladesh Journal of Scientific and Industrial Research, 32 (1): 148-152.

Abu–Sarra, A.F. and Abu-Goukh, A.A.1992. Changes in pectin esterase, polygalacturonase and cellulase activity during mango fruit ripening. Journal of Horticultural Science, 67(4) : 561-568.

Adinarayana, N.P. 1975. 'Ethrel', plant growth regulator for earlier and uniform ripening of mango fruits. Indian Food Packer, 29:15-19.

Ahmed, M. and Others, 1972. Induction of ripening delay in mangoes (cv.Dusheri) by gamma irradiation and refrigeration. Pakistan Journal of Science and Industrial Research, 15 (4/5):314-19.

Akamine, E.K. and Goo, T.1973. Respiration and ethylene production during ontogeny of fruit. Journal of the American Society for Horticultural Science, 98: 381 – 382.

Ali, Z.M., Armugam, S. and Lazan, H.1995. β-Galactosidase and its significance in ripening of mango fruit. Phyto-chemistry, 38: 1109-1114.

Ali,M., Farooqi,W.A. and Muhammed,A.1968. Preservation of mangoes (*Mangifera indica* L) by gamma radiation. Food Irradiation, 9(1-2):8-13.

Alves, RE., Filgueiras, H.A.C. and Almeida, A.D.S. 2004. Post-harvest ripening of 'Tommy Atkins' mangoes on two maturation stages treated with 1-MCP. Acta Horticulturae, 645: 627-632.

Anonymous, 1968-70. Annual Report, Central Food Technological Research Institute, Mysore, India.

Ashrof, M., Khan, N., Ahmed, M. and Elahi, M.1981. Studies on the pectin esterase activity and some chemical constituents of some Pakistan mango varieties during storage and ripening . Journal of Agricultural and Food Chemistry, 29: 526-528.

Baile, J.B., Young, R.E. and Olmstead, A.1954. Fruit respiration and ethylene production. Plant Physiology, 29: 168-174.

Balasubramaniam, V.M., Ilangantileke, S.G. and Wiese, K.L. 1991a. Effect of handling and storage temperature on mango firmness. American Society of Agricultural Engineers, No: 91-6026, 16pp.

Balasubramaniyam, V.M, Ilangantileke, S.G. and Wiese, K.L. 1991b. Effect of handling and storage temperature on mango firmness. American Society of Agricultural Engineers, No: 91-6524, 13pp.

Bandopadhyay, C. and Gholap, A.S. 1973a. Changes in fatty acids in ripening mango pulp (variety Alphonso). Journal of Agricultural and Food Chemistry, 21: 496 – 497.

Bandopadhyay, C. and Gholap, A.S. 1973b. Relationship of aroma and flavour characters of mango *(Mangifera indica L)* to fatty acid composition. Journal of the Science of Food and Agriculture, 24: 1497 – 1503.

Banks, N.H. 1984. Some effects of Tal-prolong coating on ripening of bananas. Journal of Experimetnal Botany, 35: 127-137.

Baqui, SM., Mattoo, A.K. and Modi, V.V. 1974. Mitochondrial enzymes in mango fruit during ripening. Phytochemistry, 13: 2049 – 2055.

Barmore, C.R. 1975. Ripening mangoes with ethylene and ethephon. Proceedings of Florida State Horticultural Society, 87 : 331 – 334.

Barmore, C.R. and Mitchell, E.F. 1976. Ethylene-pre-ripening of mangoes prior to shipment. Proceedings of Florida State Horticultural Society. 88: 469 – 471.

Ben-yehoshuva, S., Fisherman, S., Fang, D. and Rodor, v. 1993. New Development in modified atmosphere packaging and surface coatings for fruits Post-harvest handling of Tropical Fruits. Proceedings of International Conference. Chiang mai, Thailand,

Ben-yehoshua, S., Rodov, V., Fisherman, S., Peretz, J., Asuncion, Rode., La., Burns, P-Sornasrivichai, J. and Yantarasri , T. 1996. Perforation effects in modified atmosphere packaging: model and applications to bell pepper, and mango fruits. Proceedings of the Australian post-harvest Horticulture Conference Science and Technology for the fresh food revolution, Melbourne, Australia, 18-22, September, 1995: 143-162.

Bernarday S.A.P.F., Lajole F.M. and Cordenunsi, B.R. 2003. Evolution of starch and soluble sugar content during development and ripening of different mango cultivars. Ciencia e Technologia de Allmentos, 23 (Supplement): 116 – 120.

Besford, R.I. and Hobson, G.E. 1972. Pectic enzymes associated with softening of tomato fruits. Advances in Food Research, 10: 293-354.

Bhagyalakshmi, N., Prabha, T.N., Yshoda, H.M., Prasanna, V., Jagadesh, B.K., Tharanthan, R.N. and Drew, R. 2002. Biochemical studies related to textural regulation during ripening of banana and mango fruit. Acta Horticulturae, 575 (Vol. 2): 717 – 724.

Bhatnagar, H.C and Subramanyam, H. 1973. Some aspects of Preservation, Processing and Export of mango and its products. Indian Food Packer, 27(4): 33-52.

Bhullar, J.S. 1982. Ripenning of Langra mangoes with ethrel and calcium chloride. Progressive Horticulture, 14(1) : 71-72.

Bhullar, J.S., Khakhar, U.U. and Agnihotri, R.P. 1981. Studies on storage of Kinnow mandarin. Punjab Horticultural Journal, 21: 27-31.

Bhuva, H.P., Patel, J.C. and Amin, H.D. 1978. Effect of ethrel on ripening of some varieties of mango (*Mangifera indica* L) fruits. Indian Journal of Agricultural Research, 12(4): 263 – 265.

Bialte, J.B. and Young, R.E. 1981. Respiration and ripening in fruits – retrospect and prospect. In:T.Friend and M.J.C. Rhodes (eds) Recent Advances in Biochemistry of fruit and vegetables. Academic Press, London and NewYork, PP-1-39.

Boag, T.S., Johnson, G.I., Izarrd, M., Murray, C. and Fizsimmons K.C. 1989. Effects of gamma irradiation on mango Cv.Kensignton Pride I. Physiological responses to irradiation treatment as affected by fruit maturity and ripeness. Annals of Applied Biology, 93.

Brackman, A., Streif, J. and Bangerth, F. 1993. Relationship between a reduced aroma production and lipid metabolism of apples after long term controlled atmosphere storage. Journal of American Society for Horticultural Science, 118: 243 – 247.

Brady, C.J., McGlasson, P.J.A., Meldrum, S.K. and Kopaliovitch, E. 1985. Interaction between the amount and molecular forms of polygalacturonose, calcium and firmness in tomato fruit. Journal of American Society for Horticultural Science, 110: 254 – 258.

Brinson, K., Dey P.M., John M.A. and Pridham, J.B. 1988. Post-harvest changes in *Mangifera indica* mesocarp cell walls and cytoplasmic poly-saccharides. Phytochemistry, 27: 719-723.

Brown, B.I. 1984. Some effects of the post-harvest use of commercial wax emulsions on common mango. Proceedings of First Australian Mango Research Work Shop. 290 – 294.

Brown, B.I., Peacock, B.C. and Wong, L.S.1986. Physiological, chemical and quality changes in mangoes during post-harvest ripening. Proceedings of First Australian Mango Research Work Shop: 279 – 289.

Burg, S.P. and Burg, E.A. 1962. Role of ethylene in fruit, ripening. Plant Physiology, 37: 179-189.

Burg S.P. and Burg, E.A. 1967. Molecular requirements for the biological activity of ethylene. Plant Physiology, 42: 144 – 152.

Campbell, C.W. 1959. Storage and ripening of mangoes. Proceedings 19th Annual Meeting of Florida Mango Forum pp: 11 – 12.

Cardello, H.M.A.R. and Cardello, I. 1998. Vitamin C, ascorbate oxidase activity and sensory profile of mango(*Mangifera indica* L) var. Haden during ripening. Ciencia – e – Technologia – de – Alimentos, 18 (2): 211-217.

Carrillo, LA., Valdez, T.J.B., Rojas, V.R., Yahia, E.M. and Gomas, J.A. 1995. Ripening and quality of mango fruit as affected by coating with 'Semper fresh'. Acta Horticulturae, 370 : 203 – 216.

Castrillo, M., Kruger, N.J. and Whatley, F.R. 1992. Sucrose metabolism in mango fruit during ripening. Plant Science, Limerick, 84 (1): 45 – 51.

Chaimanee, P., Lertwikoon, N., Bungaruang. and Suntornwat,O.2000. Exo-polygalactuornose from ripe mango (*Mangifera indica* Linn.cv.Nam Dok Mai) Acta Horticulturae, 509 : 171 – 175.

Chaimanee, P. and Subhadrabandhu, S. 1992. Changes in cell wall hydrolase polygalacturonose during ripening in mango fruit. Acta Horticulturae. 321: 804 - 810.

Chalmers, D.J. and Rowan, K.S. 1971. The climacteric in ripening tomato fruits. Plant physiology. 48: 225-240.

Chaplin, G.R.1988. Advances in post-harvest physiology of mango. Acta Horticulturae. 231: 639 – 648.

Chaplin, G.R., Lai S.C. and Buckley, M.J. 1990. Differential softening and physico-chemical changes in the mesocarp of ripening mango fruit. Acta Horticulturae. 269: 233 – 239.

Charles, R.J. and Tung, M.A. 1993. Journal of Food Science, 38: 456.

Chikkasubbanna, V. and Huddar, A.G.1982. Ripening of Alphonso mangoes (*Mangifera indica* L) for processing. Indian Food Packer, 36: 38 – 42.

Chundawat, B.S., Gupta, O.P. and Singh, S.P.1973. Effect of chloro ethyl phosphonic acid (Ethrel) on ripening of mango(*Mangifera indica* L). Haryana Journal of Horticultural Science, 2(3) : 11 – 16.

Cocozza, F del M., Jorge, J.T., Alves, R.E, Filgueives, H.A.C. Garruti, D dos S and Perira, M.E.C.2004a. Sensory and physical evaluation of cold stored 'Tommy Atkins' mangoes influenced by 1-MCP and modified atmosphere packaging. Acta Horticulturae, 645: 655 – 661.

Cocozza, F del M., Jorge, J.T., Alves, R.E., Filgueivas, H.A.C and Pereira, M.E.C. 2004b. Respiration rate and chemical characteristics of cold stored 'Tommy Atkins' mangoes influenced by 1-MCP and modified atmosphere packaging. Acta Horticulturae, 645 : 645 – 650.

Conway, W.S and Sams, C.E. 1984. Possible mechanisms by which post-harvest calcium treatments reduce decay in apples. Phytopathology, 74 : 208 – 210.

Coombe, B.G. 1976. The development of fleshy fruit. Annual Review of Plant Physiology, 23. 207-268.

Cooper, T. and Bangreth, F. 1976. The effect of calcium and manganese treatments on the physiology, chemical composition and bitter pit development of "Cox orange" apples. Scientia Horticulturae, 5 : 47 – 57.

Cruikshank, T.A.M. and Perrin, D. R. 1964. In: Bio-chemistry of Phenolic compounds (Ed) J.D. Harborne. Academic Press. Newyork and London, p.511 – 14.

Cua, A.U. 1989. Ethylene biosynthesis in the 'Carabao' mango fruit (*Mangifera indica* L) during maturation and ripening. M.S.thesis, University of the Philippines at Los Bones College, Lagua, Philippines.

Cua, A.U. and Lizada, M.C.C, 1990. Ethylene production in the 'Carabao' mango (*Mangifera indica* L) fruit during maturation and ripening. Acta Horticulturae, 269: 169 – 179.

Dalal, V.B., Eipeson, W.E. and Singh, M.S. 1971. Wax emulsion for fresh fruits and vegetables to extend their storage life. Indian Food Packer, 25: 9 – 15.

Date, W.B. and Mathur, P.B.1959. Effect of post-harvest treatment with growth regulators on the ripening of mangoes (*Mangifera indica* L). Horticulture Advance, 3: 108 -110.

Dennison, R.A. and Ahmad, E.M. 1967. Irradiation effects on the ripening of 'Kent' mangoes. Journal of Food Science, 32: 702 – 705.

Dhaka, R.S.,Verma, M.K. and Agrawal, M.K. 2001. Effect of post-harvest treatments on physico-chemical characters during storage of mango Cv.Totapuri. Haryana Journal of Horticultural Science, 30 (1-2): 36 – 38.

Dhalla, R. and Hanson, S.W.1988. Effect of permeable coating on the storage life of fruits II. Prolonged treatment of mangoes. Indian Journal of Food Science and Technology, 23, 107 -112.

Dharkar, S.D., Savagaon, K.A., Srirangarajan, A.N. and Srinivasan, A.1966. Irradiation of mangoes. I Radiation induced delay in ripening of Alphonso mangoes. Journal of Food Science, 31: 863 – 869.

Dilley, D.R.1969. Hormonal control of fruit ripening. HortScience, 4: 11-14.

Dubery, I.A., Rensburg, L.J.van. and Schabort, T.C. 1984. Malic enzyme activity and related bio-chemical aspects during ripening of gamma irradiated mango fruit. Phytochemistry, 23(7): 1383 – 1386.

Eeden, S.J.van, 1992. Calcium infiltration as a possible post-harvest treatment to increase storage potential of mango fruit. Year book of South African Mango Growers' Association, 12 : 26-27.

Elahi, M. and Khan, N. 1973. Physico-chemical changes in some Pakistani mango varieties during storage ripening. Journal of Agriculture and Food Chemistry, 21 (2) : 229 – 231.

El-Zoghbi, M. 1994. Biochemical changes in some tropical fruits during ripening. Food Chemistry, 49: 33 – 37.

Esquerra, E.B., Mendoza, D.B. and Pantastico, E.B. 1978. Regulation of fruit ripening II. Use of perlite – $KMnO_4$ insert as an ethylene absorbent. Philippines Journal of Science, 107 : 23 – 31.

Fan, X.,Blankenship, S.M. and Mattheis J.P. 1999. 1-Methyl Cyclopropene inhibits apple ripening. Journal of American Society for Horticultural Science, 124 : 690 – 695.

Farkas K., Dharkar S.D. and Sreenivasan, A. 1972. Transporation and storage studies on irradiated Alphonso mangoes. Acta Aliment,1: 401 – 410.

Farooqi, W.A., Ahmand, M., Hussain, A and Hussain, A.M. 1974. Effect of gamma radiation on mangoes *(Mangifera indica L)* stored under different conditions. Journal of Agricultural Research, Pakistan, 12 (1): 31 – 42.

Faubion D, 2000. Could new ethylene inhibitor work on fruit? Good Fruit Grower, 50: 18.

Feng, X., Blankenship, A., Sisler, E.N. and Goren, R. 2000. Control of ethylene responses in avocado fruit with 1-methyl cyclo propene Post-harvest Biology and Technology, 20: 143-150.

Fergusan, H.B.1984. Calcium in plant science and fruit ripening. Plant Cell Environment, 7 : 477 – 489.

Fonseca, M.J de O., Cecn, P.R., Salouuo, L.C.C. and Pushmann, R.2004. Pulp and skin pigments in mango 'Haden' treated with fungicides and wax. Acta Horticulturae, 645 : 633 – 638.

Frylinck, L., Dubery, I.A. and Schabort, T.C. 1987. Biochemical changes involved in stress response and ripening behaviour of gamma-irradiated mango fruit. Phytochemistry, 26 (3): 681 – 686.

Fuchs. Y., Peris, E. and Zauberman, G. 1980. Changes in amylase activity, starch and sugars in mango fruit pulp. Scientia Horticulturae, 13: 155 – 160.

Fuchs, Y., Zangerman, G., Yanko, U. and Housky, S. 1975. Ripening of mango fruits with ethylene. Tropical Science. 17: 211 – 216.

Galston, A.W. and Kaur-Sawhney, R. 1995. Polyamines as endogenous growth regulators. In: P.G. Daries (ed) Plant-Hormones: Physiology, Bio-chemistry and Molecular Biology. Kluwer, Dordhecht.

Garcia – Estrada, R., Siller-Cepeda, J., Carrillo – Fasio, J., Muy-Rangel., M; Baej-Sanudo, R. and Marquez Zequera, M.Y. 2001. Effectde 1-Methylcyclopropeno enla presencia de *Colletotrichum gloeosporoides* Penz durante el mercadeo de frutos de mango. Fitopatologia Brasileira, 26: 462.

Garg, R.C. and Ram, H.B. 1973. A note on the effect of wax emulsions treatment on mango Cv.Lucknow Safeda. Progressive Horticulture, 5 (1): 35 – 39.

Gautam, D.M.1984. Post-harvest behabviour of 'Carabao' mango (*Mangifera indica* L) treated with pre-harvest calcium sprays. M.S.thesis, University of the Philippines at Los Banos college, Laguna, Philippines.

Goldstein, J.L. and Swain, T.1963. Phytochemistry, 2 : 371 – 383.

Gomez – Lim, M.A. 1993. Mango fruit ripening physiology and molecular biology. Acta Horticulturae, 341 : 484 – 499.

Gomez – Lim, M.A. 1997. Post-harvest physiology In: R.E. Litz (ed) The Mango. Botany, Production and Uses. CAB International, 425 – 445.

Goodenough, P.W., Prosser, I.M. and Young K.1985. NADP – linked malic enzyme and malate metabolism in aging tomato fruit. Phytochemistry, 24: 1157 – 1162.

Gowda, I.N.D. and Huddar, A.G. 2001. Studies on ripening changes in mango (*Mangifera indica* L) fruits. Journal of Food Science and Technology, Mysore, 38 (2): 135 – 137.

Grierson, D.1985. Gene expression in ripening tomato fruit. CRC. Critical Review of Plant Science, 3: 113 -132.

Guha, D. and Bhuiyan, M.A.J. 1997. Effect of ripening materials on mango. Annals of Bangladesh Agriculture, 7 (1): 73 -75.

Gupta O.P., Dhawan S.S. and Chauhan, K.S. 1979. Effect of pre-harvest application of ethrel (2-chloro – ethyl phosphonic acid) on ripening of mango. Research Reports on mango. Mango Workers Meeting, Panaji, Goa-India, 2-5 May 1979 pp. 304 – 305.

Harris, D.R., Leberry, J.A., Wills, R.B.H. and Spohr L.J. 2000. Effect of fruit maturity on efficiency of 1-Methyl cyclo propene to delay the ripening of banana. Post-harvest Biology and Technology, 20 : 303 – 308.

Hatton T.T. Jr., and Reeder W.F. 1965. Controlled atmosphere storage of Keitt mangoes. Proceedings of Carribean Region American Society for Horticultural Science, 10: 114-119.

Hatton, T.T. Jr,, Berha, L. and Wright, W.R.1961. Preliminary trials of gamma radiation on mature green Irwin and Sensation mangoes. Proceedings of Florida Mango Forum, 1961. pp: 15-17.

Hatton, T.T., Jr., Reeder, W.F. and Campbell, C.W. 1965. Ripening and storage of Florida mangoes. Marketing Research Report, U.S. Department of Agriculture, 725 : pp.9.

Hoda M.N., Yadav, G.S., Singh, S. and Singh, J.2001. Storage behaviour of mango (*Mangifera indica*) hybrids. Indian Journal of Agricultural Sciences, 71 (7): 469 – 472.

Hofman, P.G., Jobin, D.M., Meiburg, G.F., Manish, A.J. and Joyce, D.C. 2001. Ripening and quality responses of avocado, custard apple, mango and papaya fruits to 1- Methyl cyclo propene. Austrialian Journal of Experimental Agriculture, 41 (4): 567 – 572.

Hubbard, N.l., Pharr, D.M. and Huber, S.C 1991. Sucrose phosphate synthase and other sucrose metabolizing enzymes in fruits of various species. Physilogia Plantarum, 82: 192 – 196.

Huber, D.J. 1983. The role of cell wall hydrolases in fruit softening. Horticultural Reviews, 5: 169 – 219.

Huddar, A.G., Bharati, B.C. and Thimmaraju, K.R.1988. Note on extension of storage life of mango fruits by Tal prolong. Acta Horticulturae, 231: 668 – 639.

Hulme, A.C. 1971. The mango. In: A.C. Hulme(ed) The Biochemistry of fruits and their products. Academic Press, London, Vol.1, pp. 233-254.

Inyang, U.E and Agbo, A.U.1995. Effect of hot ash treatment of mango fruits on the physico-chemical changes during ripening. Tropical Science, 35: 259 -262.

Jacomino, A.P., Kluge, R.A., Brackmann, A. and Castro, P.R.C. 2002. Controle do amadurecimento e senescencia de mamao Com 1- Methyl ciclopropeno. Scientia Agriculture, 59: 303 – 308.

Jiang, Y. and Joyce, D.C. 2000. Effect of 1-Methyl cyclopropene alone and in combination with polyethylene bags on the post-harvest life of mango fruits. Annals of Applied Biology, 137(3): 321 – 327.

Jobin – Décor, M.P. 1989. 'Kensington Pride' mango colour development at various temperatures. Queensland Agricultural Journal, March – April: 108 – 109.

John, J., Subbarayan, C. and Cama, H.R. 1970. Carotenoids in 3 stages of ripening mango. Journal of Food Science, 35 : 262 – 265.

Joon, M.S., Kumar, J., Sharma, R.K. and Singhrot, R.S. 2001. Comparision of Calcium Carbide and Ethephon on ripening of mango. Haryana Journal of Horticultural Science, 30 (3&4): 181 – 182.

Joyce, D.C., Beasley, D.R. and Shorter, A.J. 1997. Effect of pre harvest bagging on fruit calcium levels, and storage and ripening characteristics of 'Sensation' mangoes. Australian Journal of Experimental Agriculture, 37:383 – 389.

Joyce, D.C and Shorter, A.J.1995. Partial pressure calcium infiltration of mangoes did not delay softening. Proceedings of the Austrilian post-harvest Horticulture Conference. 'Science and Technology for the fresh food revolution' Melbourne, Australia, 18-22, September, 1995. 1996: 261 – 265.

Jungalwala, F.B. and Cama H.R.1967. Carotenoids in mango (*Magnifera indica*) fruit. Indian Journal of Chemistry, 1: 36-40.

Kader, A.A. 1993. Modified and Controlled atmosphere storage of tropical fruits. In:B.R.Chapman, E.Highley and G.I. Johnson (eds) 'Post-harvest handling of Tropical fruits' Vol.50. ACIAR, Canberra, Australia, pp. 239 – 249.

Kader, A.A., Morris, L.L. and Lyan J.M.1973. HortScience, 8: 263.

Kalra, S.K and Tandon, D.K 1983a. Regulation of ripening of mango Cv.Mallika. Indian Journal of Horticulture, 40 (3&4): 155 – 159.

Kalra, S.K and Tandon, D.K 1983b. Ripening behaviour of 'Dashehari' mango in relation to harvest period. Scientia Horticulturae, 19: 263 – 269.

Kalra, S.K, Tandon, D.K. and Lohani, H.C. 1988. Ripening of Wax-coated Dashehari mango during storage. Indian Food Packer, 32:7 -11.

Kalra, S.K., Tandon, D.K. and Singh, B.P.1995. Mango. In : D.K. Salunkhe and D.K.Kadam(eds) Hand book of Fruit Science and Technology, Production, Composition, Storage and Processing. Marcel Dekker, Inc.New York, pp. 123-169.

Kalyanasundaram, P. and Parthasarathy, V.A. 1979. Effect of Thiobendazole and ethrel on ripening and stem-end-rot of mango Cv.Padiri. Annamalai University of Agricultural Research Annual, 7/8. 161- 164.

Kane, O. and Marcellin, P. 1979. Effects del'atmosphere control user la conservation des mangues (Varieties: Amelie and Julie). Fruits, 34: 123 – 129.

Kapse, B.M, and Katrodia, J.S. 1997. Ripening behaviour of 'Kesar' mangoes in relation to specific gravity. Acta Horticulturae, 455: 669 – 678.

Kapse, M.B., Rane D.A., Kulkarni, D.N. and Ingle, U.M. 1977. Bio chemical changes in Totapuri mango during ripening and storage. South Indian Horticulture, 25(3) : 119 -120.

Karmarkar, D.V. and Joshi, B.M.1941. Respiration studies of the Alphonso mango. Indian Journal of Agricultural Science, 11: 993 – 1005.

Ketsa, S., Chidtragool, S., Klein, J.D. and Lurie, S.1998. Effect of heat treatment on changes in softening, pectic substance and activities of Polygalacturonase, Pectin estarase and P-galactosidase of ripening mango. Journal Plant Physiology, 153 : 457 – 461.

Ketsa, S. and Prabhasavet, T., 1992. Effect of skin coatings on shelf life and quality of 'Nang Klangwan' mangoes. Acta horticulturae, 321: 764-770.

Ketsa, S. and Raksritong, T.1992. Effect of PVC wrapping and Temperature on storage life and 'quality of Nam Dok Mai' mango fruits in ripening. Acta Horticulturae, 322 : 756 – 763.

Khader, S.E.S.A. 1989. Delaying ripening by post-harvest treatment of GA_3 in mango. Indian Journal of Horticulture, 46 (4): 444 – 448.

Khader, S.E.S.A. 1990. Orchard Application of Paclobatrazol on Ripening, Quality and Storage of mango fruits. Scientia Horticulturae, 41: 329 – 335.

Khader, S.E.S.A. 1991. Effect of pre-harvest application of GA_3 on post-harvest behaviour of mango fruit. Scientia Horticulturae, 48: 317 – 321.

Khader, S.E.S.A., Singh, B.P. and Khan, S.A. 1988. Effect of GA_3 as a post-harvest treatment of mango fruit on ripening, amylase and peroxidase activity and quality during storage. Scientia Horticulturae, 36 (3-4): 261 – 266.

Kluge, R.A. and Jacomina, A.P.2002. Shelf life of Peaches treated with 1-Methyl cyclopropene. Scientia Agriculturae, 59: 69-72.

Konno, H., Yamaya, T., Yamasaki, Y. and Hetsusumoto, H. 1984. Pectic polysaccharide breakdown of cell walls in cucumber roots growing with calcium starvation. Plant Physiology, 76: 633 – 637.

Kosiyachinda, S. and Pankasemsum, T. 1990. Ripening of green-eating-type mangoes, 'Mamnang Mun'. Acta Horticulturae, 269: 161 – 168.

Krammer, G.F., Wang, C.Y. and Conway, W.S. 1991. Inhibition of softening by polyamine application in 'Golden Delicious' and 'McIntosh' apple. Journal of Americal Society for Horticultural Science, 116: 813 – 817.

Krishnamurthy, G.V., Jain, N.L. and Bhatia, B.S.1960. Changes in the physico-chemical composition of mangoes during ripening after picking. Food Science, Mysore, 9: 277 – 279.

Krishnamurthy, S. 1988. Effects of Tal-prolong on shelf-life and quality attributes of mango. Acta Horticulturae, 231: 675 – 678.

Krishnamurthy, S. and Rao, K.P.G 1982. Regulation of ripening by chemicals in 'Alphonso' mango. Scientia Horticulturae, 16: 179 – 183.

Krishnamurthy, S. and Subramanyam. H.1970a. Respiratory climacteric and chemical changes in the mango fruit, *Mangifera indica* L. Journal of the Amercian Society for Horticultural Science, 95: 333 – 337.

Krishnamurthy, S. and Subramanyam, H.1970 b. Effects of Maleic hydrazide and 2,4,5, Trichloro proponic acid on ripening and quality of mango fruit. Pesticide Science, 1: 63 – 5.

Krishnamurthy, S. and Sudhakar Rao, D.V.2001. Status of post-harvest management of fruits. Indian Journal of Horticulture, 58 (1&2): 152 – 163.

Krishnamurthy, S., Patwardhan, M.V and Subramanyam, H. 1971. Bio-chemical changes during ripening of mango fruit. Phytochemistry, 6: 2577 – 2581.

Kulkarni, S.G., Kudachikar, V.B., Vasantha, M.S., Prakash M.N.K., Prasad, B.A. and Ramana, KVR, 2004. Studies on Effect of Ethrel Dip treatment on the ripening behaviour of mango (*Mangifera indica* L) variety "Neelum". Journal of Food Science and Technology, 41 (2): 216 – 220.

Kumar, A and Dhawan, S.S 1995. Effect of post-harvest treatments on the enhancement of ripening of mango (*Mangifera indica* L) fruit Cv.Dashehari. Haryana Journal of Horticulture Science, 24 (2): 109 – 115.

Kumar, D.1998. Effect of post-harvest treatment on shelf life and quality mango. Indian Journal of Horticulture, 55 (2): 134 – 138.

Kumar, R. and Selveraj, Y.1990. Fructose 1,6- biphosatase in ripening mango (*Mangifera indica* L) fruit. Indian Journal of Experimental Biology, 28 (3): 284 – 286.

Kumar, R. and Selveraj, Y. 1988. Glucose 6 – phosphatase activity in ripening mango. Current Science, 57: 892 – 893.

Kumar, R. and Selvaraj, Y. 1989. Studies on ATP:Phospho fructo kinase in ripening mango (*Mangifera indica* L) fruit. Current Science, 58 (12): 666 – 668.

Kumar, S., Das, D.K., Mishra M., Singh, A.K. and Prasad, US. 1994. Sucrose metabolism during maturation and ripening of mango cultivars. Plant Physilogy and Bio chemistry (New Delhi), 21(1): 27 – 32.

Kumar, S., Das, D.K., Singh, A.K. and Prasad, U.S.1993. Changes in non-volatile organic acid composition and p^H during maturation and ripening of two mango cultivars. Indian Journal of Plant Physilogy, 36 (2) : 107 – 111.

Kumar, S.P., Roy R.N., Singh, C. and Jain, B.P. 1996. Effect of ripening materials on the quality and storage of mango (*Mangifera indica* L). Journal of Research, Birsa Agricultural University, 8(1): 33 – 37.

Kumar,J., Sharma,K.K.,Singh,R and Godara,R.K.1990. Increased shelf-life of Kinnow mandarin (*Citrus reticulata*) by different storage conditions and chemicals. Indian Journal of Agricultural Sciences, 60(2):151-154.

Lacroix, M., Bernard, L., Jobin, M., Milot, S. and Gagnon, M. 1990. Effect of irradiation on the biochemical and organoleptic changes during the ripening of papaya and mango fruits. International Journal of Radiation Application and Instrumentation. Part C- Radiation Physics and Chemistry, 35 (1-3): 296 – 300.

Lakshminarayana, S. 1973. Respiration and ripening patterns in the life cycle of the mango fruit. Journal of Horticultural Science, 48: 227 – 233.

Lakshminarayana, S. 1975. Relation of time of harvest on respiration, chemical constituents and storage life of mangoes. Proceedings of the Florida State Horticultural Society, 88: 477 – 480.

Lakshminarayana, S. 1980. Mango. In S.Negy and P.E.Shaw (Fds) Trocipal and Sub tropical fruits. AV Publishing Co.West port, Connecticut, pp: 184 – 257.

Lakshminarayana, S. and Subramanayam, H. 1966. Journal of Food Science and Technology, 3: 151 – 154.

Lakshminarayana, S. and Subramanayam, H. 1970. CO_2 injury and fermentative dicarboxylation in mango fruit at low temperature storage. Journal of Food Science and Technology, 7: 148 – 150.

Lakshminarayana, S., Subhadra, N.V. and Subrahmanyam, H. 1970. Some aspects of developmental physiology of mango fruit. Journal of Horticultural Science, 45: 133 -142.

Lakshminarayana, S., Krishna Prasad, C.A. and Shetty, M.S. 1974. Hot water treatment to regulate ripening and reduce spoilage of 'Alphonso' mango. Journal of Horticultural Science, 49: 365 – 371.

Lakshminarayana, S., Shetty, M.S. and Krishna Prasad, C.A. 1975. Accelerated ripening of Alphonso mangoes by the application of ethrel. Tropical Science, 17 : 95 – 101.

Lalel, H.J.D., Singh, Z. and Tan, S.C. 2004. Ripening temperatures influence biosynthesis of aroma volatile compounds in 'Kensington Pride' mango fruit. Journal of Horticultural Science and Bio technology, 79(1): 146 – 157.

Lalel, H.J.D., Singh, Z. and Tan, S.C. 2002. Aroma volatiles production during fruit ripening of 'Kensington Pride' mango. Post-harvest Biology and Technology, 27 : 323 – 336.

Lam, P.F., Ng.,K., Omar. D. and Talib, Y. 1982. Physical, Physiological and Chemical changes of Golek after harvest. Proceedings of workshop on mango and rambutan. ASEAN Post-harvest Training College, Lagura, Philippines: 96 – 112.

Lazan, H. and Ali, Z.M.1993. Cell wall hydrolases and their potential in the manipulation of ripening of tropical fruits. ASEAN Food Journal, 8: 47 – 53.

Lazan, H., Ali, Z.M., Lee, K.W., Voon, J. and Cheplin, G.R. 1986. The potential role of polygalacturonase in pectin degradation and softening of mango fruit. ASEAN Food Journal, 2: 93 – 95.

Lazan, H., Ali, Z.M. and Sani, H.A. 1990. Effects of Vapor Gard on polygalacturonase, malik enzyme and ripening of Harumanis mango. Acta Horticulturae, 269: 359 – 362.

Lazan, H., Ali, Z.M., Talkah, Z., Soh, J.S. and Schafferr, B.1993. The biochemical basis of differential ripening in mango. Acta Horticulturae, 341: 500 -509.

Lee, S.R.,Liu, H.L., Shiesh, C.C and Lee, K.E. 1998. Effect of temperature on ripening of 'Chim Hwang' mango (*Mangifera indica* L). Journal of the Chinese Society for Horticultural Science, 44 (2): 126 – 136.

Libermann, M., Baker, J.E and Sieger, M.1978. Influence of Plant hormones on ethylene production in apple, tomato and avocado slices during maturation and senescence. Plant Physiology, 60: 214 – 217.

Lizada, M.C.C. 1991. Post-harvest physiology of mango - A review. Acta Horticulturae, 291: 437 – 453.

Lizada, M.C.C. 1993. Mango. In: G.B. Seymour., J.E., Taylor and G.A. Tucker. (eds) Bio chemistry of Fruit ripening. Chapman and Hall, London, pp. 255-271.

Looney, N.E. and Patterson, M.E. 1967. Chlorophyllase activity in apples and bananas during climacteric phase. Nature, 214 (1): 1245 – 246.

Lopez, A.C., Valdez Torres, J.B., Villegas, R.R. and Yahia, E.M. 1995. Ripening and quality of Mango fruit as affected by coating with 'Semper fresh'. Acta Horticulturae, 370: 203 – 266.

Lopez, G.R. and Gomez-Lim, M.A.1992. A method for extracting intact RNA from fruits rich in polysaccharides using ripe mango mesocarp. HortScience, 5: 440-442.

Lopez, G.R. and Gomez-Lim M.A.1993. Changes in RNA and protein synthesis during ripening in mango fruit. Journal of Plant Physiology, 141 (1): 82 – 87.

Majumder, G., Modi, V.V. and Palejwala, V.A.1981. Effect of Plant Growth Regulators on mango ripening. Indian Journal of Experimental Botany, 191: 885 -886.

Malik, A.U. and Singh, Z-2003. Increased bio-synthesis of free polyamines in response to higher ethylene production during fruit ripening of mango. Australian Post-harvest Conference. Brisbane, Australia, 1-3 October 2003: 117 – 119.

Malik, A.U. and Singh, Z. 2004. Endogenous free polyamines of mangoes in relation to development and ripening. Journal of American Society for Horticultural Science, 129 (3): 280 – 86

Mann, S.S. and Dhillon, B.S. 1974. Effect of ethrel and calcium carbide on ripening of mango. Progressive Horticulture, 6 (1): 59 – 62.

Mann, S.S. and Singh R.N. 1975. Ripening of Dashehari and Langra cultivars of mango (*Mangifera indica* L) at different stages of maturity by various methods. Haryana Journal of Horticultural Science, 4 (1&2): 31 – 39.

Mann, S.S. and Singh, R.N. 1976. The cold storage life of Dashehari mangoes. Scientia Horticulturae, 5 : 249 – 253.

Marin, M.A. and Cano, M.P. 1992. Patterns of peroxidase in ripening mango (*Mangifera indica* L). fruits. Journal of Food Science, 57 (3): 690 – 692.

Mathur, K. and Sivastava, G.C. 2005. Effect of 1-MCP on malic enzyme activity and ethylene production in mango during ripening. Indian Journal of Plant Physiology, 10 (3) (NS): 273 – 275.

Mathur, P.B. and Lewis, N.F. 1961. Storage behaviour of gamma irradiated mangoes. International Journal of Radiation and Isotopes, 11 : 43 – 45.

Mathur, P.B. and Srivastava, H.C. 1955. Effect of skin coatings on storage behaviour of mangoes. Food Research, 20: 559 – 566.

Mattoo, A.K. and Modi, V.V. 1969a. Ethylene and ripening of mangoes. Plant Physiology, 44 : 308 – 310.

Mattoo, A.K. and Modi, V.V. 1969b. Biochemical aspects of ripening and chilling injury in mango fruit. Proceedings of International Conference on Tropical and Sub tropical fruits, London, pp111 – 115.

Mattoo, A.K. and Modi, V.V. 1970a. Partial purification and properties of enzyme inhibition from unripe mangoes. Enzymologia, 39: 237 – 247.

Mattoo, A.K and Modi, V.V. 1970b. Citrate cleavage enzyme in mango fruit Biochemical and Biophysical Research Communications, 39: 895 – 899.

Mattoo, A.K., Modi, V.V. and Reddy, V.V.R. 1968. Oxidation and carotenogensis regulation factors in mango. Indian Journal of Biochemistry, 5: 111 -114.

Mattoo, A.K., Murata T., Pantastico, P.B., Chachin, K., Ogata, K. and Phan, C.T. 1975. Chemical changes during ripening and senescence. In: E.B Pantestico(ed) Post-harvest physiology: Handling and Utilization of Tropical and Sub tropical Fruits and Vegetables. AVI publishing, West Port. 130 -156.

Mc Cready, R.M. and Mc Comb, E.R. 1954. Pectic constituents in ripe and unripe fruits. Food Research, 19: 530 – 535.

Mc Lanchlan R.L., Johnson, G.I., Mitchell, G.E. and Wills, P.A. 1990. Irradiation of Kensington Pride Mangoes. Acta Horticutlurae, 269: 469 – 476.

Medlicott, A.P. 1985. Mango fruit ripening and the effects of maturity, temperature and gasses. Ph.D.Thesis. The polytechnic, wolverchampton (CNAA).

Medlicott, A.P., Bhogol, M and Reynolds, S.B. 1986a. Changes in peel pigmentation during ripening of mango fruit (*Mangifera indica* L Var. Tommy Atkins). Annals of Applied Biology, 109: 651 – 656.

Medlicott, A.P., N'Diaye, M. and Sigrist, J.M.M-1990a. Harvest Maturity and Concentration and Exposure time to Acetylene : Influence on Initiation of Ripening in Mangoes. Journal of American Society for Horticultural Science, 115 (3): 426 – 430.

Medlicott, A.P. and Reynolds, S.B. 1988. Harvest maturity effects on mango fruit ripening. Tropical Agriculture (Trinidad), 65(2) : 153 – 157.

Medlicott, A.P., Reynolds, S.B., New, S.W. and Thompson, A.K 1988. Harvest maturity effects on mango fruit ripening. Tropical Agriculture, 65(2): 153 – 157.

Medlicott, A.P. Reynolds, S.B and Thompson, A.K, 1986b. Effect of temperature on the ripening of mango fruit (*Mangifera indica* L Var.Tommy Atkins). Journal of the Science of Food and Agriculture, 37 : 469 – 74.

Medlicott, A.P., Sigrist, J.M.M., Reynolds S.B. and Thompson, A.K: 1987. Effects of ethylene and acetylene on mango fruit ripening. Annals of Applied Biology, 111 (2): 439 – 444.

Medlicott, A.P., Sigrist, J.M.M and Sy, O.1990b. Ripening of mangoes following low temperature storage. Journal of American Society for Horticultural Science, 115 (3): 430 – 434.

Medlicott, A.P. and Thompson, A.K. 1985. Analysis of sugar and organic acid in ripening mango fruit (*Mangifera indica* L var. Keitt) by high performance liquid chromatography. Journal of the Science of Food and Agriculture, 36: 561 – 566.

Mendoza, D.B. 1981. Developmental physiology of 'Carabao' mango (*Mangifera indica* L). fruits. Ph.D. Thesis, University of the Philippines at Los Banos College, Lagura, Philippines.

Miller, W.R., Spalding, D.R and Hale, P.W. 1986. Film wrapping mangoes at advancing stages of post-harvest ripening, Tropical Science, 26: 9-17.

Mir, N.A., Curell, E., Khan, N., Whitaker, M and Beandr, R.M 2001. Harvest maturity, storage temperature and 1-MCP application frequency alter firmness retention and chlorophyll fluorescence of 'Redy Chief Delicious' apples. Journal of American Society for Horticultural Science, 126: 618 – 624.

Mitcham, E.J. and Mc Donald, R.E. 1991. Cell wall Modification during Ripening of 'Keitt' and 'Tommy Atkins' Mango fruit. Journal of American Society for Horticultural Science, 117 (6): 919 – 924.

Mitra S.K. 2008. Harvesting, Packaging, Transportation and Marketing of Tropical Fruits in West Bengal, India: Present Practices and Future needs. Acta Horticulturae, 768: 301 – 306.

Mitra, S.K. and Baldwin, E.A. 1997. Mango In: SK Mitra (ed) Post-harvest physiology and storage of Tropical and Sub tropical fruits. CAB International, 85 – 122.

Mizuta, T. and Subramanyam, H.1973. Changes in pectic and cellulosic constituents in Alphonso and Pairi mangoes during post-harvest ripening. Advances in Food Research, 21: 223 – 305.

Modi, V.V. and Reddy, V.V.R 1967. Carotenogenesis in ripening of mangoes. Indian Journal of Experimental Biology, 5: 233 – 235.

Morga, N., Lustre, A.O., Tunac, M.M., Balagot, A.H. and Soriano, M.R. 1979. Physico-Chemical changes in Philippine 'Carabao' mangoes during ripening. Food Chemistry, 15: 225 – 234.

Muda, P., Seymour, G.B., Errington, N. and Tucker, G.A. 1995. Compositional changes in cell wall polymers during mango fruit ripening. Carbohydrate Polymers, 26 (4): 255 – 260.

Mukherjee, P.K. and Srivastava, R.B. 1979. Increasing the storage life of mango (*Mangifera indica* L) by lowering the critical temperature. Progressive Horticulture, 10 (4): 63 – 69.

Musa, S.K. 1974. Preliminary investigations on the storage and ripening of 'Totapuri' mangoes in the Sudan. Tropical Science, 16: 65 – 73.

Nagaraj,P., Ramana, K.V.R., Prasad. B.A, Mallikarjunaradhya, S., Patwardhan, M.V., Anantha Krishna, S.N., Rajpoot I.N.C. and Subramanyam L.1984. Effect of calcium carbide on ripening and quality of Alphonso mangoes. Journal of Food Science and Technology, 21(5): 278 – 282.

Nandini, P.V. and Oommen, B.E. 2002. Quality Parameters of selected Mango cultivars. Journal of Food Science and Technology, 39 (4): 379 – 383.

Narayana, C.K., Roy, K. and Pal, R.K. 1991. Use of high molecular weight high density polyethylene (HM.Film) in transit of mango (cv.Baneshan). Acta Horticulturae, 291: 489 – 492.

Narayana, C.K. and Singh, B.P. 1998. Chilling injury and its effect on ripening, bio-chemical and enzymatic changes in mango. Indian Journal of Horticulture, 53(3): 166 – 170.

Nasrijal, N.H. 1993. Effect of storage temperature on pectin depolymerization and, softening of mango fruit. B.Sc. Thesis, Deportment of Botany. Universitte Kebangsaan , Malaysia.

Neeraj., Joon, M.S. and Bhatia, S.K. 2002. Effect of plastic packaging on Bio-chemical parameters of Fruits during Storage. Haryana Journal of Horticulture Science, 31 (1&2): 1-7.

Negi, S.S. 2000. Mango production in India. Acta Horticulturae, 509: 69 -77.

Nguyen, H., Mc Conchie, R., Hofmon, P., Smith, L., Stubbings, B and Adkins, M, 2002. Effect of ethylene and ripening temperatures on the skin colour and flesh characteristics of ripe ' Kensington Pride' mango fruit. Acta Horticulturae, 575 : 635 – 642.

Nour, A.A.M. 1978. The influence of pectic substances on textural properties of mango fruit. Sudan Journal of Food Science and Technology, 10: 1-3.

O'Hare, T.J. 1995. Effect of ripening temperature on quality and compositional changes of mango (*Mangifera indica* L) cv.Kensignton. Australian Journal of Experimental Agriculture, 35(2):259-263.

Ola. T.M. 1978. Fast or slow ripening of mangoes. Phulippines Farmers Journal, 20 (5): 15

Ollveira Neto, O.C. 2002. Fisiologia da maturacao de tres cultivares de bananas tratadas com 1 – metil ciclopropene e armazenades sob at mosfera modificada. Centro de Ciencies Agrarias / UFPB, Areia 112p. (M.Sc.Dissertation).

Pal, R.K. 1998a. Ripening and rhelogical properties of mango as influenced by ethrel and calcium carbide. Journal of Food Science and Technology, Mysore, 35(4): 358 – 360.

Pal, R.K. 1998b. Influence of ethrel and calcium carbide on respiration rate, ethylene evaluation, electrolyte leakage and firmness of 'Dashehari' mango (*Mangifera indica* L). Indian Journal of Agricultural Sciences, 68(4): 201 – 203.

Palejwala, V. A., Amin, B., Parikh, H.R. and Modi, V.V. 1989. Role of abscissic acid in the ripening of mango. Acta Horticulturae, 231: 662 – 667.

Pandey, M., Srivastava, G.C. and Prasad, N.K. 1998. Physiological changes Associated with ripening in two mango varieties. Indian Journal of Plant Physiology, 3 (2) (NS): 94 – 96.

Pandey, R.M. and Singh, R.N. 1976. Chemical regulation of ripening in mango. Indian Journal of Horticutlure, 33 (3-4): 209 – 211.

Pantastico, E.B., Lam, L.R., Ketsa, S., Yuniarti, Y. and Kosittrakul, M. 1984. Post-harvest physiology and storage of mango. In: Mango fruit development, post-harvest physiology and marketing in ASEAN. (Eds). D.B. Mendoza, and R.B.H Wills. ASEAN Food Handling Bureau, Kaula Lumpur, Malaysia. pp. 41 – 43.

Paramar, P.B and Chundawat, B.S. 1988. Effect of various post-harvest treatments on the physiology of Kesar mango. Acta Horticulturae, 231 : 679 – 684.

Parikh, H.R and Modi, V.V. 1990. Requirement of protein synthase during ripening of abscisic acid-treated mangoes. Current Science, 59: 468 – 470.

Passera, C. and Spettoli, P. 1981. Effects of Benzyl amino purine on mango fruit ripening. Food Chemistry, 7: 195 – 201.

Peacock, B.C. and Brown, B.I. 1984. Quality comparison of several mango varieties. Proceedings of First Australian Mango Research Workshop. Pp 334 – 339.

Peacock, B.C., Murray, C., Kosiyachinda, S., Kositrakul, M. and Tansiriyakul, S. 1986. Influence of harvest maturity of mangoes on storage potential and ripe fruit mango. ASEAN Food Journal, 2: 99 – 103.

Prabha, T.N., Veena, B.A. and Patwardhan, M.V. 1987, Metabolism of organic acids in Modified Atmosphre stored Ripening Mangoes (*Mangifera indica* L). Journal of Food Science and Technology, 24(3): 111-115,

Pruthi, J.S. 1968. Advances in Standardization, Marketing, Packaging and Transportation of Horticultural Produce both for internal and external markets. Proceedings of International sympotium on Sub tropical and Tropical Horticulture: 147.

Purwoko, B.S., Kesmayanti, N., Susanta, S. and Nasution, M.Z. 1998. Effect of polyamines on quality changes in papaya and mango fruit. Acta Horticulturae, 464 – 510.

Ram, H.B., Singh, R.V., Singh, S.K. and Joshi, M.C. 1984. A note on the effect of ethrel and hot water dip treatment on the ripening and respiratory activities of mango variety 'Dashehari'. Journal of Food Science and Technology, 21: 92 – 94.

Rama Sarma, G.B., Rao, S.D. and Hakim, D.N.1946. Carotenoid pigments of Badami mango fruit. Biochemistry Journal, 40 : 657.

Reddy, Y.V. and Srivastava, G.C. 1998. Malic enzyme and cyanide – resistant respiration in mango during ripening. Indian Journal of Plant Physiology, 3(3): (NS): 185 – 187.

Reddy, Y.V and Srivastava, G.C. 1999. Regulation of cell wall softening enzymes during ripening of mango fruit. Indian Journal of Plant Physiology, 4(3) (NS) : 194 – 196.

Reddy, Y.V. and Srivastava, G.C. 1999a. Ethylene biosynthesis and respiration in mango fruits during ripening. Indian Journal of Plant Physiology, 4(1) (NS): 32-35.

Reddy, Y.V. and Srivastava, G.c. 1999b. Regulation of cell wall softening enzymes during ripening of mango fruits. Indian Journal of Plant Physiology, 4(3): 194-196,

Reddy Y.V. and Srivastava G.C, 2001. Ethylene biosynthesis and respiration during ripening in mango cultivars. Indian Journal of Plant Physiology, 6(4)(NS) : 361 – 364.

Reddy Y.V. and Srivastava, G.C. 2003. Superoxide dismutase and peroxidase activities in ripening mango (*Mangifera indica* L) fruits. Indian Journal of Plant Physiology, 8(2): 115 – 119.

Rodov, V., Ben-yehoshua S., Fishman, S., Gotlieb, S., Fireman, T. and Fang, D.Q. 1993. Reducing decay and extending shelf-life of bell peppers and mangoes by modified atmosphere packaging. Post-harvest handling of Tropical Fruits. Proceedings of International Conference, Chianghai, Thailand.

Roe, E. and Brummer, J.M, 1981. Changes in pectic substances and enzymes during ripening and storage of 'Keitt' mangoes. Journal of Food Science, 46: 186 – 189.

Rosa, B.N., Aharoni, D., Feygenberg, O., Aharmi, N., Keynan, A. and Pesis, E. 2001. Effect of modified atmoshphere packaging on mango ripening. Acta Horticulturae, 553 (2): 607 – 609.

Roy, B.N., Biswas, S., Saha, S.K. and Pramanick, S. 1980. Effect of hotwater with captan on the storage and quality of mango fruits (*Mangifera indica* L) Cv.Langra and Safdour Pasand. South Indian Horticulture, 28 (1): 5-7.

Roy, S., Shaha, A.K. and Chettopadhyaya, T.S. 1972. Storage trials on mango fruits. Proceedings of 3rd International Symposium on Tropical and Sub tropical Horticulture: 80.

Roy, S.K. and Joshi, G.D, 1988. An approach to integrated post-harvest handling of mango. Acta Horticulturae, 231 : 649 - 652.

Roy, S.K. and Pandey, R.M. 1983. Effect of existing method of harvesting on ripening and storage of mango (*Mangifera indica* L). Silver Jubilee Celebration Seminar on Plant Physiology, Session I (Abstract) 1-22

Sadana, J.C and Ahmad, B. 1949. Metabilosim of the carotenoid pigments of mango during the development of the fruit. Indian Journal of Medical Research, 37 : 193 –196.

Sadasivam, R., Muthuswamy, S., Sundara Raj, J.S. and Daniel, S.D. 1970. Experiments with waxing of mango fruits in common storage. Madras Agricultural Journal, 57 (11) : 694 – 699.

Sahni, C.K. and Khurdiya, D.S. 1989. Physico-chemical changes during ripening in 'Dashehari', 'Chausa', 'Neelum' and 'Amrapali' mango. Indian Food Packer, 43 (5): 5-10.

Sahni, C.K. and Khurdiya, D.S. 1992. Effect of calcium carbide and ethrel on ripening of 'Amrapali' mango. Proceedings of National Symposium on Maximising and Sustaining crop and Animal Productivity by Modern Techniques, Varanasi, 14 – 17 Oct. Journal of Nuclear Agriculture and Biology, 21(3): 1097 – (Abstract No.064).

Sahni, C.K., Khurdiya, D.S and Dalal, M.A.1994. Effect of ripening aids on the quality of Amrapali mango. Maharashtra Journal of Horticulture, 8(1): 72 -77.

Salunkhe, D.K. and Desai, B.B. 1984. Mango In: Post-harvest Biotechnology of Fruits. Vol.I CRS Press, Boca Raton, Florida, USA, pp 77-94.

Sampaio, V.R. 1981a. Comparasion of methods for ripening mangoes. Anais da Escola Superior de Agricultura "Luiz de Queiroz", 38 (1): 85 -91.

Sampaio, V.R. 1981b. Effect of 2-chloro ethyl phosphonic acid on mango ripening. Anais da Escola Superior de Agricultura "Luiz de Queiroz", 38 (1): 93-98.

Santos, E.C.D., Silva, S.D.M., Mendonca, R.M.N., Santos, A.F.D., Silveria, I.R.B.S. and Silva, L.P.D. 2004. Influence of 1-Methyl cyclopropene on ripening and conservation of tree–dropped mango fruit. Cv.Rosa. Acta Horticulturae, 645: 573 – 579.

Sarker, S. and Munshi, A.A. 1981. A study on the content and inter conversions of organic acids of mango (*Mangifera indica* L) at various stages of fruit development. Bangladesh Journal of Agricultural Science, 8: 69 -75.

Saucedo – Arias, LJ. 1997. Quoted by MA Gomez – Lim 1997 In: RE-Litz(ed). Mango.

Selveraj, Y. 1989. Studies on enzymes involved in the biogenesis of lipid derived volatiles in ripening mango (*Mangifera indica* L) fruit. Journal of Food Bio chemistry, 12: 289 – 299.

Selveraj, Y. and Kumar, R. 1989a. Studies on Fruit softening enzymes and poly phenol oxidase activity in ripening mango (*Mangifera indica* L) fruit. Journal of Food Science and Technology, 26 (4): 218 – 222.

Selveraj, Y. and Kumar, R. 1989b. Pyruvate kinase activity in ripening mango (*Mangifera indica* L) fruits. Journal of Food Science and Technology, 26 (4): 228 – 229.

Seleveraj, Y. and Kumar, R. 1994. Enzymatic regulation in ripening mango fruit. Indian Journal of Horticulture, 51: 316 - 323.

Selveraj, Y., Kumar, R. and Pal, D.K 1989. Changes in sugars, organic acids, aminoacids, lipid constituents and aroma characteristics of ripening mango (*Mangifera indica* L) fruit. Journal of Food Science and Technology, 26(6): 308 – 313.

Selverajah, S., Bauchot, A.D. and John, P. 2001. Internal browning in cold-stored pineapples is suppressed by a post-harvest application of 1-Methyl cycloprpoene. Post-harvest Biology and Technology, 23: 167 – 170.

Sergent, E., Scheffer, B., Lara, S.P. and Wills, L.E. 1993. Effect of ethephon on mango (*Mangifera indica* L) fruit quality. Acta Horticulturae, 341: 510 – 517.

Sethi, V. and Maini, S.B. 1989. Appropriate Technology for reducing post-harvest losses in Fruit and Vegetables. Indian Food Packer, 43 (2): 42 – 56.

Seymour, G.B., N'diaye, M., Wain wright, H. and Tucker, G.A.1990. Effects of cultivar and harvest maturity on ripening of mangoes during storage. Journal of Horticultural Science, 65 (4) : 479 – 483.

Seymour, G.B., Wain wright, H. and Tucker. G.A. 1989. Cell wall changes in ripening mangoes. Aspects of Applied Biology, 20: 93 – 94.

Shanmugavelu, K.G. and Selveraj, P. 1974. Effect of ethrel on ripening of mango. Madras Agricultural Journal, 61 (10/12): 983 – 987.

Shanmugavelu, K.G., Selveraj, P., Veeramah, L. and Chittrai Chalvan, R. 1976. Effect of ethephon on the ripening of fruits. Progressive Horticulture, 8 : 89 -96.

Shashirekha, M.S. and Patwardhan, M.V. 1976. Changes in amino acids, sugars and non volatile organic acids in a ripening mango fruit (*Mangifera indica* L Badami variety) Lebensmitted Wissen chaft Technologie, 9 : 369 – 370.

Silva, S de M., Santos, E.C dos; Santos, A.F dos., Silvejra, I.R.B.S., Mendonca, R.M,.N and Alves, R.E. 2004. Influence of 1-Methyl cycloprpoene on post-harvest conservation of exotic mango cultivars. Acta Horticulturae, 645 :563 – 572.

Singh B.P. and Narayana, C.K. 1998. Response of 'Green Keeper' on ripening and storage of 'Langra' mango (*Mangifera indica* L). National Seminar on New Horizons in Production and Post-harvest Management of Tropical and Sub-tropical Fruits, New Delhi, 8-9th December 1998 (Abstract), p. 52.

Singh, B.P., Tandon, D.K. and Kalra, S.K. 1993. Changes in post-harvest quality of mangoes affected by pre-harvest application of calcium salts. Scientia Horticulturae, 54(3): 211 – 219.

Singh, D.B. and Chundawat, B.S, 1991. Post-harvest treatments on ripening changes and quality of mango fruit Cvs."Kesar' and 'Amrapali'. Acta Horticulturae, 291 : 472 – 478.

Singh, K.K., Kapur, N.S. and Mathew, P.B. 1954. Further studies on cold storage of mangoes. Indian Journal of Agricultural Sciences, 24: 137 – 148.

Singh, S.K. and Mathur, P.B. 1952. Ripening of mangoes . Current Science, 21: 295.

Singh, U.R., Pandey, I.C., Tripathi, B.M. and Upadhyay, N.P. 1979. Artificial Fruit Ripening in mango. Punjab Horticultural Journal, 19 (1&2): 139 – 141.

Singh, V.K. and Mishra, M. 2003. Molecular physiology in Hi-tech Horticulture: A New Parradigm in Improvement of Fruit crops. Chapter 13. In:P.K, Jaiswal and Singh, R.P (ed) Plant Genetic Engineering. Vol.6 Improvement of Fruit Crops. Sci Tech Publishing . LLC. USA. pp 245-291.

Singh, Z. and James, J. 2001. Effects of post-harvest application of Ethephon on Fruit ripening, quality and shelf life of mango under modified atmosphere packaging. Acta Horticulturae, 553 (2): 599 – 603.

Singh, Z., James, J. and Nair, S. 2001. Packaging materials affect physiological weight loss, fruit colour and quality of mango during storage. Acta Horticulturae, 553 (2) – 603 –606.

Singh Z., James, J. and Tom, S.C. 2000b. Effects of different surfactants on calcium uptake and its effects on fruit ripening, quality and post-harvest storage of mango under modified atmosphere packaging. Acta Horticulturae, 509: 413 – 417.

Sisler, E.C. and Serek, M.1997. Inhibitors of ethylene responses in plants at receptor level: Recent developments. Physiologia Plantarium, 100: 577 – 582.

Sisler, E.C., Serek, M. and Dupille, E. 1996. Comparision of cyclopropene, 1-methyl cyclopropene and 3,3-dimethyl cyclopropene as ethylene antagonists in plants. Plant Growth Regulation, 10: 169 – 174.

Smith, S.M. and Stow, J.R. 1984. The potential of a sucrose ester-coating material for improving the storage and shelf life qualities of 'Cox's Ornage Pippin' apples. Annals of Applied Biology, 104 : 383 – 391.

Smith, T.A. 1985. Polyamines. Annual Review of Plant Physiology, 36: 117 – 143.

Sornsrivichai, J., Anusadorn, P., Oogaki, C. and Gemma, H. 1989. Storage life and quality of mango (*Mangifera indica* L cv.Keaw Sawaey) fruits stored in seal packaging by plastic films and under low pressure at different temperatures. Japanese Journal of Tropical Agriculture, 339(1): 6-17.

Sornsrivichai, J,, Gomolmaused, D., Boonyakiat, U.J., Boon-long, P. and Gemma, H. 1992. Seal packaging by plastic film as a technique for limiting fungal decay of mangoes. Acta Horticulturae, 296 : 23-32.

Spalding, D.H. and Reeder, W.F.1986. Decay and acceptability of mangoes treated with combination of hot water, imazalil and Y-radiation. Plant Disease, 70: 1149 – 1150.

Spalding, D.K. and Reeder, W.F.1977. Low Pressure (Hypobaric) storage of mangoes. Journal of American Society for Horticultural Science, 102: 367 – 369.

Spalding, D.H. and von Windeguth, D.L. 1988. Quality and decay of irradiated mangoes. Hort Science, 23 (1): 187 – 189.

Srinivasan, A., Thomas, P. and Dharkar, S.D. 1971. Physiological effects of gamma radiation on some tropical fruits. In: Disinfestation of fruit by irradiation. Proceedings of a Panel on the use of Irradiation to solve Quarantine problems in the International Fruit Trade, Hawai, December, 1970. International Atomic Energy Agency, Vienna: 113 – 124.

Srivastava, H.C. 1967. Grading, storage and marketing. The Mango : a Hand Book. Chapter 5. Indian Council of Agricultural Research, New Delhi.

Srivastava, L.N. and Mukherjee, P.K. 1961. Ripening of mango variety Langra at different stages of maturity under different conditions. Annual Report of Horticultural Research Station, Sahranpur, 81 -87.

Srivastava, R.P. 1961. Studies on artificial ripening of mango (*Mangifera indica* L) Allahabad Farmer, 35 (2): 1 -65.

Stead, D.E. and Chithambo, G.S.G. 1980. Studies on the storage of tropical fruits in polyethylene bags. Luso: Journal of Science and Technology (Malawi): 3-9.

Subramanyam, H., Gouri, S. and Krishnamurthy, S.1976. Ripening Behaviour of Mango Fruits graded on specific gravity basis. Journal of Science and Food Technology, 13: 84-86.

Subramanyam, H., Krishnamurthy, S., Subhadra, N.H., Dalal, V.B., Randhawa, G.S. and Chacko, E.K. 1972. Studies on low temperature breakdown in mangoes. Annual report. Central Food Technological Research Institute, Mysore, 1967 – 1972.

Subramanyam, H. and Moorthy, N.V.N. 1973. Control of spoilage and ripening in mango fruit by Zineb and sodium diethyl di thiocarbomate, Pesticide Science, 4: 25 -31.

Subramanyam, H., Moorthy, N.V.N., Dalal, V.B. and Srivastava H.C.1962. Effect of a fungicidal wax coating with or without growth regulator on the storage behaviour of mangoes. Food Science (Mysore), 11: 236 – 239.

Subramanyam, H., Moorthy, N.V.N., Subhadra, N.V. and Muthu, M. 1969. Control of spoilage and inhibition of ripening in Alphonso mangoes by Fumigation. Tropical Science, 11: 120 – 125.

Subramanyam, H. and Sebastion, K. 1970. Effect of succinic acid, 2,2 – dimethyl hydrazide on carotone development in Alphonso mangoes. HortScience, 5: 160.

Tahir, M.A. and Malik, A.A. 1977. The activity of pectic enzymes (Pectin esterase, polygalacturonase) in various mango products. Pakistan Journal of Scientific Research, 29: 71 -74.

Tandon, D.K. and Kalra, S.K. 1983. Changes in sugar, starch amylase activity during development of mango fruit Cv.Dasheri. Journal of Horticultural Science, 58: 449-452

Tandon, D.K. and Kalra, S.K. 1997. Ripening of early harvested Dashehari mangoes with ethrel along with polyethylene packaging. Indian Food Packer, 51 (4): 11 – 12.

Tandon, D.K. and Kalra, S.K, 1995. Effect of post-harvest ethrel treatment on ripening of early season mango fruits. Proceedings of the National Seminar on Post-harvest Technology of Fruits, held at UAS, Bangalore. India. 7-9 August 1995 pp. 249 – 253.

Tandon, D.K. and Kalra, S.K. 1998. Ripening of Dashehari mangoes of different maturities with ethrel and hot water at Qurantine temperature. National symposium on Mango Production and Export. June 25 -27, 1998. Lucknow, India (Abstract), pp.81.

Tandon, D.K. and Kalra, S.K. 2001. Ripening early harvested mangoes with ethrel. Indian Horticutlure, 46(2): 32 – 33.

Teaolia, S.S., Singh, R.D. and Maurya, V.N. 1968, Studies on maturity standards for *Mangifera* indica L. vari Lavgra. Indian Journal of Horticulturae, 25(1-2): 24-30.

Thomas, P. 1975. Effect of post-harvest temperature on quality, Carotenoids and Ascorbic acid content of Alphonso mangoes on ripening. Journal of Food Science, 40 (4): 704 – 706.

Thomas, P. 1993. Irradiation of Fruits, In: K.L Chadha and O.P. Pareek (Eds). Advances in Horticulture Vol 4. Fruit Crops. Part 4: 1811 – 1833. Malhotra Publishing House, New Delhi.

Thompson, A.K. 1971. The storage of mango fruits. Tripical Agriculture (Trinidad) 48 : 63 – 70.

Tirmazi, S.I.H. and Wills, R.B.H, 1981. Retardation of ripening of mangoes by post-harvest application of calcium. Tropical Agriculture (Trinided), 58 : 137 – 141.

Tucker, G.A. and Grierson, D.1987. Fruit ripening. In : The biochemistry of plants, A Comphrehensive Treatise (ed) D.D. Danies .Vol. 12 Academic Press, 265 – 318.

Tucker, G.A. and Seymour, G.B. 1991. Cell wall degradation during mango fruit ripening. Acta Horticulturae, 292. 454 – 460.

Ulrich, R. 1970. Organic acids. In: The Biochemistry of Fruits and Their Products (ed) A.C.Hulme, Acadmic Press, London. pp.269 - 304.

Upadhyay, N.P. and Tripathi, B.M. 1985. Post-harvest changes during storage and ripening of Gaurjeet mango (*Mangifera indica* L) fruits. Progressive Horticulture, 17(1): 25 -27.

Urbain, W.M. 1986. Food Irradiation, Academic Press, New York.

Valmayor, R.V. 1972. The Philippine mango Industry – its problems and progress. Acta Horticulturae, 24 : 19-23.

Valverd, E., Hernandez, R.L. and Saenz, M.V. 1986. Effect of ethephon and calcium carbide on ripening and quality of mango fruits. Agronomia Costarricease, 10 (1-2): 153 – 163.

Vanlely veld, L.J. and Smith, J.H.E. 1979. Physiological factors in the maturation and ripening of mango (*Mangifera indica* L) fruit in relation to the Jelly – seed, a physiological disorder. Journal of Horticultural Science, 54: 282 – 287.

Vazquaz – Salinas, C. and Lakshminaryana, S. 1985. Compositional changes in mango fruit during ripening at different storage temperatures. Journal of Food Science, 50 (6): 1646 – 1648.

Veloz, C.S, Torres, F.E. and Lakshminarayana, S. 1977. Effect of refrigerated temperatures on the incidence of chilling injury and ripening quality of mango fruit. Proceedings of Florida State Horticultural Society, 90: 205 – 210.

Verma, R.A., Tripathi, M.P. and Srivastava, R.K. 1986. Studies on development of carotenoids during ripening of mangoes (cv.Dashehari). Progressive Horticulture, 18 (2): 39 -44.

Wang, T.T. and Shiesh, C.C. 1990. Fruit growth development and maturity indices of "Irwin" mango in Taiwan. Acta Horticulturae, 269: 189 – 196.

Wangchai, K., Gemma, H., Iwahori, S and Uthaibutra, J. 2000. Endogenous polyamines in 'Nam Dok Mai' mangoes with different ripening stages and its relation to chilling injury during storage. Acta Horticulturae, 509: 429 – 436.

Wardlaw, C.W. and Leonard, E.R, 1936. The storage of West Indian mangoes, Low Temperature Research Station, Imperial College of Tropical Agriculture, Trinidad, P. 1-47.

Wills, R.B.H., Tirmazi, S.I.H. and Scott, KJ, 1977. The use of calcium to delay ripening of tomatoes. HortScience, 12: 551 – 552.

Yang, S.R. and Hoffman, N.E. 1984. Ethylene biosynthesis and its regulation in higher plants, Annual Review of Plant Physiology, 35: 155 – 189.

Yantarasri, T., Ben-yehoshua, S., Rodov, V., Kumpuan, W., Uthaibutra, J and Sornsrivichai, J. 1995. Development of perforated modified atmosphere package for mango. Acta Horticulturae, 398: 81-91.

Yantarasri, T., Uthaibutra, J., Sornasrivichai, J., Kumpaun, W., Sardsud, V. and Kane-chum, N.1993. Modified atmosphere packaging by perforated polymeric film and its effect on physical properties of mango fruit. Postharvest Handling of Tropical Fruits. Proceedings of International Conference, Chaing Mai, Thailand.

Yoneya, T., Nip, W.K., Wei, P.S. and Cai, T. 1990. Physico-chemical parameters of post-harvest ripened mangoes from Hawaii. Acta Horticulturae, 269 : 291 – 297.

Yuen, C.M.C., Tom, S.C., Joyce, D. and Chettri, P. 1993. Effect of post-harvest calcium and polymeric films on ripening and peel injury in Kensington Pride mangoes. ASEAN Food Journal, 8(3): 110 -113.

Yuvaraj, K.M. and Ughreja, P.P. 1998. Effect of post-harvest treatments on ripening changes and storage life of mango (*Mangifera indica* L) fruits Cv. Kesar. National Symposium on Mango. Lucknow. 25 – 27 June, 1998 (abstract) 80p.

Zambrano, J. and Manzome, J. 1995. Effect of post-harvest calcium application on mango ripening and storage. Fruits, Paris, 50(2): 145 – 152.

Zeng, Y., Pandey, M ., Prasad, N.K and Srivastava, G.C. 1995. Ripening associated changes in enzymes and respiratory activities in three varieties of mango (*Mangifera indica* L). Indian Journal of Plant Physiology, 38 (1): 73 -76.

Zhou, Y.C., Tang, Y.L. and Tan, X.J. 1994. Changes of major carotenoid contents during ripening of mango (*Mangifera indica* L) cv. Zihna. Journal of Tropical and Sub-tropical Botany, 2(2): 77 – 79.

Zhou, Y.C., Tang, Y.L., Tom, X.L. and Guo, J.Y. 1996. Effects of exogenous ABA, GA_3 and cell wall degrading enzyme activity. Acta phytophysiligica Sinica, 22(4): 421 – 426.

CHAPTER 5

Preservation and Processing

R. Srihari Babu

Mango fruit is largely consumed as fresh fruit and is, therefore, less processed. Noverthless one thing is certain that mango is available in more processed form than any other tropical fruit. Mango is seasonal and not grown in all places of the world. But one likes to have mango throughout the year, which is rather difficult. In order to make this delicious fruit available throughout season and in all places, they are sent from production centres to consumer centres. However, because of its high perishability, short storage life and susceptibility to chilling injury it can not be made available in all places throughout the year as fresh fruit. But can be made available in preserved/processed products, because of their extended shelf life, lack of fear of quarantine pests, diseases and pesticide residues, throughout the year and in all places.

Due to the expansion of area under mango in several countries, the production increases many folds in the future. Even allowing for sizeable post-harvest losses, an increased demand (due to population increase) from domestic as well as export markets an over-supply of produce is anticipated in the future. To prevent this, export of fresh fruit is suggested. However, in near future export markets may not absorb all the production and some countries may become self-sufficient in production and may not require imports. One of the remedies suggested to such a situation is to process the mangoes into different value added products. The change in the life style and food habits also demand ready to serve/eat fruit products in several countries including India.

Processing is the best way of utilizing surplus production, particularly during the period of seasonal glut. It is an attractive alternate for export markets because it over-comes the quarantine barrier in the importing of fresh fruit due to fear of introduction of exotic and unwanted pests and disease and pesticide residues in the fruit. Preservation also enables economic utilization of surplus production at times, which otherwise may go waste and thus the level of consumption of these protective fruit is enhanced. Processing can also utilize fruit that are unsuitable for the fresh fruit market, because of skin blemishes, wind scaring, sapburn or mild skin diseases, as well as oversize, undersize or misshaped fruits. Even dropped fruits can also be utilized. Processing can also help in utilization of waste by making pectin from peel, starch and oil from seed kernel. (See chapter: Utilization of mango waste material).

Processing also directly or indirectly provides employment for several million people, reduces post-harvest losses, helps in establishment of subsidiary industries for storage, package and transportation and assures better utilization of resources.

Mango is an unique and a versatile fruit crop in the sense that at every stage of growth and development one or the other product(s) can be processed, which is unparalled by any other fruit. As said above, not only sound and harvested fruits are processed, but also dropped fruits can be utilized profitabely for processing. Mangoes are processed at both raw/(green-unripe) and ripe stages for conversion into a wide range of products as shown below:

(i) **Raw/green-unripe fruit:** Pickles, chutney, slices in brine, dehydrated slices, dried powder (amchur), dried mango flakes, raw mango beverage etc.

(ii) **Ripe fruits:** Slices in syrup, pulp (puree), juice, squash, nectar, RTS beverages, syrup, jam, fruit leather, concentrate, strained baby foods, toffies, cereal flakes, wine, frozen products (slices, puree, syrup, whole fruit) blends etc.

GREEN MANGO PRODUCTS

Preservation and processing of raw, green-unripe mangoes is an old practice in mango growing countries. The raw/unripe green mango fruits having hard pulpy matter, with developed stones, because of their acidic taste are used for making several traditional food items (pickles, chutneys, amachur, etc). Instant mango pickles, drum-dried mango powder and raw mango beverage base are the recent developments. The following are some of the important products processed and preserved from green unripe mangoes.

Green mango jam

Fruit jams are generally prepared from ripe fruits. However, it is possible to prepare jam with green mangoes also. The leftover pulpy portion remained on the cloth, after the preparation of green mango beverage can be used profitably in the preparation of green mango jam, as its bye-product, which will further lower down the production cost. Essentially the preparation of green mango jam is same as the normal preparation of jam from ripe fruits with the following recipe (Mallik, 1976).

Recipe for green mango jam

Green mango pulp	-	2.5 kg
Sugar	-	2.5 kg
Water	-	1.5 litres
Citric acid	-	30 g

Raw mango slices

Raw mango slices are traditional mango products since time immemorial. Raw green-unripe mangoes are preserved in the form of slices in brine/dry salt for later use in the preparation of pickles, chutneys and amchur.

Fibrous varieties free from terpene or paraffin off-flavour are ideally suited for preserving slices in brine, since table/grafted varieties free from fibre become soft and meshy during storage (Hussain, 2004). All those varieties which are suitable for making pickles, chutney and amechur are also suitable for making raw mango slices. The raw material for making mango slices constitutes fresh unripe green but fully developed fruits.

Methods of preparation

Raw mango slices are made and preserved either by dry salting or in brine. Conventionally green/unripe mango slices are preserved by adding 15-20% salt, draining the liquid formed therein and replacing it with fresh salt. A few methods of preparation of raw mango slices were suggested as follows:

Anand and Johar (1958a) suggested a method of steeping the freshly cut slices in equal volume of 10% salt solution (brine) to which 200 ppm sulphurdioxide (SO_2) is added. After 20 hours at 25 °C the slices are drained out and replaced with 5% salt solution containing 200 ppm SO_2 in a barrel and closed with a lid. For complete filling of the empty space in the barrel, 5% salt solution with 200 ppm SO_2 is added through the hung hole in the barrel and the hung hole is fixed tightly.

In the method, suggested by Daisy (1966), the peeled fruits after washing are cut into small pieces (slices) of different thickness and length (3.5 cm × 0.3-0.6 cm thick), common salt is added @ 4.4 kg per 47.7 kg of slices. The salted slices are packed in wooden barrels containing little water and the product is left to stand until the slices are saturated with the resultant brine (6-8 days). The surplus brine is then drained off and the slices are packed into wooden barrels and stored until their use later.

Bhatnagar and Subramanyam (1973) gave another improved method. In this method slices are kept in 10% brine containing 200 ppm SO_2 to which black mustard powder (0.5%) in 10% brine or sodium benzoate (0.1%) in 20% salt solution for longer storage of slices is added.

Dry salting (15%) results in greater softening and browning of slices. Steeping the slices in 5% salt, 1-2% acetic acid and 0.1% calcium chloride solution was suggested to overcome this problem (Sethi, 1991). Slices are cured by dry salting until tissue salt concentration reached more than 10-20% in the method suggested by Daisy (1966). The cured slices are subjected to lactic acid fermentation at 28-30 °C for 12 days. Slices containing 13-15% salt and fermented for 120 days give most desirable product with high lactic acid and total acid content. Dry salt preservation of raw mango slices results in change of colour and softening, leading to textural loss and microbiological spoilage, which can be overcome by a method, which ensures good retention of colour, texture, overall quality and reduction in spoilage during storage under ambient conditions.

The presence of SO_2 in the dried product is essential to preserve natural colour, flavour and palatability of the product and to prevent nutritional losses like ascorbic acid during drying and storage. To get maximum retention of SO_2 in raw mango slices, sulphitation should be done for 30 minutes in less than 3% potassium metabisulphite (KMS) concentration, when it is sundried and in less than 2% KMS for cabinet drying (Khedkar and Roy, 1983).

Addition of 0.2% KMS without salt gives a good constitution of dehydrated raw mango slices, but its colour and texture would be inferior in comparision to slices treated with both KMS and salt (Mehta and Tomar, 1982).

Acceptability of the dried and dehydrated slices is dependent on its reconstitution property, which is primarily determined by the cellular structure of the slices. Shrinkage of cut surface of slice occurs during dehydration and leads to cell wall repture. The rapidity with which water is removed from the material determines its textural quality. For getting an acceptable texture, water must be removed at a faster rate (Matz, 1962).

Raw mango slices are dried either in sun or cabinet drier. From time immemorial green mango slices have been used after sundrying. It takes about 16 hours exposure to sun at 35-40 °C for complete drying of the slices (Teaotia *et al* 1987). Dhabade and Khedkar (1980d), compared sundrying with cabinet drying of green mango slices and observed that the rate of drying was much more faster in cabinet drying than drying in sun. In cabinet drying it took 10 hours to dry the slices, while 15 hours were required in case of sun drying.

Packaging and storage

Wooden barrels were earlier used for packing and storing of raw mango slices. Wooden barrels besides being very expensive, heavy and occupying large space, develop leakage during storage and transport. Therefore, it is desirable to pack the drained raw mango slices in appropriate plastic films for their transport to save the cost of packing material, packing and transport.

Methods in vogue for preservation (storage) of green mango slices comprise of keeping these slices in layers in vats intercepted by varying quantities of powdered salt. The slices thus preserved become shrivelled, losing all the natural characteristics like colour, flavour and texture during storage. Sometimes due to evaporation of water from the vats, the entire mass of slices forms a hard stone like mass, which is difficult to crush (Johar and Anand, 1951). The usual practice prevalent in the processing factories to preserve green mango slices for use in off season, is to store them in wooden barrels with powdered salt. Slices stored thus in brine showed highest acidity on 6th day of storage. Subsequent use of these preserved slices depends on the concentration of salt in the slices and development of microbes. The slices in 40° salo-meter brine were soft after 2 months storage and were not fit for use in the preparation of chutney. Slices in 60° and 80° salo-meter brines showed fairly constant content of acidity with least bacterial counts. Slices in 60° salo-meter brine were crisp and retained the natural colour even after 6 months storage. By this method, it is possible to retain better texture, colour and flavour of mango slices in storage. However, storing of raw mango slices in lower concentration of salt facilitates desalting before conversion into chutney and also helps to save large quantities of common salt (Johar and Anand, 1951).

In order to store the raw mango slices for a longer time (4-5 months), Srivastava (1967) suggested smearing of the slices with salt (6.8%) till the salt concentration reaches 15-20% and development of aerobic yeast followed by draining of brine and packing them in wooden barrels, adding further quantity of salt. An improved method consists of steeping the slices in 10% brine solution containing 200 ppm SO_2 for 20 hours at 25 °C for primary salting and 5% powdered salt with 200 ppm SO_2 for storage. Raw mango slices without either KMS or salt treatment turned black in colour within a month of storage, while slices treated with salt (2.5%) alone became brown after 3 months storage. Raw mango slices treated with 5% salt and 0.1% KMS gave best results on dehydration and received highest marks on reconstitution, on conversion into products such as pickles, chutney etc. followed by treatments with 2.5% salt and 0.1% KMS after a year of storage (Mehta and Tomar, 1982). Slices dip-treated in 1.5% KMS solution and common salt were preserved for over 2 months in polyethylene pouches (Kalra and Tandon, 1983).

AMCHUR

Amchur (amchoor) is a dried mango fruit powder and is a commercial mango product of North India. It is prepared by grinding dried green mango slices. Of late, certain Telangana districts of Andhra Pradesh are also manufacturing amchur from unripe green mango fruits of seedling trees.

Amchur is a highly nutritious product as it contains all the composition of raw fruits, but has little calorific value. The powder is a rich source of carbohydrates, acids, minerals and vitamin-C. Mango powder is the best option to maintain overall fruit contents without loosing its nutritive value, natural aroma and rich vitamin content.

Uses: Amchur is mainly used as a souring agent, as a substitute for tamarind for flavouring vegetable curries. It finds use as a condiment for acidifying and flavouring a variety of culinary preparations. Sundried slices of raw mango slices (Khatai) or the ground powder (amchur) are extensively used in North Indian culinary preparations to impart characteristic sour taste and aroma. It is also an essential ingredient of many spice blends, curry powders and chat masalas. Commercially, it is employed in the manufacture of pickles, chutneys and sauces. It is also used for seasoning of ready-to-eat (RTE) processed snacks like potato chips, chudwa mix, beverages etc (Kesarwani *et al* 2000). The curry prepared from amchur had better overall organoleptic score compared to the curry from tamarind and cocum (Dabhade and Khedkar, 1980e).

Suitable varieties: Acedic varieties with higher recovery of slices and less tendency towards browning are ideal ones for making amchur. Mostly seedling types are used and proved better than commercial table varieties such as Dashehari, Bombay Green, Langre etc. Good quality amchur was prepared at Hissar from 'Ramkela' variety of mango (Pruthi and Teaotia, 1987).

Raw material

Both fresh green un-ripe mangoes and fallen fruits constitute the raw material. From time immemorial mango fruit drops are being sundried as peeled slices for use as such or after powdering. However, all fruits that drop at various stages are not useful for making amchur. Drops of post-setting are small in size and weight, have high phenolics and low acid content and so unsuitable. Fruits dropped in the 22nd metereological week and 2-3 weeks before harvest, because of size and weight are suitable for the preparation of amchur. Physico-chemical studies of dropped fruits clearly revealed the constant trend in size, weight etc. after 7th week onwards indicating their suitability for making amchur (Dabhade and Khedkar, 1980b). However, fruits at 9-10th week after endocarp hardening were most suitable for preparing amchur as they had high acidity and starch and low sugar and phenolics. Availability of higher content of pulp from the fruit is another consideration for selection of correct stage of development for making amchur (Dabhade and Khedkar, 1980c).

Methods of preparation

Amchur is prepared either from dried slices (from sound as well as fallen fruits) or from dehydrated flakes. Harvested or droped unripe mango fruits are first washed, peeled and cut into pieces of uniform size. These slices are placed in a white muslin cloth, dipped in boiling water for 2 minutes, removed and cooled in cold water. Then the slices are treated with KMS solution (1%), twice the weight of slices by steeping to reduce the moisture content to 3% or less. When dried in a cabinet drier at 55 °C $\pm$ 5 °C, 10 hours and in sun 15 hours time is required. The dried slices (sun-dried or otherwise) are ground to get amchur in powder form (Nigam *et al* 2007).

Green mango powder is prepared from flakes in an atmospheric double drum drier. This can form the base material for preparation of green mango drink and mango chutney (Gangopadhyay *et al* 1976).

Amchur is generally prepared during summer (April-May) when raw mango slices can be sun-dried in open (being cheap) and powdered, before the onset of monsoon, otherwise the product gets spoiled due to high humidity during rains, coupled with hygroscopic nature of the slices. In the absence of sun or during other periods and also to get fine quality powder, the slices are dried using different devices viz., solar drier, cabinet drier and double drum drier. Drying in solar drier is slower than in cabinet drier. Drying should be faster for rapid and economical drying (Dabhade and Khedkar, 1980a). However, in spite of slow drying, slices dried in a solar drier will be of high quality in respect of colour, flavour and keeping quality as compared to sun-drying (Nigam *et al* 2007). Use of blanching, sulphering and mechanical dehydration gives a product with better colour, nutrition, storability and less microbiological problems (Singh, 2007). Blanched slices dry earlier than unblanched ones in cabinet as well as in sundrying. However, rate of drying is little faster in un-blanched pieces, probably due to difference in the initial moisture content of the pieces subjected to drying. Faster rate of drying in blanched samples in cabinet drier as well as in sun drying may be due to the opening of tissues permitting, easy and fast removal of moisture from the pieces (Dabhade and Khedkar, 1980a).

Though pre-treatments like blanching and sulphitation cause losses (by leaching) of sugars, acids and ascorbic acid during preparation of amchur, they are essential steps in its preparation. Blanching helps to inactivate the enzymes, thereby protecting the product from enzyme catalyzed reaction and sulphitation helps to retard the oxidation reaction like non-enzymatic browning and better retention of ascorbic acid during storage (Dabhade and Khedkar, 1980c).

Storage and Packaging

Amchur is hygroscopic, hence its proper storage is very essential. It could be kept for about 6 months at room temperature (12-25 °C) without significant change in quality. It is preferable to store amchur in a cool, dark and dry places (RH < 70%). The optimum RH for storage of amchur is 40-43%. Mould growth occurs at 95% humidity with in 8 days of storage.

The powder can be packed in open top sanitary cans or foil laminate pouches. When packed in glass jars and in polyethylene bags, slight darkening of slices takes place with in 3 months (Teaotia *et al* 1987).

Quality

Great variation exists in the quality of amchur between market samples and authentic samples and also among different methods of preparation and varieties as well as pre-treatments. The mango powder prepared from 'Totapuri' and seedling mangoes with blanching and sulphitation had a better overall organoleptic score compared to other treatments and market samples. Market samples of amchur were free from sodium chloride, ascorbic acid or SO_2 and showed wide variation in water and alcohol insoluble material and non-enzymatic browning and tannin content and two types of insects and three strains of *Aspergillus* (Usha and Anand, 1981). Teaotia *et al* (1987) reported wide variation in the quality of market samples, that they were rich in sodium chloride, crude fibre, reducing and total sugars and lower acidity than the authentic lab samples.

An evaluation of quality of sundried mango slices (Khetai) and ordinary amchur, showed that Agmark brand and the control samples were superior in quality compared to the locally packed brand with relatively lower levels of moisture, total ash, acid, insoluble ash and higher non-volatile either extract. The Agmark brand received higher aroma, texture, appearance and acceptability scores. Control samples showed lower scores for texture, which were reflected in poor acceptability, high values for acid-insoluble ash. The locally packed brand samples showed a trend towards adulteration with substances like sand/grit/stones. Sand/grit/stones are common adulterants of mango powder (Trivedi *et al* 1990). Association of 3 genera of fungi within Agmark brand (*Aspergillus* sp., *Fusarium* sp and *Pennicllium* sp.) and 5 genera of mycoflora (*Aspergillus* sp., *Fusarium* sp., *Pennicillium* sp., *Rhizopus* sp. and *Sclerotium* sp.) in control samples were detected (Kesarwani *et al* 2000).

The differences in the levels of the chemical constituents such as moisture, proteins, and total ether extractants in the different samples observed were attributed to differences in variety, maturity, harvesting, seasonal variation, drying or packaging condition. Extreme variation in the quality of market samples of amchur and their poor hygienic standards reflect on the faulty methods adopted, for its manufacture. The market samples contained dust and sand, different types of fungi and insects. To avoid this, slices should be dipped in potassium metabisulphite (1%) for 1-2 minutes, before drying them in sun on raised plotfarms, preferably in polyethylene enclosures to avoid contact with dust and insects. Storage of dry slices (meant for amchur) in polyethylene lined gunny bags and evaluation of suitable packing of finished product, can improve the quality of the product and go a long way in boosting the production of amchur (Maini and Anand, 1985).

Chutney

Chutneys and pickles are important food products prepared both in homes as well as commercially. Chutney is a typical product of India. Mango chutney or sauce is a kind of spicy jam and a product prepared from peeled, sliced or grated unripe or semi-ripe mangoes by cooking them with salt, sugar, spices and vineagar to a thick jam consistency. They are appetizers. Mango chutney is an important food product exported from India.

Suitable varieties

Fibrous varieties are not considered suitable for making chutneys. Seedling varieties are more preferred as they contain more acidity. However, 'Totapuri' and 'Fazli' cultivars are ideally suited for the preparation of chutney in India.

Raw materials

Unripe/semi-ripe mangoes are utilized for its preparation. Chutney prepared from mature, but unripe fruits has good colour and flavour. Fresh as well as brined and peeled mango slices are used for its preparation. Both types of raw material are used by the industry. Frozen and cold stored (–18 °C) gratings from unripe mangoes with addition of salt (2%) can also be used (Singh, 1970).

Types of chutneys

Broadly there are two types of mango chutneys viz., (i) sweet chutney and (ii) hot chutney. However, the preparation of both is almost similar, except that more spices and less sucrose are used in hot chutney.

Methods and recipes

The method of preparation of chutney is similar to that of jam except that spices, vineagar and salt are added. A number of recipes were suggested for both sweet and hot chutneys.

Sweet mango chutney

This is made either from fresh or brined slices. The following recipes have been recommended for making sweet mango chutney (Table 5.1).

Table 5.1 Recipes for sweet mango chutney

Ingredient	Lal Singh and Girdharilal (1998)	Nigam *et al.* (2007)
Peeled mango slices	0.9 kg	1 kg
Sugar	0.9 kg	1 kg
Salt	56.6 kg	45 g
Red chilli powder	14.0 g	10 g
Mixed spices (Cardamom, Cinnamon, Cumin etc. in equal quantities)	28.3 g	10 g each
Chopped garlic	6.0 g	15 g
Chopped onion	28.3 g	50 g
Green ginger chopped	113.4 g	15 g
Clove		5 Nos
Vinegar	113.4 ml	70 ml

The fresh fruits are cut into slices and cooked in water until they become sufficiently soft. The slices are mixed with sugar and salt and cooked. The quality of a chutney depends on its cooking, which should be done for a long time at a temperature below the boiling point. Spices are coarse ground and tied in a cloth bag and placed in the slices in a boiling pan at the start of cooking. Cooking is done to the consistency of jam. To ensure proper thickening, cooking is done over moderate heat without a lid. Chopped onion and garlic are added at the start to mellow their strong flavour. Vinegar is added just before the final stage of cooking. At the final stage, the spice bag is removed and the resultant product is packed into pre-sterilized bottles and sealed air-tight. In the preparation of sweet chutney sugar is added in 3 spells in the form of a thick syrup of 75 °C brix on successive days, which avoids shrivelling of the slices. Raisins, carrots, almonds and dates can also be added depending upon preference (Nanjundaswamy, 1997).

Mango chutney from dehydrated slices

Mehta and Tomar (1982) suggested the following method of preparation of mango chutney from dehydrated slices. Dehydrated mango slices are dipped in one litre of boiling water which swells to one kilogram in weight in 2 hours. These slices are further softened in about 250 cc of water by boiling for 5 minutes and the ingredients are added as per the recipe and cooked to jam consistency. The mixture of slices, sugar and salt are allowed to sweat so that the sugar and salt dissolve in the juice exuding from the slices. Spice bag is dipped in the mixture and heating is continued with occasional stirring till the materials thicken sufficiently. After removing the spice bag, vineagar is added and heating is continued till a product of the jam consistency is obtained. After a slight cooling, the product is filled into clean, sterilized, wide mouthed glass bottles. After further cooling, the bottles are sealed air-tight.

Mango chutney prepared with dried/dehydrated slices had better overall organoleptic rating compared to chutney prepared from conventionally brined slices.

Receipe

S.No.	Ingredient	Quantity
1.	Dehydrated slices	400 g
2.	Sugar	1 kg
3.	Salt	30 g
4.	Red chilli powder	25 g
5.	Spice powder	30 g
6.	Water cooked onion, chopped	25 g
7.	Softened ginger slices in boiling water or pressure cooker	25 g
8.	Glacial acetic acid (Vinegar)	20 ml

Instant mango chutney

Chutneys prepared using conventional methods show browning, discolouration, loss of viscocity, which can be avoided by certain pre-treatments of mango slices. Ejilearassane *et al* (2001) standardized a receipe for preparing instant mango chutney mix from 'Neelum' and 'Desi' variety of mango with a pre-treatment of dipping the slices in 5% salt plus 1.5% potassium metabisulphite (KMS) for 30 minutes prior to drying for powder preparation, which is utilized for making instant chutney by mixing mango powder along with spices and condiments (in powder form), sugar, salt etc. Dehydration of the chutney was better in hot water (50 °C) as compared to ordinary water (26 °C) due to quick softening of powder, which led to better absorption of water, resulting in higher rehydration ratio. The instant chutney prepared from pre-treated mango powder scored highest when it was rehydrated in hot water due to the presence of SO_2, which exhibited protective effect on retention of quality parameters during drying. The loss in ascorbic acid and SO_2 was minimum, and non-enzymatic browning was less in pre-treated samples than controls.

PICKLES

Pickle is one of the important items of food all over the world in general and is an important part of Indian menu particularly. The preservation of food in common salt or in vineagar is pickling. It is one of the most anicient methods of preserving fruits and vegetables. Pickles are good appetizers and add to the palatability of a meal. They stimulate the flow of gastric juice and thus help in digestion. Pickles give life to otherwise insipid food and perhaps a meal is not considered complete without a touch of some kind of pickle to it (Shinde *et al* 2004).

Pickles have been a product of household since ancient times, but now have become one of the major item of food industry. Pickles occupy an important place in our traditional exports. Pickles from mango and lime are claimed to be national pickles.

Kinds/types of pickles

Different kinds of pickles are made from raw mangoes like (i) oil pickles, (ii) oil less pickles, (iii) pickles in vinegar, (iv) pickles in citrus juice or brine, (v) sweet mango pickles. Of the above, pickle in the oil is the most common one. However, in general, Indian pickles are classified as salt pickles or oil pickles.

Pickling is the result of fermentation of lactic acid bacteria, which are generally present in large numbers on the surface of fresh vegetables and fruits. These bacteria can grow in acid medium and in the presence of 8-10% salt solution, while the growth of a majority of undesirable organisms is inhibited. Lactic acid bacteria are most active at 30 °C, so this temperature must be maintained in the early stages of pickle making.

Suitable varieties

In India, mango pickle is prepared from the local seedling mangoes, which differ widely in several aspects as the mango is a highly heterozygous type. Seedling types are preferred for making pickles and chutneys because of their higher acidic nature and low sugars. Though both fibrous and non-fibrous types are utilized, preference is more for fibrous types. These characteristics help in retaining the texture of slices and their keeping quality for longer periods. High acidity inhibits the growth of spoilage organisms, besides retaining the texture (Narayana, 1976).

Each region has its own pickle varieties, which are popular and reputed for pickles in the region. Unfortunately, there is no uniformity in shape, size, flavour, acidity etc. in different seedling types of mango used for pickling. So the pickles made from seedling types vary much besides methods of preparation. Not much work was done in identifying suitable genotypes for pickling. Similarly almost no breeding work was done in this direction. Recently Shinde *et al* (2004) identified a hybrid (Neelum × Alphonso) exclusively for pickle purpose and named it as "Konkan Ruchi" on the basis of regular bearing habit, higher yield, large fruit size, maximum acidity, best organoletpic score for pickle colour, flavour and texture. The most suitable Indian cultivars are "Ramkela',

'Amlet', which also mature late in the season and the fruits are less affected by diseases like anthracnose, bacterial blight etc. However in South India, 'Totapuri' is largely used for making commercial pickles.

Raw material

Both young and unripe green fruits are utilized for preparing pickles and chutneys. Fresh unripe, green mangoes, preserved slices, less acid type fruits (sweet) and cold stored fruits are used for making pickles. Highly sour (5-6% acidity) thick skin with slight/high fibre content, round fruits and high mesocarp pulp : stone ratio fruits are used.

The selection of the right type of mango fruit for making pickles is the most important requirement, because the quality of the pickles depends on the quality of raw material. The following characteristics should be considered while selecting a mango fruit for pickling.

Fruit size: Fruit size in terms of weight and the resultant number of pieces is an important factor. The yield of pieces depends on the fruit size. As a rough guide, a large fruit size (> 250 g) was considered as a suitable size for pickling by Narayana (1976), while small sized mango fruits were also considered suitable by Das and Hota (1977). A fruit giving at least 4 standard pieces is preferable. High pulp weight is desirable. The flesh should be about 4 times the weight of stone (Narayana, 1976).

Fruit skin colour

Colour of the fruit skin is important because much of the consumer's appeal depends upon it. Green skin coloured fruits are preferable (Bhatnagar and Subramanyam, 1973., Narayana, 1976., Das and Hota, 1977).

Skin thickness: Thickness of the fruit skin/peel is also an important character. Skin should be thicker. However, pickles are prepared from peeled or unpeeled mature fruits.

Flesh colour: The fruits which exhibit white flesh colour are more preferable because they indicate unripe condition, high starch and high acidity, which are most desirable characters for acceptance of pickles.

Fibre in the pulp: The tissue holding capacity depends upon the extent of fibre in the pulp. High fibre content in the flesh is better for holding the tissue during storage, while the pieces having low fibre/no fibre get softened early and do not store for long, hence mango genotypes having very low/no fibre content are not suitable for pickling (Gangopadhyay *et al* 1976., Kocher and Sharma, 1992). However, in some areas fibreless fruits are preferred while in others fibrous fruits are desired for pickling.

Acidity: The content of acidity of a fruit is the single most important character that decides the selection of raw material for pickling. The mango fruits for making pickle should have high acidity (5-6%), as they make best pickles with better storage quality. However, mango pickles can also be made from sweet mangoes (< 1%).

Flavour: Varieties which have a characteristics harsh flavour should be avoided for pickling.

Thus, fresh, firm, fully matured, but unripe green mangoes having as much flesh as possible around the stone, high acidity (5-6%) should be selected for pickling (Sastry *et al* 1975).

Recipes

There are numerous recipes and recipe combinations of pickle preparation, because of regional variation in taste. The recipes also vary with type/kind of pickles prepared. The many kinds of recipes vary mainly in the proportions and kinds of spices used in their preparation. One basic recipe for the preparation of pickles in oils is as follows (Daughty, 1995).

Mango pieces	–	250 g
Salt	–	60 g
Mustard powder	–	30 g
Chilli powder	–	20 g
Turmeric powder	–	2-4 g
Fenugreek seeds	–	2-4 g
Bengal gram seeds	–	2-4 g
Gingelly oil	–	20-30 g

Typical recipe for mango pickle in oil (Nigam *et al* 2007)

Mango pieces	–	1.0 kg
Salt	–	150 g
Fenugreek powder	–	25 g
Turmeric powder	–	15 g
Red chilli powder	–	10 g
Cloves	–	8 nos
Black pepper, cumin, large cardamom and fennel powder – each 15 g		
Asofetida	–	2 g
Mustard oil	–	350 ml

Recipe for sweet – mango pickle

Mango slices	–	1.0 kg
Salt	–	120 g
Red chilli powder	–	1 g

Cumin	–	1 g
Cardamom	–	1 g
Fennel seed	–	1 g
Cinnamon	–	1 g
Coriander powder	–	1 g
Cloves	–	6 Nos
Jaggery/Sucrose	–	500 g

Methods of preparation

The methods of preparation of mango pickle are many and are different in different parts of India as well as world, according to taste and type of pickle. It will be a difficult task to mention all the methods and varieties of pickles in any book. However, essentially, the method of preparation includes selection of suitable mango (variety) fruits washing them, slicing, removal of stones, treating with powdered salt, exposing to sun for a specified period, mixing spices and condiments and packing and storing. Some of the typical methods of preparation of different types of pickles are presented below:

***Oil pickles*:** These pickles can be prepared from raw mangoes as detailed below:

Wash and cut fully mature unripe fruits into slices (pieces) and mix with powdered salt. Keep it in the sun in polybags or glass jars until green colour disappears from the slices. Grind ginger and garlic into a fine paste in vinegar. Separately make powder of other spices. Heat the oil to a smoking point and cool it. Add ginger and garlic paste and spice powder to oil and boil while mixing continuously to avoid charring. Add salt to cured mango slices and boil again until the whole mass is well mixed. Stop heating and add about 10 ml of glacial acid if required/preferred and small quantities of cinnamon, clove and cardamom, cool and pack into suitable containers (Gowda, 1995).

***Oil-less pickle*:** Oils are extensively used to preserve the pickles, but their increasing cost is making these pickles prohibitive. Further, oil-less pickles will certainly be a boon for human health. Gupta (1998) has developed a technology for the preparation of oil-less pickle from the fruits of 'Ramkela' cultivar, in which:

Unripe mature green fruits are peeled and cut into small pieces with a stainless steel knife, to which salt and red chilli powder are added. Asafoetida is added at 1% concentration. After adding different ingredients (spices etc.), the pieces are mixed thoroughly and kept in the sun for reducing the moisture content. The pieces are packed in glass jar and again exposed to sun with lose lid for two days, with occasional stirring for uniform mixing. Then the pickle is stored in cool and dry place at ambient temperature ($30^{o} \pm 5$ °C).

***Avakai*:** This is a special and important mango pickle, exclusive to Andhra Pradesh (India), which is a mango-sesame (gingelly) oil-based pickle. The method of its preparation is as follows:

***Selection of raw material*:** Fresh, firm, fully mature, but unripe green mangoes without surface blemishes, in which the stone has fully developed, with high content of flesh around the stone are selected for making 'avakai' pickle. The flesh should be at least 4 times the weight of seed. Injured/damaged fruits in any manner and those that have fallen on ground should be discarded and never be used, because they are heavily contaminated with microorganisms.

***Transportation and storage*:** The harvested mangoes must be suitably packed, preferably between beds of hay to prevent damage to the fruit in transit. Adequate provision must be made for cross ventilation during transit as well as in storage to prevent undue buildup of temperature. On arrival at the factory, they may be stored for a brief period only. In any case, it is undesirable to store mangoes (for pickles) for more than 4 or 5 days, because in storage, the fully matured fruits will tend to ripen and a number of physico-chemical changes occur, making them unsuitable for pickling.

***Preparation of mango fruits for pickling*:** The fruits on arrival must be given a gentle, but thorough wash, by soaking them in clear water overnight to remove the dust etc. on fruit surface and field heat. The water is drained out and the surface of the fruit is dried by wiping with a clean piece of cloth/paper knop kins.

***Cutting the fruit into slices*:** Sharp, clean and specially designed cutting knives made of steel are used for cutting the mango fruits into slices (Photo 5.1). It is desirable to cut the fruit to a sharp angle by single stroke, so that the damage to the tissues of the fleshy portion is avoided. If the tissue gets crushed too much the pieces tend to became soft and mashy in the pickle on storage. It is desirable to cut a fruit into 4 to 12 pieces of uniform size (depending on the fruit size) ensuring that every piece has a piece of the shell of the stone intact. Pieces without shell become very soft and mashy in storage. The pieces of kernel and the thin papery filament should be completely removed from the cut pieces. All the damaged pieces should be discarded.

***Preparation of avakai*:** All the dry powdered ingredients are first mixed in correct proportion as per the recipe. A sufficient quantity of sesame (gingelly) oil is mixed with powder mix to obtain a wet, but not too pasty mass. The cut pieces of mango are now mixed with the wet mass roughly in equal proportion (Photo 5.2). The total mixture is filled into smooth impervious porcelain jars and covered. After 3-4 days, there will be a considerable reduction in volume and depending on the acidity of the fruit used, the whole mass turns more fluid. At this stage the entire mass is mixed uniformly with wooden spoons. If the mass is too dry, still more pieces of mango are added and mixed. On the other hand, if it is too fluid more of the dry mixture of powder is added to bring it to a smooth pasty consistency. If needed a little more oil may also be added just enough to cover the mass to avoid exposure to atmosphere and resultant microbial spoilage. Then the pickle is filled in clean and dry porcelain jars.

***Aging*:** After preparation, aging of the pickle is essential to have a constant equilibrium of the various ingredients like salt, acid blend, drained weight, good colour, texture and flavour between the gravy and the fruit pieces to give an overall effect of an attractive mango pickle.

***Instant pickles*:** This is a new development in pickling. A process for making instant pickle has been developed to dry the salt cured slices either in the sun or in a cross flow air drier, mix with spice mixture and pack to produce an instant pickle mix. On addition of water and storing over night a good quality pickle can be obtained next day (Negaratnamma, 1987).

Ingredients and their role/purpose in pickles

Several ingredients, spices and condiments etc. are used in the preparation of mango pickles (see recipes). Each one serves an useful purpose in imparting flavour, taste, quality and extending the storage life of the pickle. The different ingredients used/added in pickles are described briefly below.

Photo 5.1: Knife for cutting mangoes into pieces for pickle and cut pieces

Photo 5.2: Mango Avakaya (pickle) after preparation

***Salt*:** Salt (sodium chloride) is added in all the pickles, mainly for preventing the spoilage and extension of storage life. Salt improves the taste, flavour and hardness of the tissues and controls fermentation. Salt content of 15% or above prevents microbial spoilage. But the salt meant for making pickles should be free from extraneous impurities (black specks, gunny fibre etc.). It must be dried thoroughly before mixing in the pickle. After drying it should be powdered and passed through a 40 or 50 mesh sieve for uniformity of particle size. The osmotic effect of the salt helps improve the texture of raw mango slices.

***Oils*:** Different types of oils viz., rape seed (Sarson/mustard)/refined groundnut/sesame (gingally) oils are used in pickles. From the nutrition point of view as well as to improve the storage capability of pickles, oils are used. Oil should be fresh, pure and free from rancid odour. In North India, mustard seed oil, while in South India, sesame oil are used for making pickles.

***Mustard flour*:** It is widely used in traditionally prepared pickles. Keeping quality, taste, flavour and colour of the pickles are maintained with the addition of mustard powder. Fresh brown mustard seed free from foreign matter should be used. Seed is first dried in sun and powdered and sieved through a 40-mesh sieve. Seed can also be dried artificially, but the temperature should not exceed 55 °C. Otherwise, the natural enzyme responsible for the liberation of the pungent volatile (Allyisothiocynate) is destroyed. A small quantity of turmeric powder with mustard flour may be added for extending the storage life of pickle. Pickling with mustard flour yields a very good pickle with attractive colour, firm texture and characteristic mango flavour (Sastry and Krishnamurthy, 1974).

***Turmeric powder*:** To prevent microbial spoilage, turmeric powder is added. Clean, good quality turmeric rhizomes of bright colour, free from infestation and artificial colouring agents, should be dried and ground to a fine powder and used.

***Chilli powder*:** Red chilli powder is added to impart hotness to the product. Fresh, sound, dry chillies of bright red colour, free from infestation and foreign matter (strings, leaves etc.) should be used without stalk. They should be dried, either in sun or artificially to a moisture content below 10%. The dried chillies are ground to a powder of 40 mesh.

***Others*:** Depending on choice of flavour, taste etc., others like fenugreek, cumin, cardamom, cinnamom, coriander powder, black pepper, asafoetida etc. are used in making mango pickles. Some people prefer to add powdered jaggery, while others add whole dried bengalgram seeds. Addition of jaggery/sugar to the pickle prepared from sweet mangoes, improves colour, texture and taste of the pickle. Addition of small amount of sugar along with spices to salt pickles produces a good taste blend. Among spices fennel seed (Saunf) and onion seed (Kalaungi) are distinctive to North India, while mustard powder and asofoetida are exclusive for South India.

General Requirement of Pickles

Pickles should be safe for consumption. They should have attractive appearance, with appealing colour and desirable texture. They should retain the mango flavour for as long a period of storage as possible, should be highly palatable; should not have any kind of disagreeable flavour, should be cheap enough to be in the easy reach of all sections of people.

Factors affecting the quality of mango pickles

The quality of mango pickle is influenced by the type of raw material, stage of fruit maturity, fruit acidity, type of ingredients added, method of preparation etc.

***Type of raw material*:** Though mango pickles can be prepared from stored, waxed and fresh fruits, better quality pickles can always be made from freshly harvested fruits (Sastry and Krishnamurthy, 1974).

***Stage of fruit maturity*:** According to Sastry *et al* (1975), the life of the growing fruit (cv. Amlet) could be devided into 4 distinct stages;

(a) Enlargement stage,

(b) early stage of maturity;

(c) maturity stage and

(d) the stage of advanced maturity.

Of which mango fruits of early maturity stage are excellent in all respects of the preparation and storage of pickles. During this stage the fruit reaches almost maximum size and weight; fruits are still green, with white flesh; stone hardened, but the shell is sufficiently soft for cutting; peel will be thinner; moisture (%) in flesh, peel and kernal is sufficiently high; total and reducing sugars continue to accumulate in flesh, peel and kernel; acidity continues to be constant in pulp while it decreases in peel and kernel; ascorbic acid decreases to half in flesh; respiration rate decreases, but remains constant at the end of the period. Pickle prepared from fruits of this stage is good in colour, texture and taste with predominant raw mango flavour. Fruits of enlargement stage are not suitable because of low pulp, thick peel, soft stone, white flash, high astringency, high moisture content, high acidity and ascorbic acid. Fruits of maturity stage are unsuitable because pickle develops fruity odour in place of characterstic raw mango flavour, with yellowish tinge of flash colour. Fruit of advanced maturity is not useful for pickling because this lacked not only characteristic raw mango flavour, but also the texture of the fruit is too soft to stand longer in storage.

Thus, the physico-chemical characteristics and pickling showed that the mangoes of early maturity (8-10 weeks) can be considered as the optimum stage of maturity for pickling in case of cv. Amlet. However, the period required to reach this optimum stage of maturity for pickling differs with variety. 'Neelum' fruits reached its optimum stage of maturity in 6 weeks, while 'Totapuri' took 9 weeks (Sastry and Krishnamurthy, 1975), 'Totapuri' and 'Neelum' fruits reach the optimum maturity between April last week and May first week, when they can be harvested for pickling.

According to Ismail *et al* (1986) the best stage of fruit maturity for pickling is just after the endocarp starts hardening in Andhra Pradesh.

***Acidity*:** Acidic fruits of mango are mostly used for pickling. Good quality pickles can be obtained from fruits having 5-6% acidity. Pickles from medium acidity mangoes (3-5%) are poor in texture and flavour with a fermented smell. Mangoes having less than 1% acidity make poor pickles in all respects of quality (Sastry *et al* 1975).

***Storage of raw material*:** The environment in which raw material is stored prior to pickling also affects the quality of pickle in many ways and the duration of storage exerts influence on the quality. Handling of raw material also influences the pickling and its quality. The quality of pickles can be improved by packing the products in food quality polyethylene bags and keeping in lacquered cans.

Problems

Spoilage of pickles in storage is the most important problem faced by the pickle industry. The spoilage of pickles in storage is due to several reasons. Unhygienic conditions of manufacturer, use of raw materials of poor and unsound quality and in-correct proportion of various ingredients often tend to deterioration of pickles in storage. It is not uncommon to find packed pickles with visible growth of fungus on the surface. Mashy texture, fermented or other undesirable smell, souring, objectionable colour are all common problems, one comes across in pickles. Some of these are due to microorganisms (fungal, bacterial, yeast). Moulds (fungi) in salt pickles can be prevented by adding benzoic acid (250 ppm). Heavy spoilage through bulging and bursting of cans, due to yeast fermentation can be prevented by initially heating the pickle, cooling and packing thereafter (Jain *et al* 1973).

Pickle in brine being highly corrosive, develops pin holes in tin cans, besides raising the tin and lead contents of the pickle. Placement of polyethylene bag barriers in these cans is partially helpful. Excessive fermentation of pickles due to *Drosophilla* sp can be prevented by application of mustard oil, camphor, asafoetida or clove oil to the lid of the container (Majumdar *et al* 1954).

Control of spoilage in pickles

Pickles are subject to physiological as well as microbial spoilage (fungi, bacterial and yeast). Both yeasts and moulds are often found associated with mango pickles and slices (Johar and Anand, 1951., Soumithri *et al* 1963a, b., Subba Rao and Johar, 1962). Spoilage in pickles is prevented by the addition of spices and condiments, oils, organic acids and chemical preservatives.

***Addition of spices and condiments*:** Both salt and spices act as preservatives besides improving texture, flavour and appetizing properties. Cloves in pickle (2%) play a preservative role. Clove, Cinnamom and Gaynine pepper were more effective (Prasad and Joshi, 1929). However, most of the spices, except cloves and cinnamom had little or no inhibitory effect on the mould. However, some of these spices themselves carry infection, for instance chillies (Thangmani *et al* 1975). For their disinfection, methods like irradiations, vaccum fumigation and thermal exposure were suggested (Pruthi, 1987).

***Oils*:** Oil pickles are particularly susceptible to oxidative changes by which acid and peroxide values tend to increase affecting the flavour during storage (Das, 1965). The oil to be used should be fresh, pure and free from rancid odour and its acid value should be below 4 to obtain better preservative action (Narayana, 1976). Based on the stability of different oils in pickles, Krishnamurthy *et al* (1957) arranged them in the ascending order: coconut, groundnut, sesame, refined groundnut oil. Combination of crude coconut and groundnut oil, however, maintained the pickles in an acceptable state even after 12 months storage at 27 °C.

***Organic acids*:** Among the organic acids, acetic acid was more effective against moulds and yeasts than lactic or citric acid and the addition of salt increased the inhibitory action of these acids (Anand and Johar, 1958b).

***Chemical preservatives*:** Among chemical preservatives, SO_2 proved most effective followed by sodium benzoate, sorbic acid and sodium propionate. Yeasts were most resistant to these preservatives than moulds. By increasing the concentration of salt in mango pickle, the corresponding levels of preservatives to control the spoilage organism could be considerably reduced (Anand *et al* 1959). A pickle preservative emulsion containing acetic acid, orange peel oil, brown mustard and turmeric powder was developed, which in combination with salt (9%) at 10% level kept well mango pickle free of moulds and yeasts for over a year, while without emulsion the pickle was susceptible to spoilage even in the presence of salt (14%) (Soumithri *et al* 1963a, b).

RAW MANGO SQUASH (PANNA)

Raw mango squash popularly known as 'Panna' is a traditional product of green mangoes in North India. It is usually prepared and consumed fresh. It is a favourite drink during summer months. It provides an instant energy and has smothering effect during hot weather. It also provides protection from sunstroke and is a good appetizer. It has therapeutic values. The slices added in the beverage act as antioxidants and increase the flavour and storability of the product, and exhibit antimicrobial properties.

***Methods of preparation*:** Different methods of preparation of Panna have been developed with different recipes. Mallik (1976) described the preparation of green mango squash (Panna) as follows:

Fully developed green mangoes are washed, sliced longitudinally and stones are removed. Then the slices are placed in 3 % salt solution and boiled in equal amount of water for 24 to 30 minutes, for making the pulp soft, adding requisite amount of citric acid. The pulp then is separated from the peel and strained through a coarse muslin cloth or by using a stainless steel sieve of 2 mm mesh. The strained juice is kept covered until the solution and mint extract is prepared. The measured quantity of sugar, salt and water are taken in another container. Citric acid, powdered cumin (Zeera) and black pepper are added and boiled. Then it is strained through a fine muslin cloth, when still hot to hot crystallization of sugar or thickening of juice, which delays straining. Fresh mint leaves (100 g) are washed, crushed with small amount of cold boiled water and strained thrice through a fine muslin cloth. The strained sugar solution and mint extract are mixed with green mango juice and stirred thoroughly. The product is filled in sterilized hot bottles, cooled and sealed either by lac or sealing wax. The filled bottles are rinsed with a wet cloth, dried, labeled and stored in a cool, dry place. The beverage thus prepared will be diluted three times and served as ready to serve (RTS) drink.

Keeping in view the nutritional and therapeutic value of raw mango squash Tandon *et al* (2004) developed a procedure to prepare raw mango squash without supplementing artificial colour and flavour, that can be stored for longer time without loss in colour, aroma and taste, as follows:

Wash the hard green mangoes of seedling cultivar and fresh mint leaves thoroughly in running water; boil the fruits with equal amount of water for 30 minutes or until thy are softened; extract pulp by hand squeezing the fruit and remove the peel and stone; homogenize the squeezed pulp in a blender; add salt (50%) to the pulp, fixed amount of fresh mint leaves (10%), roasted cumin seed powder (2%) and citric acid (3.5%), add table and black salt with a ratio of 60 : 40; homozenise the mixture thoroughly in a blender and filter through a stainless steel or nylon sieve to remove the fibrous material, add 500 ppm sodium benzoate as preservative to the prepared product (Panna); fill the product in clean, sterilized bottles and store them under ambient conditions (up to 9 months). The product can be served in 1:8 dilution with chilled water. The excellent colour, aroma and taste of the product, retained up to 9 month was attributed to the addition of mint leaves and cumin seeds. Further, higher sugar content (25%) produced better product with both salt concentrations.

Sirohi *et al* (2003) developed a whey-based mango herbal beverage containing different concentrations of mint leaves, which increased organoleptic quality and acceptability. The product showed increase in TSS, acidity and reducing sugars, on storage, for 30 days at 7 ± 1 °C.

Conventionally, Panna is prepared by boiling raw green mangoes under steam (15 p.s.i.g) for 20 minutes, peeled, and passed through a pulper and pulp is blended with salt, sugar, citric acid, and a spice mixture to prepare a beverage concentrate which has a shelf life of about 120 days at room temperature. The concentrate, upon dilution with 2-6 volumes of water gives a highly acceptable beverage.

However, Panna can also be prepared from baked raw green mangoes having fully developed stones (Thi *et al* 1989). In this technique, green mangoes without stones are backed in a muffle furnace at 200 °C for 25 minutes or at 300 °C for 20 minutes. The backed mangoes are then peeled by hand and passed through a stainless steel sieve of 20 mesh. Baking of mangoes yields higher pulp than boiling. The yield of pulp increases with increase in baking temperature from 200-500 °C. Pulping is slightly difficult in boiled mangoes due to adherence of pulp with the peel. Pulping as well as extraction of pulp are easy with baking of raw mangoes. Higher baking temperature increases hydrolysis of polysaccharides from the peel, resulting in increase in pulp percentage. Panna prepared from baked fruits at 200°-300 °C was organoleptically most acceptable.

Recipes

A number of recipes are available according to the required taste in the country for the preparation of 'Panna'.

Recipe I (Mallik, 1976)

Green mango slices - 2.5 kg

Sugar - 4.0 kg

Water	-	4.0 litres
Salt	-	200 g
Mint leaf extract	-	50 g
Powdered cumin seeds	-	25 g
Powdered black pepper	-	10 g
Citric acid	-	50 g
Preservative (KMS)	-	4 g

Recipe II (Kalra *et al* 1995)

Mango slices	-	1.0 kg
Sugar	-	1.6 kg
Water	-	1.6 litres
Salt	-	80 g
Cumin seed	-	10 g
Black pepper	-	4 g
Citric acid	-	20 g

However, the above two recipes are modified version of 'Panna'. In the traditional method, sugar is not added. Pandey *et al* (1995) tested 4 types of recipes for the preparation of 'Panna'.

Recipe A

Raw mango pulp (20%)	-	400 g
Water	-	1.4 litres
Common salt (1.25%)	-	25 g
Black salt (0.25%)	-	5 g
Roasted cumin seeds powder (0.5%)	-	10 g
Asafoetida powder (0.01%)	-	0.2 g
Black pepper powder (0.2%)	-	2 g
Red chilli powder (0.15)	-	3 g
Mint leaf extract (TSS – 20)	-	100 ml
Coriander leaf extract (TSS – 30)	-	100 ml
Edible green colour (0.05%)	-	0.1 g
Citric acid (0.1%)	-	2 g

Recipe B

Recipe A + lemon juice (50 ml) instead of citric acid

Recipe C

Recipe A + 5% sugar

Recipe D

Recipe B + 5% sugar

Organoleptic evaluation of the end products showed that 'Panna' prepared using lemon juice without sugar (recipe-B) was the best in respect of colour, flavour, consistency and taste.

***Problem*:** Formation of gel at the base of the bottle is often observed during storage of 'Panna' and this can be overcome by Pectolytic enzymatic treatment of the pulp (Anon, 1985-86).

***Quality*:** Processing time influences the quality and appeal of the product after 9 months storage. Processing time of 10 minutes was sufficient to maintain the quality and appeal of the product up to 9 months of storage. Processing for 15 minutes had an adverse effect on the quality of the product, while processing time of 5 minutes allowed mould development (Pandey *et al* 1995).

***Storage*:** Under ambient conditions, 'Panna' can be kept in storage for more than a year (Nanjundaswamy, 1997), although, ideal storage period was claimed to be 9 months (Tandon *et al* 2004).

Mango chips

Chips are usually prepared from starch-rich raw materials like potato, banana and tapioca. However, chips can also be prepared from low-starch fruits like apple and pineapple, which have been gaining popularity recently in the foreign markets, since besides being attractive, they serve as a source of vitamins, minerals and fibre (Pompen Torezan *et al* 2004). Attempts have been made in Indonesia, to make chips from mango, using the off-grade fruits of cv. Arumanis (Suhardjo and Santoso, 2000) and in Brazil (Pompen Torizan *et al* 2004).

Method of preparation

The procedure of making chips from raw mangoes is as follows :

Harvested fruits are peeled, sliced (10 ± 2 mm thickness) soaked in saturated calcium hydroxide solution for 10 minutes, dipped in sugar solution (4.5%) and dried in sun or modified oven or electric drier to 4.0 to 5.5% moisture. The dried chips are packed in 150 gauge plastic bags and stored at room temperature (Suhardjo and Santoso, 2000).

Factors influencing the quality of chips

The best quality chips should have the characteristics like low fat content, good texture, preservation of nutrients and aggreable sensory characteristics. The quality of chips (colour, taste and crispiness) depends on the fruit maturity at harvest, drying methods and pre-treatments.

Fruit maturity: Colour and taste of chips are influenced by fruit maturity at harvest. Best quality of chips (cv. Arumanies) were obtained from fruits aged 85 days after full bloom. Consumers prefer bright yellow colour of the chips and this can be achieved from fruits harvested 80-95 days after full bloom. Older fruits showed higher sugar content, but firmness, acid content and vitamin C content will be low in the resultant chips. Low fruit firmness in fruits at 90 and 95 days after full bloom causes difficulties in slicing the fruit. Younger fruits on the other hand, give stronger acid taste, so less preferable. Fruits aged 95, 90 and 85 days after full bloom give chips with better taste. The chips, produced from 90 and 95 days of fruit maturity are broken as the fruits were too ripen. The crispness of chips, however, is not influenced by stage of fruit maturity at harvest. Fruits of 85 days old make better shaped chips compared to other two maturities.

Pre-treatments: The chips finally should have low oil and moisture contents. This could be achieved by some pre-treatments, which reduce the initial moisture content of the raw material before drying. Osmotic dehydration has been widely used as a pre-treatment in moisture conservation processes of fruits and vegetables (Torrgiani and Bertolo, 2001). Osmotic dehydration consists of removal of water from solid foods, like pieces of fruits and vegetables, by their immersion in a hyper-tonic dehydrating solution. The drying force, for water removal is the concentration gradient between osmotic solution and the intracellular fluid with the cell wall acting as a semi-permeable membrane (Shi *et al* 1995). The process causes two principle flows, which are simulataneous and contrary, in flow of water from the food to the solution and simultaneous transfer of solute from the solution into the food (Saurel *et al* 1994). Osmotic dehydration carried under vacuum was more efficient and was more adequate since it resulted in less solute uptake. However, osmotic dehydration carried at atmospheric pressure confirred better quality parameters on the chips, such as low moisture content. The chips pre-dehydrated at atmospheric pressure were more preferred with respect to all the attributes except aroma. Water loss from the chips and oil uptake during frying process and handling after osmotic dehydration depend on the thickness of the chips. A thickness of 3 mm was more adequate without the removal of the peel, since this condition showed a satisfactory water loss together with low oil uptake. In 2 mm thickness, chips without skin showed greater oil content, which is undesirable, difficult to handle after osmotic dehydration due to weight loss during the process resulting in very thinned chips and consequently greater loss of material (Pompen Torrzam *et al* 2004).

Drying methods: Out of the three different methods of drying viz., electric drier (at 60, 70, 80 and 90 °C), modified oven and sundrying tested, drying in electric drier at 80 °C resulted in best quality chips in terms of colour, taste and quality. Chips dried in electric driers and modified oven were better compared to sundrying. Water content of chips dried in sun was higher than those dried in electric drier and modified oven. Chips produced by drying in modified oven showed more brown colour, compared to others. Hot air temperature (70°-100 °C) within modified oven was unstable. Chip's colour from electric drier at 60°, 70° and 80 °C were white coloured compared to others and were more preferred by consumers. Sun-dried chips were not preferred since they were dirty- coloured, probably due to differences in water content of the chips. Taste of chips, produced in electric drier at 90 °C and modified oven exhibited a slight burnt taste, while sundried chips had poor taste due to high moisture content. Water content of the chips affects their crispness, less the moisture, more will be the crispness, it is one of the determining parameter for acceptability of chips by consumers. Such preferred crispness was achieved by drying in electric drier at 70°-90° and modified oven due to lower moisture content.

On the basis of the above, it can be concluded that the best fruit maturity and best drying method for production of best preferable quality of chips are : fruits harvested at 85 days after full bloom and drying in electric drier at 80 °C (Suhardjo and Santoso, 2000),

GREEN MANGO FLAKES

The technique (method) for preparing green mango flakes from unripe green mango fruits has been standardized (Srivastava, 1998a). The method consists of heating green mango fruits in boiling water for 15 minutes; peeling and pulping in a pulper of screen size of 30 mesh. The pH of the pulp is adjusted to 5.0 and heated to 80-85 °C. During heating, corn starch and bicalcium phosphate are added to improve the free flowing character of the finished product. The final mixture is then dried in an atmospheric chrome-coated steel drum-drier, heated by steam at 50 psig and drum rotating for about 15 seconds. The finished product is packed in polyethylene-aluminium foil laminates and sanitary cans and stored at room temperature for 6 months without discolouration and was highly satisfactory in respect of colour, flavour and texture, compared to the material packed in HDPE bags (Photo 5.3).

The green mango flakes can be used for preparation of beverages and amchur (Srivastava, 1998b).

PRODUCTS FROM RIPE MANGOES

Ripe mangoes (fully mature and post climacteric ones) with full flavour and colour development can be processed as follows :

1. ***Frozen mango products:*** slices in syrup, pulp and beverage base.
2. ***Canned products:*** slices in syrup, pulp, juice and nectar.
3. Beverages and blends.

4. ***Dehydrated products:*** Fruit bars, cereal flakes, powder, strained baby food, toffies etc.
5. ***New products:*** Canned mango slices in juice, concentrates, aroma concentrates, low viscosity and low pulp containing mango beverage base, aseptic bulk packaging of pulp and concentrate, saturated products etc.

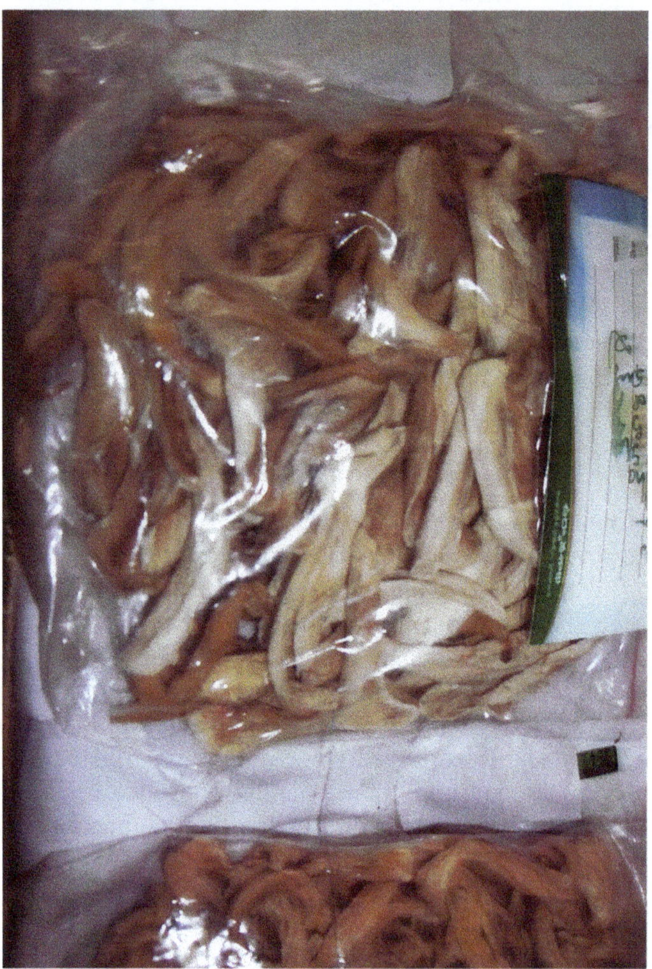

Photo 5.3: Packed dried mango slices.

BEVARAGES

Mango beverages are nutritious, easily digestable, highly refreshing and invigorative and are very popular market drinks. In tropical countries like India, fruit juices and beverages play important role in summer months, due to their thirst quenching property apart from their nutritive value.

NON ALCHOLIC BEVERAGES

Mango beverages are two types viz., non-alcoholic and alcoholic beverages. Non-alcoholic beverages include juice, squash, nectar and RTS drinks. Alcoholic beverages include wine, vermouth and vinegar. The general composition of non-alcoholic drinks of mango is given in Table 5.2.

Table 5.2 General composition of non-alcoholic mango drinks.

S.No.	Beverage/drink	Puree (%)	TSS (°B)	Acidity (%) as citric acid
1.	RTS beverage	10	15	0.3
2.	Nectar	20	15	0.3
3.	Juice	35-45	15-20	0.3
4.	Squash	25	45	1.2
5.	Syrup	25	72	1.5

Mango Squash

Squash is a non-alcoholic beverage of fruits. It essentially consists of fruit pulp, sugar, water and food additive/preservative. Mango squash is a popular beverage in mango production regions. Juicy varieties of mango are preferred for making mango squash. Ripe mangoes are utilized for making squash.

The method of preparation of mango squash is as follows (Srivastava, 1967):

Fully ripe fruits are washed well in fresh water and four vertical slits are given to each fruit with a stainless steel knife to facilitate pulping. The cut fruits are then passed through a pulping machine to separate the peel and the seed. The extracted pulp is used for making the squash. Sugar and citric acid are dissolved in hot water separately. The sugar syrup is clarified and passed through a muslin cloth and cooled. Next the syrup is added to the mango pulp and mixed thoroughly to which KMS is added and thoroughly mixed. The prepared squash is filled into sterile glass bottles and sealed. The above squash is prepared as per the following recipe:

Mango pulp - 60 kg
Sugar - 90 kg
Water - 50 lit
Citric acid - 2 kg

Besides the above ingredients, Gowda (1995) advised to include orange yellow edible colour to get attractive appearance to the product. Contrary to the suggestion of Srivastava (1967), Nigam *et al* (2007) suggested heating of the whole mixture to 76°-78 °C followed by the addition of 25-30 g of citric acid. At the time of use, 3 parts of water is added to one part of squash and served.

The squash quality is evaluated on the basis of pH, titrable acidity, soluble solids, ascorbic acid and specific gravity. Two problems are encountered in mango squash

(i) presence of air bubbles (which is a source of quick deterioration) and

(ii) separation of soluble solids (giving an undesirable appearance) (Daughty, 1995).

Mango juice

Mango juice is one of the non-alcoholic beverages obtained from ripe fruits. It is generally prepared by mixing equal quantities of fruit pulp (puree) and water and adjusting the total soluble solids (TSS) and acidity for good taste (12-15% TSS and 0.4-0.5% as citric acid).

Juicy varieties are most suited for juice making. Varieties like 'Alphonso', 'Arka Puneet', 'Arka Anmol' (Gowda, 1995) and 'Sukul' (Awasthi and Pandey, 1979) are suitable for making juice.

The process of preparation of juice is as follows :

Freshly harvested fully ripe juicy mangoes with good flavour are selected for juice preparation. They are thoroughly washed in running water, peeled with a stainless steel knife and pulped in a pulping machine. The pulp is collected in a stainless steel or aluminium tank and the juice is prepared by adding sugar and citric acid. Sugar is dissolved first in water, clarified with milk, strained through a filter cloth to remove impurities. The ingredients are mixed thoroughly and passed through the pulping machine, fitted with stainless steel sieve having required (8-9 mm) perforation. The juice is collected in a tank of stainless steel or aluminium. The juice is then passed through a heat exchanger and heated to a temperature of 82.5°–85 °C and filled hot in sterilized glass bottles. The bottles are crown-corked for 10 minutes. After sterilization, the bottles are cooled and stored in a cool, dry place.

To improve the flavour of juice, pulp from a good table variety, possessing rich and pleasant flavour can be blended up to 25-50% (Srivastava, 1967). The final product may have 25-45% pulp with TSS of 20° Brix and 0.3% acidity (Gowda, 1995).

Mango Nectar

Mango nectar is a ready to serve (RTS) beverage like mango juice. It is a very nutritious and a commercially popular product. Mango nectar contains 15-20% sugar and acidity properly adjusted to give 15° brix and 0.3% acidity as citric acid (Bhatnagar and Subramanyam, 1973).

Nectar is prepared from varieties: 'Dashehari' 'Neelum', 'Amrapali', 'Totapuri', 'Baneshan' etc. of which 'Dashehari' is the best cultivar. The procedure for nectar preparation is similar to that of juice. Nectar is prepared from either pulp from dehydrated ripe mango slices or frozen pulp, spray-dried mango puree or stored fruits and even from unripe mango fruits. Nectar prepared from mango concentrate is comparable to nectar prepared from fresh pulp.

Nectar is prepared with pulp from dehydrated ripe mango slices after reconstitution in equal amount of water and grinding and making it in the ratio of 4:3 as pulp and sugar. It contains 20% pulp with sugar and acidity adjusted to give 15°-16° brix and 0.23% acidity as citric acid. Such product claimed better acceptability in respect of colour, flavour, texture and overall quality. The recipe for nectar is as follows (Sagar and Khurdiya, 1998).

Dehydrated ripe mango slices	-	120 g
Sugar	-	90 g
Citric acid	-	1.0 mg
Water	-	390 ml

Nectar made from the frozen (–23 °C) Hawaii mango puree had better flavour than that made from unheated pulp (Brekke *et al* 1968). Nectar prepared from stored/frozen pulp is comparable to nectar prepared from canned pulp but developed off-flavour at the end of 14 months storage at 18 °C, which could be removed either by pasterusing the pulp before freezing or heating the frozen pulp before use (Ramana *et al* 1984). The nectar from spray-dried mango puree was considered to be of bright colour, but lacking flavour compared with nectar from deep frozen pulp (Baldry *et al* 1976).

Fruits stored for 6-12 days depending on the variety can be used for making nectar. Nectar prepared from fruits stored for longer period showed a decline in flavour. Carotenoids and ascorbic acid showed maximum stability in storage under cold storage at 5 °C (Sahni and Khurdiya, 1989). Nectar prepared from unripe fruits will have higher viscocity in the initial period than the one prepared from ripe fruits. Thinner viscocity in ripe fruit nectar was due to the degeneration of pectin and starch during ripening. Flavour improved with gradual ripening, mainly due to the formation of esters, which reflected in the nectar from ripe fruits (Sahni and Khurdiya, 1989).

The quality of mango nectar depends on variety, stage of maturity at harvest, ripening periods, storage periods of ripened fruits. At present nectar is made from different varieties and as a result there is no uniformity in colour and flavour of the product. 'Amrapali' pulp was superior to that of 'Totapuri' with respect to total carotenoids and colour. The brix, acidity and pH were higher in Amrapali than 'Totapuri'. 'Amrapali' nectar was highest in density, but lowest in viscocity, whereas nectar from 'Totapuri' was more viscous than that from 'Amrapali' (Khurdiya, 1993). Fruit maturity at harvest as well as the ripening period have a considerable bearing on the quality of processed products especially in terms of flavour, though the chemical composition of the juice is largely dependent on the combined influence of genetic regulatory mechanism as well as physical, chemical and biological environments. Further, more emphasis is needed in the selection of the fruit, particularly with regard to the stage of ripeness, as the quality of nectar is a function of stage of ripening of fruit (Sahni and Khurdiya, 1989). However, judicious blending of varieties considerably improves the quality of nectar compared to individuals. The cultivars 'Chausa' and 'Bombay Green' are not ideally suited for making nectars individually. Their, quality can be considerably improved by blending with 'Dashehari'.

Pulp settles during storage of mango nectar. This can be avoided with 0.5% corboxy methyl cellulose (as a stabilizer), which also improves quality of nectar (Altaf *et al* 2006).

ALCOHOLIC BEVERAGES

Mango in general contains high amount of TSS, proteinous substances, vitamins and minerals, thus making it a suitable media for the growth of wine yeasts (Kulkarni *et al* 1980). 'Sahakara Sura' an alcoholic beverage from mango and spiced liquor was among the several inebriating drinks used in ancient India between 300 BC to 75 AD (Omprakash, 1961). In tropical countries, higher temperature and lack of practice of cold storage lead to faster ripening and huge losses, which can be minimized by converting them to wine and vinegar. Chadha (1987) suggested processing of substandard mango fruit into quality wines for improving the economic condition of the grower.

A number of mango varieties were screened for dry wines (Kulkarni *et al* 1980., Onkarayya *et al* 1982) and dessert and Madeira style wines (Onkarayya and Singh, 1984). 'Langra' was found to be the best suited one for making dessert wine. Similarly varieties like 'Fazli', 'Chausa' and 'Dashehari' were suitable for 'Sweet dessert' and 'Madeira' style wines, whereas 'Totapuri' is suitable for making 'Madeira'. 'Totapuri which is less acceptable for table purpose, is cheap and so used for wine making. Mild flavour and colour of this variety are ideal as the resulting wine had a pleasant light golden yellow colour and a distinct mild fruity flavour with moderate ester content in the wines.

Akinbela *et al* (1994) from Nigeria described procedure for making wine from ripe fruits of mango as follows :

Over ripe fruits are washed, peeled and pulp is separated manually and blended for 10 minutes after adding sodium metabisulphite (0.5% 30 ml). Fibres in the mango juice are removed using a screen of 120 μm. Wine was produced by reconstructing 1.5 litre of juice with 4 litres of distilled water. The diluted juice is ameliorated to 25% sugar content with sucrose and 0.0175% (w/v) ascorbic acid. It was then sieved through a Canadian standard sieve No. 35, transferred into 5 litre fermentation tanks, inoculated with 50 ml of 20% (w/v) of yeast (*Saccharomyces cervisiae*) which has been already activated at 37 °C in 25% (w/v) sucrose solution and allowed to ferment at room temperature (28 $\pm$ 2 °C) for about 40 days. The resulting wine was fined using gelaten (1%) and tannin (1%). The wine prepared from over ripe fruits is packed in polyethylene pouches at a lower cost than bottling, with out loss in quality.

In India, mango wines are prepared by inoculating mango pulp with yeast *Saacharomyces cervisiae* var *ellipsoidum* (Paul and Tewari, 1996).

Madeirization

Dry wines and fortified wines prepared from some varieties of mango (Totapuri) lack attractive colour and body. Strong aroma in a few varieties often mask the bouquett. The lacunae in the above wines led to attempt to improve colour and other organoleptic qualities through a process called 'Madeirization'(Onkarayya *et al* 1982). 'Madeirization' is a process of baking wines in bottles in an incubation at 50 °C for 24 days. After baking the bottles are transferred to 18 °C for storage. Madeirization, after fortification with sugar and alcohol, improved colour, body and bouquet of wines of most varieties (Onkerayya and Singh, 1984). Madeirized wines are more acceptable than corresponding dry and dessert wines.

The madeirization process as above required longer period. The baking period was reduced by adding an antioxidant (ascorbic acid) (Onkarayya, 1986). Ascorbic acid was mixed with mango pulp, bottled and heated in BOD incubator at 50 ± 2 °C for 7 days. After baking the bottles were cooled to 18 °C. The 'madeirized wine' did not require aging, as ascorbic acid helped in aging rapidly upon madeirization. This process reduced titratable acidity and volatile acidity, increased pH, volatile esters, aldehydes, colour brightness and organoleptic quality of dessert wines. The colour intensity and brightness of dry wines were better than that of dessert wines. Wines from 'Raspuri' showed maximum colour intensity and brightness, while those of 'Totapuri' the least. Madeirization increased the colour and brightness of all dessert wines. Variation in colour and brightness among the different varieties screened, was considered to be due to inherent differences in pigment composition of the varieties. Sensory evaluation of wines indicated that wines from 'Mulgoa' and 'Dashehari' are not acceptable as table wines due to high astringency. Dessert wines from 'Totapuri', 'Raspuri' and 'Langra' possessed acceptable taste. Madeirization improved organoleptic characteristics of the 5 dessert wines. Improvement was mainly in colour and appearance, bouquet, freedom from acetic odour and development of a characteristic 'madeira' flavour (Onkarayya, 1986).

Vinegar

Acetic acid, obtained through the process of fermentation is known as 'Vinegar', which literally means sour wine. During the process of fermentation raw materials undergo alcoholic and aceteous fermentation (Ebner, 1982). In vinegar production, alcohol is quantitatively oxidized to acetic acid by strains of '*Acetobactor'*. During vinegar production, *Acetobactor* shows sensitivity towards absence of oxygen, hence requires continuous aeration (Tewari *et al* 1991).

Production of mango vinegar from green mangoes by cell immobilization was standardized using *Acetobactor aceti* (Garg and Tandon, 1995). Vinegar was produced from mango pulp by the twin processes of fermentation and oxidation using yeast (*Saccharomyces cerevisiae* var. *ellipsoidus*) and bacteria (*A. acete*). The yeast was recycled to improve fermentation rate, while bacteria was immobilized on wood shavings for semi-continuous vinegar production. Raw vinegar was raked at 10 ± 2 °C for

2 months, then filtered and pasteurized before bottling. The final acidity of vinegar will be 5.2% and vinegar was yellow in colour with pleasant fruity flavour, with high acceptability (Garg *et al* 1995).

Vermouth (Aromatic Wine)

Aromatized wine from mango is designated as 'Vermouth'. Herb and spice mixture formulae to produce dry and sweet vermouth with mild aroma and flavour has been developed (Onkarayya, 1985). Composition of mango vermouth is given in the Table 5.3.

Table 5.3 Chemical composition of mango vermouth made from Cv. Banganapalli fruits (Onkarayya, 1985)

S.No.	Parameter	Mango vermonth	
		Dry	Sweet
1.	Colour at 420 nm	0.42	0.68
2.	pH	3.40	3.42
3.	Total acidity (g. tartaric acid / 100 ml)	10.59	0.59
4.	Volatile acidity (g. acetic acid / 100 ml)	0.09	0.07
5.	Alcohol (v/v / 100 ml)	17.00	17.20
6.	Total Aldehydes (ppm)	15.80	26.40
7.	Total phenols (%)	0.06	0.07
8.	Organoleptic score	13.00	15.50

Carbonated mango beverage

Three mango-based carbonated beverages were prepared, containing respectively mango pulp 5, 7 and 10%; sugar 9.07, 11.7 and 13.5%; TSS, 10, 13 and 15% and acidity 0.1, 0.15 and 0.2%. All the three drinks contained carbondioxide (3% gas volume). Chemical composition, shelf life and acceptability of drinks were examined at 7 days interval for 28 days and then at 30 days interval for 210 days. Apart from vitamin C content, no significant change in colour, TSS, acidity, pH, preservative, gas volume, fat, protein and minerals content of the drinks was observed during storage. Taste pannel assessment showed differences in consumer acceptability of the beverages. The beverage containing 7% mango pulp, 11.7% sugar, 13% TSS and 0.15% acidity was preferred (Islam *et al* 1990).

FREEZING PRESERVATION

Freezing is one of the oldest and best methods of food preservation, yielding long term shelf life with retention of nutritional and sensory properties of the food. It is an excellent method of preservation of foods in general and fruits and vegetables in particular. Freezing preserves the flesh colour, flavour and texture and keeps the produce in a state close to the fresh produce. It is rated as the best techniques available for food preservation since it maintains the natural properties by reducing post-harvest changes and microbial deterioration without any influence on the original qualities. In actual practice, freezing slows down, but does not stop the physico-chemical and bio-chemical changes. Some fruits which can not be preserved by other methods can be preserved by this technique, since it is a quick, convenient, effective and easy method.

Mangoes can be frozen as whole fruits and as pulp or slices in sugar syrup or as dry slices coated with sugar. Frozen mango puree (pulp) can be used for making soft drinks, ice-creams, fruit salads etc. The frozen dried and powdered pulp can be used for preparation of juice-mix powder. The freeze-dried pulp also had very good export potential (Gowda, 1995). Whole mango fruits frozen at –30 °F in polyethylene bags were good for sucking after thawing (KrishnaPrakash *et al* 1986).

Mango beverage prepared from 'Badam' and 'Totapuri' cultivars pulp with 60% added sugar and 1% acid and packed in foil laminated pouches and stored at –18 °C has retained colour and flavour even after one year storage period with yeast (*Escherichnia coli*) (Nanjundaswamy *et al* 1988). Nectar prepared from freeze stored pulp was comparable to the nectar prepared from canned pulp (Ramana *et al* 1984). Sweetened puree containing 42% soluble solids can be preserved by canning and freezing with sucrose and high fructose (Avena and Luh, 1983).

Freezing process

In the freezing process, the temperature of the food items is reduced until the thermal centre reaches –18 °C to –20 °C with the consequent crystallization of water present in the material. Crystallization of water during the process decreases water activity of the foods, reducing chemical and biochemical reactions and microbial growth (Begona *et al* 2006).

Fruits of good edible quality, fully matured and evenly ripened should be selected for freezing preservation. However, it is advisable to ripen the fruits artificially for achieving uniform ripening. Ripe mangoes with firm flesh having fully developed colour and flavour, are washed, peeled with stainless steel knives and cut into longitudional slices/ cubes of convenient size for handling and eating. The stone with left over flesh is fed to a pulper and the pulp so obtained is frozen separately. The slices/cubes are then filled in a can/glass jar and fortified sugar is filled in the container, leaving enough head space for the expansion of the packed product during freezing. After sealing the lid, the product is frozen, using one of the different methods (see below). It takes a couple of hours to freeze the product. Freezing should be done as quickly as possible, otherwise in slow freezing, big pointed crystals are formed in the cells, which puncture the cell walls. Such product becomes unproductive and loses its identity and eating quality. The frozen product is then stored at 0 °F for up to 9-12 months (Srivastava, 1967., Singh, 1974, 1977).

Freezing mango slices in syrup

Firm and ripe mango fruits are peeled, sliced (as detailed above) and dried and put in moisture/vapour proof packages with dry sugar or lemonade. The fruit slices are covered with 40-50° Brix syrup, containing citric acid and ascorbic acid and frozen at –30 °C before storing at –18 °C. Pre-soaking of slices in calcium chloride (2%) solution improves texture. Slices frozen in 40° Brix syrup, retain better flavour than those frozen in 20° Brix syrup. Freezing and storage below freezing temperature decreases vitamin C. Air-blast freezing or plate freezing at –40 °C gives better product recovery.

Cultivars like: Amrapali, Pairi, Padiri, Mulgoa, Baneshan, Dashehari, Haden and Irwin are suitable for making frozen slices.

Methods of freezing

Different methods of freezing viz., air-blast freezing, plate freezing, cryofreezing, individual freezing etc. are adopted for freezing mango fruits and its products. The choice of freezing methods depends on the nature and preparation of food (fruit) material.

In general, cryogenic freezing results in faster rate and less moisture loss during freezing. However, exposing the material to cryogenic medium causes internal stress buildup during the process of freezing leading to cracking of frozen material (Marin *et al* 1992). However, the process is laborious and expensive.

Air-blast freezing and plate freezing at – 40° to – 45.6 °C were suggested for obtaining better quality finished product (Issac, 1986).

Pre-treatments

To avoid certain problems of freeze preservation of mango and its products, it is suggested to adopt some pre-treatments (use of citric acid, ascorbic acid, heat treatment etc.) before putting them in storage.

Pretreatments for pulp/puree

Mango pulp can be deep frozen without loss of nutrients, colour and flavour. Heating the pulp is necessary to preserve the quality of mango puree. Hawaii mango puree that has been heated to 90 °C, cooled to 35 °C before filling into tin with polyethylene line and frozen at – 23 °C kept well in storage (Brekke *et al* 1968). Heating of the pulp to 95 °C inactivates the heat resistant enzymes and destroys the spoilage organisms and extends storage life of pulp at – 18 °C.

Pasteurization prior to freezing overcomes the developments of off-flavour during storage (Avena and Luh, 1983). Mango pulp frozen (in a plate freezer at – 40 °C) without pasterurization developed off-flavour during storage, which could be overcome by heating the pulp before freezing. Alternately, the pulp may be pasteurized and filled hot (Ramana *et al* 1984).

Mango pulp with added sucrose (20%) remained in good condition for more than a year at – 18 °C (Bhatnagar and Subramanyam, 1973). Ascorbic acid and citric acid – added to the pulp helps to retain the colour, flavour and aroma during storage (Bhatnagar and Subramanyam, 1973., Ramana *et al* 1984). Mango pulp frozen with added ascorbic acid is similar to the canned pulp with regard to colour, consistency and flavour even after prolonged storage, while the mango pulp frozen without added ascorbic acid was inferior (Ramana *et al* 1984).

Pre-treatments for slices

Preliminary studies by James (1987) indicated that for satisfactory freezing of mango slices, use of 40° Brix syrup containing citric acid (0.1%), is optimum from the point of colour and flavour. But the texture of frozen product is not satisfactory after 3 months storage as the product is slightly mashy and disintegrated after thawing at 4 °C for 17 hours in a refrigerator. Pre-soaking of slices in calcium chloride solution (2%) was helpful in firming the slices during storage (Baldry *et al* 1976). Between the treatments, slices in 15° Brix syrup containing citric acid plus calcium chloride (1%) were firm, but were slightly bitter in taste, though they were acceptable. Mango slices were pre-treated in citric acid and ascorbic acid solution (to prevent discolouration) and packed in cans with sugar syrup (20-50° Brix) and frozen at –30 °C. During storage at –18 °C, colour, flavour and texture even after 12 months were retained well. Slices frozen in 40 °C Brix syrup had more flavour than those in 20° Brix syrup. Rate of freezing and addition of ascorbic acid did not influence the quality of the frozen slices (Singh, 1977).

MANGO (PUREE) PULP

Ripe mangoes are processed into pulp (puree) for making products such as beverages (nectar; juice, squash), jam, jelly and dehydrated products. The puree can be preserved and used in the off-season. Storage allows supply of raw materials during the remainder of the year, when fresh mangoes are not available. Processing of mangoes into puree allows a more economical means of storage of fruits compared with the cost of storing the finishing products, except in case of dehydrated products and provides more orderly processing during peak availability of fresh mangoes. Frozen mango puree can be utilized as a fruit spread for flavouring ice cream or for preparing fruit punch by dispensing 1.0 - 1.5 quarts puree in one gallon of lemonade. Frozen pulp with ascorbic acid, reduces the losses in ascorbic acid, β-carotene, chroma and viscosity to a bare minimum when stored at –18 °C in 14 months.

Raw material and varieties

Over ripe and fibrous mango fruits are used for preparing puree. Bombay Green, Dashehari, Langra, Chausa, Totapuri, Badami and juicy varieties are suitable for making puree.

Method

Puree can be made from whole or peeled fruits. Peeling is ordinarly done by hands using knives, which is time consuming and expensive. Hence it can be avoided. Steam and LYE peeling have been developed. A simplified method involves the exposure of whole mangoes to atmospheric steams for 2.0 to 2.5 minutes in a loosely covered chamber, followed by transfer to a stainless steel tank (Singh, 2007). The steam softened skins allow the pulping by a power-stirrer, fitted with a saw-toothed blade, mounted below a propeller blade. Then the pulp is removed from the seeds by a continuous centrifuge. The pulp material is next passed through a puddle-pulper, fitted with a 0.084 cm screen to remove fibre and reminants of peel (Singh, 2007) (Photos 5.4 to 5.7).

Addition of citric acid and ascorbic acid helps in the retention of colour and flavour of mango-puree (Bhatnagar and Subramanyam, 1973).

Photo 5.4: Washing of mango fruits prior to extract of mango pulp

Photo 5.5: Peeling and removal of mango stones for extraction of pulp

Photo 5.6: Mango cut slices entering mango pulper

Photo 5.7: Mechanical extraction of mango pulp

Packaging and Storage

Mango puree can be frozen, canned or dehydrated and stored in wooden barrels for later use in the off-season. In all the cases, heating of pulp is necessary to preserve the quality of the puree. In one process, puree is pumped through a plate heat exchanger and heated to 90 °C for 60 seconds and cooled to 35 °C before being filled into 30 lb tins with polyethylene liners and frozen at –23.5 °C. In another process, pulp is acidified to pH 3.5, pasterurized at 90 °C, hot filled into 6 kg high density bulk polyethylene containers, pre-sterilized in the boiling water. The containers after sealing are cooled in water and frozen at – 23.5 °C. Storing puree in polyethylene containers makes it possible to avoid high cost of cans (Photoes 5.8 to 10).

In places, where pulp/puree is going to be used for manufacturing jams and squashes, wooden barrels are generally used for storage. First the pulp is acidified adding citric acid (0.5-1.0%), then heated to boiling, cooled and finally SO_2 is added (1000-1500 ppm) to the pulp, before filling into barrels for later use (Singh, 2007).

Photo 5.8: Filling of mango pulp into cans

Photo 5.9: Sealing of filled cans

Photo 5.10: Packing of filled and sealed cans

For long term storage and transport two methods of dehydration can be applied viz., foam-mat drying and spray drying (Baldry *et al* 1976).

Preliminary investigations have indicated the possibility of packaging of mango fruit juices or pulpes and their products in flexible polypropylene pouches (300 gauge) after treating with KMS (Ghosh *et al* 1981).

Mango Bar/Leather

Mango fruit leather (mango sheet, mango slab, mango jelly, ampapad, amwat) is an age old mango product of the Indian sub-continent. It is a dried mango pulp in sheets mixed with a proper amount of sugar/jaggery and acid. Mango fruit bar is a modified product of traditional mango papad/leather, which is prepared by mixing mango pulp and jaggery in 3:1 ratio and sun-dried to 10% moisture level. Mango bar is a fruit confectionary product and is used in a variety of ways. It finds special place in space food, antarcatian food items and combat rations. It is consumed as a confectionary product or cut into pieces and incorporated into ice creams as fruit chunks. It is also cut into small pieces and blended with calculated amount of water to get a RTS mango juice. Mango bar is a rich source of β-carotene (vitamin A) and good source of minerals (iron, phosphorus and calcium). The product has great scope as an export product in the over seas market. Mango leather/bar is prepared as a cottage industry exclusively in Andhra Pradesh, West Bengal and Uttar Pradesh in India. In Andhra Pradesh, mango nurseries to obtain seednuts for raising rootstocks, carry mango leather manufacturing as a side business.

Raw material and varieties

Generally sound fruits are used for making mango leather. However, smaller, ill-shaped fruits, that do not fetch good price as fresh fruits and discarded on the farm as well as at the markets are also utilized. Besides ripe fruits, mango puree (pulp) is also used for making the bar (Mathur *et al* 1972), cull fruits can also be used. To spread the season, cold-stored fruits are also used.

An account of its low cost, meaty pulp and longer keeping quality 'Totapuri' cultivar is mostly preferred for mango leather (Thandra) in Andhra Pradesh. Kolanka Goa is another cultivar which is used in the state. "Alphonso" is also utilized to some extent, but its high cost discourages the industry to go for it. Other cultivars like Baneshan, Bombay Green and Dashehari are also employed. The cultivar, having good retention of carotenoids and less browning with high organoleptic score is preferred. Out of the six commercial cultivars tested 'Badami', 'Banganapalli' and 'Totapuri' gave fruit bar of soft texture, deep yellow to orange red colour with characteristics mango flavour (Nanjundaswamy *et al* 1976).

Methods of preparation

The method of preparation of mango leather/bar is quite simple and involves mainly mixing the extracted pulp with suitable ingredients and drying the pulp in sun or in a dryer using cross – flow air drying technique to around 15% moisture. However, there are more than one method of preparation of mango fruit bar.

Method followed in Andhra Pradesh (India)

Thandra (mango leather) making is done by sundrying in the state. Fruits are first ripened in paddy straw and palmyrah leaves. After washing the ripened fruits with ordinary water, the fruits are cut longitudinary into 3-4 pieces and the peels and seed stones are separated (sometimes the pulp is squeezed manually without slicing the fruits). Pulp adhering to them is scooped out with a sea-shell. Powdered jaggery or sugar is added to this pulp in wooden tubs in the ratio of 1:2 or 1:4. This mixture is force-filtered through a bamboo sieve. Mats, made from closely woven bamboo or wild date leaves (1–1.25 × 1–1.5 m) are used for sun-drying the pulp. These are spread in sunny spots either on the ground or on plat forms (1.25 cm heigh). Both home compounds and protected places on the farm are used. A thin coating of edible oil is smeared on the mats to facilitate easy removal of dried product. After pouring on the mats the pulp is uniformly, spread on the mats with hand. After the first layer has dried up, second one is spread on the first layer. Likewise 3-6 spreadings are made in a day between 10 AM and 4 PM, which are the most sunny hours. Every evening mats are staked (5-10) and covered with palmyrah leaf umbrella or tarpaulins to protect from dew. If sticky, a white cloth is used to separate two thandras. More sugar is added to top layers. After attaining required thickness (1.50-4.0 cm) thandra is cut into convenient slabs and redried before storage. The dried product has a light yellow-amber colour and possesses a delicious taste (Rameshwar, 1979) (Photoes 5.11 to 5.13).

Photo 5.11: Cutting of mango fruits for extraction of pulp from mango fruits for making mango Tandra (Leather)

Improved methods

Traditional method of sundrying the pulp for making fruit leather/bar is slow, unhygienic, yields a dark brown product, and carries dust, insect eggs, very sticky, gets spoiled fast and does not store long and suffers from certain problems and is not possible during rains and the losses are high during rainy season. To overcome these demerits and problems, a few improved methods were developed.

An improved process for the preparation of 'ampapad' was developed at Central Food Technological Research Institute, Mysore (India) and Patented (Siddappa *et al* 1953) in which the edible portion of the fruit is steeped in acidified sugar solution over night before pulping and drying the following day for improvement of colour, flavour and taste of the finished product.

Srivastava (1967) described a general method of making mango bar as follows:

The extracted pulp is strained through a 50-60 mesh stainless steel sieve and heated to 80 °C, mixed with 5% sugar, 0.05% citric acid and 0.1% KMS. The pulp is then spread in a thin a layer on trays having a tinned surface to avoid corrosion and smeared with glycerine or edible fat. The trays are kept in driers and when the first layer is dried another layer of pulp is put over it and dried in the same manner. This process is repeated until a sufficiently thick layer (1.5-3.0 cm) of the dried pulp is obtained. Then it is cut into pieces (slabs) of suitable size and shape and wrapped in butter paper or alkathene paper and finaly packed in glass jars.

Mathur *et al* (1972) prepared mango bar from ripe mango puree, through 3 stages of drying (at 80 °C for 1 hour, followed by 2 hours at 70 °C and 2-3 hours at 65 °C) in a cross-flow-air-drier, while Nanjundaswamy *et al* (1976) and later Rao and Roy (1980b) used simple stage drying at 60° or 70 °C using sugar addition to the puree, to raise the TSS content and reduce the drying period. Mango bar thus prepared was soft in texture, attractive orange yellow in colour with characteristic mango flavour. Addition of pectin (1%) and liquid glucose (6%) to the pulp gave a product of smooth texture (Mathur *et al* 1972., Rao and Roy, 1980a).

Scientists at Indian Institute of Horticultural Research, Bangalore (India) developed two methods of preparation of mango bars.

In the first method, the fully ripe fruits ['Alphonso'] are washed, peeled and pulp is extracted by passing through a pulper-cum-finisher. The extracted pulp is heated to 80 °C for 30 minutes. Sugar (20%), citric acid (0.3%) and KMS (7000 ppm) are added individually or in various combinations to the pulp. The prepared pulp is spread as thin layers on stainless steel plates, smeared with butter/glycerol and dried in a tray drier at a temperature of 65 °C. The finished product is cut in the form of slabs of suitable sizes and packed in polybags of 400 guage.

In the second method, mango pulp is mixed with sugar (20%) and citric acid (0.3%) and heated at 80 °C till the pulp mixture reached 40° Brix. The mixture is cooled and KMS (7000 ppm) is added and dried in sheets in shade or tray drier. The dried sheets are cut in the form of slabs of suitable sizes and packed in polybags. The fruit bar thus prepared from 'Alphonso' fruits was of best quality having good colour, texture and flavour (Gowda *et al* 1995).

Photo 5.12: Spreading of mango pulp on mats for drying in sun

Photo 5.13: Lifting of dried mango pulp from mats

Nanjundaswamy (1991, 1997) described two methods of mango bar preparation.

In one method, ampapad is prepared from ripe mangoes by mixing mango pulp and jaggery in 3:1 ratio, spread on a bamboo mat and sundried to about 10% moisture level. The dried product is cut and packed in cellophane paper.

In another method, fruit bar is prepared by mixing pulp with a calculated amount of sugar (sucrose) heated to 80 °C, cooled and then mixed with KMS. The pulp is then spread on a tray and dried in a cross-flow air-drier at 70 °C until the moisture is below 15 %. The dried sheet is cut into pieces of suitable size and wrapped in cellophane paper.

In the method of making fruit bar at Central Institute for Sub-tropical Horticulture, Lucknow (India), homogenized mango pulp is mixed with KMS (2 g/kg pulp) and spread, on trays smeared with oil and kept for drying in solar dehydration/sun (Photo 5.14). After drying of one layer another layer is spread over it and allowed to dry. This process is repeated as per the desired thickness. Finally, the sheets are cut into pieces and wrapped in butter paper/polythelene sheet/cellophane sheet (Nigam *et al* 2007).

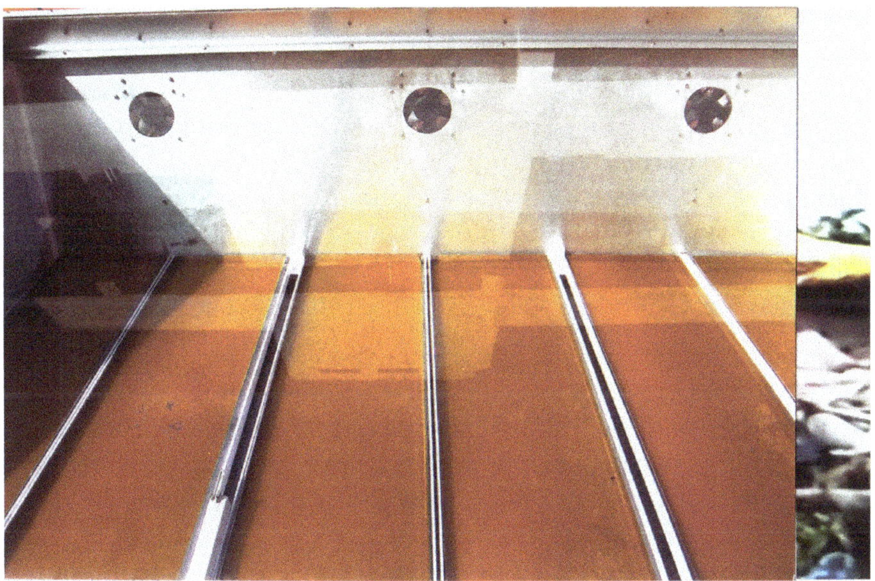

Photo 5.14: Improved method of mango leather making

Mango leather prepared by traditional method of sundrying lacks good colour, flavour and texture and needs improvement in these aspects. Heikal *et al* (1972) improved the flavour and texture of the product by the addition of citric acid and pectin. Addition of sugar improved the colour, while the addition of citric acid improved the flavour of the finished product. However, the addition of sugar increased the drying rate, as compared to the pulp without addition of sugar. Addition of pectin improved the texture of the product but pectin at higher levels impaired the flavour by masking the original flavour.

Ingredients

Sugar/jaggery, citric acid and potassium metabisulphite (KMS) are the common ingredients in mango leather/bar. The addition of these ingredients facilitates the drying and improves the product quality.

Sugar: Sugar is added to increase the sweetness of the product as well as to act as a preservative. Sugar increases the drying time, yield and content of the sugar and moisture of the final product (Gowda *et al* 1995).

Citric acid: This is used to impart a fresh-fruit 'Tanginess". It provides uniform acidity in the product, resulting in improved palatability. It is added to bring down the pH level of the product and to inactivate un-desirable enzymes causing browning.

Potassium metabisulphite (KMS): Addition of KMS provides considerable reduction in non-enzymatic browning and improvement in colour of the product (Gowda *et al* 1995). Further, KMS retards microbial spoilage (Mir and Nath, 1993), however, losses of SO_2 occur during drying (Mir, 1990).

Blends

The consistency of the pulp of the mango varieties, used for making fruit bar is not quite ideal, as thicker the consistency of the pulp, better is the texture of the final product. Thinner pulp can be mixed with banana at 3:1 ratio to get a good quality mixed fruit bar. This can be further improved by adding 1.5%, pectin (w/w) and 6% liquid glucose (w/w). Amongst different fruit combinations tried, mango plus pineapple was quite promising with respect to flavour, texture and storage stability (Muthur *et al* 1972).

Use of pectin in the preparation of ampape, is a costly proposition. Attempts were made to replace the use of pectin by adding guava pulp which is a rich source of natural pectin and vitamin C. The blended product obtained by the combination of mango and guava is a good sources of natural carotenoids from mango and of vitamin C from guava. Mango pulp with 20% guava gave a better sheet in drying followed by the one with 15% guava pulp. The sheet with 20% guava pulp was superior in colour, flavour and texture. Retention of ascorbic acid was higher in guava blended mango sheets after dehydration (Hemarker *et al* 2000).

Fruit bars generally lack protein and fat. To enrich them in these two, they are often blended with others rich in these. Supplementation of mango pulp with soybean slurry improved the taste, texture and overall quality of the product. A nutritious fruit bar can be prepared by blending mango pulp with soybean in the ratio of 70 : 30, having 14.5% moisture, 11-35% protein and 50 mg per 100 g ascorbic acid, which was the best in respect of sensory qualities like flavour, texture and taste without 'Beany' flavour in the final product (Chauhan *et al* 1997).

Quality

An ideal mango leather/bar should have not more than 15% moisture and a TSS : acid ratio of 25 : 05. However, these quality parameters are influenced by different factors like methods of preparation, drying time, thickness of the pulp, relative humidity in storage etc.

Methods of preparation: Among different methods of preparation, drying in shade or tray driers gave better quality product in terms of better colour, texture and flavour leading to higher overall acceptability as compared to sun and solar drier drying. Sun-dried mango leather often develops brown colour (Rao *et al* 1961). Good colour of the shade and tray drier drying may be due to higher amount of SO_2 retention in fruit bar and good flavour may be due to low temperature drying in shade while it was due to controlled temperature and relative humidity in tray drier (Gowda *et al* 1995). Freeze drying at –20 °C or –40 °C resulted in considerable disruption of the cellular structure, but it remained normal when dried at –19 °C after reconstitution (Ramamurthy and Bongirwar, 1979).

Drying time: Drying time was maximum under shade and minimum in tray drier. Sun drying and solar drier are intermediate. Moisture content was more in shade drying followed by tray drier drying, and it was lowest in solar drier due to over drying. Non-enzymatic browning was significantly very low in case of shade and try-drier products. Reducing sugar content was higher in solar drying due to over drying. SO_2 was best retained in shade and tray driers, due to low temperature/controlled temperature drying (Gowda *et al* 1995).

Thickness/Consistency of the pulp: Variety of mango and thickness of the pulp show definite influence on the quality of the fruit bar. Thicker the consistency of the pulp, better is the texture of the final product (Nanjundaswamy *et al* 1976). In case of thinner mango pulp, the consistency can be improved by mixing mango pulp with the pulp of other fruits like banana and pineapple (see above).

Relative humidity (in storage): Mango fruit leather/bar is a hygro-emission product and keeps well in the range of 50-60% relative humidity (RH) without any noticeable change in colour, flavour and texture. Below 50% RH, the product losses its moisture content progressively and becomes hard in texture, although the colour and flavour are good. At 70% RH and above the product absorbs moisture rapidly and the colour and texture are affected much, but without much effect on flavour (Nanjundaswamy *et al* 1976).

Ingredients: Sensory evaluation of the mango bars revealed that colour was better with KMS. The texture and flavour were better with sugar, while falvour was best with the addition of sugar, citric acid and KMS. The mango leather without the addition of ingredients was inferior in quality.

The ideal ratio of TSS : acid for getting a better product is 25:05 (Rao and Roy, 1980b). Addition of pectin at 0.5 to 0.75% improves the texture of the dried product to the desired level (Mathur *et al* 1972).

Problems

Main problems in the preparation and storage of mango leather/bar are : non-enzymatic browning, crystallization and stickiness.

Non-enzymatic browning: When the fruit bar/leather is made without any additives and preservatives, it exhibits non-enzymatic browning. With the addition of citric acid and KMS, non-enzymatic browning can be lowered. It increases with increase in storage temperature (Gowda *et al* 1995).

Crystallization of sugar: This is a common problem associated with fruit bar/leather during storage. Sugar crystals form on the surface of the product, firmly sticking within product and are not freely removable (Photo 5.15). Crystallization is due to the function of packing materials, but may be due to the alteration in solid-liquid phase equilibrium of mango bar/leather with respect to storage condition (Chandrasekhar and King, 1971). It could be retarded by addition of liquid glucose. Addition of invert sugar also prevents crystallization, but it makes the dehydrated sheet very sticky and subsequent handling difficult.

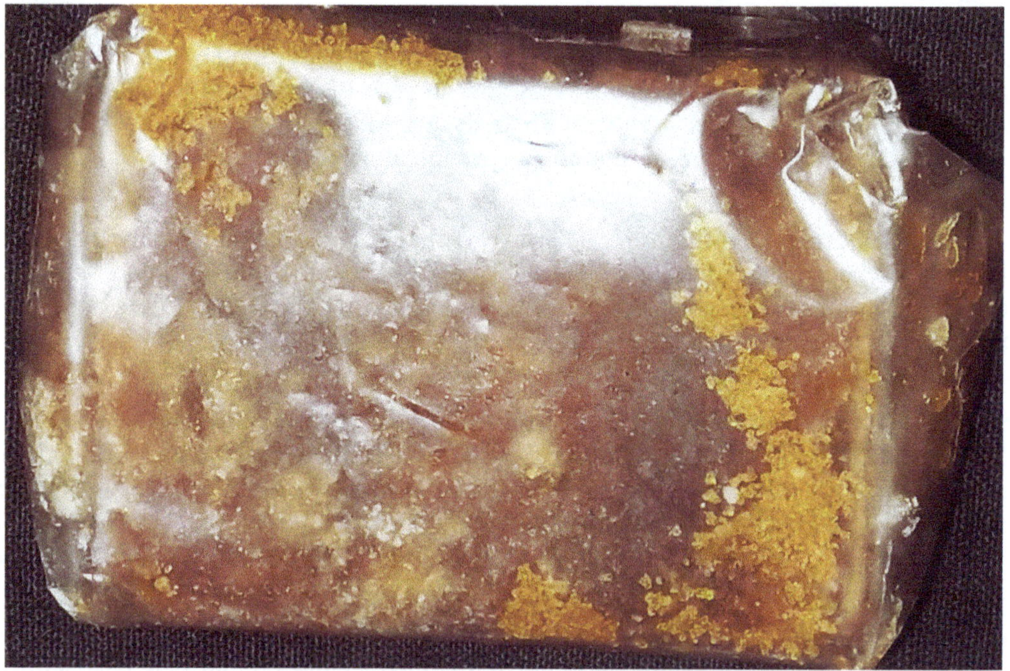

Photo 5.15: Crystallization in packed mango leather

Stickiness: The dehydrated sheets of fruit pulp were found to stick to the trays in spite of applying glycerine to the trays, due to uneven drying, as it was difficult to spread the pulp uniformly on aluminium trays (Mathur *et al* 1972). Mango bar is slightly sticky, but does not constitute a major draw back. However, it should be avoided. Effect of addition of starchy carboxy mthyl cellulose, sodium alginate, pectin etc. to pulp and dusting the surface of fruit bar with sugar were tried to prevent slickiness unsuccessfully. However, secondary drying of the cut pieces of bar/leather for a few hours appears to be helpful in reducing the stickness (Nanjundaswamy *et al* 1976).

Package and storage

Mango fruit bars/leathers, after removal from the drying, are cut into pieces of 5×5 cm size having a thickness of about 0.3 cm and arranged in 3 layers make up blocks of about 0.9 cm thickness with weight between 25 and 28 grams. Each block is separately wrapped in butter or polyethylene or cellophane paper. Unit pack consisting of two blocks is filled in printed cellophane bags (15×8 cm size) (Singh, 2007). Such packages are stored in air tight tins (Nanjundaswamy *et al* 1976) or cartons (Singh, 2007). Storage in tins will be for one year at ambient temperature (25°-28 °C) and about 6 months at 37 °C (Nanjundaswamy *et al* 1976). For longer storage life at accelerated conditions, mango bars must be packed in aluminium-foil, laminated pouches where, minimal changes in moisture, SO_2 and carotenoid pigments occur (Nandanasabhapati *et al* 1993).

Mango fruit bars/leathers should be placed in a cool dry place for storage. The ideal moisture to have storability lies between 65 and 75% (Hemarkar *et al* 2000). At high RH, the product has a tendency to gain moisture and become mouldy while at low RH, it looses moisture and becomes hard.

CANNING OF MANGO PRODUCTS

Among the different methods of preservation of fruits and vegetables, canning is the better option than freezing and chemical methods of preservation. Canning is a method of preservation, in which the product of fruits and vegetables are preserved in sugar syrup. Canning involves the application of high temperature to destroy all microorganism in the preserved product plus air-tight sealing to prevent recontamination. The degree of heat and length of time of heating vary with the type of fruit and vegetables and the kinds of microorganism that are present in the products.

Mangoes are canned as whole fruits, pulp, slices, cheeks, quarters or shoulders and to a limited extent as cubes/discs. Ripe mangoes are supplied to Defence Forces in the form of canned slices in sugar syrup. Mango pulp/puree is canned for long term storage and marketing. Canned pulp finds application in a variety of products in food industry such as beverages, ice-cream, yoghurt, milk shacks etc. The development of mango industry has increased the demand for canned products of mango. Canned mango pulp also has good export possibilities. In fact, among the different preserved and processed products of mango, canned pulp occupies the top most position from India.

Canning of whole mangoes

Indians residing in foreign countries have a craving for mangoes which could be eaten out of hand, by sucking in typical Indian style. This created a demand for canned whole mangoes outside the country. Mehta *et al* (1971) described the method of canning of whole mangoes as follows :

The fruits are thoroughly washed in clean water. The stem-end-of the fruits is cut off with a sharp stainless steel knife and the acrid juice (sap/latex) is pressed out. The prepared fruits are then filled in 401 × 411 plain or lacquered cans at 5 fruits in each can. Sugar syrup (25 °B) with 0.3% powdered citric acid for clarification and maintenance of final acidity in the range of 0.2 – 0.3% in the end product, is boiled and cans are filled leaving 1.3 cm head space. The filled cans are exhausted in a water bath for 10 minutes till the temperature in the centre of the can reaches 82 °C. The cans are then immediately sealed hot and processed at a temperature of 100 °C in boiling water for half-an-hour and cooled in running water, wiped dry and stored at room temperature (15-42 °C).

Canning of mango pulp

The mango beverage industry has great demand for canned mango pulp/puree.

Varieties: Varieties which have low/no fibre and high quantity and quality of pulp are best suited for making canned pulp. Totapuri and Neelum from South India, Dashehari, Fazli, Hasina, Chausa and Langra from North India and Alphonso from Western India were adjudged as suitable for pulp making and canning (Anon, 1997).

Essential requirements for canned pulp

1. Quality of raw material is of vital importance.
2. Brix need to be maintained at 16°-18°
3. The pH value of the finished product should be around 4.
4. Pasteuration of filled and sealed cans is essential.

Method of canning pulp: The ripe mangoes are washed, peeled or cut manually before extraction of the pulp. Two pulpers are used to obtain the mango pulp in finished form. The first one removes the pulp from the stones and/or peel and at the same time, pulp is sieved through 1/8" or 1/16" sieve, while the second one (of smaller size) is used to obtain pulp of uniform texture and to remove pieces of peel that may be coming out of first pulper in which sieves of (1/16" or 1/32") are used. The pulp is next standardized to

14-10° Brix and 0.5-0.6 % acidity as citric acid by adding sugar syrup and citric acid respectively and pH of 3.8-4.0 before heating to 85 °C. No preservative is used in canned mango pulp. Pulp is hot-filled into cans and sealed, followed by processing for 45 minutes at 100 °C, cooled, labelled and stored. However, when the pulp is filled in polyethylene jars for own use later as a base material for beverage and jam making, KMS is used as a preservative. After standardization, the pulp is heated within a range of 85-95 °C in a jacketed Kettle. The pulp is filled hot in pre-heated cans. Then after adjusting the weight of can, sealed and kept in a crate for retorting. Sterilization of cans is done in a vertical retart using steam or boiling water as heating medium for 20-40 minutes. The cans are cooled in cold water and stored (Kumbhar, 1992).

Packaging and Storage of canned mango pulp: Usually tin cans of variable size (A10 and A12) are used for packing canned mango pulp. Alternately wooden barrels (Siddappa and Krishnamurthy, 1963), high density polyethylene containers (Ambadan and Adsule, 1979) and flexible pouches (Ghosh *et al* 1981., Krishnamurthy *et al* 1982) are recommended. Kalra (1982) stored canned mango pulp in plain and flexible polyethylene containers with SO_2 (100 ppm) under cold (0-4 °C) and room temperature (30°-35 °C) conditions for 6 months. The pulp preserved much better under cold storage conditions with less changes in reducing sugars, titrable acidity and vitamin C.

Canning of mango slices (Photo 5.16)

For canning slices, fully matured and firm ripe or semi-ripe fruits are needed. The fruits should have crisp and succulent texture, pleasing flavour and attractive natural colour. The stone should be small. Fruit must be uniformly ripe, keep long on storage and retain texture on processing. Mango varieties with fine texture and strong aroma are preferred (Alphonso, Pairi, Padiri and Mulgoa). Other varieties such as Dashehari, Totapuri, Banganapalli and Fazli are less preferred because of their mild flavour and pale colour of the pulp. In Karnataka state (India), Baneshan, Suvarnarekha and 5/5 Raspuri X Langra were found to be most suitable for canning as slices (Murthy *et al* 1984). The varieties Baneshan, Langra and Dashehari have a good pulp, and slice yield while Bombay Green and Chausa have poor pulp and less slice yield. Colour and flavour of Dashehari are the best among the above varieties. However, texture of Baneshan scored highest in canned and frozen slices during storage. The overall assessment showed that Dashehari was best as canned and frozen slices followed by Baneshan, Langra, Bombay Green and Chausa after 9 months of storage. Mallika was also found to be the best for canning as slices compared to other cultivars in respect of texture and flavour. However, industry preferred mostly 'Totapuri', a less expensive fruit as compared to others especially. 'Alphonso' and also due to its high pulp content and mild flavour, which is more acceptable in the export market (Ramteke *et al* 1999).

Photo 5.16: Canned mango slices in a glass bottle

Method of preparation: Siddappa and Bhatia (1956a), Srivastava (1967), Gowda (1995) and Nigam *et al* (2007) have described the method of canning of mango slices in sugar syrup, which is prepared by dissolving 200 grams of sugar and 1 gram of citric acid in 800 ml of water for one kilo of slices.

Firm fully ripe fruits of suitable variety are thoroughly washed and the stem-end along with the neighbouring tissues is removed. Then they are peeled uniformly with a stainless steel knife, taking care that the green portion of the fruit is not left over, as it would turn browning during processing. The peeled fruit is kept in 2% common salt solution to avoid enzymatic browning. After peeling, this fruit is sliced longitudinally. Then the slices are placed in plain cans and the slices are covered with sugar syrup of 30°-40° Brix containing 0.25-0.5% citric acid. About 225 cc of syrup is required for 500 grams of butter cans and 350 cc for A ½ cans. The filled cans are then exhausted for 15 minutes at 100 °C in boiling water, to drain out air from the fruit tissue, to prevent discolouration as well as loss of flavour during storage. For exhausting, the filled cans are placed in a large kettle or open tank containing boiling water in such a way that some space (5 cm) at the top of the can remains above the level of the water. The kettle is covered and heating is continued till the centre of the can attains a temperature of 80°-82 °C with in 5-6 minutes. After exhausting, the cans are removed and the top is sealed. The sealed cans are placed in a large pan containing hot water in such a way that 2.5 -5.0 cm of each can remains below the level of water after heating. The bottom of the can is covered with a piece of thick clean cloth to act as an insulator and also to prevent breakage of glass jars, when used. The processing in boiling water (100 °C) for 500 grams cans should be done for 12 minutes and for A 2 ½ cans and glass jars (pint size) for 15 minutes. Immediately after processing the cans are cooled to avoid stack burning and browning of material. However, glass jars should not be exposed to cool surface water or air drought. After cooling the cans/jars should be stored in a cool dry place (Srivastava, 1967). Spin pasteurization is preferred and cans can be stored at 25°-30 °C for one year. β-carotene content did not change during canning and ambient storage for 200 days in cvs. Langra and Chausa, while ascorbic acid decreased by 20% after canning and another 33% during storage.

Gowda (1995) described almost similar process as above, except that he used sugar syrup having 50° Brix and 0.5% citric acid and advised to leave about 0.6 cm head space in the can, with processing in boiling water, for 15-30 minutes. Nigam *et al* (2007) recommended the addition of 350 ppm SO_2 besides 0.3% citric acid to sugar syrup and sealing with wax.

Canning mango juice

After pulping, hot sugar syrup, citric acid and water are added to make juice as per specification. The mango juice is then heated to 95 °C in a steam-jacketed vessel. The hot juice is filled in pre-sterilized cans and sealed. The retorting is done for 20-25 minutes for A-2 size cans. The remaining steps are similar to the method of making canned mango pulp (Khumbar, 1992).

Out of the IIHR hybrids, 'Arka Puneet' and 'Arka Anmole' were suitable, while 'Arka Arun' was not suitable for canning juice in comparison with 'Alphonso' and 'Totapuri' (Gowda *et al* 1994). 'Alphonso' was most suitable for making canned juice, due to its very good colour, consistency and flavour, followed by 'Padiri' 'Banganapalli'. 'Kesar' and 'Langra' gave canned juice comparable to 'Totapuri' (Gowda and Ramanjeneya, 1995).

Problems

The following problems are commonly encountered in the processing and storage of mango canned products.

Mashiness: This is a serious problem in which the canned product looses its texture. This can be avoided by using firm and semi-ripe fruits. Calcium treatment of slices was also suggested. Dipping the slices in a 0.1% calcium solution for 15 minutes was effective, but it adversely affected the flavour during storage (Ranganna *et al* 1961) and taste (SuryaPrakash Rao *et al* 1968).

Loss of flavour: In canned 'Dashehari' mangoes, the characteristic flavour of the mango fruit is not retained in the product, due to losses during processing and subsequent storage

(Saha *et al* 1976). Addition of acorbic acid to the syrup at various levels gave better acceptability for flavour, through the retention of flavour since, it is an efficient oxygen scavenger and reducing agent. Further, it is a natural constituent of the mango fruit. It is comparatively cheap and easy to incorporate into the product. A dosage of 100 mg/100 g syrup was found to be ideal for the best retention of flavour in mango canned products (Saha *et al* 1976).

Bulging of cans: Often the cans, in which mango pulp is canned, develop bulging during storage, resulting in their rejection. Anareobic microorganisms, which produce tinny bubbles of gas are responsible for this problem. Swelling of 'Alphonso' mango pulp was caused by the facultative anaerobic, spore formers (*Bacillus licheniformis)* as well as by obligate anareorbic spore former (*Clostridium sporogenes*). *B. licheniformis* raises the pH of the nutrient bath from 5.0 to 9.5 after 30 days of inoculation at 37 °C. The pH elevated by *B. licheniformis* has been reduced by acid formed from glucose and other sugars by *C.sporogenes*. The bulging of cans due to the presence of above spoilage organisms can be prevented by thermal process, to destroy them (Azizi and Ranganna, 1993).

Tin-Poisoning: Tin content in canned products is of serious concern regarding the acceptability of the product. Large quantities of canned mango products exported to overseas markets are rejected because of high tin-pick up, causing considerable loss to processing industries and exporters. Tin content and iron content in excess of 200 ppm and 20 mg respectively impart metallic taste to the product (Usharani *et al* 1992).

Various factors influence the tin-pick-up in canned products. Processing variables associated with canning such as variation in head space, oxygen in the head space, vacuum, insufficient exhaust, filling temperature, cooling and storage conditions of the cans adversely effect corrosion of the tin plate containers (Mahadevaiah *et al* 1975a).

Different types and sizes of cans are used for canning mango products. Polyster lacquered cans are better for canning mango juice in comparison with expoxy phenolic lacquer with reference to taste and flavour. Mango nectar canned in lacquered cans presented off-flavour and slight bitter taste (Mahadevaiah and Gowramma, 1979). Hot dipped tin plates are better than electrolytic tin plate for canning mango nectar. Electrolytic tin plate with tin coating of less than one 1b per base box is not suitable for canning mango produce. Corrosion was more in big-sized cans (401 × 411) with mango nectar (Mahadevaiah *et al* 1975c).

Time lag between sealing and processing, high cooling water temperature, presence of nitrate above 5 ppm and increase in acidity accelerate corrosion (Mahadevaiah *et al* 1975c).

Corrosion was less with increase in the concentration of mango pulp in the juice. Increase in the concentration of acid accelerated corrosion. Among different fractions of mango pulp only organic acids such as citric acid or maleic acid were highly responsible for corrosion in nectar during storage. β-carotene, however, did not contribute to corrosion (Mahadevaiah *et al* 1975a).

Tin corrosion was more in pulp and nectar prepared from unpeeled mangoes. The mango pulp contains gallic acid and allergic acid of which the former acts as an accelerator of corrosion (Mahadevaiah *et al* 1975a). Tin content in the nectar prepared from stored pulp was 50% less than in the case of nectar from fresh pulp during storage of 12 months at 37 °C (Mahadevaiah, 1976).

Tin corrosion in canned products of mango varies with the variety (Mahadevaiah, 1976., Adsule and Roy, 1974a., Tripathi *et al* 1985). Mango nectar prepared from 'Banganpalli' variety showed less corrosion than that from 'Badami' and 'Raspuri' varieties (Mahadevaiah, 1976). High viscosity in 'Banganapalli' mango nectar inhibited corrosion. Appreciable variation in the extent of tin discolouration was observed in canned pulp from four different varieties (Dashehari, Langra, Safeda and Chausa) tested during storage. The tin-pickup increased linearly with increase in storage period in all the varieties. At the end of storage period of 12 months, 'Safeda' showed higher amount of tin pick-up (275.8 ppm) followed by 'Chausa' (259.76 ppm) and 'Langra' (247.27 ppm), with the lowest being in 'Dashehari' (245.22 ppm). The high level of tin pick-up in 'Safeda' was due to the initial acidity level in this variety (0.17% as citric acid). Mahadevaiah *et al* (1975a) observed that the higher level of acidity resulted in accelerating the degree of tin corrosion in canned product. Adsule and Roy (1974a) suggested that higher amount of ascorbic acid in 'Langra' might have served as an anti-oxidant and counteracted the accelerating effect of higher acidity on dissolution of tin. However, acidity is not the single factor influencing the tin pick-up. Polyphenolic compounds, some enzymes, Paroxidase and Pseudo-peroxidase might play some role in enhancing the rate of tin pick-up. Varieties having low acidity and high ascorbic acid are best suited for canning as they can be stored for longer time, retain natural quality, remain acceptable and show minimum tin corrosion. 'Dashehari' is one such variety (Tripathi *et al* 1985).

Due to development of metallic taste and huge losses to the exporter as a result of rejection of exported canned products, it is highly essential to prevent/inhibit tin corrosion in canned products. Among the different additives (ascorbic acid, glucose oxidase, gelatin, agar-agar, cystein and several gums and amino acids) tested, cystein was best in checking tin corrosion in mango products (Mahadevaiah *et al* 1969). Addition of carboxymethyl cellulose or gelatin and sodium alginate showed beneficial in reducing this corrosion in canned mango nectar (Mahadevaiah *et al* 1969).

De-aeration of mango juice before filling into cans helped in reducing the extent of corrosion during storage (Mahadevaiah *et al* 1975b). Exhausting of filled cans before sealing and reduction in processing time helps in reducing the extent of corrosion (Mahadevaiah, 1976).

BULK PRESERVATION OF MANGO PUREE

Since the mangoes are seasonal and available for a short period and to run the processing unit for as long period as possible, the industry preserves the mango pulp/puree in bulk for later use for the manufacture of other processed products like jam, squash, fruit bar etc. as the conventional process of canning in cans has become expensive.

Method of preparation of mango pulp/puree

Fully ripe mangoes are washed, peeled and cut into slices for easy extraction of pulp. These slices are homogenized into pulp, which is filtered through a sieve to remove the fibres. The homogenized pulp is heated to 76-78 °C and 2 g citric acid and 2 g of KMS are added per one kg of pulp. In sweet varieties 3 g of citric acid per kg of pulp is added and the pulp is heated to 80-85 °C. Addition of ascorbic acid to the pulp is helpful in better retention of colour, flavour and carotene (Subbaiah, 1961).

Bulk preservation of mango pulp/puree: On small scale either cans or glass jars are utilized for preservation of mango pulp/puree. However, on large scale mango pulp is bulk preserved in wooden barrels using SO_2 (1000 ppm) for later use (Siddappa and Krishnamurthy, 1963). In India, stainless steel silos (capacity 10 metric tones) are being used for storing such sulphited pulp at ambient storage conditions (Najundaswamy *et al* 1988).

A thermal process for bulk packing of mango pulp in HDPE containers has been developed by which it is possible to save 30-40% of cost of packaging without sacrificing the quality during storage of 6 months (Shrikande *et al* 1976). In this method, the required quantity of citric acid (0.5-1.0%) is added to pulp in order to lower the pH to 3.8-4.0. The pulp is then heated to 85 °C and cooled at 30 °C. A calculated amount of KMS (1000-1500 ppm) is added to the cooled pulp, so that pulp contains 1000 ppm SO_2 finally. The prepared pulp is filled into the pre-cleaned and sterilized barrels, sealed air-tight and stored for later use (Bhatnagar and Subramanyam, 1973).

MANGO JAM

Mango jam is the conventional product of India. However, it is not very popular. Mango jam is prepared from both pulp and slices. It either alone or in combination with other fruits is prepared commercially.

Method of preparation

Fully ripe mangoes having characteristic flavour and good aroma are washed, peeled and pulp is extracted with the help of a pulping machine. The pulp is then treated with an equal quantity of sugar to 65-70° Brix. Citric acid is added at the end to get 0.6-0.7% acidity in the final product and to prevent crystallization and also change sugar : acid blend to improve the taste. Pectin (0.4-0.5%) is added to improve the set. Prepared jam is filled hot in sterilized containers, sealed air-tight, sterilized again in boiling water for 5-10 minutes, cooled and stored (Srivastava, 1967).

Jam is also prepared from either fresh or dehydrated slices. The fresh mangoes are washed, peeled and cut into small slices with a stainless steel knife. The pulp from the slices is treated with an equal quantity of sugar to 65-68 °C Brix and citric acid is added at the end to get 0.6-0.7% acidity in the final product. The mango jam thus obtained is highly acceptable with respect to colour, flavour and texture (Sagar and Khurdiya, 1998).

Recipe

Ripe mango slices	-	120 g
Sugar	-	120 g
Citric acid	-	9.6 g

The amount of sugar required should represent at least 60% of the weight of the mangoes prepared (Daughty, 1995). The cooking of pulp/slices is done in two stages.

In the first stage sugar (70%) and lemon juice (for citric acid) are added and stirred well during the entire cooking until 55° Brix is reached.

In the second stage of cooking, the remaining quantity of sugar and lemon juice are added and stirred well during the entire cooking until 67-68° Brix is reached.

The jars are filled while the mixture is still hot. During the operation of filling, jars must be stirred with the handle of a wooden spoon in order to eliminate the air that has entered the jars. The jars are closed with screw-tops, cooled and labelled before storage-(Dauthy, 1995).

MANGO CONCENTRATES

Preservation of single strength juice/pulp is not economical. Therefore, it is advantageous to concentrate them. It is an important export item and its demand is increasing day by day. Mango concentrate has several advantages. Concentration of pulp by reducing bulk, results in economy of handling, packing, storage, transportation and distribution of costs. However, the main disadvantage is the loss of flavour.

Mango concentrate prepared by drying mango pulp (puree), is used as a base for making beverages etc.

Methods

Mango pulp is concentrated by four different methods :

(i) Evaporation of pulp, (ii) Separation of serum, which is concentrated in reverse osmosis process and mixed with residual pulp, (iii) by drum drying the mango pulp and (iv) freeze concentration.

Method I: The extracted pulp is concentrated in a forced circulation evaporator under vacuum up to 30°–35° Brix, which facilitates easy removal of the concentrate from the evaporator (Nanjundaswamy and Gowramma, 1979., Ramteke, 1987). Aroma recovery during or before concentration enables effective restoration of mango aroma. The stripped juice and aroma concentrates can be mixed together to obtain full flavoured concentrate

(Ramteke *et al* 1999). The concentrated pulp retained high values of ascorbic acid, β-carotene and total carotene and a fairly good amount of volatile flavour. This method is now adopted on commercial scale in India (Ramteke, 1987., Nanjundaswamy, 1987, 1991).

Method II: In this method (split process) mango pulp is treated with pectic enzymes (cellulose, pectinase etc.). The serum is separated from fibrous pulp residue, concentrated and remixed with fibrous pulp residue and finally homogenized to obtain 40° Brix concentrate (Gupta and Girish, 1988). Treatment of mango pulp with a 20% enzyme (Cellulase and pectinase) is optimum to obtain serum yield of 70% for concentration. The serum so separated and having 15% total soluble solids which when blended with the pulp in the ratio of 1: 2 yields a concentrate having 45% total sugars. The loss of aroma is least in serum-pulp method of juice concentration and the product can be stored in aluminium pouches at 10 °C for 11 months.

Method III : In this method, mango concentrate is prepared by drum drying of mango pulp at 45 psig steam pressure, drum speed of 4 rpm and drum clearance of 0.3 mm. The TSS in the pulp is raised to 95% in the moist dry mass representing 5-fold concentration. The partial loss of flavour in this processes can be restored by mixing one part of dry mass with 2 parts by weight of fresh mango pulp in a kneading machine to obtain mango concentrate of 45-48% TSS and 1.02% acidity. Concentrate, thus prepared and canned retains 79% ascorbic acid and 92% β-carotene and fairly good amount of flavour (Nanjundaswamy, 1984).

Other methods of making mango concentrate include, freeze concentration, reverse osmosis and ultra filtration, that are claimed to be alternate processes by which deteriorative effects can be eliminated. However, these processes are too capital and energy intensive to be economically feasible for small scale operation (Thijseen, 1974).

RIPE MANGO POWDER

Mango powder from ripe fruits is a good source of carbohydrates, vitamins and minerals. It is the best option to maintain overall fruit contents without loosing the fruit's nutritional value, natural aroma and vitamins. Powder has many advantages.,

(i) it can be stored for longer time,

(ii) it does not require a cold chain throughout the journey to the consumer,

(iii) it is considered as a very healthy product and can be used by all age groups for taste and as a rich source of vitamin A and C and minerals like Ca etc.,

(iv) powder can be used either as a base for making beverages or as flavouring agent,

(v) it can be used in the preparation of tasty drinks and healthy dishes like milk shakes, juice mix, sharbat, ice-cream, flavours, food items, biscuits, toffies, burfi etc.,

(vi) it can be used as a raw material when fresh fruits are not available,
(vii) it requires less space for storage,
(viii) more economical in comparision with bulk pulp,
(ix) it is stable, shows reductions in the bulk and weight of the product (Sagar and Khurdiya, 1999).

Methods of preparation

Fully ripe mangoes are washed, peeled and pulp is extracted through a pulper. The extracted pulp is dried to make powder. Several methods have been utilized to dry mango pulp with or without additives to produce a free flowing powder.

Freeze drying: Freeze drying of mango pulp requires about 12 hours to reduce the moisture content to 1%. Addition of sucrose to the pulp increases the shelf life of powder, with little loss in flavour. Freeze drying results in a product with high retention of flavour and nutrients. Production of mango powder by freeze drying is costly and prohibitive (Ammu *et al* 1976., Subramanyam *et al* 1976).

Foam-mat drying: In this method, the pulp is whipped to form heat stable low density foam using a foam inducer (Baldry *et al* 1976). Glyceride monostearate (1%) is a good foaming agent for mango pulp. It is mixed with equal quantity of sugar. The foam is extended in the form of strips on the sieves and dried in hot air. Drying is accomplished in a cabinet drier at 70 °C with cross flow drying technique for an hour to a moisture level of 2%. The product is yellowish in colour and reconstitutes instantly, but the flavour is slightly affected and has poor sensory qualities (Cooke *et al* 1976., Jayaraman *et al* 1976). Foam-mat dried mango pulp loses 80% vitamin C content and is less acceptable than the spray-dried powder.

Spray drying : The spray drying process of drying mango pulp produces powder of good colour, without mango flavour (Baldry *et al* 1976). Incorporation of sugar, milk solids, glycerol monosterate and sodium alginate to mango pulp results in good quality spray-dried powder with one year shelf life, when made with the following recipe (Sharma *et al* 1974).

Mango pulp	-	35 kg
Sugar	-	10.5 kg
Concentrated buffalo's milk (30% solids)	-	32.6 kg
Cream (40% fat)	-	14.0 kg
Glycerol monostearate	-	200 g
Sodium alginate	-	200 g

Sweetened mango powder

It is a sweetened product prepared by vacuum puff drying technique using a vacuum shelf drier. Mango powder from sweetened mango pulp can be used as a base for preparation of beverages. It can be made according to the following standardized method.

Equal quantities of mango pulp and cane sugar are mixed together and calculated amount of KMS is added to this mixture, so that mixture may contain about 200 ppm SO_2. The material thus prepared is subsequently spread evenly on aluminium trays, keeping the tray load at the optimum rates of 8-10 kg per square metre. Then the trays are transferred to the vacuum shelf-drier and dried at 65 °C temperature and under 25"-29" vacuum for 8 hours (Nanjundaswamy *et al* 1976).

Package and storage

Mango powder is hygroscopic and requires moisture proof packing. Generally mango powder is packed in tin containers and stored at room temperature (25°-28 °C). Retention of ascorbic acid and β-carotene is high and the products stored at room temperature were acceptable even after 8 months, whereas products stored at 37 °C were acceptable only up to 4 months of storage. Further, the products stored at 37 °C developed non-enzymatic browning, while such browning has not developed at room temperature (Nanjundaswamy *et al.*, 1976). Ripe mango powder, packed with nitrogen either with 400 gauge LDPE or 260 gauge ALPE packaging material can be safely stored up to 6 months with better retention of colour, flavour and texture (Sagar and Khurdiya, 1999).

PRESERVATION BY DRYING/DEHYDRATION

Fruits and vegetables contain more than 75% water and are therefore, highly perishable and due to action of microorganisms, they get spoiled quickly, unless they are preserved/stored properly. They can be preserved well over a long time under dried condition by the removal of water from them. The removal of water from fruits and vegetables and other food products before preservation and marketing is referred to as "drying", "dehydration" or "evaporation". The former term is generally used for the process of drying under the influence of non-conventional energy sources like sun and wind. On the other hand, 'dehydration' means the process of removal of moisture by the application of artificial heat under controlled continuous temperature, humidity and air-flow.

Drying is an oldest method of preservation of fruits, vegetables and other food products and still in use in many parts of the world, as it is most economical method of preservation. It is achieved using solar energy, until the turn of the twentieth century and mechanical drying began to replace it thereafter. Large quantities of fruits are dried in the sun in different parts of the world since ancient times. The modern methods of drying, however, assumed greater importance in recent times due to quickness, quality and hygienic results. In the modern methods drying is done in such a way that the nutritive value, natural flavour and characteristic quality of the fresh material are retained after drying till consumption.

The main objective of drying/dehydration irrespective of the method(s) is the removal of free water from fruits and vegetables and other food products, to the extent where microorganisms fail to survive and multiply. Reduction of water content also controls the biological forces which act up on fruits etc. facilitating the preservation of these perishables. The other objective is to improve their keeping quality by reducing the water activity and enzyme activity (Setty *et al* 1975).

Advantages of drying/dehydration

Dried products have certain advantages over those preserved by other methods. They are lighter in weight and do not require refrigeration either in storage or transport. Dried products are economical to handle, transport and distribute, require less storage space. Cost of preservation/processing is very low, as less labour and no sugar are required.

Disadvantages

However, if they are kept at higher temperature and have a high moisture content (because of their hygroscopic nature) they turn brown quickly in storage.

Advantages of dehydration over sundrying

(i) Dehydration is much faster than sundrying.

(ii) Dehydration requires less or no floor and fewer trays.

(iii) Dehydration is useful in areas where fruits, vegetables etc. are available during a particular season (like rainy season) and useful during off-season.

(iv) Dehydration is done under hygeienic condition.

(v) Dehydration, like sundrying does not depend on weather.

(vi) The colour of the dehydrated products remains uniform due to uniform drying temperature.

However, dehydration is a complex one since it involves simultaneous exchange of moisture and heat and requires costly equipment as compared to sundrying.

Methods of Drying/Dehydration

During the course of time, several methods, particularly in case of dehydration have been developed. The choice of the method depends on type of raw material (fruits, vegetables etc.), condition of the end-product, operation costs and the returns realized. The different methods of drying/dehydration are detailed briefly in the following pages.

Sun-drying: In this method, fruits etc. or their pieces are spread in a singe layer on a hard surface or on a pucca floor or mats, made of bamboo etc. or loaded on metal trays and exposed to open sun and allowed to dry by the action of sunlight and heat until the food articles are about two thirds only. Then they are removed to shade and staked or covered at the spot till all the later stages of drying which proceed slowly are over. In sun-drying there is no possibility of control of temperature and humidity.

The main advantages of this method are its comparative cheapness of drying and commercial scale use even at the village level. The disadvantages include its complete dependence on sunlight, relatively hot and dry climate without rainfall during and immediately after harvesting period. Since sundrying depends on the above, which are not under the control of man, production of uniform and high quality product can not be expected. Some over drying, contamination by dust and dirt and infestation of insects in the finished product occur. Sundrying does not allow the products to dry below 15-20% moisture in the product. As such the resulting product will have a limited shelf life. Sun-drying is slow and the products develop deep brown colour. Large scale sundrying also needs large areas of land. Further, sundrying is limited to periods of dry and summer. Sun-dried foods become sore, due to the growth of microorganism, which convert carbohydrates in the food to acids.

To avoid some of the above defects of sundrying, improvements have been made. The improvement includes a hot box (a solar drier). The hot box process improves the product quality by reducing contamination by dust, dirt, insect infestation and human interference. It reduces the drying time to half that is required in open sundrying. The finished product would be more appealing with better flavour than dried in the open sun. An indirect solar drying process in which the product is exposed to heated air, rather than to direct sun was also developed. Use of hot air reduces relative humidity of the air, there by leading to drying at a reasonable rate. Solar drying with Chimni was claimed to improve the quality of finished product (Pawar *et al* 1988).

Dehydration methods

Foam-Mat drying: This process consists of essentially of incorporating small amounts of suitable foaming agents (Glycerol-mono-stearate, egg albumin, groundnut protein isolate, gaur gum, carboxy methyl cellulose etc.) into the fluid food material, so that they could be whipped to a low density foam and then dried on a mat or trays, at relatively low temperatures in an ordinary forced air-circulation drier. The material can be dried in a relatively shorter period and the dried product reconstitutes readily. The method is considerably cheaper than other methods, leaving sun-drying.

The foaming agent is first dissolved in a small quantity of water and is added to the fruit juice concentrate/pulp, mixed in Waring Blender and then foamed in a Hobart mixer fitted with a wire mesh. The foamed material is extruded into strips 3"-4" diameter and the strips are spread on stainless steel trays. Next trays are arranged in a cabinet drier, being maintained at 65°-70 °C in such a manner that the strips are parallel to the flow of air. The material is dried in 30 to 60 minutes depending upon its nature. The dried strips are scraped from the trays and filled into air-tight containers as they are hygroscopic. The strips could be dried crisp in 30 to 45 minutes to a moisture content of 1.5 to 2.0% (Vijaya, 2000).

This technique can be successfully employed in the use of a variety of fruit pulps, concentrates etc. but primarily used for liquids which are pre-foamed by whipping, adding a low level of an edible whipping agent to liquids that do not whip readily.

Drum drying: This method is employed in the preparation of dried mango powder, mango flakes, tomato flakes, strained baby foods etc. Products to be dried are allowed to flow on to the surface of two heated stainless steel drums rotating in opposite directions with little clearance between them. The product dries on the drum and the dried product is scrapped off by blades.

Its main drawback is that, the severity of heat pre-processing can produce undesirable cooked product and aroma in the dried product. The drum dried products are also extremely hygroscopic, therefore, the use of in-package desiccant is recommended during storage of the product.

Freeze drying: This (lyphilization) is a new and very advanced system of drying employed for food preservation and gives a very good quality product. The advantages are :

- Retention of shape and flavor,
- Quick reconstitution, acceptable taste,
- Non-requirement of refrigeration,
- Reduction in weight of the product (Malhotra, 1965) and
- Savings in packaging and transport cost.

However, it has one problem – requirement of cold chain from industry to house hold. Hence, the plant and machinery cost, including the cold storage unit is high (Sharma, 2008).

In this method, the material such as fruit concentrate is first frozen at –20° to –70 °C (in order to freeze the moisture in the product), on trays in the lower chamber of freeze drier and the frozen material is dried, in the upper chamber under the conditions of high vacuum. The material is directly dried by sublimation of ice without passing through the intermediate liquid stage. The dried product is highly hygroscopic and reconstitutes easily. The rate of cooling, evaporation and degree of vacuum are some of the major factors which have to be carefully controlled for satisfactory freeze drying.

The freeze-dried mango produced a very good quality product on rehydration proving freeze drying as the best method of dehydration where rehydration is to be done for consumption (Vasundhara *et al* 1985). The quality of freeze-dried mango slices depends on the rate of freezing and method of freeze drying. Improved texture is associated with faster freezing rates (Wolford *et al* 1967). Rapid freezing in liquid nitrogen resulted in less damage and consequently better texture than in conventional freeze drying (Brown, 1967). A decrease in the rate of freezing caused an increase in fluid loss and greater loss of textural properties in freeze-dried mango slices (Ramamurthy and Bongirwar, 1979). The freezing rate alters the rehydration and organoleptic properties of the freeze-dried mango slices. Rapid freezing preserves the integrity of tissue to a greater extent than slow freezing. Freezing temperature is another critical factor affecting the cell structure. Freezing at temperatures of –20° or –40 °C results in considerable disruption of cellular structure whereas it is minimal at –19.6 °C. Slices frozen quickly at –19.6 °C showed firmer structure as well as higher water holding capacity, than the slices frozen by slow (–20 °C) and moderate (–40 °C) freezing (Ramamurthy and Bongirwar, 1979).

Freeze-dried products can be kept for months, however, residual moisture content in the freeze-dried products is of critical importance for extension of their storage life (Martinez and Labuza, 1968). The temperature also plays a very important role, as the hygroscopic equilibrium of the product is influenced by the temperature (Palnitkar and Heldman, 1971). The results of a study confirm the influence of freezing rate and drying temperature on the hygroscopic behaviour of freeze-dried 'Alphonso' mango slices. Quick frozen (–196 °C) samples had a lower adsorption rates than slow (–20 °C) and moderate (–40 °C) frozen samples (Ramamurthy and Bongirwar, 1987).

Accelerated Freeze drying: This method is an adoptation of the freeze – drying technique originally developed for drying fish, meat etc. Now the method is extended to preservation of fruits, vegetables etc. In this method, the pieces of material to be dried are kept pressed between two perforated or wire mesh trays, inside a cabinet freeze drier. As the material dries, the bulk of the pieces is gradually reduced by decreasing the clearance between the two trays hydraulically with the result that the dried material retains its normal shape and regains it when rehydrated (Girdharilal *et al* 1998).

Dehydro-freezing: In this method the raw material is first dried partially (about 50% of the original weight) and then frozen and thus it is slightly different from the usual freeze-drying technique (Girdharilal *et al* 1998). The quality of hydrofrozen product is equal to that of products frozen without primary drying.

Cabinet drying: In this method, food product to be dried in a cabinet drier, is placed on trays and moved into a drying compartment where the product is exposed to the drying air. Fresh air enters the cabinet, it is drawn by the fan, through heater coils and then blown across the food trays to exhaust moisture. Screeners filter out any dust that may be in the air. The air passess across and between the trays. Cabinet-driers are usually meant for small scale operations. They are comparatively inexpensive and easy to set in terms of drying conditions.

Drying/Dehydration of mango products

The different methods of drying/dehydration that are employed for the preservation and processing of different mango products like raw mango powder (amchur), raw mango slices, mango leather/bar, ripe fruit slices etc were described under individual products. A detailed notes on drying/dehydration of mango slices and pulp by different methods are presented below:

Drying/Dehydration of Mango slices: Canning has become a costly proposition due to increasing cost of tin-cans, sugar etc. and the canned products are bulky in weight, difficult to transport and pose certain problems. In view of the above, particularly in case of supply of mango to defence forces, drying/dehydration was suggested. Drying/dehydration of mango slices is one of the most feasible and simple method. The dehydrated slices are lighter in weight and can be eaten directly. The dried mango slices can be used as such or made into pulp and/or powder with appropriate techniques. Dried/dehydrated slices have a potential for export besides serving as energy ration for defence forces (Jayaraman *et al* 1976). Mango slices can be dried/dehydrated by different methods like, sun-drying, cabinet drying, deep freezing, cross-flow air drying, through-flow air, of which the last one is the most efficient one.

Sun-drying: Freshly harvested fruits are ripened artificially and washed, peeled and cut into slices longitudinally. These slices are sun-dried as detailed earlier. After sufficient drying they are packed, sealed and stored in a cool dry place for further use as such or for conversion into pulp or powder.

Cabinet drying: Because of certain demerits and limitations of sun-drying, other methods of drying/dehydration are adopted. One of them is cabinet drying, which was standardized (Sagar, 1995., Sagar and Khurdiya, 1996., 1997) as follows.

Fully matured mango fruits (Cv. Dashehari) were artificially ripened in news paper sheets in plastic crates at room temperature (33-35 °C). After ripening, the fruits were peeled manually and longitudinal slices (6 from each fruit) were cut. The slices were dipped in equal amount of 70°Brix sugar solution, containing KMS (0.1%) and heated immediately for 2 minutes at 90 °C and allowed soaking in the same solution over night, to ensure complete immersion. Next day these slices were drained and loaded on aluminium trays (1.05 × 0.45 cm size) and dried in the cross flow hot air at a flow rate of 1.2-1.8 metres per second at 58°-60 °C in a cabinet drier to a moisture level of 5%.

Drying slices in cabinet drier took only 10 hours as against 15 hours in sun-drying and resulted in better retention of ascorbic acid (Dabhade and Khedkar, 1980d). Blanching of slices (in boiling water) reduced the drying period. Dehydration of slices after sulphitation in KMS (1-4%) and common salt (5%) helped in better retention of SO_2 during storage and the retention of SO_2 increased with the use of higher concentration of salt (Mehta and Tomar, 1982).

Cabinet dehydrated mango slices can be reconstituted by soaking them in water (1:4 ratio) for 4 hours (Mehta and Tomar, 1982). Reconstitution ratio of cabinet-dried slices was higher than that of sundried slices, possibly due to less rupturing of cells during drying in cabinet drier (Khedkar and Roy, 1980).

These dehydrated slices can be converted into powder in a powder mill and sieved in 30 mesh sieve. For converting into pulp, the dehydrated slices are soaked in equal amounts of water and wet ground and this pulp can be used for making jam, nectar, chutney and toffies (Sagar, 1995).

Good quality mango slices could be prepared from fully matured fruits up to 6th day of harvest after ripening at room temperature. These dehydrated slices contain 18.7% reducing sugar, 60.23% total sugars and 42.03% sucrose (Sagar, 1995).

Dehydro-canning: Mathur *et al* (1973) made studies on dehydro-canning of mangoes for the dual purpose of improving texture and flavour of mango slices, using unsuitable varieties and accommodating more slices in a can and standardized the method as follows:

Fully ripe fruits were soaked in potassium permanganate ($KMnO_4$) solution (0.1%) for 30 minutes, rinsed in running tap water, peeled and cut into longitudinal slices, soaked in different solutions for one hour. Then dried in a cross-flow air-cabinet drier at a velocity of 180 metres per minute and tray-load of 6.1 kg per square metre for 60 minutes at 28°-30 °C. The dried slices were then filled in tin cans (7.7 × 116 mm size), covered with 25° Brix sucrose syrup, containing citric acid (0.2%), exhausted, sealed, processed for 15 minutes in boiling water (100 °C) and cooled.

To improve the flavour quality of dehydro-canning product (the slices), the mango pulp (15%) can be added to the sugar syrup. Addition of pulp also increases the drained weight percentage. The drained syrup (with 15% pulp, 20° brix and 0.3% acidity) can be utilized as a RTS beverage (Khurdiya and Roy, 1986c). Dehydro-canning of mango slices improves the texture and flavour of canned slices. However, browning of the end product during storage is one problem encountered in using this method. Soaking of slices in di-sodium EDTA (0.1%) for one hour prior to dehydration prevents browning. Soaking in other solutions, though prevented browning, affected the texture of the slices adversely and caused blackening of can and imparted soapy taste to the product. Soaking the slices in ascorbic acid solution (0.1%) alone caused more browning of the end-product. Besides preventing browning of the product, di-sodium EDTA extended the storage life of the product to 2 years without metallic taste etc. and retained acceptability up to that period (Mathur *et al* 1973). Use of pre-treatments (sulphitation, addition of ascorbic acid and citric acid) are also useful to avoid browning of the dehydrated product. Dehydration of mango slices after steeping in 40° Brix sugar syrup in the presence of SO_2 (3000 ppm), ascorbic acid (0.2%) and citric acid (1%) gave the best product with good retention of ascorbic acid and carotenoids up to 7 and 55% respectively. In the dehydration of ripe mango slices, combined use of ascorbic acid and citric acid was superior than the use of ascorbic acid alone at higher syrup strength or in the presence of SO_2 at lower sugar concentration or without sugar. Use of SO_2 in sugared slices without addition of ascorbic acid and citric acid was able to check the browning of slices and in maintaining the taste and flavour irrespective of pre-treatments. The quality of dehydrated slices, prepared without SO_2, with or without ascorbic acid were unsatisfactory and turned brown (Teotia *et al* 1976).

De hydro-canned mango slices are to great advantage of Defence Forces, due to savings in transport and economy of imported tin plate used (Mathur *et al* 1973).

Deep freezing: Mango slices can be frozen to give, after thawing, firm slices of acceptable quality. The pre-soak and subsequent pasteurization are not desirable. Similarly addition of ascorbic acid is not necessary. Sugar percentage in covering syrup can be reduced from 40 to 20% progressively. The reported poor texture in thawed slices after deep freezing (Saleh *et al* 1960), could be improved by the inclusion of calcium chloride in the covering syrup (Netto *et al* 1971., Baldry *et al* 1976). However, freeze drying is expensive, limiting its commercial application.

Quality of dried/dehydrated mango slices

Quality of the finished product depends principally upon the quality of raw material prior to drying/dehydration, which is closely related to the post-harvest ripening process and to the ripening stage of fruits at processing time. The effect of post-harvest – ripening method and optimum ripening stage, using calcium carbide and ethrel on the quality of dried mango slices was studied (Mahayothee *et al* 2002) and the following conclusions were drawn.

(i) The quality of dried slices was strongly influenced by the maturity stage of raw material (i.e. fruits), which was related to the taste and colour of the product,

(ii) the suitable ripening stage of mangoes was 2 or 3 days after ripening at 25 °C irrespective of ripening method (agent) having a sugar : acid ratio of 14 to 20,

(iii) mango slices, comparatively rich in β-carotene can be produced without preservation and pre-treatments (blanching/dipping in sugar syrup) by using suitable fruit of appropriate stage of ripening.

Dehydration of mango pulp

Dehydration of mango pulp in double drum-drier is as follows (Gangopadhyay *et al* 1976).

Green mangoes are boiled in water for 15 minutes, peeled and the pulp is extracted in a pulper. Then it is dried in a 6″ × 8″ wide chrome coated steel drum-drier at 4 rpm, treated by steam at 50 lbs/square inch pressure for 25 seconds.

Drum-dried mango pulp contained 5.1% total carbohydrates and 252 mg ascorbic acid per 100 grams. Drum-dried mango pulp is highly acceptable as regards colour and flavour up to 6 months of storage at room temperature in polyethylene, Al-foil laminated and sanitary cans.

Osmotic dehydration

Drying of fruits, vegetables etc. by the use of conventional driers and/or vacuum driers does not allow to keep the original colour, flavour and texture of the fruits, vegetables etc. Freeze-drying on the other hand, produces products of good quality along with better storability, but expensive for commercial application. Thus the above points to the need of developing alternative comparable processes using controlled heat-transfer application

(Bongirwar *et al* 1971). Osmotic dehydration has emerged as one of these techniques, showing promise for partial dehydration of food material (Bongirwar and Sreenivasan, 1977).

Osmotic dehydration is a novel and alternative approach of dehydration of fruits, vegetables etc to reduce the water content as well as to improve the quality of final product and has received greater attention in recent years as an intermediate step in drying of several fruits and vegetables.

To minimize deterioration, application of heat should be minimum, for which osmotic dehydration is the best option. Prolonged heat treatments result in loss of subtle flavour, leading to decrease in acceptability of the product. Minimum handling and heat treatment thus, become the most imperative factors, that are involved in osmotic dehydration.

Osmotic dehydration involves two stages:

(i) removal of water (35-45%) using sugar syrup (for fruits) and brine solution (for vegetables) as osmotic agent and

(ii) subsequent dehydration in an accelerated drying, where moisture content of the material is further reduced to 15% to make the product shelf stable.

The inclusion of osmotic process in conventional dehydration has two major objectives: quality improvement and energy savings. The quality of osmotically dehydrated product is near to fresh fruits in terms of colour, flavour, texture and taste. Osmotic process, in addition to the above, improves product storability, nutrient retention and prevention of microbial spoilage during storage (Bawa, 2007).

Osmotically dehydrated products can be used in ready to eat type of foods – ice-creams, fruit salads, cakes, bakery product etc. Osmotic dehydration is utilized on industrial scale in Thailand, Philippines and Malaysia to produce dried fruits mainly from pineapple, papaya, mango and banana which are exported to Europe, USA and Japan.

Principle of osmotic dehydration

Osmotic dehydration involves the principle of osmosis for removal of water (drying/dehydration) from the material. The principle involved in this process is that water diffuses from dilute solution to concentrated solution, through a semi-permeable membrane until the concentration equilibrium is attained. This technique/principle is used to remove water from cells of fruits, vegetables etc. The cell sap with the above materials for all practical purposes of osmotic dehydration is considered as low (dilute) in concentration, as compared to the sugar syrup/brine solutions; the common osmotic agents. The cell membrane is semi-permeable, but more selective for water and acid than solute. When the pieces of fruits, vegetables etc. are immersed in concentrated sugar/salt solution a difference of osmotic pressure is set up with a flow of water starting from the material to the outside (syrup/brine) in order to equalize the osmotic pressure on both sides. By this process the product gets dehydrated without the application of heat and water can be removed to the extent of 50% of the initial weight of fruits, vegetables etc.

Thus, the main principle involved in osmotic dehydration is the product dehydration of fruits etc. by osmosis, at comparatively lower temperature in the production of dehydrated food products, which would be very near to original material from flavour, colour, texture and taste point of view.

Advantages

1. Osmotic dehydration is a cheap and simple method of preservation of tropical fruits, when combined with air/vacuum drying, it becomes cheaper than freez drying.
2. It requires minimum handling and heat exposure and thus provides products of high nutritive value and storage stability.
3. Drying period is cut down to a large extent so that heat damage to colour and flavour is minimized.
4. Osmatically dehydrated product on reconstitution is identical to fresh fruit in colour, flavour and texture.
5. Energy consumption is much less.
6. Simple equipments are required.
7. It prevents enzymatic and oxidation browning as the fruit slices are surrounded by sugar, thus making it possible to retain good colour with little or no use of SO_2.
8. Removal of acid from and uptake of sugar by the fruit pieces give a sweeter product than conventionally dried product.
9. The storage life of osmotically dried products is greatly enhanced.
10. The textural quality of products is better after reconstitution.

Disadvantages

1. The reduction in acidity level reduces the characteristic taste of some products, which can be overcome by adding fruit acid to the solution.
2. Loss of fruit acid and the tendency of syrup treated fruits (due to drying at a very low moisture content) to become rancid after storage of several weeks at low temperature is another disadvantage of the osmotic dehydration.

Process of osmotic dehydration

The process of osmotic dehydration involves 3 steps as follows (Tiwari, 2005., Sharma, 2008):

(i) ***Pre-processing:*** Raw materials (fruits, vegetables etc.) are washed with hot/cold water to remove dust, dirt, pesticides etc from the surface. They are graded, peeled, sliced and blanched.

(ii) ***Osmotic Dehydration:*** Sugar/salt solution of desired brix/concentration is prepared first. Slices of fruit, vegetables etc. are immersed in this sugar/salt solution and spread on trays and dried at specified temperature for a specified period of time. After the expiry of the time, slices are separated from the solution by hand equipment and are put in the centrifuge machine for separating the surface moisture from the slices.

(iii) ***Terminal Dehydration:*** In the osmotic dehydration process water is removed partially only. Therefore, for further removal of water from the osmotically dehydrated products and to obtain increased shelf-life, other methods of drying are employed. Air-drying after osmotic treatment is commonly used in case of tropical fruits (Toreggiani, 1993), however, osmotic air dried fruits are unstable during storage (Ponting, 1973). Instead vacuum drying (Moy *et al* 1978) and freeze drying techniques can be used. Flavour retention is better in case of osmo-vac (Osmotic + vacuum) drying than the freeze drying. Osmo-vac dried fruits are very stable in storage as the moisture content is quite low (1-3%) and retain their high quality at least for one year at room temperature. For economic reasons, use of solar drying or biofuel drying in rural areas can be better alternative.

Partially dehydrated slices are placed in vacuum drier for further removal of moisture. Drying in a vacuum drier takes at low temperature (50°-55 °C), so deterioration in colour, flavour, taste and nutrient value is arrested. Heating medium may be steam/hot water which allows saving of energy and also preserves naturalness of produce (Sharma, 2008).

The osmotically dehydrated product is soft in texture, sweetest (due to high sugar: acid ratio) with good retention of flavour. The product requires soaking in cold water at ambient temperature for 30 minutes prior to use to get an acceptable texture (Jayaraman *et al* 1976).

Packaging

Dried products are generally hygroscopic and absorb moisture from the surrounding atmosphere and lead to spoilage. To prevent this, the dried products must be packed suitably. Osmotically dehydrated mango products remained acceptable up to 6 months at room temperature in paper foil polyethylene laminated (PFL) pouches (Jayaraman *et al* 1976). However, Al-foil laminated polyethylene pouches are suggested as ideal packing materials (Sagar and Khurdiya, 1997).

Storage

The storage stability of osmotically dehydrated product varies from 6 months to 12 months depending on the type of fruit. Osmotically dehydrated papaya product remained stable up to 6 months storage at room temperature (Ahmed and Chaudhary, 1995). On the other hand, the osmotically dehydrated banana products could be preserved up to 12 months or more depending upon the storage conditions and packaging materials used (Bongirwar and Sreenivasan, 1977).

Application of osmotic dehydration in mango

The osmotic dehydration technique has been applied to many fruits like, apple, apricot, banana, cherry, citrus, grapes, guava, mango, papaya, peach, pineapple, plum, raspberry etc and vegetables like beans, carrot, melons, onion, peas, potato, sweet potato and tomato (Chaudhuri *et al* 1993). In mango osmotic dehydration was applied to both green and ripe mangoes (Jackson and Mohammed, 1971., Hope and Vitatle, 1972., Teotia *et al* 1976., Moy *et al* 1978).

From time immemorial green mango slices have been in use after sun-drying. Though sun-drying is cheaper, it has its own limitations (see sun-drying). So alternate methods were sought. It was found that osmotic dehydration is the best one for dehydration of mango pulp and slices (see previous pages). Mango cereal flakes are prepared from mango pulp by osmotic dehydration (Girdharilal *et al* 1998). The general process of osmotic dehydration in case of mango slices is as follows :

The ripened mango fruits are cleaned, washed, peeled and sliced. The slices are dipped in known concentration of sodium metabisulphite (SMS) for 15-20 minutes at room temperature. Next slices are subjected to osmosis in known concentrated sugar syrup (70° Brix) at pre-determined temperature for specified period in osmotic dehydration tanks. Immediately after this period, the slices are dipped in sulphite solution, drained and washed quickly under a water spray to drain the syrup. The washed slices are subjected to vacuum dehydration so that moisture content level in fruit slices, comes down to 8 % for safe storage. The dried products are stored suitably (Baldry *et al* 1976., Jayaraman *et al* 1976., Moy *et al* 1978., Nanjundaswamy, 1984).

Pre-treatments: Raw material is subjected to pre-treatments before osmotic dehydration. The usual pre-treatments (blanching and sulphitation) are essential steps in processing particularly green mangoes. Blanching helps in inactivating the enzymes to protect the product from enzymatic catalysed reaction and sulphitation helps to retard the oxidation reactions like non-enzymatic browning and better retention of ascorbic acid during storage (Teaotia *et al* 1987). However, heavy leaching losses of acids, sugars and ascorbic acid were observed during blanching in boiling water and sulphitation of raw mango slices. The loss of ascorbic acid was more during blanching than in sulphitation, steam blanching. Sulphuring (exposure to sulphur fumes @ 10 g/kg/hour) is a good alternative to water blanching in order to avoid heavy losses (Teaotia *et al* 1987).

Different concentrations of KMS and common salt were tried in various pre-treatments before osmotic dehydration of mango slices. The slices treated with KMS (0.1%) and common salt (5%) gave the best product on dehydration and on preparation of products such as chutney, pickle, powder etc (Mehta and Tomar, 1982).

INTERMEDIATE MOISTURE MANGO PRODUCTS

Intermediate moisture foods (IMF) are partially dehydrated foods with moisture content in the intermediate range (20-25%) and are sterilized by using suitable additives to keep water activity low at safe levels (0.6-0.85) from the angle of microbial spoilage. The concept behind the development of such foods is that, one need not dehydrate foods to 5-10% moisture level dictated by microbial stability (Vijayaraghavan, 1980). Further, there will be substantial reduction in drying and reconstitution time and better retention of original flavour and texture, compared to conventional air-dried food.

The principle involved in the preservation of IMFs is lowering the water activity by addition of solutes such as glycerol, sucrose, glucose etc and retarding the microbial growth by addition of antimicrobial agents such as potassium sorbate and potassium metabilsulphite (Mishra and Tomar, 2000).

Intermediate moisture technology has been successfully applied to develop a number of fruit products from pineapple, mango, guava, apple, papaya, banana etc. The technology involves osmotic dewatering, giving a shelf life of 6-8 months under ambient condition (Nanjundaswamy, 1991).

IM fruit products have some potential advantages over fully dehydrated or canned fruits to meet the special military feeding situations. They have better colour, flavour and texture compared to air-dried foods (fruits) and need no reconstitution prior to consumption; keep long without disintegration. Further, fruit slices need not be packed in free syrup as is essential for canning, thereby saving considerably in weight and space. IM mango slices due to their special advantages, could substitute canned dehydrated mango slices for consumption as such or as component of such convenient mixes of fruit salads or fruited curds (Yoghurts) in combination with other IM fruits like guava, pineapple, banana and apple.

Method of preparation

The ripe mangoes are washed, soaked in potassium permongante ($KMnO_4$) solution for 15 minutes, washed thoroughly, peeled and cut into longitudinal slices (1.5 cm thick and 4.0 cm long) or cubes. The IM mango slices were made by the immersion of equilibration processes, wherein the raw fruit slices were blanched and equilibrated in a solution composed of glycerol (42.33% v/v), sucrose (42.33% v/v), water (14.84% v/v), potassium sorbate (0.48% w/v) and KMS (0.1% w/v) and adding the fruit slices to the stock solution (fruit : solution = 1:24), pre-heated at 95 °C, keeping the fruit slices immersed in the solution at 90 °C for 3 minutes and then quickly cooling the mixture at room temperature. The well immersed fruit slices were allowed to equilibrate in the solution over night in a refrigerator and then drained thoroughly over stainless steel wire mesh. The drained slices are dehydrated at 60 °C for 2 hours (Jayaraman *et al* 1976., Ramanuja and Jayaraman, 1980).

By the above process, an IM product of excellent flavour and good colour and texture comparable to the fresh fruit was obtained. It was moist and soft enough to be eaten as such without any further rehydration or preparation.

Packing and Storage: IM mango slices (cv. Badami) packed in cans or PFL pouches were acceptable for more than 9 months at room temperature with retention of carotene even after 18 months storage. However, there was gradual loss of β-carotene, with subsequent fading of orange red – colour and onset of browning and off-flavour at later stage. Samples stored at 37 °C remained acceptable only up to 4 months with a retention of 35% β-carotene. Browning did not increase during 9 months storage. Storage in HDPE showed drastic loss of carotene and flavour as compared to storage in cans and PFL pouches. The water activity in the product (0.76-0.78) is good enough to prevent bacterial growth, while sorbate prevented the growth of yeasts and moulds. The control treatment took 6 hours to reach 2.3%, while osmotically treated slices took 4 hours to reach 1.6% moisture level. The treatment which contained lower amount of glycerol took more time for drying than samples which contained higher amounts of glycerol. Samples treated with glycerol were also found to retain more moisture than those of control and osmotically dried products. Frutis under control had more acidity than others and were more dried up to 12.3% moisture (Mir and Nath, 1993). Organoleptic evaluation showed that mango slices which were soaked in 68° Brix sucrose syrup containing 2% glycerol were liked most on draining and after partial dehydration were also better in flavour and texture over osmotically dried slices (Ramanuja and Jayaraman, 1980). However, storage of IM mango slices below 75.5% relative humidity was conducive to the retention of colour, flavour and texture (Jayaraman *et al* 1976).

HIGH MOISTURE MANGO PRODUCTS

Hurdle technology is a recent one for preservation and extension of food products. This approach combines several factors (hurdles) at sub-inhibition concentration that can effectively control microorganism in refrigerated foods. Common hurdles include physical heat treatment, a slight water activity, a lower pH and preservatives, which when used together interact synergistically, against microbes, which however, individually not lethal to pathogenic microbes (Kaur and Kapoor, 2000).

Hurdel technology has been extended for preparing high moisture mango products. The process includes, peeling and cutting of ripe mangoes into slices, blending them in saturated steam for 1-2 minutes, cooling in water at 20 °C and immersing in sucrose syrup containing citric acid, sodium benzoate and sodium metabisulphite until the water activity of slices is 0.93, pH 3.3 and slices contain 1400 ppm sodium benzoate and 150 ppm of sodium metabisulphite. The slices are then packed in bottles or in polyethylene pouches. The shelf life of high moisture mango slices was about 4 to 8 months at 35 °C (Alzamura *et al* 1993).

STRUCTURED MANGO

Structured foods are manufactured from several individual food components to give a specific type of product. They are designed foods which have been engineered to meet a requirement. The requirement could be extended shelf life, the need for a novel product or the need to solve a processing problem (Glicksman, 1985). Classic examples of structured foods are 'bread' and 'ice-cream', neither of which exists in nature as bread or ice cream, but are structured from individual components.

Mango fruits are also used for making structured food. Structured products are generally prepared using alginates as the structuring agents. Alginates possess the ability to form textures similar to real fruit (Glicksman, 1976). Using mango pulp, sucrose, citric acid, sequestarants (sodium citrate or sodium hexametaphase alginate), calcium phosphate and water it is possible to prepare structured mango cheeks that resemble natural mango. They can be incorporated into yoghurt, ice-cream and confectionary (Britnell, 1991).

The alginate structured mango fruits being developed include:

(i) a high quality structured mango cheek that closely resemble the natural fruit,

(ii) structured mango pieces that can be incorporated into other foods, and

(iii) structured intermediate moisture mango that can be used as substitute for dried mango.

Structured mango products have advantages over their conventionally processed mango counterparts due to their structural integrity. They offer a wide range outlets for mangoes. The products can be tailor made to suit the subsequent processes on the needs of the consumer. However, there is a drawback in structured mangoes in that they have activities, which come in the way of their acceptability as in case of other products.

Although research on structured mango has not progressed much, results to date indicate that suitable products can be developed. Research is needed to develop a structured mango product having texture that is more representative of the fruit (Britnell, 1991).

MANGO CEREAL FLAKES

Preparation of fruit products for breakfast is a novel concept and involves blending of mango pulp with small amounts of wheat flour and sugar at a fixed pH, drying of the blended and homogenized mass in an atmospheric double-drum drier at a steam pressure of 50-60 psig; speed rotation kept at 2 or 3 rpm and 0.35 mm clearance between the drums.

Mango cereal flakes can be used as a breakfast item or a shake food. When taken with milk, it serves as a good breakfast food (Jain *et al* 1957., Lal *et al* 1960). It is a nutritious food with proximate composition of mango solids (49%), wheat flour (29%) and added sugar (22%) (Srivastava, 1967).

Mango flakes are prepared both from whole fruits and stored pulp. The detailed method of preparation of mango cereal flakes is as follows (Srivastava, 1967).

The fruits are washed in running cold water, peeled and passed through a pulper, fitted with a 30-mesh to extract pulp, which can be used directly or after preservation. The pH of the pulp is adjusted to 5.0 by the addition of citric acid and then heated to 80-85 °C for 5 minutes in a stainless steel steam-jackted kettel. After heating the pulp is mixed with wheat flour and cooked with 3 times its weight of water at 75°-80 °C for 2 minutes. The other ingredients according to the following recipe are added and mixed with cooked mass. Pectin is added through a homogenizer or a fruit pulper to get an uniform mix, corresponding to a mixture of 65-70% and refractometer solids of 27-32%. The product which dries in the form of a crisp continuous sheet is broken into smaller flakes and dried in an atmospheric double drum-drier heated by steam at 4.22 kg per square centimetre and revolving at 2-3 rpm, with the clearance between the rollers being 0.35 mm (Srivastava, 1967., Bhatnagar and Subramanyam, 1973).

Recipe

Mango pulp	-	45.5 kg
Glucose	-	2.59 kg
Sugar	-	2.59 kg
Calcium carbonate	-	46 g
Sodium bicarbonate	-	91 g
Pectin	-	27 g

The quality of flakes is dependent on the colour, and flavour of the pulp used in their preparation. The main quality constituent of mango cereal flakes is β-carotene which ranges from 2572 mg (Safeda Lucknow) and 9488 mg (Alphonso) per 100 grams depending on the variety (Bhatnagar and Subramanyam, 1973).

The product is very hygroscopic, therefore, it should not be exposed to air. The flakes are packed directly from the drum-drier in sanitary cans or 300 gauge polyethylene bags which hold about 225 g of flakes. The bags are properly sealed and stored in 18 litre friction – top tins in cool and dry rooms. Alminium foil bags can also be used for packing. It is preferable that packaging of flakes is done in a special room fitted with de-humidifier and having comparatively low RH (Srivastava, 1967). The critical relative humidity of the product is 45% and the flakes packed in foil laminated pouches have one year shelf life (Padmini *et al* 1963).

STRAINED BABY FOODS

In infant feeding programmes, strained baby foods based on fruit and vegetables assumed greater importance. Strained baby foods from mango such as strained mango pulp, strained mango custard and dehydrated mango custard were successfully prepared in India, by incorporating small quantities of skim milk, pulp and corn (maize) starch to facilitate drying. Products thus prepared retain more ascorbic acid and β-carotene (Korula *et al* 1960). Strained baby foods can also be prepared by blending with apple, beans, peach, papaya or apricot and powdered milk (Anon, 1990).

Method of preparation

Ripe mangoes are washed in cool water, peeled and pulped in a pulper, then passed through 40-60 mesh sieve to remove the fibre and to this pulp, ingredients as per the following recipe are added and the blend is heated for a short time in a steam-jacketted pan to cook the corn-starch. Next the mix is passed through a homogenizer or fed directly to a double–drum drier working at 4.22 kg steam pressure and revolving at 2-3 rpm. Small quantities of skim-milk powder (5 g/100 g pulp) and corn starch (2.5 g/100 g pulp) are incorporated to facilitate drum drying. The dried material, which is in the form of a thin continuous band, is further dried in a cabinet drier to reduce the moisture content to 1.0-1.5% and then ground to a granular powder and packed for storage (Srivastava, 1967).

Recipe

Fruit pulp (Sieved)	-	200 kg
Skim-milk powder	-	2.2 kg
Sugar	-	2.0 kg
Corn starch	-	1.0 kg
Sodium bicarbonate	-	Calculated amount
Egg yolk	-	1.0 kg
Lime juice	-	1.0 kg
Cream	-	Small quantities
Salt	-	Optional

Sodium bicarbonate is added to reduce acidity of the fruit pulp to 0.25% (as citric acid) and pH to 5.6 to 5.8. Sugar is added to get the desired blend. Drum-dried powder retains more ascorbic acid than canned product and rich in β-carotene (6000 – 12000 μg/100 g) (Siddappa and Ranganna, 1961a,b).

The finished product is hygroscopic and should be packed with an in-packing desicant (Bhatnagar and Subramanyam, 1973). The mango custard powder can be packed in hermatically sealed-lagger – top cans and stored in a cool dry place. It can also be packed in cans under nitrogen atmosphere and stored well for one year at 5 °C and for 6 months at 37 °C (Siddappa and Ranganna, 1961a,b).

ASEPTIC PROCESSING AND PACKAGING

The concept and application of aseptic processing and packaging technology originated in United States of America more than half-a-century around 1960 for milk and other liquid foods as an alternate technology for conventional, canning, freezing etc. which result in over cooking (in canning) and costly (in both cases). Aseptic processing and packaging is attractive because of two reasons viz., (i) it does not require refrigeration and (ii) the container is reusable/disposable. The later is important because it eliminates costs incurred in returning the containers to the processers for reusing (Chan and Cavaletto, 1982).

Aseptic processing and packaging is a process of packing pre-sterilized and cooled products into pre-sterilized packages under sterile conditions. It involves heating the product to a high temperature in a closed system for a few seconds, then force-cooling to room temperature. While the high level of heat removes/destroys all microorganism its short duration causes less damage to nutrients and flavour compounds of the product than traditional pasterurization and canning. The techniques used for sterilization of the product in this method are-Ultra high temperature (UHT) and high temperature short term (HTST), direct heating by steam injection and indirect pre-heating. In all these cases, the product receives sterilization temperature for a short time and immediately cooled before filling the container under aseptic conditions, there by reducing the chances of over cooking of the product (Chakravarty and Narsimhan, 1983). Because of the shortened times and high temperatures, aseptic processing results in products with high nutrient retention and excellent sensory qualities (Pruthi, 1993). The basic principle of the process is that the porous cover of the container prevents contamination of the sterilized food till the cover is removed in the sterile area, at the same time the porous cover allows entry of steam necessary for sterilization of the food. The technique utilizes steam sterilization of foil and porous cover wrapped food, followed by vacuum/sterile inert gas pouching. Here in film wrapping allows steam sterilization at the same time wards off contaminants till pouching and does not allow microbial growth in side the pouch (Patwardhan, 1999).

Aseptic processing technology has been used in food industry for many years. It is suitable for low and high acid products eliminating processing contamination and allows transportation and storage at ambient temperatures without refrigeration. Various types of products such as juices, ketchups, pastes, sauces, soups and concentrates of fruits and vegetables can be packed aseptically and stored at ambient temperatures for considerable time. Tropical fruit products for trans-ocenanic shipments from under developed areas to industrially developed markets are the candidates for adoption of aseptic processing, since refrigerated transport and storage are not needed with this system (Chan and Cavalleto, 1982).

Advantages

1. Aseptic processing is much faster than traditional canning, since unlike in canning, products are flash treated for few seconds at a much high temperature.
2. In view of the fastness, the turnover capacity is much higher than canning. The maximum turnover capacity of canning is one tonne per hour, while it is 3-4 tonnes per hour in aseptic processing system.
3. It requires less energy to heat and sterilize the product.

4. It saves energy costs, as there is no refrigeration.
5. It saves handling costs (loading and unloading)
6. Transport and storage costs are also less.
7. It saves storage space, particularly when tetra pack system is followed.
8. It allows the use of wide variety of containers made from materials like plastics, cartons, metalized film and polythene film, which can not be used in conventional packaging, because they will be destroyed during thermal processing.
9. There is no restriction on the size of package.
10. Packaging costs, up to 50 pre cent over other methods are saved.
11. Most important is the possibility of recycling the packages.
12. Aseptic packages also burn cleanly and are an excellent fuel in modern waste-to-energy incinerators.
13. It provides better product quality, compared with canned products etc.
14. It allows better retention/improvement in colour, flavour and texture in the product.
15. It reduces the loss of nutrients in the finished product.
16. It minimizes the degradation of heat sensitive products.
17. It enhances food safety, as flash exposure to high temperatures destroys all the microorganisms.
18. It eliminates the use/need for preservatives.
19. Aseptically packed food products are shelf stable.
20. The products do not need any special environment in storage, especially at the retail level.

Limitations

1. It requires a large capital investment, to the tune of Rs. 50 lakhs, while it is about Rs. 20 lakhs in canning.
2. There is loss of shape and form of particulates in a product during its processing.

Though, the bulk aseptic processing is a capital intensive and high technology process, yet it turned out to be highly economical and is being used conventionally in the processing industry. One of the greatest advantages is that fruit industries can be established for the manufacture of beverages etc. by using aseptically packed pulp in bulk containers without handling fresh commodities and/or in the absence of these raw materials. In this way lots of fruits and vegetables grown in inaccessible areas can be utilized.

Problems

The problems of aseptic processing include:

Heat induced colour and flavour changes, sensitivity to oxidation, sensitivity of cell structures to shearing and the abrasion of machine parts and seals due to the presence of fibre and seed particles (Woodhams, 1982). Browning discoloration, decrease in viscosity of pulp and off-flavour development are the problems in storage at ambient conditions (Eipeson *et al* 1987). Other problems encountered are micrbiological safety and retension of desired product quality. By suitable design of the product sterilization process, right choice of packaging material, packaging closure head space etc. the above problems can be overcome (Astrom, 1990).

Steps in aseptic processing and packaging

Aseptic processing and packaging involves 3 steps:

(i) sterilization of the product,

(ii) sterilization of packaging material and

(iii) maintenance of sterility during filling and sealing operations.

Methods of sterilization: There are a number of sterilization methods available based on physical or chemical processes. In most cases the packaging machinery is sterilized by steam or hot air before processing. The oldest sterilization method in use is exposure of a product to high temperature for a specified period of time, using superheated steam or hot air. This is highly reliable approach, but seriously limits the choice of packing material. This is the major reason, why chemical sterilization or radiation sterilization are adopted. Chemical methods or combination of chemical and thermal sterilization methods can also be used. Low concentration of hydrogen peroxide with ultra-violet light has shown synergetic sterilizing effect (Patwardhan, 1999). However, the most important pre-sterilization method in aspectic packaging is the use of hydrogen peroxide. Sterilization is carried out by using a combination of hydrogen peroxide and heat. Hydrogen peroxide is sprayed into the packages or packages are immersed in its bath. The only precaution necessary is to ensure absence of residual hydrogen peroxide in the packages.

Prevention of contamination: The critical part of the aseptic packaging process is the prevention of reinfection during filling and sealing. The most common way of keeping the filling chamber free from contamination is to use filtered, sterile air at slight over pressure (Astrom, 1990). De-aeration is necessary in closed system of aspectic processing to remove the entrapped air in fruits and vegetables and should be done before sterilization. Air-removal results in improved product quality and an extended shelf life (Pruthi, 1993).

Packaging materials/containers

The packages used in aseptic packaging technology vary from the traditional metal cans and glass bottles to the flexible and semi-rigid containers based on thermo-plastics or combination of thermo-plastics and paper board and metal foil. The type of the container used is influenced by the nature of the contents, the cost of the container and the attitude of the consumer towards the product (Astrom, 1990). The specific features required for aseptic packaging material/containers are:

- Suitability for pre-sterilization and ability to allow and maintain hermit sealing.
- For selection of suitable packages, both economic and marketing factors must be taken into account in addition to the technical aspects of the product (Patwardhan, 1999).

Aseptically preserved foods can be packed in rigid or semi-rigid and flexible packages. Rigid unit type are made of glass, tin plate, aluminium, plastic and paper board combinations. Flexible forms comprise plastic films, laminates co-extruded structures, coated paper and aluminium foil and paper board combinations for retail marketing (Kumar, 1984) (Photo 5.17).

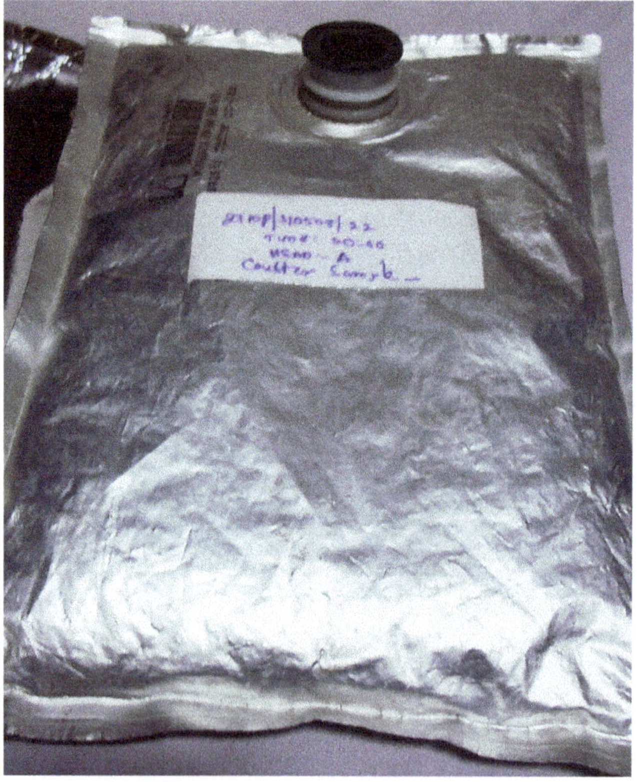

Photo 5.17: Aseptic packaging of mango pulp

Metal and glass containers are costly, but provide good protection to the packed product due to high barrier properties and reliable hermatic seal. The high thermal resistance of these materials also make it possible to use traditional physical methods for sterilization with steam or hot air (Patwardhan, 1999). Paper board laminates and plastics can reduce packaging costs considerably and may be used in rational form-fill seal packaging equipment. Laminated bags, made of polyethylene/aluminium foil/ polyethylene liner and polyethylene/metallised polyster/polyethylene/polyethylene liner of 20 kg and 200 kg capacities are used (Nanjundaswamy, 1987). Polypropylene bags (300 gauge) are used for thermal packing of mango pulp and mango syrup. On the contrary, pouches made of polyester/polypropylene, poly cell, paper foil, polylaminates and aluminium-collapsible tubes were satisfactory for mango jam (Anandaswamy, 1981). Mango pulp is aseptically packed commercially in aluminium foil laminated bags of 2200 litres capacity 'bag-in-box' or 'bag-in-drum' concept for export markets (Ramteke *et al* 1999) (Photo 5.18).

Photo 5.18: Aseptic processing of mango pulp in bags

Bulk containers are generally 'bag-in-box' type packages made of a flexible bag with an opening, contained in a rigid transportation container made of corrugated or solid fibre board, ply wood or collapsible metal boxes (Kumar, 1984).

Aseptic processing and packaging in mango

Of late aseptic processing and packaging is being increasingly used particularly for export of mango pulp in different parts of the world including India, Philippines, United States of America and Australia.

Process: Briefly, the extracted pulp is sterilized at temperature as high as 48 °C for a short time (45-60 seconds), cooling it to ambient conditions and filling aseptically in suitable containers and packing, sealing and finally storing.

Packaging and storage: In India, aseptically processed mango pulp is aseptically packed commercially in aluminium foil laminated bags of 200 litre capacity for export market. The aseplically packed mango pulp is stored for about 2-3 months at 37 °C. However, quality of the product decreases in respect of colour and flavour in storage. The shelf life of the packed product is influenced by the choice of the packaging material, package design, closures and head space, prevention of moisture and gases as well as the choice of the primary material layer in contact with the product (Astrom, 1990). In flexible packages, owing to oxygen permeation, products have a shorter life than in the metal cans. New barrier films have drastically reduced this problem (Pruthi, 1993). However, shelf life of the packed product remains a limitation for aseptically packaged products particularly in flexible packages (Nelson, 1984).

BLENDS

Blending is one of the most important operations in Food Technology. It is one of the oldest techniques having origin in the ancient art of cooking and brewing. Blending is defined as the intimate intermixing of two or more physically or chemically dissimilar groups of matter, resulting in some pre-defined state of intimacy of complete uniformity. The blending of fruit drinks is an economic requisite to utilize some fruit varieties for processing which may not have otherwise favourable characteristics including cost and product preparation. Secondly, the objective could be to supplement appearance, nutrition and flavour. Thirdly, one could simply think of new product development through blending. Forthly, it could be to utilize lesser known fruits profitably. Two or more fruit juices/pulp may be blended in various proportions for preparation of nectar, RTS drinks etc. (Tripathi *et al* 1992., Saxena *et al* 1996., Attri *et al* 1998., Deka *et al* 2001).

Reasons for blending

The chemical composition of many fruit juices is not balanced from an organoleptic point of view and to avoid chemical alteration such as neutralizing or increasing the acidity, juices of varying compositions are blended together to provide the desired balance of sweetness and sourness. Strong aromatized juice can well be appreciated only on dilution. The subsidiary ingredient can be either water or juice. Use of water or syrup in the process of blending greatly decreases the buffer strength of the juice. Hence, for all practical purposes intensely flavoured drinks should be diluted with bland juices. Juices counteract the acids of one another and regulate Brix/acid ratio. By blending it is possible to bringout the latent flavour of the ingredient juice and the resultant blend due to multiplicity of flavour and colour, is outstandingly unique. Juices are blended to give rise to 'complete juice'. Different vitamins and minerals of the juice may be supplemented by the others in mix. Hence, blends are more invigorating and nutritious.

Blending of fruit juices has some advantages like, providing a variety of new flavour, better nutritional balance and help in utilizing less used fruits.

Mango blends

Mangoes can be blended with a number of other fruits (guava, papaya, pineapple etc.), other types of products as well as different varieties of mango for product preparation as follows:

Mango + Mango blends

Though there are a number of varieties in mango all are not suitable for processing. The reasons may be many like low yields, lower edible pulp percentage, poor pigmentation etc. Some of the varieties having high yield potential may not be preferred for processing because of the poor colour of the pulp (eg. Bobbili Punasa, Banganapalli).

Srivastava (1998a) mixed the pulp from two varieties viz., 'Banganapalli' and 'Dashehari' and claimed improved RTS drink as compared to that made from original cultivars. Kalra *et al* (2001) blended the pulp of 'Bobbili Punasa' and 'Banganapalli' with 'Dashehari' mango pulp, (1 : 1) and found that the blended beverages were superior to that of pure 'Bobbili Punasa' and 'Banganapalli'. Thus prepared RTS beverage, containing 15% pulp, 15° Brix and 0.3% titratable acidity when preserved in glass bottles at room temperature (38°-42 ^{o}C) showed a slight increase in viscosity, stability in colour and general decline in the sensory evaluation scores, after 6 months of storage. Blends of both the varieties with 'Dashehari' performed better and were acceptable during the entire period of storage. However, though the colour declined, the sensory quality of blends indicated better performance of 'Bobbili Punasa' as compared to 'Banganapalli'. The original colour of the 'Bobbili Punasa' has improved when mixed with 'Dashehari', but declined gradually by the end of storage period. The average flavour score of pure beverage of 'Bobbili punasa' declined more as compared to that of its blends, mainly due to their better appearance during 6 months storage. Likewise the flavour and taste scores of pure 'Banganapalli' beverage dropped more rapidly than its blends with 'Dashehari' because of their better colour perceptive. The deterioratin of pure 'Dashehari' beverage was similar to that of other beverages of the two varieties. The microbial load of yeast, mould and bacteria was negligible throughout the storage period of 6 months. The enzymatic browning showed little change during the storage. Stability in total carotenoid content was also observed. The above study indicated that preparation of blends from such varieties having poor colour and other inferior quality traits with fruits of better quality characteristics could be an economical proposition to boost their market demand (Kalra *et al* 2001).

However, sometimes due to blending, interaction of anthocyanin, tannins or other components of fruits may lead to browning, turgidity etc. (Spayd *et al* 1984). Under such conditions improvement in the market quality of fruit beverage may be obtained by supplementation with limited edible artificial colour.

Girdharilal *et al* (1998) advised blending of pulp of poor flavour and dull colour with the pulp of bright colour and strong flavour to get superior quality flakes from the former pulp. Thus, flakes prepared from blended pulps of 'Badami' and 'Raspuri', 'Badami' and 'Totapuri' and 'Lucknow Safeda' and 'Dashehari' in the ratio of 1:2 were superior to flakes prepared from pulp of 'Raspuri', 'Totapuri' and 'Lucknow Safeda' alone.

Mango + Pineapple blend

Begum *et al* (1983) made a pineapple and mango pulp mixture blend in different ratios, having good consumer acceptability. Of late, spice beverages are gaining importance and popularity in the market in the form of fruit drinks/squashes/appetizers etc. The blending of fruit juices with various spice mixtures improves taste, aroma and nutrition of the resultant beverage. Keeping this in view, Deka *et al* (2005) prepared mango + pineapple spiced RTS beverage from 'Dashehari' mango and 'Kew' pineapple (85 : 15) having 10° brix, 0.2% and 0.006% cardamom spice drops.

Mango + Papaya blend

As compared to mango, papaya (*Carica papaya*) is cheaper and has greater proportion of pulp and colour. So to reduce the cost of premium varieties (Alphonso, Dashehari etc.) of mango, papaya can be blended to prepare beverages to improve their economic feasibility and marketability. Kalra *et al* (1991) successfully investigated the possibility of blending mango (Totapuri, Banganapalli, Dashehari and Chausa) and papaya (cultivar not indicated) in different proportions and their successful storage for one year in glass bottles under ambient conditions (20°-36 °C). The study indicated that 25-33% papaya could be incorporated in mango pulp without affecting the quality and acceptability of the blended beverage. Acidity and TSS did not change during storage. The papaya beverage (15% pulp) had double the contents of vitamin-C (8 g/100 g) as compared to pure mango beverage and its preservation was 80% after one year of storage. Yellowness of the blend remained stable during storage. This means that any proportion of papaya pulp may be blended with mango pulp without/minor effect, but from the organoleptic score it seems that 25-33% papaya in 'Dashehari' mango would be the best. The beverages from 'Totapuri' mango and its blends with papaya had the second lowest yellowness index

(Y.I) among the four cultivars and it remained stable during storage. 'Banganapalli' + papaya blend showed similar trends to that of 'Totapuri' + papaya blend. This blend has good appearance and intensity of colour with slightly higher Y.I than that of 'Totapuri' blends. Pure beverage of 'Chausa' mangoes and blends with papaya had the least Y.I of all the mango cultivars, hence the addition of papaya to 'Chausa' can not be recommended. However, organoleptic score showed that good blend could be prepared with long storability with 'Dashehari' and 'Chausa' blended with 25-33% papaya. The study also indicated that addition of papaya pulp did not suppress Y.I. of mango blends. Although the mango + papaya blend had higher content of sucrose, there was no browning or loss of colour during storage, due to lack of polyphenol oxidase and the presence of non-oxidisable phenolics in the fruit pulp (Prabha and Patwardhan, 1986).

Mango + Grape blend

Juice of 'Perlette' grapes grown in Punjab and Haryana has poor consumer appeal because of higher acidity and very weak flavour. This juice may be blended with other suitable juice/pulp profitably. Saxena *et al* (1996) blended grape and mango (3:1) having heighest sensory quality. The RTS beverage thus prepared was light brown in colour, with aroma domination masking the mild flavour of grapes.

Mango + Soybean blend

Mango juice is rich in carbohydrates, minerals, vitamin-C, starch, pectin, carotenoids, but lacks in proteins and fats and some essential amino acids. Soybean, on the other hand, is an excellent source of good quality proteins (40 – 42%) and fat (18-20%) and holds a good promise for developing proteinacious food products. The mango pulp can be fortified with protein from soybean for development of protein enriched product. Chauhan *et al* (1998) prepared a protein rich mango beverage by blending mango (cv. Amrapali) pulp with soy-protein (Yellow variety) isolate in different proportions. The beverage consisting of 60% mango pulp and 40% protein isolate with TSS of 16° Brix (raised with sugar syrup), 0.42 % acidity, 6.8% protein and 0.67% fat and score of 92% was the best. As the ratio of protein isolates to the mango pulp increased, the acidity, reducing sugars, ash, fibres increased while pH values decreased. This blended product had substantially higher protein and fat contents than those of control with good flavour, taste, body and overall quality and could be stored for 6 months without any change in constituents.

Mango + wheat + sugar blend (Mango-wafer)

Nanjundaswamy *et al* (1976) developed a novel method of preparation of wafer like product from a blend of mango pulp, fine wheat flour and sugar by adopting foam-drying technique. The dried product was crisp in texture, light yellow in colour, with mild mango flavour and pleasant taste and very rich source of β-carotene, ascorbic acid and minerals.

It can be consumed directly as snack in breakfast with milk or in the preparation of pudding etc. The product has 40% sugar and 18.8% starch. It can be packed in 300 gauge polyethylene pouches and stored in 4 gallon tins with tight lids for 3 months at ambient temperature.

The recipe followed and method of preparation of the product is almost similar to that of mango cereal flakes, except for slight modifications, namely, addition of required amount of water to the product, so as to make it amenable for foam production. Of 3 foaming agents tried, glycerol monostearate (GMS), dried egg albumin and fresh egg white, GMS was the most suitable as reducing agent and a whipping period of 20 minutes was optimum. After adding the foaming agent (GMS) to the feed material at the optimum concentration of 0.75% (on fresh weight basis), the mixed mass is subjected to the whipping in a Hobart mixture and whipped for 20 minutes to get good foam with optimum foam density. The foam thus produced is subsequently spread on stainless steel wire mesh trays and transferred to a cross flow air-drier and dried at 70 °C for a minimum drying period of 6 hours to reduce the moisture content below 2%. The dried product is cut into proper size and packed and stored at ambient temperature.

Mango + Milk blend

Mango pulp is not considered to be a nutritiously rich food, as it lacks proteins and fat. Milk is an excellent source of protein (2.9-5.0%) and fat (2.5-6.1%), but is deficient in iron, copper and vitamin C. On the other hand, mango pulp is a good source of carbohydrates; vitamin C and in-organic potassium. As such, blending of mango pulp with milk would produce a nutritionally rich, ready to serve beverage, commonly known as "Mango shake" which is a popular delicious beverage in North India as a refreshing drink, particularly during summer months.

Mango + milk blend is prepared both from fresh ripe mangoes (Hassan and Ahmed, 1998) and preserved pulp (Grewal and Jani, 1982., Yadav *et al* 2005).

Fully ripe, fresh fruits of mango (cv. Dashehari) were washed under tap water to remove all dirt, dust and adhering chemicals. Then peeled by a knife and pulp was extracted in a lab-size pulper. The pulp was heated to 85 °C for 3-4 minutes, hot-filled into cleaned glass bottles and sealed and stored under refrigerated conditions till further use. The beverage was prepared according to FPO specifications (Girdharilal *et al* 1998). A definite value of milk (15%) was added to the mango pulp at different levels (30-45%). The TSS of the beverage was maintained at 18° Brix by addition of sugar after adding and mixing all the raw materials. Blending was done in a blender for 3 minutes to make a homogenous mass. It is stored in 300 ml sterilized glass bottles and crown corked and thermally processed for 30 minutes at 88°–90 °C. Then cooled and stored at room temperature (25-40 °C).

The consistency of the blended beverage declined during storage of 6 months, due to decrease in viscosity of product, particularly at high room temperature due to development of slight brown discolouration. Changes in quality attributes of the product after 3 months storage were observed. However, the blended product was acceptable up to 6 month storage after thermal processing (Hassan and Ahmed, 1998).

The method of preparation of mango + milk blend adopted by Grewal and Jain (1982) is as follows:

Cream separated buffalo milk was blended with preserved mango pulp (cv. Chausa) and preserved in sterilized glass bottles. The acceptability of thus prepared beverage seemed to depend on its pH. The excess acidity in the blend was neutralized with sodium bicarbonate or sodium hydroxide to render it suitable for use as raw material in the preparation of milk-mango shake. At pH 6.5 the acceptability of the product was maximum. Further increase in pH from 6.5 to 7.5 decreased the acceptability. At pH 8.0 the acceptability was further decreased due to development of an off-odour (soapiness) in the beverage. The sodium-ion concentration rather than pH, affected the flavour acceptability. The added alkali was useful in improving the flavour, through increased level of sodium-ion concentration by neutralization of excess acidity. The finished product was acceptable with regard to its colour and flavour. The product was sterilized at 116 °C for 30 minutes to enhance the shelf life and stored in glass bottles without any loss in consumer acceptability.

Yadav *et al* (2005) used malted milk powder (a nutritious food stuff for growing children) for the preparation of mango-milk shake. Mango pulp (20% concentration) was added to the malted milk to impart a characteristic mango flavour and was dried at 60 °C by tray drying in hot-air-oven and in vacuum-oven with phosphorus pentoxide as desicant and 600 mm Hg vacuum. The products prepared by both above methods were of desirable quality. However, the product dried in vacuum-oven had superior quality. The product was more porous with better retention of colour and flavour and showed good solubility in cold as well as hot water. The vacuum drying was better with little sedimentation. The product retained its original quality for more than 30 days at room temperature with mango flavour and appealing.

However, commercial flavoured milk-mango shake is prepared by mixing whole or partially skimmed milk with commercial mango fruit flesh at 3-15% by volume, keeping the ratio of the two components at 1 : 5 (Korn, 1983).

Mango + Whey Blend

Whey is a major by product of dairy industry obtained in the preparation of paneer, chhano, shrikand and cheese. It contains 10% milk proteins and a rich source of carbohydrates, minerals, lactose and water soluble vitamins. It is also rich in calcium, phosphorus, essential amino acids and most importantly water soluble proteins (Parekh, 1993., Ghosh *et al* 1995). Whey possesses preventive and curative elements and is especially used to treat a wide variety of ailments such as arthritis, enamia and liver complaints (Jelen, 1992). Mango, on the other hand, is the richest source of α-carotene (precursor of vitamin A) and vitamin C and minerals like calcium, phosphorus, iron (Gopalan *et al* 2004). Disposal of whey is a serious problem of environmental pollution due to the presence of high organic matter. Hence, development of any process for its economical utilization would be a great benefit to the dairy industry (Sahu *et al* 2005).

A whey-based mango fruit juice blended RTS beverage by using whey, fruit juice, sugar and citric acid was developed (Saravana Kumar and Manimegalai, 2001). It was packed in sterilized bottle (200 ml) and pasterurized in the boiling water for 30 minutes, cooled and stored in refrigerator (5°-7 °C). The blended product had moderate yellow colour, mango flavour, highly acceptable taste and overall acceptability. The RTS beverage was safe for consumption as it had negligible number of bacteria, fungi and yeast. However, it has poor storability, as the colour, flavour, taste and overall acceptability decreased with increasing storage period.

Addition of herbal distillate such as lemon grass (*Cymbopogan flexosus*) with its germicidal, medicinal and flavouring properties will improve the quality of whey-based mango product. The lemon grass distillate also acts as a good appetizer and at the same time makes the product more palatable and acceptable. The process of making a RTS beverage with mango, whey and lemon grass distillate is as follows (Sahu *et al* 2005).

Whey based mango herbal beverage was prepared using the method reported by Gagrani *et al* (1987), Singh *et al* (1994) and Skider *et al* (2001). The paneer-whey was centrifuged to separate fat and then boiled for 15 minutes at pH 4.6 for precipitation of whey protein. The precipitated whey (32%) was heated to dissolve sugar (8%). Thereafter, mango pulp (12%) was mixed thoroughly with above mixture and the whole mixture was filtered. Then lemon grass distillate (1.5%) was added. After the addition of lemon grass distillate, the beverage was filled into glass bottles and sealed by crown corking. Bottle sterilization was done at 121 °C for 10 minutes. Bottles were cooled at room temperature and stored under refrigerated conditions (6 $\pm$ 1 °C). The beverage prepared from 12% mango pulp, 8% sugar, 1.5% lemon grass distillate, 48% water and 32% whey was quite acceptable. Apart from imparting therapeutic properties to the beverage, lemon grass also improved the shelf life and palatability of the beverage. Quality of the product was well within the acceptable limit up to 2 months of storage at 7 $\pm$ 1 °C, due to antimicrobial and antioxidative properties of lemon grass distillate.

Blended mango nectar

In mango nectar preparation, often more than one variety in combination is used. A combination of 'Dashehari' and 'Langra' (Roy *et al* 1972) and 'Totapuri' and 'Amrapali' (Khurdiya, 1993) in equal proportions gave best nectar. 'Amrapali' (Dashehari × Neelum), rich in carotenoid content, with deep orange-red flesh and without fibre, is an excellent mango variety for juice and beverage production (Khurdia and Roy, 1988). On the other hand, 'Totapuri' variety is poor in colour, has mild flavour and moderate in juice content. 'Amrapali' is superior to that of 'Totapuri' with respect to total carotenoids and colour. The °Brix, acidity and pH are also higher in 'Amrapali' pulp than in

'Totapuri'. The 'Amrapali' nectar possessed highest carotenoids to that from Totapuri and the highest in density, but lowest in viscosity, while, 'Totapuri' is more viscous. So to improve 'Totapuri' mango nectar, 'Amrapali' is blended. Nectar prepared from the pulp of 'Totapuri' and 'Amrapali' at 50 : 50 ratio was superior in colour, carotenoid contents, viscosity and sensory quality to the nectar prepared from 'Totapuri' alone or the blended nectar with 'Amrapali' at 75 : 25 ratio. The density decreased while viscosity increased with increasing proportion of 'Totapuri' in the blended nectar. The nectar (20% pulp, 15°Brix and 0.3% acidity) from blended pulps showed that 'Amrapali' had greatly influenced the quality of 'Totapuri' mango nectar in respect of carotenoid contents, 'viscosity', hunter colour values and sensory quality. Therefore, blending of 'Amrapali' and 'Totapuri' pulps at 50 : 50 ratio is the best option to get best nectar (Khurdiya. 1993).

Caution

As a rule the 'base' should be kept at least 50% and possibly not more than 75%. It is highly desirable to blend the juices as early as possible so that the two contrasting 'sets' of constituents may 'merge' together and give a smooth-mature taste. After blending it should be dehydrated, flash-pasteurized, filled into lacquered cans and finally cooked.

Blends prepared out of syrup or water can not be labeled as 'juices' but must be called 'Cocktails' or 'Ads' or some other designation which signifies beverage but not juice.

NEW PRODUCTS OF MANGO

Conventional type of mango products have been developed to a considerable extent, but the present need is to develop new and novel products to utilize the mangoes successfully and to minimize post-harvest losses and make mango orcharding more remunerative. The international market is demanding newer products; is becoming more and more competitive and each other mango producing country is also entering the market of processed products. Therefore, the research organization and the mango processing industry should attempt the development of newer products in the country, so as to retain the monopolistic position in processed products in the global market. The development and availability of newer products not only boost the mango processing prospects, but also help in full utilization as well as enhance the capacity of the existing processing units. Thus, there is a great need and scope for the introduction of newer products of mango in the markets. A few attempts have been made in this direction in India. Sagar and Khurdiya (1998) developed some new products and standardized the preparation of mango toffee, mango custard powder, mango lassi, mango shake etc.

Mango Toffee

Mango toffee is a fruit confectionary product, by which a new taste, new flavour and new texture are achieved by the use of natural ingredients like raisin, coconut etc.

The mango fruit pulp is first concentrated in a steam-jacketed kettle to about a third of its original volume. The other ingredients (sucrose, skim milk powder and hydrogenated fat) are then mixed and cooking is continued to final weight equal to about 1.2 times that of fruit pulp taken. Flavouring materials are added at this stage. The cooled mass is transferred to a level and smooth-surfaced stainless steel tray, smeared previously, lightly with fat/grease into a thin sheet of 1.6 cm thickness. The solid sheet after cooling and setting for 2 hours is cut into suitable size toffies and dried at 50°-55 °C to a final moisture content of 5-6%. Finally the toffies are wrapped in tissue paper or in cellophane paper and packed in air-tight jars (Sagar and Khurdiya, 1998).

Recipe

Fruit pulp	-	53 kg
Sugar (sucrose)	-	30 kg
Glucose	-	4 kg
Skim milk powder	-	5 kg
Vanaspathi	-	5 kg
Essence and colour of choice	-	Traces

Colour of choice

Mango toffee can also be prepared from mango slices, by mixing with milk (1:1) with following recipe. Toffies, made thus, had higher score in respect of colour, flavour and texture (Pruthi, 1992).

Recipe

Dehydrated ripe mango slices	-	120 g
Boiled and cooked milk	-	120 g
Glucose	-	9 g
Skimmed milk	-	18 g
Vanaspathi	-	10 g
Essence and colour of choice	-	Traces

Mango custard powder

This is a rich source of nutrition for children. It is made into a thin paste in water and served to children, if necessary with milk and sugar. The detailed procedure for preparation of mango custard powder was outlined in APEDA (1997) monograph on mango. The mango custard powder is prepared from ripe mango powder after mixing with water in the ratio of 1:5 and it gained highest organoleptic score in respect of colour, flavour and texture. The recipe of the custard powder is as follows:

Recipe (Sagar and Khurdiya, 1998)

Mango powder	-	120 g
Water	-	600 g
Skimmed milk powder	-	10 g
Sugar	-	20 g
Corn starch	-	5 g
Lime juice	-	4 g
Cream	-	6 g
Salt	-	6 g

The dried mango custard powder packed in cans under nitrogen can keep well for one year at 5 °C and 6 months at 37 °C (Siddappa and Ranganna, 1961a, b).

Mango Lassi

Mango lassi is a delicious and popular drink during summer months. It is prepared with fresh ripe mango powder and curd and other ingredients as per the following recipe (Sagar and Khurdiya, 1998).

Recipe

Mango powder	-	120 g
Curd	-	400 g
Salt	-	2 g
Sugar	-	27 g
Ice cubes	-	50 g / glass

The curd, mixed with cold water in the ratio of 1:2 is an excellent medium for making mango lassi. The mango lassi made with mango powder and curd at a ratio of 3:10 was highly acceptable in respect of colour, flavour and texture followed by lassi prepared from mango powder and curd at a ratio of 2:5 and 1:5 (Sagar and Khurdiya, 1998).

Mango burfi

Burfi is a traditional dairy product in which milk is used. It retains its quality for considerably long period at room temperature, due to its low moisture content and higher concentration of sugar. Conventionally orange, coconut, wood apple etc. are used in the preparation of burfi. The use of mango pulp for making burfi is a recent development (Wankhade *et al* 2007). Mango burfi containing 35% mango pulp was superior with peculiar flavour. However, increasing the mango pulp beyond 35% did not improve acceptability, besides being more costly. The superior mango burfi (35% mango pulp) contained 80.19% total soluble solids, 21.56% fat, 12.46% protein and 0.79% acidity and has good nutritive value. Mango burfi is a rich source of vitamin A, almost as rich as butter and has fair amount of vitamin C.

Mango yoghurt

Mango yoghurt is an excellent substitute for the high caloric dessert pudding which contains sufficient amount of curd with cream. Nutritionally it could be a sufficient diet for growing children, who usually develop an aversion towards milk.

Mango yoghurt is prepared with mango pulp and boiled and cooled milk in the presence of starter culture (curd). A drum-dried mango pulp and curd (1:2) was better with its colour and flavour. Mango yoghurt prepared from mango pulp and milk in the ratio of 3 : 10 was highly acceptable in respect of flavour and texture.

In an attempt to standardize the manufacture of high fat dessert of yoghurt with added pulp, Balasubramanyam and Kulkarni (1992) found that yoghurt made from milk, 10 % fat, with 8% sugar and homogenized at 200 bars was of excellent quality. Addition of mango pulp up to 4% of the milk content marginally improved the flavour characteristics. Addition of higher amount of mango pulp adversely affected the flavour, body and texture of yoghurt, irrespective of homogenization at high pressure (>200 bars).

Mango ice-cream

This is a non-conventional product. It is prepared by mixing ripe mango powder and milk. The ice-cream made from structured mango pieces modified with neutral ingredients has been discussed by Britnel (1989). A highly acceptable mango ice-cream containing mango powder and milk (3:10) was prepared, having better colour, flavour and texture with the following recipe (Sagar and Khurdiya, 1998).

Recipe

Mango powder	-	120 g
Milk	-	400 g
Sugar	-	50 g
Vanaspathi	-	8 g

Mango puree can also be used for making mango ice-cream (Bray, 1981).

Mango shake

Mango shake is a kind of mango based beverage mixed with milk. It is a delicious and popular drink during summer months. Mango shake is prepared by addition of mango powder in sweetened, boiled and cooled milk. The mango shake thus prepared, mixed in equal amount of water to the sweetened milk in the ratio of 1:1 was superior in colour, flavour and texture followed by the products made in the ratio of 1:10 and 3:1 (mango powder : milk). The following two recipes were proposed for making mango shake (Sagar and Khurdiya, 1998).

Recipe I

Mango powder	-	120 g
Water	-	120 g
Sugar	-	20 g
Milk	-	240 g

Recipe II

Mango powder	-	120 g
Water	-	400 g
Curd starter	-	Traces
Sugar	-	40 g
Salt	-	3 g
Fruits (small pieces)	-	5-10

Mango shake prepared using spray-dried product comprised of mango, milk and sugar solids and sterilized in glass bottles had good acceptability (Grewal and Jain, 1982).

Mango butter

It is prepared by mixing mango puree with 20-30% nut paste, by diluting the mixture with one-third water and heating at 90 $\pm$ 5 °C for 1 to 2 hours. It is then passed through a stainless steel press to get an uniform texture (Mortean and Danvin, 1980).

Spray-dried mango milk powder

It is prepared from mango fruit solids, milk solids and sugar in a spray dried form. It readily reconstitutes into pleasant fruit-milk drink or as an ingredient in ice-cream, puddings etc. Preliminary laboratory trials indicated that a beverage with 20% total solids consisting of 4% mango solids, 10% MSNF and 6% sugar showed maximum acceptability (Sharma *et al* 1974). The process of making spray-dried mango-milk powder is as follows:

Ripe mangoes are washed, peeled and cut into small pieces and pulped in a high speed mixer. The concentrated skim (buffalo) milk and cane sugar of ISI grade are well mixed and pre-heated to 50 °C followed by the addition of glycerol mono-stearate and sodium alginate in a ratio of 1 : 1 at the rate of 1% of the total solids of the mix. The heated mix is filtered and homogenized at 65°-70 °C. The homogenized mixture is pasteurised at

65.5 °C for 30 minutes in a hot water tank followed by cooling at 10 °C. Then the mango pulp is mixed with the cooled concentrated milk and filtered through a muslin cloth. The filtered mango-milk mix is dried in an anhydro-spray-drier to 2.5% moisture. The dried mango milk powder is collected in polyethylene bags, cooled, packed in pre-sterilized tin-plated cans, sealed and gassed twice with nitrogen at interval of 24 hours (Sharma *et al* 1974).

CHANGES DURING PROCESSING

Mango fruits under go changes in physico-chemical and nutritional characteristics during processing as well as during storage. Because of these changes, the colour, texture, flavour, taste and chemical composition of the processed products are different from those of fresh fruits. These changes influence the quality of the end products as well as their shelf life. The changes differ from product to product and also from variety to variety of mango. While, some changes are beneficial and required for the quality improvement and retention and extension of storage life of the product, others are detrimental to the quality and extension of storage life. A brief account of the changes that occur in mango products during processing and the probable reasons for them are presented below:

Carotene, precursor of Vitamin A is an important nutrient constituent of food items including mango fruit and its retention in the food is of great importance in view of the nutrition. Destruction of β-carotene in foods during processing and storage is well known. Therefore, a high stability of carotene in processed products is of special interest to both the processor and consumer. A loss of 30-40% of carotene during drying of the fruit pulp in the preparation of mango leather was observed (Das *et al* 1955). β-carotene showed little change in frozen mango pulp during canning and storage at ambient conditions for 200 days of storage (Garden and Tremazi, 1966., Nanjundaswamy *et al* 1966). Khurdiya and Sagar (1988) reported loss in total carotenoid pigments with a decrease in pH in mango pulp. These pigments are susceptible to light and temperature (Kapur, 1974., Khurdiya and Roy, 1986a). Acidity had no marked effect on the carotene pigments and as they increased, there was an appreciable loss in the total carotenoids in the mango pulp (Sahni *et al* 1994). The loss of pigments with increased acidity is due to the isomerization of pigements. Pruthi *et al* (1956) found that the retention of carotene and ascorbic acid was excellent in canned products. Siddappa and Bhatia (1956a) also observed retention of carotenoids and ascorbic acid during canning, while Bose and Das (1958) reported losses in carotenoids in canning and storage. Retention of carotenoids in dehydrated mango slices was higher (Amin and Bhatia, 1962., Sagar and Khurdia, 1997). Ramana *et al* (1984) observed better retention of carotens by fortification of ascorbic acid. In contrast to the above, Sahni *et al* (1994) observed that the total carotenoids content did not show any appreciable variation with the change in the ascorbic acid content in the pulp.

Ascorbic acid is an important nutrient constituent of mango, however, being sensitive to oxidation and heat, is largely affected during canning (Kapur, 1975). Its retention depends upon the variety and the processed product. In case of 'Raspuri' mango the retention of ascorbic acid was about 78.0%, whereas in case of 'Badami' mango, the retention was slightly less (65.2%). The percentage loss of ascorbic acid was maximum in 'Totapuri' (57.61) as compared to seedling mangoes (55.65%) (Dabhade and Khedkar, 1980c). Considerable losses of ascorbic acid occur during drying as compared to blanching. There was a heavy loss of 92.98% of ascorbic acid during drying of fruit pulp in the preparation of mango leather (Das *et al* 1955). Ascorbic acid content decreased by 20% after canning and by another 30% during storage of 200 days at ambient conditions (Garden and Tremazi, 1966., Nanjundaswamy *et al* 1966). Ascorbic acid decreased to an extent of 70-80% in case of pulp, 55-60% in case of slices and juices (Kapur *et al* 1985). About 50% of ascorbic acid is lost in canned slices of different mango varieties (Teaotia and Awasthi, 1966). Mango varieties which initially had higher amount of ascorbic acid showed better retention in the pulp than those which had initially less content (Nanjundaswamy *et al* 1966). Ascorbic acid showed a gradual decline in both the sugar syrup and sugar plus citric acid treated slices (Sagar and Khurdiya, 1997). Being a natural antioxidant, extent of retention of ascorbic acid in the end products is an important index in determining the shelf life of the product (Luh *et al* 1958). The retention of ascorbic acid in the pulp increases with increase in the acidity (Sahni *et al* 1994). The retention of ascorbic acid was better in cabinet-dried mango pieces than in the sun-dried ones (Dabhade and Khedkar, 1980c). During freezing of mango slices, ascorbic acid and reducing sugars decreased by 30 to 50% and carotenoid pigments by 15 to 30% (Adsule and Roy, 1974b).

Kapur (1975) observed a gradual decline in acidity in canned slices as well as in canned pulp (Table 5.4) and attributed it to the leaching of the water soluble acid in water, its involvement in inversion of sugars and reaction with the tin of container. The rate of increase in acidity on processing and storage was much higher in immature fruits than that of fully matured fruits (Kapur *et al* 1985). The acidity level in dehydrated slices was higher in control samples and lower in sugar syrup treated ones but without citric acid and decreased with increase in sugar concentrations during heating (Sagar and Khurdiya, 1997).

Kapur (1975) observed decline of reducing sugars in canned slices and pulp during steeping in salt/water for canning and explained the decline due to leaching in brine/ water during preparation and equilization of the solutes between solid and liquid phases. However, earlier Nanjundaswamy *et al* (1966) observed increase in the percentage of reducing sugars in processed mango pulp of different varieties. Changes in non-reducing sugars were due to an account of equilization of the soluble solids in the liquid and solid phases in slices, whereas in case of pulp the increase in non-reducing sugars was due to the partial concentration involved in hot-filling and sterilization (Kapur, 1975). In a further study, Kapur *et al* (1985) observed an increase in TSS, reducing and non-reducing sugars in canned pulp while slices and juice showed slight and insignificant variation on processing and storage. The reducing and total sugars in the sugar syrup treated slices without citric acid were higher than those in untreated and those with citric acid dehydrated slices (Sagar and Khurdiya, 1997).

Percentage of pectin and starch decreased on processing (canning) and storage due to conversion of proto-pectin into water soluble pectin during heat processing and storage (Post mayer *et al* 1956).

There was a decrease in titratable acidity and TSS and almost 50% reduction in vitamin C content of freeze dried mango pulp. The moisture, total fructose, reducing and total sugars were not affected during one year storage of pulp. During sterilization (95 °C) reducing sugars showed greater increase, but at lower temperature (75° and 85 °C) sterlization, the increases were modest (Kalra and Tandon, 1985a).

Mango wine is prepared both from ripe and over-ripe fruits by fermentation using yeast. Mango wine prepared from over-ripe fruits shows changes in pH, acidity, TSS and reducing sugars. It is seen from the Table 5.5 that all the parameters have shown decrease in wine (Akingbala *et al* 1994) (Table 5.4 and 5.5).

Table 5.4: Changes in chemical constituents of mango (Cv. Dashehari) during canning (Kapur, 1975)

Step	Slices				Pulp			
	Acidity (%)	**Reducing sugars (%)**	**Non-reducing sugars (%)**	**Ascorbic acid (mg/ 100 g)**	**Acidity (%)**	**Reducing sugars (%)**	**Non-reducing sugars (%)**	**Ascorbic acid (mg/ 100 g)**
Fresh	0.36	3.63	12.60	19.60	0.36	3.63	12.60	19.60
After preparation	0.32	3.40	11.83	17.80	0.26	3.34	11.39	16.80
Exhausting /filling	0.24	3.27	12.50	17.60	0.18	3.40	11.30	15.60
Sterilization (Canning)	0.23	2.70	14.83	10.40	0.17	3.67	14.17	9.60

Table 5.5: Changes in physico-chemical characteristics of juice and wine of over-ripe mangoes (Akingbala *et al* 1994)

Character	Juice	Wine
pH	5.00	3.70
TSS (Brix)	19.80	5.00
Titratable acid as citric acid (%)	0.42	0.38
Total reducing sugars (%)	18.80	0.12

CHANGES DURING STORAGE OF PROCESSED PRODUCTS

A number of physico-chemical, enzymatic and microbial changes occur during the storage of preserved and processed products of mango. Extent, magnitude and kind of changes and the component changes vary from product to product. The length of storage period, type of storage, storage environment in which the products are stored are the other determining factors which influence the changes as well as quality of products and their acceptability and shelf life. Microorganisms in the product also contribute their might to the changes. Most of these changes lead to deterioration in quality and nutritive value of the product during storage. The different changes that occur in storage of processed products are presented below :

Pickles

At the end of storage period of 8 days at room temperature and low temperature, there was an increase in the percentage of moisture in the fruit pieces in pickles, reducing and total sugars increased during storage, while the percentage of acidity decreased. Vitamin C content increased up to a period of 6 days after which it declined at room temperature. On the other hand, the moisture content of pickles at low temperature was constant, which indicated slow rate of metabolic changes in the product (Sastry and Habibunnisa, 1979).

Mango Leather/Bar

Storage conditions, activity and residual sulphur di-oxide (SO_2) content of the product affect the quality of mango bar. These parameters influence the extent of changes in reducing sugar, acidity, carotenoids, SO_2 and non-enzymatic browning (NEB) in storage (Rao and Roy, 1980b). Three types of mango bars (plain mango bar; mango + dessicated coconut powder bar and mango + soy protein concentrate bar) were stored for 90 days at –18 °C, at ambient temperatures; 27 $\pm$ 3 °C (65% RH) and 38 $\pm$ 1 °C (92% RH) (Mir and Nath, 1993). Moisture, acidity and reducing sugars increased in all types of mango bars during storage. Reduction in total and free SO_2, total carotenoids and β-carotene and an increase in NEB were observed. Losses of carotenoids and NEB were more in unsulphited bars than in sulphited bars. Storage decreased the overall acceptability and textural characteristics. The deterioration changes were minimum in mango bars stored at –18 °C. The changes were of lower magnitude in sulphited bars. Consequently, the NEB in sulphited bars was lower than in unsulphited bars. Sulphited samples darkened during storage due to increase in NEB, which was attributed to concomitant loss of total and free SO_2. Total carotenoids and β-carotene content decreased during storage in both sulphited and unsulphited bars and the losses were greater at higher temperatures, which could be due to non-oxidative changes or oxidative changes. Such changes alter the product colour from red-orange to pale-yellow orange and lower the flavour and nutrient value of the product (Eskin, 1979).

Satyavati *et al* (1962) related the comparatively less browning in 'Bombay Green' and 'Dashehari' leathers to higher carotenoids in these varieties, which served as a protective agent against browning. Browning was less at higher temperatures in the sheets of 'Bombay Green' and 'Dashehari' (Rao and Roy, 1980 b). More browning in 'Baneshan' leather was attributed to higher tannin content in its pulp than in the above two varieties (Adsule and Roy, 1974 b). Greater retention of β-carotene over the total carotenoids in sulphited samples was due to the protective action of SO_2 (Rao and Roy, 1980 b). Textural characteristics of mango bars changed during the storage period of 90 days. Hardiness, springiness, gumminess and chewiness of mango + soy protein concentrate bars were higher than those of other two types of bars. Presence of desiccated coconut powder in the bars prevented compactness of the bars with lower values for textural characteristics. All the values decreased significantly during storage and the changes were greater at higher temperature. A significant interaction between formulation and storage time as well as the storage time and temperature were exhibited. Storage at higher temperature caused greater loss in sensory scores and the changes in sensory parameters were similar to those observed in chemical and textural attributes. The changes in sensory parameters were minimum at –18 °C (Mir and Nath, 1973). In soy-protein enriched mango fruit bar, the moisture content increased during storage period at room temperature (Chauhan *et al* 1997) due to absorption of moisture from the external environment or due to chemical changes such as browning (Karel, 1975). Similarly the reducing sugar and acidity increased during storage. The increase in reducing sugars may be due to acid hydrolysis of sucrose (Rao and Roy, 1980b). Increase in acidity may also be due to formation of sulphurous acid from SO_2, ascorbic acid degradation or hydrolysis of pectin (Ammu *et al* 1977). NEB was less in protein enriched mango bars than in normal mango bars. Higher NEB in control samples was due to oxidation or volatalisation of food constituents (sugar, protein, pectin, lipids) (Biolin and Boyle, 1972., Ech Koff and Okos, 1986).

Mango nectar

Storage temperature showed appreciable effect on the ascorbic acid, carotenoids and reducing sugars in nectar. Losses in ascorbic acid were more rapid at ambient temperature. Similarly carotenoids were lower at room temperature storage, while viscosity, pH, acidity and TSS remained constant. Hydrolysis of non-reducing sugars to reducing sugars resulted in their rapid increase at ambient temperature, being accelerated by high temperature and acidity (Nanjundaswamy *et al* 1966., Kapur *et al* 1985., Sahni and Khurdiya, 1989). During storage of 'Amrapali' nectar and 'Taimuria' RTS drink, TSS, acidity and viscosity remained unchanged, whereas reducing sugars increased, but ascorbic acid reduced (Jain *et al* 1997). Nectar with 20% pulp, 15° Brix TSS and 0.3% acidity from 'Dashehari' variety of mango with the addition of certain stabilizers (Carboxymethyl cellulose, gumghatti, gum acacia and sodium alginate) was stored under ambient conditions for 5 months (Altaf *et al* 2006). Carboxy methyl cellulose treated ectar showed minimum changes in reducing and total sugars, titrable acidity and SO_2 and had higher organoleptic score. However, TSS fluctuated during storage. Titratable acidity increased during 5 months storage as a result of degradation of pectic substances in the nectar. Stabilizers affected the content of acidity and their concentration. Maximum acidity was found with gum acacia and sodium alginate due to the presence of manuronic acid in sodium alginate and D-Glycuronic acid in gum acacia (Ranganna, 1986). Reducing sugars increased in nectar during storage following the breakdown of sucrose etc. (Gowda and Ramanjaneya, 1995., Altaf *et al* 2006). Degradation of insoluble polysaccharides like hemicellulose and oligosaccharides continuously increased total sugars, while the total SO_2 decreased continuously. Sensory score showed a gradual decrease during storage (Altaf *et al* 2006).

Mango squash

During storage, mango squash showed increase in acidity and sugars. The increase in sugars was due to gradual inversion of non-reducing sugars to reducing sugars by hydrolysis. TSS remained unchanged in the pulp (Palaniswamy *et al* 1974). Mango squash prepared with the addition of SO_2 maintained its quality and acccptability up to 8 months, whereas pasteurized RTS and nectar lost their acceptability after 2 and 4 months storage respectively (Rabbani and Singh, 1988).

Mango juice

The colour, flavour and taste of the mango juice faded gradually as the storage period advanced. Their retention capacity varied with variety. In this respect 'Sukul' was the best and its juice did not show discolouration up to 12 months. 'Lucknow Safeda', 'Yakuti' and 'Gilas' varieties retained the characters up to 6 months only. Highest concentration of ascorbic acid was retained in 'Lucknow Safeda' and the least in 'Yakuti'. After 12 months storage 57-63% ascorbic acid was retained in the mango juices (Awasthi and Pandey, 1979).

Canned Juice

Viscosity declined in canned juice during storage. The decrease was more in viscous juice (Totapuri x Mulgoa) and least in thinner juice (Alphonso, and Totapuri x Alphonso hybrid). The decrease in carotenoid content was more pronounced in varieties having initially high carotenoid content (Alphonso and Totapuri x Alphonso hybrid). NEB increased with increase in storage period (Ambadan *et al* 1988).

Canned pulp

In canned mango pulp reducing sugars increased during storage as a result of concomitant decrease in non-reducing sugars (Patil, 1990., Ranjana, 1991., Gowda *et al* 1994., Gowda and Hudder, 2004). TSS increased slightly up to 6 months and thereafter declined. The slow breakdown of starch into sugars during storage might be the cause of increase in TSS of canned mango pulp up to a certain period of storage, while the later decline was due to its utilization for bio-chemical processes. During storage vitamin C was lost to an extent of 25-45% due to decomposition by hydrolysis producing fructose (Adsule and Roy, 1974b). However, ascorbic acid is retained in canned and frozen pulps, more so in the former one, which is probably due to

(i) heat intervention of oxidative enzymes during processing and

(ii) tin in the cans – acting as a protectant to ascorbic acid (Ramana *et al* 1984).

Retention of carotenoids in canned products is of considerable importance in view of the fact that the β-carotene is very easily destroyed in the food stuff including canned products during storage (Gowda and Hudder, 2004), which may bring about undesirable changes in the colour and nutrition of the product. A number of factors influence the storability of carotenoids, water activity, presence of antioxidants, storage conditions like light, temperature, type of containers used, duration of storage etc. Sudhakar and Maini (1994) investigated the effects of various biochemical components, other additives and different storage conditions on the stability of carotenoids in the sulphited canned mango pulp. A gradual decrease in carotenoid pigments of the pulp occurred during storage at pH 4.5 and 5.0, but the colour was bright yellow at pH 2.5 and 3.0 compared to that of high pH value. The carotenoids were relatively more stable at lower concentrations of sugars as compared to higher concentrations which may be due to high moisture level at lower concentrations of sugar in the pulp, that exerts a protective effect against carotenoids. At higher levels of SO_2, carotenoids were more stable due to retention of more free SO_2 during storage at higher concentrations of added SO_2. Ascorbic acid and antioxidants (butylated hydroxyl amisole) helped in better retention of carotenoids of pulp. Ascorbic acid as an antioxidant protected the loss of carotenoids during storage. Packaging materials also played a role in the retention of carotenoid pigments in the stored pulp. Reduction of carotenoids was higher in the pulp packed in glass containers and with least surface area exposed to air and lower amount of oxygen coming in contact with the pulp, thereby reducing the oxidation of carotenoids. Higher losses of carotenoids at low temperature in PVC and HDPE containers was due to the permeability of containers to oxygen. Higher temperature coupled with higher permeability of pack led to less stability of the pigments in the pulp stored at room temperature and 40 °C

(Maini *et al* 1984., Sahni and Khurdiya, 1989., Sudhakar and Maini, 1994). The carotenoids of sulphited pulps of different varieties of mango decreased gradually during storage and losses were higher in 'Rataul' and lower in 'Dashehari' during 4 months of storage at room temperature. Maximum retention of carotenoids was in the pulp of 'Neelum' followed by 'Totapuri' and 'Chausa'. Reduction in flavour of canned pulp during storage was attributed to the loss of flavour compounds, which are highly volatile (Gowda and Hudder, 2004).

Canned mango slices

Acid content and reducing sugars showed increase in canned slices, while ascorbic acid content was much reduced during storage in all the varieties. Heating and processing have destroyed ascorbic acid (Pruthi *et al* 1956). Similarly Palaniswamy *et al* (1973) while evaluating the suitability of certain mango varieties for canning, observed loss of ascorbic acid and slight increase of TSS, acidity and sugars in canned slices during storage. However, Kapur (1975) in a similar evaluation of some varieties for canning observed no changes in TSS content in canned slices stored at room temperature (not mentioned), 37 °C and 55 °C. After two months of storage, acidity declined as a result of its involvement in reaction with sugars and tin surface of the cans, particularly at higher temperatures. During storage, the content of reducing sugars increased following the inversion of non-reducing sugars. The rate of inversion was more rapid at higher temperature. Changes in sugar, acid and ascorbic acid were accompanied by changes in quality factors and condition of the containers, which largely determined the shelf life of the product. Ascorbic acid showed gradual loss. The product acquired brown colour. All the above changes took place more rapidly at higher temperature storage. At room temperature, canned mango slices retained colour, flavour and taste up to 70 days, whereas deterioration in the characteristics occurred within 50-60 days at 37 °C and 25- 50 days at 55 °C. The colour of the canned slices changed from yellow to light brown and finally to deep brown colour (Kapur, 1975) on account of changes in sugars or ascorbic acid (Wilson, 1928). Development of the brown colour in the canned slices, according to Kapur (1975) was closely accompanied by the change in flavour and taste, which finally determined the shelf life of the product.

Mango pulp

Mango pulp is often stored for later use in the preparation of several processed products such as nectar, juice, jam etc. During this storage period several changes occur in the pulp. 'Dashehari' mango pulp stored with SO_2 (1000 ppm) in wax-sealed glass containers for a year, showed increase in reducing sugars four fold and in pentose by 1.6 times, with a decrease in SO_2 to one-fourth of the initial level. Non-reducing sugars were gradually converted into reducing form. Moisture, titratable acidity, pH and TSS varied within 10% in the pulp at the end of one year storage (Kalra and Tandon, 1985b).

Kalra (1982) studied the changes during storage (3 months) of mango pulp, aseptically processed and packed in glass and HDPE containers at ambient (30-35 °C) and cold temperature (0-4 °C) in case of three varieties (Langra, Chausa, Dashehari) and observed little differences between the varieties in respect of moisture, colour and flavour, whereas acidity, reducing sugars and tannins showed increase in their levels at room temperature. Tannins contribute to the astringency of the pulp. Rise in the tannins in storage at room temperature indicated unsuitability of mango pulp. There was general decrease in protein and vitamin contents during storage irrespective of the temperature, at which the pulp was stored. Between the two temperatures tested, preservation of aseptically processed pulp with lesser alterations was better at low temperature than at room temperature.

Freezing preservation of pulp

During storage, ascorbic acid and reducing sugars decreased by 30-50%, while carotenoid pigments decreased by 15-30% in mango pulp preserved by freezng at –5 ± 2 °C (Singh, 1970., Krishna Prakash *et al* 1986). Ascorbic acid, total carotenoids, colour and viscosity of pulp decreased in the frozen pulp of 'Totapuri' mango during prolonged storage of 14 months at –18 °C (Ramana *et al* 1984). Loss of ascorbic acid in frozen control pulp was faster than in pulp fortified with ascorbic acid and frozen or canned due to higher activity of oxidative enzymes in the control pulp. Ascorbic acid in the fortified pulp acted as an oxidant. The decrease in viscosity in frozen pulp may be due to the activity of the pectin degrading enzymes.

Frozen slices

Total sugars and TSS remained unchanged during storage of frozen slices, which could be due to complete absence of osmotic differences at –12.2 °C. The acidity (as citric acid) did not show much variation in frozen slices. The ascorbic acid as well as carotenoids decreased during storage due to presence of residual oxygen (Andrabi *et al* 1956), Colour and flavour of the frozen slices scored more up to 6 months and thereafter remained constant. However, texture of frozen slices scored less (Adsule and Roy, 1974b).

RTS beverages

The acceptability of fruit drinks/beverages is very much dependent on their physico-chemical proeperties and their retention during storage. Changes in colour and volatile flavouring compounds during storage determine the quality of the product. The RTS beverages from mango, stored in white and amber coloured bottles for 6 months under 3 different storage environments viz., ambient temperature (12.5-36.0 °C), cool chamber (10.0-29.6 °C) and low temperature (4 ± 1 °C) showed a gradual decrease in sensory quality, acidity, ascorbic acid and tannins. Retention of ascorbic acid was more in beverages stored in amber coloured bottles under low temperature environment. TSS, reducing sugars, total sugars and NEB showed an increasing trend during different storage environments (Deka *et al* 2005).

Wine

The main change in fruit wines during storage is the "oxidative browning". Some of the wines develop deep brown despite the use of SO_2 and following standard enological procedures for their preparation. Such wines can not be stored at lower temperatures. Different factors influence the susceptibility of the wines to oxidative browning (Berg and Akiyoshi, 1956., Caputi and Peterson, 1965., Simpson, 1982).

Initial colour of wines does not show a positive correlation to browning, though colour after heating is positively correlated with internal colour primarily followed by anthophylls and phenolics (Lakshminarayana, 1980). Increase in colour at 420 nm after heating is considered as an indicator for browning susceptibility.

Total phenolic content of mango wines is fairly highly correlated with browning. The other components of wine like reducing sugars, ethanol, aldehydes, pH, total titratable acidity and volatile acidity are not correlated with either increase in colour or colour after heating. The wine with lower pH is more resistant to browning (Onkarayya and Rao, 1988). On the basis of normal distribution of phenolics, varieties with less than 812 mg per litre, such as 'Bombai', 'Dashehari', 'Fazli' and 'Langra' are least susceptible, those varieties with 1550 mg per litre like 'Mallika', 'Seroli', 'Bappakai' and 'Elamandala Thiyyamamidi' are most susceptible to oxidative browning (Onkarayya and Rao, 1988).

The pH of the wines from over-ripe mango fruits decreased while the acidity increased with age in polyethylene pouches. The greater increase in total acidity of wine in polyethylene pouches may be due to gradual introduction of oxygen from the air into the wines (Amerine *et al* 1980) resulting in a series of oxidation reactions which converted ethanol and lower alcohols to acids and aldehydes (Rankine, 1965). The increase in ester concentration with time might be due to increased reactivity of alcohols and acids (Amerine *et al* 1980). The wine exhibited sweet-swelling with characteristic grape wine aroma. However, the taste was harsh. This could be due to high level of sodium metabisulphite used during fermentation, the addition of tannic acid during fining the acidity of the wine, or combination of these factors. A longer storage period may ameliorate the harsh taste of the wine (Akingbala *et al* 1994).

Intermediate moisture mango products

Partially dehydrated foods with intermediate level of moisture (20-50%) followed by sterilization to keep water activity at safe levels (0.6-0.85%) and thereby avoid microbial spoilage are referred to "intermediate moisture products". Samples of such product of mango (cv. Badami) packed in different containers, stored for more than 9 months at room temperature, showed retention of carotene even after 18 months storage. There was a gradual loss of β-carotene with subsequent fading of orange-red colour and onset of browning and off-flavour at later stage of storage. Browning did not show significant increase during 9 months storage. Samples packed in HDPE suffered more drastic carotene and flavour loss than corresponding samples in cans and PFL pouches and became unacceptable after 4 months at room temperature (Jayaraman *et al* 1976). Slight decrease in acidity was observed after 6 months storage of the product treated with glycerol. The reducing sugars content was twice in the product just after drying in comparison to soaked slices. Reducing sugars increased slightly during storage as a result of acid hydrolysis of sucrose (Mir and Nath, 1993).

Amchur

Raw mango powder (amchur) prepared from unblanched, blanched, sulphured and sulphited, sun-dried and cabinet-dried pieces were packed in polyethylene bags and paper bags for storage during which time the following changes in the powder were noticed. The storage period as well as packing material influenced the changes in the chemical composition of powder and in the incidence of browning of the product (Dabhade and Khedkar, 1980c). A rapid increase in moisture content was noticed during storage period of 6 months at room temperature in both types of bags, which was attributed to high humidity in the atmosphere as a result of rains at the time of packing and the highly hygroscopic nature of the product. The loss or gain in moisture from the powder was mainly due to the permeable nature of bags. The gain was comparatively more in paper bags than in polyethylene bags, since paper bags are more permeable to moisture. Decrease in acidity of powder during storage could be due to disappearance of SO_2 from the product and the trend of decrease was more in paper bags than in polyethylene bags. Unblanched control sample had more acidity than blanched, sulphited and sulphured ones. Total and reducing sugars increased in both types of bags, greater being in paper bags due to more rapid hydrolysis of polysaccharides and their subsequent inversion to reducing sugars. Greater leaching losses of sugar in sulphited than in control lots of mango powder resulted in lower amount of sugars in treated lots. Starch content decreased throughout the storage period, irrespective of package due to hydrolysis of starch. The loss of ascorbic acid was rapid in sun-dried, paper packed powder samples than cabinet-dried powder packed in polyethylene bags. Higher retention of ascorbic acid in cabinet-dried product may be due to rapid drying in cabinet drier. Ascorbic acid was more in cabinet-dried powder and packed in polyethylene bags after 6 months storage period. NEB showed increase both in sun-dried and cabinet-dried products during storage and was inversely proportioned to SO_2 treatment. Greater browning in sun-dried powder was due to the longer exposure of the pieces to atmosphere during drying. High moisture content in the paper bags resulted into more browning than in polyethylene bags and acceleration of browning. Thus, during storage, cabinet-dried, sulphured and sulphited product had better retention of ascorbic acid and less browning as compared to sun-dried and control lots, with a progressive decrease in ascorbic acid, acidity and starch and increase in both reducing and total sugars.

Spray-dried mango milk powder

Studies on changes during a storage period of one year at room temperature and at controlled temperature of 30 $\pm$ 1 °C indicated changes in the spray-dried mango-milk powder. There was no perceptible change in colour, flavour and reconstitution of the product, but there was a progressive increase in the free fat content, volatile free fatty acids, acidity and optical density values. The colour of the powder increased from the beginning at room temperature and it was still darker at 30 $\pm$ 1 °C during one year storage. The change in colour may be due to the interaction of reducing sugars, lactose and fructose (from fruits) with NH_2 groups of the proteins. However, the overall quality of the powder remained acceptable at the end of one year storage at either of the storage temperatures (Sharma *et al* 1974).

Blends

Mango + pineapple spice blended beverage (a RTS drink) stored in white and amber coloured bottles for 6 months at ambient temperature (12.5-36.0 °C), cool chamber (10.0-29.6 °C) and low temperature (4 $\pm$ 1 °C) showed a gradual decline in sensory quality, acidity, ascorbic acid and tannins. Retention of ascorbic acid was more in amber coloured bottles under low temperature. TSS, reducing and total sugars and NEB showed increasing trend with increase in storage period.

The sensory quality of the beverage decreased linearly over a period of 6 months. TSS increased linearly with advancement of storage period due to conversion of polysaccharides to simple sugars. The blended beverage showed less decrease in acidity in coloured bottles, than in white bottles. Ascorbic acid decreased continuously during storage, but its retention was more in amber coloured bottles under low temperature storage as the total carotenoid content decreased. However, the retention of carotenoids was higher than that of ascorbic acid as a result of higher storability of carotenoids at higher temperature, compared to ascorbic acid. Retention of carotenoids was higher at low temperature storage than at other temperatures. Loss of carotenoids was due to oxidation of tannins. Flavour also decreased during storage. The tannin content of the blend decreased linearly, more in white bottles than in amber coloured bottles. Irrespective of the storage conditions and container NEB showed linear increase. However, NEB was less in amber coloured bottles and at low temperature storage. Non-enzymatic reaction of organic acids with sugars or oxidation of phenols led to increase in NEB in storage. Terpenyl acetate, the major volatile flavouring compound of the blend decreased during storage (Deka *et al* 2005).

Mango milk shake is a favourite beverage drink during summer season. However, it does not stand storage much. During 6 months storage, there was decrease in pH, TSS and viscosity (Gowda and Ramanjeneya, 1995). The decrease in viscosity was the result of break down of large sugar molecules to smaller ones and the development of slight brown discolouration. Quality attributes have changed after 3 months storage (Hassan and Ahmed, 1998).

REFERENCES

Adsule, P.G. and Roy, S.K. 1974a. Tin pick-up in mango products of different varieties. Indian Food Packer, 28(4) : 51.

Adsule, P.G. and Roy, S.K. 1974b. Studies on some important commercial varieties of mango of North India in relation to canning, freezing and chemical preservation of pulp. Journal of Food Science and Technology, 11 : 257-260.

Ahemed, J. and Chaudhary, D.R. 1995. Osmotic dehydration of papaya. Indian Food Packer, 49 : 5-11.

Akinbala, J.O., Oguntimein, G.B., Olunlade, B.A. and Aina, J.O. 1994. Effects of pasteurization and packaging on properties of wine from over-ripe mango (*Mangifera indica*) and banana (*Musa acuminata*) juices. Tropical Science, 34 : 345-352.

Alzamura, S.M., Tapia, M.S. and Welli, A.A.J. 1993. Application of combined methods of technology in minimally processed fruits. Food Research International, 26 : 125-130.

Ambadan. and Adsule, P.G. 1979. Bulk preservation of mango pulp in high density polyethylene (HDPE) containers. Indian Food Packer, 33(6) : 34-37.

Ambadan., Adsule, P.G., Iyer, C.P.A. and Subramanyam, M.D. 1988. Suitability of some mango hybrids for processing for export. Acta Horticulturae, 231 : 776-781.

Amerine, M.A., Berg, H.W., Kenku, R.E., Ough, C.S., Sringleton, V.L. and Webb, A.D. (eds) 1980. The technology of wine making. 4th Edition, West point, Connecticut, AVI Publishing Company.

Ammu, K; Krishnappa,K.G., Subramaniam, V., Sharma, T.R. and Nath,H. 1976. Freeze drying of mango pulp. Food Technology, Australia.28: 339-341.

Ammu, K., Radhakrishna, K., Subramanian, V., Sharma, T.R. and Nath, N. 1977. Storage behaviour of freeze dried fruit juice powders. Journal of Food Technology, 12 : 541-554.

Anand, J.C. and Johar, D.S. 1958a. Improved method for commercial preservation of raw mango slices. Food Science, 7(6) : 150-151.

Anand, J.C. and Johar, D.S. 1958b. Action of certain organic acids on spoilage organisms in mango pickles. Journal of Science and Industrial Research,17A : 16.

Anand, J.C., Soumithri, T.C. and Johar, D.S. 1959. Prevention of fungal spoilage in mango pickles with chemical preservatives. Indian Journal of Agricultural Science, 29 : 181.

Anandaswamy, B. 1981. Preliminary studies on suitability of flexible packaging materials for packing processed food products. Indian Food Packer, 35(2) : 56-60.

Andrabi, M.H., Magor, N.G., Pruthi, J.S. and Girdharilal, G.A. 1956. Effect of storage on the nutrients of Badami and Raspuri mangoes. Indian Food Packer, 10(11) : 9.

Anonymous, 1985-86. Raw mango beverage base formation. Annual Report, CFTRI, Mysore (India), 13, p. 53.

Anonymous, 1990. Mango in India : Production, Preservation and Processing. Extension Bulletin. Central Food Technological Research Institute, Mysore, India, p. 20.

Anonymous, 1997. Mango Processing For Value Addition. Farm Digest. July – September, 1997, 2-7.

APEDA. 1997. Ministry of Commerce, Government of India, New Delhi.

Astrom, A. 1990. Aseptic Packaging. Indian Food Industry, 9(6) : 23-29.

Altaf, A., Singh, D.B. and Tandon, D.K. 2006. Studies on effect of stabilizers on the quality of mango nectar. Progressive Horticulture, 38(2) : 208-213.

Attri, B.L., Lal, B.B. and Joshi, V.K. 1998. Physico-chemical characteristics, sensory quality and storage behaviour of sand pear juice blended with temperate fruit juice / pulp. Indian Food Packer, 52 : 36-42.

Avena, R.J. and Luh, B.S. 1983. Sweetened mango puree preserved by canning and freezing. Journal of Food Science, 48 : 406-410.

Awasthi, R.K. and Pandey, I.C. 1979. Screening of mango varieties for juice production. Indian Food Packer, 33 : 40-43.

Azizi, A. and Ranganna, S. 1993. Spoilage organisms of Canned Acidified mango pulp and their Relevance to Thermal processing of Acid Foods. Journal of Food Science and Technology, 30(4) : 241-245.

Balasubramanyam, B.Y. and Kulkarni, S. 1992. Standardization of manufacture of high fat Yoghurt with natural fruit pulp. Journal of Food Science and Technology, 28(6) : 389-390.

Baldry, J., Breag, G.R., Caygill, I.C., Cook, R.D., Ferber, C.F.M. and Lalitha, K. 1976. Alternate methods of processing mangoes. Indian Food Packer, 30 : 56-62.

Bawa, A.S. 2007. Perspectives and prospects of food technology. Indian Food Industry, 26(6) : 7-17.

Begum, J.A., Shams-Ud-Din, M. and Nural Islam, M. 1983. A study on the shelf life and consumer's acceptability of mixed squash prepared from pineapple juice and mango pulp. Bangladesh Journal of Science and Industrial Research, 18 : 48-54.

Berg, H.W. and Akiyashi, M. 1956. Some factors involved in browning of Wine. American Journal of Enology and Viticulture, 7 : 1-7.

Bhatnagar, H.C. and Subramanyam, H. 1973. Some aspects of preservation, processing and export of mango and its products. Indian Food Packer, 27 : 33-52.

Biolin, H.R. and Boyle, F.P. 1972. Effect of storage and processing on sulphurdioxide in preserved fruit. Food Product Development, 7 : 82-88.

Bongirwar, D.R., Pendharkar, M.B. and Sreenivasan, A. 1991. Proceedings of the Ist National Seminar on Heat and Mass Transfer. Published by Indian Institute of Technology, Madras, 14 : 1.

Bongirwar, D.R. and Sreenivasan, A. 1977. Studies on osmotic dehydration of banana. Journal of Food Science and Technology, 14(3) : 104.

Bose, A.N. and Das, K.J. 1958. Journal of Science and Industrial Research, 174 : 202-205.

Brekke, J.E., Cavaletto, C.G. and Stafford, A.F. 1968. Mango Puree Processing. Hawaii Agricultural Station, Technical Progress Report No. 167.

Bray, F. 1981. Use of tropical fruits in ice cream production. Indian Alimentari, 20(3) : 171.

Britnell, P.M. 1989. Development of structured mango slices from puree. Proceedings of 3rd International Symposium on mango held in Darwin N.T. Australia. Old Department of Primary Industries, Australia (Abstract), p. 99.

Britnell, P.M. 1991. The development of structured mango products. Acta Horticulturae, 291 : 554-562.

Brown, M.S. 1967. Journal of Science, Food and Agriculture, 18 : 77.

Caputi, A.Jr. and Peterson, R.G. 1965. The browning problem in wines. American Journal of Enology and Viticulture, 16 : 9-13.

Chadha, B.S. 1987. Improvement in heat resistance and ethanol tolerance of *Saccharomyces cerevisiae* strains. M.Sc. Thesis, PAU, Ludhiana, India.

Chakravarty, S.C. and Narasimhan, K.R. 1983. Aseptic Packaging of Food Products. Indian Food Packer, 37(1) : 57-62.

Chan, H.T.Jr. and Cavaletto, C.G. 1982. Aseptically Packaged Papaya and Guava puree : Changes in Chemical and Sensory Quality during Processing and Storage. Journal of Food Science, 47 : 1164-1169, 1174.

Chandrasekhar, S.K. and King, C.J. 1971. Solid-liquid phase equilibria in multi component aqueous sugar solutions. Journal of Food Science, 36 : 699-700.

Chaudhuri, A.P., Kumbhar, B.K., Singh, B.P.N. and Narain, M. 1993. Osmotic dehydration of fruits and vegetables – A Review. Indian Food Industry, 12(1): 20-27.

Chauhan, S.K., Lal, B.B. and Joshi, V. K. 1997. Preparation and evaluation of protein enriched mango fruit bar. Indian Food Packer, 51(5) : 5-7.

Chauhan, S.K., Lal, B.B. and Joshi, V.K. 1998. Development of a protein rich mango beverage. Journal of Food Science and Technology, 35(6) : 521-523.

Cooke, R.D., Breag, G.R., Ferber, C.E.M., Best, P.R. and Jones, J. 1976. Studies on mango processing I. The foam mat drying of mango (Alphonso cv). Puree. Journal of Food Technology,11 : 463-473.

Dabhade, R.S. and Khedkar, D.M. 1980a. Studies on Drying and Dehydration of Raw Mangoes for Preparation of Mango Powder (Amchur) Part-I. Physico-chemical composition of Raw Mangoes During Growth and Development. Indian Food Packer, 34(3) : 3-17.

Dabhade, R.S. and Khedkar, D.M. 1980b. Studies on Drying and Dehydration of Raw Mangoes for Preparation of Mango Powder (Amchur). Part II Mango Drops and their Assessment for physcio-chemical composition. Indian Food Packer, 34(3) : 18-31.

Dabhade, R.S. and Khedkar, D.M. 1980c. Studies on Drying and Dehydration of Raw Mangoes for Preparation of Mango Powder (Amchur). Part III. Leaching losses in Raw mango pieces during blanching and sulphitation. Indian Food Packer, 34(3) : 32-34.

Dabhade, R.S. and Khedkar, D.M. 1980d. Studies on Drying and Dehydration of Raw Mangoes for Preparation of Mango Powder (Amchur). Part-IV. Drying and Dehydration of Raw Mango pieces. Indian Food Packer, 34(3) : 35-42.

Dabhade, R.S. and Khedkar, D.M. 1980e. Studies on Drying and Dehydration of Raw Mangoes for Preparation of Mango Powder (Amchur). Part-VII. Organoleptic Evaluation of Raw Mango Powder. Indian Food Packer, 34(3) : 55-59.

Daisy, F.K. 1966. Indian Food Packer, 20(6) : 21.

Das, D.P., Jain, N.L. and Girdharilal, 1955. Bulletin, Central Food Technological Research Institute, Mysore, India, 7 : 157.

Das, L. 1965. Studies on mango, turnip pickles. M.sc. Theins. Indian Agricultural Research Institute. New Delhi, India.

Das, R.C. and Hota, B.N. 1977. Some indigenous mango varieties of Orissa. Orissa Journal of Horticulture, 5(1) : 38.

Daughty, M.E. 1995. Fruit and Vegetable Processing. International Book Distributing Company, Lucknow, 197-198, 318-319.

Deka, B.C., Sethi, V., Prasad, R. and Batra, P.K. 2001. Application of mixtures methodology for beverages from mixed fruit juice/pulp. Journal of Food Science and Technology, 38 : 615-618.

Deka, B.C., Sethi, V. and Saikia, A. 2005. Changes in quality of mango-pineapple spiced beverage during storage. Indian Journal of Horticulture, 62(1) : 71-75.

Ebner, H. 1982. Vinegar, In : Prescott and Dunn's Industrial Microbiology. G. Reed (Ed.). 4th edition. AVI Publishing Company, West Port, CT pp. 302-734.

Echkoff, S.R. and Okos, M.R. 1986. Kinetics of sulfite reactions in corn grains. Journal of Agriculture and Food Chemistry, 34:239-245.

Eipeson, W., Ramteke, R.S. and Nanjundaswamy, A.M. 1987. Unpublished data. Quoted by Nanjundaswamy, A.M. 1991. Mango processing : Present status and Future Outlook. Acta Horticulturae, 291:525-545.

Ejilearassane, K., Sethi, V. and Singh, G. 2001. Quality evaluation of Instant Dehydraed Mango Chutney during storage in different containers. Indian Food Packer, 55(6):151-154.

Eskin, N.A.M. 1979. Plant Pigments, Flavours and Textures. The Chemistry and Biochemistry of selected compounds. Academic Press, London.

Gagrani, R.L., Rathi, S.D. and Ingle, U.M. 1987. Preparation of fruit flavoured beverage from whey. Journal of Food Science and Technology, 24:93-96.

Gangopadhyay, H., Chaudheri, D.R. and Mukherjee, S. 1976. Dehydration of green mango pulp in double drum drier. Indian Food Packer, 30(5):51-54.

Garden, S.A.H.Z and Tremazi, S.A. 1966. Study of the effect of canning and storage on various constituents of mangoes. Pakistan Journal of Science Research 18:198-201.

Garg, N. and Tandon, D.K. 1995. Standardization of Immobilization Technique for *Acetobacter aceti* for the production of Mango Vinegar. Haryana Journal of Horticulture Science, 24(3-4) : 243-244.

Garg, N., Tandon, D.K. and Kalra, S.K. 1995. Production of mango vinegar by immobilized cells of *Acetobacter aceti*. Journal of Food Science and Technology 32(3) : 216-218.

Ghosh, K.G., Nirmala, N., Krishnappa, K.G., Parameshwar, P.M., Barker, H. and Vijaya Raghavan, P.K. 1981. Preservation of fruit juices and pulp in flexible pouches. Indian Food Packer, 35(2) : 50-54.

Ghosh, S., Kenowjia, S.K. and Singh, S. 1995. Recent advances in whey permeat based sports drinks. Beverage and Food World, 22(1) : 24-26.

Girdharilal, Siddappa, G.S. and Tandon, G.L. 1998. Preservation of Fruits and Vegetables. Indian Council of Agricultural Research, New Delhi, India.

Glicksman, M. 1976. Hydro colloid utilization in fabricated foods. Cereal Foods World, 21 : 17-26.

Glicksman, M. 1985. Food gels – the role of alginates. Food Technology, New Zealand. 20(10) : 78-85.

Gopalan, C., Ramasastry, B.V. and Balasubramanian, S.C 2004. Nutritive value of Indian Foods. National Institute of Nutrition, Hyderabad, India.

Gowda, I.N.D, Ramanujaneya, K.H., Iyer, C.P.A., Subramanyam, M.D. and Dinesh, M.R. 1994. Physico-chemical and processing quality of four new mango hybrids in comparision to two commercial cultivars. Journal of Food Science and Technology, 31 : 385-388.

Gowda, I.N.D. 1995. Processed Products of Mango. Lalbagh, 39(3-4) : 44-47.

Gowda, I.N.D., Ambadan and Ramanjaneya, K.H. 1995. Studies on Mango Fruit Bar Preparation. Indian Food Packer, 49(2) : 17-21.

Gowda, I.N.D. and Hudder, A.G. 2004. Investigation on Processing Quality of some Mango varieties, Hybrids and Their blends. Journal of Food Science and Technology, 41(2) : 154-159.

Gowda, I.N.D. and Ramanjeneya, K.H. 1995. Evaluation of some mango varieties for their suitability for canned mango juice. Journal of Food Science and Technology, 32 : 323-325.

Grewel, K.S. and Jain, S.C. 1982. Studies on the feasibility of incorporating preserved mango pulp in the preparation of milk-mango shake. Indian Food Packer, 36(4) : 95-97.

Gupta, G.K. 1998. Standardization of concentration of additives for development and processing of oil-less mango pickles. Indian Food Packer, 52(3) : 15-20.

Gupta, T.R. and Girish, K. 1998. Studies on mango pulp filtration for concentration by pectolytic enzyme. Indian Food Packer, 42(6) : 7-14.

Hassan, M. and Ahmed, J. 1998. Physico-chemical and Sensory characteristics of mango milk beverage. Indian Food Packer, 52(2) : 32-35.

Heikal, H.A., El-Sanafirt, N.Y. and Shooman, M.A. 1972. Some factors affecting quality of dried mango sheets. Agriculture Research Review, 50 : 185-194.

Hemarkar, A.K., Tomar, M.C. and Singh, U.B. 2000. Studies on Blending of Guava Pulp with Mango pulp for dehydration (Mango-Guava sheet). Indian Food Packer, 54(4) : 45-47.

Hope, G.W. and Vitale, D.G. 1972. Osmotic dehydration : A cheap and simple method of preserving mangoes, bananas and plantains. Contribution 175 Food Research Institute, Canadian Department of Agriculture, Ottawa, Canada.

Hussain, S.Z. 2004. Raw Mango Products - An Overview, Land Bank Journal, 43 (1): 11-23.

Islam, M.N., Begum, J.A. and Shamsuddin, M. 1990. Studies on carbonated beverage based on mango pulp. Bangladesh Agriculture Science, 17(2) :169-172.

Ismail, S., Rameshwar, A. and Rao, M.N. 1986. Maturity standards for fruits of pickle mangoes. Journal of Food Science, 23 : 105.

Issac, A.R. 1986. Freezing of mango slices. Food Technology in Australia, 38(9) : 383-384.

Jackson, T.H. and Mohamed, B.B. 1971. The Shambat process : New development arising from the osmotic dehydration of fruits and vegetables. Sudan Journal of Food Science and Technology, 3 : 18.

Jain, N.L., Lal, G. and Murthy, G.V.K. 1957. Further studies in the preparation and uses of mango cereal flakes. Indian Journal of Horticulture, 14 : 172-178.

Jain, N.L. Lal, G. and Siddappa, G.S. 1973 (unpublished data) quoted by H.C. Bhatnagar and H. Subramanyam 1973. Some aspects of preservation, processing and export of Mango and its products. Indian Food Packer, 27(4) : 33-52.

Jain, V., Tiwary, B.L. and Sharma, H.G. 1997. Evaluation of Early Maturing Mango Varieties for the preparation of Beverages as Nectar and RTS. Orissa Journal of Horticulture, 25(1) : 40-42.

Jalen, J.M. 1992. Whey, Cheeses and beverages. In: Whey and Lactose Processing. Elsevier Applied Science, London.

James, C.P. 1987. Effect of pre-treatment on the quality of frozen mango slices. M.Sc. (Food Technology) Dissertation. CFTRI, Mysore, India.

Jayaraman, K.S., Ramanuja, M.N., Govardhanan, T., Bhatia, B.S. and Nath, H. 1976. Technological aspects of use of ripe mangoes in the preparation of some convenience food for Defence Services. Indian Food Packer, 30(5) : 76-82.

Johar, D.S. and Anand, J.C. 1951. Note on the Microbiology of Brined Mango Slices. Indian Journal of Horticulture, 8(4) : 45-8.

Kalra, S.K. 1982. Preservation of mango (*Mangifera indica* L.) pulp. Progressive Horticulture, 14(4) : 205-208.

Kalra, S.K. and Tandon, D.K. 1983. Salt treated raw mango slices in polyethylene pouches. Indian Food Packer, 37(6) : 101-107.

Kalra, S.K. and Tandon, D.K. 1985a. Preservation of mango pulp by freezing. Indian Food Packer, 39(5) : 23-27.

Kalra, S.K. and Tandon, D.K. 1985b. Physico-chemical changes in Mango pulp during ambient storage in glass containers. Journal of Food Science and Technology, 22 : 350-353.

Kalra, S.K., Tandon, D.K. and Garg, N. 2001. Improved Beverages from mangoes with poor coloured flesh. Indian Food Packer, 55(5) : 70-72.

Kalra, S.K., Tandon, D.K. and Singh, B.P. 1991. Evaluation of Mango-Papaya Blended Beverage. Indian Food Packer, 45 : 33-36.

Kalra, S.K., Tandon, D.K. and Singh, B.P. 1995. Mango. In : Hand book of Fruit Sciences and Technology (Ed) D.K. Salunkhe and S.S. Kadam. p. 123-168. Narcel Dekkar Inc. New York, 123-168.

Kapur, K.L. 1974. Studies on biochemical changes in mango during growth and ripening. Indian Food Packer, 39(5) : 60-67.

Kapur, K.L. 1975. Varietal suitability of mangoes from Uttar Pradesh for canning. Indian Food Packer, 29(3) : 33-35.

Kapur, K.L., Verma, R.A. and Tripathi, M.P. 1985. Effect of maturity and processing on quality of pulp, slices and juice of mango. Cv. Dashehari. Indian Food Packer, 39(5) : 60-67.

Karel, M. 1975. Water activity in food preservation. In : Fennema (Ed.) Principles of Food Science. Part II. Physical Principles of Food Preservation. Marcel Deckkar, Inc. New York, : 255-258.

Kaur, C. and Kapur, H.C. 2000. Minimal Processing of Fruits and Vegetables. Indian Food Packer, 54(6) : 156-164.

Kesarwani, M., Kohli, U., Raghuvamshi, R.S., Chauhan, G.S. and Dwivedi, T.S. 2000. Quality Evaluation of Raw Dried Mango Slices, Khatai and Mango Powder, Amchur sold in the retail market. Journal of Food Science and Technology, 37(5): 449-454.

Khedkar, D.M. and Roy, S.K. 1980. Histological Evidence for the Reconstitutional property of Dried/Dehydrated Raw Mango Slices. Journal of Food Science and Technology, 17(6) : 276-277.

Khedkar, D.M. and Roy, S.K. 1983. Absorption and Retention of Sulphurdioxide in raw mango slices during drying and dehydration. Journal of Food Science and Technology, 20(1) : 38-39.

Khurdiya, D.S. 1993. Composition and quality of Nectar prepared from Blended Pulps of Amrapali and Totapuri mangoes. Journal of Food Science and Technology, 30(2): 139-140.

Khurdiya, D.S. and Roy, S.K. 1986a. Studies on ripening and canning of mangoes. Indian Food Packer, 40(1) : 40-45.

Khurdriya, D.S. and Roy, S.K. 1986b. Studies on canning of mango slices in covering syrup containing mango pulp. Indian Food Packer, 40 : 52-54.

Khurdiya, D.S. and Sagar, V.S. 1988. Effect of pH and temperature on carotenoid pigments in mango pulp. Beverage and Food World. July-September 1988, pp. 19-20.

Kocher, G.K. and Sharma, K.K. 1992. Fibre content and its composition in commonly consumed Indian vegetables and fruits. Journal of Food Science and Technology, 29(3) : 187-188.

Korn, H. 1983. Flavoured milk of commercial quality made with fruit or fruit pieces. German Federal Republic Patent Application. DE 31, 40111, Al.

Korula, S., Joseph, K., Rao, M.N., Indiramma, K. and Siddappa, G.S. 1960. Supplementary value of mango custard to milk diet. Annu. Biochem. Exp. Med, 20 : 65-72.

Krishnamurthy, K., Pantulu, A.J., Swaminathan, M. and Subramanyam, V. 1957. Studies on the stability of oils in pickles. Food Science, 6 : 133.

Krishnamurthy, S., Krishna Rao, K.P.G. and Onkarayya, H. 1982. Storage of mango pulp in bulk and consumer packs. Indian Food Packer, 36(1) : 32-38.

Krishna Prakash *et al* 1986. Unpublished data. Quoted by Nanjundaswamy, A.M.. 1991. Mango Processing : Present Status and Future outlook. Acta Horticulturae, 291 : 525-545.

Kulkarni, J.H., Singh, H. and Chadha, K. L. 1980. Preliminary screening of mango varieties for wine making. Journal of Food Science and Technology, 17(4) : 218-220.

Kumar, K.R. 1984. Aseptic Packaging materials and Packaging forms. Indian Food Industries, 38(3) : 95.

Kumbhar, B.K. 1992. Processing of mango in the Industry – A Scenario. Indian Food Industry, 11(6) : 33-37.

Lakshminarayana, S. 1980. Mango. In : Tropical and Sub-tropical Fruits – Composition, Properties and Uses (Ed. S. Nagy and P.E. Shaw), AVI, West Port, CT, p. 184.

Lal, G., Siddappa, G.S. and Tandon, C. L. 1960. Preservation of Fruits and vegetables. ICAR, New Delhi, pp. 60-70 and 94-114.

Lal Singh and Girdhari Lal 1998. Quoted by Giridharilal *et al.* 1998. In. Preservation of Fruits and Vegetables. Indian Council of Agriculture Research, New Delhi. pp. 239.

Luh, B.S., Leonard, S.J. and Marsh, G.L. 1958. Objective criterion for storage changes in tomato paste. Food Technology, 12 : 347-351.

Mahadevaiah, M. 1976. Factors affecting Tin pick up in canned mango products. Indian Food Packer, 30(5) : 92-94.

Mahadevaiah, M. and Gowaramma, R.V. 1979. Suitability of lacquered cans for canning mango juice. Journal of Food Science and Technology, 16 : 114-115.

Mahadevaiah, M., Gowaramma, R.V., Eiperson, W.E., Anandaswamy, B. and Sastry, L.V.L. 1975a. Studies on the internal corrosion of tin-plate with mango (*Mangifera indica* L.) juice. Indian Food Packer, 29(1) : 5-12.

Mahadevaiah, M., Gowramma, R.V., Eipeson, W.E. and Sastry, L.V.L. 1975b. Internal corrosion of tin plate containers in canned mango nectar. Journal of Science, Food and Agriculture, 26 : 821-833.

Mahadevaiah, M., Gowaramma, R.V., Eipeson, W.E. and Sastry, L.V.L. 1975c. Influence of tin plate variables on the internal corrosion of tin containers with mango and orange products. Journal of Food Science and Technology, 13 : 17-23.

Mahadevaiah, M., Gowaramma, R.V., Setty, G.R., Sastry, M.V., Sastry, L.V.L. and Bhatnagar, H.C. 1969. Studies on variation in tin content in canned mango nectar during storage. Journal of Food Science and Technology, 6 : 192-196.

Mahayothu, B., Muhbauer, W., Neidhar, S., Leitenberger, M. and Carla, Q. 2002. Non-destructive determination of maturity of Thai mangoes by near infrared spectroscopy. Paper presented at 7th International Mango Symposium. Recife, P.E. Brazil, 22-27th September.

Maini, S.B. and Anand, J.C. 1985. Improved Fruit and Vegetable Products. Indian Horticulture, 30(2) : 10-13.

Maini, S.B., Diwan, B. and Anand, J.C. 1984. Use of high molecular weight high density (HMHD) film pouches for processing and storage of mango and tomato pulps. Journal of Food Science and Technology, 21 : 410-412.

Majumdar, S.K., Muthu, M. and Pingale, S.V. 1954. Protection of mango pickle from insect damage. Indian Food Packer, 8:8-13.

Malhotra, S.S. 1965. Freeze Drying of Foods. Indian Food Packer, 19(1) : 15-16.

Mallik, S.C. 1976. Recipe and Marketing of Green Mango Beverage and Jam-I. Indian Food Packer, 30(5) : 37-39.

Marin, M.A., Cano, P. and Fuster, C. 1992. Freezing preservation of four Spanish mango cultivars. Labensum Unters. Forsch, 194(6) : 566.

Martean, P. and Dannin, A. 1980. Process for extraction of mango butter and product derived from said process. French Patent Applied, 2 : 437.

Martinez, F. and Labuza, T.P. 1968. Rate of deterioration of freeze-dried salmon as a function of Relative Humidity. Journal of Food Science, 33 : 241-247.

Mathur, V.K., Das, S.A., Jayaraman, K.S. and Bhatia, B.S. 1972. Preparation of fruit bars for use in combat rations. Indian Food Packer, 26(2) : 33-35.

Mathur, V.K., Siddaiah, C.H., Bhatia, B.S., and Nath, H. 1973. Studies on dehydro-canning of mangoes. Indian Food Packer, 27 : 15-18.

Matz, S.A. 1962. Food Texture, A.V.I. Publishing Company Inc. West Port, Connecticut, 1962: 212.

Mehta, G.L. and Tomar, M.C. 1982. Dehydration and utilization of raw mangoes. Indian Food Packer, 33(6) : 52-57.

Mehta, G.L., Tomar, M.C., Agnihotri, B.M and Tripathi, S.R. 1971. Canning of whole Mangoes. Progressive Horticulture, 3(3) : 49-56.

Mir, M.A. 1990. Development and evaluation of fortified mango bars. Ph.D. Thesis, G.B. Pant University of Agriculture and Technology, Pantnagar, India.

Mir, M.A. and Nath, N. 1993. Storage changes in fortified mango bars. Journal of Food Science and Technology, 30(4) : 279-282.

Mishra, J. and Tomar, M.C. 2000. Studies on the preparation and storage of Intermediate Moisture Ripe Mango Slices of Dasheheri cultivar. Indian Food Packer, 54(2) : 61-64.

Moy, J.H., Lau, N.B.M. and Dollar, A.M. 1978. Effects of sucrose and acids on osmotic dehydration of tropical fruits. Journal of Food Processing and Preservation, 2 : 131.

Murthy, G.V.K., Giridhar, N and Raghuramaiah, S. 1994. Suitability of some mango varieties for processing. Journal of Food Science and Technology, 21(1) : 21-25.

Nagaratnamma, 1987. Unpublished data quoted by Nanjundaswamy, A.M. 1991. Mango processing. Present Status and Future outlook. Acta Horticulturae, 291: 525-545.

NandanaSabhapathi, A.M., Srivastava, A.N. and Natraju, S. 1993. Storage studies on mango bar in flexible packaging materials. Indian Food Packer, 47 : 5-8.

Nanjundaswamy, A.M. 1984. Studies on preparation of dehydrated products from tropical fruits and their pulps. Ph.D thesis, University of Mysore, India.

Nanjundaswamy, A.M. 1987. Developments in technology for processing of mangoes. A chapter on mangoes – A review (Eds) RT. Prinsley and T. Tuker, Common Wealth Science Council, London.

Nanjundaswamy, A.M. 1991. Mango processing : Present Status and Future Outlook. Acta Horticulturae, 291 : 525-545.

Nanjundaswamy, A.M. 1997. Processing. In: The Mango, Botany, Production and Uses. (Ed.) R.E. Litz. CAB International, Willingford, p. 515.

Nanjundaswamy, A.M. *et al* 1988. Private communication. Quoted by Nanjundaswamy, A.M. 1991. Mango: Processing: Present Status and Future outlook. Acta Horticulturae, 291 : 525-545.

Nanjundaswamy, A.M. and Gowaramma, R.V. 1979. Suitability of lacquered cans. Journal of Food Science and Technology, 16 : 114-115.

Nanjundaswamy, A.M., Rahim, A. and Siddappa, G.S. 1966. Proximate composition and suitability for preparation of some important varieties of mangoes grown in Salem. Indian Food Packer, 20 : 15-21.

Nanjundaswamy, A.M., Setty, G.R. and Saroja, S. 1976. Studies on the development of newer products from mango. Indian Food Packer, 30(5) : 95-103.

Narayana, B.S. 1976. Some Technological Aspects of the Preparation of Popular varieties of Green Mango Pickles. Indian Food Packer, 30(5) : 40-44.

Nelson, P.E. 1984. Outlook for aseptic bag-in-box packaging of food products for remanufacture. Food Technology, 38(3) : 72-73.

Netto, A.G., Blunroth, E.W. and Lazzarini, L.C. 1971. Proceedings of International Congress of Refrigeration, 3 : 265.

Nigam, S., Bhatt, D.K. and Jha, A. 2007. Different products of mango : The king of fruits. Processed Food Industry, July 2007 : 32-40.

Omprakash, 1961. Food and Drinks in ancient India. Munisharma Manoharlal Oriented Book Sellers and Publishers, New Delhi.

Onkarayya, H. 1985. Mango Vermouth – a new alcoholic beverage. Indian Food Packer, 59 : 40-45.

Onkarayya, H. 1986. A Rapid Modeirization Process to Improve Mango Dessert wines. Journal of Food Science and Technology, 23 : 175-177.

Onkarayya, H. and Rao, G.S.P. 1988. Oxidative browning of mango wines – A mathermatical model to predict browning susceptibility. Acta Horticulturae, 231 : 736-743.

Onkarayya, H. and Singh, H. 1984. Screening of mango varieties for dessert and madeira-style wines. American Journal of Enology and Viticulture, 35 : 63.

Onkarayya, H., Singh, H. and Chadha, K.L. 1982. Madeira style mango wines. Screening of mango varieties. Proceedings of Ahara International Conference of Association of Food Scientists and Technologists (India), pp. 160-163.

Padmini, T., Narayanaswamy and Murthy, H.B.N. 1963. Packaging and storage studies on mango cereal flakes. Food Science (India), 12 : 315-317.

Palaniswamy, K.P., Muthukrishnan, C.R. and Shanmugavelu, K.G. 1973. Studies on the evaluation of certain mango varieties of Tamilnadu for canning. Indian Food Packer, 27 : 9-14.

Palaniswamy, K.P., Muthukrishnan, C.R. and Shanmugavelu, K.G. 1974. Studies on the evaluation of certain mango varieties of Tamilnadu for pulp and squash. Indian Food Packer, 28(27) : 5-9.

Palnitkar, M.P. and Heldman, D.R. 1971. Journal of Food Science, 36 : 1015.

Pandey, D.C., Tomar, M.C. and Singh, U.B. 1995. Preparation, Processing and Storage Studies on Raw Mango Panna. Indian Food Packer, 49(6) : 5-9.

Parekh, J.V. 1993. What is new in whey processing. Food Technology, June-August 1993, 8-11.

Patil, R. 1990. Evaluation of mango (*Mangifera indica* L.) Cvs. Alphonso, Ratna, Pairi and Kesar fruits for physico-chemical composition, storage and processing. M.Sc. Thesis, Konkan Krishi Vidyapeth, Dapoli, Ratnagiri, India.

Patwardhan, S.G. 1999. Aseptic Packaging of particulate Food Products. A Review. Indian Food Packer, 53(6) : 72-80.

Paul, N. and Tewari, H.K. 1996. Preparation of fruit wine from mango variety Dussehri. Research and Development Reports, 13(1&2) : 55-61.

Pawar, V.N., Singh, N.I., Dev. D.K., Kulkarni, D.N. and Ingle, V.M. 1988. Solar drying of white onion flakes. Indian Food Packer, 42(10) : 15-28.

Pompon Torezan, G.A., Favaret, P.C., Pallet, D. and de Menezes, H.C. 2004. Use of a Combined Process of Osmotic Dehydration and Deep-Fat-Frying to obtain mango chips from the cultivar Tommy Atkins. Acta Horticulturae, 645: 285-291.

Ponting, J.D. 1973. Osmotic dehydration of fruits – Recent modification and applications. Process Biochemistry, December : 18-20.

Post Mayer, H.L., Luh, B.S. and Leonard, S.J. 1956. Characterization of pectin changes in free stone and cling stone Peaches during ripening and processing. Food Technology, 10 : 618-625.

Prabha, T.N. and Patwardhan, M.V. 1986. Endogenously oxidizing polyphenols of mango, sapota and banana. Acta Alimemtarie, 15(2) : 123-128.

Prasad, H. and Joshi, M.V. 1929. Preservative values of spices used in pickling raw fruits in India. Agriculture Journal of India, 24 : 402.

Pruthi, J.S. 1987. Microbiological contamination in spices and their decontamination - A hygienic necessity. Souvenir, National Symposium on Spice Industry. Present Scenario, Problems and Prospects. AFST (India) Delhi Chapter, New Delhi, pp. 2-11.

Pruthi, J.S. 1992. Simple innovations in mango processing technologies. Indian Food Packer, 46(5) : 45-55.

Pruthi, J.S. 1993. Innovations in Fruit Processing. In : K L Chadha and O.P. Pareek (Eds). Advances in Horticulture vol. 4. Fruit Crops. Part 4, 1876-1911, Malhotra Publishing House, New Delhi.

Pruthi, J.S., Lal, G., Dhoopeshwarker, G.L. and Nagar, N.G. 1956. Fruit and Vegetable preservation industry in India. Central Food Technological Research Institute, Mysore, p. 243.

Pruthi, J.S. and Teaotia, M.S. 1987. Studies on physico-chemical composition, packaging and formulation of quality standards for raw mango powder (Amchur). Indian Cocoa, Areacanut and Spices Journal, 11 : 22-28.

Rabbani, A. and Singh, I.S. 1988. Evaluation of Local sucking mango varieties for Beverage Industry. Acta Horticulturae, 231 : 715-720.

Ramamurthy, M.S. and Bongirwar, D.R. 1979. Effect of Freezing Methods on the quality of Freeze Dried Alphonso Mangoes. Journal of Food Science and Technology, 16(3) : 234-236.

Ramamurthy, M.S. and Bongirwar, D.R. 1987. Influence of Processing variables on the Hygroscopic Behaviour of Freeze dried Alphonso Mango Slices. Indian Food Industry, 6(3) : 122-124.

Ramana, K.V.R., Ramaswamy, H.S., Prasad, B.A., Patwardhan, M.V. and Ranganna, S. 1984. Freezing Preservation of Totapuri mango pulp. Journal of Food Science and Technology, 21 : 282-285.

Ramanuja, M.N. and Jayaraman, K.S. 1980. Studies on the preparation and storage stability of intermediate moisture banana. Journal of Food Science and Technology, 17 : 183.

Rameshwar, A. 1979. Tandra (Mango Leather) industry in Andhra Pradesh. Indian Food Packer, 33(2) : 11-12.

Ramteke, R.S. 1987. Characterisation and storage behaviour of aroma constituents recovered from tropical fruits. Ph.D Thesis, Mysore University, Mysore, India.

Ramteke, R.S., Vijayalakshmi, M.R. and Eipeson, W.E. 1999. Processing and Value Addition to Mangoes. Indian Food Industry, 18(3) : 155-162.

Ranganna, S. 1986. Handbook of analysis and quality control for fruit and vegetable products. Second edition Tata McGraw Hill Publication Co.Limited, New Delhi.

Ranganna, S., Sastry, M.V. and Siddappa, G.S. 1961. Confructa, 1 : 5.

Ranjana, P.J. 1991. Investigation on varietal evaluation of mango (*Mangifera indica* L.) for processing. Ph.D Thesis, University of Agricultural Sciences, Bangalore, Karnataka, India.

Rankine, B.C. 1965. Factors influencing the pyruvic acid content of wine. Journal of the Science of Food and Agriculture, 16 : 394-398.

Rao, P.V.S., Rao, S.K and Rao, B.K. 1961. Preliminary studies on the dehydration of mango pulp. Andhra Agricultural Journal, 8(5) : 248-253.

Rao, V.S. and Roy, S.K. 1980a. Studies on dehydration of mango pulp I standardization for making mango sheets/leather. Indian Food Packer, 34 : 64-71.

Rao, V.S. and Roy, S.K. 1980b. Studies on dehydration of mango pulp II. Storage studies on mango sheets/leather. Indian Food Packer, 34 : 72-79.

Roy, S.K., Singh, R.N. and Singh, R. 1972. Studies on evaluation of some mango varieties of North India for processing as nectars. Indian Food Packer, 26 : 5-8.

Sagar, V.R. 1995. Studies on cabinet drying and storage of ripe mango slices. Ph.D Thesis, Indian Agricultural Research Institute, New Delhi.

Sagar, V.R. and Khurdiya, D.S. 1996. Effect of ripening stages on quality of dehyrated ripe mangoes slices. Indian Food Packer, 50(5) : 15-21.

Sagar, V.R. and Khurdiya, D.S. 1997. Studies on dehydration of Dashehari mango slices. Indian Food Packer, 51(1) : 5-8.

Sagar, V.R. and Khurdiya, D.S. 1998. Improved products from ripe mango (cv. Dashehari). Indian Food Packer, 52(6) : 27-30.

Sagar, V.R. and Khurdiya, D.S. 1999. Effect of packaging material and mode of pack on the quality of ripe mango powder. Journal of Applied Horticulture, Lucknow, 1(2) : 115-117.

Saha, N.K., Agarwal, P.C. and Rodriquez, R. 1976. Effect of Ripeness Level, Storage Period, Processing conditions and Ascorbic acid on Flavour retention in canned Dashehari Mango. Indian Food Packer, 30(5) : 87-91.

Sahni, C.K. and Khurdiya, D.S. 1989. Effect of ripening and storage temperature on the quality of mango nectar. Indian Food Packer, 43 : 5-10.

Sahni, C.K., Khurdiya, D.S. and Dalal, M.A. 1994. Studies on quality and carotene profile of mango pulp as influenced by Temperature and Additives. Indian Journal of Horticulture, 51(3) : 229-233.

Sahu, C., Patel, S. and Chaudhary, P.L. 2005. Technology for manufacturing of whey based mango-herbal (lemon grass) beverage. Journal of Food Science and Technology, 42(5) : 421-424.

Salah, M.A., El. Said., Amer, M. and Shehata, S.G. 1960. Suitability of some Egyptian varieties of mango for preservation by freezing. Agricultural Research Review, 38 : 369-379.

Saravana Kumar, R. and Manimegalai, G. 2001. Technology of Whey Based Mango Fruit juice blended RTS. Beverage. Processed Industry, October, 2001 : 23-24.

Sastry, M.V. and Habibunnisa. 1979. Studies on Indian Pickles – Part V. Storage of Raw mangoes for pickling. Indian Food Packer, 33(6) : 10-18.

Sastry, M.V. and Krishnamurthy, N. 1974. Studies on Indian pickles : preparation and storage of pickle from mangoes. Indian Food Packer, 28(1) : 32-44.

Sastry, M.V. and Krishnamurthy, N. 1975. Studies on Indian pickles – IV, Physico-chemical variation of mangoes (Amlet variety) and studies of pickles from them. Indian Food Packer, 29(3) : 55-61.

Sastry, M.V., Nagaratnamma, K. and Habibunnissa, 1975. Studies on Indian pickles. Part III, physico-chemical variation during growth of mangoes (Amlet variety) and storage studies of pickles from them. Indian Food Packer, 29 : 39-54.

Satyavati, V.K., Bhatia, B.S., Sastry, L.V.L. and Lal, G. 1962. The influence of some carotenoid pigments on non-enzymatic browning in some food products. Indian Journal of Applied Chemistry, 25 : 66.

Saurel, R., Raoult-Wack, A.L., Rios, G. and Guilbert, S. 1994. Mass transfer phenomana during osmotic dehydration of apple I. Fresh Plant tissue. International Journal of Food Science and Technology, 29 : 531-542.

Saxena, A.K., Teaotia, M.S. and Berry, S.K. 1996. Studies on the development of grape–mango and grape-pineapple beverage blends. Indian Food Packer, 50(4) : 26-29.

Sethi,V. 1991. Preservation of raw mango slices for use in pickles and chutneys. Journal of Food Science and Technology, 28 (1) : 54-56.

Setty, G.R., Reddy, K.B.S.M., Saroja, S., Chikkappaji, K.C. and Nanjundaswamy, A.M. 1975. Developments in Dehydration Techniques of Fruits and Vegetables. Indian Food Packer, 32(1) : 115-126.

Sharma, S.C. 2008. Fruit and Vegetable Processing. Agribusiness and Food Industry, 5(4) : 50-53.

Sharma, S.P., Bhanumurthi, J.L. and Srinivasan, M.R. 1974. Studies on the production and storage behaviour of spray dried mango milk powder. Journal of Food Science and Technology, 11(4) : 171-174.

Shi, X.O., Fito, P. and Chirato, A. 1995. Influence of vacuum treatment on mass transfer during osmotic dehydration of fruits. Food Research International, 28: 445-454.

Shinde, A.K., Wagha, R.G., Joshi, G.D., Waghmare, G.M. and Shivsagar, P.J. 2004. Pickle Purpose Mango variety – Hybrid – 4 (Konkan Ruchi). Indian Food Packer, 58(2) : 54-55.

Shrikande, A.J., Srirangarajan, A. and Nadakarni, G.B. 1976. A thermal process for bulk packing of mango pulp. Indian Food Packer, 30 : 63-67.

Siddappa, G.S. and Bhatia, B.S. 1956. Bulletin, Central Food Technological Research Institute, 5 : 236-238.

Siddappa, G.S., Bhatia, B.S. and Lal, G. 1953. Indian Patent No. 49441. CFTRI, Mysore, India.

Siddappa, G.S. and Krishnamurthy, G.V. 1963. Bulk storage of mango pulp. Punjab Horticulture Journal, 3(2/4) : 300.

Siddappa, G.S. and Ranganna, S. 1961a. Strained Baby Foods : I Proximate, mineral, vitamin composition of some canned fruit and vegetable pulps and mango custard. Food Science (India), 10 : 27-36.

Siddappa, G.S. and Ranganna, S. 1961b. Strained baby foods : 2 – Drying of strained mango pulp and custard. Food Science (India), 10 : 37-40.

Sikder, B., Sarkar, K., Roy, P.R. and Ghatakur. 2001. Studies on shelf life of whey based mango beverages. Beverage and Food World, 28 : 53-54.

Simpson, R.T. 1982. Factors affecting oxidative browning of white wine. Vitis, 21 : 233-239.

Singh, K.K. 1970. Studies in Frozen Storage of Some Fruits. Climate Control, 3(3) : 33-36.

Singh, K.K. 1974. Refrigerated storage of Mango in India. Climate Control, 7(4) : 33-39.

Singh, K.K. 1977. Refrigerated storage and freezing of Mango fruit in India. Punjab Horticulture Journal, 17(1-2) : 6-12.

Singh, N.P. 2007. Fruit and Vegetable Preservation. Oxford Book Company. Jaipur, India, pp. 232-236.

Singh, S., Ladkani, B.C., Kumar, A. and Mathur, 1994. Development of whey based beverages. Indian Journal of Dairy Science, 47 : 586-590.

Sirohi, D., Patel, S., Choudhary, P.L. and Shan, C. 2003. Studies on preparation and storage of whey based mango herbal (Pudina) beverage. Paper presented at 5th IFCON-2003. Innovative food technologies and quality system – strategies for global competitiveness, held at CFTRI, Mysore (India), 5-8, December 2003, Abstract No. TC 27, p. 170.

Soumithri, T.C., Subba Rao, M.S. and Johar, D.S. 1963a. Studies on brine pickles : Part I. Salt and acid ratio in Indian pickles and the incidence of microbial spoilage. Food Science, 12 : 374.

Soumithri, T.C., Subba Rao, M.S. and Johar, D.S. 1963b. Studies on brine pickles. Part II–Isolation and identification of yeasts from pickles. Food Science, 12: 377.

Spayd, S.E., Nagel, C.W., Hayrynen, L.D and Praker, R. 1984. Colour stability of apple and pear juice blended with fruit juices containing anthocyanin. Journal of Food Science, 49(2) : 411-414.

Srivastava, H.C. 1967. Grading, Storage and Marketing. In : Mango – A Hand Book. Indian Council of Agricultural Research, New Delhi, pp. 99-149.

Srivastava, J.S. 1998a. Comparative study of RTS drinks prepared from Dashehari and Banganapalli mangoes. Indian Food Packer, 52(2) : 38.

Srivastava, J.S. 1998b. Mango processing industries – A sceneraio. Indian Food Packer, 52(6) : 43-51.

Subbaiah, K. 1961. Quoted by Nanjundaswamy, A. M. and Mahadevaiah M. 1993. In : Fruit processing. In : K.L. Chadha and O.P. Pareek, 1993 (Eds) Advances in Horticulture. 4, Fruit crops. Part 4, 1856p. Malhotra Publishing House, New Delhi.

SubbaRao, M.S. and Johar, D.S. 1962. Spoilage in pickles. Food Science and Technology, 4 : 51.

Subramanian, V., Sharma, T.R. and Nath, H. 1976. Freeze dehydration of mango pulp as ready to serve beverage. Indian Food Packer, 30 : 70-75.

Sudhakar, D.V. and Maini, S.B. 1994. Stability of carotenoids during storage of mango pulp. Journal of Food Science and Technology, 31(3) : 228-230.

Suhardjo, Y. and Santoso, P. 2000. Determination of Drying Method and Fruit Maturity in Preparing chips of Mango. cv. Arumanis. Acta Horticulturae, 509 : 515-521.

Suryaprakash Rao, P.V., Prasad, P.S.R.K., Giridhar, N. and Nageswara Rao, G. 1968. Evaluation of mango varieties of Andhra Pradesh for processing as Slices, Nectars and Juices. Indian Food Packer, 22(3) : 11-17.

Tandon, D.K., Dikshit, A. and Kumar, A. 2004. Raw mango squash (Panna concentrate) A good health drink. Processed Food Industry, July 2004 : 13-15.

Teaotia, M.S., Manan, J.K. and Saxena, A.K. 1987. Green mango processing – A review. Indian Food Packer, 41(6) : 75-87.

Teotia, S.S. and Awasthi, R.K. 1966. Indian Food Packer, 20(3) : 10-14.

Teotia, S.S., Mehta, G.L., Tomar, M.C. and Garg, R.C. 1976. Studies on dehydration of tropical fruits in Uttar Pradesh-I. Mango (*Mangifera indica* L.). Indian Food Packer, 30(6) : 15-19.

Tewari, H.K., Manwaha, S.S., Gupta, A. and Khanna, P.K. 1991. Quality vinegar production from juice of sugarcane var. Co-64. Journal of Research, Punjab Agricultural University, 28 : 77-84.

Thangamani, R.R., Mattada and Sankaran, R. 1975. Microbial contamination in spices. Indian Food Packer, 29 : 11.

Thi, H., Chan, W., Pal, R.K. and Roy, S.K. 1989. Studies on extraction of pulp and development of beverages from green mangoes. Indian Food Packer, 43(3) : 27-34.

Thijseen, H.A.C. 1974. Fundamentals of concentration processes – In : Advances in pre-concentration and dehydration of foods (Ed.) A. Spincer, Applied Sciences Publishers Limited, London, p. 13.

Tiwari, R.B. 2005. Application of Osmo – Air dehydration for processing of Tropical Fruit in Rural areas. Indian Food Industry, 24(6) : 62-69.

Toreggiani, D. 1993. Osmotic dehydration in fruit and vegetable processing. Food Research International, 26 : 59-68.

Toreggiani., D. and Bertolo, G. 2001. Osmotic pre-treatments in fruit processing : chemical, physical and structural effects. Journal of Food Engineering, 49 : 247-253.

Tripathi, V.K., Jain, S.P., Ram, G.B. and Singh, S. 1985. Studies on Tin dissolution in canned mango pulp of different varieties during storage. Journal of Food Science and Technology, 22(2) : 147-150.

Tripathi, V.K., Lyndgsh, K. and Singh, S. 1992. Studies on blending of pineapple juice with different ratios of guava juice for preparation of RTS beverage. Progressive Horticulture, 24 : 60-65.

Trivedi, S., Nagar, R. and Nagar, P.N. 1990. Adultarants detected in five species of Jaipur (Rajasthan). Indian Journal of Nutrition and Dietitian, 27 : 331-335.

Usha, K. and Anand, J.C. 1981. Quality aspects of commercial raw mango powder (Amchur). Journal of Food Science and Technology, 18(5) : 208-210.

Usha Rani, M., Rama Rao, S.N. , Girija Bhai, R. and Nagaraja, K.V. 1992. Studies on quality standards of Indian Commercial Pickle. Indian Food Packer, 46(4) : 27-33.

Vasundhara, T.S., Siddaiah, C.H., Phanindra Kumar, H.S. and Jayathilakam, 1985. Improvement of certain organoleptic qualities of freeze dried fruit juices. 5th Indian Convention of Food Scientists and Technologists, New Delhi, 3-5, April 1978, 2-2 (Abstract).

Vijaya, K. 2000. Text book on Food Storage and Preservation. Kalyani Publishers, Ludhiana.

Vijaya Raghavan, P.K. 1980. Convenience foods for defence personnel. Proceedings of Seminar on Convenience Foods, 20-21 December, Bombay.

Wankhade, S., Akhere, S.B., Morey, K. and Pawar, R.V. 2007. Use of mango pulp in the preparation of Burfi. Crop Pollution and Production, 3(2) : 83-84.

Wilson, C.P. 1928. Industrial Engineering and Chemistry 200 : 1302-1306.

Wolford, L.R., Ingerlsbic, D.W. and Boyle, F.P. 1967. Proceedings of XIIth International Congress on Refrigeration, 3 : 459.

Woodhams, D.J. 1982. Aseptic processing of viscous and particulate products In : Aseptic processing and packaging. University of Auckland, Newzealand.

Yadav, B.S., Ritika and Singh, R. 2005. Preparation of Malted milk powder with mango flavour. Indian Food Packer, 59(5) : 61-62.

CHAPTER 6

Varietal Suitability for Processing of Mango

K. Vanajalatha

More than 1000 named varieties and hybrids exist in India and the number may go over 2000 in the entire world. However, all varieties are not suitable for preservation and processing. Only a few are found suitable and are utilized for processing. There may be multiple reasons for this limitation, as most of the mango varieties have lower yields or lower edible pulp percentage and cultivated in negligible area, thereby not readily available and in large quantities and confined to a particular region or other unfavourable quality characteristics. Further, suitability of a variety varies with the kind of the product processed i.e. each product requires specific quality parameters. Some varieties may be suitable for making more than one product, while others for making a particular product. The suitability of a variety varies with its maturity stage and ripening for making different products. Both processing qualities and production capacity are important for evaluation of various mango varieties for processing industry. Quality components and production capacity vary from region to region, year to year, in the same region/garden and in the same varieties, their age, the care bestowed by the grower etc. Computed data for production capacity and quality components of 20 varieties of mango are presented in Tables 6.1 and 6.2 based on the reports of various mango research workers' reports presented at Mango Research Workers' meeting held in 1979 at Panaji, Goa (Singh and Dhawam, 1983).

Table 6.1 Production capacity of some mango varieties

Varieties	Yield per		Pulp per		Total soluble solids per	
	Tree (kg)	Hectare* (Qtls)	Tree (kg)	Hectare* (Qtls)	Tree (kg)	Hectare* (Qtls)
Alphonso	45.5	45.5	35.7	35.7	6.6	6.6
Amrapali	27.0	27.0	20.2	20.2	4.6	4.6
Badami	88.0	88.0	41.5	41.5	9.1	9.1

Table Contd...

Varieties	Yield per		Pulp per		Total soluble solids per	
	Tree (kg)	Hectare* (Qtls)	Tree (kg)	Hectare* (Qtls)	Tree (kg)	Hectare* (Qtls)
Bombay green	37.0	37.0	27.8	27.8	5.0	5.0
Dashehari	27.5	27.5	18.8	18.8	3.2	3.2
Dilpsand	42.3	42.3	20.8	20.8	3.8	3.8
Fazli	28.0	28.0	15.5	15.5	2.1	2.1
Kesar	27.7	27.7	11.2	11.2	1.8	1.8
Langra	60.2	60.2	53.8	53.8	10.5	10.5
Mallika	41.0	41.0	30.8	30.8	7.5	7.5
Mulgoa	51.0	51.0	26.2	26.2	5.6	5.6
Neelum	62.4	62.4	36.5	36.5	6.7	6.7
Pedda rasam	54.0	54.0	28.1	28.1	4.6	4.6
Pairi	21.5	21.5	15.7	15.7	2.3	2.3
Rumani	73.0	73.0	29.2	29.2	4.1	4.1
S.B. Chausa	85.0	85.0	48.6	48.6	8.6	8.6
Swarna Jahangir	94.0	94.0	54.9	54.9	9.5	9.5
Totapuri	77.2	77.2	56.9	56.9	7.9	7.9
Vanraj	34.4	34.4	23.5	23.5	3.6	3.6

* 100 trees / ha at 10 × 10 m spacing

Table 6.2: Quality components of some mango varieties

Variety	Pulp (%)	TSS (%)	Acidity (%)	Remarks
*Alphonso	78.5	18.4	0.08	Stable flavour, but soft flesh
Amrapali	74.8	22.8	0.12	Excellent in β-carotene content
*Badami	47.2	22.0	0.18	Good canning quality, but low pulp content
*Banganapalli	50.2	19.5	0.10	Good canning quality
*Bombay green	75.2	18.0	0.17	Excellent flavour
*Dashehari	68.2	16.8	0.16	Flavour unstable
*Dilpasand	49.2	18.5	0.23	Low pulp content
*Fazli	55.3	13.5	0.21	Inferior quality
Kesar	40.4	16.5	0.33	Low pulp content
*Langra	89.4	19.5	0.17	Lacks intense colour

Table Contd...

Variety	Pulp (%)	TSS (%)	Acidity (%)	Remarks
Mallika	75.2	24.5	-	Excellent fruit quality, keeps well
*Mulgoa	51.3	21.4	0.14	Good canning quality
Neelum	58.4	18.3	0.32	Disagreeble texture
Pedda rasam	52.2	16.4	0.21	Fibrous flesh
*Pairi	72.9	14.4	0.38	Low productivity
*Rumani	40.0	14.0	0.06	Disagreeable texture
*S.B. Chausa	57.2	17.6	0.22	Lacks intense colour
*Swarna Jahangir	58.4	17.3	0.18	Higher productivity
*Totapuri	73.8	13.8	0.05	Inferior quality
Vanraj	68.2	15.2	0.19	Excellent flavour and colour

* Average of three reports

Basic requirements for processing in mango

1. High and regular yield of high quality fruits.
2. Readily available at reasonable price.
3. Bigger fruits with small stones, to give higher pulp yield (> 60%).
4. Good sugar: acid blend.
5. Preferably fibre-less and firm texture, with good pulp colour (high content of carotenoids).
6. Good flavour, which should be stable in processing and storage.
7. Free from fruit disorders like spongy tissue, internal necrosis or internal breakdown and/or from insect pests like fruit fly and store weevil and diseases.
8. High keeping quality.
9. Amenable for processing and preservation.
10. Less demand for table purpose.

Selection criteria

In many advanced countries processing industry takes into account total yield of pulp/juice and TSS/tree or hectare/acre for selecting a variety for processing. In India these are not adopted. The selection of varieties for processing in India has been mostly based on the fruit size, pulp percentage and organoleptic score (Kalra *et al* 1994). The prime quality parameters in the selection of a suitable variety for processing include colour, flavour, texture and nutritive value, which depend on the physico-chemical composition of the fruits which in turn depend on the variety, fruit maturity, place of cultivation etc.

However, complete physico-chemical and nutritional characterisation and the significance of these quality parameters visa-a-visa in respect of processing has not been clarified.

Many factors are involved in selecting the variety for processing. Internal fruit quality and production capacity of a variety, which cut the cost of production are of paramount importance, but not adopted in India.

Physical parameters

1. ***Yield:*** Fruit yield in terms of weight per tree or unit area and regularity in bearing are considered. The variety should be a high and regular yielder. `Nisar Pasand' and `Banazir' although produce bigger fruits, being medium in bearing are not suitable. Shy bearing tendency and low-bearing tendency are not commercially viable for processing. 'Bangalora' (Totapuri), 'Neelum' and 'Saheb Pasand' with high degree of regular bearing and prolificity are considered as potential processing varieties in spite of their undesirable characteristics (Kalra *et al* 1994).
2. ***Fruit size:*** Size of the fruit is an important criterion, since mango fruit is generally and mostly peeled manually in the processing industry. Smaller sized fruits (< 200 g) are not considered for processing for economic reasons. 'Gaurjeet', 'Gulabkhas', 'Amin Khurd', 'Asodia Deoband' and 'Bariellywala', though heavy bearers, because of smaller sized fruits are not suitable for processing. Similarly medium bearing varieties (Amin Khura, Amin Buland Bagh) have smaller fruit size, hence not suitable for processing (Kalra *et al* 1994).
3. ***Fruit colour:*** External colour of the mango fruit is of significance in fresh fruit market both in the internal and overseas business, but has no relevance in processing industry. On the other hand, internal colour of the fruit is considered important. The internal colour of the fruit is mainly due to the presence of carotenoids. The varieties like 'Alphonso', 'Amrapali', 'Dashehari' etc. develop good internal colour, hence considered to be highly suitable for processing. However, varieties like 'Bangalora', 'Mallika', 'Saheb Pasand' etc. do not develop good colour, hence not suitable from the view point of colour. However, they can be processed by blending with other varieties having higher colour. High amount of carotenoids gives deep yellow colour to the product and provides more vitamin A (Gowda and Hudder, 2004).
4. ***Pulp:*** The pulp percentage is an important factor for determining the suitability of a variety for processing. Varieties with less than 60 per cent pulp (Asodia Deoband, Bariellywala, Gaurjeet etc.) are not suitable, but those having more than 70 per cent pulp are suitable. However, varieties having 65-67 per cent pulp can also be considered provided other parameters are satisfactory. Higher pulp content reduces the cost of production even in spite of higher cost of raw material. The percentage of pulp depends upon the size of the fruit as well as the size of the stone. Fruits having smaller stones provide higher pulp percentage irrespective of

fruit size. Bulky fruits with higher percentage of pulp and good flavour have a good advantage and economy for processing (Kalra *et al* 1982). Thickness of the pulp is important for getting thicker body in the processed product. The texture of the pulp should be fine, smooth, but not gritty or powdery (Chandrag *et al* 1988., Kalra *et al* 1994).

5. ***Fibre:*** Fibre in the pulp is a negative trait for processing particularly for making juice, nectar, squash, slices etc., but is needed for pickles. So varieties producing fruits with fibre are suitable for making pickles and those producing fibreless fruits are suitable for making other processed fruit products with ripe fruits. Fibrous varieties are not suitable for RTS beverages due to possible interference of fibres in smoothness of drinks (Kalra *et al* 1982).

B. Chemical parameters

1. ***Total soluble solids:*** Varieties having higher content of total soluble solids (TSS) are suitable for processing. Higher TSS means less requirement of sugar for preparation of juice, nectar, squash, jam etc. However, TSS beyond 14 per cent reduces the organoleptic score of beverages (Shetty *et al* 1978., Gowda and Hudder, 2004).
2. ***Acidity:*** This is another important parameter, considered in the selection of suitable varieties for processing. Acidity has greater commercial utility during juice preparation, pickling or preservation. Varieties having high acidity are suitable for making pickles, chutneys, amchur etc. Acidity beyond 1.5 per cent reduces the organoleptic score of squash (Shetty *et al* 1978).
3. ***Ascorbic acid:*** Though the presence of ascorbic acid (Vitamin C) contributes to the nutritive value of the fruits and fruit products it is not considered as important, since it is lost during processing and storage (Kalra *et al* 1994).

Organolepetic characters: The studies of Kalra *et al* (1982) revealed that organoleptic evaluation is the main criterion for judging the suitability of a variety for processing, because the composition of the fruit pulp varies with variety, stage of maturity and agro-climatic condition of the area of production.

The chemical composition at best can provide general guidelines for selection of a particular variety for processing. Nevertheless chemical composition has great importance in processing for adjusting TSS and acidity of the final product besides being useful in assessing the nutritive value of fruits and products.

GREEN MANGO PRODUCTS

Varieties suitable for pickles and chutneys

Varieties for preparing pickles and chutneys should have higher acidity and fibre content and low sugars. These characters help to retain the texture in slices and in storage.

Varieties producing fruits with thicker peel, white flesh and good fruit size with higher pulp weight are preferred for pickle making. In general local seedling types having high acidity are used for making pickles and chutneys. Each district/region has its own preferred types/varieties for making pickles. For example, in Andhra Pradesh varieties like 'Jalal' (Photo 6.1), 'Tellagulbi,' 'Hamlet,' 'Acharpasand' are used for pickling in homes, while Totapuri is mostly used for pickling commercially. 'Totapuri' and 'Fazli' are ideally suited for making chutney. 'Chandrakaran' and 'Pulimong' are the best pickle varieties of Kerala (Satyavathi *et al* 1972) and 'Carreira Goa' and 'Amlet' are the popular pickle varieties in Karnataka (Singh and Dhawan, 1983). 'Amlet' variety is exceptional in maintaining its high acidity content even at fully ripe stage of 8-10 days storage at room temperature. Recently, from a survey of pickle genotypes, an ideal genotype, christened as 'Konkan Ruchi' was identified in Maharashtra (Tidke *et al* 2006). It is a regular, high yielder, having large fruit size, maximum acidity and best organoleptic score for colour, flavour and texture in pickle (see also 'pickles' in chapter 5). In North India variety 'Ramkela' is used for making mango pickles (Photo 6.2).

Photo 6.1: Jalal (pickle variety)

Photo 6.2: Ramkela (pickle variety)

Amchur

This is normally prepared from immature raw green fruits from seedling types and dropped fruits having high acidity. Though there is a fairly good demand for this product, no particular variety has neither been identified nor screened so far. Acidic types with high recovery of slices and less tendency for deterioration are ideal for amchur preparation.

RIPE MANGO PRODUCTS

Varieties suitable for beverages

Mango fruit beverages include: nectar, squash, juice and RTS drinks. Varieties for making mango beverages should have good consistency, flavour and colour without terpentine odour. Traditionally beverages are prepared from different varieties and as a result there is no uniformity in colour, flavour and taste of the product. This emphasizes the need for selection of a suitable variety for making beverages. The choice of most suitable variety for making mango beverages has been the subject of investigation of many researchers in India as well as other countries. 'Swarna Jahangir', 'Dashehari', 'Bombay Green' and 'Vanraj' have excellent juice colour, hence they have better potentiality for beverage industry. These can also be used to blend the juice of those varieties which lack intense colour such as 'Chausa' variety (Singh and Dhawan, 1983). Out of the seven sucking varieties (small fruited, highly firbrous types, which cannot be easily sliced), 'Safeda Jauhari' and 'Gaurjeet' showed desirable characteristics, such as attractive colour, rich flavour, high content of TSS and vitamin A for beverage (RTS drinks, nectar and squash) processing. Between the two 'Safeda Jauhari' was more suitable (Rabbani and Singh, 1988). Commercial varieties of North India like 'Langra' and 'S.B Chausa' lack intense colour and 'Dashehari' has no stable flavour, hence not suitable for making beverages, unless blended with pulp having more colour and flavour. 'Sulkul' and 'Sindhuri', because of higher acidity are not suitable.' Alphonso', 'Pairi', 'Padiri', 'Banganapalli', 'Langra', 'Carbao' and 'Pico' are suitable for making beverages (Rabbani and Singh, 1988). The demand for fruit beverages is largely based on their nutritive value, flavour and colour, which depend directly on the physico-chemical characters of the variety i.e., the ultimate quality of mango product largely depends on the selection of suitable variety(s).

Varieties suitable for making nectar

Mango nectar is an important nutritious beverage and is very popular among the masses. Several workers studied suitability of different varieties and hybrids for making quality nectar. Mango hybrids from Kodur (Andhra Pradesh, India) were evaluated for their suitability for making nectar by Surya Prakash Rao *et al* (1968b). According to them, hybrids – '2/2 Neelum × Banganapalli', '4/3 Banganapalli × Alumpur Baneshan' and '2/7 Chinna Suvarna Rekha × Neelum' were satisfactory for making nectar as they posses good aroma, flavour and textural characteristics among the nine hybrids. Out of the four commercial table varieties and three hybrids evaluated for nectar, all except 'Bangalora' and 'Khader' varieties and 'Neelum × Yerramulgoa' hybrid were suitable. 'Baneshan' and 'Neelum' varieties and the hybrids 'Bangalora × Alumpur Baneshan' yielded nectar of very good quality, considering their consistence, aroma and flavour and overall sensory impressions. The nectar from '2/2 Neelum and Banganapalli' hybrid possessed attractive deep orange colour, soft and silky texture and pleasant mouth feeling properties. The smooth and melting texture of the flesh of this hybrid mango seemed particularly conducive for the production of nectar. The development of a terpeny odour and a pungent taste during storage marred the quality of nectar from 'Baneshan' and 'Khader' varieties and '2/4 Neelum' and 'Yera mulgoa' hybrid.

Among the four mango varieties of North India, 'Dashehari' was adjudged as the best variety for nectar making, as it had high edible pulp, high carotenoid content and excellent flavour, while others were not suitable. Though 'Chausa' had maximum content of edible part (67.6%), it contained less content of carotenoids, lost flavour during processing, had poor colour and good body. 'Langra' had higher vitamin-C and high acidity but had poor colour and good body and flavour, however, because of high acidity and poor colour it was not suitable for nectar. Similarly 'Bombay Green' was also unsuitable because it lacked good colour, good flavour and had minimum edible part (55.5%) (Roy *et al* 1972).

Some table varieties (Alphonso, Dashehari etc.,) are best suited for making good quality nectar, but are very expensive, which could be prohibitive for the processing industry. Hence, Kalra *et al* (1982) screened 20 less priced varieties against 'Dashehari' and found 'Alifaila', 'Deltia' and 'Prabhasankar' as the best varieties after 'Dashehari', while 'Mithwa Malda', 'Baramasi Dhanipur' and 'Firdus Pasand' were good varieties for nectar preparation. Among the 20 varieties cv. 'Hurrhad' had high percentage of pulp (79%), but showed very less score for other characteristics. Similarly 'Lazat Bakshaw', 'Hamlet' and 'Eruvadi Bangalora' though liked slightly to moderately, could be processed with advantage because of higher pulp content. 'Bangalora', a big fruit (> 200 g) with 78 per cent pulp content exhibited poor organoleptic score, 'Barbacha' and 'Bappa Kai' varieties produced the lowest quality nectar.

Krishnamurthy *et al* (1984) found that 'Baneshan', 'Navaneetham', 'Goa bunder', 'K.O.7' and 'Sharbatigadi' were good for nectar in Karnataka. Nectar from 'Baneshan' and 'Navaneetham' retained their consistency, flavour and taste even after one year storage at ambient temperature followed by 'Goabundar' and 'K.O.7'. Except for dull colour, nectar from 'Kurnuli Mulgoa' had very good consistencey, pleasant flavour and taste. Nectar from 'Jahangir', 'Suvarnarekha', 'Roayl Special', 'Rumani × Neelum', 'Neelum × Bangalore'. 'Samar-e-Bahist x Dashehari' hybrids had mild flavour. 'Totapuri' was not found suitable for making nectar as it developed unpleasant flavour during storage.

Based on the pulp yield, colour, flavour and overall acceptability of the canned nectar and juice 'Dashehari' and 'Tamburia Dasheri' were found to be good and comparable to 'Alphonso'. From colour and flavour point of view 'Dasheheri' and 'Mallika' were excellent, while others were fairly good. Next preferred varieties were 'Malda Fazli', 'Lucknow Fazli', 'Chausa' and 'Golbhadiya safeda'. 'Lucknow Fazli' gave inferior quality nectar and juice (Chakraborty *et al* 1991).

Kalra *et al.* (1994) explored the possibility of utilizing some non-commercial varieties for making nectar. From the processing point of view of nectar, varieties like 'Benazir', 'Gaurjeet', 'Gulabkhas', 'Nisar Pasand' and 'Saheb Pasand' and hybrid 'Mallika' were suitable. However, 'Gaurjeet' and 'Gulabkhas' with smaller fruits were not acceptable for processing. 'Banazir' and 'Nisar Pasand' although have bigger sized fruits, but being medium bearers, do not satisfy the needs of processing industry as the supply would be less and questionable and unreliable. However, these varieties with higher carotenoid content can be used for blending with others having low carotenoid content like 'Bangalora'. 'Saheb Pasand', on the other hand, being a fairly regular bearer can be a potential processing variety. It is an East Indian variety, having high TSS (30 °C), bright coloured pulp and firm texture. It not only gives an excellent beverage but also yields good slices with stable colour during storage (Yadav *et al* 1987).

Sharma *et al* (1995) investigated the suitability of certain polyembryonic and monoembryonic mango cultivars for making nectar. These workers claimed 'Kurukkan' a polyembryonic variety as best for nectar in comparison with commercial varieties followed by 'Dashehari' (mono-embryonic). However, 'Dashehari' could not be placed at par with 'Kurukkan' because of its poor body and un stable flavour. On the other hand, 'Kurukkan' produced both good body and better flavour though the colour was not good. The nectar from 'Alphonso', 'Baneshan' and 'Totapuri' were at par with that of 'Dashehari'. The good body of nectar from 'Kurukkan' was due to greater difference between brix and total sugars compared to other varieties, in case of 'Chausa' and 'Langra', resulting in more accumulation of water soluble polysaccharides (gum or mucilage) (Roy *et al* 1972).

Among the two cultivars (Kesar and Taimuria) and two hybrids (Amrapali and Mallika), 'Amrapali' was best suited for making nectar as it scored highest sensory score (89.6%) and was graded as Grade I quality followed by 'Kesar' and 'Taimuria', which were rated as Grade – II quality. The highest score for 'Amrapali' nectar was due to its richness in colour, taste and flavour (Jain *et al* 1996). Khurdiya and Roy (1988) also reported 'Amrapali' as the most suitable and best variety for nectar making. In another study, out of the six early maturing varieties 'Kurukkan' was considered as the best, since it scored highest organoleptic score for nectar, in spite of lower pulp content (54.08%) (Jain *et al* 1997).

Varieties suitable for making juice

Sucking verities are considered good for juice production as compared to table varieties, because of their abundance and consistency of the juice. Among the different sucking varieties of Uttar Pradesh (India), on the basis of bigger fruit size, lower peel and stone contents, high juice recovery (53.7%), thin consistency, attractive colour (lemon yellow) fine taste and pleasant flavour and better acceptability, 'Sukul' a juicy variety from Basti (Uttar Pradesh) was claimed as best suited for juice preparation on commercial scale and its juice can be stored for one year without quality loss (Awasthi and Pandey, 1979., Singh and Dhawan, 1983). In spite of their better productivity for juice, 'Chausa' and 'Rumani' are not suitable for juice because 'Chausa' lacks intense colour and 'Rumani' has disagreeable odour (Singh and Dhawan, 1983). Of the 5 cultivars screened, 'Safeda Jauhari' was rated best for quality of the pulp, organoleptic score and storage quality of juices and squashes (Singh and Rabbani, 1986). 'Amrapali' and 'Langra', because of their deeper colour have greater potentiality for preparation of juice (Khurdiya and Roy, 1988., Gowda and Ramanjaneya, 1995). Out of the 14 diverse cultivars of West Bengal (India) based on sensory evaluation and biochemical constituents, 'Langra', 'Dashehari', 'Sitabhog', 'Rani Pasand', 'Amrapali' and 'Himsagar' were judged suitable for juice making (Chakraborty and Debnath, 2004).

Varieties suitable for squash

Based on the organoleptic evaluation, 'Alphonso', 'Olour', 'Neelum', 'Khudadad', 'Peter', 'Rasam' and 'Surangadi' were suitable for squash making (Palaniswamy *et al* 1974). Satyavati *et al* (1972) were also of the opinion that 'Olour' a polyembryonic variety was the best suited one for squash among different varieties of Kerala (India).

Varieties suitable for RTS drinks

Among the 6 early varieties of mango, 'Amin' was best suited for making RTS drink, as it scored highest organoleptically (Jain *et al* 1997). 'Taimuria' gave best and Grade I quality RTS drink having most attractive colour and flavour (Jain *et al* 1996).

Varieties suitable for mango wine

Preliminary studies on mango wine making indicated that all the varieties are not suitable for wine making. To make wine, the pulp of mango varieties screened was subjected to alcoholic fermentation using yeast (*Saccharomyces crevisiae var. ellipsoideus*) for 7-10 days at 22 °C. Wines made from 'Fazli', 'Langra' and 'Chausa' were best having golden yellow colour with brilliant appearance and good body. Although alcohol content in the wine from 'Fazli' was low (8%) its good sugar and acid blend of the final product, made it more acceptable. Wines from 'Suvarnarekha' had a high acidic taste. The titratable acidity and volatile acidity were too high in the wines from 'Lucknow Safeda', while alcohol in them was only 5 per cent. Sweet wine made from 'Dashehari' was also good, possessing characteristic flavour due to abrupt arresting of the fermentation through addition of mango brandy, which resulted in the retention of some fruit sugars and flavours. However, due to its complete dryness, it was not acceptable (Kulkarni *et al* 1980).

Varieties for Blends

Since all the mango varieties do not possess similar processing characteristics, lack in one or more characteristics and most of them are lost during various processing operations and storage and to make the differences good among the varieties to produce ideal beverages, blending of varieties was conceived. In general blends produce better results in comparison with individuals. In case of 'Chausa' considerable improvement of flavour in nectar was achieved by blending with 'Dashehari'. The combination of 'Dashehari' and 'Langra' followed by 'Dashehari' and 'Chausa and 'Langra' and 'Bombay Green' was best (Roy *et al* 1972). To get an ideal nector, 'Kurukkan' (for good body and aroma) and 'Dashehari' (for good colour) were blended in different proportions. Most of the combinations were good in comparison with individuals. Among them 'Kurukkan' and 'Dashehari' (50:50) was best (Sharma *et al* 1995).

The RTS drink from the combined pulp of 'Banganapalli' and 'Dashehari' (50 : 50) was the best and scored highest sensory marks as compared to the one from either of these alone, both from fresh as well as stored pulp (Srivastava, 1998).

Varieties suitable for canning

Mangoes are canned as whole fruits, pulp, slices and juices; the later three are the main export items from India. Canned mango products are prepared from different varieties showing different quality attributes in the end products and variations in the sensory score and acceptability. The best mango cultivars identified for canning are: 'Badami' (Girdharilal, 1956), 'Himsagar' (Bose and Das, 1958), 'Dashehari' (Jain, 1961., Teaotia and Awasthi, 1966), 'Alphonso' (Nanjundaswamy *et al* 1966) and 'Baneshan' and 'Bangalora' (Surya Prakash Rao *et al* 1968a). Among the hybrids 'Neeleshan' (Neelum × Baneshan), 'Mallika' (Dashehari × Neelum) and 'Amrapali' (Neelum × Dashehari) are best for canning. In Karnataka (India), 'Totapuri' and 'Neelum' are the main varieties used for canning juice and pulp and 'Alphonso' for canned slice production (Krishnamurthy *et al* 1984).

Varieties suitable for canning whole fruits

Varieties 'Gaurjeet', 'Sipia' and 'Lucknow Safeda' were suitable for canning whole fruits in syrup, because of their smaller size fruits. The colour and flavour of 'Gaurjeet' were acceptable but its price is rather high and availability is low, because it is confined to a particular area and the area under this variety is comparatively small. Though 'Sipia' is suitable, it is not attached much value, as it has less attractive colour and its taste is much inferior to other varieties (Mehta *et al* 1971).

Varieties suitable for canning pulp

The desirable characteristics of a variety for canning pulp are:

1. higher percentage of pulp recovery,
2. maximum pulp weight,
3. smaller size stone,
4. low moisture content, as higher moisture content is detrimental for distant transport and storage.

According to Iyer *et al* (1991) the basic requirements for canning pulp in varieties are: high yielding, high pulp to stone ratio, good sugar: acid blend, good pulp colour, free from spongy tissue, internal necrosis, fruit fly and stone weevil and fibre. Varieties having low fibre content and high quality of pulp are best suited for canning pulp. Mainly 'Totapuri', 'Badami', 'Banganapalli', 'Dashehari', 'Fagli' and 'Hasina' are used for canning of pulp presently in India (Chakraborty *et al* 1991). However, from time to time a number of varieties are tested for their suitability for making canned pulp in India.

With regard to colour and taste of the canned pulp, 'Olour' was best followed by 'Cripe', 'Neelum', 'Priyor', 'Muvandan' and 'Sundri', in order of preference. 'Olour' had the maximum β-carotene (2865 μg/100 g pulp) content, 10°Brix and 0.48 per cent acid and its canned pulp was bright orange yellow in colour, with characteristic pleasant mango flavour and taste. Canned pulp from 'Cripe' and 'Priyor' was yellow in colour, with good flavour and taste. 'Neelum' and 'Priyor' are quite suitable for canning pulp with good yellow colour and pleasant flavour and taste. Muvandan had the least β-carotene content with dull yellow colour and insipid flavour and taste, hence unfit for canning. Thus 'Olour', 'Neelum' and 'Priyor' were best for canning as pulp in Kerala (Satyavati *et al* 1972).

Paliniswamy *et al* (1973) screened 20 varieties for their suitability for canning in Tamilnadu and found that 'Alphonso', 'K.O.8'., 'Baneshan' and 'Mulgoa' were best for texture, flavour, clarity of syrup etc. in the canned pulp product followed by 'Olour', 'Peter', 'Alumpur Baneshan', 'Bangalora', 'Rumani' and 'Eruvadi Rumani' while others were not suitable for canning.'Dashehari' and 'Safeda' were adjudged as good for canning, while 'S.B. Chausa', being fibrous was unsuitable for canning (Kapur, 1975).

'Navaneetham' variety from Andhra Pradesh was considered ideal for making canned pulp, in view of maximum retention of flavour, thick consistency, even after one year of storage at ambient temperature and acceptability. 'Baneshan', 'Goa Bundar', 'Royal Special' varieties and 'Neelum × Baneshan' (Neeleshan) hybrid were also acceptable for making canned pulp in that order (Krishnamurthy *et al* 1984). Among the 4 varieties (Alphonso, Banganapalli, Neelum and Totapuri) and 2 hybrids (Mallika and Amrapali), 'Alphonso' and 'Amrapali' were best suited. Though the size of 'Amrapali' fruit is small (photo. 6.13) its pulp yield was on par with that of 'Alphonso' due to its thin peel and also due to the absence of spongy tissue problem unlike in 'Alphonso'. In sensory evaluation of canned pulp, 'Amrapali' and 'Alphonso' scored maximum for colour, 'Totapuri' for highest consistency, which is desirable to get thicker product; 'Alphonso' for highest flavour and best taste closely followed by 'Amrapali'. In consumer acceptability 'Alphonso' was best followed by 'Amrapali' (Gowda and Hudder, 2004). In Cuba, mango varieties viz., 'Lopez', 'Samfelip,' 'Santacruz' and 'Samora' were suitable in view of their high pulp yield and flavour, colour and aroma (Finera *et al* 1995).

'Totapuri' and 'Neelum' have mild flavour and poor colour (due to low content of carotenoids) while 'Alphonso' and 'Amrapali' have excellent flavour and higher colour (due to higher content of carotenoids), but suffer from spongy tissue and relatively smaller sized fruits respectively. Further, 'Alphonso' fruits are expensive, while 'Totapuri' fruits are comparatively cheaper. Therefore, it was investigated how best the quality and pulp colour of 'Alphonso' and 'Amrapali' can be improved by blending with the others and to find out a suitable variety for blending 'Totapuri' and 'Neelum'. For blending 'Totapuri' was very ideal with either 'Amrapali', which has viscous pulp or 'Alphonso', since it improved the viscosity of these two varieties besides improving taste and flavour to a great extent. 'Totapuri's thick pulp was responsible for the higher viscosity of other varieties. Blending the pulp of 'Totapuri' or 'Neelum' either with 'Alphonso' or 'Amrapali' improved their colour. The quality and marketability of high yielding and regular bearing, low–priced 'Totapuri' pulp can be improved by blending with 'Alphonso' or 'Amrapali' at 50 : 50 strength. 'Totapuri' and 'Amrapali' blends are mutually beneficial, since consistency of 'Amrapali' is poor, while colour and flavour of 'Totapuri' are poor (Gowda and Hudder, 2004).

Varieties suitable for canning slices

Although large number of fleshy varieties are available, all are not equally suitable for canning of mango fruit in the form of slices. Varieties for canning as slices should have pleasing flavour, crisp and succulent texture, attractive colour, smaller stone, uniform ripening, free from spongy tissue, internal necrosis, fruit fly and stone weevil and fibre and should retain texture and flavour on processing and subsequent storage of the canned product. The varieties should maintain clear syrup during processing and storage. The clarity of the syrup depends upon fruit texture at canning. Varieties which retain shape and texture of slices at canning provide clean syrup (Palaniswamy *et al* 1973). In fact texture retention is the main criterion for canning slice. Several Indian varieties like 'Alphonso', 'Bangalora', 'Beneshan', 'K.O.7', 'K.O.8', 'K.O.16', 'Mulgoa', 'Chausa','Safeda', 'Pairi', 'Mallika', 'Amrapali' etc. were screened for their suitability for canning as slices and were adjudged as suitable. Generally, varieties like 'Alphonso', 'Pairi', 'Padiri' and 'Mulgoa' are preferred. 'Dashehari' ranked first among all the varieties studied followed by 'Bombay Yellow', 'Bride of Russia' and 'Neelum' in order of preference (Teaotia and Awasthi, 1966). 'Himasagar' of West Bengal was best followed by 'Langra' and 'Fazli' (Bose and Das, 1958).

However, among the Indian mangoes screened 'Alphonso' was the most suitable for canning slices in syrup. Others like 'Dashehari', 'Totapuri', 'Banganapalli' and 'Fazli' yielded canned products with mild flavour and colour, hence not suitable (Siddappa and Bhatia, 1956). Among the 13 varieties evaluated 'Dashehari' was the best variety for canning in view of texture, flavour, colour, clarity of syrup and recovery of slices followed by 'Bombay Yellow', 'Bride of Russia', 'Neelum' and 'Bangalora'. The percentage of 'A' grade slices was higher in the case of 'Dashehari' (65%) which was due to its thin and lower percentage of peel and stone. Others like 'Mohanbhog' and 'Langra' also showed quite high percentage of good slices, but their texture was not very firm, which is a disqualification in canning. Slices of 'Dashehari' were of attractive yellow colour with pleasant flavour. 'Bombay Yellow' and 'Bride of Russia', although produced good and firm slices, but were poor in other traits. Among the late matured varieties 'Neelum' performed well for canning (Teaotia and Awasthi, 1966).

Satyavati *et al* (1972) screened grafted as well as seedling types of mango in Kerala and adjudged 'Priya' and 'Neelum' (grafted ones) as best followed by 'Sundri' and

'Cripe' (grafted ones). The former two varieties had low fibre content and their canned slices were firm in texture. 'Muvandan' and 'Olour' (seedling types) with high fibre content (> 0.5%) were unfit. However, irrespective of their suitability, the canned slices of all the varieties kept well during 6 months storage at room temperature (25-30 °C) with no visible corrosion of the cans, in which the slices were packed and stored.

Based on visual and organoleptic tests, after one year storage, 'K.O.8', 'Alphonso', 'Mulgoa' and 'Baneshan' were adjudged as good for canning followed by 'Olour'; 'Peter', 'Alumpur Baneshan', 'Eruvadi Rumani', 'Bangalora' and 'Rumani' in respect of flavour, texture and taste. The crisp succulent texture with pleasing flavour and attractive natural colour of the slices were well preserved even after one year. Others were not suitable as they failed to give the product of acceptable quality either due to disintegration of the slices or due to fibrous nature (Palaniswamy *et al* 1973).

Similarly based on the overall assessment of four mango varieties of North India. 'Dashehari' was the best for canned and frozen slices followed by 'Baneshan', 'Langra', 'Bombay Green' and 'Chausa' after 9 months of storage. 'Dashehari' maintained its colour and flavour better in canned and frozen slices compared to other varieties. Among the varieties, 'Baneshan' had better texture in both canned and frozen slices during storage (Adsule and Roy, 1974).

Though 'Dashehari' satisfies most of the requirements of a good canning variety, its characteristic flavour is not retained in the end product due to the influence of certain factors like VRS value, storage period, processing conditions, addition of ascorbic acid to the syrup (Saha *et al* 1976).

Among the 17 fleshy mango varieties 'Safeda Malihabad' was superior for canning as it retained most of its original qualities (colour, texture and flavour) during canning and storage along with clear syrup. Next in order of suitability were 'Anopan Budulka Kelwa', 'Langra Dudhia' and 'Himsagar', which gave fairly good product. 'Pairi' was satisfactory in caning, but due to its higher acidity, the product was poor in taste and flavour, so not preferred for canning. Similarly, though 'Bangalora' gave the highest yield of slices (62.7%), it was adjudged as unsuitable because of its mild flavour and lower content of carotenoids (Awasthi and Pandey, 1980).

During the past 50 years several hybrids were developed from different centres in India, with a main objective of obtaining a regular and prolific bearer. Some of these hybrids were evaluated for their suitability for canning as slices and also compared with existing commercial varieties. Out of the 9 mango hybrids (Kodur) evaluated for canning slices in syrup, the hybrid (7/1) 'Bangalora × Himayuddin' alone was found promising as per the requirements, norms of preparation, processing and storage (Surya Prakash Rao *et al* 1968a). Only 2 of the 5 commercial varieties of Andhra Pradesh (India) viz., 'Beneshan' and 'Khader' and 3 hybrids of Kodur viz., '2/4 Neelum × Yerramulgoa', '2/2 Neelum × Banganpalli' and '4/3 Bangalora × Alumpur Baneshan' studies were satisfactory. The crisp, succulent texture, the pleasing characteristic flavour and attractive natural colour of the slices were well preserved even after a storage period of one year in both 'Baneshan' and 'Khader' at ambient temperature (25-37 °C). 'Rumani' × 'Neelum' and 'Bangalora' × 'Alumpur', 'Baneshan hybrids had fibrous pulp, which presented disagreeable texture in the end product. 'Bangalora' fruits and the hybrids of '2/14 Neelum × Yerra mulgoa' and '2/2 Neelum × Banganpalli' presented a very soft and melty texture pulp resulting in a soft and marshy canned product. Canned slices of the other hybrids studied viz., '11/13 Chinna-Suvarnarekha × Jehangir', '3/16 Neelum × Alampur Baneshan' and '3/7 Neelum × Himayuddin' were not suitable even for nectar and puree, due to terpeny odour and taste. Canned 'Banglora' slices developed a disagreeable camphorous odour during storage (Surya Prakash Rao *et al* 1968b).

Nine table varieties, 4 juicy varieties and 5 hybrids of Andhra Pradesh were tested for their suitability for canning as slices, juice and nectar (Krishnamurthy *et al* 1984). Out of which 'Baneshan', Suvarnarekha' and '5/5 Raspuri × Langra', were suitable for canning as slices, as their slices had a good flavour and texture. Slices of 'Kuroli Mulgoa' were good, but had dull yellow colour while those from 'Intermax', 'K.O.7' and 'Neelum × Baneshan' had become meshy during processing. 'Goa Bundar', 'Rumani × Neelum', 'Jehangir', 'Saman-e-Bahist × Dashehari' and 'Neelum × Baneshan' were fibrous and not suitable for canning. Further, 'Samar-e-Bahist × Dashehari' hybrid and Dashehari × Mahimooda developed off-flavour during processing itself. Slices of Jehangir developed uneven white patches and softness, 'Baneshan' and 'Raspuri × Langra' retained their firm texture, colour, taste and flavour during storage period of one year, while 'Suvarnarekha', 'Goa Bundar' and 'Rumani × Neelum' hybrid retained their texture and flavour only.

According to Khurdiya and Roy (1986a), 'Mallika' was best compared to other cultivars in respect of texture and overall organoleptic evaluation of canned slices followed by 'Amrapali', 'Dashehari' and hybrid '16/5'. 'Mallika' gave slice yield of 53 per cent against 40 per cent in others, along with good texture and flavour. Due to its poor flavour and texture hybrid '16/5' was not acceptable as canned slices.

Ambadan *et al* (1988) evaluated mango hybrids of I.I.H.R, 'Bangalore' for canned slices suitability and found that the hybrid 'Banganapalli × Alphonso' as suitable and rated the hybrid as superior to 'Banganapalli' and only slightly inferior to 'Alphonso', which has smaller fruits with larger stone than the hybrid. Fruits of this hybrid were medium in size, but with lesser stone percentage, resulting in more yield of slices, less carotene content than 'Alphonso'. But the sensory quality of the hybrid was more or less at par with that of 'Alphonso' indicating that this hybrid is very close to 'Alphonso' in quality.

Varieties suitable for canning juice

Juicy varieties of Andhra Pradesh (India) viz., 'Peddarasam', 'Cherukurasam' and 'Panchadara Kalasa' are quite suitable for juice canning. The attractive colour and characteristic pleasing flavour of these varieties are well retained during processing and storage (Suryaprakash Rao *et al* 1986b). 'Badami' was best suited for canning juice in Tamil Nadu (India) followed by 'Padiri' (Paliniswamy *et al* 1973) and in Karnataka (India) (Gowda and Ramanjaneya, 1995). 'Padiri' performed similar to 'Alphonso' in respect of flavour and consistency, but scored less in colour due to lower carotenoids content (Iyer *et al* 1991). Canned juice from 'Kuroli Mulgoa' and 'K.O.7' had good consistency, flavour and taste, but was unattractive due to dull yellow colour while the canned juice from 'Suvarnarekha' and 'Dashehari' and 'Rumani × Neelam' hybrid had mild flavour. 'Navaneetham', 'Baneshan', 'Goa Bundar', 'Royal Special', 'Hyder Sahib', '9/4 Neelum × Baneshan' hybrids were the best for making canned juice. Others were not suitable as they developed disagreeable flavour after processing (Krishnamurthy *et al* 1984).

Four hybrids of Bangalore viz., 'Arka Aruna', 'Arka Puneet', 'Arka Anmol' and 'H-51' were screened for their suitability for canned juice against 'Alphonso' and 'Totapuri', the main varieties used for processing in Karnataka (Gowda *et al* 1994). All the hybrids yielded marginally higher pulp and had lower peel, stone and fibre contents. Carotenoid content of all hybrids was markedly lower than that of 'Alphonso' but higher than that of 'Totapuri' except in case of 'Arka Aruna'. Vitamin-C was lowest in all hybrids except in 'H-51', 'Arka Aruna' had higher fruit weight, volume and less fibre than all other hybrids and varieties included in the study. 'Arka Aruna' and 'H-51' were not suitable as they scored less in colour, consistency and flavour and were not acceptable as compared to others. In terms of colour, consistency and flavour, 'Arka Puneeth' and 'Arka Anmol' were better than the other two hybrids, but scored less than 'Alphonso' due to lower content of carotenoids, but comparable to 'Alphonso' in flavour and consistency. 'Totapuri' and 'H-51' performed poorly in sensory evaluation due to poor colour, low consistency and moderate flavour. From the above, it is evident that hybrids 'Arka Puneeth' and 'Arka Anmol' were best suited for canning juice production, comparable in quality to 'Alphonso'.

Among the eleven mango varieties from different regions of the country, 'Alphonso' and 'Padiri' were ideally suited for the preparation of canned juice in that order, in respect of acceptability, colour, consistency and flavour. Others viz., 'Banganapalli', 'Kesar', 'Langra' and 'Totapuri' performed moderately due to lower content of carotenoids, but did well with regard to flavour and consistency (Gowda and Ramanjeneya, 1995) (Table 6.3)

Table 6.3 Mango varieties and hybrids, best suitable for canning in different states of India

State	Variety (s) and Hybrid (s)	Product (s)	Reference
Andhra Pradesh	Bangalora, Baneshan, Juicy varieties (Peddarasam, Cherukurasam, PanchadaraKalasa)	Canned pulp Canned slices Canned juice	Surya Prakash Rao *et al* (1968a) Krishnamurthy *et al* (1984) Krishnamurthy *et al* (1984)
Kerala	Olour, Neelum and Priyor	Canned pulp	Satyavati *et al* (1972)
Karnataka	Alphonso, Totapuri, Neelum Alphonso, Padiri,	Canned pulp Canned juice	Girdharilal, 1956 Gowda and Ramanjeneya, 1995
Tamilnadu	Alphonso, K.O.8 Baneshan, Mulgoa, Padiri	Canning Canned juice	Palaniswamy *et al* (1973)
Maharashtra	Alphonso	Canning	Jain, 1961
Uttar Pradesh	Dashehari Mallika, Amrapali	Canning Canning	Teootia and Awasthi, 1966 Khurdiya and Roy, 1986 a,b
West Bengal	Himsagar	Canning	Bose and Das, 1958

However, most of the above varieties and hybrids do not meet all the requirements of canning. For example, 'Alphonso' has an excellent and stable flavour, but it is an irregular bearer and has limited adaptability, besides developing spongy tissue and expensive fruit. 'Totapuri' and 'Rumani' have good productivity and regular bearing tendency, but have poor quality pulp. 'Langra' lacks intense colour. Flavour of 'Dashehari' is unstable during processing and storage. 'Bombay Green' has excellent flavour but has low productivity. 'Badami' and 'Banganapalli' seem to be better choice for canning industry, an account of their firm pulp, excellent flavour, smaller stone with good keeping quality and optimum productivity and weak tendency to alternate bearing in case of 'Baneshan'. Among hybrids, 'Amrapali' and 'Mallika' both have excellent flavour and fruit quality and are regular bearers, but productivity is less. If these two hybrids can be adopted to high density orcharding, their yields can be increased and then they become potential for processing industry (Singh and Dhawan, 1983).

Varieties suitable for jams and fruit bars

Pulpy varieties are considered ideal for preparation of jams and fruit bars. However, not much work was done to screen the suitable varieties for these products. 'Totapuri' is mostly used for making bar/leather, followed by 'Kolanka Goa' in Andhra Pradesh. 'Alphonso' is also used to some extent, but its high cost comes in the way. 'Bombay Green' and 'Dashehari' also are used to some extent. Recently, 'Malcorada and Xavier' from Goa have been found suitable for making jam. 'Langra' and 'Totapuri' have high productivity for pulp, but lack intense colour and if these two varieties are blended with pulp having intense colour such as 'Alphonso', 'Dashehari' and 'Amrapali', they can be utilized in jam and fruit bar industry. Non-firbours varieties with compact texture, otherwise not suitable for table purpose, canning, pulp or beverages are useful for preparation of preserves and crystallized fruit at a slightly unripe stage (Bhatnagar and Subramanyam, 1973).

Varieties suitable for mango cereal flakes

According to Girdharilal *et al* (1960) suitable varieties for preparation of mango cereal flakes are – 'Badami', 'Raspuri', 'Totapuri', 'Neelum', 'Padiri', 'Malihabad Safeda', 'Lucknow Safeda' and 'Dasehari' as well as combination of some of these varieties. Their studies have shown that the best quality flakes could be prepared from

(i) 'Badami'; mango pulp alone,

(ii) Badami pulp blended with 'Raspuri', 'Neelum' or 'Padiri' pulp in the ratio of 2 : 1,

(iii) 'Dasehari' mango pulp alone,

(iv) 'Dasehari' pulp blended with the pulp of 'Lucknow Safeda' or 'Malihabad Safeda' in the ratio of 1 : 2.

Flakes from 'Badami' pulp had maximum orange yellow colour and flavour, whereas 'Padiri' and 'Neelum' varieties yielded products with attractive colour and mild flavour. Flakes from 'Raspuri' were full yellow in colour, with satisfactory flavour. Flakes from 'Totapuri' had very light yellow colour and poor flavour. Flakes from 'Dashehari' pulp had maximum orange yellow colour and flavour followed by 'Safeda Malihabad'. Flakes from 'Lucknow Safeda' had dull yellow colour and mild flavour.

Varieties suitable for frozen products

Frozen products of mango (pulp and slices) are most popular and have bright future for export. Varieties with firm and higher content of pulp like 'Totapuri', 'Badami' and 'Alphonso' should be tried for making frozen slices and 'Mulgoa', 'Langra', 'Chausa' and 'Swarna Jehangir' for frozen pulp (Singh and Dhawan, 1983). Singh (1974) screened 'Alphonso', 'Pairi', 'Padiri' and 'Mulgoa' for freezing in the form of slices. Among the Indian varieties, 'Alphonso', 'Baneshan' and 'Dashehari' were better suited (Adsule and Roy, 1974., Cooke *et al* 1976).

Vasquez-Caiceda *et al* (2004) have summarized the requirements of different mango products as follows:

(i) The production of puree, concentrates and beverages requires specific characters where sugar content (TSS) and sugar : acid ratio together with an intense yellow orange colour and pleasing flavour,

(ii) On the other hand, production of dried, frozen or canned slices requires firmer fruits, where colour is not of less importance,

(iii) For production of puree and nectar soft fruits with low TSS/acid ratio accompanied by intense yellow orange colour and a pH below 4.5 are required;

(iv) Mango slices for canning, freezing and drying require firmer texture, whereas other properties such as high activity of oxidative enzymes detrimental to dried and frozen products should be taken into account,

(v) High fibre content may be a hinderance in fresh fruit consumption, but not so in case of slices, where fibres are removed by pulping.

Characteristics and suitability of mango varieties and hybrids for processing

Although many commercial and non-commercial varieties and hybrids were tested for their suitability for processing different types of products, only few stood the acid test. From the published information, most important characteristics of certain mango verities and hybrids are presented below along with their suitability for processing different products.

Alphonso (Photo 6.3): It is one of the finest Indian mangoes and rated as the best both domestically and internationally as a table variety. It is the leading variety of export from India and is in great demand in both domestic and overseas markets. It is a dual purpose variety for table as well as processing. It is well known for high quality and keeping quality. It is very expensive than any other commercial variety in India. The fruit has an attractive shape and colour (Photo 6.3). The pulp is firm and fibreless, juice is moderate to abundant. The flesh (pulp) is rich in colour due to high content of carotene and has pleasing flavour due to high number of (50) flavouring compounds and their proportion (Selveraj *et al* 1989). It has excellent sugar/acid blend. It is high yielder, but exhibits alternate bearing tendency, which affects its regular availability. It suffers from 'Spongy tissue' which reduces the yield of pulp (Iyer *et al* 1991). It has limited adaptability, best quality 'Alphonso' is mainly confined to Ratnagiri and Sinidurg districts of Maharashtra (India). It is very expensive which increases the cost of production of processing products, so being uneconomical it is not much preferred by the processing industry. The variety is considered to be the best one for processing in general and in particular for canning purposes as it yields higher pulp and for nectar and juice concentrates (Girdharilal, 1956., Jain, 1961., Nanjundaswamy *et al* 1966). Incidence of spongy tissue makes the variety unfit for slice production (Krishnamurthy *et al* 1984).

Bangalora (Totapuri) (Photo 6.4) : 'Bangalora' is a well known commercial variety of South India for high yielding potential and regular bearing habit. It is by far the most largely processed mango fruit in India, although it has medium fruit quality in terms of colour, flavour and taste. Fruits are medium to large (2-3 fruits/ kg; 200-800 g per fruit). Peel is thick and adhering, pulp yield is high (70-80%), firm, fibreless, yellow to light yellow due to low carotenoid content, mild flavour, insipid taste to moderately sweet and less acidic, juice scanty, viscosity is high due to presence of large amount of water insoluble solids (Gowda *et al* 1994., Gowda and Ramanjeneya, 1995., Gowda and

Hudder, 2004). The stone is large, papery thin, but occupies small area in the pulp. Fruit withstands transport, rough handling in transit, poor packing etc. Keeping quality is excellent. Fruit is less expensive than most of the commercial varieties. However, it is highly susceptible to stone weevil. Despite the above, it is ideally suited for pickles and chutneys, canned pulp, slices and fruit bar, but not good for nectar. The high preference of 'Bangalora' by processing industry for processing is mainly due to abundant availability, low price, high keeping quality, high pulp content and its versatile nature in processing into different products.

Banganapalli (Baneshan) (Photo 6.5): It is a leading commercial variety of Andhra Pradesh and neighbouring states of India. It is a high yielder and fairly regular. It is mostly consumed as a fresh table variety. Fruit is large (2-3 fruits weighing in a kg, often weighing 500-1000 g per fruit). Skin is thin, golden yellow in colour, pulp firm, fibreless, yellow, with pleasant flavour and sweet taste, moderately juicy, but viscosity is lower. Keeping quality is good, best suited for making pulp, for canning in the form of slices/cubes (Surya Prakash Rao *et al* 1968b) and good for nectar (Krishnamurthy *et al* 1984).

Photo 6.3: Alphonso

Photo 6.4: Banglora (Totapuri)

Photo 6.5: Banganapalli (Baneshan)

Bombay Green (Photo 6.6): It is one of the early maturing varieties of North India. It is a high yielder, but tends to be an alternate bearer and highly susceptible to mango malformation. Fruit is medium in size, thick skin and green in colour; pulp firm to soft, fibreless. Lacks colour, flavour and juice; pulp content is less (< 55%) hence not considered good for processing products like nectar.

Dashehari (Photo 6.7): This is one of the most popular and largely cultivated varieties of North India. It is well known for its excellent fruit quality. It is a high yielder, but an alternate bearer. It satisfied most of the requirements of good processing variety in general and for canning in particular, but expensive for the processing industry. Peel medium thick and peels easily and yellow in colour, pulp firm, fibreless, attractive yellow, rich in vitamin A and C (Singh and Rabbani, 1986), stone bigger (about 25%) and peel content is also high (about 20%) as a result, yield of edible pulp is less (about 56%); tastes very sweet and juice moderately abundant. However, flavour is unstable, not retained in the processing and storage. But considered best suited for canning slices (Bose and Das, 1958., Jain, 1961., Teaotia and Awasthi, 1966) and nectar (Roy *et al* 1972). It has low moisture content, which is considered as one of the reasons for its good keeping quality (Sanyal *et al* 1991).

Photo 6.6: Bombay Green

Photo 6.7: Dashehari

Photo 6.8 Himsagar

Himasagar (Photo 6.8): This is a choice variety of Bengal, a heavy and regular yielder. Fruit medium in size with good fruit and keeping qualities. Pulp firm, fibreless, yellow with delightful flavour, sugar/acid well blended and very sweet and abundantly juicy. It is considered best suited for canning (Bose and Das, 1958).

Langra (Photo 6.9): After Dashehari, this is the most important and popular commercial variety of North India. It is a high yielder, but an alternate bearer. Fruits medium in size, with thick greenish, yellow colour. Pulp firm to soft, fibreless, lemon yellow, but not intense (Rabbani and Singh, 1988), rich in vitamin C and titratable acidity, juice moderately abundant; flavour is good, with good body. It is best suited for making wine (Kulkarni *et al* 1980), but not suitable for nectar, because of high acidity and poor colour (Roy *et al* 1972). Canned products of Langra could be stored only for short period of 4 months (Siddappa and Rao, 1953., Bose and Das, 1958).

Photo 6.9: Langra

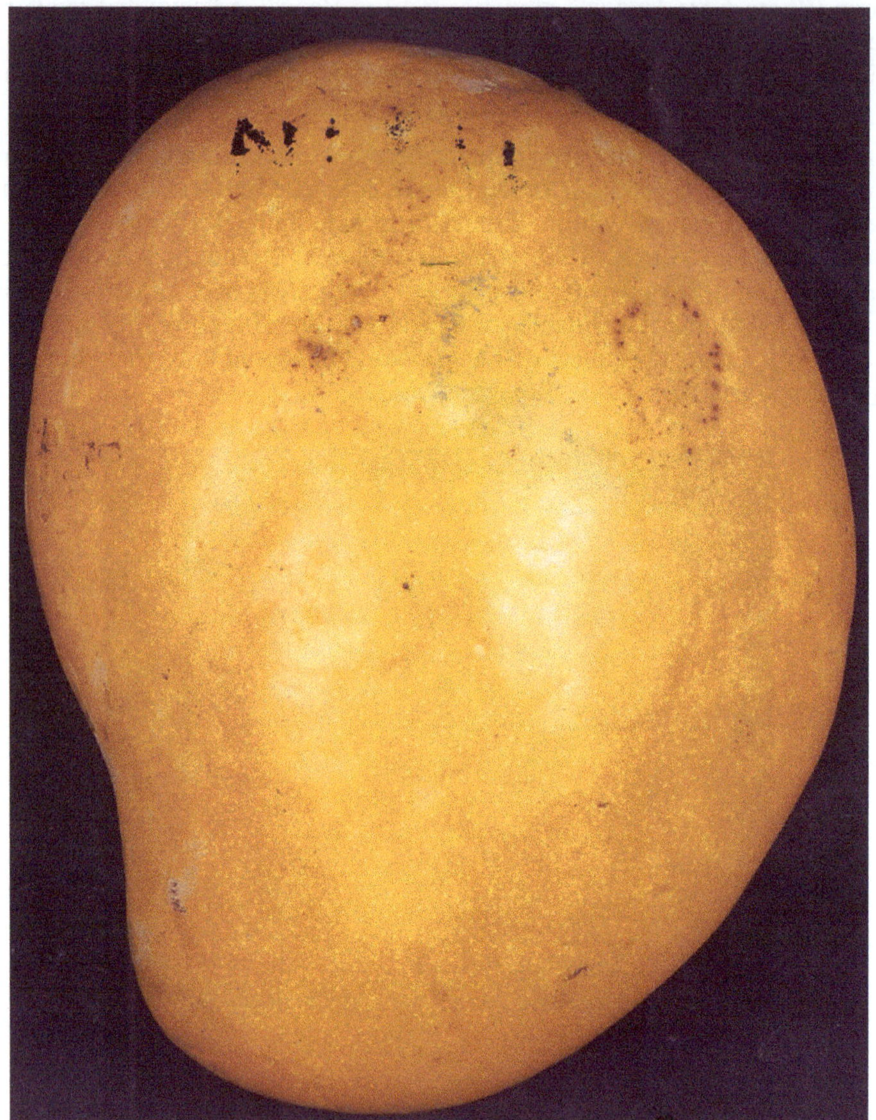

Photo 6.10: Neelum

Neelum (Photo 6.10): This is one of the important commercial variety of South India, well known for its high yielding and regular bearing habit. Fruits small to medium; peel medium thick and yellow; pulp fibreless, flavour good, light coloured, sweet, juice moderately abundant, good keeping quality. It is susceptible to stone weevil. Because of lack of colour, it is not suitable for making beverages and canned products. Pulp yield is also less (< 60%), because of thick peel and larger stone.

Padiri (Photo 6.11): This is one of the important varieties of Tamil Nadu (India). It is an early, heavy and regular yielder, hence its availability is good. Fruits medium to large with smaller stone and hence provides higher pulp percentage. Pulp has high TSS and high sugars and low acidity and good colour. It has good processing quality and scored high organoleptically. Iyer *et al* (1991), considered this variety good particularly for canned juice and can be used for creation of variation in the product.

S.B. Chausa (Photo 6.12): This is one of the best late maturing varieties of North India, but an alternate bearer. Fruits medium in size but with maximum edible pulp (> 65%). Pulp is fibrous, sweet, but lacks colour due to less content of carotenoids. It looses flavour during processing. Hence, not suitable for canning, besides being fibrous (Kapur, 1975) and for juice due to poor colour (Singh and Dhawan, 1983), however, best suited for making wine (Kulkarni *et al* 1980).

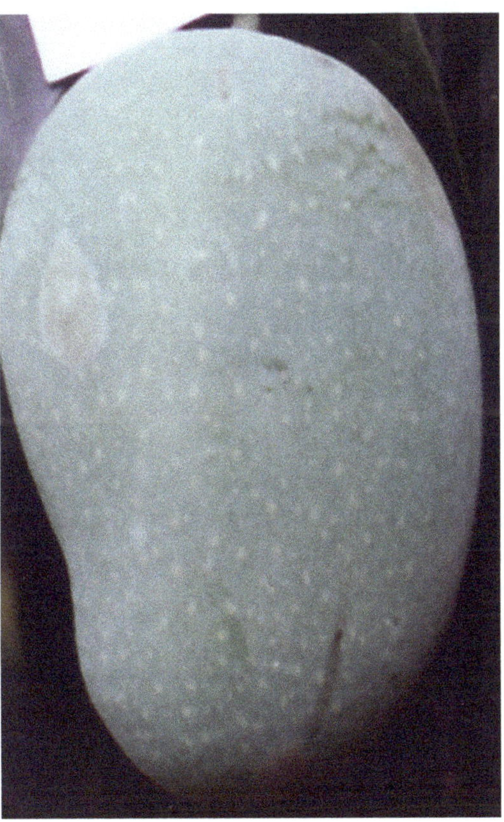

Photo 6.11: Padari (Padiri)

Photo 6.12: S.B. Chausa

HYBRIDS

Amrapali (Dashehari × Neelum) (Photo 6.13): This is a prolific and regular bearer particularly in North India. The fruits are small to medium in size. Pulp fibreless, very sweet, less acidic and deep orange red in colour due to high carotenoid content (2-3 times higher than parents) (Gowda and Hudder, 2004). Though fruits are smaller, its pulp yield is on par with that of 'Alphonso', due to thin peel and also due to the absence of spongy tissue (Khurdiya and Roy, 1988). Because of highly attractive colour 'Amrapali' is best suited for processing highly coloured nectar (Khurdiya and Roy, 1988), good for juice (Khurdia and Roy, 1988) and canning in the form of pulp (Gowda and Hudder, 2004). Being rich in carotenoids it can enhance the colour and nutrition of beverages prepared from other varieties like 'Totapuri', 'Neelum' etc., on blending.

Mallika (Neelum × Dashehari) (Photo 6.14): This is also a heavy and regular bearing hybrid. The fruits are large (300-320 g each). Peel is tough, thick and easily separable like 'Dashehari'. Pulp yield is high (76.8%) and is firm, tender and juicy with little fibre, deep yellow to orange in colour, sweet (Kalra *et al* 1994); lowest in flavour due to less number of flavour compounds and their composition; consistency of pulp is moderate (Gowda and Hudder, 2004). It is good for making nectar (Chakraborty *et al* 1991).

However, it may be preferred as a table variety rather than a processing one, because of its sauvery pulp characteristics and fibrous nature, large fruit size combined with firm flesh and good quality of fruits (Kalra *et al* 1994). 'Mallika', from the point of higher cheek yield (53%), texture and flavour is considered best for dehydration (Khurdiya and Roy, 1986b).

Totapuri* × *Alphonso: This hybrid is from Indian Institute of Horticultural Research, Bangalore. Fruit size is longest, stone is thinnest, pulp is fibrous, has more carotene content and hence more attractive. Though the hybrid had many of good points like bigger fruit size, higher carotene content, with better sensory score, because of its fibrous pulp, it is not considered for processing by the industry. Further, the hybrid is more susceptible to fruit fly (Ambadan *et al* 1988).

Totapuri* × *Mulgoa: This hybrid is from Indian Institute of Horticultural Research, Bangalore. It is suitable for canned juice and a suitable substitute for 'Totapuri'. It has mild 'Mulgoa' flavour (Ambadan *et al* 1988).

Conclusion

The demand for mango fruits and processed fruit products both for domestic and export consumption is ever increasing during the last few decades. It is hoped that the demand will increase further. Out of the 1000 or more varieties in India, hardly 20-25 varieties are under cultivation and only a few selected varieties are commercially identified for processing like 'Alphonso', for export and 'Bangalora' (Totapuri) for processing. However, due to one or another reason, they cannot be dependable entirely. Suitability of lesser known mango varieties for processing was investigated earlier and several varieties were reported suitable without reference to the yield potential and availability. With the increasing demand as well as awareness of health hazards, there is urgent need for widening the number of varieties for export as well as for processing and markets. So far identified non-commercial varieties were not brought under cultivation on large scale to meet the demands of processing industry. Therefore, an awareness in this direction among the fruit growers should be created by extension methods. Simultaneously, the commercial cultivation of these non-commercial varieties, that are suitable for processing should be encouraged through subsidies and contract farming with by – back facility to avoid total dependency on one or two varieties and thus provide constant regular supply of produce, which is a must for sustainable fruit industry.

Photo 6.13: Amrapali

Photo 6.14: Mallika

Many processing industries are now on the lookout for a few more processing suitable varieties, which would add variety to the products in the market, in addition to minimize difficulties in getting the raw materials. However, it is not enough to identify the suitable varieties for processing only, but the area under these identified varieties should be enhanced to make their availability in abundance, through the ways and means indicated above.

Preliminary studies on wine making from certain mango varieties indicated that all the varieties of mango are not suitable for wine making (Kulkarni *et al* 1980). Therefore, it was suggested to screen a still large number of varieties before attempting to produce wine on large scale commercially.

REFERENCES

Adsule, P.G. and Roy, S.K. 1974. Studies on the physico-chemical characteristics of some important commercial cultivars of North India in relation to canning and freezing of slices. Journal of Food Science and Technology, 11(6) : 269-273.

Ambadan, Adsule, P.G., Iyer, C.P.A. and Subramanyam, M.D. 1988. Suitability of some mango hybrids for processing for export. Acta Horticulturae, 231 : 776-781.

Awasthi, R.K. and Pandey, I.C. 1979. Screening of sucking mango (*Mangifera indica* L.) varieties for juice production. Indian Food Packer, 33(4) : 40-43.

Bhatnagar, H.C. and Subramanyam, H. 1973. Some aspects of Preservation, Processing and Export of mango and its products. Indian Food Packer, 27(4) : 33-52.

Bose, A.N. and Das, K. 1958. Journal of Science and Industrial Research, 17A: 202-205.

Chakraborty, V.I. and Debnath, S. 2004. Screening of Mango cultivars for processing. Indian Food Packer, 58(5) : 71-74.

Chakraborty, S., Bhist, H.C., Agarwal, M.D., Verma, L.N., Shukla, I.C., Madaiah, N., Krishnaprakash, M.S., and Nanjundaswamy, A.M. 1991. Studies on varietal screening of mangoes of Uttar Pradesh for their suitability for production of canned nectar, juice and pulp. Indian Food Packer, 45(5) : 49-56.

Chaudry, M.A., Ahmad, M. and Khan, I. 1988. Technological studies on mango squash : Cultivar effect and cost economics. Nucleus – Karachi, 25(1-2): 7-9.

Cooke, R.D., Alliso, C.J., Baldry, J., Caygill, J.C., Ferber, C.E.M and Lalitha, K. 1976. Deep Freezing of mango slices. Journal of Food Technology, 11 : 475-484.

Finera, R.M., Casah, C., Batista, A.R. and Cerezal, P. 1995. Evaluation of mango cultivars for pulp production. Alimentaria, 32(260) : 29-31.

Girdharilal. 1956. Proceedings of symposium on Fruit and Vegetable preservation industry in India, pp. 67-74.

Girdharilal., Krishnamurthy, G.V., Jain, N.L. and Bhatia, B.S. 1960. Suitability of different varieties of mangoes for the preparation of mango cereal flakes. Food Science, 21-123.

Gowda, I.N.D. and Hudder, A.G. 2004. Investigation on Processing Quality of Some Mango Varieties, Hybrids and Their Blends. Journal of Food Science and Technology, 41(2) : 154-159.

Gowda, I.N.D. and Ramanjeneya, K.H. 1995. Evaluation of some mango varieties for their suitability of canned mango juice. Journal of Food Science and Technology, 32(4) : 323-325.

Gowda, I.N.D., Ramanjaneya, K.H., Iyer, C.P.A., Subramanyam, M.D. and Dinesh, M.R. 1994. Physico-chemical and processing quality of four new mango hybrids in comparison to two commercial cultivars. Journal of Food Science and Technology,31 : 385-388.

Iyer, C.P.A., Subramanyam, M.D. and Dinesh, M.R. 1991. 'Padiri' A promising mango variety for processing. Indian Food Packer, 45(1) : 13-15.

Jain, M.L. 1961. Chemistry and Technology of mango. Review in Food Technology, 3 : 131-162. Published by Association of Food Science and Technology, Mysore, India.

Jain, V., Tiwary, B.L. and Sharma, H.G. 1997. Evaluation of Early maturing mango varieties for the preparation of Beverages as Nectar and RTS. Orissa Journal of Horticulture, 25(1) : 40-42.

Jain, V., Tiwary, B.L., Sharma, H.G. and Saxena, R.R. 1996. Evaluation of late maturing mango varieties for the preparation of beverages as nectar and RTS. Indian Food Packer, 50(3) : 9-11.

Kalra, S.K., Tandon, D.K., Singh, H. and Chadha, K.L. 1982. Assessment of some mango cultivars for nectar preparation. Progressive Horticulture, 14(4) : 220-224.

Kalra, S.K., Yadav, I.S. and Sinha, G.C. 1994. Screening of Some North Indian Non-commercial Mango varieties for processing. Indian Journal of Horticulture, 5(2) : 136-140.

Kapur, K.L. 1975. Varietal suitability of mangoes from Uttar Pradesh for canning. Indian Food Packer, 29(3) : 33-35.

Khurdiya, D.S. and Roy, S.K. 1986a. Studies on ripening and canning of mango. Indian Food Packer, 40 : 15-18.

Khurdiya, D.S. and Roy, S.K. 1986b. Studies on canning of mango slices in covering syrup containing mango pulp. Indian Food Packer, 40(2) : 50-54.

Khurdiya, D.S. and Roy, S.K. 1988. Studies on processing quality of mango hybrids developed at IARI. Acta Horticulturae, 231 : 709-714.

Krishnamurthy, G.V., Giridhar, N. and Raghuramaiah, B. 1984. Suitability of some mango varieties for processing. Journal of Food Science and Technology, 21 : 21-25.

Kulkarni, J.H., Singh, H. and Chadha, K.L. 1980. Preliminary Screening of mango varieties for wine making. Journal of Food Science and Technology, 17(4) : 218-220.

Mehta, G.L., Tomar, M.C., Agnihotri, B.N. and Tripathi, S.R. 1971. Canning of whole mangoes. Progressive Horticulture, 3(3) : 49-56.

Nanjundaswamy, A.M.; Rahim, A., and Siddappa, G.S. 1966. Proximate composition and suitability for preservation of some important varieties of mangoes grown in Salem. Indian Food Packer, 20(3) : 15-21.

Palaniswamy, K.P., Muthukrishnan, C.R. and Shanmugavelu, K.G. 1973. Studies on the evaluation of certain mango varieties of Tamilnadu for canning. Indian Food Packer, 27(3) : 9-14.

Palaniswamy, K.P., Muthukrishnan, C.R. and Shanmugavelu, K.C. 1974. Studies on the evaluation of certain mango varieties of Tamilnadu for pulp and squash. Indian Food Packer, 28(2) : 5-9.

Rabbani, A. and Singh, I.S. 1988. Evaluation of local sucking mango varieties for beverage industry. Acta Horticulturae, 231 : 715-720.

Roy, S.K., Singh, R.N. and Singh, R. 1972. Studies on evaluation of some of mango varieties of North India for Processing as Nectars. Indian Food Packer, 26(5) : 5-8.

Saha, N.K., Agarwal, P.C. and Rodriguez, R. 1976. Effect of ripeness level, storage period, processing conditions and ascorbic acid on flavour retention in canned Desehri mango. Indian Food Packer, 30(5) : 87-91.

Sanyal, D., Roychaudhury, R., Dhua, R.S. and Mitra, S.K. 1991. Study of some mango cultivars of West Bengal. Indian Food Packer, 45(2) : 29-33.

Satyavathi, V.K., Bhat, A.V., Varkey, A.G. and Mookerji, K.K. 1972. Studies on the suitability of different varieties of mango of Kerala for processing. Indian Food Packer, 26(4) : 8-12.

Selveraj, Y., Kumar, R. and Pal, D.K. 1989. Changes in sugars, organic acids, aminoacids, lipid constituents and aroma characteristics of ripening mango (*Mangifera indica* L.) fruit. Journal of Food Science and Technology, 26 : 308-331.

Sharma, H.G., Jain, V. and Tiwary, B.L. 1995. Performance of polyembryonic and monoembryonic mango cultivars for processing as nectar. Orissa Journal of Horticulture, 23(1&2) : 65-69.

Shetty, M.S., Malathi, H.N. Nanjundaswamy, A.M. and Potty, V.H. 1978. Mango – An industrial profile. Central Food Technological Research Institute, Mysore (India).

Siddappa, G.S. and Bhatia, B.S. 1956. Food Research, 21: 163-165.

Siddappa, G.S. and Rao, G.R. 1953. Tin content of some important Indian canned mangoes and other fruits. Indian Journal of Horticulture, 10 : 49.

Singh, I.S. and Dhawan, S.S. 1983. Potentiality of various fruits for processing industry. Indian Food Packer, 37(3) : 47-55.

Singh, I.S. and Rabbani, A. 1986. Evaluation of mango (*Mangifera indica* L.) cultivars for beverage processing. Narendra Dev Journal of Agricultural Research, 1(1) : 56-60.

Srivastava, J.S. 1998. Comparative study of RTS drinks prepared from Deshehari and Banganapalli mangoes. Indian Food Packer, 52(2) : 38-40.

SuryaPrakash Rao, P.V., Chalapati, G., Bhaskar Rao, K. and Sudarsana Rao, K. 1968a. Evaluation of some mango hybrids for canning. Andhra Agricultural Journal, 15(1) : 10-17.

SuryaPrakash Rao, P.V., Prasad, P.S.R.K., Giridhar, N. and Nageshwar Rao, G. 1968b. Evaluation of mango varieties of Andhra Pradesh for processing as slices, nectar and juice. Indian Food Packer, 22 : 18-22.

Teaotia, S.S. and Awasthi, R.K. 1966. Studies on varietal suitability on mango fruit by canning. Indian Food Packer, 20(3) : 10-14.

Tidke, G.A., Desai, U.T., Kshirsagar, D.B., Garad, B.V. and Shirsath, H.K. 2006. Assessment of Local Mango (*Mangifera indica* L.) seedling types for pickling characters. Orissa Journal of Horticulture, 34(1) : 13-21.

Vasquez, Caicedo, Al., Neidhart, S. and Carle, R. 2004. Post-harvest ripening behaviour of Nine Thai mango cultivars and their suitability for industrial applications. Acta Horticulturae, 645 : 617-625.

Yadav, I.S., Sinha, G.C., Rajan, S. and Kalra, S.K. 1987. 'Saheb Pasand' a potential mango cultivar. Progressive Horticulture, 19(3/4) : 316-317.

CHAPTER 7

Microbiology of Processed Products

B. Neeraja Prabhakar

One of the causes of spoilage of processed products is invasion of microbes. Because of high acidity, high sugar content, and low pH in mango fruits and their products, they get spoiled readily due to microbes. Microorganisms on the surfaces of freshly harvested fruits include not only those of the normal surface flora, but also those from soil and water. If the surfaces are moist or damaged, growth of some microorganisms occurs between harvesting and processing or consumption. Fungi, yeast and lactic acid bacteria are the major spoilage microorganisms (Table 7.1). Among these, yeasts are fast growing and cause fermentation spoilage quickly (Ethiraj and Suresh, 1988). The nature and type of microorganisms that cause spoilage depend upon the chemical composition, preventive techniques and sanitation observed during the pre-and post-harvest handling, processing, packing and storage conditions. In case of juices and juice concentrates non-spore forming microbes play a dominant role. Improper harvesting and improper post-harvest handling result in the injury to the surface layer, making the fruits more prone to attack of spoilage of microorganisms. During transportation, mechanical damage may increase the susceptibility of fruits to decay and growth of microbes. Utensils and gunny bags used in the processing industry also contribute to high microbial incidence and loads. Increase in microbial load in raw mango slices during storage was due to uncleaned barrels. Contamination of fruits from the field and their nonremoval before processing, under-processing, inadequate quantities of chemical preservatives are some of the causes of spoilage of mango processed products in storage. Unhygienic conditions in the industry and storage places, lack of proper packaging also lead to spoilage of processed products. Microbial spoilage of different processed products of mango are detailed below:

Pickles

Pickles made out of mango, lime *etc;* are traditional food items, in which vinegar, salt, or oil or both are added for preservation. However, inspite of great care taken, in the preparation of pickles, they often get spoiled. The spoilage of pickles includes, alteration in flavor, colour, and texture of fruit pieces such as softening, blackening, slime production, ripeness, mashy texture, unpleasant odour, bubbling etc. In pickles, the colour changes from yellow to black, texture from soft to slimy to lumpy and mashy and unpleasant sour earthy or oily odour develops during spoilage.

Microorganisms are mainly responsible for spoilage of pickles. Microbial spoilage of pickles is attributed to inadequate concentration of added salt. Pradhan *et al* (1985a) found that commercial pickle samples in Pune city (India) had a salt content of less than 18 per cent, which resulted in spoilage. Using artificial inoculation of salt tolerant strains of *Aspergillus niger,* Kanekar *et al* (1989) concluded that 15 per cent salt concentration as optimum to prevent fungal spoilage in the unfermented mango pickle. Oil is another ingredient which is used in the preparation of pickles. Oil covers the surface of the pickles as a layer and prevents exposure of the pickle to atmosphere, there by prevents spoilage by aerobic microorganisms. Inadequate addition of oil does not prevent the incidence of microorganisms, that results in spoilage. The spoiled pickles had maximum oil content (35%) which indicates that oil alone cannot inhibit microbial spoilage. The degradation of edible oil, may be another important physiological action of microorganisms causing spoilage in the form of rancidity development in pickles. Ground nut oil has no preventive role against microorganisms, but provides physical barrier against microbial infection. Fungal spoilage is not arrested by the native acidity of the pickles (4.8%), since *A. niger* tolerates 12 per cent citric acid (Pradhan *et al* 1985a) and *Penicilium sp* and *Candida utilio* are tolerant to 10 per cent citric acid (Pradhan *et al* 1985b). Tolerance to acidity seems to be another important property of microorganisms to cause spoilage.

Several number of microorganisms were isolated from spoiled pickles. Pradhan *et al* (1985a) isolated 48 microorganism from the spoiled pickles of Pune city, of which 27 were bacteria, 16 were yeasts and 5 were fungi. Among these, *Bacillus sublitis, B.brevis,* and *Micrococcus lutens* were the most predominant bacteria, while *A. niger*, and *Penicillium* sp. were the most predominant moulds and the predominant yeast was *Saccharomyces sp* followed by *Hansenula sp, Candida sp* and *Torulopsis.* Thus, the spoilage in commercial pickles involved bacteria followed by moulds and yeasts. Even though bacteria were isolated from spoiled pickles, surface growth of moulds made the products unacceptable. Pradhan *et al* (1985b) isolated salt acid and oil tolerant microorganisms from spoiled pickles.

Amchur

Spoilage of amchur seems to be due to little care bestowed for their hygienic preparation, as it's preparation is concentrated in small scale sector as cottage industry, manual peeling, slicing of the fruit and exposure of slices to dust, insects *etc* (Teotia *et al* 1987., Teotia and Pruthi, 1987). Usha and Anand (1981) observed three strains of *Aspergillus* in market samples of amchur *viz., A. niger var. tieshen, A. candidus and A. flavoum* and presumed that these moulds were either derived from the raw material and persisted through the process of dehydration or gained access to the product during storage.

Mango Pulp

Any contamination during the preparation of pulp may persist and lead to the spoilage of the subsequent products made from the pulp. The sources of microbial contamination may be fruit surface, water, additives, processing equipment, packing, containers, storage environment etc (Garg and Kalra, 2002).

Raw pulp contains high microbial load, which could be due to composite raw material, good and partially damaged fruits and to containers from processing equipment and personnel (Ethiraj and Suresh, 1988). Spoilage of mango pulp due to *Bacillus megaterium* and *B.cereus* was reported (Sridhar and Shankhpal, 1986). Bacteria out numbered yeasts in both unwashed and washed fruits. Flesh from both washed and unwashed fruits contains yeasts, of which *Kloeakera sp* , *Hyphopichia sp* and *Candida* sp were the major ones. Species of *Candida, Kloeakera* and *Khyveromyces* were the predominant yeasts in the unheated raw mango pulp whereas heated pulp did not show the presence of any yeast (Ethiraj and Suresh, 1988).

Heating of mango pulp having a pH less than 4.0 to 95 °C would destroy the spoilage organisms (Ramana *et al* 1984). The higher load of microbes (bacteria, yeast and mould) can be reduced by sterilization of pulp at 75 °C, which could further be reduced to negligible limits during freezing storage (Kalra *and* Tandon, 1985a).

Mango Jam

Spoilage in mango jam occurs in the form of surface microbial growth and fermented odour. In a survey of mango jam samples, the highest incidence of *Micrococcus lutens* was observed followed by *Bacillus coagulam, Aspergillus niger, Penicillium sp.,B. laterosporus* and *Staphylococcus saprophyticus* which were sugar and acid tolerant (Kanekar *et al* 1992).

Mango Leather

This is a popular mango confectionery product prepared by drying the pulp in open sun with or without sugar, mostly by small/cottage scale industries. Garg *et al* (1993) assessed the microbial quality of mango leather from markets. Besides bacteria, insect larvae and adult insects were detected in the samples. The poor quality of commercial leather samples might be due to unhygienic conditions and lack of other quality controls during processing, drying and storage of leather.

Raw mango slices in brine

Unripe mango slices are usually preserved with dry salt or in brine for conversion into pickles or chutney. Johar and Anand (1951) noticed yeast and bacteria in 10 per cent brined mango slices sold in the market. Transportation of cured slices in gunny bags increases the microbial load by 12-15 times. Further, increase in microbial load by eight times was observed when the slices were stored in wooden barrels (Maitra *et al* 1979). Raw dry mango slices and mango powder from the market were contaminated with different fungi, during harvesting, drying and storage (Manjari *et al* 2000). *Aspergillus, Penicillium, Rhizopus, Fusarium* and *Sclerotia* fungal genera were isolated from raw dried mango slices.

Ripe mango slices in syrup

Garg *et al* (1995a) studied the microbial load of stored ripe mango slices preserved in 20° or 40° Brix syrup with or without calcium and initially with no microbial count. However, after 6 months of storage the spoilage was higher in slices preserved in sugar syrup of 20° Brix than that of 40° Brix. *Bacillus sp* was pre-dominant bacterium followed by *Thermoactinomyces sp, Asperigillus niger, A. restrictus, A. pencilloides, Rhizopus sp, Eurotium sp, Saccharomyces bailli* and *S.bisporus.*

Beverages

Kalra *et al* (1991) observed the presence of live yeast, mould and bacteria in different samples of mango beverages collected from different packaging houses. Suresh *et al* (1982) isolated 15 morphologically different groups of yeasts from fermented mango juice, while WarnaSuriya *et al* (1985) isolated 36 strains of yeasts belonging to six genera. Heat resistant fungi, *viz., Pacilomyces variotic, A. tamari; A. flavus* and *A. ochraceus* were found in 27 per cent mango juices, the last being pre-dominant (Obita and Ogwuanyi, 1995). The presence of bacterium *viz., Leuconostoc oenos* and some yeasts in spoiled canned mango juice was reported (Ethiraj and Suresh, 1985). The isolation of yeasts along with bacterium suggested that spoilage could have been due to either under-processing or the ingress of contaminated cooling water through the cans. Sometimes microbial growth may occur if significant delay takes place in between heating and filling of containers. Yeasts like *Kloeckara, Pichia, Candida*, and *Hypopichi sp* are known to be associated with fresh mango juice (Suresh *et al* 1982).

Prevention/Control of Microbial Spoilage

Presence of certain inhibitory substances like phenols and natural defence mechanisms of fruits like pH and phyto-toxins of the product are helpful in warding off the microbial incidence in processed products (Sharma and Desai, 1978), but still spoilage occurs. Hence, as in other cases, prevention is better than cure in this case also. Prevention/control of microbial spoilage in mango processed products primarily depends upon good manufacturing practices and one of the most effective ways is the application of "Hazard Analysis of Critical Control of Points" (HACCP). It requires establishment of hazard, identification of critical point, effective monitoring follow up and evaluation (Underwood, 1995). For a food process, it is necessary to have the knowledge of control of the microbial quality of the mango fruits, its processing environment and the packaging components. This also requires the validation of all the processes stages designed to destroy both the pathogens and the spoilage organisms and the efficacy of the preventive systems (Gould, 1995). In developed countries, microbial quality monitoring is an essential practice to be observed by all the processors. In India, this practice is however, not seriously followed. It is time that Indian Food Processing Industry must immediately adopt the concept of clean room processing. The advantages of such practice shall be of long-term perspectives both for domestic as well as international trade. Therefore, maintenance of clean room environment, good management and personal training are necessary and essential for prevention of spoilage of processed products.

Sharma and Desai (1978) described different methods of preventing/controlling microbes in processed products of fruits and vegetables as follows:

Decontamination

This mainly consists of washing of fruits and vegetables using chlorinated water or borax solution. Detergents may be added to facilitate the removal of dirt and microbes. Mould growth can be removed using non-ionic detergent solution. Mere washing of the fruits and vegetables in running water also reduces the surface micro-flora considerably. The pesticide residues on mango fruits were reduced to 66-68 percent in case of dimethioate and fenthion as against 21-27 percent in case of fenvalerate and cypermethrin simply by washing treatment i.e., dipping in water for 10 minutes. The peeling of the fruit pericarp (skin) dislodged pesticide residues to an extent of 100 per cent in all the cases of pesticides (Awasthi, 1993).

Coating techniques

The shelf life of fresh fruits and vegetables can be extended significantly by coating their surfaces with mineral oil, waxol or talprolong or frutox and such other materials, which obstruct the entry of microbes. Often anti-microbial agents such as fungicides like benlate or benomyl or bavistin are incorporated in the coating materials for more effective control of microbial attack. Commonly bananas and citrus fruits are preserved by such treatment.

Microbistatic Techniques

Chilling

Fruits and vegetables can be preserved by cooling promptly after receipt from the field and keeping at low temperatures (chilling). The chilling can be done using cold water, ice, cool-air or mechanical refrigeration or by vacuum cooling. However, certain commodities have an optimal storage temperature well above freezing point, below which these commodities suffer from chilling injuries. In chilling storage due consideration for relative humidity of the atmosphere should be given. Too low relative humidity results in desiccation and very high relative humidity favours the growth of spoilage organisms.

High humidity near to saturation is required for bacterial growth; yeasts require 90-92 per cent relative humidity and moulds grow at 85-90 per cent relative humidity.

Freezing

Fruit pulps, juices *etc.* are more commonly preserved by freezing. Microorganisms associated with frozen products remain viable and multiply during thawing. Under high frozen condition, the growth of microbes is prevented almost entirely, but not killed. Scalding or parboiling or blanching prior to freezing invariably lowers the microbial load. Quick freezing is always preferred as slow freezing results in large ice-crystals which damage the cells. Quick freezing is a method of freezing a commodity within 30 minutes and can be accomplished by:

(i) Direct immersion of a commodity in a refrigerant,

(ii) immersion of commodity package in refrigerant at –17 °C to – 45 °C,

(iii) frigid air blast at –17 °C to –34 °C,

(iv) contact plate freezing.

Drying

Drying can be accomplished by sun-drying, mechanical drying, or freeze drying. Drying by any of these methods reduces the water activity in fruits and vegetables to a level where microbes cannot multiply. However, the organisms may not be necessarily killed during the drying process. Pulps and juices that are dehydrated by spray, drum or freeze drying are first subjected to pasteurization to eliminate pathogens. However, depending on the condition and rate of drying especially sun drying, some proliferation of micro-organisms may occur. Yeasts are often associated with several dry fruits (dates, figs, raisins etc).

Preservatives

These are chemicals which when added to food tend to prevent or retard deterioration thereof, but do not include common salt, sugar, vinegars, spices or oil from spices. Organic acids and their salts, benzoic acid, propionic acid etc., may be added or developed in processed products to prevent them from deterioration and spoilage.

Microbicidal Techniques

Heat

Heat kills microbes by way of coagulating proteins and inactivation of enzymes. The heat treatment, necessary to kill organisms varies with the kind of microorganisms, its state, environment and pH of the product. Thermal process schedule for low-acid products ensures destruction of *Clostridium botulinum,* capable of producing exotoxin, the consumption of which is fatal. Thermal process schedule, evolved on the basis of the inactivation of heat resistant enzyme systems such as peroxidase and pectin esterase which cause undesirable changes, renders the canned products microbiologically safe (Nath and Ranganna, 1977). The processing industry is making use of the thermal processes evolved by Siddalinga *et al* (1983) for the canning of the fruit pulps including mango pulp and this should continue in case of products having pH of 4.0 or lower. However, the products having pH higher than 4.0 but lower than 4.6 should be given thermal process adequately to destroy *Bacillus licheniformis* with the exception where *B. coagulant* spoilage is likely to occur (Azizi and Ranganna, 1993).

Blanching

Scalding or parboiling or blanching briefly by hot water or steam before processing serves to wash the commodities, sets their colour, softens the tissues to aid packaging, helps de-aerate and kills some microorganisms.

Pasteurization

This is a type of treatment of fruits and vegetables and food material adopted to kill the pathogens and some non-spore formers by applying heat in a flash process (72 °C per 5 seconds) or by a holder process (63 °C per 30 minutes). Thermoduric cells and spores, however, survive in this treatment.

Sterilization

This is used to make fruit and vegetable products completely free from microbes and their spores. The temperature required is as high as 100-120 °C for few to several minutes. Actual time and temperature combination need to be worked out for particular product depending upon its nature and load and nature of microbes involved.

Irradiation

Highly penetrating radiation like Y-rays from cobalt or cesium sources or accelerated electrons effectively kills the microbes in foods. However, high doses of radiation tend to harm the quality attributes of the food besides killing the microbes. Still, gamma irradiation can be used to preserve fruits and vegetables by combination with heat. A dose of 65 Krad and heat treatment of 60 °C/30 minutes can preserve dry fruits like cashew nuts and raisins effectively. Heat and radiation combination treatments preserve mangoes and bananas effectively. Against microbes, micro-wave radiations are now being used for quick heating of foods and killing of microbes thereby.

Prevention/Control of microbial spoilage of mango processed products

Pickles

Microbial spoilage in mango pickles can be effectively checked by addition of salt and spices. Salt concentration of over 20 per cent is recommended to prevent microbial spoilage (Prasad and Joshi, 1929), though 15 per cent salt can serve the purpose (Kanekar *et al* 1989). Microorganisms tolerate salt up to 20 per cent. Maximum tolerance to salt at higher levels was exhibited by A. *niger,* A.*fumigatus* and *Pencillium sp* (Pradhan *et al* 1985b). Black mustard powder at 1 per cent level of the mango slices was very effective in checking mould growth. A preventive emulsion prepared from acetic acid (5 g), brown mustard powder (16 g) or orange peel oil (0.2 g), turmeric powder (2 g), gum- acacia (8 g) and water (100 g) is very efficacious at low salt and/or acid level in checking microbial spoilage without altering the colour, flavour and taste of the product (Subba Rao *et al* 1963). Since heat sterilization effects the texture of pickles, heat cannot be used for prevention of spoilage in mango pickles (Hussain, 2004).

Pulp

Washing and peeling of the fruits reduced the microbial load from the fruit surfaces, while homogenization of pulp increased it (Garg *et al* 1995 c). The microbial population decreased markedly after thermal processing of pulp at 78 °C, however, a tailing population of yeasts, moulds and bacteria still survived, which are known to be heat resistant. The spores of *Bacillus sp.* in thermal preserved mango pulp could survive 5 hours of heating at 100 °C and 110 °C. Thermal death time studies indicated that heating at 110 °C for 15 minutes destroyed all spoilage microorganisms, but processing to such a high temperature may adversely affect the quality of the product (Garg *et al* 1995b). However, Ethiraj and Suresh (1988) claimed that heating the mango pulp to 88-90 °C in Steel-Jacketted Kettle completely eliminated all the microorganisms in freshly extracted pulp.

All the yeasts isolated from 'Totapuri' during processing failed to survive against the heat treatment given at 60 °C for 20 minutes except, *Pichia membranaefaciens* whereas *Saccharomycodes ludwigii* which showed slightly higher degree of tolerance to preservatives was checked at 60 °C for 5 minutes. Therefore, mild heat treatment and use of preservatives at lower level would be helpful in checking the growth of yeasts in mango pulp during storage.

Chemical preservatives are also helpful in checking the microbial spoilage. However some organisms may be resistant to chemical preservatives and spoil the product. *Pichia membranae faciens* is a benzoate resistant yeast isolated from spoiled 'Totapuri' mango pulp, but it showed less tolerance to KMS and potassium sorbate (Ethiraj and Suresh, 1988). Sodium benzoate (500 ppm) inhibited all the yeasts isolated from 'Totapuri' mango pulp during processing, except *Saccharomycodes ludwigii* while, potassium sorbate and KMS (both at 500 ppm) inhibited all the yeasts. Combination of SO_2 and powdered salt treatment preserves the mango slices for over three months without any spoilage (Kalra and Tandon, 1983) (Table 7.1).

Table 7.1 Pre-dominant microflora identified in the market drinks (Garg and Kalra, 2002)

Bacteria	**Yeast**	**Moulds**
Acetobactor sp	*Candida sp*	*Aspergillus sp*
Bacillus sp	*Saccharomyces cerevisiae*	*Botrytrichum sp*
Coryneb bacterium sp		*Paccilomyces sp*
Macro coccous sp		*Lasquicellum sp*
Serratia sp		*Spheridium sp*
Streptomyces sp		*Neurospora sp*
		Pencillium sp

REFERENCES

Awasthi, M.D. 1993 Decontamination of Insecticide residues on Mango by washing and peeling. Journal of Food Science and Technology, 30 (2): 132-133.

Azizi, A. and Ranganna, S. 1993. Spoilage organisms of Canned Acidified Mango pulp and Their Relevance to Thermal processing of Acid foods. Journal of Food Science and Technology, 30(4): 241-245.

Ethiraj, S. and Suresh E.R. 1985. A note on the occurrence of *Leuconostoc oenos* as a spoiling organism in canned mango juices. Journal of Applied Bacteriology, 53(3): 239-242.

Ethiraj, S. and Suresh E.R. 1988. Studies on microorganisms associated during processing of mango. Acta Horticulturae,. 231:731-735.

Garg, N. and Kalra, S.K. 2002. Microbiology of Processed Mango Products. Processed Food Industry, 5(6): 15-18.

Garg, N. and Kalra, S.K. and Tandon, D.K. 1993. Quality Evaluation of market raw mango powder and mango leather. Beverage and Food world, 20(3):13-14.

Garg, N., Kalra, S.K. and Tandon,D.K. 1995c. Microbial load of stored mango slices. In: Phal Samskarana. (eds.). G.S.V. Raghavan., V.Orsat., H.S. Ramaswamy; T.N. Tulsidas and S.Subramaniyan. Olympiad Press, Bangalore,pp.99-101.

Garg, N., Tandon,D.K and Kalra, S.K. 1995b. Determination of microbial load during various stages of mango processing for pulp. Beverage and Food world, 22(3): 14-15.

Garg, N., Tandon, D.K., Kalra, S.K. and Splittstoesser, D.F. 1995c. Heat resistance of a new isolate of *Bacillus sp.* from heat preserved mango pulp. Paper presented at 36th Annual conference of Association of Microbiologists of India, held at Hissar, Nov. 8-10. 1995.

Gould, G.N. 1995. Biodeterioration of foods and an overview of preservation in food and dairy industries. International Biodeterioration and Biodegradation, 267-277.

Hussain, S.Z. 2004. Raw mango products - An overview. Land Bank Journal, 43(1): 11-23.

Johar, D.S. and Anand, J.C. 1951. Note on the Microbiology of Brined Mango Slices. Indian Journal of Horticulture, 8(4): 45-8.

Kalra, S.K. and Tandon, D.K. 1983. Salt treated raw mango slices in polyethylene pouches. Indian Food Packer, 37(6): 101-107.

Kalra, S.K. and Tandon, D.K. 1985 Preservation of mango pulp by freezing. Indian Food Packer, 39(5): 23-27.

Kalra, S.K ., Tandon, D.K., Garg. N. and Singh, B.P. 1991. Quality evaluation of some market fruit drinks. Indian Food Packer, 44 (1) : 48-52.

Kanekar, P., Sarnaik, S, and Joshi, N. 1992. Sugar and acid tolerant microorganisms causing spoilage in mango jam. Journal of Food Science and Technology, 29(5): 278-280.

Kanekar, P., Sarnaik S., Joshi, N. Pradhan, L. and Godbole, S.H. 1989. Role of salt, oil and native acidity in the preservation of mango pickle against microbial spoilage. Journal of Food Science and Technology, 26(1): 1-3.

Maitra, R.K., Choudhary, D.R. and Bose, A.N. 1979. General survey on the quality of cured mango slices. Indian Food Packer, 33(6): 52-57.

Manjari, K., Kohli, U., Raghuvamshi, RS., Chauhan, G.S. and Dwivedi, T.S. 2000. Quality evaluation of Raw Dried Mango slices, Khotai and Mango powder (Amchur) sold in the Retail Market. Journal of Food Science and Technology, 37(5): 449-454.

Nath, N. and Ranganna, S. 1977. Determination of thermal process of Canned Mandarin orange segments. Journal of Food Science and Technology, 14: 113-119.

Obita, J.A.N. and Ogwuanyi, J.O. 1995. Heat resistant fungi in Nigerian heat processed fruit juices. International Journal of Food Science and Technology, 30(5): 587-590.

Pradhan, L., Kanekar, P. and Godbole, S.H. 1985a. Microorganisms Isolated from Spoiled Mango Pickles. Journal of Food Science and Technology, 22(1): 17-20.

Pradhan, L., Kanekar, P. and Godbole, S.H. 1985b. Microbiology of Spoiled Mango Pickles. Tolerance to Salt, Acidity and Soil of the Microbes Isolated from Spoiled Mango pickles. Journal of Food Science and Technology, 22(1): 339-341.

Prasad, H.H. and Joshi, N.V. 1929. The preservative value of spices used in pickling raw fruits in India. Agriculture Journal of India, 24:202.

Ramana, K.V.K., Ramaswamy, H.S., Prasad B.A., Patwardhan, M.V. and Ranganna, S. 1984. Freezing preservation of Totapuri mango pulp. Journal of Food Science and Technology, 21. 282-285.

Sharma, A. and Desai, S.R.P. 1978. Microbiology of Fruits and Vegetables. Indian Food Packer, 32(5): 31-39.

Siddalinga, S.B., Padwal, R.A, and Rangnna, S. 1983. Determination of thermal process schedule for canned mango, papaya and guava pulp. Acta Alimentaria, 14(4): 331-342.

Sridhar, P. and Shankapal, K.V. 1986. Bacterial spoilage of canned mango pulp and green garden peas. Indian Journal of Microbiology, 26(1/2): 39-42.

Subba Rao, M.S., Soumitri, T.C., Johar, D.S. and Subramanyam, V. 1963. Studies on brine Pickle. Part-III. Preservative emulsion for pickles. Food Science, 12: 381.

Suresh, E.R., Onkarayya, H and Ethiraj,S. 1982. A note on the yeast flora associated with fermentation of mango. Journal of Applied Bacteriology, 52: 1-4.

Teotia, M.S., Manan, J.K and Saxena, A.K. 1987. Green Mango Processing – A Review. Indian Food Packer, 41(6): 75-86.

Teotia, M.S. and Pruthi, J.S. 1987. Techno-economic Aspects of 'Amchur' Manufacture. Indian Food Packer, 41(6): 26-30.

Underwood, E.1995. Good Manufacturing Practices- A means of controlling Biodeterioration. International Biodeterioration and Biodegradation, 449-457.

Usha, K. and Anand, J.C. 1981. Quality aspects of commercial raw mango powder (Amchur). Journal of Food Science and Technology, 18(5): 208-210.

WarnaSuriya, D., Liyanage, A.W., Weerawansa, G.G., Atuada, P.K. and Jayatissa, 1985. Isolation and characterization of yeasts of some fruits and fruit products of Sri Lanka. Journal of National Science Council of Sri Lanka, 13(1): 71-75.

CHAPTER 8

Storage of Processed Products

M. Madhavi

Mango (*Mangifera indica* L.) fruits can be preserved and processed in to different kinds of products for value addition. However, the processed/preserved products are not only for instant/current use but also for longer periods of usage, particularly during the lean period of the crop and in places, where fresh fruits are not available, until the products are exhausted or spoiled and unfit for consumption. The products are invaded by different microorganisms like yeasts, moulds and bacteria leading to their spoilage and shortened shelf life and decline in nutritive value and consumer acceptability. On long storage certain biochemical changes also occur to the disadvantage. Therefore, after preparation, the products need to be stored/preserved suitably to avoid spoilage, extend their shelf life and retain their nutritive value and acceptability. Different types of packages and methods of storage are employed for storage of processed products, such as, thermal preservation, sulphiting, addition of preservatives etc. Packaging and storage go hand in hand. Storage of processed products is influenced by the kind of package material used besides others. During storage, the stored products exhibit certain changes in biochemical composition and also some problems.

Amchur

`Amchur' preparation is confined to the mango season only (March-May), hence it should be preserved to make it available during the rest of the year. 'Amchur' is prepared by different methods and the method of its preparation influences its storage. Raw mango powder from unblanched, sulphured and sulphited, sun-dried and cabinet dried slices are stored in different packaging materials and at room temperature (27 $\pm$ 5 °C) for 6 months. Cabinet dried, sulphured and sulphited products have better retention of ascorbic acid, and less browning as compared to sun-dried and unblanched lots (Dhabade and Khedkar, 1980). `Amchur' could be stored for 6 months in friction-top-tins at room temperature (12-25 °C). There were no significant changes in quality, except slight decrease in acidity after 6 months storage. However, the product showed slight darkening in polyethylene bags and glass jars within 3 months of storage (Pruthi and Teotia, 1987).

'Amchur' being a dry product, its storage depends upon its moisture content as well as on relative humidity of storage atmosphere. Mango powder with a moisture content of less than 1% was acceptable even after one year's stage at 37 °C (Ammu *et al* 1976). Products having more than 2% moisture were stable and acceptable under ambient condition for one year, but those stored at 37 °C, got spoiled with in 3-4 months of storage (Subramanyam *et al* 1976). Products stored at room temperature and at 37 °C showed slight increase in acidity, which declined with the advent of browning of the product. Fading of colour, associated with decrease in β-carotene occurred throughout the storage period.

At room temperature loss of carotenoids did not result in off-flavour but at 37 °C off-flavour was observed before 50% β-carotenoids was destroyed (Subramanyam *et al* 1976). Dehydrated powder having less than 1% moisture packed in cans could be stored at ambient temperature (24-28 °C) as well as at 37 °C and was acceptable up to one year (Subramanyam *et al* 1976). Dhabade and Khedkar (1980) reported that optimum relative humidity for storage of raw mango powder (amchur) was 43% and mould growth occurred at 95% relative humidity within 8 days of storage. Deterioration in colour and texture and mould infection occurred when the product was stored at 70% relative humidity (Teotia *et al* 1987). Therefore, dry 'amchur' should preferably be stored in a cool, dark and relatively dry place having less than 50% relative humidity for extending its storage life.

Pickles and Chutneys

Pickles and chutneys are marketed/stored in various types of containers. During storage a number of chemical changes occur in the product, which make the product either unfit for consumption or sometimes to rejection on aesthetic grounds.

Mango oil pickle was stored at ambient temperature (30 $\pm$ 5 °C) for 6 and 12 months period in 5 types of packages. It was observed that ascorbic acid, sugars, proteins, phosphorus, etc, decreased with increase in storage period. On the other hand, there was increase in reducing sugars. The progressive decrease in protein was due to degradation of proteins or leaching of protein molecules from the tissue. The loss of phosphorus might be either due to its utilization for different metabolic processes or its leaching to the suspension medium. The pH, browning and rancidity increased. Maximum rancidity (melanin aldehyde) was in samples packed in LDPE bags. HDPE containers gave stronger off-flavour. The pickles stored in A 2½ cans had no bad smell and retained better colour, but the product was very soft and mashy in texture. Pickles stored in glass jars were superior in quality and scored highest marks (58%) after storage of one year. Other samples were not satisfactory due to strong rancid odour and dull colour. Irrespective of the type of containers, browning showed increasing trend during storage. Browning was minimum in cans and maximum in LDPE bags and friction top containers. Pickles stored in glass jars and HDPE containers showed moderate browning. Hence, storing of pickles in LDPE bags and friction top containers is undesirable (Verma *et al* 1986).

The sensory scores of mango pickle decreased for appearance, colour, taste, texture, consistency and acceptability, due to changes in the quality of fat and raw materials, particularly spices or due to increase in peroxide content or browning when stored for 6 months (Thakur and Arya, 1990).

Pulp (Puree)

Fruit juices and pulp are generally preserved either by heat or by addition of chemical preservatives such as potassium metabisulphite, sodium metabisulphite/sodium benzoate and packed in tin cans and glass bottles. However, the exorbitant prices of tin plates had necessitated the use of plastic containers for packaging and storage of these mango products. Further, for domestic markets, less expensive packs (HDPE containers, glass bottles, polyethylene pouches and polypropylene pouches) are used, whereas for export purposes cans are used, inspite of being expensive.

Preservation by sulphitation : The mango pulp is often preserved in bulk quantities by the addition of chemical preservatives like SO_2 in the form of potassium metabisulphite (KMS) or sodium metabisulphite (SMS) followed by lower temperature storage processing, which preserves the delicate mango flavour. SO_2 acts as a preservative against the growth of moulds and also bleaches the colour of the product and makes the colour of the product uniform and attractive. SO_2 helps in better retention of carotenoids and flavour during storage of mango pulp (Sethi and Maini, 1991). However, there are some constraints in the use of sulphitation like development of residual flavour of the product (Contractor, 1975). Sulphitation at high temperature (32-35 °C) was less effective due to high pH (4.55 of the pulp) (Kalra and Chadha, 1984).

Canning of mango pulp being uneconomical for bulk storage, alternate packaging types have been used such as wooden barrels, glass bottles, HDPE containers, polyethylene and polypropylene pouches etc., and found to be suitable for storing sulphited mango pulp. Sulphited (1000 ppm SO_2) mango pulp was stored for 12 months in wooden barrels without deterioration in quality (Siddappa and Krishnamurthy, 1963., Ambadan and Adsule (1979). Besides conventional cans, glass and HDPE containers were found suitable for storing mango pulp (Kalra, 1982), particularly in sealed glass containers, mango pulp could be stored for over 2 years, while flexible pouches were good only for a shorter period of storage (Kalra and Chadha, 1984). Kalra and Tandon (1985) found that glass and HDPE containers were satisfactory for storing 'Dashehari' mango pulp, heated to 75-76 °C and preserved with SO_2 (1000 ppm) for a year.

Glass bottles were more satisfactory at ambient temperature with least changes in moisture, acidity, TSS, tannins, proteins and pH and without microbial spoilage as compared to PVC and HDPE pouches both at room temperature and low temperature for storage of mango pulp with SO_2. The pulp was organoleptically acceptable up to one year in glass bottles, 7-8 months in PVC bottles, 2 months in HDPE and HMHD pouches (Sethi and Maini, 1991). Storage of pulp without addition of any preservative had shown that hot pulp filled at 80 °C could stay 12 months in glass bottles at low temperature as compared to 8 months in HDPE containers, which was attributed to the non-permeability of glass bottles (Santha *et al* 1982). The mango pulp stored with 1000 ppm SO_2 in bulk glass containers, with glass lids and wax-sealed under ambient conditions showed a sharp increase in reducing sugars and freedom from microorganisms during the 12 months of storage. Thus the glass containers were superior for storage of mango pulp in bulk quantities (Kalra and Tandon, 1985).

Sulphited (500 ppm SO_2) fresh mango pulp could be stored for 12 months at room temperature (25°–28 °C) and for more than a year at low temperature in HDPE bottles (Santha *et al* 1982). Mango pulp without suphitation could be preserved for 6 months after thermal process and bulk packing in HDPE containers (Shrikhande *et al* 1976). Hot pulp of mango without any preservative, filled at 80 °C could store well for 8 months in HDPE containers at low temperature (5 °C). Cold pulp with 750 ppm SO_2 was in good condition for 4 and 6 months at room temperature and low temperature respectively in HDPE containers (Santha *et al* 1982).

Mango pulp in polyethylene and polypropylene pouches can be kept in good condition for 2-8 months depending upon the storage temperature and the treatment used. At 42-45 °C the shelf life was one month, with browning. In the frozen condition, the pulp could stay for more than 6 months without browning (Santha *et al* 1982). However, these flexible polypropylene pouches were unsuitable for long storage for mango pulp with sulphitation (350 ppm SO_2). But still higher levels of SO_2 may improve the preservation of mango pulp in such flexible pouches (Kalra and Chadha, 1984).

Fresh mango pulp of 'Amrapali' with 500 ppm of SO_2 could be preserved for more than 9 months in glass bottles and 7-8 months in PVC bottles at ambient temperature and 12 and 10 months at low temperature respectively (Sethi, 1995). The fresh pulp of different mango varieties could be stored up to one year with 500-1000 ppm SO_2 treatment at ambient temperature in HDPE and glass bottles, while at low temperature it could be kept well for more than 12 months (Santha *et al* 1982., Kalra and Chadha, 1984., Sethi and Maini, 1985).

Mango pulp could be stored in good condition for 7-8 months depending on the plastic container and the treatment adopted. The shelf life of mango pulp was minimum in flexible pouches and maximum in glass bottles. The increase in shelf life of pulp in glass bottles was due to its non-permeability to moisture and gasses, which helped in better retention of SO_2 in the stored pulp. Glass bottles were more suitable than plastic bottles regarding the retention of carotenoids and reduction of non-enzymatic browning (Sethi, 1995). PVC bottles improved the preservation of mango pulp by better retention of SO_2 which in turn improved quality and shelf life of the product (Sethi, 1995).

To obtain more benefit, often two or more preservatives are employed for preserving mango pulp. Fresh mango pulp with SO_2 alone or in combination with sodium benzoate had a shelf life of 10-12 months at room temperature (25°-35 °C) in glass bottles, 4-8 months in PVC and 2 months in HDPE pouches and 45 days in HMHD pouches (Sethi and Maini, 1991). Shelf life of stored mango pulp at room temperature decreased with decrease in concentration of SO_2 and sodium benzoate by 2 to 4 months in glass and PVC bottles respectively. However, there was no change in the duration of shelf life of pulp stored in HDPE pouches. At low temperature (1°-3 °C) the pulp had more than year shelf life in glass bottles. Pulp stored with SO_2 and sodium benzoate showed more non-enzymatic browning of pulp in HDPE containers as compared to PVC and glass bottles. Off-flavour developed in pulp stored in HDPE containers after 2 months and 4 months in PVC bottles containing sodium benzoate. Combination of sodium benzoate and SO_2 was equally effective for the retention of colour, but there was a slight increase in non-enzymatic browning and decrease in carotenoid content and flavour during storage (Sethi and Maini, 1991). With combination of SO_2 and sodium benzoate (each 250 ppm), the pulp remained good for 2 to 6 months, but thereafter showed browning in all the containers with perceptible off-flavour, resulting in non-acceptability of the pulp. Up to 2 months of storage, both KMS and KMS + sodium benzoate were equally effective with respect to retention of carotenoids and flavour of the pulp at different temperatures, but later on up to 5 months, KMS alone was better in the retention of carotenoids irrespective of storage temperature (Sethi, 1995). Addition of sodium benzoate along with KMS showed better storage life of pulp in PVC bottles at room temperature, but not at low temperature.

To prevent the microbiological spoilage of pulp, cooking (heating) of the pulp is generally suggested. However, heating of the pulp was not adequate to completely inhibit bacteria although yeast and moulds were completely inhibited. This appears to be due to the presence of heat-resistant bacteria in the pulp. In such cases addition of preservatives to the pulp is suggested. Hot pulp with either SO_2 (250 ppm) or potassium sorbate (50 ppm) was found adequate to inhibit the growth of bacteria, yeast and moulds during storage. However, cold pulp with either of the preservative at 500 ppm each, was spoiled early. Untreated mango pulp was spoiled with in 24 hours as a result of 100 fold increase in microorganisms. However, addition of potassium sorbate alone or with SO_2 resulted in browning the pulp with in a week or two of storage at room temperature in HDPE containers and glass bottles as compared to poly-ethylene and polypropylene pouches (Sethi and Maini, 1991).

The extension and spoilage of sulphited product depend on the existence of free form of SO_2 in the product. Free form of SO_2 in turn is influenced by pH and sugars of the pulp, concentration of SO_2, additives, storage temperature, type of container and surface area exposed to air (Sudhakar and Maini, 1997). Their study showed that the retention of free form of SO_2 was highest at pH 3.5 to 4.0 of the pulp. Addition of ascorbic acid (0.01-0.03%) markedly increased the retention of free form of SO_2, as a result of antioxidative properties of ascorbic acid, which prevented the oxidation of SO_2 to SO_4. The retention of free form of SO_2 decreased with an increase in the concentration of tannins. The retention of free form of SO_2 increased with increase in SO_2 concentration from 250 to 1000 ppm in the pulp (Bose and Lodha, 1960). The retention of SO_2 in the pulp packed in glass bottles was higher as compared to that of the other containers during storage at different temperatures. The low retention of SO_2 in PVC and LDPE containers was ascribed to their permeability to gases. The retention of free form of SO_2 in the pulp decreased with increase in storage temperature irrespective of packing containers and the loss of free SO_2 could be due to the effect of high temperature on retention of SO_2 (Adsule and Anand, 1977., Maini *et al* 1984). Mango pulp packed in containers with least surface area exposed to air and stored at cold temperature had maximum SO_2 in the free form in the pulp due to rapid loss of SO_2 to atmosphere when the exposed area was more (Santha *et al* 1982., Sudhakar and Maini, 1997). From the above, it was concluded that for longer storage of mango pulp, it should have higher ascorbic acid content, low reducing sugars, tannins and pH (3.5-4.0) and stored at low temperature and the exposed area should be more (Santha *et al* 1982., Sudhakar and Maini, 1997).

Dried / Dehydrated pulp: Mango pulp is dried by different methods for storage and these methods influence the storage life of the stored pulp. Drum-dried green mango pulp was highly acceptable as regards to colour and flavour after 2, 4 and 6 months of storage at room temperature in polyethylene bags, Al-foil laminates and sanitary cans, respectively, but those stored in HDPE pouches were less satisfactory. In all cases, ascorbic acid content fell by 20% after 9 months of storage (Gangopadhyay *et al* 1976).

Frozen pulp: Frozen mango pulp, with or without pre-treatments is supplied to armed forces. Homogenized puree packed in different cans or waxed paper containers sealed and frozen at 20 °F could be stored at 0 °F for 18-24 months without effect on the colour, flavour and taste (Singh, 1977). Sterilized mango pulp could be stored at –5 °C $\pm$ 2 °C for one year satisfactorily without any spoilage. However, there was a slight decrease in titratable acidity and TSS and 50% reduction in Vitamin – C content. The moisture, total fructose, reducing and total sugars were not much affected during one year of storage. Pre-heat treatment of pulp did not allow development of any spoilage microbes and consequently there was no spoilage of frozen pulp.

Pasteurized pulp: The pasteurised pulp in 300 gauge polypropylene flexible pouches could be stored for 3 months at 5 °C and at ambient temperature, but only 2 months at 37 °C. The quality and shelf life of the pasteurized pulp were comparatively better with SO_2 (Ghosh, 1981) and without addition of synthetic colouring and flavouring agents (Ghosh *et al* 1982).

Mango pulp, pasteurized for one month at 88 °C-92 °C, filled into sterilized pouches, evacuated, heat sealed and chilled at 4 °C and stored at 18 °C –20 °C for 4 months showed no sensory deterioration and microbial spoilage. However, pasteurized pulp could be stored for a year at –12 °C in flexible pouches with good retention of 50-90% carotenoids, 11-18% ascorbic acid and little changes in total sugars and acids (Kalra *et al* 1995).

Changes in pulp during storage

Several physico-chemical changes occur in the pulp of mango during storage under the influence of containers, storage temperature and the duration of storage. Pulp stored in glass bottles showed minimum changes in chemical composition during storage, as compared to PVC and HDPE pouches both at room temperature and low temperature. Organoleptic evaluation of the stored pulp showed that both glass and PVC bottles were acceptable with regard to retention of colour and taste (Sethi and Maini, 1991). Changes in quality parameters such as total carotenoids, non-enzymatic browning, SO_2 in mango pulp stored in HMHD film pouches were faster at 40 °C than at 0 °C. Storing the pulp with SO_2 in the dark was as good as keeping the same at 0 °C (Sethi and Maini, 1991).

The pulp stored in HDPE containers showed higher brix, more acidity, reduction in total sugars, increase in reducing sugars and decrease in total carotenoids during 12 months storage at room temperature with decrease in SO_2 concentration. The pulp stored in glass bottles showed minimum changes. Similar changes were observed in the pulp stored in polyethylene and polypropylene pouches during the storage of 4 months. Viscocity of pulp also showed change during storage (Santha *et al* 1982).

Kalra and Chadha (1984) observed rise in pH, protein content, and tannins and decrease in total sugars and soluble solids, no appreciable loss of moisture or Vitamin C, but complete loss of SO_2, during storage of mango pulp in different containers. The pulp under ambient conditions exhibited an increase in reducing sugars and a decrease in TSS and total sugars. There was a slight increase in pH of stored pulp without concurrent decrease in titratable acidity and no loss of Vitamin C. Soluble proteins increased but more so under ambient temperatures. Tannins rose strongly at room temperature but in cold storage they increased up to 2 months and then declined. There was no deterioration in colour or organoleptic quality of the pulp or the juice prepared from the stored pulp (Kalra and Chadha, 1984). The pulp stored in PVC bottles showed high TSS, reducing sugars and total sugars and decrease in total carotenoids and total SO_2 at ambient temperature, as compared to glass bottles in which the above changes were less. Inspite of the above, glass bottles were claimed more suitable than PVC containers for long term storage of mango pulp at both room as well as low temperature due to less permeability of glass to moisture and gases (Sethi, 1995).

LEATHER/BAR

Mango leather was traditionally stored very tightly in wooden boxes using *Bauhina* dried leaves for lining for getting equilibrium of moisture in the product (in India) for about a year, during which time colour may become slightly darker (Rameshwar, 1979).

The present method of packing mango bar in cellophane, polypropylene materials etc., does not ensure the required storage life. Out of the different packaging materials (Paper, Al-foil, LDPE, HDPE and metalised polyster) evaluated, Al-foil based materials were found to be suitable for long term storage of mango bar (Nadanasabhapathi *et al* 1993). Mango bar packaged in various above packing materials and stored at room temperature (26 °C-32 °C) and at 37 °C with 90% relative humidity showed insignificant changes in moisture content, SO_2 content and carotenoid content, which did not affect the acceptability of the product. The total sugars and reducing sugars content did not change during storage. In all other packaging materials, sugar was crystallized on the surface of the mango leather. This phenomenon (crystallization of sugar) was not observed in the samples stored at 40 °C. At this temperature, after 5 months of storage, the sensory evaluation showed that the product was not acceptable due to development of undesirable odour as a result of crystallisation of sugar in the product (Chandrasekhara and Judson-king, 1971). The residual SO_2 content of mango bar in various above packaging materials was of same order, but at 40 °C, the loss was more as compared to the storage at room temperature and 37 °C and hence acceptable. On storage at room temperature and 37 °C, carotenoid content decreased, but it was completely lost when the product was stored at 40 °C in all the above packaging materials (Nadanasabhapathi *et al* 1993).

Method of preparation of mango bar also influences its storage life. Mango bar prepared by dehydration at 65 °C stored for longer periods (81 days) than that was sun-dried (68 days) (Sunita Singh *et al* 2005). Mango bar obtained by cross-flow drying technique (at 70 °C for 20-22 hours) could be stored for one year at 50-60% relative humidity (Nanjundaswamy *et al* 1976).

The storage life of mango bar depends on the moisture content of the product as well as the relative humidity of the storage atmosphere. The ideal moisture for storage stability was found to be 15% or a little more with relative humidity between 63 and 70%. The optimum relative humidity for storage of mango sheet, theoretically was 70.5% when the initial moisture content in the product was 17.95% (Rao and Roy, 1980).

SLICES

Raw mango slices: Raw mango slices during the season are preserved as salted, dried/ dehydrated ones for use during off season to make pickles, chutneys etc.

Salted slices: Traditionally raw mango slices are preserved in powdered salt in alternate layers in gunny bags. But such slices can store for less than 3 months only. Dry salting makes them soften and such slices show browning (Sethi, 1991). Dry salting is not good with regard to colour and texture.

Johar and Anand (1951) recommended the storage of raw mango slices in 15% brine solution and reported that 60° salometer brine retained slices crisp with natural flavour up to 6 months. Later, Anand and Johar (1958) suggested an improved method of storing raw mango slices for two weeks in 10% brine solution containing 200 ppm SO_2 for primary salting for 20 hours followed by addition of 5 % powdered salt with 200 ppm SO_2 for storage and finally filling the empty space in the wooden barrel with 5 % brine solution containing 200 ppm SO_2 and sealing the barrel.

Raw mango slices could be preserved up to one year in an acidified chemical solution containing salt (5%), acetic acid (1.2%) and KMS (0.1%) with good retention of colour and texture. Addition of calcium chloride (0.5%) helped in retention of texture of slices during subsequent storage in glass jars at room temperature (30 $\pm$ 5 °C). Retention of nutrients was better in this technique (steeping preservation) than dry salting (Sethi, 1991).

However, brined slices showed leaching loss in acidity, sugars and ascorbic acid during storage. Vitamin–C completely disappeared after 3 months of storage and after 6 months storage, the brined slices could not stand and showed deterioration (Khedkar and Roy, 1988).

Raw mango slices were treated with salt saturated-and half-saturated brine solution with or without KMS treatment and were stored in LDPE pouches. Combined SO_2 and salt treatment preserved the slices for over 3 months and SO_2 alone for 70 days. The mango slices without SO_2 were spoiled in a week's time, with mould growth in saturated-/half-saturated brine solution. Only powdered salt treatment could preserve the slices for more than 5 months and along with SO_2, there was an improvement in the preservation of slices. Slices dipped in saturated-/half-saturated brine solution containing KMS (1.5%) for 5 minutes were spoiled within 40 days storage (Kalra and Tandon, 1985).

The moisture content of slices declined slightly during storage due to osmosis (Maitra *et al* 1979a). TSS increased about 3 times, which was attributed to salt absorption and adsorption. Reducing sugars did not change much. Acidity decreased by 15 to 20% during the first week of storage due to leaching during treatment and also due to lactic fermentation and thereafter remained more or less constant. Ascorbic acid was totally destroyed after 2 weeks as a consequence of oxidation reaction (Maitra *et al* 1979b), total soluble carbohydrates increased 3-fold during storage in most of the treatments tried, but decreased in powdered salt treatment (Kalra and Tandon, 1983). During storage, loss of acids, sugars and Vitamin-C was observed in slices preserved in brine solution due to leaching, being water soluble. Consistent decrease in SO_2 was also observed. Salt content of the slices in the brine was more or less uniform with slight increase after 6 months of storage period.

Dried/Dehydrated slices: Raw mango slices meant for chutney making are traditionally preserved either with dry salt or brine solution. Such preserved slices with dry salt turn greyish black, soft and get shrivelled (Anand and Johar, 1958), while those preserved in brine solution show leaching losses of the important constituents during storage (Pruthi and Bedekar, 1963). In either method, packaging of the treated slices is done in wooden barrels, which besides occupying vast space, also adds the freight charges. An alternate method of preservation of slices is, therefore, sought, which should provide better quality slices, occupy less space and be amenable for easy and less expensive transport. In this context drying and dehydration of raw mango slices were suggested.

The sun-dried and cabinet dried slices with different doses of SO_2 (0.5-2.5% KMS) were stored at room temperature (30 $\pm$ 5 °C) in 400 gauge polyethylene bags (Khedkar and Roy, 1988). During storage, the cabinet-dried and sulphited slices had better retention of SO_2 and ascorbic acid and less browning as compared to sun-dried and control lots. Among different SO_2 levels, slices treated with KMS at 1.5% level for 30 minutes and dried in a cabinet drier had scored maximum organoleptically and gave best chutney. Retention of SO_2 was more in cabinet-dried slices than in sun-dried ones. In general, sulphited, cabinet-dried slices had better reconstitution ability than sun-dried and control ones. Decrease in reconstitution ratio can be attributed to texture degradation of the slices, which resulted in less absorption of water during reconstitution. Retention of ascorbic acid was proportional to SO_2 concentration during drying and storage, being better in cabinet-dried slices. However, increase in the retention of SO_2 resulted in increase of non-enzymatic browning and it was inversely proportional in both sun-dried and cabinet-dried slices. The slices of control treatment showed more browning than sun-dried slices due to rapid loss of SO_2 and longer exposure of slices to atmosphere during drying. Higher acidity in control samples could be due to blanching and not sulphitation and hence leaching losses during sulphitation were avoided. Increase in sugars in slices during storage was due to hydrolysis of starch (Khedkar and Roy, 1988).

Moisture did not increase much after a month's storage in unsalted slices, when packed in 250 gauge polyethylene bags and placed in friction top tins at room temperature, but it enhanced in salted slices during storage due to its hygroscopic nature. Unsalted slices, however, turned black after a month's storage, while the salted slices (2.5%) without KMS became brown after 3 months storage (Mehta and Tomar, 1982).

Cabinet-dried, sulphited mango slices can be stored for 6 months in polyethylene (100 gauge) pouches for making `amchur' (Dhabade and Khedkar, 1980).

Ripe mango slices

Osmotically dried slices: Keeping the osmotically dried ripe mango slices between 64.8 and 75.5% relative humidity is conducive to the retention of colour, flavour, taste and texture of the product. High humidities stimulated moisture pickup, resulting in mouldy growth within one to four weeks depending on the level of relative humidity. Similarly, storage at lower humidities was also detrimental to the quality of the product, as there was moisture loss and the slices become gradually hard (Amitabh *et al* 2000).

Dehydrated ripe mango slices: For the storage of dehydrated ripe mango slices, 260 gauge Al-laminated polyethylene pouches were found to be better as compared to 400 and 200 gauge LDPE pouches with respect to colour, flavour, texture and overall quality for 6 months. The slices packed in 200 gauge LDPE pouches gained more moisture as compared to 400 gauge LDPE and 260 gauge Al-laminated polyethylene pouches depending on the permeability of packing material. Browning was more in the samples packed in 200 gauge LDPE pouches as compared to 400 gauge LDPE and 200 gauge Al-laminated polyethylene pouches, which was due to higher moisture content and inadequate SO_2 levels in the packed dehydrated slices (Sagar and Khurdiya, 1999).

Canned ripe mango slices: For dessert purposes, ripe mango slices are often canned in sugar syrup in cans. During storage in cans, acidity and non-reducing sugars are reduced, while reducing sugars increased. The rate of these changes was more rapid at higher temperature and markedly affected the quality and shelf life of the canned products. Samples stored at room temperature did not show deterioration or develop off flavour while slices stored at higher temperature developed hydrogen smell with discolouration and spoilage (Dhopeswarkar and Magar, 1955., Kapur, 1975). Storage quality of `Badami' mango slices in sugar syrup in A 2 ½ tin cans was satisfactory up to 12 months of storage at room temperature and 9 months at 37 °C (Mahadevaiah *et al* 1977).

Frozen ripe mango slices: Ripe mango slices are frozen in cans/glass jars and then stored at –18 °C, where they could be kept well for about one year (Srivastava, 1967). Singh (1977) suggested polyethylene, cellophane or plio-film bags as packing material instead of cans/glass jars. Such frozen slices can be held at 0 °F for 12-18 months without the loss of natural flavour, colour, taste and texture.

Intermediate moisture slices: Keeping these types of slices below 75.5% relative humidity was conducive to the retention of colour, flavour and texture of the slices, having 2.5% moisture (Mishra and Tomar, 2000).

Minimally processed products: These products such as slices are highly perishable, requiring control of storage temperature and environment as well as sanitation to maintain the quality (Watada *et al* 1996). The shelf life of minimally processed products (slices) based on visual appearance, agreeable aroma was 14 days in polyethylene tenephthalate trays (PET), while in other packages it was 11 days only. The titratable acidity in chunks (slices) packaged in polypropylene cups or PET trays was reduced during storage and the colour changed from light yellow to dark yellow. The best packages for storing the minimally processed mango slices were the cups (Donaden *et al* 2004). The colour changed gradually from pale yellow in the product packaged in cups, however, there was no such change in those packaged in plastic bags (Rathanapanone and Watada, 2000., Donadone *et al* 2004). Rattanapanone and Watada (2000) observed that the TSS content, in slices was not affected by temperature, atmospheric composition of storage system or storage temperature. However, Allong *et al* (2000) observed decrease in titratable acidity and ascorbic acid with storage time in the chunks of mango stored at 10 °C and 5 °C.

Fresh cut slices: Fresh cut mangoes have a potentially wide market demand, however, browning, translucency and loss of firmness limit their quality and shelf life. Cut surface of the slices/cubes acts as an ideal place for microbial growth (Soliva-Fortuny and Martin-Bellosa, 2003). The following symptoms limit the shelf life of fresh cut mangoes;

(i) tissue darkening,
(ii) faster ripening,
(iii) development of a 'glossy appearance',
(iv) surface desiccation,
(v) loss of tissue firmness and
(vi) loss of weight.

Therefore, a suitable storage system, avoiding the above and maintaining good quality for longer time is required. Extended post-cutting life of fresh cut mangoes was reported with low temperature, atmospheric modification, ethylene absorbants, calcium treatment, anti-browning agents, antioxidants and their combination (Abe and Watada, 1991., Wright and Kader, 1997., Rattanapanone *et al* 2001).

Low temperature storage: A storage temperature of 5 °C was more effective than 10 °C in delaying ripening and inhibiting microbial growth, while preserving sensory quality of fresh cut slices of mango. However, the response to low temperature depended on the stage of maturity of fruits at the time of slicing. Half-ripe (HR) and firm-ripe (FR) mango fresh cut slices held at 10 °C showed unacceptable microbial counts, with minimum changes in sensory quality at HR stage of ripeness. There was a decline in sensory quality of mature green (MG) mango slices after 2nd day storage at both temperatures (5 °C and 10 °C). On the other hand, HR and FR mango slices were acceptable after 8 days of storage. Therefore, MG fruits are unsuitable for fresh cut purpose because although ripening appeared to take place during storage, sensory quality of cut slices after 2-4 days at both temperatures has suffered. HR fruit slices held at both temperatures were of an acceptable quality for 6 days longer than MG fruit slices at both temperatures. Thus, the optimal ripeness for preparing fresh cut slices was HR and FR (Allong *et al* 2000).

Fresh cut mango slices lost weight when stored at 7 °C (Maciel *et al* 2004). According to Watada *et al* (1996) fresh cut slices have a large surface area without any skin thus have potential for losing substantial amount of weight. Weight loss was maximum in LDPE film, which possesses excellent barrier property to water vapour and good permeability to oxygen (Wills *et al* 1989).

Although the shelf life of fresh cut mango slices can be partly extended by low temperature (< 5 °C), further extension of shelf life (approximately 5 days) makes distribution and marketing difficult as they get spoiled (Rattanapanone *et al* 2001., Izumi *et al* 2003).

Storage by atmospheric modification: Since extension of storage life of fresh cut mango slices more than 5 days posed problems (above), other means in combination with low temperature were tried. Such treatments include low O_2, enhanced level of CO_2, calcium chloride and browning inhibitors.

In general, lowering the level of O_2 in storage atmosphere is beneficial in maintaining quality and extending shelf life of fresh produce. Low O_2 atmosphere has minimal effect on the physico-chemical changes in fresh cut slices of mango during storage. The flesh became translucent in air and slightly browning in low O_2 atmosphere (Rattanapanone and Watada, 2000). Reports from the industry indicated that shelf life of mango cut slices may be extended to as much as 6 months by combining low O_2 and enhanced CO_2 concentration and organic acids. Chantana Warangoon and Kader (2000) reported that a combination of low O_2 and enhanced CO_2 and calcium chloride can increase shelf life to 15 days at 5 °C in case of mangoes.

Reduced O_2 (2.5%) was effective in controlling tissue discolouration and development of a 'glossy' appearance while calcium chloride (3%) was partly effective at controlling tissue darkening and loss of firmness (Souza *et al* 2006). CO_2 (5-40%) and citric acid had little positive effect on shelf life of cut slices, but promoted tissue softening. A combination of O_2 and calcium allowed slices a storage life of 15 days at 3 °C (Chantana Warangoon and Kader, 2000., Souza *et al* 2006). Desiccation of slices would be minimized if slices are stored in modified atmospheric packs having high humidity and low air movement; however, the loss of firmness limits the shelf life.

Despite low O_2 atmosphere's inhibiting effect on darkening and glossy tissue development, visual appearance of slices was still observed to decline during storage due to gradual development of surface desiccation as reflected by weight loss of slices. However, surface desiccation is a less timing factor in slices packed commercially in MAP, due to high humidity environments and negligible air flow. The reduction of slice firmness by high CO_2 is not similar in all cases (Poubol and Izumi, 2005), but varied with variety (Rattanapanone *et al* 2001). TSS remained relatively constant over the storage period and was unaffected by the storage temperature (Leon *et al* 2000., Rattanapanone and Watada, 2000., Donadone *et al* 2004., Souza *et al* 2006). Application of organic acids (citric or ascorbic acid) failed to exert significant effect on visual appearance, firmness and TSS in fresh cut slices of mango (Maciel *et al* 2004., Souza *et al* 2006). However, citric acid (2%) slightly increased titratable acidity initially, probably due to the uptake of citric acid when the slices were immersed in citric acid solution (Souza *et al* 2006). In contrast to calcium, slices treated with citric acid lowered tissue firmness. Combination of citric acid and calcium doubled weight loss. Souza *et al* (2006) suggested osmotic effect in this. Calcium also reduced titratable acidity in the treated slices (Tovar *et al* 2001.,

Souza *et al* 2006). Browning and water soaking, as limiting factors affecting visual appearance of fresh cut mango slices stored under normal O_2 levels, due to reduction of O_2 concentration during storage period were reported (Rattanapanone and Watada, 2000., Chantana Warangoon and Kader, 2000., Rattanapanone *et al* 2001., Izumi *et al* 2003., Poubol and Izumi, 2005., Souza *et al* 2006).

Treatment with 1-MCP and others: Ethylene has an undesirable effect on the quality of fresh cut fruit slices. Its accumulation in the packages of fresh cut slices can be detrimental to their quality and shelf life. Therefore, removal of accumulated ethylene from the storage environment is essential to extend the shelf life of fresh cut fruits, for which an ethylene inhibitor is needed.

1-Methylcyclo propene (1-MCP) is a recently developed inhibitor of ethylene action (Sisler and Serek, 1997). Ripening and senescence of intact fruits and fresh cut slices of fruits have been delayed by 1-MCP action in different fruits (Golding *et al* 1998., Jiang *et al* 1999., Kim *et al* 2001., Vilas-Boas and Kader, 2007., Mao *et al* 2007). It was effective in delaying further ripening of ripe fresh banana (Pelayo *et al* 2003) indicating its potential commercial value to slow the changes associated with loss of quality and to extend the shelf life of fresh cut slices of fruits (Vilas-Boas and Kader, 2006).

1-MCP can be applied to whole mangos prior to cutting or directly on slices after cutting, but the effect differs. Application to whole mangoes prior to cutting maintained freshness and colour of fresh cut pieces, but did not delay ripening (Plotto *et al* 2003). Direct application of 1-MCP on slices delayed their softening and darkening.

The treatment with 1-MCP (1 $\mu l/l$) in 6 hours had the best quality fresh cut mango slices in terms of firmness, during storage at 10 °C. Softening and browning were delayed for 9 days by direct application on fresh cut slices. Respiration rate of slices was not influenced, whereas ethylene production was affected. Application of 1-MCP at lower dose (0.5 $\mu l/l$) for 24 hours and at higher dose (1.0 $\mu l/l$) for 6 hours decreased ethylene production by fresh cut mango slices stored at 5 °C on the 4th day of storage. Neither chemical ($CaCl_2$ 1% w/v) nor 1-MCP applied on intact fruit (before slicing) affected the firmness of fresh-cut mango slices stored at 5 °C. However, when applied directly to the slices 1-MCP effectively delayed the softening of cut slices for 9 days. Chemical dip and 1-MCP applied directly to fresh-cut slices delayed their colour darkening during 9 days of storage at 5 °C (Vilas-Boas and Kader, 2007).

Treatment of fresh cut mango slices with combination of D-isoascorbic acid, 4-hexyl resorcinol and potassium sorbate and storage at 10 °C for 14 days resulted in best maintenance of colour, visual appearance and fewer symptoms of browning and decay (Gonzalez-Aguilar, *et al* 2000). However, the chemical dip decreased the respiration rate of the slices of mango (Vilas-Boas and Kader, 2007). Calcium chloride has been successfully used in fresh cut slices of mango to maintain appearance and to extend shelf life (Mootoo, 1991).

Pre-treatment with Ethanol vapour: Exposure of fresh cut slices of mango to ethanol vapour for 24 hours extended the shelf life, but was detrimental to flavour, while short time exposure was not detrimental, but was not effective in extending the shelf life significantly (Plotto *et al* 2003). On the other hand, whole fruit treatment with ethanol vapour maintained higher firmness, visual quality and lower sugar levels in fresh cut slices (Plotto *et al* 2003). However, the treatment with ethanol vapour resulted in undesirable flavour notes in fruit tissue, without decay of slices after 2 weeks of storage at 7 °C. More than 20 hours exposure of fresh cut slices results in the development of off-flavour in the product. At shorter exposure times (8-10 hours) ethanol vapour could be used as a safe microbial control (Plotto *et al* 2006).

Hot water treatment initiates the activity of ripening – related enzymes, while ethanol inhibits them. The inhibitory effect was more pronounced when mangoes were ethanol – treated sooner after hot water treatment (Plotto *et al* 2006).

BEVERAGES

Both alcoholic and non-alcoholic beverages are prepared from mangoes. The latter includes nectar, juice, squash, RTS drinks etc. They being very nutritious and refreshing have become very popular, particularly in summer.

Fruit juice

Fruit juices were usually packed and preserved in tin cans. However, due to their high cost and possibility of tin corrosion and with the advent of plastic packaging materials, the tin cans were discontinued.

'Totapuri' mango juice packed in tin-plate containers was acceptable up to 12 months at ambient temperature of 25-30 °C (Naresh *et al* 1983., Gowramma *et al* 1988) and 9 months at 37 °C (Gowramma *et al* 1988). The storage behaviour of the juice depends on the extent of presence of fruit pulp in the juice. Mango juice having more pulp than nectar showed better storage in tin-plate containers (Gowramma *et al* 1988). White-lacquered-cans based on polyester, both imported and domestic, were suitable for packing mango juice without loss of flavour for 12 months of storage at ambient temperature and 9 months at 37 °C and was acceptable accordingly. However, juice packed in lacquered cans had slight lacquer flavour after 9 months storage (Gowramma *et al* 1988).

Pouches made from a laminate of polyster/polyethylene were not suitable for packing and preserving mango juice. In plastic flexible pouches of above packaging material, there was considerable amount of loss in β-carotene content and ascorbic acid as compared to juice packed in tin-cans. Outer coverings of yellow and black polyethylene protected β-carotene and ascorbic acid very slightly. However, after 3 months of storage, mango juice in laminated pouches of polyster/polyethylene and without secondary packaging was unacceptable, since it had lost its original taste and flavour (Gowramma *et al* 1988).

Percentage of ascorbic acid loss, non-enzymatic browning and titratable acidity increased with storage period in all packaging materials (polyethylene films, polyethylene tetraphthelate (PET) or plastic bottles and transparent glass bottles) stored at 6 °C, 26 °C and 34 °C in 8 weeks. However, pH of the juice decreased with storage time and soluble solids remained constant. Higher percentage of ascorbic acid loss, browning index and titratable acidity occurred in juices packed in polyethylene films than in PET and glass containers, while the same increased with storage temperature. However, titratable acid decreased with increased storage temperature and soluble solids remained constant throughout the storage period of 8 weeks (Alaka *et al* 2004).

Mango nectar

Mango nectar is a RTS beverage, much like mango juice, but containing more sugar. The storage quality of mango nectar in plain electrolytic tin plates was very good for one year at ambient temperature (25-30 °C) and 6 months at 37 °C. Mango nectar packed in different types of indigenous hot-dipped tin plate cans and stored at ambient temperature (25-30 °C) was satisfactory even for 15 months of storage but at higher temperature (37 °C) after 12 months storage, it developed metallic taste; having the tin content beyond permissible level. Hot-dipped tin plates were better than electrolytic tin plates for storing mango nectar due to high tin-coating weight and lesser ATC value compared to electrolytic tin plate. The organoleptic quality of mango nectar was satisfactory in both (A 2½ and A 5½) size canes after 12 months at room temperature, but the tin content was less in A 5½ cans as compared to A 2½ cans. At 37 °C after 9 months storage the tin content was more in A 2½ cans (Mahadevaiah *et al* 1975, 1976).

PANNA

Raw mango panna is a very popular drink in North India. It kept well in glass containers up to one year at room temperature (22°-35 °C) without change in taste and appeal when diluted for consumption (Singh *et al* 1999).

RTS drink

Mango ready to serve (RTS) drink prepared from combined pulp of 'Dashehari' and 'Banganapalli' cultivars could be stored for 6 months at ambient temperature without any adverse change in all quality parameters and organoleptic evaluation (Srivastava, 1998).

REFERENCES

Abe, K. and Watada, A.E. 1991. Ethylene absorbent to maintain quality of lightly processed fruits and vegetables. Journal of Food Science, 56 : 1589-1592.

Adsule, P.G. and Anand, J.C. 1977. Evaluation of rigid thermoplastic container, for packing of some fruit and vegetable products. Indian Food Packer, 31(6) : 16-22.

Alaka, O.O., Aina, J.O. and Falado, K.O. 2004. Effect of storage condition on the chemical attributes of Ogbomoso mango juice. European Food Research and Technology, 218(1) : 79-82.

Allong, R.D., Wickham, L.D. and Mohammed, M. 2000. The effect of cultivar, fruit ripeness, storage temperature and duration on quality of fresh-cut mango. Acta Horticulturae, 509 : 478-494.

Ambadan. and Adsule, P.G. 1979. Bulk preservation of mango pulp in high density polyethylene (HDPE) containers. Indian Food Packer, 33(6) : 34.

Amitabh, Singh, R.D. and Tomar, M.C. 2000. Studies on osmotic dehydration of some varieties of ripe mangoes grown in Uttar Pradesh. Indian Food Packer, 54(3) : 66-68.

Ammu, K., Krishnappa, K.G., Subramanian, V., Sharma, T.R. and Nath, H. 1976. Freeze drying of mango pulp. Food Technology. Australia, 28 : 339-341.

Anand, J.C. and Johar, D.S. 1958. Improved method for the commercial preservation of raw mango slices. Food Science, 7 : 150.

Bose, A.N. and Lodh, S.B. 1960. Storage loss of sulphur dioxide in ripe mango pulps preserved with bisulphites. J.Proc.Aus.Chemi, 32 : 9-13.

Chandrasekhara, S.K. and Judosonking, C. 1971. Solid-liquid phase equilibria in multi component aqueous sugar solutions. Journal of Food Science, 36 : 699-670.

Chantana Warangoon,S. and Kader, A.A. 2000. Calcium treatments and modified atmospheres extend post-cutting life of mango cubes. Institute Of Food Technologists Annual Meeting 10-14th June 2000. Dallas, Texas, USA, paper 65 A.24 (Abstract).

Contractor, J.D. 1975. Fruit Juices and Beverages. Seminar on 'Fruits and Vegetables' held at New Delhi, 25-27, April. 1975.

Dabade, R.S. and Khedkar, D.M. 1980. Studies on drying and dehydration of raw mangoes for preparation of mango powder (Amchur). Part-VI. Changes in chemical constituents of Raw Mango powder during storage. Indian Food Packer, 34(3) : 48-34.

Dhopeshwrkar, G.A. and Magar, N.G. 1955. Journal of Science and Industrial Research, India, 14C, 27-31.

Donadon, J.R., Durigan, J.F., Teixeira, G.H de A., Lima, M.A. and Sarzi, B. 2004. Production and preservation of fresh cut 'Tommy Atkins' mango chunks. Acta Horticulturae, 645 : 257-265.

Gangopadhyay, H., Chaudheri, D.R. and Mukherjee, S. 1976. Dehydration of green mango pulp in double drum drier. Indian Food Packer, 30 (5) : 51-54.

Ghosh, K.G. 1981. Preservation of fruit juices and pulp in Flexible pouches. Indian Food Packer, 35(2) : 36.

Ghosh, K.G., Nirmala, N., Krishnappa, K.G., Parameshwaraiah, P.M., Borker, H. and Vijayaraghavan, 1982. Preservation of fruit juices and pulps in flexible pouch. Indian Food Packer, 36(4) : 23-26.

Golding, J.B., Shearer, D., Wylie, S.G. and Mc. Glesson, W.B. 1998. Application of 1-MCP and propylene to identify ethylene-dependent ripening processes in banana fruit. Post Harvest Biology and Technology, 14 : 87-98.

Gonzalez – Auguilar, G.A., Wang, C.Y. and Buta, J.G. 2000. Maintaining quality of fresh-cut mangoes using anti-browning agents and modified atmosphere packaging. Journal of Agriculture and Food Chemistry, 48 : 4204-4208.

Gowramma, R.V., Mahadevaiah, M. and Naresh, R. 1988. Storage behaviour of mango products in different types of containers. Acta Horticulturae, 231 : 782-794.

Izumi, H., Nagatoma, T., Tanaka, C, and Kanalayanarat, S. 2003. Physiology and quality of fresh cut mango as affected by low O_2, controlled atmosphere storage, maturity and storage temperature. Acta Horticulturae, 600 : 833-838.

Jiang, Y., Joyce, D.C. and Macnish, A.J. 1999. Extension of the shelf life of banana fruit by 1-Methyl-cyclopropene in combination with polyethylene bags. Post harvest Biology and Technology, 16 : 187-193.

Johar, D.S. and Anand, J.C. 1951. Note on the microbiology of brined mango slices. Indian Journal of Horticulture, 8(4) : 45.

Kalra, S.K. and Chadha, K. L. 1984. Studies on the suitability of polypropylene pouches for packing mango pulp. Journal of Food Science and Technology, 21(5) : 317-318.

Kalra, S. K. and Tandon, D.K. 1983. Salt-treated Raw Mango Slices in Polyethylene Pouches. Indian Food Packer, 37(6): 101-107.

Kalra, S. K. and Tandon, D.K. 1985. Physico-chemical changes in mango pulp during ambient storage in glass containers. Journal of Food Science and Technology, 22(5) : 350-353.

Kalra, S.K., Tandon, D.K. and Singh, B.P. 1995. Mango. In : Hand book of Fruit Science and Technology – Production, composition, storage and processing (Ed) D.K. Salunkhe and S S Kadam. Marcel 'Dekker Inc., New York, Basal, Hongkong.

Kapur, K.L. 1975. Varietal suitability of mangoes from Uttar Pradesh for canning. Indian Food Packer, 29(3) : 33-38.

Khedkar, D.N. and Roy, S.K. 1988. Storage studies in dried and dehydrated raw mango slices. Acta Horticulturae, 231 : 721-730.

Kim, H.O., Hewett, E. W. and Lallu, N. 2001. Softening and ethylene production of Kiwi fruit reduced with 1-Methyl cyclopropene. Acta Horticulturae, 553 : 167-170.

Leon, D.M., Pesis, J.C. and Garcia, H.S. 2000. Physiology of 'Tommy Atkins' mango slices stored in modified atmospheres. Institute of Food Technologists Annual Meeting, 10-14 June, 2000, Dallas, Texas. Paper 65A-26 (Abstract).

Maciel, M. I. S., Lima, V.L.A.G., Melo, E.A., Prazeres, F.G., Lima, D.E.S., Souza, A.M. and Lima, M.S. 2004. Quality evaluation of minimally processed mango. Cv. 'Espada'. Acta Horticulturae, 645 : 261-265.

Mahadevaiah, M., Eipeson, W.E., Gowramma, R.V. and Sastry, L.V.L. 1977. Assessment of Rourkela Electrolytic Tinplate for canning Fruit and Vegetable products. Indian Food Packer, 31(2) : 5.

Mahadevaiah, M., Gowramma, R.V., Eipeson, W.E. and Sastry, L.V.L. 1975. Internal Corrosion of Tinplate Containers in canned mango nectar. Journal of Food Science and Agriculture (UK), 26(6) : 821-833.

Mahadevaiah, M., Gowramma, R.V., Eiperson, W.E. and Sastry, L.V.L. 1976. Influence of tinplate variables on the internal corrosion of tin plate containers with mango and orange products. Journal of Food Science and Technology, 13(1): 18-21.

Maini, S.B., Diwan, B. and Anand, J.C. 1984. Use of high molecular weight high density (HMHD) film pouches for processing and storage of mango and tomato pulps. Journal of Food Science and Technology, 21(6) : 410-412.

Maitra, R.K., Chaudhuri, D.R. and Bose, A. N. 1979a. General survey on quality of cured mango slices. Indian Food Packer, 33(6) : 52-57.

Maitra, R.K., Chaudhuri, D.R. Mitra, S.N. and Bose, A.N. 1979b. Lactic fermentation of green mango slices in lime. J. Instn.Chem.India. 51(4) : 156-158.

Mao, I., Wang, G. and Que, F. 2007. Application of 1-Methylcyclopropene prior to cutting reduces wound responses and maintains quality in cut Kiwi fruit. Journal of Food Engineering, 78 : 361-365.

Mehta, G.L. and Tomar, M.C. 1982. Dehydration and utilization of raw mangoes. Indian Food Packer, 36(6) : 75.

Mishra, J. and Tomar, M.C. 2000. Studies on the preparation and storage of Intermediate moisture ripe mango slices of Dashehari cultivar. Indian Food Packer, 54(2) : 61-64.

Mootoo, A. 1991. Effect of post-harvest calcium chloride dips on ripening changes in 'Julie mangoes'. Tropical Science, 31 : 243-248.

NadanaSabhapathi, S., Srivastava, A.N. and Nataraju, S. 1993. Storage study on mango bar in flexible packaging materials. Indian Food Packer, 47(4): 5-7.

Nanjunadaswamy, A.M., Shetty, G.R. and Saroja, S. 1976. Studies on development of newer products from mango. Indian Food Packer, 30 : 85-88.

Naresh, R., Gowramma, R.V., Mahadevaiah, M. and Ananda Swamy, B. 1983. Suitability of tin free steel container for Canned Processed Food Products. Indian Food Packer, 37(3) : 56.

Pelayo, C., Vilas-Boas, E.V., de Be., Baviechou, M. and Kader, A.A. 2003. Variability in responses of partially ripe bananas to 1-Methyl cyclopropene. Post-harvest Biology and Technology, 28 : 75-85.

Plotto, A., Bai, J., Baldwin, E.A. and Bracht, J.K. 2003. Effect of pre-treatment of intact 'Kent' and 'Tommy Atkins' mangoes with ethanol vapor, heat or 1-Methyl cyclopropene on quality and shelf life of fresh cut slices. Proceedings of Florida State Horticultural Society, 116 : 394-400.

Plotto, A., Bai, J., Nawciso,J.A., Brecht,J.K. and Baldwin, E.A. 2006. Ethanol vapour prior to processing extends fresh-cut mango storage by decreasing spoilage, but does not always delay ripening. Post-Harvest Biology and Technology, 39: 134- 145.

Poubol, J. and Izumi, H. 2005. Shelf life and microbial quality of fresh cut mango cubes stored in high CO_2 atmospheres. Journal of Food Science, 70 : 69-74.

Pruthi, J.S. and Bedekar, S.V. 1963. Studies on varietal trials in salting or brining of raw mango slices for subsequent use in Chutney and Pickle. Punjab Horticulture Journal, 3 : 264-271.

Pruthi, J.S. and Teotia, M.S. 1987. Studies on the physico-chemical composition, packaging and formulation of quality standard for raw mango powder (Amchur). Indian Cocoa, Arecanut and Spices Journal, 11(1) : 21-28.

Rameshwar, A. 1979. Tandra (Mango leather) industry in Andhra Pradesh. Indian Food Packer, 33(2) : 11-12.

Rao, V.S. and Roy, S.K. 1980. Studies on dehydration of mango pulp II storage studies on mango sheet/leather. Indian Food Packer, 34(3) : 72.

Rattnapanone, N., Lu, Y, Wu, T. and Watada, A.E. 2001. Quality and microbiological changes of fresh cut mango cubes held in controlled atmosphere. HortScience, 36: 1091-1095.

Rattanapanone, N. and Wateda, A.E. 2000. Respiration rate and respiratory quotient of fresh cut mango (*Mangifera indica* L.) in low oxygen atmosphere. Acta Horticulturae, 509 : 471-478.

Sagar, V.R and khurdiya, D.S. 1999. Indian Food Packer, 53(1).

Santha, K., Rao, K.P.G. and Onkarayya, H. 1982. Storage of mango pulp in bulk and consumer packs. Indian Food Packer, 36(1) : 32-38.

Sethi, V. 1991. Preservation of raw mango slices (var. Neelum) for use in pickle and chutney. Journal of Food Science and Technology, 28(1) : 54-56.

Sethi, V. 1995. Storage study of mango pulp from hybrid 'Amrapali' in different containers. Indian Food Packer, 49(2) : 5-10.

Sethi, V. and Maini, S.B. 1985. Suitability of different types of containers and preservatives for storing mango pulp. Paper presented at II International Symposium on Mango held at Bangalore, May 20-24, Abstract No. 8-17.

Sethi, V. and Maini., S.B. 1991. Studies on storage of mango pulp. Indian Journal of Horticulture, 48(3) : 228-231.

Shrikhande,A.J., Srirangarajan, A.N. and Nadkarni,G.B. 1976. A thermal processing for bulk packaging of mango pulp. Indian Food Packer, 30(5):65.

Siddappa, G.S. and Krishnamurthy, G.V. 1963. Bulk storage of mango pulp. Punjab Horticulture Journal, 3(2-4) : 300.

Singh, K.K. 1977. Refrigerated storage and freezing of mango fruit in India. Punjab Horticulture Journal,17(1-2) : 6-12.

Singh, S.S., Tomar, M.C. and Singh, U.B. 1999. Preparation and storage studies on raw mango panna concentrate. Indian Food Packer, 53(6) : 63-66.

Sisler, E.C. and Serek, K.M. 1997. Inhibition of ethylene responses in plant at the receptor level. Physiologica Plantarium, 100 : 577-582.

Soliva-Fortuny, R.C. and Martin - Belloso, O. 2003. New advances in extending the shelf life of fresh cut fruits - A Review. Trends in Food Science Technology, 14 : 341-353.

Souza, B.S. de., O' Hare, T.S., Durigan, J.F. and Souza, D.S. de. 2006. Impact of atmosphere, organic acids and calcium on quality of fresh cut 'Kensngton' mango. Post-harvest Biology and Technology, 42 : 161-167.

Srivastava, H.C. 1967. Grading, Storage and Marketing. In : Mango : A Hand Book. Indian Council of Agricultural Research, New Delhi. 99-149.

Srivastava, J.S. 1998. Comparative study of RTS drinks prepared from Dashehari and Banganapalli mangoes. Indian Food Packer, 52(2) : 38-40.

Subramanyam, V., Sharma, T.R. and Nath, H. 1976. Freeze dehydration of mango as pulp and as a ready-to-serve beverage. Indian Food Packer, 30(5): 70-75.

Sudhakar, D.V. and Maini, S.B. 1997. Status of sulphite during storage of mango pulp. Indian Food Packer, 51(5) : 14-16.

Sunita Singh, Arajariya, D. and Singh, D. 2005. Effect of thermal treatment on destruction of microorganisms in mango bar and its shelf life. Indian Journal of Microbiology, 45(4) : 295-300.

Teotia, M.S., Manan, J.K. and Saxena, A.K. 1987. Green Mango Processing – A Review. Indian Food Packer, 41(6) : 75-86.

Thakur, B.R. and Arya, S.S. 1990. Packaging requirements and stability of fried wheat snacks. Journal of Food Science and Technology, 27(2) : 76.

Tovar, B., Garcia, H.S. and Mata, M. 2001. Physiology of pre-cut mango II. Evolution of organic acids. Food Research International, 34 : 705-714.

Verma, R.A., Tripathi, V.K., Tripathi, M.P. and Singh, S. 1986. Studies on the effect of packaging materials on mango pickle during storage. Indian Food Packer, 40(6): 22-25.

Vilas – Boas, E.V. de Be and Kader, A.A. 2006. Effect of atmospheric modification, 1-MCP and chemicals on quality of fresh-cut banana. Post-Harvest Biology and Technology, 39 : 155-162.

Vilas – Boas, E.V. de Be and Kader, A.A. 2007. Effect of 1- Methyl cyclopropene (1-MCP) as softening of fresh-cut Kiwi fruit, mango and persimmon slices. Post-Harvest Biology and Technology, 43:238-244.

Watada, A.E., Ko, N. P. and Minott, D.A. 1996. Factors affecting quality of fresh-cut-horticultural products. Post-Harvest Biology and Technology, 9 : 115-125.

Wills, R. B.H., Mc Glasson, W.B., Greham, D., Lu, T.H. and Hall, E.G. 1989. Post-harvest and introduction to the physiology and handling of fruit and vegetables. B.S.P. Professional and Book, London, 59 and 142.

Wright, K.P. and Kader, A.A. 1997. Effect of slicing and controlled atmosphere storage on the ascorbate content and quality of strawberries and persimmons. Post-Harvest Biology and Technology, 10 : 39-48.

CHAPTER 9

Quality Control of Processed Products

Veena Joshi

'Quality' is a measure of the degree of excellence or a degree of acceptability by the consumer. It is the standard, how good a material is as measured against similar produce. The International Organisation for Standardization (1S0) defines `quality' as the operational techniques and activities that are used to fulfil requirements of quality. In operational terms it is, classifying, sorting or grading the produce/product according to their condition of excellence, value or worth. Quality is a combination of characteristics that have significance in determining the degree of acceptability of the product to a consumer. Quality of a food refers to the complete characteristics which differentiate individual units and enable determination of the degree of acceptability by the consumer. Thus `quality' is defined in various ways. However, quality is easier to perceive but difficult to define.

The production of high quality food products has been an important goal ever since man first began to produce food. The food industry should be aware of the consumers' image of food quality and should strive to produce quality products. The expectations of consumer with regard to quality have been changing from time to time. Of late, there is a world-wide trend towards more stringent customer expectations with regard to quality. Therefore, continuous improvements are often necessary to achieve and sustain good economic performance.

Quality Characteristics of Fruits for Processing

For fresh marketing purposes, the characteristics of primary importance are: size, attractiveness, maturity, organoleptic quality (flavour, texture, etc.,) nutritive value and freedom from infections. When the same fruit/vegetable is to undergo processing other properties assume more importance like colour, flavour and texture. Obviously fruit/vegetable of poor quality can not assume a good quality processed product. The quality parameters that consumers look in a processed produce are: wholesomeness and safety, nutrition, appearance, texture and flavour, trust and confidence and marketability of the product. These characteristics are dependent on the physico-chemical composition of raw material, which in turn depend upon the pre-harvest factors (environmental, cultural practices) and post-harvest factors (temperature, relative humidity, respiration, post-harvest handling, post-harvest changes etc).

Quality characteristics are important since they are related to the total yield of a finished product and are therefore, major consideration in processing. In many countries there are precise specifications of various characteristics of products intended for processing. The overall quality of a food product may be broken into component characteristics – colour, texture, flavour, nutritional value, freedom from harmful microorganism and undesirable substances (Ranganna, 1975). Quality characters are also classified as external (colour, form, firmness etc.,) and internal ones (sugar, acid etc., ripening and storability). The quality characteristics of a product may be divided into 3 major categories as follows (Kramer and Twigg, 1973).

I. Sensory characteristics

These include appearance (colour, size, shape and defects); texture and flavour (taste and odour) which the consumer evaluates with his senses. These are generally experienced by all and referred to as organoleptic or sensory properties, since they are perceived by the senses. The sensory tests are of two types:

(i) the difference test, and

(ii) the performance test.

In the first one, the object is to distinguish minor differences (ex. sweetness in mango etc.,) and in the second one to indicate preference for a particular sample. The two types of tests are combined to judge the quality or acceptability of a processed product.

(a) Appearance (Eye-appeal judged by sense of light): The overall eye appearance of a fruit product is more important than taste and odour and may determine acceptance or rejection without a trial testing. Because of the excellent appearance of the product, the consumer is drawn and tempted to purchase the product even before he tasted it. Appearance of the product, therefore, deserves much consideration in food processing.

1. ***Colour:*** Surface colour is of importance for the fresh produce market, while internal colour (of the flesh) is important for the processed products. Colour increases the attractiveness of fruits/vegetables. It is also maturity index and is associated with flavour, texture, nutritive value and wholesomeness. Green colour is an indication of insufficient maturity/ripeness of fruits. Colour of the fruits on ripening should be full and uniform. Carotenoids, chlorophyll and anthocyanin pigments are responsible for colour of fruits and vegetables. Carotenoids produce yellow colour and play an important part in the colour of canned mangoes, citrus and pineapple. The red and violet colour in fruits/vegetables is due to anthocyanin and this gradually passes into the syrup

or brine used in canning. Certain fruits and vegetables like guava, litchi, banana and broad beans may turn pink during processing due to the presence of leuco-anthocyanins. Pigment determination, though used as a measure of colour quality of processed products, is rarely used for measuring the colour of fresh produce even when intended for processing (Ranganna, 1975).

The colour requirement of fruits and vegetables used for quick-freezing differs markedly from those needed for canning, since in quick-freezing, there is little change of chlorophyll to phaeophylum, no marked change in leuco-anthocyanins and little migration of anthocyanin from fruit to syrup. Colour and appearance are extremely important quality attributes in unblanched frozen, cut fruits because they are subject to enzymatic browning when thawed.

The visual colour depends on the spectral distribution of light reflected from it. Light reflected from coloured object can be divided into 3 components: termed as hue, value and chroma. 'Hue' is the predominant wave length, which determines whether the colour is red, green, yellow etc, 'value' **(Y percent)** refers to lightness or darkness of colour and 'chroma' refers to the intensity of the colour (Ranganna, 1975).

The instruments used for colour measurements are: Hunter colour difference meter, Spectronic-20 (reflectance measurement), Loviband Tinto meter and Spectrophotometer (Mehta *et al* 2002). To denote colour, reference is made to colour dictionaries/charts. The colour of the product can also be measured by comparing to defined colour disks or tiles (Ranganna, 1975).

The measurement of surface colour presents certain problems. When the unit is large, such as mango, papaya, guava, apple, pumpkin, etc., individual units have to be measured. When the unit size is small such as strawberry, grapes, phalsa etc., a representative sample can be measured. Kramer and Twigg (1973) suggested that colour measurement, limited to lightness only is a good indicator of end product. Colour quality in those instances where lightness is the dominant colour quality of the processed product as well.

(b) ***Size and shape:*** The importance of size and shape of fruits and vegetables is often under-estimated. They make important contribution to the appearance of fresh produce and processed products. To determine the suitability of fruits and vegetables for processing, grading of fruits and vegetables into various size and shape categories is one of the first steps in packaging and processing operation. Size grading is done mainly to facilitate succeeding operations such as cutting, peeling or blanching to obtain uniformity in the product and to provide consumers with the preferred size. Shape of the raw materials sometimes determines the suitability for processing. To reduce losses during mechanical trimming and handling the shape of the fruit/vegetable should readily lend itself to such operations. Many tropical fruits often have inconvenient shape or size and thus present problems of handling, while processing. ex: mango, papaya, guava etc. Size and shape can be measured manually with simple scale, vernier callipers, micrometers, planimeters and machines, which measure weight, diameter and length (Mehta *et al* 2002).

Size, shape and wholeness make important contribution to the appearance of the fresh material as well as processed products. Wholeness refers to degree of whole and broken and mashy pieces, for Ex. canned mango slices may have a few soft and mashy pieces in the can. In the packaged product, container label and the pattern of arrangement contribute to the appearance of the product (Ranganna, 1975).

(c) ***Defects:*** Defects in the raw material lower the grade of the product. Defects are imperfections due to the presence of something that detracts from perfection. Fruits and vegetables may be attacked with fungal infection either before harvest in the field or after harvest. Some fruits may exhibit physiological defects like soft rot, jelly seed, spongy tissue in mango, granulation in citrus etc. Fruits may be infested with insect pests. If adequate care is not taken, the finished product may contain insect fragments, eggs or larvae or faecal matter (ex. stone weevil in mango). Mechanical damage may occur due to hail storms in the field or during harvesting, handling and transport, which manifests as bruises, scars, cuts, lesions, splits etc. Such defects might be either major or minor depending upon the end product. Examination of a random sample for specific defects, determines whether they are critical, major or minor and help in deciding to accept or reject the lot. Examination for fungal infection involves either visual or cutting the material at random. The use of highly amplified light transmitting instrument for the purpose was proposed (Birth and Norris, 1965). In addition to visual observation in the field and before processing, it is a good practice to carry out occasionally micro-analytical examination for insect fragments etc., as a part of quality control procedure (Ranganna, 1975). In the acceptance inspection of the raw material, it is of utmost importance to determine the limiting attributes which determine the ultimate quality of the product. Defects in processed fruit products besides the above also include deformities due to hereditary abnormality and extraneous matter which might be harmless, but not a part of the edible portion, such as peel in mango slices etc (Ranganna, 1975).

(d) **Texture:** Texture qualities of food are those which one can feel with finger, the tongue, the palate or the teeth. The range of texture is considerable and the departure from an expected texture is a quality defect. It involves both hand feel (firmness, softness, juiciness) and mouth feel (chewiness, fibrousness, grittiness).

The instruments used for measuring textural equality are: Succulometer, tenderometer and pressure meter (Ranganna, 1975., Mehta *et al* 2002). The principle mainly consists of resistance to force (Ranganna, 1975).

(i) **Pressure tester:** This is a very light and portable instrument, having a spring scale with mechanical rod, rounded at the bottom attached to a spring. This is used to measure the maturity and texture of fruits. It measures rapidly, but the measurement is not precise.

(ii) **Mechanical thumb:** To overcome the problem of penetration and variability in results obtained, a mechanical thumb is used. It consists of a sieve attached to the base of pressure tester spring.

(iii) **Fibre pressure tester:** In this case the penetration end of the pressure tester is replaced with the blade of specified thickness.

(iv) **Tenderometer:** It consists of lower and upper sets of bars. With rotation of upper bars, tenderness of sample (ex: peas) is judged with this.

(v) **Shear press:** It consists of a basic unit capable of applying force under controlled rates per unit time and the force extended can easily be judged.

(e) **Sugar: Acid Ratio:** From the stand point of taste appeal, particularly in the case of juices, sugar:acid ratio is an important criterion. Soluble solids content indicates the maturity of the raw material. It is measured by a hand refractometer, with the assumption that the soluble solid content consists entirely of sucrose. Brix hydrometer is used for determining the soluble solids content in juices. Acidity, which is a measure of sourness, is important from the stand point of flavour and wholesomeness of the processed products. The titratable acidity indicates the sources, while pH is more important from the processing point of view (Ranganna, 1975).

(f) **Flavour:** The term 'flavour' is generally used to denote both taste and aroma (odour) in combination. Flavour is defined as the totality of sensation, aroma and taste, coupled when present with warmth, cold and mild pain. According to another definition 'flavour' is a complex sensation comprising taste, odour, roughness, smoothness, hotness or coldness, pungency or blandness. Odour or aroma is a complex sensation and is the most important factor in flavour. Taste indicates sweetness, saltyness, bitterness, sourness and astringency. Smell may be fragrant, acidic and burnt, due to enzymatic, physiological and chemical changes.

In case of fruits, one is mainly interested in the aroma, which in most cases is due to the presence of volatile odoriferous compounds. In fruits, the skin and peels carry most of the flavour (ex: citrus), while in case of other fruits (ex: mango, guava), flesh is the main contribution to the characteristic aroma. Flavours of fruits are multi-component systems carrying 5000 or more constituents. Changes in the raw material, maturation and ripening, handling, processing and storage conditions affect the flavour of fruits and their processed products (Teotia *et al* 1987).

Gas chromatography, Spectrometry and objective measurements are used to determine flavour (Mehta *et al* 2002). Unlike colour and texture, flavour is difficult to assess objectively. Since, flavour is a physiological sensation, it is by definition subjective and therefore, is not amenable for measurement. Basic characteristics of taste, like sweetness, saltiness, sourness, and astringency can be determined objectively, but subtle odour characteristics are difficult to measure. On the contrary some progress was made in the dilution of off-flavours like determination/estimation of volatile acids, amines and succinic acid and enzymes' activity as indication of off-odour/off-flavour in stored fruits and vegetables. However, gas chromatographic methods of measuring flavour, remained largely as research and quality control tools. When it comes to consumer quality acceptance, there is still no substitute to sensory methods by a panel consisting of trained individual/groups (Ranganna, 1975).

II. Hidden characters

These are the characters which a consumer can not evaluate with his senses. Examples include, nutritive value of raw material, presence of harmless adulterants and presence of toxic substances (Mehta *et al* 2002).

(i) **Nutritive value:** Nutritional quality characters include specific vitamins, minerals etc., that can be assessed by chemical, instrumental or biological methods. The details regarding nutritive value of mango fruits are presented in Chapter I. Higher the nutritive value of the raw material, the more nutritious product is incidentally prepared and preferred.

(ii) **Toxicity:** Various chemical compounds (pesticides, growth regulators, etc.,) are employed extensively in fruit and vegetable production. Edible tissues may accumulate amounts of persistant pesticides, particularly those belonging to chlorinated hydrocarbon group, even beyond permissible limits. The chemical residues lead to bitter or mashy flavour in the processed products and pose a health hazard to the consumer. Therefore, only after assessing the chemical residues, one can determine to use that raw material for processing.

III. Qualitative characteristics

These characteristics (crop yield, yield of end product) are also considered as attributes of food quality, since they form a part of total evaluation of the product (Mehta *et al* 2002).

(a) **Crop yield:** Good yields, pest and disease resistance of the crop, and freedom from physiological disorders in the raw material (fruits) reduce the cost of production and processing and help in producing quality products.

(b) **Yield of finished product:** Cost of raw material per unit weight of finished product is another important consideration in processing. It is calculated by determining the amount of product yield per unit weight of raw material. The ratio of the weight of raw material to the weight of pre-packaged finished product is called the overall "Shrinkage ratio". The higher the ratio for a given product, the greater will be unit cost of the finished product. Naturally low 'shrinkage ratios" are desired. Factors, important in determining overall "shrinkage ratios' include : dry matter content of the raw material, peels, stone size and shape of the fruit (small and irregular shaped 'fruits' have greater peeling, slicing costs and losses and rejects for poor colour, odour and composition (in mango). In the processing of juice concentrate, the solid content and yield of juice are equally important and determine the cost of the finished product.

IV. Additional quality characteristics

Apart from the above, two other important ones, not always apparent to sensory observations are (i) sanitary quality and (ii) keeping quality of raw material as well as that of the products.

(i) **Sanitary quality:** This refers to the freedom of raw material as well as processed product from infection and infestation. Sanitary quality is usually measured by bacterial, yeast, moulds and insect fragment counts etc. in the raw material as well as processed material. Raw material borne infection and infestation reduce consumer confidence in the safety of product, which is considered low/inferior in quality.

(ii) **Keeping quality:** The finished product should be able to store for prolonged storage till disposal/consumption without quality deterioration. To determine keeping quality or storage stability, accelerated storage tests like extremes of temperature, humidity or other variables (O_2, CO_2) are used to hasten development of quality defects in a short time. However, such methods/tests may be chosen with considerable care since an extreme temperature or other variables frequently will alter the pattern of quality determination.

V. Others

Cuts, cracks and other deformities on surface of the fruits and vegetables lower the quality and consequence acceptability of the produce. Such defects lead to increased chances of infection and wastage and ultimately may create risks and poor products as well as increased cost in processing and preservation.

QUALITY CONTROL

Food is defined under the law as any article used for food or drink by man or animal and is interpreted to include substances added to food which may have no nutritive value, what so ever, ex : artificial colouring, condiments, spices, flavour and preservatives. Food quality control is generally defined as the regulation by laws of food manufacture, distribution and sale, in order to prevent health hazards and fraud to the consumer. Thus, it becomes a criminal offense to sell (deliberately or in any other way), adulterated, filthy or contaminated food. There are three main aspects of the application of food quality controls:

(i) Moral,

(ii) Commercial and

(iii) Legal.

The application of control on food quality puts a moral responsibility on the food manufacturer towards the consumer. The commercial aspect requires that standard quality products are economically produced. In increasing competitive markets it is important that producers exercise a stringent degree of quality control on what they produce. The legal view point demands that the quality of products confirm to national and international standards. Fruit and vegetable industries, which produce very large quantities of products, have a special responsibility to ensure that their products are both wholesome and safe as well as successful in the markets. The control of food quality by law leads to:

(i) Improved quality of production.

(ii) Achievement of greater consumer satisfaction.

(iii) Promotion of quality consciousness.

(iv) Increased consumption and sales.

(v) Employment opportunities for scientific and technical personnel.

(vi) Avoidance of controversy and litigation in marketing at the nation and international levels.

(vii) Promotion of national and international trade.

(viii) Provision of the means of the intelligent comparison of prices in relation to quality and grade.

(ix) Greater confidence in the minds of consumers.

Quality control should start right from the field during production of crop in relation to the intended use. Quantity and timing of irrigation; amount, source and timing of manures and fertilizers and pesticides etc. have profound influence on the quality and yield of end-products.

Quality control within the fruit processing industry demands constant vigilance at all stages in processing, so that the necessary adjustments can be made at the appropriate time. The specific responsibility of quality control is to ensure that the systems used produce a standard product with the acceptable quality in respect to nutrition, purity, wholesomeness and palatability. In other wards, the purpose of quality control is to ensure production of products which meets defined standards within the accepted tolerance at a cost, which is competitive with the market for which it is desired and the price at which it will sell. Thus, the precision and the accuracy of the specification are the two important requirements of quality control.

Quality control involves

(a) sampling

(b) raw material control at

   (i) grower's level

   (ii) crop production level,

   (iii) harvest level and at

   (iv) processing level,

(c) inspection of finished product and

(d) score cards for assessment of quality etc.

Quality control involves operational techniques and activities aimed both at monitoring a process and at eliminating causes of unsatisfactory performance at different stages in order to result in economic effectiveness. Thus, quality control involves procedures and systems used to ensure that all aspects of manufacturing processes are regulated so that the quality of the final product is assured (Sardana and Vaidya, 1994). Once a product has gone through the manufacturing process, little/nothing can be done to alter the quality or to overcome the defects/results. Effective control over raw materials and process minimizes the efforts needed in the elimination of the finished product and consequent rejection of the product if not up to the desired standard (Ranganna, 1975., Mehta *et al* 2002).

QUALITY CONTROL PROCEDURES

1. **Sampling:** A representative sample is drawn and used, since it is not possible to examine each unit. Sampling must enable rejection of lots sampled if they do not pass rigidly established quality levels. In such cases simplified procedure (sampling by attributes) may be used. For example, raw materials may be sampled for critical defects which would render the finished product unsaleable, such as moulds, yeasts and bacterial count of raw material which could not be entirely eliminated from the finished product. In the sampling, errors must be established for each commodity at the point of the section in terms of standard deviation.

2. **Raw material control:** According to the food industry "raw material' is anything purchased/used by the manufacturer for direct or indirect use in food (fruits, vegetables etc.,) processing. Raw materials include: food ingredients (fruits, vegetables etc.,) water and packaging materials. The quality of raw materials required varies according to the material, the intended processed product and the standard qualities desired in the end-product.

The quality of a processed fruit/vegetables product ultimately depends upon the quality of the raw material that is used to make the product. As such the quality of raw material goes back to the tree stock in case of fruits, growing conditions and practices and weather conditions, which influence the quality of raw material. The degree of maturity and ripeness when picked and the method of harvesting are important from the point of processing. Quality control at harvest level is practised to test the stage of maturity in order to determine the exact time at which average quality of the lot harvested will be optimum for processing (Mehta *et al* 2002). Appropriate time to harvest the fruit depends on variety, location, weather, ease of removal from the parent tree and the purpose for which it is being harvested. In general, fruits for canning should be slightly less ripe than for table purpose as processing causes further softening. Fruits, when ripe are of high quality, as they would be larger, more sweet, less sour and/or succulent. Over ripe fruits become soft and mashy on processing (Ranganna, 1975).

Fruits do not ripen uniformly and have a short period of optimum maturity. Hence, it is essential to time the harvest so that the grower and the processor can benefit. The harvesting season can be adjusted according to the need of the processor, by staggering the date of flowering (ex. vegetables), by using varieties of different harvesting seasons (early, mid and late) and by getting raw material from different growing areas (Ranganna, 1975).

Quality control of raw material should be exercised during the production stage of the crop. Irrigation, application of manures and fertilizers, pesticides etc., have significant effect on the quality and yield of the end product. Fruits and vegetables intended for fresh market when irrigated towards the end of the growing season help in maximising the yield (by weight). On the contrary, if the crop is intended for processing a concentrated or de-hydrated product, irrigation towards the end of the growing season should be reduced or entirely suspended to increase the weight of the solids and reduce the cost of concentration or dehydration (Kramer and Twigg, 1973).

Raw material inspection includes tests for genuineness, composition, freedom from contamination and conformity with official standards. Once the raw material quality standards are established, regulatory checks are carried out to produce the standardized products. After the selection and control of sample batch a trail run is carried out which gives a pre-view of the end product. All control tests are run on the sample and any adjustments in processing procedure are made wherever necessary. The shelf life run is examined for sign of corrosion due to acidity or alcohol content of the raw materials used. Approval for processing of raw material should be given only after all quality specifications on sample run have been met rigidly (Mehta *et al* 2002).

Before buying raw material in bulk, processing industries generally examine a buying sample to make sure that it confirms to the factory's specifications. This is very important especially if samples have to be obtained from different sources. It is also necessary to examine the representative sample of any subsequent purchase/delivery to make sure that the bulk is up to approved factory specifications. Failure to check conformity to the specification may lead to the acquisition of unusable quantities of raw materials.

3. **In-plant quality control:** To maintain the quality of a processed product, a systematic in-plant control programme is essential. This programme is dependent upon the quality demanded and the price, which the customer can afford to pay. In processing plant to meet the quality specifications and market demands, quality control testing must start with i) acceptance, sampling and testing of raw material, ii) process control and iii) examination of finished products.

   (i) ***Acceptance inspection:*** This is the most important part of the quality control system. At this point the processer has the chance of rejecting unacceptable raw material. It would also help to determine the processing method to get the end-product of desired quality. The raw materials, which make up the product may be classified into major and the minor. For example, in the preparation of mango pickle, mango cut pieces is the major raw material and other raw materials (ingredients) which make up the product like salt, oil, spices etc. are the minor raw materials. The raw materials used should be tested in relation to their contribution to product quality. Raw materials which deteriorate during prolonged storage should be analysed at intervals to ensure that on preparation, they yield product of desirable quality, besides providing special storage facilities to avoid their deterioration. Parameters have been developed for testing the quality of fresh fruits and vegetables (Ranganna, 1975).

   (ii) ***Process control:*** During the actual processing, careful attention should be paid to the processing procedure. The process should be checked whether standardized procedures are followed, whether ingredients are used in correct proportions, whether accurate methods of preparation and mixing are employed, whether containers being used are sound and clean, whether standardized processing times and temperature are followed, to make sure the desired results are achieved. The production staff should be acquainted with the quality specifications, established by technical staff of the processing industry. The technical staff should also evaluate the production samples for compliance of the specification.

   It is vital that quality control tests run continuously and concurrently with a 24 hour production schedule. Failure to maintain a round the clock check can lead to rejection of raw material as substandard, as it does not meet the quality specification. This is a very costly inspection, but liaision between the quality control and production department is of paramount importance.

Intermediary product samples may be tested on routine basis to establish that specific targets of quality are being achieved. The desired composition, consistency, colour and concentration are checked and ensured. Rapid, on the spot analyses are to be used to achieve prompt results so that appropriate information can immediately be supplied to the concern. This rapid communication of information is very important in process control. Where processing controls are not properly employed, ex: during dehyeration, the quality of the product may be seriously impaired, as evident as change in shape or structure cracks, case hardening, browning reaction etc.

Satisfactory hygienic conditions should also be maintained during processing in order to prevent microbial contamination. For the routine microbial control of the processing factory microbial counts on utensils, equipment, working surfaces, walls and floors are to be regularly carried out and the results are recorded for immediate action. In addition, periodic inspection of the plant should be made by trained inspectors to make sure that adequate hygiene standards are maintained.

Process control must relate to the result of processing, to the test performed on raw materials and meet the standards of the finished product. The manufacturing processes, more so, where a batch process is adopted, may cause considerable variation in the quality of the finished product. It is, therefore, essential to instal suitable testing procedures during the process.

In planning, a process control, the sequence of steps in the process should be drawn preferably with the help of the flow diagram. Alternate processes and materials should be considered at this stage. The critical points in the process which bring about alteration that reflect in the quality of the finished product should be defined and alteration is concentrated on this. The critical points for processing of canned mango slices, for example, are

1. grading of fruits for size and maturity,
2. peeling,
3. slicing,
4. sorting of slices,
5. can filling,
6. syrup composition and temperature filling,
7. ex-hausting,
8. can closing,
9. processing and
10. cooling.

(iii) ***Inspection of finished product:*** Inspection or examination of the finished product is necessary to ascertain to what extent the desired quality specifications have been met/achieved. If the finished product is faulty, nothing can be done, except rejection. The points of consideration are:

1. the external appearance of can/pack,
2. net weight,
3. vacuum,
4. drained weight,
5. internal appearances of the can,
6. number, colour and texture defects, if any and taste and flavour of syrup, in case of canned products.

The ability to withstand storage can also be confirmed on the finished product. For example, in case of canned products, careful examination is made of the external conditions of the can for distinctive signs. A can with both ends being concave is said to be `flat' and is considered to be good because it means that the vacuum in side is high enough to maintain the ends in a concave conditions.

Cut-out analysis may be carried out to know the product's chemical composition and to check that the composition is confined to meet legal requirements. In case of dried products, samples are to be examined for a blemish count. The dried product is regularly checked for its reconstitution value to enable the correct cooking instructions to be supplied on the package for the ultimate user.

At the end of the production line, packets are weighed for their correct weight and volume as specified. Tests are also performed to check certain physical properties such as crispness, colour, viscosity and texture, which are related to the palatability and acceptability of the product. The keeping qualities of foods, which depend upon sugar, salt or acid for their preservation are checked to ensure that they will keep well under the condition to which they are most likely to be exposed. Both the contents and containers are to be examined for faults, rancidity, development of microorganisms, loss of colour and flavour. Microbial examinations are carried out to check whether proper hygienic procedures have been followed and whether the finished product is safe to keep and to eat/drink.

The ultimate criterion for desirability in the product is checked by sensory evaluation by both expert sensory judgement as well as market testing.

(iv) ***Sensory evaluation:*** After physical, chemical and microbiological examinations have been performed on a finished product, with satisfactory results, the product is considered ready for distribution/sale, but it should be done only after its palatability or sensory quality is assessed. Palatability or sensory quality is of great importance to both processors and consumers. To the processors a palatable product ensures sales, because good palatability attracts consumers; to the consumer, palatability satisfies his aesthetic and gustatory senses. Sensory quality is a combination of different senses of perfection, which come into play in choosing and eating a food item. The principal sensory properties which affect the palatability of food are: appearance, texture and flavour.

The desired physical and chemical tests to measure differences in the sensory qualities of foods, alone are not adequate to give the required information on their acceptability by the consumer. Human judges, therefore, have to be employed. Measurement of the relative palatability of a food product is carried in two ways:

(i) by obtaining the judgement of experts and

(ii) by testing the performances of a sample of the public for whom the product is intended, also known as "market testing".

(i) ***Expert Sensory Judgement**:* Sensory evaluation (acceptance measurement) is generally performed by a panel of judges, who are trained for the purpose in order that their sensitivity and consistency are well established by repeated tests. These judges after testing and tasting allot marks for each parameter. The average of these marks will be considered for judging the quality and acceptability of the finished product. The responsibility of a trained panel of judges is generally to look after inline quality of the final product, process development and to a limited extent preliminary acceptance testing.

(ii) ***Market testing:*** This is carried out to obtain the preference of a sample of the public for whom the product is intended. An untrained panel, made up of number of men and women selected to be representative of the population to be served, is used after ascertaining their natural emotional reaction to the selected food product. However, such survey though required and beneficial in promoting the product, is time consuming and costly. They are, therefore restricted to products selected through a series of laboratory tests and presented in their marketable form. Data from these surveys (consumer survey) are analysed statistically to determine the acceptability or rejection of the product in question. However, obtaining a representative sample of consumers and the interpretation of market survey data are very difficult.

(iii) ***Score cards for assessment of quality:*** As mentioned earlier, it is easy to perceive quality than to define. Based on this, sensory quality was developed. For assessment of quality of finished products score cards are developed to be used by the panel of expert judges and consumers. Each country has developed its own score cards. The Indian Standards Institution (ISI) specifications for scoring mango products are as follows (Table 9.1):

Table 9.1: ISI Specifications for scoring of mango products

Code No.	Product	Grade I				Grade II			
		Colour	Texture and Uniformity of size	Taste and Flavour	Absence of defects	Colour	Texture and Uniformity of size	Taste and Flavour	Absence of defects
IS: 2867-1964	Canned mangoes	15	19	15	26	13	16	13	23
IS:3880-1966	Canned mango pulp	15	15b	30	15	13	13b	26	13

b. Denotes consistency

Packaging

Packaging has multipurpose functions for food products. The primary purpose of a manufacturer is to protect the food product, to keep it in good condition and to preserve the flavour until it reaches the consumer. It is essential, therefore, that a suitable form of packaging is choosen for a finished product. The package must be capable of protecting the product from thermal changes, humidity variation, hazards of rough handling and infestation and contamination by rodents and insects. It must be kept under hygienic conditions at all times. The material used for packaging must not contaminate the product and must be effective in preventing the product from deterioration.

Labelling

After packaging, the product should be labelled suitably. Labelling must reflect on the quality of product. Effective labelling should be clear and informative. The name and addresses of the manufacturer, the list of important ingredients and additions and the net weight/volume of the product, manufacturing date, expiry date and instruction for use, time within which it should be used best, should be declared on the label. A good and attractive label is an aid to the successful marketing of the products (Mehta *et al* 2002).

Storage

The quality control must be extended and excercised to storage to protect the product from post-processes contamination to keep it in sound condition and to preserve its flavour and wholesomeness until it reaches the consumer. Therefore, finished products must be stored under optimum and appropriate conditions. High moisture, high storage temperatures, light and air/oxygen, are all determined to stored foods in general and to dehydrated foods in particular. Dehydrated foods are generally hygroscopic and easily pick up moisture from their surroundings during storage followed by mould and insect infestation. It is important that dehydrated foods are stored in a cool, dry and dark places to prevent the deteriorative effects of high storage temperatures, light, air and humidity. Since most manufactured foods can be stored for many years under optimum and appropriate conditions, it is important to establish the conditions and storage periods that afford optimum balance between the cost of storage and the changes in quality of stored products. Warehouse facilities are generally provided close to industries for the storage of both raw material and finished products. Careless handling should be avoided. Quick and efficient transportation need to be provided for the distribution of the product to the market.

Quality Control Parameters for Different Mango Products

I. For Juices/Concentrates/Beverages

(a) **Total solids** are determined by Vacuum oven drying at 70 °C and should be (%) 100- M.C.%

(b) **Total soluble solids:** These are determined by brix hydrometers or with hand refactometer.

(c) **Acidity/pH:** Usually acidity is determined by titration with standardized alkali using phenolphthalein as an indicator. The pH should be around 3-4.

(d) **Sugar: acid ratio**: This is the ratio of degree of Brix to one gram of anhydrous citric acid in 100 grams of juice/concentrate.

(e) **Pectic substance**: It is determined as calcium pectate.

(f) **Tannins:** Tannins are determined by titration with (i) $KMNO_4$ or (ii) indigo caramine solution.

II. Dehydrated products

(a) **Moisture:** Moisture in the product is determined by drying the sample in vacuum oven at 70 $\pm$ 1 °C at a pressure of 50 $\pm$ 10 mm and expressed in percentage.

(b) **Enzyme test:** The acidity of catalase and peroxidase are tested qualitatively.

(c) **Bulk density:** This is determined by filling the sample into a container of specified volume and then weighed.

(d) **Per cent reconstitution:** It is done to determine the dehydration of the product. It tells about the success of the dehydration process.

(e) **Rehydration ratio:** It is the ratio of weight of rehydrated product to the weight of dehydrated product.

(f) **Coefficient of rehydration:** This differs from product to product and process to process being used.

(g) **Per cent water in rehydrated material:** It is given by the formula:

$$\frac{\text{Drained weight of rehydrated material-Dry matter content in sample taken for rehydration}}{\text{Drained weight of rehydrated sample}}$$

III. Canned Products

(a) **External appearance:** Body dents, leakage around seam, condition of ends etc.

(b) **Weight of can:** It is noted as gross weight.

(c) **Vacuum or pressure:** Vacuum or pressure in side the can is determined by using a guage which indicates both vacuum and pressure.

(d) **Internal quality parameters**

(i) ***Inspection of contents:*** Appearance of the surface, under-filling and over-filling, cloudiness in syrup or brix are noted after opening of can.

(ii) ***Drained weight:*** Weights of contents after draining the syrup over a circular sieve are determined.

(iii) ***Net weight of contents:*** This is measured by subtracting tare weight of can from gross weight.

(iv) ***Internal condition of can:*** This is observed for any evidence of corrosion, blistering or removal of lacquering.

(v) ***Condition of syrup:*** Colour, clarity and flavour of syrup are noted.

(vi) ***Soluble solids:*** Determined by using hand refractometer.

(vii) ***Acidity:*** Determined by titration against standard alkali using phenolphthalein as indicator.

(viii) ***Defects in canned product:*** Here, all types of extraneous matter, blemishes, non-uniformity in grading etc., are noted.

(ix) ***Colour of the product:*** Pale and dull appearance makes the product unattractive and so not preferred by consumers.

(x) ***Texture:*** The product should have a good texture and be free from porosity.

(xi) ***Biological examination:*** In canned products, presence of bacteria, total counts and particular microorganism are determined. For this canned product is incubated at 37 °C and 55 °C for 7 and 10 days respectively to check any growth of bacterial spoilage/bulging.

IV. Pickles

The following are the quality control parameters in case of pickles. (Usha Rani *et al* 1992).

1. Moisture content,
2. Salt concentration,
3. pH,
4. Acidity,
5. Total oil,
6. Drained oil,
7. Fluid portion,
8. Insect fragments,
9. Trace elements,
10. Heavy metals and
11. Free fatty acids.

The microbial condition of the product governs the safety and keeping quality. Advances made in thermal processing have rendered the canned foods absolutely safe from food poisoning. Contamination may occur from raw materials, equipment, water, working staff in the processing industry etc. The construction of the buildings, walls, floors and ceilings should be such that they are not able to harbour microbial growth. Cleaning schedule must be drawn up to maintain absolute sanitary conditions. There should be adequate toilet and washing facilities. The measures taken should eliminate or at least reduce to a minimum the possible sources of contamination with microorganisms. The measures should also eliminate possible sources of insect infestation and rodent contamination.

Responsibilities of Quality Control Unit

The responsibility for quality control in food processing industries is generally delegated to a person or a department, depending on the size of the factory, whose specific responsibilities include the following:

1. Standardizing procedure for sampling and examination of raw materials.
2. Development of test procedures.
3. Establishment and implementation of quality standards for fresh and processed products.
4. Setting up preventive quality control methods for in-plant liaison between manufacturing section and test laboratories.
5. Examination/inspection of finished products.
6. Storage control.
7. Recording and reporting.

8. Dealing special problems, including attendance to consumer complaints, by locating their cause and eliminating them.
9. Research and development into new products.
10. Packaging and storage.

QUALITY ASSURANCE

Quality assurance is defined as the provision or proof or evidence that the requirements for quality have been met. ISO defines it as all the planned and systematic actions necessary to provide adequate confidence that a product service will satisfy given requirement for quality. Thus it is a system designed to prevent, detect and correct product defects that would cause customer dissatisfaction. Quality assurance is concerned with survey, audits and the evaluation of the quality factors that affect the specifications in section and customer acceptance of a particular product. It also involves a continuing evaluation of adequacy and effectiveness of the quality of control programme. Although quality assurance and quality control may be a bit confusing, quality assurance is a broad programme that tries to anticipate and prevent potential defects from occurring, while quality control attempts to detect and correct the defects in products (Sardana and Vaidya, 1994).

Parameters which are analysed during quality assurance:

Mehta *et al* (2002) have detailed various parameters that are generally analysed during quality assurance as follows:

(a) ***Moisture:*** It helps in determining maturity and tenderness of fresh fruits. It is determined by hot-air oven method or distillation method or Brabender moisture tester.

(b) ***Water insoluble solids:*** This gives information regarding texture. Fruits are considered more ripe if these solids are less. These solids are determined in per cent by amount of dried residue left after boiling and filtration.

(c) ***Alcohol insoluble solids:*** Fruits are considered fully ripe and less mature if these solids are low. The alcohol insoluble solids include some starches, cellulose, fibre, pectins and proteins. The method involves extraction of sample with 70-80 per cent alcohol, washed and dried. Wasir Blender is used in the determination of these solids.

(d) ***Total soluble solids:*** These give an indication of sweetness of the product which in turn reveals maturity and ripeness of fruits. For fruits high level of total soluble solids are desirable. These are determined by using Abbel hand refractometer.

(e) ***Acidity:*** It indicates flavour as well as wholesomeness of the product. It is determined by titration method or as pH.

(f) ***Sugar: Acid Ratio:*** This is evaluated to determine the taste of the fruit product. Total sugars and titrable acidity are calculated to derive at the ratio. For each product, there is a desirable ratio, ex: for apple products 40:1, for orange juice optimum acceptable ratio is 10 : 1.

(g) ***Density:*** It is highly correlated with the total solid contents and measure of density can frequently replace total solid determination. Density of the product can be measured by flotation method.

(h) ***Bulk density:*** It gives an indication of fluffiness, compatability or over-run. Freeze dried products would have higher bulk densities than conventionally dehydrated products.

(i) ***Succulence:*** It is closely associated with moisture or TSS. Hence, it may be used to provide the information on state of maturity. It can be checked by an instrument "succulometer".

(j) ***Fibre:*** It represents all the tough tissue, which are not readily comminuted by normal chewing.

(k) **Percentage of recovery**: This is related to the state of maturity or ripeness of crop when harvested or processed.

The Need for Quality Control and Quality Assurance

Many studies on quality evaluation of various products of mango from the commercial centres and markets have indicated that the products are of poor quality; do not meet the specific standards, have no uniformity in either number of ingredients used or in their preparations, not made with uniform methods, contain excess or low quantity of various ingredients, contaminated with microbes, insects, rodents etc. (Usha and Anand, 1981., Teotia *et al* 1982., Usha Rani *et al* 1992). Further, the present population, both national and international has become more health consciousness and demand high quality products without reference to the prices and are willing to pay more for a quality product, which is safe and hazardless for consumption.

Food safety is attracting a lot of attention, because of free trade and microbial safety, which are the pre-requisite for long term survival of trade and for entry into competitive global markets. The products, free from preservative chemicals and pesticide residues are in greater demand. Techniques that produce safe and hygienic goods at minimum cost and without compromising on environment safety parameters are getting priority.

Further, doors are open for export of products and the global market demands, high quality, healthy, safety and hazardless products. So for constituent production of a satisfactory produce at a minimum cost, there is a need for quality control unit in every processing industry (Saradana and Vaidya, 1994). In order to be successful a processing unit/industry should offer products that:

(i) Meet a well defined need, use or purpose.

(ii) Satisfy customer's expectations.

(iii) Comply with applicable standards and specifications.

(iv) Comply with statutory (and other) requirements of the society.

(v) Are made available at competitive prices.

(vi) Are provided at a cost which will yield a profit.

A well organized and properly functioning quality control systems will:

(i) Improve product quality.

(ii) Reduce defective items and rejects.

(iii) Maintain uniform quality.

(iv) Improve plant sanitation and house keeping.

(v) Decrease consumer complaints.

(vi) increase consumer satisfaction.

(vii) Improve employee awareness and boost morals.

(viii) Minimise cost of production etc.

(ix) Ensure food safety and quality.

Sanitation and Health Problems in Quality Control

Ranganna (1975) has enumerated the following sanitation and health problems in quality control by processing industries.

In the commercial processing/manufacturing of fruit and vegetable products, it is not only necessary to manufacture the products that would satisfy the consumer but also safe and hazardless to his health. The production of each product is affected by the conditions of cultivation, time of harvest, condition of fruit and vegetable at harvest (maturity/ripeness), susceptibility to diseases, pests and their incidence etc, which in turn affect the quality of ultimate product. Fruits and vegetables may be contaminated with pesticide residues which might have toxilogical significance or affect the sensory characteristics of the product. Additives are being used either to improve the product quality or to preserve the product. The permitted lists of these additives are being constantly revised on the toxicological data. Manufacturers should know and keep abreast of these changes. In the manufacture of products meant for export, appropriate additives permitted in the importing country must 'only be used'.

The packaging material used should not only meet the packaging needs but also safety requirements. For example, paraffin wax is generally considered as inert, safe and free from any hazard after ingestion. However, it was shown that some of the paraffin waxes may be contaminated with well known corcicogens such as 3, 4 - benzopyrine (Shubik *et al* 1962). This shows the utmost importance of an appropriate quality control check for any unsuspected health hazards.

QUALITY STANDARDS

India, being a signatory to WTO agreement, has an obligation on its part to follow universal product quality standardization through CODEX particularly whenever products are exported to global markets. This standard is basically meant for observing the hygienic practices in the food processing industry right from the farming and growing aspects of the raw material until it reaches the consumer in a processed form. Quality standard is a technical document used as the basis for commercial transaction. The standards for a material are dependent on the product in which the material is used in the manufacturing process. Standards for each ingredient vary and have no universal application. However, the specifications should be realistic. The tolerance limits for each ingredient and product should be set down.

Each processing unit should have its own manufacturing standards to ensure excellence of its products. Standards laid down by the Indian Standards Institute (ISI) and trade standards are adopted on a voluntary basis to ensure acceptable quality and to prevent the lowering of standards of quality produced by the industry as a whole. The Government standards (F.P.O. and P.F.A.) are mandatory standards laid down to protect the health of the consumer and prevent deception. In addition to these, the products meant for export should meet the standards prevalent in the importing country(s). The specifications of standards should include the following (Ranganna, 1975).

(a) **Identity:** Definition of the ingredient or material, source, any variations in composition etc.

(b) **Purity and composition**

(i) **Physical:** Size, colour, texture, viscosity etc.

(ii) **Chemical:** Moisture, TSS, acidity, fat, proteins, vitamins etc.

(iii) **Microbiological** tests from the hygienic point of view.

(c) **Packing:** Size and type of package.

(d) **Transport and schedule of supply:** Quality standards play a vital role in marketing of products. In early 1960s, the Government as well as the industry felt the need to compile Indian standards for fresh as well as processed products of fruits and vegetables, specifically for those items which have an export market. As a result, standards covering fresh as well as products of fruits and vegetables, methods of making, methods of tests and guide for storage, package and transport were brought about. At the same time several Indian standards were brought out on mango products. The salient features of these Indian standards in case of some of the mango products are briefly presented below:

Quality Standards for Raw Mango Products

1. **Mango chutney (IS : 3500 – 1966)**

   Mango chutney should contain less than 50 per cent TSS. The Indian standard (IS) also specifies the requirements of fruit content, preservatives, acidity, total ash, acid-in-soluble ash, microbiological requirements including mould content and limits for metallic impurities. Chutney covered under this specification could be of two grades and score card system for grading the products has been included.

2. **Mango pickles (IS : 3501-1966)**

   The pickles covered under this standard are of 3 types viz., pickles in vinegar, pickles in citrus juice or brine and pickles in oil. Requirements specified include fluid portion, acidity, salt and metallic impurities. The product could be of two grades and score card method for grading the product has been included.

3. **Amchur (Raw mango powder)**

   No standards have been prescribed for raw mango powder by Agmark, ISI or FPO. Based on the histograph constructed on the analytical data on 35 samples, tentative standards for amchur have been suggested by Pruthi and Teotia (1987), as moisture 12 per cent maximum, acidity 12 per cent minimum, total ash 6 per cent maximum and insoluble ash 2.5 per cent maximum, crude fibre 6 per cent and total sugar 17 per cent maximum.

Quality standards for ripe mango products

1. **Canned mangoes (IS : 2867 – 1964)**

   This standard prescribes the requirements for canned mangoes in syrup. It gives the various varieties suitable for canning and styles of canned mangoes. The canned mangoes were covered under two grades. In addition to general and hygienic requirements, the requirements for vaccum, head space, drained weight, pH of cut-out syrup, specific gravity, acidity, microbiological and limits for metallic contamination have been covered. The standard includes detailed methods for grading the product based on score card system.

2. **Canned mango pulp (IS : 3880-1966)**

   The standard gives list of varieties suitable for preparation of canned mango pulp. The requirements for total sugar, organoleptic quality, vaccum, head space, degree brix, acidity, total ash, microbiological and metallic impurities have been covered.

   (i) canned mango pulp (natural) should contain not less than 12 per cent TSS. The material shall be passed through a minimum 1.5 mm mesh sieve so as to have a homogenous product. It shall be free from any cooked flavour, black spects, extraneous matter (portion of skin, fibrous matter, larvae and other insects or fragments thereof). The pulp should not show any sign of bacterial growth when incubated at 37 °C. The product shall not show any positive pressure at sea level. The variety used shall be clearly and conspicuously marked on the container.

(ii) Canned mango pulp (sweetned) should not contain less than 12 per cent TSS and a minimum acidity of 0.3 per cent as citric acid, and the other specifications are similar to that of canned mango pulp (natural).

3. **Mango squash (IS : 4936-1968)**

The specification covers requirements for additives, fruit content, degrees brix, microbiological and limits for poisonous metals.

4. **Fruit Jams, Jellies and Marmalades**

The specification gives the list of various ingredients and the limit for fruit content, in different types of jams etc. The requirements for end product include TSS, method of scoring and limits for poisonous metals.

OTHER RELATED STANDARDS

1. **Guide for storage and transport**

To provide guidelines for cold storage industry in India, a series of IS guides were prepared. IS : 7252-1974, guide for testing of physical condition in cold storages, covering definitions of terms most commonly occurring in industrial cold storage. It also includes methods for measuring physical condition inside cold stores. The guides for storage and transport of mangoes were also prepared at national and international levels.

2. **Methods of Test**

To bring about uniformity in analytical techniques and to ensure repeatability and reproductability of results, following standards as methods of tests have been issued.

(a) IS : 2860 – 1964 : Methods of sampling and tests for processed fruits and vegetables.

(b) IS : 4939 – 1968 : Methods of tests for products derived from fruits and vegetables.

(c) IS : 7254 – 1974 : Methods of test for determining preservatives in fruits and vegetables.

3. **Hygiene standards**

Consumers do not have the knowledge or means of determining the hygienic quality of mango products they purchase, but except and believe that processed products were prepared under most hygienic conditions. Unless the factory manufacturing the products is covered by the strict hygienic guidelines in respect of layout plan and personnel, the quality of the product produced can not be considered safe, however nutrition and delicious it may be. Keeping this in view, the IS has prepared IS : 6542-1972 – Guide for hygienic conditions for fruit and vegetable canning units. Standards have also been prepared for determination of coliform bacteria, plate count, yeast and mould count, bacteria responsible for food poisoning and food-borne diseases.

4. Standards for Sensory Evaluation

It is frequently asked whether human response to taste, colour etc. can be measured by commercial and instrumental means. The researchers have proved that human senses are capable of integrating sensory stimuli better. Hence, final standard of quality is human evaluation. Keeping this aspect in view, basic standards, in the field of sensory evaluation covering glossary, methodology, quality characteristics, optimum requirements, methods and evaluation cards have been prepared. A standard for the specific field of sensory evaluation of fruits and vegetables products was also prepared.

5. Standards for Labelling and Advertising

Food labelling and advertising are essential components in all food processing industries. Their purpose is to inform the consumer in an unambiguous manner all the contents of the food inside the package or container. The consumers have the right to know what does a package contain and they are not paying for an unknown quality and quantity. Further, advertising is vital for successful marketing of food products. Secondly, advertising should be truthful, realistic etc, so as not to dupe or confuse the consumer. Keeping these aspects in view, the ISI prepared the following standards.

(i) IS : 7688 (part I) – 1975 : Code of practice for labelling of pre-packaged foods. Part I. General Guidelines.

(ii) IS :7655-1975 : Code of practice for Food Advertising.

6. ISI Certification Mark Scheme

For providing practical utility of standards to the consumers, the Institution is operating ISI certification mark scheme under the powers vested in it by ISI certification mark Act 1952 (as amended in 1961). Under this scheme, licences are granted to the manufacturers producing goods in confirmity to the relavant IS. Certification Mark Scheme gives assurance to the consumers that the products marketed have been tested, inspected and certified by the National Standard Body and they may purchase with the assurance of quality. The 'Mark' is operated on a voluntary basis and the scheme is primarily responded as technical audit for ensuring the quality system employed in production.

Besides the above, different departments of GOI have promulgated a few more standards viz., FPA standards, FPO standards etc., for maintenance of standards of processed products.

GOOD MANUFACTURING PRACTICES

For production of safe, high quality and acceptable products the following good manufacturing practices (GMP) should be followed strictly.

I. Hygiene Requirements

A. Personnel

1. **Disease control :** Any person, who has an illness, open lesions, including boils, sores, infected wounds or any other abnormal source of microbial contamination must not be allowed to work in any operation in a food processing factory. Otherwise, it would result in the food, contact surface, or food packaging materials becoming contaminated.
2. **Cleanliness:** People who in direct contact with food preparation, food ingredients or surface of equipments or utensils that will contact food, must wear clean outer garments, maintain a high degree of personal cleanliness and confirm to hygienic practices while on duty; they must wash their hands thoroughly with soap and, if they are working at a job where it is necessary, they must also sanitize their hands before starting work, after each absence from the work spot and at any other time when the hands have become soiled or contaminated; they must also remove all unsecured jewelry. People who are actively handling food, should remove any jewelry that cannot be properly sanitized from their hands; it is necessary, to wear effective hair restraints, such as hairnets, caps, head bands or beared covers; operators must not store clothing or other personal belongings in food processing areas. Eating food, drinking beverages etc. must not be allowed in food processing area. All necessary steps should be taken up by supervisors to prevent operators from contaminating foods with microbes or foreign substances such as hair, perspiration, cosmetics, tobacco, chemicals and medicants.
3. **Education and training:** Sanitary monitoring persons must have the education and/or experience to demonstrate that they are qualified. Food handlers and supervisors should be trained to make them aware of the danger of poor personnel hygiene and unsanitary work habits.
4. **Supervision:** Responsibility of supervising the respective personnel to comply with all the requirements of GMP's should be assigned to an appropriate person for effectiveness.

B. Plants and Grounds

Grounds around a food processing centre, which are under the control of the centre must be free from condition such as: improperly stored equipment, litter, waste or refuse; weeds or grass close to buildings; dusty roads, yards or parking lots; inadequately drained areas-potential foot borne filth or breeding places for insects, worms, or microorganisms, inadequately operated system for waste treatment and disposal.

Plant contribution and design shall provide enough space for sanitary arrangement of equipment and storage materials; floors, walls, and ceilings must be constructed so that they are cleanable and must be kept clean and in good repair; separate by partition, location, time and other means, any operation that may cause cross – contamination of food products with undesirable microorganisms, chemicals, filth or other extraneous materials; provide effective screening or other protection to keep out birds, animals and vermin such as insects and rodents.

Adequate ventilation should be provided to prevent contamination of foods with odours, noxious fumes or vapour (including steams), light bulbs, sky lights or any other glass must be of the safety types or protected so that glass contamination can not occur in case of breakage.

II. Sanitary Operations

General maintenance

The plant and all fixtures must be kept in good condition and be maintained in good sanitary conditions. Cleaning operations must be conducted in a manner that will minimise the possibility of contaminating foods or equipment surfaces that contact food.

Pest-control

- No animals or birds should be allowed anywhere in the plant;
- Programmes must be in effect to prevent contamination by animals, birds and pests (rodents and insects).
- Insecticides and rodenticides may be used as long as they are used properly (according to label instructions).
- These pesticides must not contaminate food or packaging materials with illegal residues.

Sanitation of Equipment and Utensils

- Utensils and equipment surfaces that are in contact with food must be cleaned as often as necessary to prevent food contamination.
- Equipment surfaces that are not in contact with food should be cleaned as frequently as necessary to minimize accumulation of dust, dirt etc.

- Simple service articles such as disposable utensils, paper cups, paper towels etc. should be stored in appropriate containers. Handled, dispensed, used are disposed off in a manner that prevents contamination of food or equipment.
- Where there is a possibility of introducing undesirable microbes into food, all utensils and equipment surfaces that contact food must be cleaned and sanitized before use and following an interruption during which they may have become contaminated;
- When utensils or equipment are used in a continuous production operation, they must be cleaned and sanitized on a pre-determined schedule.
- Any facility, procedure, machine or device may be used for cleaning and sanitizing as long as it has been established that the procedure will do the job effectively.

Storage and Handling of Clean Portable Equipment and Utensils

Portable equipment or utensils which have surface that will contact foods, after cleaning and sanitizing should be stored in a manner that will protect the food contact surfaces from splash, dust and other contaminations.

SANITARY FACILITIES AND CONTROLS

Water supply: Any water that comes into contact with food or processing equipment must be safe and of adequate quality/quantity.

Sewage disposal: Must flow into an adequate sewage system or disposed off through other adequate areas.

Plumbing: Must be of adequate size and design to:

- Supply enough water to areas in the plant where it is needed.
- Properly convey sewage or other disposable liquid waste from the processing plant.
- Not to create a source of contamination or unsanitary conditions.
- Provide adequate floor drainage. Where, hosing type cleaning is done or where operations discharge water or liquid waste on to the floor.
- Insure that there is no back flow from cross connection between piping systems, that discharge waste water or sewage and those that carry water for food or food manufacturing.

Toilet facilities

- Toilets and hand washing facilities must be provided in side the fruit and vegetable processing centres with either adequate water or toilet paper.
- Toilets must be kept in sanitary and in good usable condition.
- Toilet rooms must have self-closing doors.

- Toilet rooms must not open directly into areas where food is exposed unless steps have been taken to prevent air-borne contamination (ex: double doors, positive air flow etc.).
- Signs must be posted that direct employees to wash their hands with soap or detergent after using the toilet.

Hand Washing Facilities

(a) Adequate and convenient hand-washing and, if necessary, hand-sanitizing facilities must be provided anywhere in the processing plant where the nature of employees jobs required that they wash, sanitize and dry their hands.

(b) These hand washing facilities must provide:

- Running water at a suitable temperature.
- Effective hand cleaning and hand sanitizing preparations.
- Clean towel service or suitable drying devices.
- Easily cleanable waste receptacle (waste box).
- Water control valves designed and constructed to protect against recontamination of clean, sanitized hands.
- Signs, directing employees (handling unprotected food) to wash and if appropriate, sanitize their hands before starting work after each absence from the work spot and any other time when the hands have become soiled or contaminated.

Rubbish and offal disposal must be handled in such a way that they do not serve to attract or harbour pests or create contaminating conditions.

Equipment and Utensils

(a) Equipment and utensils must be designed and constructed so that they are cleanable and will not adulterate food with lubricants, fuel, metal fragments, contaminated water etc.

(b) Equipment should be installed so that it and the area around it can be cleaned.

(c) Food contact surfaces shall be made of non-toxic materials and must be corrosion resistant.

(d) Seams on food contact surfaces shall be smoothly bonded, or maintained in order to minimize the accumulation of food particles, dirt and organic matter.

(e) Equipment in processing areas that do not come into contact with food shall be constructed so that it can be kept clean.

(f) Holding, conveying and manufacturing systems, including gravimetric, pneuematic, closed and automated systems shall be maintained in sanitary conditions.

(g) Each freezer and cold storage compartment shall have an indicating thermometer, temperature measuring or recording device and should have an automatic control for regulating temperature or an automatic alarm system to indicate a significant temperature change.

(h) Instruments and controls used for measuring, regulating or recording temperatures, pH, acidity, water activity etc. shall be adequate in number, accurate and maintained in proper conditions.

Processes and Controls

There must be an individual, responsible for supervising the overall sanitation of processing plant.

Raw materials and ingredients

(a) Must be inspected and stored to insure that they are clean, wholesome and fit for processing into human food.

(b) Must be stored under conditions that will protect against contamination and minimise deterioration.

(c) Must be washed or cleaned to remove soil and other contamination.

- Water used for washing, rinsing or conveying food products must be of good sanitary quality.
- Water must not be used for washing, rinsing or conveying, if contamination of food may result.

(d) Containers and carriers (such as trucks or railway wagons) should be inspected to assure that their conditions do not contaminate or have not contaminated raw ingredients.

(e) Raw materials should not contain levels of microorganisms that may produce food poisoning or other disease or they should be pasteurized or otherwise treated during manufacturing operation so that the product will not be adulterated.

(f) Materials susceptible to contamination with natural toxins like aflotoxins, should comply with national and international official levels before they are incorporated into the finished product.

(g) Raw materials susceptible to contamination with pests, undesirable microorganisms or extraneous materials should comply with national and international regulations, guidelines and defect action levels.

(h) Materials should be stored in containers and under conditions which protect against contamination.

(i) Frozen materials should be kept frozen. If thawing is required prior to use it should be done in a manner that prevents contamination.

Manufacturing Processes/Operations

(a) Food processing equipment must be kept in a sanitary condition through frequent cleaning and when necessary, sanitizing. If necessary, such equipment must be taken apart for through cleaning.

(b) It is necessary, to process, package and store food under conditions that will minimize the potential for undesirable microbial growth, toxin formation, deterioration or contaminations. To accomplish this it requires careful monitoring of such factors as time, temperature, humidity, pressure, flow rate etc. The object is to assure that mechanical breakdown, time delays, temperature fluctuations or other factors do not allow the foods to decompose or become contaminated.

(c) Food should be held under conditions that prevent the growth of undesirable microorganisms as follows :

  (i) Frozen foods should be maintained in a frozen state,

  (ii) Acid or acidified foods should be held in hermetically sealed containers at ambient temperatures, should be heat treated to destroy mesophyllic microorganisms.

(d) Measures such as sterilizing, irradiating, pasteurizing etc. should be adequate to destroy or prevent the growth of undesirable microorganisms.

(e) Work-in-process should be protected against contamination.

(f) Finished food should be protected from contamination.

(g) Equipment, containers and utensils should be constructed, handled and maintained to protect against contamination.

(h) Measures – sieves, traps, metal detecters, should be used to protect food against the inclusion of metal or other extraneous material in food.

(i) Food or materials that are adulterated should be disposed off in a manner that prevents other food from being contaminated.

(j) Mechanical manufacturing steps such as washing, peeling etc., should be performed to protect against contamination by providing adequate protection from contaminants that may drip, drain or be drawn into the food, by adequately cleaning and sanitizing all-food contact surfaces and by using time and temperature controls at and between each manufacturing step.

(k) Heat blanching should be done by heating the food to the required temperature, holding it at this temperature for the required time, and then either rapidly cooling the food or placing it to the next manufacturing step without delay.

(l) Filling, assembling, packaging etc., should be performed in such a way that the food is protected against contamination by :

  (i) use of a quality control operation in which the critical control points are identified and controlled during manufacturing;

  (ii) adequate cleaning and sanitizing of all food contact surfaces and food containers,

  (iii) using materials for food containers and food packaging materials that are safe and suitable,

  (iv) providing physical protection from contamination, particularly air-borne contamination,

  (v) using sanitary handling procedures.

(m) Food such as, dry mixes, nuts, intermediate moisture food and dehydrated food, that relies on the control of water activity for preventing the growth of undesirable microorganisms should be processed and maintained at a safe moisture level by

  (i) monitoring the water activity of food,

  (ii) controlling the soluble solids/water ratio in finished food,

  (iii) protecting finished food from moisture pickup by use of a moisture barrier or by other means, so that the water activity of the food does not increase to an unsafe level;

(n) Acid and acidified food, that relies principally on the control of pH for preventing the growth of undesirable microorganisms should be monitored and maintained at a pH of 4-6 or below by

  (i) monitoring the pH of raw materials, food-in-process and finished product;

  (ii) controlling the amount of acid or acidified food added to low acid food.

(o) If ice is used and comes in contact with food products, it must be made from potable water and be in a sanitary condition.

(p) Areas and equipment that are used to process human food should not be used to process non-human food grade, animal feed or inedible products unless there is no possibility of contaminating the human food.

(q) A coding system should be utilized that will allow positive lot identification in the event it is necessary to identify and segregate lots of food that may be contaminated.

(r) Records should be kept for a period of time that exceeds the shelf life of the product, but not beyond two years.

HAZARD ANALYSIS AND CRITICAL CONTROL POINTS

The other most significant practices which need to be observed in food (fruit) processing industries are "Hazard Analysis and Critical Control Points" (HACCP) and 'Sanitary and Phytosanitary standards' (SPS), which relate to the protection of the raw materials for processing industry from the risks arising from the disease causing organisms and other chemical additives. SPS measures are getting universal importance due to increased import of food products. The industry also needs a total quality management (TQM) system to ensure the standard quality management (SQM) plants at every stage of procurement, handling and processing of raw material.

Pre-processing steps – Converting Raw Foods to Ingredients

Raw materials (fruits and vegetables) generally vary in composition, colour, flavour, structure and nutrient content. Thus, a key part of describing a process is the preparation of detailed specifications for ingredients and packaging materials to ensure that final product performance and composition specifications can be met with the specified process and equipment. This is only possible if ingredients are pre-processed to the desired specification.

Thus, the processing of foods must be separated into two broad areas:

(A) Preparation of raw fruit and vegetable for further processing to consumer products or ingredients,

(B) Assembling of pre-processed ingredients to finished, consumer products.

(i) Typical operations using in converting raw fruits and vegetables to processed ingredients for packaging and preservation or prior to a further heat treatment processing areas follows:

Harvesting or gathering → Transport → Storage → Washing → size grading → peeling → size reduction → separation of waste → sorting inspection → storage.

(ii) The assembly of pre-possessed ingredients to finished consumer products include principles used to describe a process for quality control, which apply to the conversion of raw foods to processed products. The chief differences are the variation in raw product specifications and certain washing, peeling, size reduction and blanching or heat treatment steps needed to convert highly valuable raw materials to standardized ingredients.

Process Description for Quality Control

One of the most important specifications of a product is its safety in terms of microbial contamination, freedom from hazardous chemicals and absence of foreign materials (metal pieces etc.), non-edible parts (bits or woody stem material) and dirt, insect parts or other extraneous material.

The microbial safety of processed fruits and vegetables in of prime importance from a quality view point. The following analysis provides a description of a process to help ensure that microbial safety can be achieved with a minimum opportunity for failure of the finished product to meet specifications.

Process Operations

Food processing steps require detailed descriptions when microbial safety is a concern, because of possible contamination, through ingredients, packages, equipment and surroundings. Under suitable conditions of pH (acidity), temperature, moisture and nutrients, microbial growth will be rapid on processing equipment and also in the food. Thus the microbial counts can increase during each step of processing, if the process is not designed properly. One of the first requirements for the description of a process is to determine if an individual process step will increase (+), decrease (–) or result in non-change (=) in the microbial content of food undergoing the processing step. This can be determined from the chemical and physical conditions of the food and surroundings in each processing step and be recorded for each process step needed to prepare a finished product.

Individual processing steps are identified in the preparation of the canned juice drink from warehouse inventory to just prior to can labelling. Processing steps needed to prepare the package for filling and sealing are listed on the right side of the chart.

For purpose of analyses, the juice drink must have a pH of 8.8, a fill temperature of not lower than 190 °F and must leave the cooling tunnel at a temperature below 90 °F. The can fill temperature must be 190 °F or above. Regular can and lid inspections are needed to ensure that double seals are always within specifications. The pH of the system may be less critical since the fruit concentrate will bring the pH below 4.5 even without the dry mix.

Even with these requirements to control, the heat-treatment given, the products (heating at 200 °F) with a variable (0-60 minutes) hold is not sufficient to inactivate *Clostridium botulinum*. Thus the product must be cooled at 38 °F within 40 minutes to prevent any possible germination and out growth of *C. botulinum* spores.

Further, the manufacturer of the product must provide a means to ensure that the product is used within a specified time and is always kept at or below 35 °F if it is to be released to the public.

Food Processes in General

It is important to know the following information to determine the microbial safety of a product and to determine what quality control action will be needed to achieve a safe product. First, all raw material, ingredients, and packaging materials in contact with the food must be specified and if needed tested to meet specifications. Second, it is important that an element of food be observed as it enters, passes through and leaves the process.

This 'ride' should provide the following data at each process step from entry to discharge:

(i) Food temperature.

(ii) Time at each temperature.

(iii) pH

(iv) Oxygen concentration

(v) Water activity.

Further, if preservatives are required to prevent microbial growth, their addition point and concentration throughout the entire food mass should be shown. Finally for products which depend as a sealed package, one should check the seal integrity of each container produced.

STATUTATORY PROVISIONS FOR QUALITY CONTROL

For ensuring safety and health of the consumers while consuming food items, legislations have been introduced by all the Governments as a first step. Indian Government also enacted a few Acts with the following objectives (Srivastava and Kumar, 2006).

(i) To maintain the quality of food produced in the country.

(ii) To prevent exploitation of the consumer by sellers.

(iii) To establish criteria for quality of food products, since more and more foods are eaten in processed, rather than in natural forms.

Indian Government has enacted the following Acts for quality control of food items.

1. Agricultural Produce (Grading and Marketing) Act, 1937.
2. Prevention of Food Adulteration (PFA) Act 1954.
3. Fruit Products Order (FPO), 1955
4. Export (Quality Control and Inspection) Act, 1963.
5. Bureau of Indian Standards Act 1986 and

In addition to the above, Defence Ministry, GOI, prepared specifications, known as Army Corps Specifications (ACS) for their purchases of food grains etc.

Majority of these Acts are aimed at manufacturing/producing/distributing food items under hygienic conditions. Many of these legislations are meant for production and distribution of food items for local consumption, though the regulatory agencies are different.

Agricultural produce (Grading and Marketing) Act 1937 (AGMARK)

This is one of the oldest Act of Food Laws promulgated in India with the basic aim of protecting the consumer and at the same time ensuring the producer a just return for his produce and providing quality control of agriculture and allied produce and products through grading and marketing. It empowers GOI to formulate standards and to carryout grading for quality control programmes for agricultural and allied products. The Act provides compulsory standards for domestic as well as for export markets. These standards are formulated on the physical and chemical characteristics of food both the natural as well as those acquired during processing. This Act is administered by the Directorate of Marketing and Inspection with Agricultural Marketing Advisor to the GOI, Faridabad, as the Chairman.

'Agriculture Produce' includes all produce of agriculture or horticulture and all articles of food or drink, wholly or partly manufactured from any such produce. A number of grading standards of raw or processed agricultural or horticultural produce have been brought about by AGMARK. The AGMARK standards stipulate common physical and chemical parameters like, odour, size, colour extractions matter, mould and insect infestation, admixtures, bulk density, moisture level, total ash, insoluble ash etc. Keeping in view of the requirement of the sophisticated import markets, the standards would be revised from time to time.

The Directorate of Marketing and Inspection established under this act has set up a number of analytical laboratories and commodity testing centres for enforcing the quality control programme on agricultural commodities. The quality control testing is done by primary grading laboratories which are under private, cooperative and state sector and they carry out testing and grading at producers level. Their work is checked and supervised by the Regional Agmark Laboratories established at 21 centres. These laboratories are under technical control of the Central Agmark Laboratory at Nagpur. It is an appellate and apex laboratory of the Directorate (Vijaya, 2000).

The net work of Agmark laboratories test about 140 agricultural and horticultural commodities. The schedule consists of fruits, vegetables, eggs, dairy produce, tobacco, coffee, processed cereals and also lac, sunnhemp, myrobalans, wool, raisin and turpentine, essential oils, spices, honey, tapioca chips, oil cakes etc. (Vijaya, 2000).

The Directorate authorises marking of packings of agriculture products (bag, bottle, can, tin etc.,) with their logo and designation mark on payment of specified fees. This implies that the product is duly tested and found in conformity with the Directorate Standards. It is known as AGMARK standards. There are penalties for unauthorised marking with grade designation, counterfeiting grade designation mark and for selling misgraded articles (Vijaya, 2000). Any organisation or packer, desirable to use 'AGMARK' logo on the label of the sample containers can apply to the Directorate. The authorities of Directorate after inspection and ensuring that the necessary facilities like equipment, laboratory etc are available, allows them to use 'AGMARK' symbol which ensures that the quality standards laid down for that product under AGMARK have been compiled with. This gives a sort of quality assurance to the consumer and at the same time a fair return to the producer (Srivastava and Kumar, 2006).

The AGMARK standards are quality standards depending on the degree of excellence of the produce/product and are more rigid than Prevention of Food Adulteration

Standards. AGMARK standards, however, are voluntary in nature in general and the producer/manufacturer registers with the Directorate with a view to boost up his sales with AGMARK registration mark on his product. AGMARK label on a container helps the consumer to be sure of quality of product what he purchases. There is a consumer protection cover when one buys AGMARK products. That means a defective product will be replaced free of cost or money is refunded (Vijaya, 2000).

The general Grading and Marketing Rules 1937, as amended in 1988 and further supplemented by specific commodity, grading and marketing instructions facilitate implementation of the provision of Act. These standards have been made applicable to exports also (Nair, 1992).

Prevention of Food Adulteration Act 1954 and Rules 1955 (PFA Act)

This act was promulgated in 1954 and is primarily intended to check adulteration of food stuffs available in the country and those entering into market. The rules and regulations that are formulated for operation of this Act are directed to make available pure food materials devoid of adulterants and contaminants to the Indian consumers. Any food not confirming to the standards of Act is considered to be adulterated and contaminated and unfit for human consumption from hygienic angle. The rules include among others, specifications for a large number of raw and processed foods. The minimum standards for various types of fruits and vegetable products are also covered under the PFA Rules. These rules are administered by the Ministry of Health, GOI, State Governments and Union Territories through Food Inspection. The PFA requirements and FPO specifications are practically designed with each other, except that the latter are a little more elaborate as they encompass the specific field of fruits and vegetables. The Central Committee for Food Standards advises the government on the implementation of the provision of the Act/Rules. The provisions made under PFA are mandatory and it is the responsibility of the manufacturer/whole saler/vendor etc. to abide by the standards of various food commodities. Broadly the control of the quality can be categorised under three heads (Vijaya, 2000) :

(i) Enforcement,

(ii) Analysis and

(iii) Prosecution.

The PFA standards are the minimum standards of purity and are based on the agricultural and manufacturing practices followed in the country. The Act deals with the preservatives, poisonous metals, naturally occurring toxic substances, antioxidants, emulsifying and stabilising agents, flavouring agents, colouring matter and other food additives, insecticides and pesticides, solvent extracted oils and edible flours, non-alcoholic beverages, starchy foods, spices and condiments and their mixes, honey, jaggery, coffee, tea, milk and milk products, fruit and vegetable products, edible oils, cereals, backed products, sweets and confectionaries and a range of similar products. The Act deals with definition and standards of quality (Vijaya, 2000).

To enforce the provisions in the Act and Rules, GOI has set up four well equipped Central Food Laboratories and a number of laboratories at State and Union Territories level.

Fruit Products Order 1955 (FPO)

The Fruit Products Order 1955 was promulgated by the Department of Food, Ministry of Food Processing Industries, GOI, under the powers vested in the Government under the Essential Commodities Act to ensure the quality of the fruit and vegetable products. The FPO regulates the manufacture, storage and sale of fruit and vegetable products. This order controls production, distribution and quality of the fruit and vegetables products manufactured in the country as well as registration, licensing and operation of manufacturing units. Under the FPO, the Central Government has constituted an expert Committee called "Central Fruit Products Advisory Committee" to advise the Government on standards and policy matters pertaining to fruit and vegetable preservation/processing industry (Srivastava and Kumar, 2006).

The various provisions of the FPO are being implemented by the Director, Fruit and Vegetable Preservation attached to the Ministry of Food Processing Industries with the assistance of Zonal Deputy Directors, stationed at Chennai, Mumbai, Kolkata and New Delhi and large number of technical staff. Under this Act, no person can carry on the business of manufacture except under and in accordance with the terms of an effective licence, granted to him. A fee is levied on the manufacturer depending on the scale of production. 'FPO' mark is given to each processor, after the grant of licence for manufacturing fruit or vegetable products, after the inspection of the factory for hygiene and sanitation. FPO Mark and licence number are required by law to be exhibited on labels of each processed item along with other information as laid down in the FPO Rules. Continuous inspection scheme and pre-shipment inspection scheme are being exercised under the FPO especially regarding export-oriented fruits and vegetable products. The licencing officer can inspect the manufacturing unit, collect samples intended for sale and get them analysed at an authorised laboratory. Penalities such as fine, closure of the unit and even imprisonment will be levied if malafied intensions and neglect are proved in the manufacture and complying with FPO.

The processor has to manufacture fruit and vegetable products in conformity with the sanitary requirements and the appropriate standards of quality and composition specified in the schedule to this order. The processor has to comply with the specified requirement in regard to the packing, marking and labelling of containers of products. Each container should display licence number of the special identification mark, code number, indicating the lot or the date of manufacture of such product. The labels should not contain any statement, claim, design or device, which is false or misleading concerning the contents in the container (Vijaya, 2000).

The FPO specifications cover, methods of preservation/processing, permissible covers in the preparations and also the minimum quality requirements of the final products. In addition to specifications for various products, limits for poisonous metals, list of permitted and harmless food colours, limits for permitted preservatives and other additives are given in the schedules of the Act. The fruit and vegetable products which do not conform to the FPO specification are considered adulterated (Vijaya, 2000, Srivastava and Kumar, 2006).

The FPO covers all types of fruit products such as juices, pulp concentrates, squashes, nectar, aerated water containing fruit juices, bottled and canned fruits and vegetables, jams, jellies, fruit cheese, preserves, chutneys, pickles, soups, ketchup, paste, dehydrated vegetables, dehydrated onions etc. (Vijaya, 2000). Bulk products of fruit and vegetable industry relates to the products such as jam, jellies, marmalades and squashes. In these cases, the quality of the product is controlled 'on the spot' inspection for the reason that the quality of fruit can not be estimated from the 'end product' which losses its identity with respect to its starting raw materials.

Bureau of Indian Standards (BIS)

Bureau of Indian Standards (BIS), a Government body, also prescribes quality standards for agricultural products and processed foods thereof. The Agricultural and Food Products division constitutes different committees consisting of representatives of concerned Government departments, scientific institutions, processors, traders etc., who prepare quality standards mainly for processed food items. They also prepare standards for methods of analyses. It is obligatory to adopt these methods of analysis. As in case of AGMARK, ISI mark can be used by the processor after registering with them and complying with their stipulations regarding inspection and testing of their products. ISI standards are quality standards and more rigid than PFA standards (Vijaya, 2000). BIS has laid down standards for all the aspects from constitution of the manufacturing factory, raw materials and processing until its packing.

Indian standards are laid down by the ISI. These standards are arrived at by experts in the industry, research and development organisations and by those in marketing and concerned ministry. The standards are aimed at providing a good quality product to the purchaser of the product for the price he pays. These standards are not mandatory in nature, but voluntary. The ISI requires the industry if it intends to make use of their standards and allows them to print ISI mark on the packing. The ISI inspects the product during the manufacturing process to ensure that the concerned industry produces the product as per the specifications. The ISI marking on the packing indicates good quality of the product. Indian standards are available for food products in addition to industrial products etc. (Vijaya, 2000).

Army Corps Specifications

For the exclusive use of Defense personnel, the Army Purchase Organisation (APO), Food stuffs has brought out several Arm Corps Specifications which include specifications on fruit and vegetable products among others. These standards are designed with the relevant Indian Standards. Often the Indian Standards are adopted by the APO in 'toto'.

Export (Quality Control and Inspection) Act 1963

Export Inspection Council has been set up under the Export Inspection Act in order to have a check on quality of food items exported abroad. All packaged fruit and vegetable products were brought under compulsory pre-shipment inspection from January 1972. Under these rules every shipment need to be examined for conformity to specifications laid down for the respected products to be accompanied by a certificate of uniformity and quality issued by the authority concerned. The objectives of the pre-shipment inspection are to eliminate sub-standard products from being exported and to create a sense of confidence among the buyers/consumers by assuring them of the minimum quality as specified in the export standards. One important feature of the sampling and inspection plan is that no allowance is given for defectives in the sample. If any sample is found defective, the entire lot under the code is considered as sub-standard and rejected for export (Thakur *et al* 1976).

Food labelling

Food labelling is an essential component in all food processing industries. The purpose of labelling is to inform the consumers about the content of the foods inside the package or container. The consumer wants to know what a package contains, to be sure that he is not paying for false product of an unknown quality and quantity of the food in the package. This necessitated the enumeration of the ingredients, net contents and other essential points about the food in the package, on the label. It is also equally important that the declaration and claim made on the label are true and reflect the product packed (Siddappa and Tandon, 1959).

The standard has been issued in three parts to provide important information about the product to the consumer as follows (Vijaya, 2000).

Part I: Covers general guidance on labelling.

Part II: Covers claims and lays down the condition underwhich claims including nutritional dietetics and medical claims may be made. It also lays down that marketing, unsubstantial, meaning-less claims or those which arouse fear in consumers or give rise to doubts about safety of similar foods are prohibited.

Part III: Covers guidelines on labelling with respect to nutritional information. The standard lays down that any nutrient added to food or any nutritional claim for information should be declared on the label. The manner in which the information is declared is also given on the label as also the amounts of addition under which declaration is compulsory or optional for each nutrient.

Nutritional labelling has come in conjunction with food labelling in recent times to give more information to the consumers. It helps consumer to buy the foods, he exactly needs with specific nutrients. Nutritional labelling will upgrade the nutritional quality of food because the food industry has increased responsibility to determine what nutrients are actually contained in the food product they are selling. The importance of nutritional labelling can be observed in baby foods, convenient foods, therapeutic foods, genatric foods, etc. and needs to be extended to fruit and vegetable processed food products.

Labelling of marketed food products is necessary to increase the sales. Apart from attractive design, name and colour the label has to indicate some details of the product for the benefit of customers. This is made obligatory by FPO in case of all processed fruit products marketed. Besides, AGMARK standards and ISI standards are of voluntary nature. The respective organisations permit the use of AGMARK or ISI mark on the labels when the manufacturer prepares the product according to the specifications of the said organisations (Vijaya, 2000).

PROBLEMS AND APPROACHES RELATED TO QUALITY CONTROL IN PROCESSING INDUSTRIES

In order to monitor and maintain the quality of processed products, a well organized and systematic quality control programme is essential for every food processing industry. Further, in view of the demand for quality products in the coming years in both domestic and international markets, it is necessary to maintain both an optimum quality of life and balanced approach to maximum food safety. In developing countries like India, though the industry has grown to a considerable extent, still some problems are encountered for running effective quality control programmes, which need to be solved in order that food industries have a sound quality control system. Sardana and Vaidya (1994) have discussed some problems of processing industries in India and suggested approaches to overcome them.

1. **Lack of communication**

   Many processing industries in India do not have a quality control programme and where it exists, it is restricted merely to control composition of the product and in some cases microbiological quality. The results are recorded in a daily record in the laboratory, based on which the products are released or rejected. But in case of defects there is no quality improvement since the results of evaluation of finished product are not communicated to the working staff. Therefore, it is very important to communicate the results eventually with appropriate recommendations to the working staff, making the product, so that they could adjust the process if needed for the quality improvement of the product. At the same time, the top managements must have a clear idea about the company's quality control policy and top management must be involved and committed to it. Good communication between all departments/sections in a processing industry is, therefore, vital for the production of quality food products.

2. **Negligence on the part of producers**

   Some of the producers are unaware of the need for quality control of their products, who have the sole motive of making business and profits. On the other hand, some, though aware of the need for quality control of their products, they just neglect this important part of the industry and the result is that the product does not stand in the market in the long run. When the purchasing capacity of people is increasing particularly for the high quality products and at the same time the imported processed food products availability at comparatively lower prices in the market and making their presence felt, the food industry in India must give prime importance to quality control in order to stand up well in the growing competition.

3. **Lack of trained power**

   One of the bottle necks in implementing a successful control programme in food industries in India is unavailability of appropriately trained man power in this field. As a result, men from related fields or unconcerned fields are often engaged. In addition to a degree in general chemistry, to work in quality control of a food industry a person needs to have a special intensive course in food chemistry. This situation is mainly due to lack of appropriate courses in the universities and other institutes. However, of late, realising the importance of such courses many universities and institutes have introduced special food chemistry courses. This was possible due to the availability, technical know how from organisations like Central Food Technological Research Institute, National Dairy Research Institute etc. in the country.

4. **Unavailability of facilities and infrastructure and organisational set up**

   Many processing industries are not equipped with the minimum requirement of laboratories for quality analysis and quality control of their processed products. Every unit must have adequate facilities for testing the processed product. It should be made mandatory. Processing industries must have a quality control section and have minimum requirements for a laboratory for testing and monitoring. Right at the establishment stage of the industry, it should be taken care and insisted with equal importance as others.

5. **Lack of motivation and participation of staff at all levels**

   Often, it is observed that the failure to produce a quality product in a processing unit is due to lack of motivation and participation of staff at all levels from top to bottom. Cooperation and coordination between different sections of the industry are not existing. Quality control is a kind of job which one individual alone cannot perform. It is a team effort and everybody in the system must participate fully to make the output of the industry of the best quality. This demands cordial labour relations, a cooperative atmosphere, worker's participation, sense of quality control in each and everyone and above all devotion and motivation in the workers. The top management should take utmost care in this aspect.

6. **Lack of systematic approach**

   The processing industries should implement quality control systems with a systematic approach. Product specification should be well defined, testing methods should be standardized, simple statistical methods should be applied for evaluating the performance of product in average and each and every processing industry should develop its own quality control hand book or quality control manual with necessary details like terms of references, guidelines for quality control, quality functions and polices, product specifications, minimum requirement of quality and testing methods. This system should be strictly followed and a follow-up and monitoring must be carried out from time to time.

7. **National Legislation and National Standards**

   National legislation plays an important role in performing and maintaining quality control in food industries in a country. The primary aim of such legislation is to ensure the consumer safety on one hand and on the other for the manufacturer of food articles of quality and standards needed for consumer health and safety. Food laws must protect the interests of both producer and consumer. However, such laws should never be inhibitory to the development of food industry at the same time. There should be very strong legislation with regard to food industries for minimum quality requirements and it should be implemented strictly. Because of WTO, the world has become a single village. Goods are exchanged between nations. Therefore, to promote technical progress and trade, international legislations are enacted. National and International laws are the foundations on which rests the future improvements in technology and promotion of trade. The national standards should be fixed on the following :

   1. *Standards to identify*: defining and describing food products and laying down requirements for their composition, labelling etc.
   2. *Engineering standards*: describing equipment used in food industry, requirements for these equipment, terminology etc.
   3. *Analytical standards*: laying down chemical or bacteriological methods for the determination of certain constituents or organisms in food product.

Future

Hormonization of India's food standards is necessary as the WTO agreement recognises the rights of member countries to adopt measures deemed necessary to safeguard the health of their people. Thus, it is imperative that the Indian standards be gradually equated to that of the developed countries. Hormonization of standards would also obviate the need to modify the standards to meet the exports to suit many different standards of various countries which at the same time are India's competiters in global trade.

Effective participation of all stake holders, especially producers, processed food distributors and research organisations is needed in setting standards as well as guidelines for food safety initiatives that are acceptable to buyers in domestic as well as international markets. The proposed strategy should address the broader concept of risk along the entire fruit production chain. It will take into consideration the need for sustainable agricultural production system in all regions and will redirect some of the existing approaches to ensure that they meet the emerging challenges of global food safety.

In order to have overall control over the processed mango products, it is desirable to narrow the ranges which will not only control national losses, but also improve general quality of the product.

REFERENCES

Birth, G.S. and Norris, K.H. 1965. The Difference Meter for Measuring Interior Quality of Foods. USDA Tech. Bull.No.1341.

Kramer, A. and Twigg, B.A. 1973. Quality control for the Food Industry. 3rd Edn.Vol.2. The AVI Publishing Co., Inc., West Port, Connecticut.

Mehta, A., Ranote, P.S. and Bawa, A.S. 2002. Indian Fruit Processing Industry: Quality Control Aspects. Indian Food Industry, 21(1): 37-40.

Nair, P.B. 1992. Quality Control and Inspection of Spices and Spice Powder for Export. Food Digest, 15(3): 263-266.

Pruthi, J.S. and Teotia, M.S.1987. Indian Cacoa, Arecanut and Spices Journal, 11(1) : 21-28.

Ranganna, S.1975. Quality Control in Fruit and Vegetable Processing. Indian Food Packer, 29(1):83 -111.

Sardana, V. and Vaidya, R.1994. Quality Problems in Agro-industries: Challenges and Approaches. Indian Food Packer, 48(4): 29-32.

Shubik, P., Saffiotti., Linijinsky, W., Pietra,G., Ruppaport, H., Toth,B., Reha,C.R., Tomatis,L., Fieldman,R. And Ramah,H. 1962. Applied Pharmacology, 4:1.

Srivastava, R.P. and Kumar, S. 2006. Fruit and Vegetable Preservation: Principles and Practices. International Book Distributng Co. Lucknow, U.P. India.

Teotia, M.S., Manan, J.K. and Saxena, A.K. 1982.Green Mango Processing – A Review. Indian Food Packer, 41(6): 75-86.

Thakur, S.P.D., Chaudhuri, D.R., Mitra, S.N. and Bose, A.N. 1976. Quality Aspects of Processed Mango Products. Indian Food Packer, 30(5):45-50.

Usha, K. and Anand, J.C. 1981. Quality Aspects of Commercial Raw Mango Powder (Amchur). Journal of Food Science and Technology, 18: 208-210.

Usha Rani, M., Rama Rao, S.N., Girija Bai, R. and Nagaraja, K.V. 1992. Studies on quality standards of Indian Commercial Pickle. Indian Food Packer, 46 (4): 27-33.

Vijaya, K 2000. Textbook on Food Storage and Preservation, Kalyani Publishers, Ludhiana.

CHAPTER 10

Post-Harvest Diseases

R. Srihari Babu and V. Krishna Rao

Introduction

Mango (*Mangifera indica* L.) fruits suffer from several diseases both in the field (pre-harvest) and after harvest (post-harvest). Post-harvest diseases are those which manifest as rots/decays of fruits after harvest during transport, storage, marketing and ripening. Post-harvest diseases are often extensions of diseases from field which are called latent or quiescent infections. The time between infection and appearance of disease symptoms is known as latency or quiescence (Berger and Bartz, 1982). The term refers to quiescent or dormant parasitic relationship of disease which after a time changes to an active one (Verhoeff, 1974).

Mango fruits are attacked because of their low pH, high moisture and nutrient contents. During prolonged transport and storage, fruits become more susceptible to microbial changes occurring in the fruit after maturity and during ripening and decline of natural host (fruit) resistance (Prusky and Keen, 1993).

Several microorganisms including fungi and bacteria are known to cause post-harvest diseases of which fungi are predominant. As many as 45 species belonging to 21 genera of fungi are recorded causing post-harvest diseases of mango in different mango growing regions.

The post-harvest diseases as well as causal organism differ from place to place, further different pathogens cause the same disease in different places. For example stem-end-rot of mango fruits is reported to be caused by different pathogens *viz*., *Diplodia natalensis* (= *Botryodiplodia natalensis* = *Lasiodiplodia theobromae*), *Dothiorella* sp., *Phomopsis* sp.

The different post-harvest diseases of mango are as follows (Table 10.1):

Table 10.1: Different post-harvest diseases and causal organisms on mango fruits

S.No.	Post-harvest disease	Causal pathogen
1.	Anthracnose	*Colletotrichum gloeosporioides, C. acutatum, C. capsici*
2.	Stem-end-rot	*Lasiodiplodia theobromae, Phomopsis mangiferae Dothiorella dominicana*
3.	Black rot	*Aspergillus niger, A. variecolor, A nidulans, A. flavus, A. fumigatus and A. chevalieri*
4.	Botryosphaeria rot	*Botryosphaeria ribis, B. dothida*
5.	Brown rot	*Lasiodiplodia theobromae*
6.	Stem-end-soft rot	*Ceratocystis paredoxa*
7.	Pestalotiopsis rot	*Pestalotiopsis versicolor, P. mangiferae, P. glandicola*
8.	Charcoal rot	*Macrophomina phasolina*
9.	Alternaria rot	*Alternaria alternata, A. tenuissimo*
10.	Phoma rot	*Phoma mangiferae*
11.	Macrophoma rot	*Macrophoma mangiferae*
12.	Rhizopus rot	*Rhizopus arrhizus, R. oryzae, R. nigricans*
13.	Cladosporium rot	*Cladosporium herbarum, C. cladosporioides*
14.	Fusarium rot	*Fusarium oxysporum*
15.	Canariomyces rot	*Canariomyces notabolis*
16.	Rhizoctonia rot	*Rhizoctonia solani*
17.	Mucor rot	*Mucor subtilissimus*
18.	Grey mould	*Botrytis cinerea*
19.	Blue mould rot	*Penicillium expansum*
20.	Watery soft rot	*Sclerotinia sclerotiorum*
21.	Sclerotium rot	*Sclerotium rolfsii*
22.	Yeasty rot	*Saccharomyces sp.*
23.	Dry rot	*Boothiella tetrasora*
24.	Bacterial rot	*Pseudomonas mangiferae*
25.	Soft brown rot	*Hendersonia creberrima*

A majority of the pathogens of post-harvest diseases are weak and wound parasites and need wounds/injuries for access into the fruit. Some gain access through lenticels or stem-end-scar etc. (Pathak, 1980).

The known post-harvest diseases are broadly grouped into two categories *viz.*,

(i) Latent infections, which can be traced back to infections that occur during flowering, fruitset and fruit growth and development prior to harvest. Quiescent infections are characterized by infection in the field in the early stages of the fruit development, dormancy for certain period, before the resumption of fungal growth and expression of symptoms. The latent infections are not generally visible prior to or at harvest, but appear after harvest at the commencement of ripening in transit, storage and marketing under favourable conditions. e.g., anthracnose, stem-end-rot etc.

(ii) Post-harvest infections, which are mainly infected after harvest, during harvest, subsequent handling, transit, storage and marketing. e.g., soft rot, blue mould rot etc.

Post-harvest diseases are particularly severe in high rainfall areas having high relative humidity. Under dry conditions, though the diseases occur, their extent and severity are less.

The sources of post-harvest diseases will be as follows (see also sources in case of individual diseases).

1. Orchard atmosphere, trees and their different infected parts (leaves, twigs, panicles and fruits).
2. Soil, in some cases.
3. Latent infections become the source of secondary infections during transit, storage and ripening possibly by contact.
4. Atmosphere of transport vehicles, pack houses, packages, store houses and even handling persons and equipment, serve as sources. The situation gets aggrevated with repeated use of the same infrastructure several times without disinfection and using same facilities for different, particularly susceptible kinds of fruits and vegetables.

The important consequences of post-harvest diseases are as follows:

1. Reduction in nutritive value and palatability of the produce due to utilization of biochemical constituents by invaded pathogens.
2. Reduction of palatability due to production of mycotoxins by the pathogens.
3. Production of toxic metabolities by diseased tissues in response to the disease infection by the pathogens/exposure to ethylene produced as a result of infection and making them unfit for consumption and processing (Phillips, 1984., Moss, 2002).

4. Unacceptable taste of the produce/products.
5. Contamination of fruits by microbes and mycotoxins.
6. Disintegration of processed fruits by heat tolerant pectionolytic enzymes of causal pathogens.
7. Limiting the storage period and marketability and marketing life of the fruits.
8. Destruction of fruit quality.
9. Economic losses to all those involved in production and distribution of the produce to the ultimate consumer. The damage occurs at very important time i.e. when the fruits are ripening and are ready for sale. By then, harvesting, packing, transport and marketing expenses have already been added to production costs, hence economic losses are huge and high.
10. Consumers do not get sound and healthy wholesome fruit.
11. The export of fruits to global markets is impaired.

Geographical Distribution

Mango post-harvest diseases are ubiquitous/omni-present in all the mango production regions. However, the type(s), their extent of incidence, severity and their causal agents vary from region to region and variety to variety.

According to OmPrakash and Srivastava (1987) most important microorganisms responsible for post-harvest diseases of mango in India are: *Colletotrichum gloeosporioides, Diplodia natalensis, Aspergillus niger, Rhizopus oryzae, Bacillus subtilis* and *Bacterium carotovorum*. Garcha and Singh (1980) reported *Aspergillus niger* and *A. terreus* on Deshari and Totapuri and *A. flavus*, *A. fumigatus*, *A. niger* and *Botryodiplodia theobromae* and *B. ribin* on Langra in Punjab. Thakur and Chenulu (1967) surveyed fungal diseases of mango fruits in Delhi markets and found that Diplodia stem-end-rot, soft rot (*Rhizopus arrhizus*), black mould rot (*Aspergillus niger*) and *Aspergillus* rot (*A. fumigatus* and *A. nidulans*) as important post-harvest diseases. Patel and Goswami (1984) in their regular inspection of the Manek Chowk market, New Delhi recorded the occurrence of bacterial and fungal diseases on mango fruits. The fungi included were *Aspergillus niger* (black mould rot), *A. nidulans* (Aspergillus rot), *Rhizopus arrhizus* (soft rot). *Diplodia natalensis* (stem-end-rot), *Colletotrichum gloeosporioides* (anthracnose), *Botryodiplodia theobromae* (Botryodiplodia rot), while bacterial pathogen included was *Xanothomonas mangiferae* (bacterial rot). A survey of Lucknow fruit market indicated that *Aspergillus niger* was predominant among all the post-harvest fungal diseases of mango (Raoof and OmPrakash, 1983). Pandey *et al* (1983) claimed stem-end-rot (*Botryodiplodia theobromae*) as the common post-harvest disease in Uttar Pradesh. Sumbali and Badyal (1991) recorded post-harvest diseases like *Penicillium crustosum, Rhizopus oryzae, Botryodiplodia theobromae, Aspergillus niger, Alternaria alternata, Geotrichum candidum, Glomerella cingulata* on mango in Jammu. The incidence of anthracnose (*Colletotrichum gloeosporioides*) was recorded in Udaipur market (Rajasthan) (Chandra and Pathak, 1989, 1992a). In a survey of Junagadh (Gujarat) fruit markets, *Lasiodiplodia theorbromae* and *Aspergillus niger* were observed mainly causing fruit rots (Jadeja, 2000). In a survey of processing units in Junagadh (Gujarat), Lasiodiplodia rot, Aspergillus rot, Rhizopus rot, Macrophomina rot and Colletotrichum rots were recorded on Alphonso and Kesar mango fruits (Jadeja and Vishnav, 2000). A total of 17 pathogens were found in association with post-harvest spoilage of mangoes in a survey of fruit markets, godowns and canning industries in Himachal Pradesh of which *Collectotrichum gloeosporioides, Gloeosporium mangiferae, Botryodiplodia theobromae, Dathiorella robis, Pestaloliopsis glandicola, Phomopsis mangiferae, Rhizopus arrhizus, Aspergills niger* were common (Sharma *et al* 1994).

According to Santha and Rao (1983) stem-end-rot and anthracnose are the most important post-harvest diseases of mango in South India. In Andhra Pradesh, anthracnose and soft brown rot were more important post-harvest diseases of mango, where as Mucor rot was sporadic, but not severe. In a survey of fruit markets at Tirupati (Andhra Pradesh) during a fruit season, several fungi *viz*., *Aspergillus niger, A. flavous, A. terreus, Colletotrichum gloeocosporiodes*, *Rhizopus stolonifer* and *Thielaviopsis paradexa* were found to associate with storage diseases (Srikanth, 1989). In Krishna district of Andhra Pradesh, several fungi *viz*., *Colletotrichum gloeosporioides, Botryodiplodia theobromae, Rhizopus arrhizus, Pestalotia* sp, *Aspergillus niger*, *A. flavus, Penicillium purpurogenum, Macrophomina*, *Phaseolina*, *Phoma* sp, and *Penicillium* sp. were observed causing mango fruit rots, out of which *Colletotrichum gloeosporioides* and *Aspergillus niger* were common in all stages of fruit transaction (Bhuvaneswari, 1999). Fungi like *Aspergillus niger, A. flavus, Colletotrichum gloeosporioides* and *Botryodiploidia theobromae* were found to be in association with post-harvest decays of mango in Hyderabad fruit markets (Madhavilatha, 1997). Similarly, Gopi (2003) observed fungal pathogens, *viz*., *Colletotrichum gloeosporioides, Diplodia natalensis, Aspergillus niger* and *Rhizopus* sp. causing post-harvest rots in mango in the fruit markets of Hyderabad (Andhra Pradesh).

In Other Countries

The most serious post-harvest diseases of mango in Florida (USA) are anthracnose (*Colletotrichum gloeosporioides*) and stem-end-rot (*Diplodia natalensis; Phomopsis citri*) (Spalding, 1982). The main post-harvest diseases of mango in Western Australia are stem-end-rot (*Dothiorella dominicana*) (Muller and Bart, 1989). *Penicillium expansum, Botrytis cinerea, Stemphylium vesicarium* and *Mucor circinelloides* are reported as post-harvest diseases of mango (cv. Kensington Pride) for the first time in Australia (Johnson *et al* 1990). Stem-end-rot (*Dothiorella dominicana*) is the most serious post-harvest diseases of mango in South Africa (Darvas, 1993) followed by *Hendersonia creberrima* (Pelser and Lesar, 1989). Major post-harvest decay of mangoes in Philippines was due to anthracnose (*Colletotrichum gloeosporioides*) and stem-end-rot (*Lasiodiplodia theorbromae*) (Esguerry *et al* 2004). Stem-end-rot on mangoes in Nigeria was reported to be due to *Dothiorella dominicana* (Alabi, 1986). *Alternaria alternata* is an important post-harvest disease of mango in Israel (Prusky *et al* 1981).

Losses Due to Post-harvest Diseases

The losses that occur from harvest of the produce until it reaches the consumer are generally referred to 'Post-harvest losses'. Any post-harvest change in the quality of harvested produce due to biotic and abiotic causes, that prevents/alters, its intended use or decreases its nutritive value is also considered as post-harvest loss. These losses may be due to mechanical, physiological and pathological reasons. In the following pages, pathological reasons for post-harvest losses in mango are presented. For other post-harvest losses, the reader is referred to Chapter 12 in this volume.

The pathological reasons are mainly infections of microorganisms (Fungi, bacteria etc.), which cause decays/rots after harvest of the produce. The losses due to post-harvest diseases occur at all levels/stages, right from harvest till the produce is consumed in the house or utilized in the processing factories. The post-harvest losses due to post-harvest disease in fruits like mango are high in tropical countries like India.

Post-harvest disease causes both economical and quality losses which are significant and important. However, accurate estimates of economic losses are not available. The extent of economic losses due to individual pathogens (i.e., diseases) has also seldom been estimated. The economic losses incurred through post-harvest diseases might exceed that caused by field diseases, because of the investments with overall production i.e. harvesting, sorting, grading, packing, storage and transport costs, that the produce passes through from harvest until it reaches the consumer. Economic losses due to post-harvest diseases are considerably more than are often realized, because the losses increase many fold in unit volume, while passing from the field to the houses. Losses in developing countries have been estimated to be in the range of 5-15% or more of the harvest and are greater than these in developed countries, due to lack of adequate refrigeration, poor sanitation and poor and improper post-harvest handling (Eckert, 1981). World wide post-harvest losses have been estimated to be 50% of the harvested crop, which are mainly attributed to post-harvest diseases. However, loss statistics based on percentage spoilage would not reveal the full economic impact of such wastage on the industry. Decay of fruits till they reach the consumer through market processes represent a loss of large financial investment in cultivation and in post-harvest processes till the produce reaches the consumer, the combined cost of which may be many folds greater than the value of crop production. In monetary benefit, *Aspergillus* rot, *Rhizophus* rot *and Lasiodiplodia* rot resulted in a national loss of Rs. 2033.7, Rs. 3362.2 and Rs. 2674.0 millions respectively in India (Saini and Pathak, 1991).

Post-harvest losses due to decays/rots can be apportionated into losses at various stages *viz*., field, market and finally consumer. The losses due to different pathogens were broadly estimated. The losses vary according to the pathogen involved in the spoilage, mango varieties, place, level of marketing etc. The losses vary in kind, magnitude, susceptibility of hosts to the pathogen and during different stages of handling. The magnitude of losses are often substantial, but difficult to quantify and the prevailing state of supply and demand have a decisive influence on the extent of physical and monetary losses.

In the past, several workers have recorded the post-harvest losses due to post-harvest diseases with reference to the supply of the produce to markets as well as pathogen-wise, cultivar-wise and place-wise.

A. Losses during transport and storage

Long distance transport, particularly without refrigeration results in heavy losses due to post-harvest decays. Clara (1927) reported a loss of 39% mangoes in storage in the Philippines. Su (1934) estimated 8% losses in storage and transit due to stem-end-rot. *Aspergillus niger* alone was responsible for about 25% spoilage in storage (Verma and Kamal, 1951). The losses due to stem-end-rot infection were estimated at 10% in Langra variety (Chakravarthy and Srivastava, 1964). Dharamvir *et al* (1967) estimated a loss of 10-15% due to *Diplodia natalensis* and *Alternaria tenuis* in transit. On an average 17.7% mangoes were spoiled and lost due to fungal diseases only in transit, storage and marketing (Sharma *et al* 1994). Thus, the major constraint in mango export to domestic as well as global markets is the spoilage of fruits due to microorganisms during post-harvest handling, storage and transit.

B. Losses at different stages of marketing

Bhuvaneswari (1999) estimated the post-harvest losses due to post-harvest rots at field, wholesaler and retailer levels in Nuzwid mango belt of Andhra Pradesh and reported as follows :

Level	Loss (%)
1. Field level	1.46 – 3.42
2. Wholesaler	0.95 – 6.36
3. Retailer	9.33 – 15-30

The losses at field level were due to faulty harvesting and handling (Verma and Kamal, 1951., Madan, 1988., Rath and Mohanan, 1986). The lower losses at wholesaler may be due to quick disposal of the produce as he is only an intermediary in the marketing of produce (See chapter 12). On the other hand, the highest losses at the retailer level might be due to acceleration of latent infections following transport and storage at wholesale markets, which were aggravated by the open exposure of fruits at the retailer level.

C. Losses at processing factories

In a survey of markets and processing units at Junagadh (Gujarat) Jedeja and Vaishnav (2000) observed disease incidence of 10% at the markets and about 31% at processing units. The higher incidence of diseases/losses at processing units was attributed to improper handling and faulty ripening practices. Nearly 23% was the direct loss due to *Lasiodiplodia* rot alone followed by *Aspergillus* rot (about 7%) at the processing units in the above survey.

D. Losses due to different pathogens

About 25 fungal as well as bacterial pathogens cause post-harvest decays in mango, throughout the mango production centres. However, all do not occur in each and every place. Further, the magnitude of spoilage/loss differs with the pathogen and place. The information available on the subject for brevity is summerized in the Table 10.2

Table 10.2: Losses in mango due to different pathogens

S.No.	Name of the pathogen	Loss (%)	Place	Reference
1.	Multiple infection	14.4 – 20.00	Delhi	Thakur and Chenulu, 1970b
2.	Multiple infection	5.3 – 26.0	Warangal (Andhra Pradesh)	Laximinarayana and Reddy, 1977
3.	Multiple infection	10.0 – 30.0	Bangalore	Sohi *et al* 1973
4.	Multiple infection	10.0	Lucknow	Raoof and Om Prakash, 1983
5.	Multiple infection	10.0 – 30.0	Lucknow	Raoof and Om Prakash, 1983
6.	*Aspergillus niger*	4.0 – 35.0	Different parts of India	Tandon 1967b
		5.0	Gujarat	Verma and Kamal, 1951
		16.0 – 17.0	Warangal	Patel, 1972
		5.0 – 6.0		Laximinarayana and Reddy, 1977
7.	*Colletotrichum gloeosporioides*	6.0 – 15.0	Different parts of India	Tandon, 1967 b
		29.0	Bangalore	Sohi, 1975
		2.0	Warangal	Laximinarayana and Reddy, 1977
		24.0	Lucknow	Raoof and Omprakash, 1983
		29.6	Himachal Pradesh	Sharma *et al* 1994
8.	*Botryodiplodia theobromae*	8.0 – 20.0	Different parts of India	Tandon, 1967b
		26.71	Himachal Pradesh	Sharma *et al* 1994
9.	*Diplodia natalensis*	10.0	Udaipur	Chakravarthy and Srivastava, 1964., Chandra and Pathak, 1989
		2.2 – 4.2		
10.	*Rhizopus oryzae*	6.3	Delhi	Thakur and Chenulu, 1970b
		5.0 – 6.0	Warangal	Laximinarayana and Reddy, 1977
	R. arrhizus	5.5	Himachal Pradesh	Sharma *et al* 1994

Srivastava *et al* (1965) carried out a survey on post-harvest losses due to *Colletotrichum gloeosporioides* in the markets of Allahabad, Bombay, Pune, Calcutta and Kanpur and estimated the loss as 15, 13 and 12% in Dashehari, Himsagar and Rumani respectively. Pathak and Srivastava (1967b) observed 6.2% loss in Langra variety as against the 4.3% loss in Totapuri in Delhi market. The post-harvest loss of mango fruits due to different species of *Aspergillus* in Gujarat was observed to be 16-17% (Patel, 1972). In a survey of markets in Gujarat losses due to *Aspergillus niger* in case of Totapuri, indicated the greater susceptibility of Totapuri to the pathogen. Losses in Karnataka were substantial, the loss was maximum (36.7%) in Alphonso followed by Banganapalli (20%) (Madan and Ullasa, 1988). Post-harvest losses due to post-harvest decays and disorders were in the range of 5-10% (Salunke and Desai, 2001). Market survey indicated a maximum loss (50%) in cv. Neelum followed by Baneshan (34%) in Manni market in Karnataka (Prasanna Kumari *et al* 2007). Arauz and Mora (1983) recorded 43.7% losses in mango in Costa Rica markets.

ALTERNARIA ROT

Alternaria rot (Black spot and black rot) is one of the post-harvest rots of perishables like fruits and vegetables and has been increasingly reported as an important disease of various stored fruits including mango (Prusky *et al* 1980). In prolonging the storage life of perishables, infection by *Alternaria* sp. has become one of the important bottle necks. The pathogen causes post-harvest rot during ripening and storage. It becomes serious when anthracnose and stem-end-rot are well controlled (Johnson and Coates, 1993).

Geographical Distribution

The disease is ubiquitous in all the mango growing areas of the world. It has been reported from Australia, India, Israel, South Africa etc. It is the main post-harvest disease of mango in Israel (Prusky *et al* 1997).

Damage/Loss

Pre-mature fruit fall due to the *Alternaria alternata* is a field loss of mango fruits due to formation of abscission layer prematurely. The stalk-end dries, followed by shrivelling, which changes from normal green color to brown colour. Artificial inoculation of Dashehari and Safeda Lucknow fruits resulted in 10-25% loss (Pandey *et al* 1983). Besides fruits, the leaves and inflorescence are invaded by the pathogen. Panicle infection results in decrease in fruit set (Ploetz, 2004).

Symptoms (Photo 10.1)

On fruits the fungus produces latent infection which develops when fruits start to ripen. The symptoms on fruits differ with the species involved in the rot.

Symptoms of *A. alternata* include the development of small black circular water soaked spots around the lenticels, with a dark centre and diffuse borders on fruit surface. Lesions develop on fruit, concentrated around the stem-end, where the lenticels are concentrated in large numbers. The fungus develops in the lenticels and penetrates the fruit resulting in darkening of inter-cellular spaces, as a result the cells collapse (Prusky *et al* 1983). These spots grow and coalesce into a single large spot covering much of the fruit surface on the upper part of the fruit. The centre of the spot is highly sunken.

Under humid conditions, dark brown spores appear over the lesions. Initially the affected fruit is firm and the pathogen does not penetrate the pulp more than 1-2 mm until late. Later the disease progresses into the flesh, which darkens and becomes partially soft. In advanced stages, the whole fruit gets softened. Beneath the skin, reddish brown patches appear on the flesh. Conidia of pathogen develop in the lesions under moist conditions, which look as olive brown spots on the fruit surface (Prusky *et al* 1983., Om Prakash and Srivastava, 1987., Prusky, 1994., Dodd *et al* 1997., Ploetz, 2003). Symptoms of Alternaria rot are more concentrated than those caused by 'Anthracnose' with darker and hard affected tissue (Ploetz, 2004). In advanced stages, the whole fruit becomes rotten.

Symptoms of *A. tenuissima* start as small water-soaked circular brownish spots, which enlarge to form irregular patches. Reddish patches develop on the flesh below the spots on the fruit surface (Aulakh *et al* 1998).

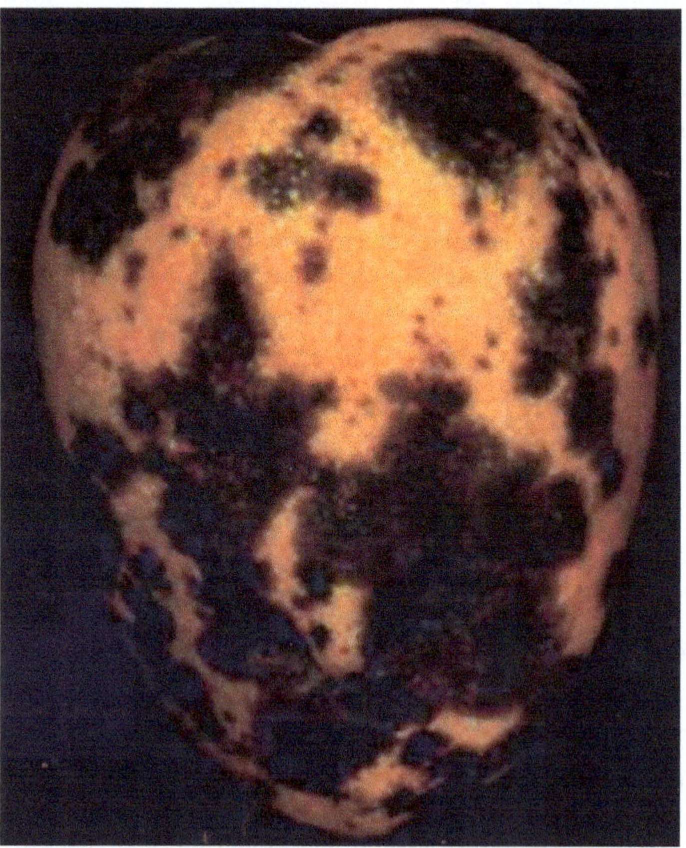

Photo 10.1: Symptoms of black spot of mango

Varietal susceptibility

Most of the commercial varieties of mango grown in Florida (USA) are susceptible to alternaria rot. Cv. 'Keiti' was more resistant than cv 'Haden' and cv. 'Tommy Atkins' (Droby *et al* 1986., Ploetz, 2004). Similarly, most of the mango cultivars being grown in Israel were highly susceptible (Prusky, 1994).

Etiology

Alternaria rot in mango is caused by more than one pathogen *viz., Alternaria alternata* (Fr. Fr) Keissal (Neergaard, 1945), *A. tenuissima* (Fr) Wittshire (Mukherjee and Bhattacharya, 1965., Aulakh *et al* 1998) and *A. solani* (Rotem, 1978). Out of the three, the first one is wide spread and is the main species that causes post-harvest decay of mango fruits (Neergaard, 1945., Domisch, *et al* 1980). Singh and Tandon (1967) and OmPrakash and Raoof (1985) isolated *A. alternata* from mango leaves causing leaf spot of mango. Srivastava *et al* (1964) isolated *A. tenuissima* from the fruit, while Yadava and Udainarayain (1970) isolated from the fallen twigs of mango. Singh and Tandon (1967) found that isolates of *A. tenuissima* from banana, guava and mango were pathogenic under artificial conditions except the isolate from papaya.

Disease Cycle

Infected leaves, twigs and inflorescences are the important sources of inoculum for fruit infection. The fungus overwinters on fallen leaves on the orchard floor and acts as a source. Thus the main source of fungal infection on fruits was the dry parts of the mango trees on the tree as well as on the orchard floor (Munishamanna *et al* 2009). Alternaria isolated from the leaves was able to infect fruits and the isolates from rotten fruits were able to infect detached mango leaves (Prusky *et al* 1983).

The conidia are air borne and are moved by run off of dew from dried or newly infected leaves within the tree canopy (Person and Hace, 1975., Ploetz *et al* 1994). In Israel, sand storm spreads the disease (Prusky, 1994).

Different paths of penetration of fungus have been described for *Alternaria.* The conidia of *A. alternata* penetrate the fruit through the lenticels at the stalk-end and/or wounds and remains latent in the fruit until ripening (Prusky *et al* 1983., Dodd *et al* 1997). However, the lenticels of mango fruit seem to be the main penetration path with further intercellular development. Isolations from the peel indicated that infection occurred at all stages after fruit set. Two week-old fruitlets showed the presence of latent infection. An assessment of latent infection in mango in a 10-week growing season indicated a continuous increase in the incidence of alternaria infection from fruitset until harvest (Prusky *et al* 1983). A significant correlation was observed between the quiescent infected area at harvest time and the incidence of black spots on mature green fruits in storage (Prusky *et al* 1983). The pathogen becomes quiescent soon after the initial stages of infection are completed (Prusky *et al* 1981., Droby *et al* 1986.). The quiescent period of *A. tenuissima* is 4 days at 17 °C (Schiffmann-Nadel *et al* 1983). When the fruits, start ripening the pathogen resumes its growth and develops intercellularly resulting in an active decay lesion and rotting (Prusky *et al* 1983). The harvested fruits gradually loose their resistance as they ripen and the pathogen resumes its growth (Eckert and Ratnayake, 1983). Infection latency was found to associate with two fungitoxic resorcinols that are present in the peel prior to ripening, but decline in concentration as ripening and post-harvest decay begin (Droby *et al* 1986, 1987).

Epidemiology

The environmental parameters, which influence the infections of Alternaria rot in the order of importance include, relative humidity, direction of surface wetness on the fruit, rainfall and temperature (Anon, 1988). Relative humidity is the dominant factor that influences the incidence of quiescent infection of *A. alternata* (Prusky, 1994). Alternaria infection is heavily dependent on moisture (Rotem, 1964, 1965). The pathogen has the ability to utilize dew for fruit infection (Prusky *et al* 1993). A film of free moisture on the host surface is essential for infection by most pathogens including *A. alternata* (Hunter and Kummoto, 1974., Rotem, 1978). Fruit infection has been related to the length of time of relative humidity and prevailing temperature during fruit growth. A minimum of 350 hours and a relative humidity of 80% are needed for quiescent infection and the infection increases with increase in duration and relative humidity. In most humid conditions in coastal regions of Israel, with 620 hours of relative humidity more than 80%, the average quiescent index area (Q1A) was 29 cm^2, while it was only 14 cm^2 under less dry conditions in the same region having 493 humid hours. In the inland valleys of Israel, with 326 hours of relative humidity over 80% disease was not detected (Prusky *et al* 1983, 1992, 1993). Prusky *et al* (1993) developed a map of average daily maximum relative humidity during fruit growth to use as a basis for predicting the incidence of the disease in Israel.

Sufficiently high night temperatures (20-25 °C) during mango fruit growth lead to infection (Anon, 1983). Under these conditions several hours of marginal relative humidity of 80-85% may facilitate the development of the disease (Rotem, 1978).

Because *Alternaria* spores are extremely resistant to heat and desiccation , they are able to utilize dew for infection and initiate infection at night permitting infection to occur under short wet periods, interrupted and dry periods (Rotem, 1978, 1981). Fruit inoculations in the laboratory with conidia at 25 °C showed symptoms of infection 3-4 days after inoculation. However, according to Prusky (1994) alternaria rots are generally significant in arid climates like in Israel.

Control/Management

The losses caused by alternaria rots can be minimized by a regular pre-harvest application of fungicides on the trees in the orchard, post-harvest fungicide treatments and accelerated ripening of fruits with ethylene (Ploetz *et al* 1994). The pre-harvest treatments reduce the quiescent infection development and the incidence of decay after harvest, where as post-harvest treatments inhibit the development of quiescent infection already present in the fruit (Johnson and Coates, 1993., Peterson, 1986., Prusky *et al* 1981).

As alternaria infection on mango fruit occurs from fruit set onwards until harvest and because the fruit growth period is relatively short (3-4 months) infection continues till harvest, continuous protectant treatments are necessary to reduce the latent infection. Any delay in taking protectant treatments after fruit set decrease their effectiveness. Assessment of latent infections aids in determination of the infection period and timing of application of protectant sprays (Prusky *et al* 1981). They developed a technique to monitor quiescent infection of *A. alternata* in mango in Israel from fruitset to harvest and found a good correlation (r – 0.92) between the pre-harvest estimates of quiescent infection and the severity of disease development after harvest. Their results indicated the possibility of using pre-harvest assessment of latent infection as a basis for a control programme of post-harvest treatments (Prusky *et al* 1981).

Maneb (2.5 g/l) sprayed four times (once in every 2 weeks) from a fortnight after fruit set reduced infection from 60 to 10% in Hadean fruits in Israel (Prusky *et al* 1981).

Three applications of the fungicide – Maneb, beginning 2 weeks after fruit set were the most effective before harvest and its efficacy decreased when the treatment was delayed. Maneb application resulted in a lower incidence of latent infection and consequently a lower percentage of infected surface after 22 days of storage (Prusky *et al* 1983).

In India, the disease could be managed by the application of Cuprosan Super D followed by Bordeaux mixture (Arya, 2004). Mukherjee and Bhattacharya (1965), used Mercinbin followed by mercerized copperoxychloride for the control of the disease. Aureo-fungin was found effective in the control of the rot (Sinha and Bilgrami, 1987).

Post-harvest Measures

The effectiveness of mancozeb (Amresh, 1997., Khannan and Subbaraja, 1998., Pandey *et al* 2000) and Carbendazim against *A. alternata* (Khannan and Subbaraja, 1998) was reported. The use of tricyclazole was suggested for the control of the Alternaria in mango. Prasanna Kumar *et al* (2006) conducted laboratory studies on the effectiveness of systemic as well as non-systemic fungicides on the growth of *A. alternata*. Tricyclazole was more effective followed by thiophanate methyl, carbendazim and benomyl in inhibiting the growth. Among 5 non-systemic fungicides, iprodine inhibited mycelial growth to the maximum followed by copperoxychloride, mancozeb and chlorothalonil. Least inhibition of mycelial growth was observed with sulphur. However, contrary to the results of Khannan and Subbaraja (1998), Prasannakumar *et al* (2006) reported that carbendazim and benomyl were not effective against *A. alternata* in laboratory.

No pre-harvest treatments are adopted in Israel to prevent post-harvest rot due to *A. alternata*. On the other hand, post-harvest treatments are applied depending on the climatic conditions where the grove is located. Fruits from groves located in dry region and where the incidence of the disease is low, are only water washed. On the other hand, fruits from groves located in humid region, where the incidence of decay is very high, are treated with 900 μg/ml prochloraz (Prusky *et al* 1983). Since, a single post-harvest treatment with prochloraz is as effective as a pre-harvest measure, this post-harvest treatment is mostly applied in Israel (Prusky *et al* 1981, 1983).

The efficacy of post-harvest dip (in hot water) treatment with iprodione (1.58/l) on mango fruits was demonstrated by a decrease in the incidence of Alternaria rot and the percentage of unmarketable fruits after harvest (Prusky *et al* 1983). However, it was cautioned that mango fruit with values around 5 should not be treated with iprodione, because post-harvest treatments will not significantly reduce disease incidence. Higher values of latent infected surface indicate the need for post-harvest treatment. However, unheated fungicide application as post-harvest treatment substantially reduces the incidence of rot spots in mango (Prusky *et al* 1983). Post-harvest dip with thiophanate methyl or prochloraz (0.01%) in hot water reduces the incidence of *A. alternata* decay.

A simple and efficient treatment for the prevention of post-harvest decay of *A.alternata* in mango was developed in Israel. It involves hot water spray at a temperature of 50-60 °C at the time of brushing the fruit (Hot water brush). Instead of 5 min. dip treatment, the new technique requires about 15-30 seconds only. The treatment significantly reduced the decay of *A. alternata* to levels similar to those reached with prochloraz and storage at 12 °C for 3 weeks. However, when the fruits were stored for longer periods of time, the fungicide was still more effective than the hot water brush (HWB) treatment (Prusky *et al* 1997). The effect of HWB at 70 °C was higher than at 55 °C on fruits, 23 to 25 days after harvest. But, prochloraz was even more effective and further reduced the incidence of black rot decay on such fruits. In 'Kent' mango fruits, both treatments were efficient in reducing the *A. alternata* decay, after 17 and 21 days after harvest (Prusky *et al* 1997).

A. alternata was well controlled by hot benomyl followed by prochloraz (Johnson *et al* 1990). Fruits should be stored at lower temperatures (6-12 °C) and marketed promptly to avoid the risk of black spot decay in mango. However, the development of alternaria rot symptoms in stored mango fruits is usually prevented by a combination of HWB with prochloraz at 225 ul/ml. This leads to eradication of latent infection and consequently prevents fungal colonization (Prusky *et al* 1999).

The finding that *A. alternata* alkalinizes, the host environnent pH (Eshel *et al* 2002), prompted the examination of possibility of modulating the pH by acid treatments as a means of reducing fungal colonization. Prusky *et al* (2006) compared the effectiveness of hydrochloric acid (HCl) alone or incombination with prochloraz in controlling the quiescent infection of *A. alternata* in Israel. Spore germination and germ tube elongation *in vitro* were inhibited by 95 and 65 per cent respectively by exposure to 1.25 mM HCl and fungal germination was completely inhibited by 2.5 mM hydrochloric acid (HCl). Application of a combination of hot water spray and HWD for 15-20 seconds followed by spraying with increasing concentration of prochlaraz at 45 to 900 ug/ml in 50 mM HCl was effective as treatment with acid alone in preventing the development of rot. The enhanced activity of prochloraz in acidification was attributed to its enhanced solubility, which resulted in an increase with fungicide active ingredient in the solution. This suggested that the combination of acid solution alone or in the presence of reduced prochloraz concentration provides a simple treatment for the control of post-harvest diseases, that alkalinize the host environment to promote the activities of pathogenicity factor and fungal attack.

The effect of Ocium leaf extract against *A. alternata* spore germination, growth, proteins and pectolytic and cellulolytic enzymes of various rot causing pathogens was reported (Pathak, 1977). Prasanna Kumar *et al* (2006) reported that Ocium leaf extract was the best in inhibiting the mycelial growth of *A. alternata* followed by neem leaf extract, while onion bulb extract was least effective *in vitro.*

ANTHRACNOSE

There are several diseases which limit the successful cultivation, production, marketing and storage of fruits and vegetables. Among these 'anthracnose' is the most important and serious disease in the field as well as after harvest. It is a commercially important post-harvest disease in many fruits and vegetable crops, particularly in humid growing areas. The incidence can reach almost 100% in crops produced under wet or humid conditions. Mango, in all the production areas suffers both under orchard (pre-harvest) and post-harvest stages from anthracnose. Fruit rot after harvest due to anthracnose is particularly a serious concern when mango is stored for a considerable period or is required to be shipped to distant markets, taking longer time in transit. It is the major constraint in mango export to the world markets.

Anthracnose invades leaves, inflorescence and fruitlets in the field and gets extended as a fruit rot following harvest. It was first reported on mango from the Philippines (Ocfemia, 1931) and in India, it was first recorded in Punjab (McRae, 1924).

Geographical Distribution

Anthracnose is the major and foremost pre-and post-harvest fungal disease of mango in all the production areas. As the most serious post-harvest disease it enjoys world wide distribution. Its occurrence was reported from India (McRae, 1924., Sattar and Malik, 1939), the Philippines (Clara, 1927., Ocfemia, 1931), Brazil (Bitancourt, 1938). South Africa (Doidge, 1932), USA (Traub and Robinson, 1938), Trinidad (Wardlaw *et al* 1939), Peru (Rada, 1939), Indonesia (McKee, 1940), Portugal (Camera and Daluz, 1941), Hawaii (Anon, 1947), Fench Guinea (Brun, 1951), Srilanka, Pakistan and West India (Pathak, 1980) and Puerto Rica (Arauz, 2000). Anthracnose in storage was reported from India by Sinha (1948) and Cheema *et al* (1950). However, it is rarely found in stored mangoes in Israel (Prusky *et al* 1983). The most common and wide spread disease of mango is anthracnose in South Africa (Doidge, 1932). Anthracnose is a major factor limting the production of mango in north coast of New South Wales, Australia (Fitzell, 1979).

In India, anthracnose is quite common in the mango production states like Punjab, Uttar Pradesh, Bihar, Assam, Kerala, Andhra Pradesh, Tamilnadu, Maharashtra, Gujarat, Karnataka etc. (Mc Rae, 1924., Uppal *et al* 1935., Stevans and Pierce, 1933., Sattar and Malik, 1939., Hayes, 1953., Gandhi, 1955., Sohi *et al* 1973., OmPrakash and Raoof, 1985 etc.).

Economic Importance and Losses

Anthracnose is very destructive in field as well as in storage. The effects of the disease are manifested in two phases ; infection of flowers and fruitlets, which leads to direct reduction in yield, while quiescent infection of fruits leads to post-harvest disease development that renders the fruits unmarketable (Jeger *et al* 1987). If wet conditions prevail during flowering the disease causes, blossom blight, which kills the panicles and no fruit sets (Fitzell, 1979). The fungus attacks both the young as well as mature fruits. Early infection of fruits results in pre-mature dropping. Mature fruits are blemished and fetches low prices in the market.

Anthracnose on fruits causes huge losses, as the infected fruits become unfit for marketing and consumption. The average loss of Alphonso fruits due to anthracnose was to the tune of 11.2% ((Dalvi *et al* 2010). The post-harvest losses vary depending upon the variety as well as area of production. Under humid/wet conditions, the incidence is more as compared to drier-arid areas. Clara (1927) from the Philippines reported 39% decay of fruits in storage. The disease spoiled 29% Alphonso fruits in Bangalore, while Bangalora (Totapuri) and Baneshan (Banganapalli) varieties showed less attack (Sohi *et al* 1973). Laxminarayana and Reddy (1977) in Warangal market (Andhra Pradesh) observed a loss of 20% in Baneshan variety, while Srikanth (1989) found a loss of 10% in Baneshan at Tirupati (Andhra Pradesh). A loss of 24% in Dashehari fruits was reported from Lucknow (OmPrakash and Raoof, 1988). Sharma *et al* (1994) recorded a loss of 29.6% in Dashehari fruits at Kangra fruit market in Himachal Pradesh.

Varietal Susceptibiilty

More than 1000 named varieties exist in India of which about 20-25 varieties are under commercial cultivation. Mango genotypes (varieties) show variation in their susceptibility to anthracnose. Bose *et al* (1973) observed that varieties - Bangalora and Baneshan showed comparatively less attack of anthracnose. Sohi (1975) observed maximum damage on Neelum, while Edward showed resistance to this disease in Kerala. Neelum is particularly susceptible to anthracnose when it is affected by 'red tip' disorder. Susceptibiity of Bangalora was more in processing units when the fruit was left for natural ripening in heaps/multiple layers. Baneshan fruits infested with thrips are more readily infested with anthracnose in Andhra Pradesh. In the field, the disease does not appear on important varieties like Chausa, Dashehari and Langra at Sahranpur, but it invaded some of the late varieties, maturing in August-September (rainy season) like Asouja Reoband, Neelum and Peter (Tandon and Singh, 1968a).

Varietal susceptibility was also reported from other mango producing countries. Anthracnose decay was more serious in 'Kent' mangoes than in 'Keitt' mangoes in Florida, USA (Spalding and Reeder, 1972). Indian 'Mulgoa' was very susceptible (Malo, 1977). 'Tommy Atkins' in Florida (Spalding, 1982) and 'Doodoothi' (Passam, 1982) showed susceptibility. Peterson (1986) observed that mango cultvars – 'Carrier', 'Carabao,' 'Florigen' and 'Saigon' were resistant and 'Kensington Pride' was moderately resistant to anthracnose. 'Fyerto' was highly susceptible (Fitzell, 1987). According to Ploetz (1994) 'Sensation' is a susceptible one and 'Haden' is highly susceptible (Campbell, 1992).

Symptoms

Descriptions of symptoms of fruit anthracnose both under natural infection and artificial inoculations are available with slight differences; this may be due to the fact that under natural conditions, more than one microorganism (both disease causing and antagonistic ones) invade the fruit, which hinder the expression of disease symptoms of one another. On the other hand, in case of artifical inoculations, the fruits are inoculated with the pathogen in question after sterilization of the fruit surface, which allows freedom to the inoculated pathogens to express symptoms clearly. Further, variations are also noticed in disease symptoms from place to place, species to species of pathogen, variety to variety of mango etc. The symptoms and damage/loss also varied with the time/stage of maturity of the fruit.

Fruits smaller than pea size on infection get aborted and are shed in large numbers. Larger fruits, that are aborted because of natural thinning or due to other physiological causes are usually mummified. These mummies are invaded saprophytically by anthranose fungus, which sporulates abundantly on them (Arauz, 2000).

Post-harvest anthracnose on fruits is characterized by the appearance and development of black spots of various forms and sizes (Photo 10.2). Lesions develop anywhere on the fruit surface, but often appear concentrated at the stem-end of the fruit. Lesions larger than 2 cm in diameter are fairly common. Lesions appear rounded brown to black in colour. Lesions of different sizes and shapes coalesce into larger ones and cover extensive areas of the fruit surface. On half-mature fruits slightly sunken black spots are produced. These lesions are initially superficial on the skin of the fruit, but at later stages, the fungus penetrates deep (8-10 mm) into the pulp. Then the pulp becomes discoloured and rots. On severe infections, some times, cracks may appear on the skin exudating affected pulp some times even on the intact fruits on the tree (Photo 10.2). In advanced stages of the incidence and under moist conditions the fungus produces acervuli and abundant orange to salmon-pink masses of conidia on the lesions (Baker, 1938., Sattar and Malik, 1939., Muirhead and Grattidge, 1984., Das Gupta and Mandal, 1989., Lonsdale, 1992., Aranuz, 2000., Sharma, 2002). Whole fruit turns blackish and shrinks within 3 or 4 days under humid conditions and the fruits rot and become unfit for consumption. On the ripening fruit the disease occurs as sunken blackish brown blotches upon which salmon buff masses of spores develop in storage (OmPrakash and Srivastava, 1987).

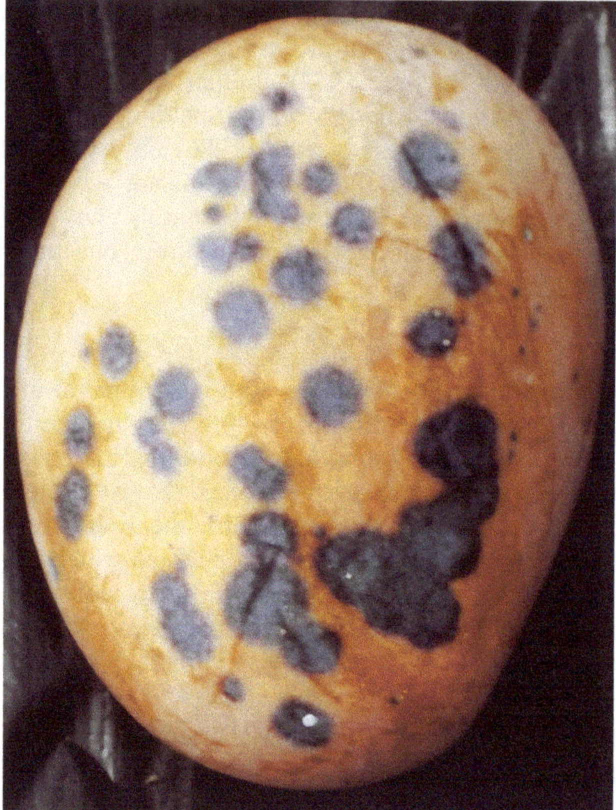

Photo 10.2: Symptoms of anthracnose

In some cases, the lesions or spots typically form "Tear star" (Tear drops) (Photo 10.3) pattern from the apical end (stalk-end) towards the distal end and on one side of the fruit, suggesting that the disease has spread through spores washed down by rain water from diseased twigs or flower stalks (Baker, 1938). Such fruits either rot on the tree or more often carry the disease and rot during transit and/or storage (Singh, 1960., Omprakash and Srivastava, 1987., Dodd *et al* 1997., Arauz, 2000., Sharma, 2002).

Fitzell (1979) compared the symptoms of *Colletotrichum gloeosprioides* var. *minor* and *C. acutatum* on inoculated leaves, panicles and fruit. On leaves and ripening fruit, the lesions developed slowly in case of *C. acutatum* as compared to that *C. gloeosprioides* var. *minor* and were restricted to dark brown flecks instead of a mere characteristic greasy spot. By comparision, *C. acutatum* did not cause defoliation of flushes and as fruit began to ripen, this fungus did not spread into surrounding cells as quickly as *C. gloeosporioides* var. *minor*. Both fungi produced abundant acervuli in lesions. *C. gloeosporioides* produces large, spreading lesions on the surface of the fruit (Muirhead and Grattidge, 1984., Lonsdale, 1993).

Photo 10.3: Tear – stained anthracnose

Smoot and Segall (1963) observed that during the early ripening period, fruits developed rot slowly from pin-point latent infection in the skin; as the fruit began to soften, the individual spots enlarged rapidly to the size of a half-dollar or larger before the fruit was ripe; severe decay resulted from the coalescence of a number of spots to form a large blackened area involving most of the fruit.

A few workers attempted to reproduce the disease symptoms of anthracnose on fruits under laboratory conditions using fungus inoculum isolated from infected fruits from markets (Smoot and Segall, 1963., Lakshminarayana and Reddy, 1977., Srikanth, 1989., Chandra and Pathak, 1992c., Madhavilatha, 1997., Bhuvaneswari, 1999., Gopi, 2003., Bagwan, 2010).

Mango fruits of Neelum variety inoculated with *C. glooesporioides* exhibited round or elliptical spots on the fruit surface, which increased in number and size and coalesced gradually to form bigger lesions. These symptoms were found similar to those of the disease fruit from market (Srikanth, 1989). Mango fruits of cv. Totapuri inoculated by anthracnose fungus produced small circular isolated sunken lesions on the outer rind of the fruit, which later turned into black or dark brown in colour. As the lesions increased in size, several lesions coalesced to form large sunken areas (Madhavilatha, 1997). Anthracnose caused by *C. gloeosporioides* produced isolated, circular sunken, light yellow lesions on the fruit surface, which later coalesced into large lesions measuring 2-4 cm in diameter turned brown to dark brown in fruits of variety 'Baneshan' (Bhuvaneswari, 1999). Chandra and Pathak (1992c) observed that on inoculation, symptoms developed only on wounded fruits and disease severity was higher on ripe fruits. On the inoculated fruits, the spots appeared at first near the stalk-end as very small brown spots that increased rapidly and became black. Often the entire surface was covered by the coalescing spots. Under moist conditions, pink spore masses appeared in the fruiting bodies that were scattered over the lesions (Bagwan, 2010). Symptoms, appeared on artificially inoculated fruits, were more or less similar to those reported earlier from naturally infected fruits (Baker, 1938., Franco, 1946., Meijiao *et al* 2005).

Etiology

Anthracnose is a fungal disease infecting leaves, inflorescenes and fruits. Mango anthracnose is caused by *Glomerella cingulata* (Stoneman) *Spauld & H. Schrenk* (anamorph *Colletotrichum gloeosporioides* (Penz) Penz and Sacc). Besides the above, *C. gloeosporioides* var. *minor* Simmonds (Fitzell and Peak, 1984) *C. acutatum* Simmonds (Freeman *et al* 1998) and *C. capsici* were also reported to infect mango fruits causing fruit rot (Arauz, 2000). The anthracnose of mango due to *C. acutatum* is common in Australia (Simmonds, 1965) and India (OmPrakash, 1990).

Studies on the isolation and identification of the pathogen on infected fruits in different parts of the world revealed that anthracnose of mango fruits was caused mainly by *C. gloeosporeioides* Penz (Small, 1924., McRae, 1924., Stevens and Pierce, 1933., Uppal *et al* 1935., Sattar and Malik, 1939., Sohi *et al* 1973., OmPrakash and Raoof, 1985., Johnson *et al* 1989a., Srikanth, 1989 etc.). In Thailand, the main pathogen of anthracnose was identified as *C. gloeosporioides* (*Glomerella cingulata*) (Johnson *et al* 1989).

Anthracnose in New South Wales, Australia is caused by *C. acutatum* Simmonds (Fitzell, 1979). However, in the field *C. gloeosporioides* var. *minor* is the dominant spices, but *C. acutatum* appears more frequently in trees that have received fungicidal sprays. *C. acutatum* was isolated from leaves, panicles and fruits showing symptoms of anthracnose. The pathogen in the field blights the flowers, young foliage, inflorescence tissue and young fruits. It also initiates fruit infection that remains incipient until ripening (Jeffries *et al* 1990).

The fungal pathogen may exhibit both asexual state (acervuli with conidia) and the sexual state (perithecia and ascospores). The fungus is heterothallie and the sexual state is not common.

Both saprophytic as well as pathogenic forms occur making the detection of pathogenic form difficult. Furthermore, in mango *C. gloeosporioides* infects foliage, inflorescence, immature and ripe fruits. During the early stages of fruit development infection is manifested as blossom blight, fruit staining and later as decay.

The pathogen can survive on leaf spots on the tree throughout the year. The fungus has a long saprophytic survival ability on dead twigs (Singh, 2000). The ability of the pathogen to survive long periods of drought is related to the presence and function of the extra conidial matrix (Louis *et al* 1988). There is no evidence to show that the fungus perpetuates through ascospores i.e. they probably have no role in the epidemics. The peritheicial stage in the fungus is not very common (Sharma, 2002). However, if this stage does not occur, then there is no doubt that this may be the major means of survival in decaying leaves/twigs/debris and for sudden and rapid development of new race of the fungus (OmPrakash and Srivastava, 1987).

Sattar and Malik (1939) studied the viability of the fungus in detached twigs and leaves and isolated the fungus from them even after two years of exposure to open weather conditions. The spores of the fungus borne in acervuli on the twigs showed 75% viability and remained viable for 14 months on diseased leaves, while on dead twigs, the disease persisted still longer periods.

Fungal species exists in isolates and these isolates show resistance to control measures particularly to fungicides. The development of resistance in microorganisms is favoured by continuous and in discriminate use of the same fungicide(s). Isolates of *C. gloeosporioides* from different areas showed resistance to different fungicides. However, it can not be ruled out that the same isolate may show resistance to each and every fungicide.

Isolates of *C. gloeosporioides* in Florida showed resistance to benomyl. The development of resistance to benomyl by isolates of mango pathogens in Florida was probably the result of repeated use of benomyl on mango trees to control anthracnose during production. Isolates of anthracnose fungus have also shown resistance to thiabendazole and thiophanate methyl. The isolates showed greater resistance to benomyl and thiophanate methyl than to thiabendazole (Spalding, 1982). Resistance to benomyl has been reported in *gloeosporioides* isolated from mango in Thailand (Farung Sang and Farung Sang, 1992).

Resistance of *C. gloeosporioides* to benzimidazole fungicides has been reported in many cases (Spalding, 1982., Astua *et al* 1994., Barquero and Arauz, 1996), but resistance to prochloraz has not been reported yet, but its possibility can not be ruled out, because of two factors, i) resistance to prochloraz in other fungi has been reported (Cavelier *et al* 1994) and ii) there is considerable variation in sensitivity to prochloraz among isolates of *C. gloeosporioides* from mango (Madrigal, 1997).

Laboratory studies of Suvarna *et al* (2009) exhibited variation in degree of sensitivity of different fungicides to the isolates of *C. gloeosporioides* from Chittoor and Kadapa districts of Andhra Pradesh (India). These studies indicated the development of resistance to carbendazim (50 and 100 ppm), thiophonate methyl (50 and 100 ppm) and propioconazole (25 and 50 ppm). Based on these studies, the screened isolates of *C. gloeosporioides* were classified as highly sensitive ones and sensitive ones.

In view of the development and possible development of resistance to fungicides by pathogens, it was advised to frequently change the fungicides in pathogen control programmes. The answer for the problem is the development of more potent fungicides such that these can be alternated in control programmes to avoid the development of resistance of mango pathogens. Availability of such more potent compounds avoids the need for exclusively sole use of a particular fungicide which is quite risky from resistance point of view (Spalding, 1982).

Disease Cycle

Disease cycle deals with source and spread of inoculum, time, mode of infection, progress and development of disease in the host tissue etc.

Source: The disease is incited by conidia of *C. gloeosporioides*. The disease appears both in the field and after harvest. The fungus perennates on the infected leaves, twigs and panicles which fall on the ground on which it can survive even as saprophyte on these debris where spores are produced under high humid conditions (Sharma, 2002).

Under field conditions the pathogen produces conidia on lesions on leaves, twigs, panicles and mummified fruits (Fitzell and Peat, 1984., Arauz and Gonzalez-Lobo, 1986). Under favourable moisture conditions in tropical areas, fresh supplies of spores are continuously produced throughout the year on intact as well as dead and fallen parts of the tree (Sattar and Malik, 1939). The different diseased parts (leaves, twigs, panicles and mummified fruits) serve as foci for infection for the succeeding season. Since the condia are formed abundantly in the tree canopy, this is considered to be the primary source of inoculum in the field (Fitzell, 1979., Johnson and Coates, 1993., OmPrakash and Pandey, 2000). There will be abundant production of conidia during flush growth and flowering of the plant (Sharma, 2002).

Spread

The spread of the disease within the mango tree in the field is by water borne conidia of the pathogen during rains (Fitzell and Peak, 1984). The conidia are rain splashed to leaves, flowers and developing fruits from source in the field causing secondary infections, thus the disease is polycyclic in these organs (Ganfotli and Davis, 1993). Development of infection in healthy fruit occurs by contact with diseased fruits (Backer *et al* 1940, Sohi *et al* 1973). However, Arauz (2000) claimed that there was no such fruit to fruit infection by contact.

Time of infection: The fruits are susceptible at all the stages of growth, but most of the infection takes place at the young stage (a week or 2 week-old fruits), which continues until they are half-grown. Infections appear from the start of blooming until the fruits are more than half-grown. Latent infections occur on the fruits while they are still green and unripe on trees, which expresses symptoms after harvest (Wardlaw *et al* 1939., Cheema *et al* 1950).

Mode of infection: After landing and getting hold on the fruit surface infection pegs of the spores enter the green and unripe fruits through pores/lenticels of the fruit skin. Most of the spores do not germinate immediately on the fruit but, remain firmily attached to the fruit as latent stage of the pathogen until the on set of ripening (Sharma, 2002).

Injury was found to be essential for infection under laboratory conditions. Wounding gave maximum disease incidence compared to rubbing of fruits. Fruits without injury did not show the symptoms (Chandra and Pathak, 1992c). Pathogenicity tests in the laboratory revealed that symptoms developed on wounded fruit one day after inoculation and 3 days later on unwounded fruits (Srikanth, 1989).

Development/Progress of the disease: Disease develops on latent infected fruit when the ripening of the fruit commences and progresses rapidly until rotting (Arauz, 2000). The infection progresses faster in wounded lesions and ripe fruits than on wounded, unripe fruits. On unwounded ripe and unripe fruits lesions developed at a slow pace (Quimio and Quimio, 1975).

The quiescence of the fungus terminates when the fruit starts ripening. This phenomena of termination of fungus quiescence has been attributed to the changes in the nutritional status of the host fruit upon ripening (Prusky and Keen, 1993)., to production of ethylene by the ripening fruit (Flaishman and Kolathukudy, 1994) or to the reduction of antifungal compounds (phenolics) in the fruit peel (Prusky, 1996). As the fruit ripens, inhibiters that maintain the quiescence state of the pathogen disappear allowing the pathogen to resume its growth. For this type of interaction quantitative rather than qualitative reactions are the general rule in the post-harvest host-pathogen interaction at the level of fungal penetration and development of appressoria (Prusky, 2009).

The mechanism involved in the onset and termination of quiescence is not clearly understood. However, regardless of the mechanism involved, quiescent infection appears to be a case of co-evolution, as it is advantageous to both pathogen and host in a natural ecosystem. In relation to the pathogen quiescent infection of climacteric fruits by *C. gloeosporioides* represents an example of pathogenic adoption to host physiology. The involvement of ethylene in the termination of quiescence strongly suggests that *Colletrotrichum* sp. must have co-evolved to develop a mechanism to use the host fruit's ripening hormone (ethylene) as a signal to resume its activity, infection process and growth (Arauz, 2000). This mechanism prevents contact of the pathogen with host tissues that have high levels of antifungal compounds. On the other hand, resistance to the pathogen in mango fruit tissue is advantageous to the host during seed development, but not afterwards, because invasion of ripe fruit by saprophytes and weak pathogen can eliminate the ripe pulp, liberating the seed for perpetuation (Prusky and Keen, 1993).

Epidemiology

Weather conditions (temperature, rainfall, humidity) greatly influence the incidence, development, severity and spread of the disease in the field. Most research on the effect of weather parameters on the *C. gloeosporioides* mango interaction was done in Australia (Fitzell *et al* 1984) and the Philippines (Dodd *et al* 1991b., Estrada *et al* 1993).

Humid and misty conditions are most favourable for the incidence (Rada, 1939). Fruit infection of anthracnose is associated with rainfall and four consecutive days of rainfall are necessary for establishment of infection (Kotze, 1978). If wet weather due to rainfall occurs during flowering the disease appears and can prevent fruit set (Patil and Pathak, 1993). Prolonged rains during the active growth and flowering results in out-breaks of the disease (Rada, 1939). The winter rains preceeding or during blossoming provide favourable condition for the spread of the disease. Late set fruits/late maturing varieties which come to harvest and ripen in July-August when the South-West monsoon rains commence as in North India, are seriously damaged by the disease (OmPrakash and Srivastava, 1987). In general, such is not the situation in South India. However, intermittent rains during fruit set, growth and development and thrips damage in South India favour the disease incidence in some years.

In the Philippines, shorter intensive rain showers were more effective than prolonged light rainfall in washing the conidia over the fruit surface, resulting in a corresponding increase in the number of conidia trapped (Dodd *et al* 1989). During light rains, conidia collected in reservoirs around the stem/fruit attachment point are subsequently dislodged under higher rainfall conditions. Presumably conidia become attached to the fruit surface by the mucilage surrounding them under wet conditions (Louis *et al* 1988).

Similarly presence of a suitable temperature is one of the most important environmental factors required for further establishment and progress of the post-harvest anthracnose. The optimum temperature for sporulation was found to be 30 °C and the fungus grows satisfactorily within the temperature range of 10-30 °C and exposure at 40 °C and above kills the spores (Sattar and Malik, 1939). Simmonds (1965) observed that *C. gloeosporioides* grew at 15-35 °C (optimum being 25°-30 °C) and growth was completely inhibited at 10 °C. The optimum temperature for infection was found to be 25 °C (Sattar and Malik, 1939). In general, infection is favoured at temperature ranging from 20 °C to 30 °C. Within this range, there is, however, considerable variation in the optimum temperature requirements for germination of conidia and appressorium formation among the isolates of the fungus from different locations, which was suggested to be due to ecolytic variations among the different locations (Dodd *et al* 1991). The optimum temperature, which favours the growth and sporulation of the pathogen in an artificial culture medium is also suitable for the advancement and spread of the disease in field and storage. It should, however, be understood that the optimum temperature favouring the progress of fruit rot can often be higher than the optimum temperature for the growth of the pathogen in artificial culture medium. This is because of the fact that higher temperature enhances the process of ripening of the fruit, resulting in increase in sugar content and decrease in acidity as well as the firmness of the fruit tissues with concomitant release of ethylene and reduction of antifungal compounds in the fruit. Although the optimum temperature for growth *in vitro* is 25 °C, a higher temperature of 33 °C favours the maximum advancement of the fruit rot by anthracnose fungus (Bhargava *et al* 1965).

Baker *et al* (1940) worked out the exact conditions for infection of mango fruits by *C. gloeosporioides* as, 25 °C and a relative humidity of more than 95% for 12 hours being essential for infection and quick development of the diseases on the fruit. High incidence of anthracnose and stem-end-rot fungal spoilage in mango may be due to the prevalence of favourable weather conditions (25-30 °C, 80-90% relative humidity) for infection in Kangra area of Himachal Pradesh, India (Sharma *et al* 1994). Similarly the higher incidence of fungal diseases in coastal area of India is due to higher humidity, favourable temperature and intermittent rains during flowering and fruiting periods of mango.

Thus humid and misty conditions, temperature of 24°-32 °C at the time of development of shoots, flowers and fruits are most favourable for the infection of the disease under field conditions.

Control/Management

The efforts to reduce the loss caused by the post-harvest disease of tropical fruits (like mango) include both pre-harvest as well as post-harvest measures. The strategies recommended to control/manage anthracnose in mango include site selection, varietal selection, cultural practices, use of protective and curative fungicides, physical methods (hot water treatment, irradiation) etc., which can be arranged broadly under two headings, *viz.*, pre-harvest and post-harvest measures. Since the disease affects the fruit and causes direct loss, the above management strategies should be followed in systematic manner.

The main objective of disease management is to keep the fruit disease free/symptom – free until it is marketed/consumed. The management strategies aims at prevention, eradication and delaying the infection progress and the symptoms expression of the disease.

A. Pre-harvest measures

Management of mango anthracnose in the fields involves cultural and chemical practices as well as site and varietal selection.

I. Cultural Practices

(i) **Site selection:** The incidence and development of mango anthracnose is dependent principally on wetness or high humidity (see epidemiology). Hence, ideally mango orchards should be established in areas with a well – defined dry season during flowering and fruiting periods to allow fruit development in conditions unfavorable for disease development. In tropics mango flowering occurs during dry season. In the entire Indian sub-continent flowering in mango occurs during dry months (December to March). However, within a given area mango trees can flower at any time during the dry season depending on factors such as maturity of the shoot (Chacko and Ananthanarayanan, 1985) and water stress (Rameswar, 1988). If flowering occurs early before the dry season is well established, flowers and young fruit can be infected. The situation will be worst when trees flower last in the dry season and fruits are at the end of their development, where flowering occurs in response to low temperature rather than water deficit (Shu and Sheen, 1987., Nunez-Elisea *et al* 1993). Advancing flowering with potassium nitrate or paclobutrazole or both is also helpful in avoiding anthracnose (Arauz, 2000).

(ii) **Use of resistant varieties:** Although all commercial mango cultivars are susceptible to anthracnose, some of them are comparatively less susceptible than others. Such less susceptible varieties may be chosen for raising mango orchards.

(iii) **Field sanitation:** Saaiman (1997) stressed the importance of orchard sanitation as a means of controlling post-harvest diseases in mango. The main aim of field sanitation is to eradicate the inoculum sources. It involves the destruction of source of inoculum. Diseased leaves, twigs, panicles and fruits that have fallen to the ground as well as on the tree should be collected and burnt at appropriate times, because the fungus has a long saprophytic survival ability on the dead parts of the tree (see disease cycle).

Tree sanitation is difficult to practice. Elimination of dry panicles and mummified fruit is time consuming. As the majority of the post-harvest diseases emanates from fields/orchards, it is logical that phytosanitation would provide a simple, cheap and effective measure to keep the incidence of post-harvest diseases at bay/low (Fitzell and Peak, 1986). However, removal of dry panicles was not found much useful in reducing the severity of anthracnose as compared to those trees where dry panicles were not removed. Leaf inoculum in such situations is probably sufficient to result in infection (Arauz, 2000). Not withstanding the above, it is always beneficial to remove the sources of inoculum from the tree as well as from the ground to minimize the disease incidence. Canopy management of the trees, by way of thinning the canopy and opening the tree to allow sun and free circulation of air is also quite helpful in reducing the inoculum source and avoidance of the diseases. However, the wide spread occurrence of the inoculum of the fungus in the orchard makes it impracticable to control the disease by pruning alone. Hence, pruning has to be supplemented by a suitable spray schedule and other measures.

(iv) **Maintenance of tree vigour:** Plant vigour plays an important role in keeping them free from the diseases. Nutritionally weakened trees are more prone to disease infections. Therefore, maintenance of tree vigour at high level is essential, through proper manuring and fertilization and irrigation to avoid the disease. Manuring and fertilization should be as per the soil and leaf analysis.

(v) **Bagging of fruits:** Landing of the rain-splashed conidia occurs on the fruit surface. Therefore, protection of fruit surface assumes significance, which can be achieved by bagging of fruits and protective chemical sprays. Bagging of developing fruits in paper bags against the incidence of diseases like anthracnose etc. and insect pests like fruit flies, stone weevils etc. has been found to be beneficial in Australia (Hofman *et al* 1997), the Philippines (Bugante and Lizada, 1997) and India (Anon, 2005). However, the practice of bagging, reduced the red colour of the fruits, which had negative impact on consumer appeal in Australia and the Philippines. The choice of bagging fruits depends upon the market conditions and disease pressure (Hofman *et al* 1997). The cost effectiveness of this practice depends on the market price of the bagged fruit (Arauz, 2000). Dodd *et al* (1997) felt that though bagging of fruits is laborious and time consuming, it is cost effective and fetches higher price, because the fruits are clean, unblemished, spot less and not chemically treated. However, the practice of bagging did not find any favour from the mango grower in India, as it is laborious, time consuming and costly and lack of awareness of the consumer towards such bagged fruits. Bagging of fruits one month before harvesting keeps them free from disease and pest infestation for about 15 days after harvesting (Anon, 2005).

(vi) **Care at and during harvesting** : The severity of post-harvest diseases induced by wound pathogens like *C. gloeosporioides* etc., is proportional to the damage inflicted to the fruit by faulty harvesting and rough handling at and during harvesting and after harvest. Chosing the most appropriate time and technique of harvesting is of utmost importance and significance in preventing the post-harvest diseases. Proper handling of the fruit at and during harvesting and after harvest is one step which helps a great deal in the prevention of post-harvest diseases.

The maturity of climacteric fruits particularly at the time of harvesting is a factor which can influence the magnitude of post-harvest losses due to anthacnose. Over mature or ripe fruits if harvested pose the problem of being prone to physical damage during harvesting, transit, staking and storage. On the contrary immature fruits, if harvested tend to get damaged being unable to carryout the desired physiological functions needed for their ripening. Both pre-mature as well as be-tated harvesting can, therefore, make the fruits more susceptible to post-harvest diseases.

Mango fruits unlike citrus fruits are delicate in nature and therefore, require careful harvesting. Suitable harvest aids should be used for harvesting – like those developed at Bangalore, Dapoli, Lucknow and New Delhi. Whatever harvesting method or tool is adopted, the basic care that needs to be exercised is avoidance of injuries, wounds or bruises since the anthracnose pathogen is a wound parasite.

Harvesting of mango fruits with pedicel (8-10 cm) is better than the conventional method of harvesting without stalk (Photo 3.1). Fruits harvested with stalks showed less decay loss and physiological loss of weight (PLW) and better quality and shelf life. On the other hand, the fruits harvested without stalks recorded more PLW and poor shelf life and more losses due to diseases. The losses due to anthracnose and stem-end-rot were also significantly higher in fruits harvested conventionally as compared to fruits harvested with stalk (Kapse, 1993). Conventional harvesting is a most unscientific and improper method, resulting in mechanical damages and bruises producing ingress to wound pathogen like anthracnose. Besides, wounds favour ethylene production and increased respiration and thereby hasten ripening and development of latent fungus if any, after harvest. If the fruits are harvested without stalk, the pathogen can get an easy entry through the stem-end, which could result in early ripening, reduced storage life and fungal spoilage (Roy and Pal, 1991). The disease reduction index of stem-end-rot and anthracnose was higher in fruits harvested with stalk than fruits harvested conventionally without stalk (Kapse, 1993., Kapse *et al* 2009).

II. Pre-harvest chemical control measures

In the modern mango cultivation, the use of chemicals (fungicides/insecticides) has become a regular process and practice particularly during flowering and fruiting periods in view of the invation of several diseases and pests. Use of fungicides of different kinds, both systemic and non-systemic fungicides has been recommended for the management of post-harvest diseases that occur in the field and carried to pack houses, storage chambers etc. after harvest as latent infections. Field sprays with suitable fungicides can prevent disease infections on fruit surfaces by checking spore germination and the formation of appressoria or deep seated infections in the lenticels or in the floral reminants. A pre-harvest fungicidal treatment is considered as an appropriate strategy in view of the occurrence of disease infections in the field, while the fruit is still on the tree and also in situations where considerable harvest injuries are anticipated. Protective sprays have been tested and are being used to prevent fruit anthracnose etc., of mango in several countries.

Baker and Wardlaw (1937), showed that fruit is liable to infections at almost any stage throughout its growth and development on the tree, although the subsequent development of the disease is post-poned until maturity and ripening of the fruit. It follows that if the fruit is protected throughout the growing season with fungicides, the incidence of anthracnose on ripening fruit can be avoided or reduced to a low level. In view of the above, sprayings should commence from the flowering stage. However, since weather conditions during April to June in India are not favourable for the infection, sprayings during this period may be omitted. The same advice may be followed in other countries during dry periods. This will reduce the number of sprays and consequently the cost. Two-three sprayings during flowering and fruit setting periods with subsequent sprays from mid-June (which is the probable time for the commencement of rains) to until harvest were effective in North India. However, under South Indian tropical situations sprays are restricted to periods of flowering and fruiting, which proved effective.

The total number of pre-harvest sprays may vary with the variety, according to their time and duration of flowering, fruiting and time of harvesting and the disease pressure. Generally 2 to 4 sprays during flowering and fruiting are adequate to keep the disease under check in countries like India. However, in extreme situations, where fruit develops completely under disease favourable conditions, up to 25 sprays of protectant and systemic fungicides have been reported (Dodd *et al* 1997). Due to constant presence of anthracnose pathogen in the field and particularly under highly congenial weather for the disease, repeated sprays are required for effective control of mango anthracnose in the field.

Pre-harvest field sprays with different kinds of fungicides in order to check the spread of disease infections, especially the latent infections have been tried successfully in several mango production countries. Similarly several kinds of fungicides like copper-based ones, dithio carbomates, benzimidazoles etc., were tested against mango fruit anthracnose in the field.

Several workers (West, 1934., Stevens, 1936., Baker, 1938., McKee, 1940., Ruehle, 1953 etc.,) were able to control anthracnose by copper based fungicides like Bordeaux mixture, Burgundy mixture, basic copper sulphate, Cuprous oxide and copper oxychloride (COC), of which Bordeaux mixture proved most effective. A 6 : 6 : 100 formula of Bordeaux mixture was recommended in Florida (USA) as a pre-harvest spray for the control of anthracnose in mango (Lynch and Mustard, 1950). But in view of its phytotoxicty on mango flowers and increase in scale and mealy bug infestation, Bordeaux mixture is not generally advised, instead Zineb, which is equally effective was recommended (Stevans, 1936). Ferbam and Ziram, which were reported to have given good control (Hendrix *et al* 1947) were inferior to Bordeaux mixture and Zineb (Tandon and Singh, 1968a). Anthracnose was fairly controlled during flowering and early fruit development by a complete spray programme using captan, copper and other fungicides (Ruehle and Ledin, 1955). Trees sprayed with Bordeaux mixture (1%), Zineb, Maneb, Blitox (0.2%) and Captan (0.3%) showed minimum incidence of latent infections of anthracnose (Conover, 1969., Tandon and Singh, 1968a). Wettable cuprocide (1.51b/100 gallons of water), tribasic copper sulphate (30 lb/100 gallons of water) and other fixed copper fungicides proved quite effective as Bordeaux mixture spray (Singh, 1968). Two sprays of Zineb/Bordeaux mixture (4 : 4 : 50) twice at flowering and then at 14 days interval until harvest were highly effective (Tandon and Singh, 1968a). Micop, Fycol, Blitox, Diathane or Blizone sprays were also effective (Lingaraj, 1969). Diathane Z-78 was highly toxic to the fungus in culture (Bose *et al* 1973). Gadre (1979) achieved good control by pre-harvest spray of Bordeaux mixture, Captan, Blitox and Difolaton.

In Australia, Johnson and Muirhead (1988) recommended weekly sprays of mancozeb (2 g/l) during flowering and then monthly until harvest. COC was suggested as possible substitute for mancozeb after flowering (Thompson, 1987). In a study, monthly application of COC was the only treatment that reduced the incidence of anthracnose on Kent mango fruits in South Africa (Saaiman, 1996). However, the efficacy of copper fungicides was reported to be lower than that of dithiocarbomates under high disease pressure (Jimenez *et al* 1989., Cordoba, 1992). Spraying of COC plus Zineb after completion of heavy showers followed by wettable sulphur (0.2%) before flowering, carbendazim (0.3%) at pea stage and Zineb (0.2%) before maturation of stone (Jadeja and Vaishnav, 2000) and Bavistin (0.1%) at 15 days interval thereafter proved effective against anthracnose (OmPrakash and Misra, 1988). In a study of fungicidal sprays, on the basis of decay reduction index, Diathane M-45 (0.2%) and Captan (0.2%) were effective on 'Neelum' and 'Totapuri' mangoes (Roy and Pal, 1991). Topsin M or Bavistin (0.1%) as pre-harvest spray gave the best control of post-harvest diseases of mango (OmPrakash, 1993).

Pre-harvest sprays of some systemic fungicides were also reported to be effective against *C. gloeosporioides* in mango. Mc Millan (1973) observed that pre-harvest application of Benomyl (0.03 g/l) was very effective against the fungus, while Pathak and Shekhawat (1977) noted that Benlate was most effective. Thompson (1987) recommended up to 2-3 sprays of Benomyl for adequate control of anthracnose. Lonsdale (1993) also showed that pre-harvest application of Benomyl was effective. Benomyl has been used in calendar based schedule usually in a mix with protectant fungicides, to delay the build up of resistance in the pathogen population (Jimenez *et al* 1989). It has been applied as an eradicant spray following infection periods (Estrada *et al* 1996). Pre harvest sprays of Benomyl combined with anti-transparent (Vapor Guard) reduced the disease incidence in storage at 21.1 °C more effectively than an application of Benomyl with NU film 1.7 (Barmore *et al* 1973). Mc Millan (1973) found a similar reduction with post-harvest anthracnose of mango cv. Irwin when the orchards were sprayed with Benomyl (50% a.i.) at 1.121 kg/ha or Maneb (80% ai) at 1.681 kg/ha, along with a surfactant. In Philippines, Dangam *et al* (1988) recommended tank mixing of a fungicide either with Benomyl or mancozeb at 5-10 g and 48 g per 20 litres of water respectively with the chemical flower inducer (KNO_3) to reduce the inoculum of pathogen load on the tree prior to flowering. Subsequent weekly applications of fungicidal spray until fruit set was also proposed to ensure fruit set and development. The current practice in Philippines is similar to this recommendation and involves about 5-6 sprays during flowering plus 1 or 2 sprays during fruit development for the effective control of the disease.

Carbendazim is another systemic fungicide which finds extensive use in the control of fungal disease in several crops including field and post-harvest diseases. Lonsdale (1993) showed that pre-harvest application was effective against *C. gloeosporioides* on mango. Sharma and Badiyala (1994) found effective control of anthracnose on 'Dashehari', 'Langra' and 'Alphonso' mango fruits with pre-harvest sprays of carbendazim. Lonsdale (1993) and Sharma and Badiyala (1994) claimed carbendazim as most effective treatment followed by Bordeaux mixture and COC. Carbendazim, tricyclazole, Benomyl and COC were more effective than Bordeaux mixture (Sharma and Badiyala, 1994).

Pordesimo and Barredo (1976) observed that spraying 'Carabao' mango orchards in Philippines with mancozeb or Karathane between flower initiation and fruit development stages resulted in fewer post-harvest fruit rots due to *C. gloeosporioides* and *Botryodiplodia theobromae*. Pre-harvest sprays of Benlate, Bavistin, Topsin-M and Baycor (each at 0.1%) at 15 days interval resulted in the lowest percentage of fruit rotting due to anthracnose after harvest (Ullasa and Rawal, 1982). Banik *et al* (1998) concluded that the systemic fungicides Benomyl and Topsin-M may be effectively used in the control of *C. gloeosporioides* and non-systemic fungicides were less effective in this respect. Lim (1980) reported that Thiophanate and Benzimidazoles were extremely toxic to anthracnose pathogen.

Fungicides with after-infection activity for post-harvest anthracnose include benzimidazoles and the imidizole prochlorez. Prochloraz has been used as a protectant or as an eradicant spray (Estrada *et al* 1996). Dithiocarbamate fungicides are highly effective for control of anthracnose on mango fruits in the field (Arauz, 2000).

Three pre-harvest fungicidal sprays of thiophanate methyl and carbendazim could manage the quiescent infection of anthracnose on Alphonso mangoes (Sharma and Verma, 2007). Most effective fungicides were tricyclazole, followed by thiophanate-methyl, carbendazim, mancozeb plus carbendazim, propinates and defenoconazole. However, mancozeb was least effective due to differences in climate at Vengurla or cultivar (Dalvi *et al* 2010).

Prevention of pre-harvest infections appears to be more logical way of saving post-harvest losses. The studies of Swart *et al* (2009b) confirmed the importance of good, pre-harvest programmes to ensure effective control of post-harvest diseases in mango. Pre-harvest application of fungicides significantly reduced the pre-harvest infection of mango fruits. Although fungicides were found to be effective as pre-harvest applications, their use as pre-harvest orchard sprays is not totally advisable, because of the possibility of emergence of resistant strains of the pathogen, in spite of alternation of fungicides. However, use of Benomyl usually in a mix with protectant fungicides delays the buildup of resistance in the pathogen (Jimenez *et al* 1989). Resistance of *C. gloeosporioides* to benzimidazole fungicide was reported (Astua *et al* 1994) but not to Prochlorez, yet the possibility exists and can not be ruled out (see etiology). Hence, it was advised not to use prochloraz in field to avoid resistance development in *C. gloeosporioides* and its strains (Arauz, 2000).

Complete elimination of all infections, however, is difficult or impossible, because the fruit is susceptible for a relatively long period. However, sprays of COC during monsoon and post-monsoon are recommended as necessary to suppress the inoculum already existing on the trees. Just before stone formation stage till 15 days before harvest, 3-4 sprays of carbendazim, prochloraz, thiophanate methyl alternatively will reduce the pre-harvest infections of anthracnose and stem-end-rot coming from the field (Anon, 2005).

The choice of the fungicides to be used for after infection sprays in the field should be determined by the target market and by the fungicides that would be used in post-harvest treatments. Field and after harvest application of the same fungicide(s) with a risk of cross-resistance should be avoided. Bavistin and Prochloraz give good after infection control, but Prochloraz is the only fungicide registered for post-harvest use in some countries. Therefore, its use in the field should be restricted to situations where no other options are available.

III. Disease Forecasting System

To avert economic losses due to diseases, to take up timely protection measures and to reduce the number of applications of fungicides, disease forecasting systems are required, in the present context of international markets and higher cost of inputs in commercial cultivation of mango. Two predictive models based on temperature and moisture requirements for infection of mango by *C. gloeosporioides* have been developed, one in Australia (Fitzell *et al* 1984) and the other in the Philippines (Dodd *et al* 1991a) which have been used in the field to time fungicide application.

In Australia, Fitzell *et al* (1984) studied the requirements of temperature and wetness duration for production of dark appressoria from conidia applied to detached young mango leaves under laboratory conditions. They estimated the infection periods based on quantification of conidia that germinated on infected leaves, duration of leaf wetness and temperature during wetness. They found that infection was positively correlated with temperature (25 °C was the highest temperature tested) and the duration of free moisture on leaf surface. By combining this relationship in the form of an infection model and utilizing orchard weather data, they were able to estimate when and at what levels anthracnose infection occurred. When high infection levels (> 50%) of conidia germinated and coincided with flush growth, severe out breaks of the disease with decreased fruitset were observed. These laboratory data were used to derive a model to estimate infection periods in orchards during flowering using the following formula.

$$Y = \text{In } p/q = 16.114 + 1.12\,T - 0.0225\,T^2 + 1.0862 \text{ In } \; t$$

where P = Percentage of conidia forming dark appressoria

T = Temperature (°C) during wetness period.

t = Wet period (hours) and

q = 100 – P

This equation was built into a microprocessor-based forecasting system to monitor and store environmental data and was programmed to estimate the percentage of conidia forming appressoria following rain (Peak *et al* 1986). This instrument called 'Mango Anthracnose Estimator' (MAE) was used to time applications of prochloraz during the mango flowering in Australia. Infection levels were estimated from temperature and wetness duration following at least 0.2 mm rain. Prochloraz (0.6 mg/ml) was applied if the model predicted that 10, 20 or 40% of conidia would form dark appressoria, which were considered as threshold levels by the authors for low, medium or high levels of anthracnose infection respectively. Depending on the threshold, use of the model resulted in a reduction of 4 to 8 fungicide sprays per season to control flower anthracnose compared with weekly sprayings, which was the commercial recommendation in Australia (Peak *et al* 1986).

A similar forecasting system for mango anthracnose, based on the laboratory studies was developed in the Philippines (Dodd *et al* 1991). It differs from the Australian model in that it was developed using combined data from fruit and leaf inoculation and it includes relative humidity in addition to wetness and temperature. Further, the emphasis in the Philippine model was on the dispersal of conidia onto the surface of developing fruit rather than on the blossom blight stage (Dodd *et al* 1991., Estrada *et al* 1996). This model was tested under field conditions for the control of post-harvest anthracnose in the Philippines. Benomyl, Prochloraz (both at a rate of 0.47 mg/ml) or triforine (0.375 mg/ml) was applied following a predicted infection level with a threshold of 40% of conidia producing dark appressoria. One-time application of benomyl or prochloraz was as effective as six calender based sprays of either fungicide (Estrada *et al* 1996).

A comparision of the two models (Australian and Philippine) has shown important differences between them, such as adaptive differences between the isolates used in Australia (Fitzell *et al* 1984) and the isolates used in the Philippines (Dodd *et al* 1991) and in the levels of infection predicted from a given combination of temperature and wetness duration. For example, at 25 °C 10% conidia would form dark appressoria in about one hour according to the Australian model and in about 16 hours according to the Philippines model. The conidia of Australian isolates germinated and formed dark appressoria at lower temperatures than those of the two Philippine isolates (Dodd *et al* 1991a,b). Such a discrepancy indicates that weather based forcasting system for mango anthracnose should not be extrapolated from one region to another and the infection criteria should be based on local data and conditions. The differences between the two models also reflects the differences in the experimental methods especially in the plant tissue used to develop the models, Tissue specificity can occur in different isolates of *C. gloeosporioides* (Ganfotli and Davis, 1987), which should be taken into consideration with due weightage in the development and application of mango anthracnose forecasting systems.

The forecasting systems can be simplified in certain tropical areas, where, the temperature of a given location does not vary considerably from one wetness or high humidity period to another and longer wetness periods. Thus, it can be assumed that infection occurs whenever there is wetness period. Therefore, it is possible to time fungicide spays based on the mere occurrence of a wetness or high humidity period. Working under this assumption, a considerable reduction in the number of fungicide sprays was achieved compared with a calendar based programme for post-harvest anthracnose control (Hord *et al* 1992).

The use of these forecasting systems to program fungicide applications based on predicted infection levels following individual wetness periods (Peak *et al* 1986., Estrada *et al* 1996) is appropriate for leaf or flower anthracnose, but not for fruit anthracnose. Because quiescent infections that result in post-harvest anthracnose on mango fruit accumulate over several infection periods of varying durations and temperatures throughout the fruit development. Therefore, disease should be forecast on the basis of accumulated infection rather than on individual wetness periods. Otherwise, several below threshold infection periods would result in enough infection to account for high disease severity. Conversely, one application of a fungicide with after infection activity could be sufficient to reduce significantly an infection level resulting from several infection periods (Arauz, 2000).

An anthracnose forecast based on accumulated infection would also be useful for optimizing post-harvest disease control. A forecast of cumulative infection could be used to decide which post-harvest treatment would be most cost effective for a given level of infection. One such approach to anthracnose forecasting would be the monitoring of latent infections to determine the need for post-harvest treatments as proposed by Prusky *et al* (1983) for *Alternaria alternata* in Israel. Several monitoring methods are possible : direct planting of sections of mango epidermis on culture media (Prusky *et al* 1983), fruit treatment with herbicide to form the rapid expression of quiescent infections (Biggs, 1995) or correlation of accumulated weather conditions or cumulative predicted infection periods with final disease severity (Rios *et al* 1999).

Forecasting systems are useful for diseases that are important but sporadic (Fry, 1982). Anthracnose forecast in the seasonal tropics would be most useful in dry seasons, when sporadic rate is possible or during transitional periods between dry and wet seasons. Once the rainy season is established, calendar based fungicide applications are the best strategy for chemical control, since conditions are usually favourable for disease development. The most advisable strategy would depend on the time of flowering of a given orchard (Arauz, 2000).

B. Post-harvest measures

Fruit anthracnose is a quiescent fungal disease appearing mainly after harvest. Storage of harvested fruits with quiescent infection is generally limited to less than 2 weeks at 55°F before decays become problem. Therefore, their protection after harvest is imperative as long as possible. Traditionally post-harvest control of mango fruit anthracnose has been aimed at prevention and reduction of the level of quiescent infection on fruits, especially when pre-harvest control methods were not undertaken or ineffectively undertaken and even if undertaken, such measures have not provided adequate protection to the fruit in the orchard. Post-harvest treatments which showed relatively successful results in controlling anthrocnose mainly include chemical (use of fungicides etc.,) physical (heat treatments, irradiation, waxing, wrapping etc.,) methods and their combinations and of late biological control methods etc. Post-harvest treatments such as regulation of ripening, storage by different methods, using host defences etc. are treated separately under "delaying of disease symptoms", besides post-harvest handling practices.

The possibility of prolonging the period of fruit resistance and delaying the onset of anthracnose development until the fruit reaches edible ripening stage was also suggested (Prusky and Keen, 1993).

Careful handling of harvested fruit is the most important step that has to be taken care at all stages to avoid wounds to the fruit that lead to quick expression of the disease and also post-harvest infections. Sorting and grading of the harvested fruit is a good farm house practice to avoid the post harvest diseases. Eliminating the fruit showing bruises, injuries or wounds and early symptoms of the disease and over ripe fruits substantially reduces the chances of diseases at later stages.

However, the use of different post-harvest treatments has gained recognition as promising methods of reducing fruit decays in mango.

I. Post-harvest chemical control measures

One of the methods of controlling post-harvest pathogens is the treatment of harvested produce with chemicals like fungicides, sterilants etc., usually by dipping or immersing the fruits in weak solution for a specified period in cold/hot water. The storage conditions no doubt suppress the growth of the fungal pathogens, but the incidence of decay appears and increases as the fruit ripens after removal from the storage. Therefore, fungicide treatments are applied to the fruits prior to placing them in storage. A lot of research efforts have been made in this direction to develop chemical treatments to limit the post-harvest diseases. A number of fungi-toxic chemicals such as borax, biphenol, imazalil, sodium-o-phenyl phenol, thiobendazole etc., have been tested and put to use with varying success. These chemicals are mostly used as dilute solutions in which the fruits are dipped before storage or as solution for washing or hydro-cooling of fruits immediately after harvest. Some chemicals are impregnated in the packaing containers, cushioning material or wrappers of the fruit.

Dipping the fruits immediately after harvest in weak solution of calcium salts controlled the post-harvest diseases like 'bitter pit' of apple caused by *Penicillium expansum* (English *et al* 1975) and soft rot of potatoes caused by *Erwinia canatovora* pv. *arroseptica* (Mc Guire and Kelman, 1986). The mechanism by which calcium reduces disease levels is not known, but it may act partially by affecting pectolytic enzymes produced by the pathogens. It has been suggested that an increase in cell wall calcium content might be regulating the activity of such enzymes (Conway *et al* 1988). In view of the above, the effect of calcium on the production/activation of the pectolytic enzymes of *C. gloeosporioides* was investigated in mango using calcium chloride and calcium nitrate. It was found that calcium activated and induced higher amounts of pectatelyase, but had no effect on the activity of pectolytic enzyme of the pathogen, nor was able to reduce its induction. It was shown that either dipping or vacuum infiltering of mango fruit with calcium chloride solution did not control anthracnose.

A few chemical sterilants were tested against anthracnose of mangoes *in vitro* (Gopi, 2003). Boric acid (10%) was very effective, compared to other sterilants, followed by potassium permanganate (2%), calcium chloride (4%) and salicyclic acid (0.2%) which did not differ from control treatment. After 20 days of storage, 50% reduction in the incidence of anthracnose by *C. gloeosporioides* was observed in fruits treated with boric acid (10%), copper sulphate (2%), potassium permanganate (2%) and solicyclic acid (0.4%). The disease appeared 7 days after storage in these treatments and 5 days after storage in other treatments and control. However, the sterilants showed least residual toxicity with the increase in storage period, hence ineffective in reducing post-harvest anthracnose decay at later stages of storage (Gopi, 2003). Hence, their use is not encouraged.

The elimination of quiescent infections is achieved commercially by chemical or thermal treatments individually or a combination of both (Mc Millan, 1987). Both systemic and non-systemic fungicides were tested. Several workers attempted the possibilities of employing fungicides individually and also coupled with hot water in laboratories *in vitro* and *in vivo* for the control of anthracnose on mango fruits. A brief account of these studies is presented below.

Benomyl and Diathane M-45 were extremely toxic to *C. gloeosporioides in vitro* (Bose *et al* 1973). Bavistin (0.1%) and Blitox (0.3%) were effective against the pathogen followed by Diathane M-45, while Aureofungin and Captan were least effective (Subramanyam, 1985). Similarly Bavistin (1000 ppm) and Benomyl (100 ppm) caused complete inhibition of mycelial growth of the anthracnose fungus *in vitro* followed by Diathane M-45, while Aureofungin – Sol (500 ppm)) was ineffective, but Captan could inhibit the growth up to about 64% (Srikanth, 1989). Complete inhibition of *C. gloeosporioides* growth was observed with Bavistin (400 ppm) followed by Captaf (450 ppm), Topsin (450 ppm), Cuman L (550 ppm) and Kavach (650 ppm) *in vitro*. Systemic fungicides lke carbandazim (Bavistin) and Thiophenate methyl (Topsin M) in general were found to be superior to non-systemic fungicides *in vitro* in checking the growth of the fungus (Banik *et al* 1998).

Soaking of mango fruits (inoculated with *C. gloeosporioides*) in Benomyl (0.1%) for 10 min. effectively inhibited the decay in storage (Quimio and Quimio, 1974b). Thiabendazole (0.1%) was highly effective in checking the anthracnose infection on fruits collected from market providing 95% decay reduction index (DRI) followed by Indofil M-45 (Diathane M-45), Carboxin *in vivo* studies, which led to the conclusion that anthracnose can be easily controlled by giving pre-inoculation dips of mango fruits for 4 min. in Thiabendazole (0.1%) (Sharma *et al* 1994). *In vivo* studies on inoculated fruits, Bavistin and Topsin M were the best in reducing the incidence of anthracnose followed by Cuman L and Captaf (Banik *et al* 1998). Post-inoculation treatment with carbendazim (0.1%) was highly effective to control anthracnose (Ullasa, 1989), which enhanced the shelf life of mangoes (Ullasa, 1991).

Benzimidazole group of fungicides are systemic in nature/action and include fungicides like Benomyl, carbendazim, thiophenate methyl etc. These were found highly effective, with or without hot water combination in warding off of post-harvest diseases including anthracnose rot of mangoes.

Benomyl (a Benzimidazole) is one of the several systemic fungicides tested and being used as a post-harvest dip treatment commercially at rates ranging from 500-1000 ppm, for the control of mango fruit anthracnose rot after harvest in storage (McMillan, 1973). The fruit dipping in Benomyl solution (500 ppm) immediately before storage drastically reduced the disease (Sohi *et al* 1973). A post-harvest dip of mango fruits in Benomyl solution (600-1000 ppm) containing Tween 40 (0.5%) for 10 min. effectively inhibited the anthracnose decay (Quimio and Quimio, 1974b). Pathak and Shekawat (1977) reported that Benomyl was effective in controlling anthracose under controlled conditions, but not at room temperature. However, Benomyl in cold water failed to check the disease effectively as compared to hot-benomyl and this failure was suggested to be due to the inability of the fungicide to penetrate the thick cuticle of the fruit to kill the deep seated hyphae (Muirhead, 1974). On the otherhand, Santha and Rao (1983) reported that Benlate (500 ppm) in cold water was effective in case of 'Alphonso' mango fruits but not in case of 'Totapuri' mango fruits and stated that the difference could be attributed to the structural differences between the two varieties like thickness of the peel, maturity of fruits, metabolic activity of the pathogen, chemical composition of the fruits etc. Several workers reported the effectiveness of Benomyl in hot water (hot benomyl). Spalding and Reeder (1972) showed that anthracnose can be controlled by hot water treatment (HWT) with benomyl at 54.4 °C for 5 min. Jacob *et al* (1973) succeeded in controlling mango anthracnose by dipping the fruits, in hot benomyl (55 °C, 5 min.), but not at room temperature in case of cold stored mangoes. Hot-benomyl treatment apparently reduced natural fruit luster, however, accelerated fruit shriveling. It was suggested that this problem can be successfully overcome by waxing the fruit immediately after harvest prior to hot benomyl treatment. Santha and Rao (1983) observed that by dipping the fruits in a suspension of Benomyl at 53 °C for 5 min, the post-harvest decay of mangoes by *C. gloeosporioides* could be controlled. Instantaneous dip of fruits in hot-benomyl appears to be the most effective means of controlling anthracnose development in harvested mango fruits (Dodd *et al* 1997). Hot benomyl treatment as post-harvest treatment offers some additional advantages like controlling stem-end-rot and soft brown rot along with anthracnose (Pelsar, 1985., Johnson *et al* 1989a., Johnson, 1994), enhancing the fruit quality etc. (Bryan *et al* 1972). In Australia, the standard PHT for control of all post-harvest diseases is hot benomyl (Ladger, 1991). In South Africa hot-benomyl dip (18/l 5 min.) is registered as PHT for local markets (Swart *et al* 2009a). In spite of the above, Benomyl is no longer permitted for post-harvest use in some countries (Arauz, 2000) because of world wide concern of human health and development of resistance to Benomyl in *C. gloeosporioides* (Spalding, 1982).

Carbendazim (Bavistin) is another systemic benzimidazole tested as PHT for controlling the anthracnose decay of mangoes with or without HWT. Carbendazim is more popular and commonly used as a pre-harvest spray, rather than PHT in India for the control of mango anthracnose. As a PHT, it has shown certain benefits like extension of shelf life of treated fruits (OmPrakash and Pandey, 2000., Kapse *et al* 2009). Mango fruits dipped in hot water (52 °C, 3 min.) together with carbendazim (0.1%) could be effectively stored for 26 days at 12 °C without any anthracnose infection (OmPrakash and Pandey, 2000). Almost all the ripening parameters were benefited by Bavistin treatment like lowering physiological loss in weight, retarding in ripening, besides freedom from anthracnose and stem-end-rot, when fruits were dipped in Bavistin at a concentration of 1000 ppm till the end of shelf life of 13 days (Kapse, 1993., Kapse *et al* 2009). Carbendazim dip (1000 ppm) was effective in reducing the incidence of anthracnose up to 50% compared to untreated control fruits, which was attributed to its systemic nature (Gopi, 2003).

Thiabendazole (TBZ) is yet another benzimidazole, with systemic action and has been registered as a PHT chemical against anthracnose in some countries (Arauz, 2000). The effectiveness of TBZ in reducing the incidence and D1 of anthracnose was documented (Spalding and Reeder, 1972., Sohi *et al* 1973., Sharma *et al* 1994). Dipping of mango fruits in TBZ solutions (1000-2000 ppm) was found to be effective in checking the anthracnose infection on harvested fruits. The fruit dipping in solutions of TBZ (1000 ppm) immediately before storage reduced the disease (Sohi *et al* 1973) and gave over 95% DRI (Sharma *et al* 1994). Instantaneous dip in TBZ (900 ppm) reduced the incidence of disease to about 9% from 29% of untreated fruits (Sohi *et al* 1973). TBZ (0.1%) was the most effective fungicide closely followed by dip in carbendazim (0.1%) providing a DRI of over 90% (Sharma *et al* 1994). Spalding and Reeder (1972) observed that dipping of mango fruits (cv. Keitt) in TBZ (0.1%) in hot water could control anthracnose.

Imazalil and prochloraz are the imidazoles tested against anthracnose in mango. Imazalil was tested and found to be effective against fruit anthracnose of mango as a PHT. Post-harvest dips of mango fruits in Funga Flor 500 EC with a.i. imazalil in hot water (53 °C) was tested in Florida and was found to be effective as PHT. Imazalil significantly reduced anthracnose from 88% in control to 38% after treatment. However, cultivars of mango showed variable response. Further, its efficacy also varied with production centres. It was not effective in Costa Rica (Cuendis, 1988), but effective in USA (Mc Guire and Campbell, 1993). This has been attributed to differences in cultural practices, such as fertilization, irrigation and/or pest control measures between the locations (Mc Guire and Campbell, 1993). It was not labeled for use on mango in Costa Rica (Arauz, 2000).

Prochloraz, of late entered the scene of post-harvest disease control programme of mango and other fruits and gained importance. Efficacy of prochloraz depends upon the extent of disease pressure. It varied from 65% under high disease pressure to 94% under moderate disease pressure (Arauz, 1995). It is used in rates up to 1000 pm in case of mango fruits shipped to European Union (Arauz, 2000). Mangoes dipped for 20 seconds in prochloraz emulsion (18 ml/10 l) at ambient temperature, comply well with local and export requirements (Nel *et al* 2003). Prochloraz very effectively controlled anthracnose on 'Kent' and 'Keitt' mangoes (Swart *et al* 2009a). However, the decision whether to apply prochloraz as a post-harvest fungicide would depend on weather conditions, treatments applied during fruit development, and the market requirements i.e. organic or conventional (Arauz, 2000). Prochloraz is effective for the control of anthracnose and complies well with the maximum residue levels (MRL) of 5 ppm set by European union, hence, in South Africa, hot dip in prochloraz at 9 ml/10 l for 2 min. was registered as PHT of mangoes for local markets (Swart *et al* 2009a). Hot water (55 °C 2 min.) having prochloraz (81.0 g a.i./100 l) was found effective in reducing fruit rot (Saaiman and Lonsadale, 1994).

Of late, many modified formulations of the recommended fungicides are introduced. Joshi *et al* (2010) evaluated the performance of the recently introduced formulations *viz.*, mancozeb (flow), copper oxychloride (water dispersible granules) and carbendazim (flow) against anthracnose in mango. The comparison revealed that mancozeb flow (0.25%) and mancozeb 75 WP (0.25%) were equally effective. Similarly copper oxychloride 50% water dispersible granules (0.25%) and copper oxychloride 50 WP (0.25%) were equally effective. However, carbendazim flow 50% was less effective when compared with its formulation carbendazim 50 WP, which was ascribed to change in the formulation.

Fludioxnil, a new generation fungicide, as dip effectively reduced anthracnose on infected mango fruits from gardens (Swart *et al* 2009a).

Often fungicide mixtures are found to be more effective than individual applications. The value of using two different, chemically non-related fungicides in a mixture each with good activity against a specific disease and some against another can be an excellent strategy for total disease control especially if concentrations can be lowered without losing efficacy in treated solutions (Swart *et al* 2009a). Fruitox and Fritox with Bavistin (carbendazim) gave maximum control of anthracnose disease [25% and 20%, respectively) over control (90%). Hence, Frutox in combination with Bavistin can be used safely for the control of anthracnose of mango under storage conditions (OmPrakash and Raoof, 1988). Swart *et al* (2009a) evaluated the efficacy of Fludioxnil plus prochloraz mixture against anthracnose in 'Keitt' and 'Kent' mangoes. It was observed that in 'Keitt' mangoes anthracnose was better controlled with a mixture of fludioxnil and prochloraz, while in 'Kent' mangoes the disease was better controlled when fruits were treated in heated fungicide solution.

In a search to find out a new method of post-harvest treatment that does not leave toxic residues in fruit, harmful to human health, a new one was identified (Ducamp-Cokin *et al* 2009). The lacto peroxide system (LPS) is a natural system in milk and has antibacterial activity. Soaking of mango fruits (cv. Keitt) in the LPS solution at 35 °C for 120 min. decreased the disease symptoms in Vietnam *in vitro* during storage under standard conditions up to 22 days. *In vitro* studies showed that LPS had antifungal effect against *C. gloeosporioides*. The growth of the fungus was strongly inhibited by LPS. The antifungal action of the LPS is due to hypothiocynate released by lactoperoxidase. LPS is safe to humans and is a good alternative to traditional synthetic fungicides such as immersion in hot (55 °C) benomyl solution, spraying with prochloraz (Prior *et al* 1992., Johnson and Coates, 1993) or HWT at 53 °C (Spalding and Reeder, 1986). Use of LPS avoids the use of chemical fungicides and also avoids high temperature treatments, thus keeping the fruits as natural as possible (Ducamp-Cokin *et al* 2009). In order to extend the shelf life of fruits, the combination of LPS with other natural antifungal components or the inclusion of LPS into a coating that would be directly applied on the fruits would be of great interest. However, further studies should be carried out to confirm the above findings and to consider its industrial application in case of mangoes and other fruits.

Often, the success of post-harvest control of anthracnose on fruits by post-harvest fungicides treatments is not complete. The performance of post-harvest fungicides can be influenced by magnitude of pre-harvest control of the disease, temperature and pH of fungicide solution, environmental factors during transport and storage, duration of treatment, concentration and formulation of toxicant, variety and age of fruits, specificity of chemical and isolate of the pathogen (Pathak and Shekhawat, 1977). When pre-harvest measures were effective in reducing the disease pressure in the orchard, the post-harvest decay could be controlled more effectively. Low post-harvest decay is associated with effective protection of fruit throughout the growing season irrespective of the post-harvest control strategy. On the other hand, poor control of post-harvest anthracnose disease is associated with poor control of the disease in the orchard and high disease pressure. Anthracnose can only be controlled effectively when spray volume and frequency of application were adequate to ensure continuous protection against infection in the orchard.

The increased use of synthetic fungicides during the last few decades, however, is not without problems. Unforeseen side effects such as toxicity to non-target organisms, development of resistance in pathogen to materials used and environmental contamination with potential to affect entire food chain, high and residual toxicity, long degradation period, effect on fruit quality, side effects on humans are some of the drawbacks of fungicides (Lingk, 1991).

II. Post-harvest heat treatments

Restrictions and bans on the use of chemical treatments, due to their higher cost, particularly their residues in the fruits and other perishables, beyond the set MRLs and the limited success of control offered by bio-agents, prompted the need to findout and develop alternate methods of controlling post-harvest diseases. This need revised interest in the use of non-chemical procedures such as cold, heat and irradiation. However, mangoes are sensitive to low temperature and irradiation has not been accepted by consumers at large, leaving only 'heat' as the most appropriate option for the control of post-harvest diseases.

Heat has fungicidal as well as insecticidal action and can be applied by humid-hot-air (Vapour heat and hot forced air) hot water and radiation. Humid-hot air is effective, but expensive, whereas the effects of radiation on mangoes are not well understood and is not fully accepted by consumers because of unfounded apprehensions that consumption of irradiated food is fatal to human health. Therefore, the choice rests on hot water treatment (HWT), which is widely used not only for the control of post-harvest diseases but also for the disinfestation of mango fruits from insect pests, as a quarantine treatment in case of exports to international markets.

Hot Water Dip Treatments

Heat treatments are promising and have been used with success in eradicating or suppressing the development of fungi on the fruit surface as well as those situated in the first few cell layers below peel, as a result of pre-harvest infection. In most cases the fruits are immersed in hot water at different, but suitable and unharmful temperatures for variable periods. HWT is most suitable for such diseases where there is wide gap in the thermal death point of the pathogen and its host and where the temperature of water does not spoil and injure the produce by scaling of skin, loss of natural colour, flavour or softening of the flesh. HWT has shown some benefits. It leaves no chemical residues on the fruit and hence could be a good option for the control of anthracnose in case of organically grown mangoes and/or for export of mangoes where no fungicides are currently labelled for post-harvest use. Besides it serves dual purposes like disinfestation of insect pests and control of post-harvest diseases. However, one drawback with HWT is, hastening of ripening. Not withstanding this, HWT is an effective and feasible method for control of fruit anthracnose of ripe mangoes, particularly when fruits are stored at lower mild temperature after the HWT.

The effectiveness of HWTs as post-harvest treatments for the control of post-harvest diseases of mangoes such as fruit anthracnose and for disinfestation of quarantine insect pests has been in vogue for many years and is an old and successful strategy in this sphere. HWT for fruit flies provides acceptable anthracnose control under low infection levels and it may not be necessary to any other control measure to fruit produced under dry conditions (Arauz, 2000).

Several workers from different mango production centres have reported success in the control of mango fruit post-harvest anthracnose with HWT, but the temperature of water and time exposure varied with the worker and area of production and cultivation of mango, with the corresponding variation in the success of control. The combination of water temperature and time of immersion of fruit required for the purpose of control of anthracnose varied with the cultivar of mango (Smoot and Segall, 1963., Tandon and

Singh, 1968b). Further, HWT is more effective on some cultivars than on others and varietal differences exist in susceptibility of cultivars to heat damage in HWT (Smoot and Segall, 1963). Also fruits can be easily damaged by over-exposure to heat irrespective of the cultivar. In view of the above, temperature and time of exposure are critical and assume greater significance. Appropriate temperature and time of exposure, therefore, should be worked out for each cultivar and production centre, which would not be harmful to the fruit in any way. Tandon and Singh (1968b) have recommended a 15 min. immersion of fruits of cv. 'Dashehari' and 'Langra' in hot water at 50 °C in India.

Pennock and Maldonaldo (1962) were the first to demonstrate the usefulness of this technique (HWT) for the post-harvest control of anthracnose of mango fruits. They recommended immersion of fruits for 15 min. in hot water (51.0 to 51.5 °C) for commercial practice, before packing and shipment. In Florida reduction of anthracnose rot by immersion of fruit of 5 cultivars in hot water (54.5 to 56.9 °C) for 5 to 7 min. was achieved. A slightly lower temperature of 51.7 °C was also effective when exposure time was extended to 15 min (Smoot and Segall, 1963). The disease could be controlled by HWT at 55 °C for 5 min. in 8 out of 9 cultivars in Florida in packing houses (Hatton and Reeder, 1964). Mangoes of cvs. 'Chausa', 'Dashehari,' 'Langra' etc., treated with hot water at 50-55 °C for 15 min showed a substantial reduction in anthracnose decay and treated fruits could be stored for 5 weeks at 9 °C-10 °C (Tandon and Singh, 1968b). The use of HWT to unhibit anthracnose in mango was described by Subramanyam *et al* (1972) and Laxminarayana *et al* (1974). HWT at 52 °C for 30 min. provided complete protection from anthracnose up to 8 days storage at ambient temperature and for 26 days at 12 °C. Short duration HWT (5-15 min.) was not effective, because of higher incidence of disease (99.33 and 86.66%) and percentage disease index (58.67 and 28.0%) at ambient temperature. The ineffectiveness of shorter duration of treatment could be due to the thermal death point of the mycelium of *C. gloeosporioides*, which is 10 min. at 53-54 °C (Quimio and Quimio, 1974a). According to Laxminarayana *et al* (1974) post-harvest spoilage could be prevented by treating the fruits with hot water at 54 °C for 5 min. in case of cv. 'Alphonso'. In cv. 'Irwin' a hot water dip for 5-10 min. at 56 °C could substantially reduce the post-harvest anthracnose rot (Chang, 1975). Akamine (1977) reported 20 min. hot water dip at 47 °C to be effective for controlling storage decay of mangoes in Hawaii. The untreated control fruits and those fruits subjected to phased exposure of hot water (36.5 °C) for 30 min. stepped-up to 46.5 °C for 22 min. showed highest severity of anthracnose, which was attributed to quiescent infection and the susceptibility of the cultivar to the disease (Fitzell and Peak, 1984).

Post-harvest fruit anthracnose has been controlled by immersing of fruits in hot water (50-60 °C) for up to 15 min. in mangoes (Rappel *et al* 1991). Hot water dip of bagged fruits (with brown paper) for 10 min. resulted in the least incidence of anthracnose in the Philippines (Bugante *et al* 1997). HWT at 46.5 °C for 45 min. significantly reduced the incidence of anthracnose in mangoes (Nyanjage *et al* 1998). HWT at 52 °C for 30 min. could very effectively controlled the disease and the treated fruits could be stored safely for 26 days at 12 °C (Omprakash and Pandey, 2000). Immersion of fruits at a constant temperature of 46 °C for 90-115 min. also checks the anthracnose as well as stem-end-rot (Singh, 2000).

The disparity observed above in temperature of water and exposure time is due to varying tolerance levels of cultivars to the treatment and variable sensitiveness of the causal pathogen i.e. *C. gloeosporioides* strains from different mango growing regions to HWT (Dodd *et al* 1997). Cultivars vary in their tolerance to hot water temperatures, therefore, should never exceed 55 °C for 5 min, when HWT is used alone. The general observation from the above review of information is that as the temperature of water increased, the exposure time decreased and vice versa. The general recommendation is that the water temperature should be between 50 and 55 °C and dipping time must be at least 5 min. when the fruits are treated with hot water alone.

Effectiveness of HWT in controlling the fruit anthracnose is influenced by the treatment temperature and duration of exposure. Heat units have no influence on the effectiveness of HWT. Further, the effectiveness of HWT varies with the amount/intensity of infection (disease pressure) and the storage regime and declines as time after harvest increases (Johnson *et al* 1997). In naturally infected fruits, when subjected to HWT at various times after harvest, the results differed. Treatment 12 hours after harvest delayed the onset of disease such that on average fruits exhibited the symptom only when almost fully yellow in case of Cv. Carabao in the Philippines. In contrast, those treated early (6 hours) or late (more than 12 hours) after harvest showed symptoms at the yellow with tip green stage. This demonstrated that the efficacy of HWT was governed by the time that lapsed after harvest (Lizada *et al* 1997).

Combination of HWT and Fungicides

Hot water dips of fruits are moderately efficient. When the levels of infection are low HWT is adequate for the control, but when the infection levels are high addition of another component like fungicides is needed for effective and enhanced control of post-harvest diseases (Spalding and Reeder, 1972., Pelser, 1985, 1987., Pelser and Lesar, 1990., Lonsdale, 1993). The elimination of quiescent infections could be achieved by a combination of thermal and chemical treatments (Mc Millan, 1987). Cold fungicides are less effective. The combination of hot water and fungicides is the most effective commercial post-harvest treatment for the control of post-harvest fruit anthracnose of mango. Even at very high levels of infection, efficacies over 95% can be achieved by applying hot water treatment and fungicides sequentially or together (Arauz, 1995). The advantages of such combination are many. The combination allows the improvement of efficacies of either component on mutual basis due to synergism. Combination treatment allows the use of lower temperature and a shorter time of exposure (Spalding and Reeder, 1972., OmPrakash and Pandey, 2000). Addition of fungicides increases the efficacy of HWT (Spalding and Reeder, 1972., Muirhead, 1976., OmPrakash and Pandey, 2000). The combination enables the concentration of fungicide and temperature of water to be reduced compared with those required when either one was applied alone, resulting in improved control of post-harvest diseases (anthracnose and stem- end-rot) and results in greater amount of marketable fruits (Spalding and Reeder, 1972). The efficacy of fungicides with hot water improves since hot water is a more efficient heat transfer medium than cold water and hot air (Shellic and Mangan, 1994). The combination treatment lowers physiological loss in weight and improves organoleptic score when stored in cool chamber as compared to storage at room temperature (Waskar and Gaikwad, 2005). Storage life of fruits is extended by combination treatments (OmPrakash and Pandey, 2000).

Various combinations were suggested like hot water with benomyl or thiabendazole (Spalding and Reeder, 1972., Tandon and Singh, 1968b), hot water with carbendazim or thiophenate methyl (OmPrakash and Pandey, 2000), hot water plus prochloraz (Saaiman and Lonsdale, 1994), hot water plus zineb (Subramanyan and Moorthy, 1973) etc.

Spalding and Reeder (1972) showed that anthracnose could be controlled by HWT at 54.4 °C for 5 min. by adding benomyl or thiobendazole fungicide. The combination allowed the use of lower temperature and a shorter time and extension of storage life of 4 weeks, as against 3 weeks of storage life with HWT alone. Hot water in combination with sodium diethyl dithiocarbamate gave best results in controlling anthracnose on mango fruits (Subramanyan and Moorthy 1973). HWT (52 °C for 15 min) with carbendazim (0.1%) allowed 26 days storage of treated fruits at 12 °C, while those treated with HWT (52 °C for 15 min) and Thiophenate – Methyl (0.1%) could be stored for 19 days only at 12 °C. Reduction of exposure time from 15 min to 2 min with the above combination did not provide protection of fruits against anthracnose. However, the former combination of HWT and carbendazim/thiophenate methyl with an exposure of 30 min showed slight shrinkage of fruit skin without adverse effect on fruit ripening (OmPrakash and Pandey, 2000). Combination treatment of hot water plus Bavistin (0.1%) was found to be the best against anthracnose and SER of mango fruits (Waskar and Gaikwad, 2005). Treating mango fruits in hot water (50 °C) with carbendazim (Bavistin 0.1%) and methyl thiophanate (Roko, 0.4%) gave effective control of anthracnose rot with higher DRI (58.34 and 62.55 respectively) and lower number of spots (6.10 and 4.22 respectively). Similarly, a combination of hot water (50 °C) with common salt (1%) for 5 min gave complete control of rot with DRI of 95 to 85 and reduction of average number of anthracnose spots per fruit to 4 as compared to zero DRI and 11.25 spots in the control fruits (Saxena and Rawal, 1998). Of late, it is recommended to dip the fruits in prochloraz (0.1%) followed by HWT to keep the fruits free from fungal infection (Anon, 2005). When HWT is combined with fungicides (e.g., benomyl or imazalil) water temperatures should be reduced to 52°-53 °C (Dodd *et al* 1997).

Problems

One limitation of HWT is that temperature required to kill or inactivate the latent pathogen is often high to cause scald injury and induction of pre-mature and abnormal ripening of treated fruits. Heat injury is caused by HWT with or without fungicides, when the treated fruits are stored at low temperature (< 13 °C). The heat injury occurs due to over exposure of fruits to higher temperatures (> 55 °C). Heat injury manifests as scalding of fruit skin, acceleration of flesh softening and change of peel colour from green to yellow, within the first week of low temperature storage. Therefore, in future the optimum temperature as well as the duration of exposure to HWT should carefully worked out for each cultivar, preferably for each mango growing region for the same cultivar. In future, any recommended HWT must be based on lack of injury to the fruit held at less than 13 °C after treatment (Smoot and Segall, 1963., Hatton and Reeder, 1964., Spalding and Reeder, 1972).

The emergence of heat resistant strains of the pathogen is another problem encountered and apprehended (Aragaki *et al* 1981). It is also apparent that under hot tropical climatic conditions many of the pathogens have evolved which are inherently resistant to high temperatures. This could easily explain some failures with HMT of mangoes in post-harvest disease eradication programme at Allahabad (India), where maximum day temperature (in shade) during pre-harvest season of mango ranges between 45° and 48 °C and the same holds good in other tropical areas (Ghosh, 1998).

Aerated Steam

This method has been found to be effective against some post-harvest diseases (Pathak *et al* 1988., Patil, 1992). It is more effective than HWT against spore germination and disease severity. In aerated steam treatment, latent heat of vapourization is transferred to the spore/fruit when steam condenses and it coagulates the enzymes of the pathogens more effectively, resulting in their inactivation or killing.

Heated Forced Air

Miller *et al* (1986) found that a heated forced air treatment at 51.5 °C for 125 min. of mango (Cv. Tommy Atkins) apart from being effective in the eradication of Caribbean fruit fly, can concurrently reduce anthracnose and SER of mango substantially.

Mc Guire (1991) studied the effect of heated forced air treatments on 3 cvs of mango and observed that the treatment at 48 °C for 150 min. not only did considerably reduce the anthracnose and SER, but also concurrently eradicated the infestation of Caribbean fruit fly. Coates *et al* (1993) successfully used heated air to control anthracnose, by raising the core temperature of mango fruit to 46 °C within 24 min. or 48 °C within 8 min. The efficacy of fruit fly disinfestation quarantine treatment as a measure of post-harvest disease control was evaluated. High humidity (> 95% RH) hot air treatment alone (to a core temperature of 46.5 °C for 10 min.) reduced the incidence of anthracnose in mangoes stored at 13 °C prior to ripening at 22 °C. A treatment consisting of high humidity hot air combined with either a heated fungicide (benomyl) or an unheated superior fungicide (prochloraz) gave complete control of anthracnose under these storage conditions. The efficacy of fruit fly quarantine treatment varied from 60-85% for elimination of anthracnose infection (Mc Guire, 1991).

Irradiation

This is a physical method employed for the control of post-harvest anthracnose incidence of mango fruits in which the harvested fruits are exposed to ionising radiation from sources like gamma rays, infer-red and ultraviolet rays. Among these sources, gamma irradiation has been most successfully used for inhibiting the growth of pathogen in side the fruits, as the rays penetrate fruits and destroy the pathogen in side even in established lesious deep in the host tissue (Spalding and Reeder, 1986). They found that fruits of Cv. 'Tommy Atkins', after exposure to gamma irradiation at 0.2 kGy or 0.75 kGy exhibited reduction in infection caused by *C. gloeosporioides* (anthracnose) and *Botryodiplodia theobromae* (SER), but the higher dose (0.75 kGy) caused browning and higher pitting of the skin. Johnson *et al* (1990) observed that gamma irradiation ranging from 0.3 kGy to 1.2 kGy reduced the incidence of post-harvest diseases of 'Kensington Pride' mango fruits, but a dose of 0.6 kGy resulted in lenticel spotting and surface discolouration of the exposed fruits. The incidence of anthracnose decreased with increasing dose of irradiation (Johnson *et al* 1989b). Higher doses, however, were phytotoxic and also affected the palatability of mango fruits (Hatton *et al* 1961., Sreenivasan *et al* 1971).

However, irradiation with gamma rays was not always successful (Lim and Khoo, 1985). Reports of radiation effects on the ripening of mangoes have ranged from no delay in 'Haden' mangoes irradiated at 1000 Gy (Akamine and Goo, 1979) to a delay of 3 days in Sensation (100 or 250 Gy) and 'Irwin' (500 Gy) mangoes (Hatton *et al* 1961). Thus low dose irradiation of mangoes can be beneficial through extension of normal shelf life of some cultivars (Spalding and Reeder, 1986). Irradiation can inhibit development of red colour, which reduces eye appeal and salability (Spalding and Reeder, 1986). Inhibition of colour development varies with the cultivar, as well as maturity and condition of treatment such as temperature and dosage of irradiation. A slight inhibition of colour development in mangoes irradiated at 250 Gy was observed (Spalding and Reeder, 1986).

The mango global export market demands the supply of high quality fruits that are free from any pests and diseases and blemishes in order to get economic returns. A post-harvest treatment (PHT) is, therefore, essential to ensure adequate control of post-harvest diseases without any effect as fruit quality and salability to meet the demand of the international markets. A PHT of mangoes consisting of 5 min. hot water dip (50 °C) followed by 20 sec. dip in prochloraz at ambient temperature is the conventional standard method to ensure adequate control of post-harvest diseases in South Africa which is a time consuming process, thus presents a bottleneck in the pack houses. As an alternate, a short wave infrared (IR) treatment was developed in South Africa and was found to be more effective against post-harvest diseases. Short wave IR radiation has long been recognized as being a highly efficient form of heating. It reduced anthracnose levels similar to commercial HWT in checking the anthracnose fungus (Arauz, 2000). The IR treatment is cheaper and faster than the HWT (Saaiman, 1993). The advantage of IR treatment lies in the enhanced speed and economy pack times. Since IR treatments only lasts for 3 min. against 5 min. of HWT, 40% more fruit can be treated per hour of active packaging for the same labour costs (Saaiman, 1993). One min. exposure of 'Kent' mango fruits to short wave IR was as effective as the commercially used 5 min. hot water bath treatment in checking the anthracnose. No differences between IR and HWT were observed in their effect on fruit quality (Lonsdale and Droamer, 1994., Saaiman, 1993). However, the effect of IR treatments rests on the time of duration of exposure of fruits. Under conditions of low disease pressure, the disease was significantly higher in fruits treated with 1 or 2 min. than on fruits treated with the conventional HWT. Fruits treated with IR for more than 3 min. showed equal efficacy to commercial standard PHT of 5 min. HWT controlled anthracnose and soft brown rot (*Hendersonia creberrima* Syd and Butt) of 7 mango cultivars in South Africa without effect on skin colour and fruit quality, but IR treated fruits showed more scalding and lenticel damage than those subjected to conventional HWT (Saaiman, 1993). However, IR exposure of more than 2 min. was phytotoxic to 'Irwin' mangoes in South Africa (Lonsdale and Droamer, 1994). This was attributed to the fact that surface temperatures of fruits were higher following treatment with IR than the HWT (Saaiman, 1993).

Ultraviolet (UV) rays is another source of irradiation. Exposure of fruits to UV radiation is considered as an effective photo-chemical threaupy to control post-harvest diseases and to maintain the quality of harvested fruits (Baka *et al* 1999). However, it appears that no work on mangoes with UV radiation was done. In a comparative study, exposure of fruits to IR radiation for 1-30 min. or to UV radiation for 10-30 min did not control fruit decay effectively (Lonsdale, 1993). The fruits exposed to IR radiation for 30 min. exhibited lower incidence of anthracnose followed by UV radiation for 20 min, IR for 5-15 min. and UV radiation for 15 min. In another, experiment, UV for 20 min. showed 26-30% disease index, 20 days after storage, compared to control treatment (Gopi, 2003).

Combinations of irradiation with HWT/fungicides showed promising results. A combination of HWT plus irradiation treatment gave complete control of anthracnose decay of mango fruits. The combination of hot water with irradiation has the advantage of leaving no chemical residues within the fruits and also providing control of fruit fly (Burditt *et al* 1981) and seed weevil (Thomas and Brodrick, 1977). Increased control of decay was possible with irradiation and HWT combination (Ahmed *et al* 1969). A combination of hot water (55 °C for 5 min.) with radiation (750 Gy) is used commercially for testing mangoes in South Africa and is reported to act synergistically in controlling anthracnose (*Colletotrichum gloeosporioides* Penz) and soft brownrot (*Hendersonia creberrima* Syd and Butt) of mangoes in addition to providing quarantine control of the mango seed weevi (*Sternochetus mangiferae* Fab.) (Thomas, 1975). A combination of hot water (50 °C, 5 min.) and 50 Krd of gamma irradiation reduced the incidence of both anthracnose (*C. gloeosporioides*) and stem-end-rot (*Diplodia natelensis*) during fruit ripening and post-climacteric stages (Thomas and Desai, 1976). Such combination was the most effective and comparable to dipping of fruits is hot water (50 °C) plus benomyl (0.1%) for 5 min. on artificially inoculated mango fruits. *In vitro* studies with isolated pathogen also corroborated the above findings and the treated fruits were comparable to untreated control fruits in respect of organoleptic qualities. Therefore, hot water dip combined with irradiation offers better prospects for delaying both ripening and control of fungal infections (Padwal Desai *et al* 1973).

Spalding and Reeder (1986) studied the effects of various combinations of hot water, imazalil fungicide and gamma irradiation on decay control and acceptability of 'Tommy Atkins' mangoes in Florida (USA). Combination treatments with radiation at 200 or 750 Gy and hot water (53 °C) or hot (53 °C) imazalil (0.1%) for 3 min. were more effective than simple treatments for control of anthracnose (*C. gloeosporioides*) and stem-end-rot (*Diplodia netalensis* and *Phomopsis citri*) of 'Tommy Atkins'. The combinations of hot water plus radiation or hot imazalil plus radiation gave the most effective decay control and the highest percentage of acceptable fruit. Unheated water plus radiation had no effect on decay or percentage of acceptable fruit. Irradiation can inhibit development of ripe skin color, which reduces eye appeal and salability of fruits. Irradiation (750 Gy) after the HWT (55 °C for min.) nullifies the accelerating effect of the HWT on the ripening process (Thomas, 1975). Later work (Brodrick and Van der Linde, 1981) suggested that the colour development was still a problem in mangoes with hot water plus radiation and when placed in cold storage at 10 °C-12 °C for 3 weeks before ripening and gassing the mangoes with ethylene before placing them in cold storage stimulated yellow colour development. It was recommended that picking mangoes should be at colour break rather than at an earlier stage to ensure proper ripening. Inhibition of colour development may also vary with cultivar as well as maturity and conditions of treatment such as temperature, duration of hot water treatment and dosage of radiation. The slight inhibition of colour development noted in mangoes irradiated at 200 Gy was eliminated when irradiation was preceeded by the HWT (Spalding and Reeder, 1986). The combination of hot imazalil with radiation was no better than hot imazalil alone or the combination of hot water with radiation. Irradiation should be done within 24 hours of the HWT to be effective. Disease control declines rapidly with longer delays (Thomas and Brodrick, 1977., Brodrick and Van der Linde, 1981).

IR radiation treatments for 1 to 5 min. followed by a 20 sec. prochloraz (81 g a.i/100 l) dip effectively controlled anthracnose and soft brown rot. However, a 20 sec. dipping of prochloraz (71 g ai/100 l) followed by 1-5 min. IR exposure gave satisfactory control of anthracnose, but did not control, soft brown rot (Lonsdale and Droamer, 1994).

Thus heat and ionizing radiation can be suitably combined to eradicate latent fungal infections of fruits for export or food processing industry. If these combination treatments are adequately developed, the use of fungicides of dubious safety at any stage could altogether be avoided.

Biological Control

Biological control of post-harvest diseases of perishables (fruits and vegetables) is of recent development and involves the use of antagonistic microorganisms (bioagents) against post-harvest diseases. This offers one of the most effective means of controlling the disease without any adverse effect on the environment, on the commodity and the consumer, in compliance with sustainable environmental issues, natural and ecofriendly (Patel and Patel, 1998). Further it is safe, non-hazardous either to environment or human consumer and cheap. It offers a reliable approach either alone or in integration with other disease management practices (Patibandla and Prasad, 2004). However, whether the pathogen in question can develop resistance against the antagonist, remains a question, though no such information is available till to date. Such possibility does not seem to exist, since the antagonists are existing since a very long time in the nature and the pathogens have not shown any resistance to the natural antagonists, rather the antagonists and pathogens coexisted in the nature, probably striking a balance. Yet such a possibility in nature can not be ruled out, particularly because of human intervention, by way of continuous use of the same antagonist against a particular pathogen.

Biological control of mango anthracnose has been reviewed by Jeffries and Koomen (1992). A number of microorganisms are reported to be antagonistic against plant pathogens including *C. gloeosporioides* in nature. For example, Prusky and Keen (1993) found a non-pathogenic strain of *Colletotrichum magna* that colonizes the mango fruit endophytically and prevents infection by *C. gloeosporioides*. The efforts of Ilag and Co-workers (unpublished) has resulted in the selection of natural isolates that can control *Colletotrichum* symptoms on mango fruits (Prusky *et al* 2004) based on

(i) lack of pathogenicity on mango plants and

(ii) the capability to decrease disease incidence in the plant and in the fruit.

One of these isolates in Philippines was *C. capsici*. Its mutant (Isole $O_{13}M_4$) was highly effective in reducing the incidence of fruit anthracnose infection of *C. gloeosporioides* irrespective of the stage of mango fruit development. Sprays of the mutant at pre-bloom stage, at the flowering stage and fruitset were equally effective (Prusky *et al* 2004).

Control of anthracnose needs an effective, cheap and environmentally friendly non-chemical method. Attempts have been made successfully in this direction. A number of microorganisms (bacteria and fungi), with *in vitro* or *in vivo* activity against anthracnose causal pathogen, *C. gloeosporioides* have been isolated and tested (Jeffries and Koomen, 1992., Koomen and Jeffries, 1993., Gopi, 2003., Udhaykumar and Usha Rani, 2010). But none have been exploited commercially in the field, possibly because of the limitations of biological control under field conditions compared with controlled environmental situation after harvest, in spite of the fact that comparatively adequate information is available on post-harvest biological control of mango anthracnose from laboratory studies. Lack of commercial production of antagonostics may be another reason for not using biological control of anthracnose commercially. And also lack of information on quiescence of anthracnose pathogen may be yet another limitation.

In an investigation with a strain of *Bacillus* sp. that exhibited *in vitro* activity against *C. gloeosporioides*, it was found that disease control *in vivo* was possible when fruits were inoculated with the bacterium 24 hours prior to inoculation with anthracnose fungus but not when fruits were inoculated first with the pathogen which indicated that quiescence phase of the pathogen was not affected by the antagonistic bacterium (Fernandoz and Arauz, 1992). It is not clear whether the quiescent structure in mango anthracnose is a dominant appressorium or a subcuticulator hypha and this information could be critical and vital to the successful post-harvest application of biological control agents (Jeffries and Koomen, 1992). If quiescence is dependent on an appressorium, an effective biocontrol agent could possibly act by antibiosis, competetion or parasitism on the fruit surface or by inducing resistance. If the quiescent phase is a subcuticular hypha, an effective biocontrol agent should produce an antibiotic substance that penetrates the cuticle or the antagonist it self should be capable of cuticular penetration. The latter mechanism appears more promising because of public concern about antibiotic production on edible commodities (Wisniewski and Wilson, 1992).

The only other example of successful post-harvest application of biological control agent for the control of mango anthracnose was the application of an isolate 558 of *Pseudomonas fluroscens* by Koomen and Jeffries (1993). The bacterium significantly reduced anthracnose development, compared with the control fruit, but the mode of action involved was not clear. Korsten *et al* (1993) found that post-harvest dips of mango fruits in *Bacillus licheniformis* gave excellent control of fruit anthracnose and stem-end-rot on 'Sensation' and 'Keitt' mango fruits at concentration of 10^7 cells/ml of water and the effects were comparable to commercial fungicide treatment (50 ul/litre of water prochloraz or benomyl 0.25 g/litre of water). However, Koomen and Jeffries (1993) achieved limited success with post-harvest dips in bacterial suspensions. UdhayKumar and Usha Rani (2010) tested the efficacy of bacterial antagonists *viz.*, *P. fluroscens* and *B.subtilis* against mango anthracnose pathogen *in vitro*. An isolate of *P. flurescens* showed maximum inhibition of spore germination, germ tube and mycelial growth of the pathogen (*C. gloeosporioides*).

Among the different biocontrol agents tested *in vivo*, *Trichoderma virens* was most effective in controlling anthracnose, followed by *T. harzianum*. The disease appeared 7 days after storage in treated fruits with *T. virens*, whereas with other biocontrol agents the disease appeared 5 days after storage (Gopi, 2003).

However, Papavizes and Lumsden (1980) were of the opinion that, in the management of post-harvest diseases of perishables, fungicides need to be used with biocontrol agents without any toxic effect on antagonists. The future research should be aimed to identify potential biocontrol agents and their compatibility with the fungicides to reduce the dependence on ecologically hazardous chemicals and improving the knowledge of antagonist-pathogen-fungicide interaction and mechanism involved therein (Suvarna *et al* 2009).

Delaying Disease incidence and Symptoms

Since quiescent infection of anthracnose and stem-end-rot resumes after harvest with the commencement of ripening of fruits, it was felt that by delaying ripening by some means, the incidence and the disease symptoms can be delayed until the fruit are disposed off by sale or consumed, as one of the strategies of managing the diseases. Pathak (1997) suggested regulation of ripening (see Chapter on ripening), waxing, coating and wrapping of fruits and different methods of storage as some of the methods for delaying resumption of disease and appearance of disease symptoms for post-harvest disease in tropical and subtropical fruits. Some of them, which are relevant and tried in case of mango fruit anthracnose are presented below:

(i) **Fruit Waxing and Coating:** Wax can be applied to fruits alone or in combination with chemicals as a protectant against biotic and abiotic pressures. Wax improves fruit visual appearance, firmness and brightness. Likewise plastic or paper wraps can be used alone or after impregnation with chemicals.

Fruits can be individually wrapped in a heat shrinkable polyethylene film that prevents the spread of the diseases as well as the transfer of spores from fruit to fruit in a container. Some advanced countries have adopted a package of wrapping fruit and vegetables to ensure better storage and limit losses due to post-harvest diseases during transit and storage. Plastic film was tried for the control of post-harvest fungal decay of mango fruits. However, mangoes individually wrapped in heat shrinkable film developed excessive decay and failed to ripen properly and became unacceptable (Spalding and Reeder, 1986). Sornsrivichaai *et al* (1989) found that the plastic wrapping extended the storage life of mangoes without increasing the disease severity, but the treatment caused tainting of the fruits.

Wax coating alone (6%) was moderately effective against *C. gloeosporioides* decay of mango fruits (Krishnaiah *et al* 1986., Srikanth, 1989). On the other hand, mango fruit treated with Bavistin (1000 ppm) plus wax (3%) exhibited a disease index (DI) of 2.67 on wounded fruits and 1.33 DI on unwounded fruits (Srikanth, 1989). In pre-and post-inoculation trials, waxol (Wax emulsion 100%) effectively reduced the severity of anthracnose (Chandra and Pathak, 1992c). Sheikh *et al* (1977) found that application of fungicidal wax emulsion during storage reduced spoilage and delayed ripening.

The development of resistant strains of *C. gloeosporioides* to benomyl and imazalil (Spalding, 1982) prompted to search for alternatives for the fungicides. The fungicide benzalconic chloride with wax controlled satisfactorily the incidence of anthracnose in fruits under ambient conditions. Therefore, benzalconic chloride with wax is a viable alternate to benomyl in the control of anthracnose. It is a contact fungicide with short residual effect. The enhanced effect of the fungicide in combination with wax may probably be due to synergistic effect with wax (Fonseca *et al* 2004).

El-Ghouth *et al* (1992) found that 'Chitosan' an ingredient of some fruit waxes reduced the growth of *C. gloeosporioides 'in vitro'* and suggested that it has potential for disease control.

(ii) Low Temperature Storage: Tropical production areas like India face high temperature during summer months, when the fruits are ready for harvest, which makes preservation of mango fruits for longer time impossible without cold storge. Temperature near 0 °C will retard the disease development. However, such low temperature storage is less suitable for tropical fruits like mango as compared to subtropical and temperate fruits, because of their sensitivity to chilling. Hence, low temperature can not be adopted to retard decay development during marketing (Hatton *et al* 1965). Bhargava *et al* (1965) reported that when mango fruits of Cv. 'Langra' were stored at 10 °C and brought back to room temperature showed chilling injury and enhanced rate of advancement of fruit rots caused *by C. gloeosporioides* and *Aspergillus niger,* so recommended 15 °C for the cold storage of mangoes for avoiding chilling injury as well as the disease incidence. Boulet *et al* (1982) observed that mango fruits stored at 4 °C, when brought to room temperature, exhibited a marked chilling injury and rapid proliferation of fungi, where as those stored at 12 °C did not show these effects. According to Singh (2000) storage of fruits should be under dry and cool conditions to avoid chilling injury and diseases.

(iii) Modified/Controlled Atmosphere Storage: Altering the levels of oxygen and carbondioxide in the storage atmosphere surrounding the stored commodity can influence disease development either by direct inhibition of the pathogen or by altering the host resistance. Modified/controlled atmosphere storage methods have found applications in extending the shelf life of harvested fruits and are basically aimed at maintaining

(i) a low temperature,

(ii) an adequate humidity,

(iii) low oxygen tension, and

(iv) low ethylene concentration around the stored fruits.

Incidentally, as a result the fruits show less decay under this storage.

Success in delaying decay of mangoes (cv. Tommy Atkins, Maya, Haden and Keitt) was achieved by storing the fruits in controlled atmosphere storage in Israel (Sive and

Resinzky, 1985). Sangchote (1989) observed that when preceded by chemical treatment, mangoes stored for one month under modified atmosphere (2% CO_2, 6% O_2, 13 °C) showed very little decay caused by *C. gloeosporioides*. Johnson *et al* (1990) have successfully employed controlled atmosphere storage of mangoes (cv. Kennington Pride) (5% O_2, 2% CO_2, 13 °C) for 26 days followed by air at 20 °C for 11 days for controlling the common storage diseases.

Since, anthracnose rot also develops from latent infections in storage, it becomes necessary to check it in storage also. HWT at 55 °C controlled mango post-harvest anthracnose during storage for 11 days at 25 °C. The addition of benomyl (500 or 1000 ppm) allowed the storage at 51.5 °C and 48.5 °C respectively without reduction of efficiency of hot water (Muirhead, 1976).

Integrated Disease Management

Fruit rot management has mostly confined to individual strategies with exceptions. Several investigators have reported better control with combined strategies (see above). Patil (1992) reported encouraging results of integrated disease management in the control of post-harvest diseases in mango. The hot water dip followed by wax plus prochloraz and *Bacillus licheniformis* treatment was found to be more effective in reducing the fruit rot in mango than HWT or the wax plus *B. subtilis* treatment. Wax, however, did not enhance the activity of *Bacillus* sp. (Villiers and Korsten, 1994). HWT combined with Bavistin and storage in low cost zero energy cool chamber could be an integrated treatment which would control both insect and fungal pests; beneficially influence the commodity quality in addition to being economically alterantive and possible solution to the problem of storage of mangoes (cv. Kesar) in India (Waskar and Gaikwad, 2005). These findings suggest that more investigations on integrated disease management (IDM) are needed to improve efficacy, applicability and economics of such IDM practices.

Integrated management of post-harvest mango anthracnose, under tropical conditions requires knowledge of the biology of the pathosystem and the technologies available for control, their economic feasibility and viability and economical acceptability.

Successful integrated management of post-harvest anthracnose of mango must take into account the following key points (Arauz, 2000).

(i) The disease system is rain moisture driven and infection criteria have been elucidated, so that they can be used to time fungicide application.

(ii) Severity of post-harvest anthracnose is the result of cumulative quiescent infections that develop after harvest as fruit commences ripening and which are subject to eradication by post-harvest treatments.

(iii) Flowering in mango can be managed so that susceptible tissue is produced during the drier months of the year, for example, in New Southwales, commercial crops are produced only in dry areas, which are unfavourable for this disease (Fitzell, 1979).

(iv) Several pre-and post-harvest control treatments are commercially available and

(v) The efficacy of the various treatments is dependent on infection level.

Combination of Pre-and Post-harvest Measures

Combination of pre-and post-harvest control measures in case of latent infections of post-harvest diseases is a better approach than either of the measures alone. Attempts in this direction were successful and many workers advocated such practice for effective control of the fruit anthracnose as well as stem-end-rot etc. A few examples wherein the combinations were tried successfully are presented below:

Both treatments i.e. pre-and post-harvest treatments in combination resulted in fewest fruits with anthracnose and high percentage of marketable fruits (Bryan *et al* 1972). They also observed intensification of red colour with pre-harvest spray of Benomyl followed by post-harvest measures. Pre-harvest spray with Bavistin (0.1%) or Topsin-M (0.1%) or prochloraz (0.1%) followed by dipping of fruits in the same solution after harvesting was found to be effective (Rawal, 1997). Kapse and Katrodia (1997) observed promising results in the control of anthracnose and stem-end-rot with pre-harvest spray and post-harvest dip in Bavistin (1000 ppm) and Diathane M-45 (2000 ppm). Such combination showed a beneficial effect on ripening of 'Kesar' mango fruits over control. Anthracnose was effectively controlled with pre-harvest spray of copper fungicides followed by post-harvest application of prochloraz (Kapse *et al* 2009).

However, adoptation of such combination is possible when the grower himself is a trader/exporter. Otherwise grower has to stick up to his pre-harvest measures and the trader/exporter has to practice post-harvest measures, whether the farmer has taken up pre-harvest sprays or not.

Future

In view of the expansion of mango cultivation to new areas with its associated pest and disease problems, expansion of global export trade, climate change and pathogen resistance, the following strategies are to be addressed in future.

(i) Finding a simple, economically feasible and cheap treatment that is environmentally friendly and does not leave toxic residues on fruit is a major challenge in the coming years.

(ii) Adjustment of flowering and fruiting to coincide dry periods particularly in areas of high humidity and wet conditions that prevail during the above stages in mango.

(iii) ***Development of Resistant varieties:*** Use of disease resistant cultivars is one way of avoiding the diseases. Although some mango cultivars are moderately resistant, none are sufficiently resistant to be produced without fungicides, particularly in humid areas (Litz, 1997). Development of resistant varieties has been suggested as a measure of managing anthracnose in mango. Wide collections of germplasm are available at different centres in the world. Thorough screening of the genotypes in the germplasm collections against the diseases for selecting suitable disease resistant clones for general cultivation and breeding new ones is needed. And it is possible to induce natural resistance in the fruits of off spring by genetic manipulation approach (see below). Incorporation of genetic code for disease resistance characters into fruit requires transformation systems. In case of mango, Mathews and Litz (1991) noted that the polyembryonic mangoes from South East Asia, that evolved under constant disease pressure were generally more resistant to diseases than monoembryonic cultivars. Stimulating soma-clonal variation would unmask useful genes, which while broadening the genetic base for resistance to mango disease pathogens, could improve productivity additionally. Therefore, work in this direction need to be initiated and intensified.

Utilisation of Host Resistance

All living organisms have built-in natural resistance to biotic and abiotic pressures. Such built-in natural resistance also exists in mango. Disease control through maintenance and utilization of host resistance is an interesting and noval approach. Prusky and Keen (1993) and Prusky (2009) dealt the resistance of fruits and its mechanism of action in their reviews.

Anthracnose fungus does not infect immature and unripe fruits unless they are injured or wounded. This is attributed to the existence of built-in natural resistance of host against the microbial attack in immanture and unripe fruits. It is also reported that this natural resistance declines towards maturity and ripening of fruits allowing resumption of quiescut infection.

Resistance to post-harvest pathogen is usually the result of several genes that interact in a way that is not well understood. Resistance to post-harvest pathogen in fruits and vegetables is described as a 'dynamic incompatability' between host and pathogen. The response of the host resistance genes to products of a pathogen's virulent genes would be prevention or retardation of pathogen's growth (Prusky, 2009). Host barriers may be imposed at various levels of fungus pathogenicity; at the level of fungal penetration, inhibition of appressoria development or inhibition of fungal colonization (Prusky, 2009).

This natural resistance has been attributed to the presence of certain antifungal compounds produced by the host. Plants produce a diverse array of secondary metabolities that are toxic to fungi or which affect fungal colonization (Verhoeff, 1974., Prusky, 1996). Fungal development in the host tissues may be restricted by host barriers (see above) or inhibited by pre-formed compounds in the host. This type of inhibition of fungal development does not involve temporarily differential gene activation, but it does involve the presence of biologically active compounds, whose activity affords protection against diseases (Ingham, 1973).

In mango fruits, two antifungal compounds have been identified *viz*., 5-(12-cis-heptadecenyl)-resorcinol and 5-pentadecyl resorcinol (Cojocaru *et al* 1986., Droby *et al* 1986., Prusky and Keen, 1993). A third compound 5-(7, 12-heptadecadie-nyl) resorcinol was also found in mango fruits (Kobiler *et al* 1998). These pre-formed antifungal compounds may directly affect the pathogen itself and delay its development until their concentrations decline to sub-fungitoxic level (Droby *et al* 1986). These compounds inhibit the penetrating hyphae from appresoria that have directly breached the host (fruit) cuticle. These pre-formed compounds or phyto-anticipants differ from the inducible phytotoxins that are synthesized from remote precursors in response to pathogen attack (Prusky, 2009).

Antifungal compounds exist in immature, unripe mango fruits and the concentration of these compounds decreases as the fruit matures and ripens. The decline in these antifungal compounds is brought by oxidative process. This decline can be prolonged to delay the onset of anthracnose development until the fruit ripens (Prusky and Keen, 1993). In avocado, the decline of the concentration of antifungal compounds was delayed by exposure of fruit to an atmosphere of 30% CO_2 for 24 hours or by treatment of fruit with an antioxidant compound, butylated hydroxyl amisole (BHA) (Prusky and Keen, 1993). In Costa Rica, post-harvest treatment with BHA resulted in reduced activity of mango anthracnose, although the level of reduction achieved was less than that obtained with a prochloraz dip (Rios and Arauz, 1997).

The antifungal compounds exist in the peel and act on hyphae from germinating appressoria (Prusky and Keen, 1993). The climacteric rise in respiration in mango occurs simultaneously with the production of ethylene and with changes in external parameters and with a reduction in fruit resistance to fungal attack (Prusky and Keen, 1993). The relationship of preformed antifungal compounds to fruit resistance is not clear, because of lack of knowledge about the distribution and concentration of inhibitors in the infection courts, the sensitivity of pathogen to inhibitors in the plant, inhibitor concentration and host resistance (Prusky, 2009).

Most of the antifungal compounds are located in the peel of the fruit (Prusky *et al* 1990., Prusky and Keen, 1995), whereas sub-fungitoxic concentrations were found in the susceptible flesh of the fruit. Specific signals originating from the pathogen could enhance the release of pre-formed compounds from storage to active sites. The amount of chemical available depends on the cultivar (in Haden being grater than that in Tommy Akins) and fruit maturity (younger growing fruits contain higher concentration than more mature ones) (Prusky, 2009).

The mechanism by which the pre-formed antifungal compounds (PAC) in mango fruits declines is not known, though it may be related to enzymatic change during ripening and senescence (Prusky, 2009).

The mechanism of resistance of unripe mango fruits to fungal attack is complex and several factors have been involved in its expression. The presence of fungitoxic amounts of PAC was correlated to fungal resistance (Prusky, 2009).

Resistance could also be induced by modulation of particular stages, such as the synthesis of 5-substituted resorcinol during the post-harvest period. This regulation can be done by the application of elicitors as post-harvest treatment and by means of synthetic activities. Induction of resistance seems to occur by a series of early events in which the pathogen and host release reactive oxygen species. Resistance of mango fruits can also be affected by modulation of the secretions of specific pectilytic enzymes of the pathogen (Prusky, 2009).

Although pre-formed chemicals have been considered as non-inducible (van Etten *et al* 1994), significant increases in the levels of fungal inhibition have been associated with exposure to various biotic and abiotic elicitors (Prusky and Keen, 1993). The level of 5-substituted resorcinol can be modulated by increasing it and/or preventing its decrease. The present aim is to prevent its decrease. Inoculation of unripe mango fruits with spores of a pathogenic fungus induced the level of the antifungal compounds (Kobiler *et al* 1998). Freshly harvested fruits exposed to a stream of 20-60% CO_2 in air for 24 hours showed no significant change in their ripening, but decay development was significantly delayed (Kobiler *et al* 1998).

Manipulation of the concentration of pre-formed and induced antifungal chemicals, in harvested fruits and vegetables is a logical approach to the development of new and improved control for post-harvest diseases and the future research should be in this direction. The relevant strategies are based on prevention of the loss of fungal inhibitory compounds or selection and/or breeding of plant genotypes that are rich in such antifungal compounds and therefore, more resistant to fungal attack (Droby *et al* 1986, Prusky *et al* 1988). The assumptions underlying this strategy are that such compounds are safe because they are natural, effective since they have evolved in nature to specifically protect the plant against pest and diseases and have enabled existing plants to survive the selection pressure of evolution. In mango fruits, the protection of host during specific periods of susceptibility by manipulation of the levels of the natural compounds is a safeway to maintain host resistance. For example, a high level of 5-substitued resorcinol can be maintained in stored mango fruits by CO_2 treatment (Prusky, 2009).

The mango grower can not, however, depend on these treatments and needs easier, safer and cheaper treatments against the diseases. An alternative to the preservation of defence compounds by application of external treatments to the host is the prevention of the development of fungal pathogenicity factors. Recently it has been reported that post-harvest pathogens of *Colletotrichum* and *Alternaria* actively, secrete ammonia at the site of infection by the fungus. Prusky *et al* (2001) and Eshel *et al* (2002) showed both in culture and in host tissues that *C. gloeosporioides* and *Alternaria alata* actively modified their environments by secreting large amounts of ammonia, resulting in alkalinization of the environment (Prusky and Yakoby, 2003). This indicates that the ability of fungal hyphae to colonize the plant tissue may depend on modulation of the pathogenic secretion system. These findings lead to the idea that neutralization of the post-harvest changes induced by the pathogen could serve as an efficient and friendly treatment for post-harvest disease control in mango fruits. Recent results indicated that acid solution reduced *A. alternata* incidence, which suggests that acidification of the host tissue can neutralize the alkalinization by the pathogen during fungal colonization (Prusky, 2009). This also suggests that the gradual fall in the acidity of mango fruit towards maturity may be contributing to the decline of antifungal chemicals and inviting the resumption of latent infections and this needs to be studied in the future. However, strategies for the control of anthracnose and/or other post-harvest disease to be developed based on the above findings, require some more time.

Mean while, one has to depend on the existing strategies. Post-harvest practices such as low temperature storage and controlled/modified atmosphere storage maintain resistance to decay by delaying the ripening process, but there are two limitations to the potential benefit of this approach in mango :

(i) mangoes are sensitive to chilling and are injured at temperatures lower than 10 °C to 13 °C depending on variety and duration of exposure to such low temperature and

(ii) once the fruits are allowed to ripen under ambient conditions, disease develops normally (see low temperature storage).

Neverthless, some progress has been made in this direction towards anthracnose control through maintaining of fruit resistance beyond the onset of ripening. However, Dodd *et al* (1997) felt that natural host resistance to anthracnose is not well developed and hence offers little opportunity to plant breeder. However, future research should be with an aim of utilising the host resistance/defenses in combating, the post-harvest disease, to avoid the dependence on external treatments.

Changes

As a result of post-harvest resumption of the latent infection the infected fruits undergo drastic structural as well as bio-chemical changes. The pathogen exerts an indirect influence by altering the normal metabolic processes of the host fruit. Apart from the reduction in aesthetic appeal of fruits, post-harvest disease also brings a substantial decrease in their food value. The taste and flavour of the fruit change to an unacceptable level (Ghosh *et al* 1966).

Post-harvest infections bring about a reduction in the sugar content of fruit as the pathogen finds the sugar as an excellent source of energy required for its growth and mycelial build up. The sucrose content in the fruit tissue, infected with *C. gloeosporioides* showed a marked decline and was not traceable after 12 days incubation. Simultneously there was also a fall in glucose and fructose content of the infected fruits. Non-reducing sugars, on the other hand, increased in anthracnose infected fruits (Bagwan, 2010). The decrease in ascorbic acid was more pronounced in the fruit infected with

C. gloeosporioides (no ascorbic acid could be detected in the tissues of the infected fruits) (Ghosh *et al* 1966).

Pathogen produced enzymes like PMG, PG and Cx *in vitro* and *in vivo*. PG had maximum activity on 9th day of inoculation *in vitro* and *in vivo* also it showed maximum activity followed by Cx and PMG (Chandra and Pathak, 1992c).

Recommendations

Management of mango post-harvest disease is mainly based on orchard sanitation, pre-harvest fungicide applications and post-harvest HWT with or without fungicides. These strategies controlled well the disease in South Africa. Use of resistant cultivars and preventive treatments with copper based or triazols are the common practices of mango growers in Brazil (Pinto *et al* 2004). In Australia (Dodd *et al* 1997) and Israel (Prusky *et al* 2001). Anthracnose is controlled with fungicides based on forecasting a certain control management schedule for the control of anthracnose depending upon the local conditions. In India mango broadly is cultivated in drier areas, except in places like Kerala and North Eastern States, where humidity is higher. So based on the local environmental conditions the following pre-and post-harvest control measures against anthracnose were recommended.

Pre-harvest Measures

1. Follow orchard sanitation by way of destruction of diseased leaves, twigs, panicles and fruits.
2. Adopt canopy management for free air flow and light after fruit harvest.
3. Prune and destroy all the dried and diseased wood if any before flowering.
4. At bud break spray and also during flowering period, fungicides: Copper oxychloride (3 g)/Carbendazim (1 g)/Mancozeb (3 g)/Ontracole (2 g) at weekly interval and at 15 days interval twice/thrice during flowering.
5. Bag the fruits one/two months prior to harvest depending on the disease pressure and environmental conditions of the area.
6. Although total suppression of infection by the above practices is difficult and not possible, the disease can be reduced considerably. The above schedule also minimizes the incidence of latent infections of fruits (OmPrakash and Srivastava, 1987).

Post-harvest measures

1. Harvest the fruits with its pedicel.
2. Follow prompt and proper post-harvest handling practices of fruits to minimize the disease.
3. Dip the fruits immediately after harvest/before storage in hot water (51 °C for 15 min.) combined with carbendazim (0.05%) thiophenate methyl (0.05%) or benomyl (500-1000 ppm) or prochloraz and store at 12 °C-13 °C.

RHIZOPUS ROT

Rhizopus rot, some times referred to as 'soft rot' is one of the soft rots of mango fruits and is an important transit and market diseases (Thakur and Chenulu, 1966). However, it is found to be serious at processing units. But it is not as common as *Aspergillus* rot and stem-end-rot, although it can be destructive if fruits are badly bruised and are transported in closely packed containers like baskets at high temperatures over long distances.

Geographical Distribution

The disease has been reported since 1945 (Sinha, 1945., Thakur and Chenulu, 1966) from different markets as well as processing units in different states of the country *viz*., Andhra Pradesh, Punjab, West Bengal, Gujarat, Delhi etc. (Thakur and Chenulu, 1966., Laximinarayana and Reddy, 1977., Garcha and Singh, 1980., Mandel, 1981., Patel and Goswami, 1984., Srikanth, 1989., Jadeja and Vaishnav, 2000., Gopi, 2003). Besides India, it was reported from Queensland, Australia (Muirhead and Grattidge, 1984).

Losses

The disease caused considerable losses in different markets as follows (Table 10.3)

Table 10.3: Losses in mango at different markets in India, due to rhizopus rot

S.No.	Market	Loss (%)	Reference
1.	Delhi	6.3	Thakur and Chenulu, 1966
2.	Warangal, Andhra Pradesh	4.6	Laxminarayana and Reddy, 1977
3.	Tirupati, Andhra Pradesh	2.5	Srikanth, 1989
4.	Ahmedabad, Gujarat	7.2	Patel and Goswami, 1984
5.	Junagadh, Gujarat	0.45	Jadeja and Vaishnav, 2000

The incidence of rhizopus rot in fruit processing units/factories may be due to faulty ripening practices adopted (Jadeja and Vaishnav, 2000).

Symptoms

The pathogen causes a typical soft rot of fruits. The tissues of the affected fruits become soft and watery. Pulp will be sour in taste. Juice of the infected fruit loses its flavour, becomes dissolved and watery in consistency with increased pH (Thakur, 1972a).

The rot can appear from any part of the fruit surface, where spores of the fungus have lodged on wounds. The disease is characterized by large water soaked areas developing on any portion of the fruit, with profuse growth of the fungus, having black sporangial bodies. Initially the symptoms appear in the form of light, brown, water soaked circular spots on the fruit surface which get covered with a first whitish growth, then black sporangial bodies of the fungus. The intensity of the fungus colour, varies according to the variety infected and time after infection. The lesions grow in a regular manner and form large circular patches. The fruit skin becomes soft and is easily peeled off from infected flesh. This type of soft decay is very common, emitting foul smell. In severe cases, water oozes out from the cracks of the skin. In latter stages fluffy white growth of the fungus occurs over the fruit, with black spore heads. The odour of the diseased fruit is peculiar and unpleasant unlike the rancid smell of *Dothiorella* rot or the putrid odour of the bacterial rot. The sporangia on the fruit surface can be seen with naked eyes and they arise from dark branched, root like structures (Rhizoids). Infected fruit gets disfigured within 2-3 days, as the disease progreses rapidly after infection (Thakur and Chenulu, 1966., Laxminarayana and Reddy, 1977., Patel and Goswami, 1984., Singh, 2000).

Small pale spots are the first symptom of *Rhizopus stolanifer* in Australia, which enlarge rapidly and extend into the flush causing the skin to peel off and black spore bodies to form (Peterson, 1986).

'Neelum' mango fruits artificially inoculated with *R. stolaniferae* showed irregular and water soaked areas and fruit was disfigured within 2-3 days of inoculation. The symptoms appeared in the form of browm lesions and discoloration later, fruits exhibited soft rot with sporulation, blakish discolouration on the peel, while the internal peel of the affected fruit turned brown, became soft and emitted fermentive odour (Srikanth, 1989., Bhuvaneswari, 1999). Under artificial inoculation of 'Totapuri' fruits in the laboratory, the symptoms were in the form of circular concentrated rings with light brown patches. The diseased spongy area drastically increased and covered with white mycelial growth. (Munishamanna *et al* 2008).

Varietal Susceptibility

'Neelum', 'Bangalora' and 'Kesar' were reported to get heavily infected with *Rhizopus* sp. in different markets of the country (Thakur, 1972a., Laxminarayana and Reddy, 1977., Srikanth, 1989., Bhuvaneswari, 1999., Jadeja and Vaishnava, 2000).

Etiology

More than one species of *Rhizopus* were claimed to incite this soft rot. Baker and Wardlaw (1937) and Rao (1968) reported *R. nigrican*, whereas Laxminarayana and Reddy (1977) reported *R. oryzae* Went and Gerlings, causing rhizopus rot on mango. Dasgupta and Bhatt (1946), Thakur and Chenulu (1970b) and Munishaamanna *et al* (2008) made *R. stolonifer* responsible for the decay of mango fruits in Delhi market and Bangalore market respectively. Studies on the isolation and identification of the causal organisms revealed that *R. stolonifer* was mainly responsible for soft rot of mango fruits at Tirupati market (Srikanth, 1989., Bhuvaneswari, 1999). *R. stolonifer* was observed causing soft rot of mango fruits in Australia (Peterson, 1986). On the other hand, Thakur and Chenulu (1966) considered *R. arrhizus* alone as responsible for the decay of mango fruits in Delhi market. Singh (2000) included *R. stolonifer*, *R. arrhizus* and *R. oryzae* as common pathogens causing soft rot in India, of which *R. arrhizus* was most common inciting a typical soft rot of mangoes. All these three pathogens have been observed causing soft rot in apple, banana, mango, tomato, potato etc. (Thakur and Chenulu, 1970b., Thakur, 1972b), *R. stolonifer* although also reported from temperate regions, is often destructive on tropical fruits like mango and papaya.

All the above 3 species are wound parasites and destructive, if fruits are badly bruised in transit, particularly when they are improperly packed.

Disease Cycle

Rhizopus sp. exist in transport vehicles, godowns in the market and processing units and cold storage rooms, which serve as sources of the disease (Kantilal and Reddy, 1981). A spoiled fruit due to the pathogen also serves as a source of infection for other healthy fruits in storage and transit by contact.

The pathogen enters the fruit surface after harvest mainly via injuries caused by improper harvesting, handling in transit and packing. Once the infection occurs, the disease spreads rapidly under favourable conditions, particularly when the fruits are in closely packed containers like baskets. Over long distances, at high temperatures the disease is more destructive. Under laboratory conditions, wounded portion of the fruit developed disease symptoms on the second day of inoculation and infected fruits showed complete rotting by 5th day. Unwounded fruits exhibited initial symptoms of brown lesions and discolouration on 4th day and by 8th day all unwounded fruits exhibited soft rot (Bagawan, 2010).

Epidemiology

Environmental factors like temperature and relative humidity play very important roles in the development and spread of the disease. Their influence also varied with the *Rhizopus* sp. involved in the disease. A few studies on the effect of these factors in the laboratory were made (Thakur, 1972b., Patil and Pathak, 1993., Bhagwan and Yeole, 2003).

Thakur (1972b) studied the influence of temperature, humidity and variety of host on the development of soft rot by 3 species of *Rhizopus* under laboratory conditions and found a wide range of temperature (7-40 °C) being effective against the disease. The time taken for decay of fruits depended upon the temperature to which the fruits were exposed. At 10°-25 °C, with in 4 days, 100% fruit decayed. While, at 20-40 °C. 100% fruit decay was observed in 20 days. At low temperature of 7°-8 °C disease development was slow and of low intensity, whereas at temperature between 20°-40 °C the intensity of infection increased and the development was very rapid. Severity of rot and percentage of spore germination were maximum at 35 °C and 100% RH. At 10 °C even after 24 hours of incubation, there was neither spore germination nor appearance of symptoms. In general, there was an increase in severity levels of rot, with increase in temperature and time (Patil and Pathak, 1993., Bagwan and Yeole, 2003). At low temperature (7°-8 °C) *R. oryzae* proved most aggressive as compared to *R. arrhizus* and *R. stolonifer*. The temperature also influenced the duration of incubation of the fungus. At low temperature, the incubation period is longer than at optimum temperature.

Dry atmosphere (0% RH) showed slight difference in the intensity of infection, because, at this level of relative humidity, the intensity of infection ranged between light and moderate levels. However, different levels of relative humidity exerted definite influence on the incubation period. At dry atmosphere (0% RH), *R. arrhizus* produced visible symptoms on inoculated mango fruits in 6 days, whereas at 80 per cent relative humidity, rotting was observed in 36 hours. *R. stolonifer* was most aggressive at 90 per cent relative humidity (Thakur, 1972b). Soft rot increased with increase in relative humidity from 30 to 100 per cent. *R. arrhizus* was most severe at highest relative humidity (100%) and least at 30 per cent. The spore germination was least at 30 per cent relative humidity and the disease was most severe at 100 per cent relative humidity (Bagwan and Yeole, 2003). A relative humidity of 50 per cent and above allowed rapid development of soft rot of moderate to heavy intensity. Rotting was earlier at 80-90 per cent relative humidity, but took a little longer time for symptom development, because of resistance of thick peel and chemical composition of the pulp. However, on the whole, it was concluded that humidity has slight influence on disease development as there was considerable infection at all the levels of humidity (0-100%) (Thakur, 1972b).

The above discussion suggests that storage of mango fruits at lower temperature (10 °C) and low relative humidity (0-30%) may help in reducing the post-harvest rots due to Rhizopus sp of mangoes (Bagwan and Yeole, 2003).

Control/Management

Treatment with 2-Aminothiazole with 2-Amino pyridine (each at 5%) was effective against the rot up to 20 days. After 20 days, however, the rotting was rapid and all the chemicals failed to protect the fruit from rotting (Thakur and Chenulu, 1970a). Tetracycline, oxytetracycline and chloramphenical were recommended for control of rhizopus rot (Sundararaj *et al* 1972). Post-harvest dip of fruits in thiophenate methyl (0.1%) in hot water (52 $\pm$ 1 °C for 30 min.) deactivated the latent infection of the disease. Bavistin (1000 ppm) plus wax (3%) was superior to all other fungicidal treatments with disease index of only 1.00 in controlling the rhizopus rot. Captan (2000 ppm), Aureofungin – Sol (500 ppm) Diathane – M-45 (1000 ppm) and Blitox (2000 ppm) were effective in controlling Rhizopus soft rot in case of inoculated unwounded fruits (Srikanth, 1989). There was no decay due to *Rhizopus* sp. in fruits dipped in carbendazim and benomyl (each at 1000 ppm) even after 20 days of storage, whereas infection of Rhizopus rot appeared 11 days of storage in control fruits. Similarly complete control of rot was observed after 20 days of storage in fruits treated with potassium permanganate (2%) and boric acid (5 and 10%). The incidence of rot was more than that of control, in the fruits treated with salycyclic acid (0.2%), calcium chloride (4 %) and copper sulphate (1 and 2%). These sterilants showed least residual toxicity with increasing storage period and were ineffective at later stages of storage. Moreover, calcium chloride showed phytotoxicity (Gopi, 2003).

Plastic wrapping of fruits effectively reduced the development of post-harvest rots including Rhizopus rot, by reducing the metabolic activities of fruit rotting fungi (Chandra and Pathak, 1992b).

Mustard, caster and paraffin oils were found to be effective against the rot. Dipping of fruits in combination of above oils each at 75 per cent concentration with soap solution (1%) was effective in preventing fruit rot infection up to 20 days (Pathak *et al* 1971).

Exposure of fruits to UV radiation (30 and 150 wattage) for 20 and 30 min. and infrared radiation (30 and 150 wattage) for 20 and 25 min. completely avoided the incidence of Rhizopus rot in exposed fruits (Gopi, 2003).

Storing of the fruits below 10 °C considerably slows down the growth of the fungus and prevents secondary infection. However, storage at temperatures below 5 °C in cold storage attracts chilling injury. Combination of wax-coating of fruits and storage at low temperature (10 °C) avoids the disease and extends the storage up to 20 days (Patel and Goswami, 1984).

Fruits treated with ocium leaf extract (10%) remained free from Rhizopus rot even after 20 days of storage. However, its infection was observed in other treatments including control between 9 and 11 days of storage at room temperature. The rot infection was more in fruits treated with onion bulb extract (5%) and ginger rhizome extract (10%) compared to control, 20 days after storage (Gopi, 2003).

Possibility of controlling Rhizopus rot in mango fruits, using bioagents was explored, without success. None of the biocontrol agents effectively reduced the Rhizopus rot and recorded higher incidence than control treatment. However, fruits dip-treated with *Trichoderma harzianum* recorded least percentage of disease index, when the disease incidence was delayed by 9 days. The disease appeared 5 and 7 days after storage in treated fruits with *T. virens* and *T. viride, in vitro* (Gopi, 2003).

Rhizopus rot disease was least severe, in mango fruits, subjected to an integrated treatment that combined aerated steam plus bavistin plus D-arginine plus Ocium leaf extract plus *Trichoderma harzianum* as compared to that in fruits subjected to individual treatments (Patil, 1992).

The following control measures have been suggested to avoid/control Rhizopus rot in mango.

1. Fallen and rejected fruits in the orchard as well as in packing sheds, godowns and storage chambers should be removed and destroyed to minimise the source of inoculum.
2. Follow careful harvesting and handling of the fruits after harvest.
3. Strict hygiene conditions in the orchards as well as in packing yards, markets, godowns and storage rooms, minimize the incidence.
4. Rapid cooling of harvested fruits by pre-cooling prior to storage and storing at lower temperature (10-12 °C) and lower relative humidity after fungicidal treatment.

5. Keep the fruits dry after harvest and during other handling processes, since low relative humidity discourages spore germination and spread of the disease.
6. Avoid tight packing, particularly with diseased and healthy fruits in the same container.
7. Avoid transporting of fruits during day time when the temperature is high.
8. Transport after disinfection of packages and vehicles.

STEM-END ROT

Stem-end rots (SER) are important, destructive and serious post-harvest diseases of mango causing huge losses during ripening, transit, storage and marketing in several mango production centres of the world. 'Stem-end rot' has been used to describe the development of disease lesions mainly at the pedicel (stalk or stem-end) of the fruit and its surrounding area after harvest. In India, it is often referred to as 'Diplodia stem-end-rot' after the name of the principal causal pathogen. The disease becomes a more prominent and problematic issue in older orchards, where anthracnose levels are reduced by either climatic factors or by pre-harvest fungicide treatments for the control of several diseases of mango. The disease affects both the domestic and export trade of mangoes.

Geographical Distribution

The SER of mango was reported from several mango growing regions of the world (Table 10.4).

Table 10.4: Geographical distribution of stem-end-rot on mango

S.No.	Country	Pathogen	Reference
1.	Mynmar (Burma)	-	Su, 1934
2.	Srilanka (Ceylon)	-	Fernando, 1937
3.	Florida (USA)	-	Ruehle and Wolfen Barger, 1949
4.	India	*Diplodia natalensis*	Cheema *et al.*, 1950
5.	Palastine	*Lasiodiplodia* sp.	Minz & Yohebeo, 1952
6.	Philippines	-	Alicbusan & Schaffer, 1958
7.	Mauritius	-	Orian, 1960
8.	Nigeria	*Dothiorella dominicana*	Alabi, 1986
9.	Australia	*D. dominicana*	Mullar and Bart 1989
10.	China	*Diplodia natalensis*	Huang & Liu, 1995
11.	Thailand and	*Dothiorella mangiferae* *Lasiodiplodia theobromae*	Sepaiah, 1986
12.	Malaysia	*Phomopsis mangiferae*	
13.	Thailand	*Diplodia natalensis*	Pavichitrasakul, 1980

Thus, stem-end-rot enjoys world wide distribution. It's wide spread in India, was reported from every mango producing state (Cheema *et al* 1950., Chakravarti and Srivastava, 1964., Srivastava *et al* 1965., Rath and Mohanan, 1986., Om Prakash and Srivastava, 1987., Patel *et al* 1985).

Economic Importance and Losses

Stem-end-rot (SER) is very destructive and assumes greater economical significance, as it causes heavy losses at the time of disposal of the produce in domestic as well as overseas markets. Losses can increase when fruits are stored for prolonged periods at low temperatures or when fruits are transported and ripened at high temperatures (> 28 °C) (Ploetz, 2004). The extent of losses depends on the causal organism, place of production, cultivar and post-harvest handling of fruits. Losses due to SER in mango were observed and estimated at field level, transit, storage and markets in different countries. A few instances of losses due to SER are furnished below (see also post-harvest losses due to post-harvest diseases in this volume).

In India, Chakravarti and Srivastava (1964) estimated 10% infection in Cv. "Langra" in Delhi market, whereas Pathak and Srivastava (1967a) observed that SER spoiled 4-6% in Indian markets, with 6.2% loss being in 'Langra' and 4.3% loss in 'Totapuri'. Dharamvir *et al* (1967) reported 10-15% fruit spoilage due to SER during storage and transport. Jagdish and Pathak (1989) observed 2.2 to 4.2% incidence of SER in Udaipur markets during May-August. In Lucknow markets, the losses were 18-20% in 'Langra', 'Dashehari', 'Chausa' and 'Lucknow Safeda' cultivars (Anon, 1975-85). The incidence of SER disease in fruits of 12 cultivars transported from different places (Ratnagiri, Ahmedabad, Delhi, Meerut and Sahranpur) to Udaipur market at termination of mango season was maximum owing to high temperature and greater relative humidity in the atmosphere (Jagdish and Pathak, 1989). Sharma *et al* (1994) observed that *Botryodiplodia theobromae*, *Dothiorella ribis* and *Phomopsis mangiferae* caused 26.7, 2.5 and 40.05% decay respectively in Cv. 'Dashehari' in fruit markets of Himachal Pradesh. Survey of fruit markets of Junagadha in Gujarat indicated a loss of 4.6% and 22.6% loss at processing factories due to *Botryodiplodia theobromae*. The higher loss of fruits due to SER at processing units was attributed to faulty and improper methods of ripening, while the lower extent of spoilage at markets was ascribed to quick disposal of the produce to destination points, instead of retaining for long at the market yards (Jadeja and Vaishnav, 2000).

Reports from other countries like Srilanka (Fernando, 1937), Mynmar (Su, 1934), USA (Ruehle and Wolfen Barger, 1949), Mauritius (Orian, 1960), Philippines (Alicbusan and Schaffer, 1958) etc., showed that SER is quite destructive on mangoes in storage and transit. SER, due to *Diplodia natalensis* was associated with mangoes to an extent of 8% in transit and storage in Mynmar. The main post-harvest disease of mango in Western Australia is SER, leading to losses exceeding 50% after 4 weeks of storage at 13 °C (Muller and Bart, 1989). An average loss of 12.5% due to SER was reported from Bangladesh (Quroshi and Meah, 1991).

SER adversely affects the export of fruit to global markets, because of its appearance during transit. Pathak (1966) observed as much as 50% damage in Indian consignments landed at the United Kingdom airport.

Early infection by *Dothiorella* sp. caused fruitlet abortion and the infections that incited post-harvest SER resulted in spoilage of fruits due to decay, making them unfit for consumption or processing. Once the disease appears the fruit is no longer edible, even slightly affected fruits are rendered unmarketable and unconsumable (Johnson *et al* 1991a).

Symptoms (Photos 10.4 to 10.9)

It is easy to recognize the disease by its unique characteristic symptoms on the epicarp at the base of the stalk (pedicel). However, the SER of mango fruits is caused by different fungi (see etiology) and accordingly the symptoms differ considerably. Irrespective of the pathogen, however, the symptoms of SER on mango include principally the discolouration of the skin at the stem end and progressing down to the rest of the fruit surface on harvested fruits when they start ripening in either storage or transit. Infection of *Diplodia natalensis* always starts from the stalk end, while in case of other fungi, infections may start additionally on the sides of the fruit (Fernando, 1937). Symptoms of SER were described from both natural infections and artificial inoculations and also due to different causal organisms.

Photo 10.4: Stem-end-rot

Photo 10.5: Stem-end-rot

According to Fernando (1937) visible infection of the disease appeared on the ripe fruits as a slight darkening of the epicarp around the base of the fruit pedicel (0.5 cm across). This infected patch extended in due course of time, with remarkable rapidity and could cover an area of 4 cm in diameter with in 24 hours. At this stage, the affected area appeared water soaked and could exhibit a slight wrinkling of the epicarp.

The first sign of the disease by the *Diplodia natalensis* is the discoloration of the base of the pedicel and surrounding area of the skin. In a few hours, the affected area enlarges to form a circular patch of 3-4 cm in diameter. The entire fruit decayed within 7 days. However, under humid conditions, the whole fruit turns completely black within 2-3 days. The affected skin was somewhat softer than the unaffected healthy portion. The pulp of the diseased fruit became light brown in colour and softened (Chakravarthi and Srivastava, 1964., Srivastava and Durgapal, 1965., Pathak and Srivastava, 1967a., Gopi 2003). The pedicel if present, becomes dry in diseased fruits. The basal part of the pedicel darkens initially followed by the discoloration of the epicarp around the pedicel. The fungus progresses when the fruit starts ripening. Few days later, black pycnidial are pushed up from the cuticle, with time. Oblong to cylindrical pycnidial spores appear. According to Pathak and Srivastava (1968), the affected skin remained firm, but decay developed in the pulp below and emitted an unpleasant odour.

Botryodiplodia theobromae causes soft rot of mango in India (Srivastava *et al* 1965). Under laboratory conditions, the infection usually started at the stem end of the fruit and very rarely at other places. It began as a water-soaked lesion around the stalk, which increased and soon covered 1/3 rd to 2/3rd of the fruit surface. At this stage, the central part of the neglected area acquires a dark brown colouration and is surrounded by a peripheral zone, the irregular margin of which assumed a water-soaked appearance. Subsequently pycnidia developed. The rot is soft in nature (Pathak and Srivastava, 1967b., Bhuvaneswari, 1999).

The symptoms of *B. theobromae* on the diseased fruits from Bangladesh markets were described by Quroshi and Meah (1991). First indication of the infection was a purple discolouration surrounding the stem end of the fruit. This extended down the body of the fruit on all sides, sometimes as a strip to the distal end of the fruit. Intially the diseased area was brown in colour and later turned to black when fully rotten. The infected fruits disintegrated by splitting on the body or at the top, through which a light brown exudation came out, which sometimes condensed on the fruit skin, especially when the fruits were rotten inside the bamboo baskets. Slits also exhibited white cottony mycelium.

Lasiodiplodia theobromae causes SER of mango in Australia (Photo 10.6). Symptoms included the development of diffuse areas of water soaked tissue that radiate from the stem end of the fruit in finger like projections, which quickly darkened and coalesced into circumpendicular lesions with crenate or lobed margins. Necrosis of the pulp remained below the cuticle (subcuticular) and penetrated all fruit flesh within seven days or less at 25 °C. Subsequently superficial mycelia appeared around the pedicel through ruptures in the skin or directly through the epidermis. A brown liquid drained out from the stem end or from surface ruptures (Johnson *et al* 1991a., Johnson, 1994).

Another causal organism of SER in Australia is *Dothiorella dominicana* (Johnson *et al* 1989a). It caused a decay of mango fruits which spreaded rapidly from the stem end to destroy the entire fruit (Photo 10.8). However, the symptoms of *D. dominicana* are virtually identifical with those of *Lasiodiploida theobromae* (= *Botryodiplodia theobromae*) on mango fruits in tropical regions (Peterson, 1986., Johnson *et al* 1989a) and by *Hendersonia creberrim* (Photo 10.7) on mango in South Africa (Brodrick and van der West – Thizon, 1975).

Varietal Susceptibility

'Langra' and 'Chausa' were comparatively more susceptible to SER in India (Pathak and Srivastava, 1967a). Out of the 20 cultivars screened, 6 cultivars (Biswanath Chatterjee, Dashehari, Lakhman Bhog, Lalphute, Piavophute and Totapuri) were moderately resistant, 2 cultivars (Krishna Bhog and Totapuri Red Small) were resistant and the rest

12 cultivars were susceptible in West Bengal (Awasthi *et al* 2004). 'Narm Dok Mai', 'Kensington Pride' and 'Davis Haden' were highly susceptible showing greatest losses due to SER in Australia (Muller and Wood, 1991). 'Okrong' cultivar was more susceptible to SER than 'Carabao.,' in the Philippines (Huang and Liu, 1995). Varieties 'Sethi Bombai', 'Langra', 'Dashehari', 'Totapuri', 'Safeda' and 'Chausa' are more susceptible (Verma and Kamal, 1951).

Photo 10.6: Lasiodiplodia stem-end-rot

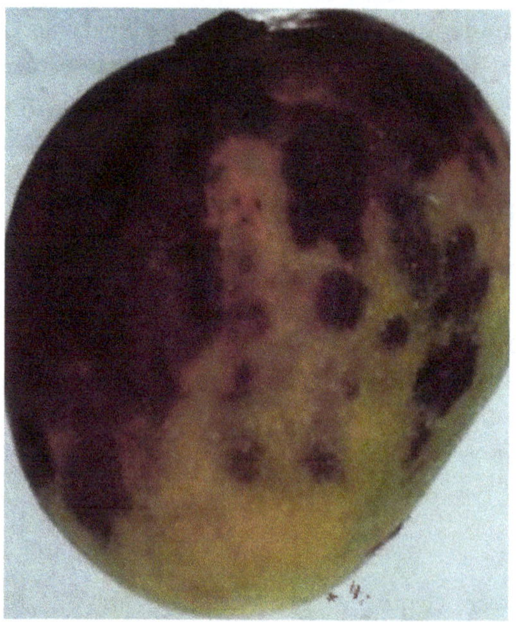

Photo 10.7: Stem-end-rot of mango caused by *Dothiorella mangiferae* (Syn : *Hendersonia creberrima*)

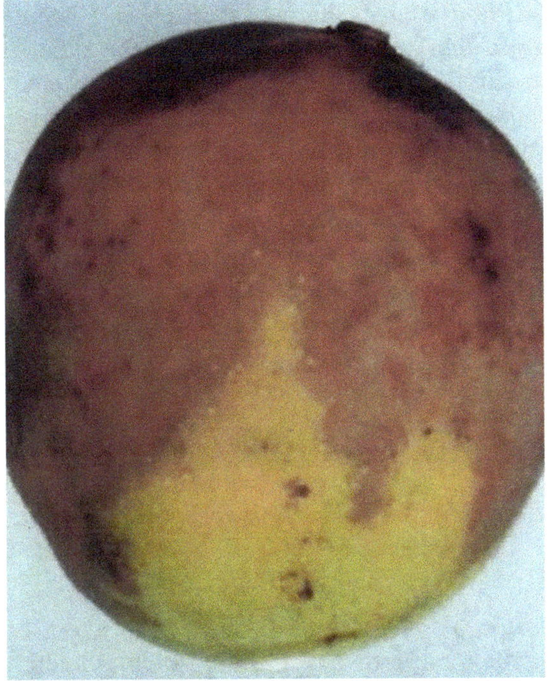

Photo 10.8: Stem-end-rot of mango due to *Dothiorella dominicana*

Changes

Biochemical changes occur in infected fruits. Infected fruits show decline in the contents of ascorbic acid (Vitamin C), non-reducing sugars, sucrose, glucose, fructose and amino acids progressively. The loss of vitamin C under pathogenesis might be due to the production of suitable ascorbic acid degenerating enzymes by the fungus alone or by the host pathogen complex. The loss might also be due to increased respiration in the diseased tissues (Srivastava and Tandon, 1966). The total absence or reduction in the amount of amino acids in the infected fruits was attributed to their utilization by the pathogen or to their degradation by enzymes or their utilization in the synthesis of proteins (Tandon, 1967a).

Etiology

The post-harvest stem-end-rots of mango are caused by a number of fungi. Different fungi are associated with SER in different regions.

In India, *Diplodia natalensis* Pole Evans and *Botryodiplodia theobromae* Pat are the principal causal agents of SER of mango fruits (Srivastava, 1972., Singh, 2000). In a survey of fruit markets in Himachal Pradesh, SER of 'Dashehari' fruits was found to be due to more than one pathogen like, *Botrydiplodia theobromae, Dothiorella ribis* and

Phomopsis mangiferae (Photo 10.9) (Sharma *et al* 1994). Cheema *et al* (1937) reported the association of *Gleosporium mangiferae* with SER and other rots in mango varieties under cold storage. The causal agent of SER according to Srivastava and Durgapal (1965) was *Physalospora rhodina* (Bark and Curt) Cooke (= *Diplodia natalensis* Pole Evans). OmPrakash and Srivastava (1987) attributed SER of mango in India to *Lasiodiplodia theobromae* (Pat) Griffi and Mauble (= *Botryodiplodia theobromae*) alone.

In Hainan Province of China, *Diplodia natalensis* predominates causing latent infection of SER of mangoes (Huang and Liu, 1995). While in South China, *Dothiorella dominicana* (most predominant), *Botryodiplodia throbromae* and *Phomopsis mangiferae* were identified as the causal pathogens of SER (Wang *et al* 1994). However, according to Johnson *et al* (1992a) *Lasiodiplodia theobromae* is the main cause of SER of mango in Asia.

Several fungal pathogens *viz*., *Phomopsis mangiferae* Ahmad, *P. citri*, *Dothiorella dominicana* Pabric and Cif, *D. mangiferae* Syd and Syd, *Lasiodiplodia throbromae* (Pat) Grift and Mauble (= *Botryodiplodia theobromae* Pat = *Diplodia natelensis* Pole) have been isolated from SER infected/spoiled fruits all over the world. (Spalding and Reeder., 1972., Jacobs *et al* 1973., Muirhead and Grattidge, 1984).

In South Africa, *Nattrassia mangiferae* is the causal agent of various mango post-harvest diseases *viz*., SER, soft brown rot, Blossom blight etc. (Saaiman and Smith, 1997). In Australia, SER of mango is the resultant of infection by *Dothiorella* sp, *Lesiodiplodia theobromae, Phompsis mangiferae* and others (Johnson *et al* 1991a., Coates *et al* 1997), while in Queensland, Northern Australia *D. dominicana* is the major cause of SER of mango (Johnson *et al* 1992b). *Botrydiplodia theobromae* predominates in Thailand (Sangchote, 1987). *Alternaria alternata, Phomopsis* sp. and *Lasiodiplodia* sp. (= *Botryodiplodia*) are the main causal pathogens of SER in Israel (Kobiler *et al* 2001).

Collaboration research between Australia, Thailand and Malaysia, however, reported that *Lasiodiplodia theobromae*, *Phomopsis mangiferae* and *Dothiorella dominicana* (Sacc) Petro and Syed, caused SER of mango in all these countries, while *D. mangiferae* Syd and Syd also occurred in Australia and Thailand (Sepaiah, 1986).

The causal agents of SER vary among different production areas of the same country/ region depending upon the environmental conditions of the location. Johnson *et al* (1991b, 1992a) and Johnson (1994) showed that host and environmental factors influence the prevalence of different species in different situations. For example, *Lasiodiplodia theorbromae* prevails in hot regions and *Dothiorella dominicana* prevails in sub-tropical and elevated growing regions, while, *Pestalatiopsis mangiferae* occurs in dry areas and *Fusicocum arsente* is predominant in cooler areas (Johnson, 1994).

Thus, the major fungi, involved in SER of mango are: *Botryodiplodia theobromae* Pat (Perfect state, *Physalospora rhodina* (Berk and Curt) Croke). *Diplodia natalensis*, Pole Evans, *Dothiorella* sp and *Phomopsis* sp. The less frequent SER fungi are:

Colletotrichum gloeosporioides, *Fusarium mangiferae*, *Pestalotiopsis mangiferae*, *Crytosphara mangiferae* and *Dothiorella longa* and *D. aromatica* which are endophytes in mango (Johnson *et al* 1992a., Slippers *et al* 2001., Jagadish and Pathak, 1989., Meah *et al* 1991).

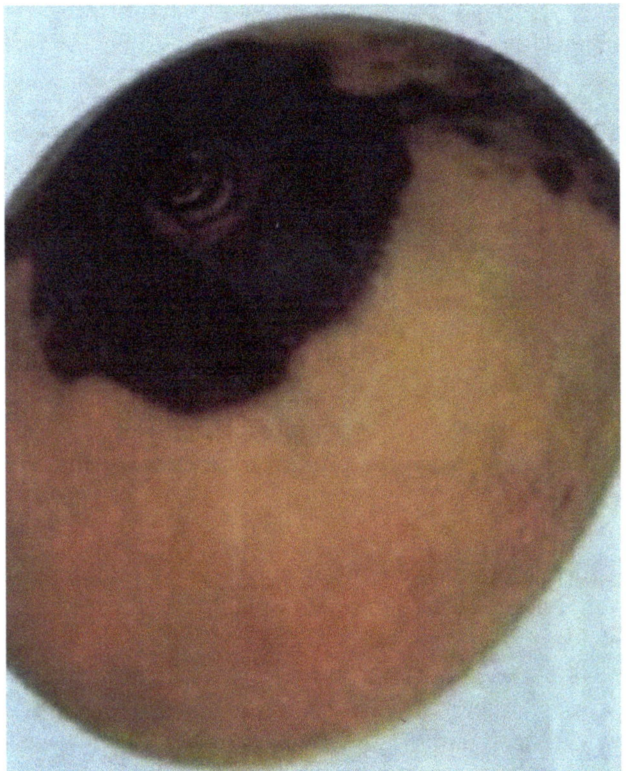

Photo 10.9: Stem-end-rot of mango due to *Phomopsis mangiferae*

Disease Cycle

Information on disease cycle of the pathogen, sources of inoculum, time and mode of infection and its progress is a pre-requisite for making a plan for control/management measures. The infection of SER fungi occurs at two different times, first from flowering to fruit set (pre-harvest) leading to latent infections and next after harvest, which express immediately after the infection that occur after harvest. Latent infections are endophytic in nature (Johnson, 1994). Accordingly the sources of fungal inoculum differ.

Sources

Both plant parts and soil were reported as sources of inoculum of pathogen. Growing fruits and their pedicels while on the tree and green fruits are not infected by the fungus, hence are not considered to serve as source of inoculum for latent infections. But bark and dead twigs of the tree by virtue of harbouring the pathogen serve as primary infection source of *Botryodiplodia theobromae* in India (Pathak, 1966., Pathak and Srivastava, 1969b). Aeroscopic studies showed that the atmosphere of the orchard with the onset of rains gets charged with the spores of the fungus causing infection (Pathak, 1998).

Dead twigs of mango found in the leaf litter in the tree basins floor are the source of *Dothiorella dominicana* in Australia (Johnson *et al* 1991a, 1993).

Pathak and Srivastava (1968) demonstrated that harvested fruits left on the bare orchard soil for long (8 hours or more) got contaminated with different fungi including SER fungi on the pedicel end. According to Johnson *et al* (1990) *Lasiodiplodia theobramae* in Australia also occurs in the soil and can infect fruits that are dropped to ground during harvesting and placed in the soil for desapping. The same situation can not be ruled out in other mango growing countries under similar situations. According to Alicbusan and Schaffer (1958) the source of primary infection of SER in the Philippines was saprophytically infected banana leaves used for packing mangoes in baskets. This may happen with mango or other leaves which are saprophytically infected. Similarly, packing yards, packing materials, vehicles used for transport, godowns and storage and ripening chambers serve as source of inoculum for harvested fruits.

Spread

In nature the pathogen is dessiminated by its spores. During rains the spores from dead twigs on the tree are washed down contaminating the fruits. Infection reaches the pedicel of the fruit from infected and senescenced inflorescences. Fruit to fruit spread also occurs after harvest in packages, godowns, storage rooms and vehicals during transit, particularly when diseased and sound fruits are interminglad (Alvarez Garcia and Lopez-Garcia 1971., Muirhead and Grattidge, 1986).

Ripe vs Unripe Fruits: Under laboratory conditions unripe fruits were not affcected by the disease, whereas ripe fruits had shown disease on inoculation. Disease, appeared earlier in ripe fruits had shown greater severity on 5th day of inoculation. On semi-ripe fruits, however, the infection appeared on 5th day of inoculation. Unripe and unwounded fruits were not infected by inoculation (Chandra and Pathak, 1989). The above situation may be explained as follows:

Pre-formed inhibition in unripe fruits retard the growth of the pathogen (Simmonds, 1963). The presence of a mixture of antifungal compounds [5-(12-cistrepa decenyl) resorcinol and 5-penta decenyl resorcinol] at fungitoxic level in the peel of unripe mango fruits might have checked the access of fungus into the fruit, which decrease in ripe fruits permitting the access and development of disease symptoms (Droby *et al* 1986). The concentration of these compounds decreases further and faster during ripening in susceptible cultivars than in resistant ones. This effect is enhanced by induction of ripening from ethylene treatment (Pathak, 1998).

Mode of Infection and Route

Different modes of infection of SER fungi were reported. Mode of infection also depends upon the pathogen involved. Infection occurs both in the field, while the fruit is still on the tree and after harvest in assembling centres, package sheds, storage and ripening chambers and also during transit.

According to Johnson *et al* (1991b, 1992b, 1993) infections of SER fungi occur prior to harvest well before fruit set and post-harvest losses do not result from such infections at flowering or fruitset because fruitlets infected at this time are aborted and shed. They also maintained that SER fungi remained as endophytes in mature stem tissues and colonize the inflorescence and ultimately reach the stem end of the fruit via peduncle and pedicel several weeks after flowering. They further suggested that the infection of fruit occurs from the endophytic pathogen already present in the peduncles, which forms an important mode/route in the infection of *Dothiorella* sp. Such stem end infections in the intact fruitlets remain quiescent until fruits start ripening.

SER fungi are wound invaders. Wounds or injuries are necessary for their access into the fruits, especially in case of harvested fruits. It was demonstrated that the only avenues of infection of the uninjured fruits by the spores of *L. theobromae* were the exposed surfaces of the pedicel or pedicel scare or the wounds on the fruits (Pathak and Srivastava, 1969b). These are also claimed as the natural mode of infection.

Fernando (1937) obtained successful infection by inoculating the fungal culture on the fresh pedicel scar of the fruits. Chakravarthy and Srivastava (1964) obtained similar results by inoculating fungal culture on the wounds near the stalk end of the fruit. Srivastava and Durgapal (1965) showed that fungal cultures could infect readily when modulated in cavities made on the pedicels of the fruit. Pathak and Srivastava (1967b, 1969b) found that the natural way of infection of the uninjured fruit by spores was the exposed surface of the pedicel or the scar when the pedicel was removed. Uninjured fruits did not show infection, even when the fruit is fully ripe. However, when injuries were made either on the sides of the pedicel or the epicarp of the fruit, the pathogen successfully showed 90 and 100% infection respectively. SER infections may also originate at injuries in other locations of the fruit (Pathak and Srivastava, 1969b). Halos and Divinagracia (1970) confirmed that *L. theobramae* invaded the mango fruit pedicel through wounds on detached mature green and ripe fruits with the invation of fruit occurring without any apparent quiescent stage. Injuries to pedicel or epicarp of the fruit may happen during harvest, after harvest, in post-harvest handling, transit, in packing, packages, storage etc., which often result in infections. However, mere rubbing of fruits with each other will not lead to the infections (Pathak and Srivastava, 1969b). In packaged cases, healthy fruits get contaminated through wounds caused at the walls of the wooden boxes on account of continuous friction, particularly when diseased fruits are packed along. Even otherwise, the source of infection being the packages, wounded healthy fruits also get infected (Dharamvir *et al* 1967). However, while Srivastava (1972) observed no secondary infection of the disease from diseased to healthy fruits, when kept together for a long time and ruled out such possibility due to lack of evidence. Johnson (1994) maintained that such spread occurs by physical contact or by exudates from decaying fruits. Fruits during desapping in soil get infected through the cut-end of the pedicel (Johnson, 1994). However, Halos and Divinagracia (1970) found that the pathogen also penetrated the cuticle directly. Thus, removal of fruit pedicel and injury to fruit have great significance in the incidence of SER of mango fruits.

Fruits with or without pedicels left uncovered in the orchard for 8 hours or more after harvest developed infections. But when fruits were covered immediately after harvest none of them showed infection (Srivastava, 1972). Similarly, fruits in which the cut ends of the pedicel were coated with a fungicidal wax immediately after harvest and exposed in the orchard remained free from infections. Unripe fruits with or without pedicels exposed in the orchards for 8 hours following harvest and later stored in cellophane bags in the laboratory showed 9 and 24% infection respectively at ripening time (Pathak and Srivastava, 1967a). Exposure of harvested and desapped fruits to the atmosphere in the orchard beneath the trees for 6 hours did not show the incidence of SER indicating that airborne inoculum was not a significant source of inoculum for SER in Australia (Johnson *et al* 1993). However, the fact that non-pedicellate fruit showed 15% greater infection than the pedicellate ones, when exposed in the orchard, suggests that all pedicellate infection may not efficiently lead to fruit infection.

D. dominicana infects mango fruit – endophytically via the fruit stalk (Johnson *et al* 1991b). Endophytic colonization of the inflorescence and pedical tissue was found to be the primary route of infection of fruit that developed into SER during ripening (Johnson *et al* 1992a). In the inoculated pedicel the pathogen was mainly confined to the vascular tissue through which it proceeded into the fruit intercellularly. Colonization of *D. dominicana* in inoculated fruits initially followed the vascular tissue or lactifers and then infected the fruit flesh. Infection of seed seemed to follow the same path i.e. down the vascular tissue present in the funicules to infect endocarp, testa and embryo. However, as there did not seem to be a direct vascular connection from the funicules to the embryo, it is not clear how the embryos are infected. Colonization of the embryo does not occur until the visible SER symptoms appeared on the fruit surface (Gosbee *et al* 1997). Infected seed produced infected seedlings, as *D. dominicana* is seed transmissible (Johnson *et al* 1993).

Nature of Infection

Field infections that occur before harvest are latent (Quiescent) in nature. The fungi infect the inflorescence early in its development and reach the pedicel several weeks after flowering and remain quiescent until the fruit commences ripening. In latent infections, the spores germinate on immature fruits and form appressoria. Some of these appressoria germinate to form infection hyphae, which mechanically penetrate the cuticle and develop to a limited extent on the epidermal layers of the fruit. However, most of the appressoria, do not germinate immediately, but remain firmly attached to the host surface as the latent stage of the pathogen (Swinburne, 1983., Pathak, 1998).

Factors: Temperature, duration of transit and storage facilities were considered as the main physical factors influencing the infection of SER pathogen in mango. A linear relationship between the incidence and the transport distance (i.e. time) was observed. The delays in the transportation aggrevated the disease. Further, transportation on rough roads, with jerks and vibrations and bumps aggrevates the disease. Fruits under such conditions are more prone to injuries and subsequent fungal infection during transit (Quroshi and Meah, 1991). During transit fruits in bamboo baskets get heated up and sweat leading to rapid development of the disease (Prasad and Sinha, 1981).

Epidemiology

Among the different epidemiological parameters, temperature and relative humidity (RH) play important and greater role in the disease incidence, its development and progress. SER mango in India is considered to be a high temperature disease (Singh, 2000).

On artificial media, temperature requirement of different SER fungi varied. *B. theobromae* grew within the temperature range of 15-35 °C with an optimum range of 30-35 °C. Optimum temperature range for the growth of *L. theobromae* was 24-34 °C, while *Diplodia natalensis* grew well at a temperature range of 20-40 °C, with optimum being around 30 °C. *Dothiorella dominicana* and *D. mangiferae* failed to grow at 40 °C and *L. theobromae* failed to grow at 10 °C, while low temperature was favoured by *B. cinerea* (Johnson *et al* 1991b., Johnson and Coates, 1993).

Maximum temperature for infection by *L. theobromae* was 31.5 °C and minimum was 25.9 °C (Pathak and Srivastava, 1967b, 1969a). The high incidence of SER fungal spoilage (up to 26.7%) observed in 'Dashehari' fruits in fruit markets of Himachal Pradesh (India) might be due to the presence of favourable weather conditions (25-30 °C and 80-90% RH) for infection. Warm and moist storage conditions favour the incidence and spread of the disease. Storage of fruit at 30 °C favoured the successful infection of *L. theobromae* in Australia over *D. dominicana*, while it was reverse at 25 °C and lower, explaining the apparent differences in the main cause of SER between tropical and subtropical growing areas (Johnson, 1992., Johnson and Coates, 1993). SER develops rapidly at room temperature.

The post-harvest quiescent period for SER pathogens was influenced by storage temperature, the pathogen involved and the mode of infection, ranging from 4 to 5 days at 22 °C and 14 days at 13 °C in fruits that were naturally infected with *Dothiorella* sp or inoculated with *D.dominicana* at stem end from 5 to 13 days at 23 °C depending upon whether the primary inoculeum was soil borne (*L.theobromae*) or endophyte (*Dothiorella* sp.) (Johnson *et al* 1993). Symptoms appeared 3-7 days after harvest at 25 °C, but can get delayed for additional 7-12 days at 13 °C at which the growth of the fungus was inhibited. Ethylene treatment prior to storage at 13 °C however, can promote SER development in storage (Ploetz, 2004).

Fungal spores generally require high humidity (80-100%) for germination to cause infection (Prasad and Sinha, 1981). Infection usually occurs when the fruits are harvested on a cloudy day and in wet weather. Conidial production is enhanced by moist and wet weather. Fruit rot is severe in areas of excess rainfall, resulting in more than 80% RH and high temperature. Late varieties, which are harvested in the months of June to August are more susceptible to rots due to prevailing high temperatures and RH at and after harvest. The incidence of latent SER would be significantly higher in areas where there would be unbroken spell of wet weather during the flowering season than would be in areas where there would be unbroken spell of dry weather (Huang and Liu, 1995). The condensate from transpiration of fruits or fruit exudates resulting from minor bruises and insect punctures provide the fungal spores with adequate humidity needed for germination and subsequent penetration into the fruit. Once the disease is infected/initiated, moisture never becomes a limiting factor inside the fruit pulp.

Harvesting methods, tree age etc. significantly influence the incidence of SER. The incidence of SER increases with the tree age. Older trees were more prone to the disease, because of dense canopy and resultant high humidity within the canopy and lack of sunlight and air-circulation. Faulty and improper harvesting methods enhance the chances of disease incidence through injuries that happen during harvesting. Fruits harvested with pedicel (8-10 cm long) and shifted immediately from orchard to packing yards never show post-harvests rots (Jadeja and Vaishnav, 1998). Water stress and defoliation favour colonization of the fungus and depress host response mechanism against endophytic SER (*D. dominicana*) in Australia (Johnson *et al* 1993). Application of copper sprays for the control of anthracnose increased the incidence of *D. dominicana* SER in Australia by reducing the competetion from other endophytic colonists (Johnson *et al* 1991a). When different SER fungi prevail at the same time/or simultaneously, *Diplodia theobromae* predominates over *F. aserculi* at high temperatures, whereas *F. aserculi* predominates at < 25 °C. Although these fungi out compete other SER colonists, *Colletotrichum glaoeo–sporioides* can depress upto 25% of *F. aserculi* infection at 13-18 °C temperature (Ploetz, 2004). Meah *et al* (1991) studied the growth and infectivity of *B. theobromae* of mango and found that the pathogen grew and sporulated best at 28 °C on media amended with ripe mango extract and dextrose or sucrose and observed that agar amended with unripe mango pulp caused a reduction in growth rate of the mycelium of the pathogen.

Control/Management

The disease control/management strategies aim at prevention and erradication of disease incidence and/or delaying the disease symptoms and progress of the disease leading to collapse of the host. Mode and time of infection and form of quiescent structures for a specific combination of penetration, under different host pathogen environment combination must be clearly elucidated and understood to target the type and timing of control measures (Pathak, 1998). Since the infections occur both in the field (prior to harvest) and after harvest, one has to think of both pre-harvest (mainly preventive) and post-harvest (mainly curative of latent infections) measures for the control/management of the disease and should be undertaken in two directions *viz.*, preventive and curative in both strategies.

Pre-harvest Measures

Pre-harvest measures include mainly preventive measures, where main objective is to prevent/reduce/eradicate the inoculum/infection of the fungus.

Orchard sanitation is one such measure suggested. The pathogenic fungus over winters in the bark and dead twigs and leaves on the tree as well as in the tree basins. Hence, their prompt and timely removal and destruction reduce the inoculum in the orchard. For more benefit, it should be followed by fungicidal sprays to further destroy the inoculum if any remained on the tree. Orchard sanitation, including removal of weeds etc., if continued regularly, timely and promptly for few years continuously may eradicate the inoculum from the garden for a longer period.

Both pre-mature as well as be-lated harvesting can make the fruits more susceptible to post-harvest infection. Therefore, the fruits must be harvested at most appropriate stage of maturity as per the prescribed maturity/harvest indices in each case of the variety. Choosing the most appropriate maturity stage, time and method of harvesting are of utmost importance in preventing the inoculum/infection.

The fruits should be harvested with pedicel using modern harvesting aids for prevention of the disease, because fruits without pedicels (stalk) are prone to post-harvest infection. The length of pedicel at harvest and subsequent incidence of SER are interrelated. Fruits with longer stalks do not usually develop the symptoms and if developed, the development of symptoms is more slow than those with short stalks. Mangoes should be harvested with 8 to 12.5 cm long stalks. Fernando (1937) suggested to reduce the length to 2.5 cm by cutting the stalk in methylated spirit and with a final dip of 1.25 cm stalk in molten wax to prevent the access for infection. However, this suggestion is cumbersome besides not being useful when the infection already has taken and advanced into the fruit. Instead, it is safer to cut the pedicel up to a length of 1.0 to 1.5 cm and dipping the cut end in molten wax or smearing the cut ends with a disinfectant/fungicide to avoid the ingress of the pathogen through the cut end after harvest.

To avoid the pre-harvest infection one can resort to pre-harvest field sprays of suitable fungicides on the trees at appropriate time. Field sprays with fungicides/chemicals can prevent germination and formation of appressoria or deep seated infections in lenticels or in the floral reminants. Therefore, a pre-harvest treatment with fungicides is an appropriate preventive strategy. These control measures should start right from blossoming in trees, as it was shown that infection of SER started during blooming and fruit set and continued until few days before harvest. Timing of the fungicidal sprays is more critical in these measures. Pre-harvest field sprays of both systemic and non-systemic ones against SER of mango were tried/tested, of which the former were found to be more effective in the field against the disease.

Three combined sprays of benomyl and captan during fruit development (June to early August) delayed the appearance of fruit rots (unspecified) and greatly reduced the amount of rot development during storage in Israel (Zauberman *et al* 1976). Kalayansundaram and Parthasarathy (1979) obtained best control of *Diplodia natalensis* with sprays of thiobendazole (0.1%). Rawal and Ullasa (1988) demonstrated that pre-harvest field sprays of carbendazim (0.1%) were highly effective in checking the post-harvest decay of mango fruits by *B. theobromae* at Bangalore, while Raoof and OmPrakash (1986) found two pre-harvest sprays of Bancop (0.2%), Thiabendazole (500 ppm), Agrimycin (1000 ppm) and Aureofungin (250 ppm) at 15 days interval before harvest were quite effective followed by Prochlaraz in reducing the disease incidence in the month of July at Lucknow. Sharma and Badiyala (1994), while working on SER of 'Dashahari', 'Langra' and 'Alphonso' varieties of mango, observed effective control of SER by 3 sprays of carbendazim (0.1%) at fortnight intervals commencing 45 days prior to expected harvest. According to Banik *et al* (1998), the systemic fungicides like Bavistin (2000 ppm) and Topsin M (2000 ppm) may be effectively used as pre-harvest sprays for the control of *D. natalensis*; the non-systemic fungicides such as Captaf, Cunean L and Kavach were less effective in reducing SER symptoms on 'Himsagar' mango fruits. Pre-harvest sprays with copper oxychloride, iprodione, or propiconizol reduced the incidence of SER in Australia (Johnson *et al* 1991a., Johnson, 1994). Pre-harvest sprays of carbendazim (0.1%) thrice at fortnightly intervals starting from 45 days before harvest lowered the disease percentage and resulted in higher decay reduction index on 'Baneshan' mangoes compared to that of thiophonate methyl, Aureofungin, Captan and Mancozeb (Narsireddy *et al* 2004).

Controlling disease of the blossom complex have the advantage of reducing post-harvest diseases especially SER as it invades the fruitlets during blooming period. The best control was obtained with fungicidal treatments applied at 5-7 weeks after bloom. Spray of 3 fungicides (azoxystribin, benomyl and carbendazim) decreased SER when applied 5 weeks after full bloom. Benomyl was also effective 7 weeks after full bloom. The rest of the two fungicides could also reduce 'anthracnose' when applied 11 weeks after full bloom. Early application reduced SER (Swart and van Broekhuzen, 2004). Swart *et al* (2009b) from South Africa, confirms the value of using fungicides during late blooming and early fruit set period in order to reduce SER. The effective fungicides differed between locations. At one site, Propiconazole (Tilt) or Ortira reduced the occurrence of SER, while at another site, Fludioxinil showed superior control of SER.

The use of trunk injected potassium phosphonate for the control of phytophthora disease is well known. Phosphonate acts as an antifungal agent through both direct and indirect effects on the pathogen (Guest and Bompiex, 1990). Trunk injection of 400 g/l of water formulation of potassium phosphonate (at 15 ml/meter of tree canopy diameter) at 'hen egg' stage of fruits reduced SER levels in fruts after storage at 22 °C for 20 days (Cooke *et al* 1997). However, injection and soil drenches with the above chemicals during flowering and fruit development did not reduce SER levels on 'Kensigton Pride' mangoes in Australia (Cooke *et al* 1997).

Post-harvest Measures

Post-harvest measures include both preventive and curative measures. Curative measures mainly include treatment of harvested fruits with different types of chemicals and different methods of application. Physical methods like hot water treatment and irradiation with or without chemicals etc. have been employed as part of post-harvest treatments of fruits for prevention as well as control of SER. Of late, use of plant oils, botanicals and biological control are increasingly evaluated with promising results.

Preventive measures: Proper harvesting of the fruits and subsequent handling in shifting to pack yards, packaging, transporting and storage are steps which help a great deal in the prevention of post-harvest infections. Sorting of the harvested fruits is a good practice for prevention of post-harvest disease infections naturally and by contact. Eliminating the fruits showing injuries, bruises and early symptoms of the disease and over ripe fruits can substantially reduce the intensity of the disease at later stages. The practice of using mango leaves, banana leaves etc., for packing or cushioning mango fruits, which may serve as a source of inoculum should be discouraged/avoided, instead the use of paddy/wheat straw, shredded paper, plastic sleeves etc. must be encouraged (Srivastava *et al* 1964). Wrapping of fruits in plastic film or in heat-shrinkable film will prevent the spread of the disease by contact in a pack or container (Barmore *et al* 1983). Disease infections can also be prevened by avoiding injuries to the pedicel or fruit skin at or during harvesting, avoiding exposure of harvested fruits for longer periods to sun and contaminated atmospheres of the orchard. Shifting of the harvested fruits in cellophane or other type of bags/sacs with in 6 hours of harvesting to the pack houses, ripening chambers etc., avoiding snapping of pedicel during post-harvest handling operations and smearing of the cut ends of stalk/pedicel scar with protetive chemicals/fungicides are also helpful in preventing the infections. To check the spread of the disease, Srivastava and Durgapal (1965) suggested to sort out those infected fruits which show characteristic browning of the pedicel base and internal discolouration of the pedicel tissue prior to transport, packaging, ripening and storage, to minimize the losses by disposing such fruits in the local markets and consuming them on priority basis in the houses. Further, if the infection has not advanced up to the fruit pulp in infected portions, the pedicel need to be removed to save the fruit from further infection and rotting. Since the growers/pre-harvest contactors, wholesalers and other middle men involved in marketing of the fruits, have no chance to take up such preventive measures, as the disease manifests only at or after ripening, the retailers have to take up the above preventive measures to reduce their losses and also by selling the ripened fruits on a prirority basis (Srivastava and Durgapal,

1965). However, an intelligent mango grower can also take up above preventive measures before disposal of the fruits to markets/middlemen to gain confidence and better returns.

Curative Measures

Fungicidal treatments: Fungicides alone as dip treatments of harvested fruits were evaluated for the control of SER mangoes. A perusal of the available information indicates that different causal agents needed different fungicides for their control. It was observed that efficacy of fungicides varied with the fungal species involved in the disease and also the location of mango growing.

Post-harvest dip treatment of mango fruits with aureofungin (500 ppm), an antibiotic, prolonged their post-harvest life by 18-20 days against infection of *Diplodia* stem-end-rot, while control fruits rotted with in 2-3 days *in vitro* (Dharamvir *et al* 1967). On the other hand, Pathak *et al* (1971) reported failure of aureofungin to control SER. It is a common practice in vogue to use news papers as padding/wrapping material in the wooden boxes during transportation (Photo 2.5) which is believed to contaminate the packed fruits with diseases. The use of aureofungin as spray for the treatment of padding or wrapping material inhibited the growth of *Diplodia natalensis 'in vitro'* for 20-25 days. The combination of post-harvest dip and use of aureofungin treated padding or wrapping material in the wooden boxes at the time of packing the produce was effective against *Diplodia* SER in mango. And this combination of treatment will help the growers and traders to reduce losses by such post-harvest disease especially during transit (Dharmvir *et al* 1967).

Benomyl is used for post-harvest treatment of mangoes in many countries and has the highest sensitivity to *D. natalensis*. Under laboratory conditions, benomyl, carbendazim and thiophenate methyl were able to inhibit the growth of *D. natalensis* (Mandal and Das Gupta, 1985). OmPrakash and Misra (1988) were of the opinion that *D. natalensis* could be controlled by benomyl, carbendazim, thiobendazole, brine or hot water at 50 °C. Linear growth of *D. natalensis* was completely inhibited at 400 ppm Bavstin *in vitro* followed by Captaf (450 ppm), Topsin-M (450 ppm), Cuman L (600 ppm) and Kavach (650 ppm). Bavistin and Topsin M were the best in reducing the incidence of *D. natalensis* followed by Cuman L and Captaf, when the fungicide treated fruits were artificially inoculated *in vitro*. Pathogenic effect of Captaf and Karach however, remained fairly high (Banik *et al* 1998). In the pre-inoculation treatments, thiabendazole (500 ppm) completely controlled Diplodia rot of mangoes (Pathak *et al* 1971). The disease incidence was delayed in carbendzim treated fruits, where in first appearance of disease was noticed 14 days after storage against 11 days after storage in fruits treated with benomyl and in untreated control fruits (Gopi, 2003).

Prochloraz gave good control of SER due to *Lasiodiplodia theorbromae* and *Phomopsis mangiferae* in Florida (Spalding, 1982). Muller and Bart (1989) could get control of *L. theobromae* in harvested fruits for up to 4 weeks by prochlaraz dip (0.025%) at an ambient temperature of 31 °C followed by storage at 13 °C. Hence, it was recommended to use prochlaraz as a fungicidal dip treatment of mangoes for the control of SER because, it is easy to apply and does not require water heating equipment and is registered for use on mango in Western Australia. Muirhead an Grattidge (1986) found that *Dothiorella dominicana* was controlled by benomyl, but not by prochloraz. Prochloraz and calcium chloride, however, did not control *D. dominicana* SER in Australia (Johnson *et al* 1989b).

In laboratory, fruits dipped in thiobendazole (0.1%) showed higher reduction (90.7%) in SER incidence followed by SOPP (Sodium orth phenyle phenate) (61.1%) and Indofil-M-45 (60.30%), but was less effective than carbendazim (0.1%). Therefore, it was concluded that SER by *B. theorbromae* could be easily controlled by dipping fruits for 4 min. in carbendazim or thiobendazole (Sharma *et al* 1994). Dipping of mango fruits in Diathane-M-45, thiobendazole and potassium metabisulphite was effective (Das Gupta and Mandal, 1989). Dipping of fruits in potassium metabisulphite (KMS), dodine (scorbic acid 50% flowable) and roval (iprodine) did not prevent SER, but increase in concentration of fungicide lowered DRI linearly. The efficacy was more pronounced with SOPP and KMS (Sharma *et al* 1994).

SER of mango can also be kept undercheck by ethylene dibromide fumigation (Spalding *et al* 1976).

All the chemical sterilants, except boric acid (10%) recorded higher disease incidence. SER was higher in fruits treated with calcium chloride (4%), copper sulphate (8%) and potassium permanganate (2%) (Gopi, 2003). The sterilants showed least residual toxicity with the increase in storage period, hence ineffective in reducing the disease at later stages of storage. Phyto-toxicity was also observed with calcium chloride (4 and 8%) (Tirmazi and Wills, 1981., Gopi, 2003).

Thus, for the control of fruit infections, fungicides are usually applied as a post-harvest dip treatment. However, latent as well as established infection at the time of harvest are difficult to eradicate by mere post-harvest protective fungicides etc.

Heat Treatments

Traditionally decay control in perishables during storage, transit and marketing has primarily relied upon fungicidal application and manipulation of storage conditions. However, in view of public concerns about the health associated with fungicidal treatments and possibility of development of resitance in phatogen, alternate, non-chemical control measures were explored (Johnson and Sangchote, 1994). The efforts resulted in the development of heat therapy, as a non-chemical means of controlling post-harvest diseases. Heat theraupy may be applied as hot water dip (Pennock and Maldonaldo, 1962., Lizada *et al* 1986), vapour heat [Coates *et al* 1993], heated dry air (Lurie, 1998) or as hot water rinse and brushing (Fallik *et al* 1996., Prusky *et al* 1999). The treatment is applied for a relatively short time (few seconds/minutes), since the target pathogen is usually found on the surface of the fruits or in the first few layers under the fruit skin (Schirra *et al* 2000). Heat and radiation treatments are meant for killing or breaking the quiescent pathogens, thereby enhancing the shelf life of perishables.

Dipping of 'Carabao' mango fruits in hot water (53 °C) for 10 min. effectively reduced the occurrence of SER in the Philippines (Quimio and Quimio, 1974a). Hot water treatment (HWT) of 'Tommy Atkins' mango fruits reduced SER (*D. natalensis* or *Phomopsis citri*) and doubled the percentage of acceptable fruits in Florida (Spalding and Reeder, 1986). A similar hot water dip at 55 °C for 5 min, reduced the SER caused by *D. dominicana* in Australia (Muller and Bart, 1989). Immersion of 'Keitt' and 'Tommy Atkins' mango fruits in hot water (46.1 °C) for 65 min, lowered the incidence of SER and anthracnose (Sharp and Spalding, 1984). However, heat injury was frequently observed in hot water treated fruits (Rappel *et al* 1991). Not withstanding the above heat injury, the only post-harvest treatment used commercially for decay control in mangoes in Florida is immersion in or drenching with hot water (53 °C) for 3 min. (Spalding and Reeder, 1986).

The time of immersion/treatment in hot water differed with the causal pathogen of SER. An immersion time of 5 min. was found to be sufficient to completely kill the inoculum of *Phomopsis* sp at a temperature of 25 °C. *L. theobromae* and *D. dominicana*, however, required an immersion time of 8 and 18 min. respectively at 53 °C for complete killing of their mycelia. Perhaps *L. theobromae* become more deeply embedded in the mango stem-end tissue and therefore, more protected from heat than *D. dominicana* hence required more time of immersion.

Of late, heated forced air treatment has gained some importance in the control of post-harvest diseases of mango. A heated forced air treatment at 51.5 °C for 125 min. of 'Tommy Atkins' mango fruits, apart from being effective in the eradication of Caribbean fruit fly, could concommittanly reduced the incidence of anthracnose (*Colletotrichum gleosporioides*) and *Botryodiplodia* stem-end-rot infections substantially (Miller *et al* 1986., Mc Guire, 1991). However, the hot air treatemtn alone gave no control of SER (*Dothiorella* and *Lasiodiplodia*) in mangos stored at 13 °C prior to ripening. A supplementary hot fungicide treatment was required for acceptable control of SER in cold stored fruits (Golan, 2001).

Vapour heat treatment (VHT) was also attempted for control of post-harvest diseases including SER in mango. VHT clearly gave control of SER in the artificially inoculated fruits, but not SER of naturally infected fruits. This difference in response to VHT, might be reflecting the differences in the colonization of mango stemend tissues in the artificially – inoculated (where the remaining fruit surface in addition to stemend was inoculated) and naturally infected fruits. However, VHT in Australia provided inadequate control of SER of mango (Coates *et al* 1997).

There has been variable success with HWT in controlling SER, which might partly be due to differences in the causal agents and their strains (Spalding *et al* 1988., Esguerra and Lizada, 1990., McGuire, 1991). Isolates of SER causing fungi showed variation in response to VHT. Isolates of *D. dominicana* were found to be the most resistant and that of *Phomopsis* sp. were the least resistant. VHT (at fruit surface temperature of 46.5 °C for 10 min.) more effectively controlled SER caused by *D. dominicana* than that caused by *L. theobromae* (Coates *et al* 1993).

All heat treatments (hot-water dip, forced air and vapour heat) resulted in a reduction of SER incidence in 'Carabao' mango fruits in the Philippines. However, VHT alone did not control the rot completely. HWT with VHT was effective (Esguerra *et al* 2004). Heat treatments may kill or reduce the population of beneifical antagonists present on the fruit surface and thereby favour the colonization of decay causing pathogen(s). Aerated steam treatment was more effective than HWT against spore germination and disease severity (Pathak, 1997).

HWT plus Fungicides etc

However, heat treatments alone were considered less effective control measures against post-harvest rots (Ploetz *et al* 1994). HWT does not protect fruits against reinfection by the microorganisms. HWT may be more effective against SER etc., in combination with others like, fungicides, irradiation etc. Addiiton of various fungicides has been observed to incrase the effectiveness of the HWT for disease control (Muirhead, 1976., Spalding and Reeder, 1972). It has also been observed that by combining 2 or 3 different treatments, due to their synergism, the efficacy can be improved in controlling the diseases.

Dipping of mangoes (green, hard and pre-climacteric) soon after harvest in warm (43 °C) borax solution (6%) for 3 min. could reduce SER during ripening at room temperature. The hot borax treatment also accelerated the rate of ripening and improved the skin colour of mangoes (cv. Alphonso) on ripening (Thomas and Dalal, 1968).

The combination of hot water and benomyl has proved promising and best in several mango growing regions (Dodd *et al* 1991a). In Queenstand, benomyl at 52 °C water temperature gave good control of *D. dominicana* (Muirhead and Grattidge, 1986). Post-harvest immersion of fruit in hot water (50-55 °C) combined with benomyl (850 g a.i./litre of water) for 5-10 min. was shown to control SER (Dodd *et al* 1991a). Dipping of mango fruits (cv. Keitt) in benomyl (0.1%) in hot water at 70 °F for 2 min. was, however, ineffective against *B. theobromae*, but dipping in hot water plus benomyl at 130 °F for 5 min. was effective, indicating that the efficacy of benomyl depends on the water temperature as well as duration of treatment (Spalding and Reeder, 1972). Hot benomyl provided good control of SER caused by *D. dominicana* and *L. theobromae* at 52 °C for 5 min. in Australia (Johnson *et al* 1990, 1991a., Rappell *et al* 1991., Sangchote, 1991., Johnson and Coates, 1993). Good control of *D. natalensis*, for up to 13 days after treatment was achieved with hot water plus benomyl in Brazil, without ill effects on the fruit external apperance (Sampaio *et al* 1981). 'Kensington Pride' mango fruits can be stored for up to 4 weeks after dipping in benomyl (0.050%) at 50 °C water temperature with significant reduction in losses from *L. theobromae* as compared with control ambient water treatment (Muller and Bart, 1989). In view of the above results hot benomyl was recommended for the control of SER of mango in Australia (Sangchote, 1989), but its use is not allowed in various other countries (Ploetz *et al* 1994). However, there are inherent difficulties in the use of hot benomyl, such as heat damage and benomyl stripping by mango sap (Wells *et al* 1989., Rappel *et al* 1991). Dipping of mango SER (*P. mangiferae*) affected fruits in hot benomyl (0.05%) at 52 °C water temperature for 5 min. controlled the disease but resulted in scalding on fruits to an extent of 2.6% (Muller and Bart, 1989). Severe scalding occurred when fruits were dipped in hot benomyl at 54.4 °C (Spalding and Reeder, 1972). Muirhead (1976) however, did not observe scalding of fruits at 51.3 °C. Mullar and Bart (1989) showed that benomyl (0.05%) at 50 °C water temperature gave good control of SER (*Phomopsis mangiferae* and *D. dominicana*) for 2 weeks without scalding on the fruit. The addition of gaur gum to hot benomyl improved control of SER, but did not improve marketability of fruits (Spalding and Reeder, 1986).

In Queensland, hot prochlaraz was not effective (Muirhead and Grattidge, 1986), while it gave partial control of SER (Dodd *et al* 1991a). 'Kensingtan Pride' mangoes could be stored for up to 4 weeks after dipping in prochlraz (0.025%) at 31 °C (Muller and Bart, 1989). Prochlaraz showed better control of SER, particularly when it is heated, inspite of lowering the dosage (Swart *et al* 2009a). Prochlaraz combined with thiobendazole showed a very little effect on SER. Johnson *et al* (1990) felt that a dual treatment of hot benomyl followed by prochlaraz was necessary for long term control of SER caused by *D. dominicana* or *L. theobromae*, although hot water plus benomyl was providing good control of SER caused by above fungi during short term stages.

Thiobendazole (0.01%) dip in hot water at 70 °F for 2 min. was ineffective, but at 130 °F with 5 min. dip was effective (Spalding and Reeder, 1972). A dip of fruits in thiobendazole (0.01%) at 50 °C proved less effective against SER (*L. theobromae*) in Queensland (Mullar and Bart, 1989). Sharma *et al* (1994) also found that thiobendazole (0.1%) was less effective against Diplodia SER as compared to carbendazim. However, carbendazim was partially effective against SER, especially in combination with prochloraz.

Tilt (propiconazole) in hot water controlled SER of mango (*D. natalensis*) most effectively (75-80%) at lower dosage (250 ppm) without affecting the fruit quality. Tilt showed not only protective action, but also curative action against the pathogen *in vitro*. The next best treatment was hot benomyl which had protective action 90-100% without curative effect. Mixtures of benomyl with guazatine or imazalil were more effective than individual fungicides. Tecto was least effective. Hot Imazalil was poor in both protective and curative action. Hot solution of benomyl mixed with tilt or Imazalil showed better results than the fungicides applied separately. Tilt + benomyl was the best (Huang and Liu, 1995).

Mango fruits treated with hot water at 52, 55 and 58 °C plus Captan (0.5%) for 5 min. showed reduction of losses due to SER (Roy *et al* 1980).

Fludioxonil (a new generative fungicide) applied as hot dip (300 or 600 ppm), could control SER effectively in South Africa. At higher doses it also showed some efficacy against anthracnose (Swart *et al* 2002). In a later study fludioxonil proved effective in reducing SER in either cold or hot water treatment. It can therefore, be a valuable fungicide for control of SER on mangoes exported from South Africa (Swart *et al* 2009a).

Irradiation: Attempts, made to control SER disease of mangoes by irradiation using Gamma or Ultra Violet (UV) radiation were not successful at different doses and durations (Pathak and Khandalwal, 1969., Gopi, 2003). However, combination of irradiation with hot water and fungicide proved fruitful. Padwal Desai *et al* (1973) hoped that hot water dip combined with irradiation would offer better process of delaying fungal infection, besides ripening. Combination of hot water dip (50 °C, 5 min.) and 5 Krad of gamma rays reduced the incidence of SER during ripening and even during post-climacteric stages (Padwal Desai *et al* 1973). The efficacy of this combination was tested against artificial inoculation and it was found that the combination treatment was most effective and comparable to dipping of fruits in hot water (50 °C) plus benomyl (0.1%) for 3-5 min. *In vitro,* studies with isolated pathogen also corroborated the above findings and combination treated fruits were comparable to untreated fruit at their best in respect of organoleptic qualities (Thomas and Desai, 1976). Gamma irradiation (1000 Gy) which is aimed at eradication of fruitfly infestation of mango fruits, improved the efficacy of hot water against both SER and anthracnose (Spalding and Reeder, 1986).

Combination treatments with radiation at 200 Gy and hot water (53 °C) plus hot imazalil (0.1%) at 53 °C for 3 min. were more effective than single treatment for the control of anthracnose (*Colletotrichum gloeosporioides*) and SER (*Diplodia natalensis* or *Phomopsis citri)* of Tommy Atkins mango fruits in Florida (Spalding and Reeder, 1986). However, unheated water + radiation (750 Gy) or unheated imazail + radiation were ineffective against SER (Spalding and Reeder, 1986). Hot-benomyl treatment immediately followed by irradiation provided effective control of SER (*D. dominicana*) during short term storage at 20 °C. For 15 days. Satisfactory control of SER was achieved during long term controlled atmosphere storage with hot benomyl treatment followed by prochloraz and irradiation. Effects of fungicide or irradiation treatments were additive to HWT. Fungicide or irradiation individually were ineffective (Johnson *et al* 1990).

Disease Control by Oils

Fixed oils act both as protectants and therapeautants in preventing fungal decay of perishables. Many workers earlier reported oil emulsion as effective control agent against post-harvest decays of fruits and vegetables (Aulakh and Grower, 1969., Grower and Aulakh, 1968., Sumbali and Mehrotra, 1980).

Paper wrapping impregnated with oils, organic compounds (Ekert and Kolbezon, 1962), copper compounds (Butler, 1957) and ionized wraps (Tomkins, 1936) provided satisfactory results.

Mustard and castor oils check the rot completely at 75% concentration as pre-and post-inoculation dips *in vitro*. While linseed and Mohua oils could check 96% and 95.85% fruits respectively as pre-inoculation dips (Pandey *et al* 1983).

Healthy fruits were preserved from infection of SER by wrapping the fruits in paper impregnated with mustard oil emulsion (50%) and castor oil emulsion (75%) etc. Linseed and mustard oils showed maximum prevention of decay (93.2% and 95.7% respectively) as compared to fruits wrapped with untreated papers. All the fruits treated with mustard oil emulsion developed better colour, treated fruits were firm and had a pleasant flavour, while in case of Mahua oil, a slight unpleasant smell was experienced. It is thus evident that mustard and castor oil emulsions, either directly applied or impregnated in wrapping paper could protect the mango traits from SER (Pandey *et al* 1983).

Fruit coating with mustard oil, groundnut oil, castor oil, coconut oil and vegetable oil were effective in the management of *B. theobromae* SER of 'Kesar' mango fruits. Except mustard oil, however all other oils produced phytotoxic effects as fruit skin. Mustard oil treated fruits retained their original colour up to 14 days without the incidence of diseases (Jadeja and Vaishnav, 1998). Mustard oil is free from human health hazards and did not encourage the fungicidal resistance in the causal pathogen (Chandra and Pathak, 1989).

In pre-inoculation tests, mustard oil was effective up to 5 days and as post-inoculation treatment it was effective up to 8th day of inoculation. Pathak *et al* (1971) and Rao of and OmPrakash (1983) reported the effectiveness of mustard oil against Rhizopus rot and Black mould rot.

Waxol

Waxol immersion was more effective in controlling Diplodia SER disease of post-harvest mango fruits (Chandra and Pathak, 1989).

In pre-as well as post-inoculation tests waxol was most effective treatment on 5th day of inoculation. A big advantage of using waxol could be that the treatments do not carry any hazard to human health. Further it does not encourage the fungicides resistant strains of pathogen (Chandra and Pathak, 1989).

Botanicals

All the plant extracts tested at all concentrations were not effective in reducing SER on mango fruits. In most of the treatments, the incidence of the disease was on par with the control except in onion bulb extract (10%) and ginger rhizome extract (5%). In these treatments, significant increase in DDI over control was observed. The disease appeared between 7 and 9 days after storage at room temperature; except in the treatments of onion bulb extract (10%) and garlic bulb extract (10%) and Ocium leaf extract (5%) wherein the disease appearance was noticed from 5th day onwards (Gopi, 2003).

Biological Control

The high cost of agrochemicals, development of fungicidal resistance in target pathogen and hazardous effects of fungicides on the ecosystem and harmful residues of the fungicides in food led to the development of alternate methods of control of diseases, of which one is "biological control". Biological control agents are well known to avoid the adverse effects and effective against the post-harvest pathogens without being hazardous to environment and leaving harmful residues in the food. Control of SER resulting from pre-harvest infection is difficult with post-harvest treatments. Post-harvest losses in mango due to post-harvest diseases have been reduced mainly through fungicides and to a lesser extent through management practices. However, the development of pathogen resistance to fungicides and harmful effects of fungicides to human health and environment led to search for alternate control methods of post-harvest diseases. Biological control has emerged as one effective strategy to combat major post-harvest decays of fruits including SER.

Patil (1992) reported the antagonistic nature of *Trichoderma* sp. against fruit rot incited by *L. theobromae* and *Rhizopus arrhizus* in mango. Moreno and Paringbataon, (1995) reported on the biocontrol of *Diplodia* SER of mango by the antagonistic activity of fungus – *T. viride*. Its anlagonistic activity was enhanced by higher concentration at lower pressure of the pathogen.

Disease incidence was reduced following treatment with yeast. Moreno *et al* (1997) and Nigam and Sharma (2000) successfully utilized yeast isolates, *Candida maritime* and *Debryomyces hansenii* respectively against mango SER.

Bacillus sp which are dominant colonizers of the phylloplane and fructoplane, control latent infections on mango (De Jagar *et al* 2001). Spray applications of *Bacillus* on fruits have produced promising results to check latent infections carried from the orchard. The effectiveness of these *Bacillus* isolates can be increased through manipulation of their mode of application (Janisiewiez and Kortsun, 2002). Okigbo and Osuindo (2003) found that *B. subtilis* isolated from soil under mango tree, inhibited *L. theobromae* by 61% on agar plates *in vitro*. The effect of fructo plane *Bacillus* sp in reducing SER disease of mango fruits was investigated (Jadeja and Bhatt, 2008). Four *Bacillus* sp isolated from mango fruit (cv. Kesar) surface were tested against *L. thiobromas* on artificially inoculated mango fruits, using two methods of injury (Cork wound and Pin prick). In cork wound method bacterial application proved better and *Bacillus* isolate B was most effective. Fruit rot reduction under pin prick injury method was higher than cork-wound method. The application of any of 4 *Bacillus* sp, isolated from fruits resulted in more than 50% reduction of natural incidence of SER (Jadeja and Bhatt, 2008).

However, none of the bioagents (fungi, yeast, bacteria) was effective in controlling SER of mango on inoculation in the laboratory. In all the treatments, the disease appeared 9 days after storage as in the case of untreated control (Gopi, 2003).

Storage

Stem-end-rot can also be avoided by adopting suitable and different methods of storage. Low pressure storage, cold storage and modified atmosphere storage are aimed not only to extend shelf life of perishables, but also at preventing the perpetuation and spread of the diseases. However, mere refrigeration of infected fruits helps little as the disease continues to advance at low temperature, but at a slightly reduced pace (Pathak and Srivastava, 1968). SER becomes a serious problem in untreated fruits after 3-4 weeks of storage at 13 °C (Spalding and Reeder, 1972).

The success in delaying decay of mangoes (cvs. Tommy Atkeins, Maya, Haden and Keitt) by using controlled atmosphere (CA storage) has been recorded in Israel (Sive and Resnizky, 1985). When preceded by chemical treatment, mangoes stored for one month under modified atmosphere storage (2% CO_2, 6% O_2, 13 °C) showed very little damage due to SER and anthracnose decays (Sangchote, 1989). During long term storage, controlled atmosphere storage of 'Kensington Pride' mangoes (5% O_2, 2% CO_2, 13 °C) for 26 days followed by air at 26 °C for 11 days in controlling the common storage disease was successful with hot benomyl followed by prochloraz. These treatments also achieved the highest level of marketeable fruits in Australia (Johnson *et al* 1990). Where SER, due to *L. theobromae* predominates, storage at 13 °C offers potential to reduce the decay and resultant losses (Johnson *et al* 1991a). When fruits are placed in cold storage for longer periods, they must be treated with a fungicide (Ploetz, 2003). SER (*L. theobromae*) was controlled in harvested mango fruits up to 4 weeks by fungicidal dipping followed by storage at 13 °C in Australia (Muller and Bart, 1989).

Combination of Pre-and Post-harvest Measures

Since the disease originates in the field and extends after harvest, it was thought wise to combine pre-harvest and post-harvest measures to control the disease. Both types of treatments i.e. pre-harvest sprays of fungicides and post-harvest dips were successful. Pre-harvest application of copper oxychloride and post-harvest application of prochloraz as dip treatment reduce SER in mango. Pre-harvest and post-harvest dip treatment with Bavistin (1000 ppm) and Diathane-M-45 (2000 ppm) were eqally good against SER (Kapse *et al* 2004). Dipping of fruits (cv. Baneshan) from trees sprayed with fungicides in cold water was found to be an effective technique for the control of post-harvest diseases of mango fruits. Pre-harvest application of carbendazim (0.1%) thrice at fortnight intervals starting from 45 days before harvest followed by fruit dip at harvest, resulted in lower disease percentage and higher decay reduction index (RDI) of *D. natalensis* as compared to that of thiaphenate methyl, aureofungin, captan and mancozeb (Narsi Reddy *et al* 2004).

Recommendations

Based on critical studies on stages and mode of infection, following preventive and management measures are suggested in India for avoiding SER of mango fruits (Pathak and Srivastava, 1968., Pathak, 1970., 1973., Srivastava, 1972).

Pre-harvest Measures

1. Adopt orchard sanitation strictly by destruction of sources of inoculum.
2. Adopt pre-harvest field sprays of appropriate fungicides like carbendazim, thiophenate methyl or copper oxychloride.
3. Harvest fruits at appropriate stage of maturity promptly and timely; avoiding harvesting immature and over mature fruits.
4. Use modern harvesting aids to avoid injuries during harvesting of fruits.
5. Harvest of fruits on a clear, cool and dry day.
6. Harvest fruits with sufficient length of the pedicel (8-10 cm).

Post-harvest Measures

1. Take care to prevent snapping of the fruit pedicel at any stage of handling the harvested fruits, smear the cut end/pedicel scare with appropriate disinfectants/ fungicides to avoid the ingress of pathogen.
2. Cover the harvested fruits, to avoid exposure to sun and contaminated orchard atmosphere and shift the fruits from the orchard to pack houses/ripening houses immediately after harvest.
3. Avoid injuries to fruits at all stages of handling.
4. Avoid contact of fruits with soil either during harvesting or desapping.
5. Disinfect the packing yards, packages, cushioning or wrapping materials, ripening chambers, vehicles to avoid infections.
6. Avoid using mango leaves etc. as packing/cushioning materials to avoid infections from them.
7. Cover the fruits individually with plastic covers etc. to avoid spread of the disease by contact.
8. Adopt post-harvest dips of fruits in hot water (52 ± 1 °C) supplemented with carbendazim or thiophenate methyl for 15 min or borax at 43 °C for 3 min.

BLACK MOULD ROT

Black mould rot (black rot, Aspergillus rot) though reported from other mango production countries, is of particular importance in India, where it causes huge losses in long distance transit involving high temperature and rough handling of fruits (Srivastava *et al* 1965., OmPrakash and Srivastava, 1987). Black mould rot of mango fruits incited by *Aspergillus* sp., a fungal pathogen., is a common disease both in the field and storage conditions (Chakravarti, 1982). Banerjee *et al* (1934) first reported this fruit rot due to *Aspergillus* sp. in India, while Dasgupta and Bhatt (1946) reported as latent infection carried from the field. Verma and Kamal (1957) made a first report of this disease in nature from Lucknow and it was predominant among all the storage fungal diseases of mango in Lucknow markets (Raoof and OmPrakash, 1983). It is a more common and serious disease during transit, storage and marketing processes.

Geographical Distribution

Besides India, the black mould rot (BMR) was reported on mango fruits from the Philippines (Clara, 1927), Australia (Baker and Wardlaw, 1937) and Venezuela (Arya, 2004). In India, it was reported from all the mango growing states, especially from Andhra Pradesh and Karnataka (Sohi *et al* 1973).

Losses

Huge losses due to BMR in mango have been reported in India. Long distance transport of South Indian mango cultivars (Baneshan, Totapuri etc.,) to North Indian markets by rail or road without care resulted in heavy losses of fruits (Raoof and OmPrakash, 1984). Laxminarayana and Reddy (1977) estimated a loss of 6.5% of fruits due to *Aspergillus niger* in Warangal markets, while Raoof and OmPrakash (1983) observed 10% loss in Lucknow market. Survey of different mango markets indicated huge losses, while survey of processing units in Gujarat over two years indicated 7% incidence of black rot in mango (Jadeja and Vaishnav, 2000).

The BMR was responsible for about 25% loss in several mango varieties (Verma and Kamal, 1951). Raoof and Om Prakash (1983) observed 10% loss in case of 'Dashehari'. The disease caused 20-22% loss in 'Totapuri' and 'Langra' varieties in Allahabad and 35% loss in 'Safeda Lucknow' variety at Lucknow (Arya, 2004).

Symptoms (Photo 10.10)

Disease symptoms due to *Aspergillus* sp. have been described both from orchards/ markets as well as under artificial inoculation conditions in the laboratory. Symptoms show variation due to different species of *Aspergillus* involved.

Usually the basal part of the fruit is most affected but cases have been observed when other regions like stem-end, but never at the tip of the fruit (OmPrakash and Srivastava, 1987). Stem-end-rot due to *Aspergillus niger* is quite distinct from the rots caused by other organisms. The disease also appears on bruises/injuries in which case rot can start at injured portion of the fruit. The incidence is further aggrevated by improper handling of fruits during harvesting, post-harvest, transit and storage levels. Frequently fruits are infected while they are still on the tree due to wounds. Thus mango fruits may be infected before or after harvest. Semi-mature fruits are more susceptible to decay during storage than fully matured fruits. Fruits are more prone to decay during storage (Pathak, 1980).

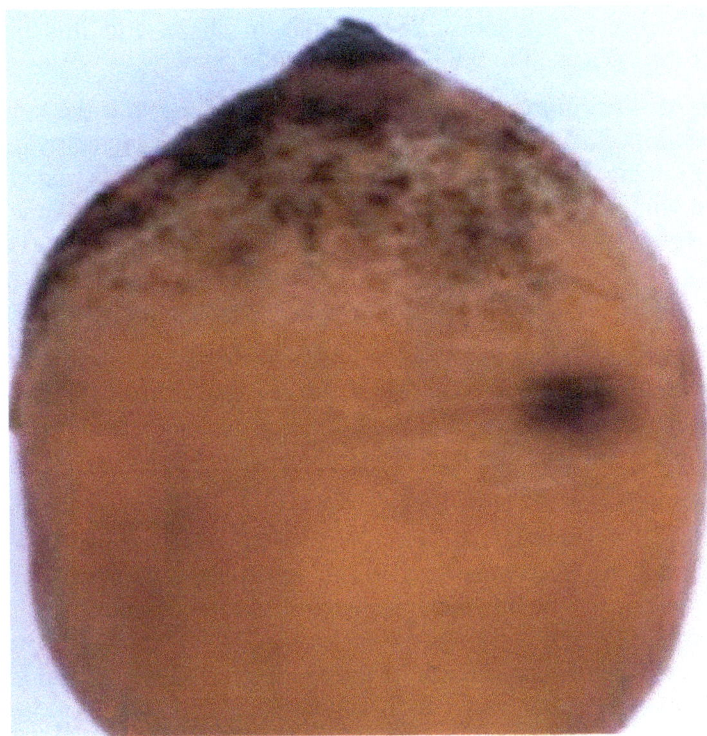

Photo 10.10: Black mould rot

In India, the fruits infected by *Aspergillus niger* externally show characteristic yellowing at the base (stalk-end) with hazy, dull gryeish spots, which later coalesce into large black necrotic areas of 1.0-1.5 cm diameter. These spots show copious growth of black mould with depressed lesions and soft mesocarp (flesh). In the beginning light brown circular spots develop on the fruit. In 3-4 days, the infected portions become depressed and later the fruit shrinks and the infected portion of the flesh becomes soft underneath the lesions, while remaining portion is quite hard. Sclerotial bodies are produced on mature lesions, which are usually white and later turn into dark brown. In advanced stages, copious growth of fungal colonies develop and the diseased fruits emit foul odour (OmPrakash and Srivastava, 1987). Besides softening into pulpy mass, the mesocarp assumed brown colour and showed the presence of hyphae and spores. Sangchote (1987) from Thailand also reported similar symptoms with *Aspergillus niger* infection from harvested fruits collected from different locations.

Sometimes, small spots also appear near the stemend and become large and dark black when the fruit ripened completely. At severe stage of the disease, entire surface of the fruit is covered by small black spots with black spore mass (Bagwan, 2010).

Besides causing stem-end-rot, *Aspergillus niger* also causes later rot, tip-end-rot or rot at any other places. In these cases, the place of lesions depends on the position of injury on the fruits. Sometimes, the spots are irregular in shape and this might be due to different types of injury caused to the fruit (Arya, 2004).

The infection of *Aspergillus flavus* also starts at stemend, but the side of the fruit is rarely involved. Similar to BMR caused by *A. niger* the fruit infected by *A. flavus* showed characteristic yellowing with irregular, dull greyish spots, which developed into dark necrotic areas showing copious growth of black mould with depressed and soft mesocarp. In advanced stages, rotting sets with characteristic foul odour (OmPrakash and Srivastava, 1987).

Fruits collected from markets, affected by *Aspergillus niger* (BMR) and *A. flavus* (black blotch) produced larger sunken, light brown necrotic areas on the outer surface of the fruits. The internal tissue became soft and emitted foul smell. Black conidial heads of *A. niger* and light yellow green conidial heads of *A. flavus* were observed on the infected fruits (Laxminarayana and Reddy, 1977., Bhuvaneswari, 1999).

Similar symptoms were developed on mango fruits collected from fruit markets of Tirupati (Srikanth, 1989), Hyderabad (Madhavilatha, 1997., Gopi, 2003) and Krishna district, Andhra Pradesh (Bhuvaneswari, 1999). Inoculation of mango fruits incited the symptoms as yellowish patches, which increased in a regular manner. Later the infected fruits became soft and were covered with black mass of spores. Mostly the basal part of the fruit was affected, while other parts of the fruit were hard and unaffected (Srikanth, 1989., Madhavilatha, 1997., Bhuvaneswari, 1999., Gopi, 2003).

Mango fruits infected with BMR on inoculation with *A. niger, A.flavus* and *A. terreus* exhibited spore mass in the sunken regions with foul smell (Srikanth, 1989). The symptoms were similar to those described by Srivastava *et al* (1965), Laxminarayana and Reddy (1977), Pathak (1980) and OmPrakash and Srivastava (1987), both under market and laboratory conditions.

According to Arya (2004), the disease first appeared as a large brown coloured patch on the fruit surface at the stalkend. The intensity of colour of infected area depended upon the variety of the fruit and the stage of the disease. The lesions grew in a regular manner and formed a large circular spot encircling the stalk. After 3 or 4 days, the infected region became sunken and at an advanced stage, it crumpled. The fruit rot was soft in nature. Black conidial heads of the causal organism became distinct after 5 or 6 days of initiation of the symptoms. The stem-end-infection resulted in pre mature fruit drop.

Changes

Black mould rot infection caused rapid decrease in ascorbic acid content in diseased fruits of 'Dashehari' and 'Langra' (Ghosh *et al* 1966) and disappearance of sucrose in inoculated fruits (Tandon, 1970). The development of black spots due to *A. flavus* was accompanied by high levels of citric acid, total acidity and sucrose. A 3-fold decrease in total sugars was observed in infected fruits. These changes were associated with a considerable decrease in amylase and invertase activities. Decrease in the invertase activity and utilisation of the reducing sugars for the fungal growth might have reulsted in a decrease in the content of reducing sugars in the infected fruit tissues (Majumdar and Modi, 1980). Succinic acid and an un-identified organic acid appeared in *A. niger* infected fruits (Singh, 1968., Bagwan, 2010). Calcium content was significantly reduced in all the mango varieties (Kesar, Alphonso, Baneshan, Dashehari and Local) due to inoculation with isolates of *A. niger*. The level of decrease varied with the variety and isolates. The decrease in calcium content was higher 'in Dashehari', 'Alphonso' and 'Kesar'. The decrease in calcium content was attributed to its utilization by the pathogen as a substrate (Gadgile and Chavan, 2009).

Health Hazards

Black mould rot diseases pose a threat to the public health as the pathogen, secretes some mycotoxins on the fruit surface, which cause various ailments in human beings and animals. The mycotoxins commonly encountered after *Aspergillus* infection are aflatoxins and tremorganic toxins. Aflatoxins are responsible for liver cancer in tropical countries like India. Tremorganic toxins produce marketed body tremors followed by convulsive seizures, which are fatal. Even refrigeration can not prevent the production of aflatoxns and tremorganic toxins in the infected fruits (Chakravarthi, 1982).

Varietal Susceptibility

Sinha (1948) observed that *A. niger* fungus was capable of infecting fruits of 'Bombai' and 'Safeda' varieties of mango and out of the two 'Bombai' was more susceptible at all stages of maturity. Whereas, *A. nidulans* preferred 'Safeda' while 'Bombai' was resistant. Varieties 'Dashehari', 'Amin', 'Mohanbhog', 'Tukkhaja', 'Gola', and 'Safeda' are usually affected, the last two are particularly susceptible (Verma and Kamal, 1951). *Aspergillus* rot was observed on 'Alphonso'; 'Pairi' and 'Totapuri' in Gujarat (Patel, 1972) of which 'Totapuri' was highly susceptible (Patel 1972., Pandey and Mohammad, 1976). Cultivars like 'Sethi,' 'Bombai,' 'Langra,' 'Dashehari,' 'Totapuri,' 'Safeda' and 'Chausa' were susceptible to infection by *A. niger*. Infection on 'Langra' was more severe than on others (Bhargava and Singh, 1975). 'Krishnabhog' varietywas severely damaged by *A. niger* in the orchards of Krishnanagar in West Bengal. Besides being badly rotten, the infected fruit has become mummified on account of loss of water. Such fruit remained attached to plants and thus served as source of inoculum during the ensuing season. It is thus clear that aspergillus rot persists in the garden and appears quite early and damages the fruit in orchards as the pathogen remains dormant and becomes responsible for post-harvest decay at a later stage (Arya, 2004).

Garcha and Singh (1980) from Ludhiana, recorded the association of *A. niger, A teru* on 'Dashehari' and 'Totapuri' and *A. flavus*, *A. fumigatus* and *A. niger* with 'Langra'.

Etiology

Black mould rot in mango is caused mainly and mostly by *Aspergillus niger* Van Tiegh, but other species of *Aspergillus* were also isolated from the decaying fruit, whose pathogenicity, however, was not established. Several workers reported `*A. niger'* as the causal organism of black mould rot in mango in India (Banerjee *et al* 1934., Das Gupta and Bhatt., 1946., Sinha, 1948., Patel, 1972., Majumdar and Modi, 1980., OmPrakash and Srivastva, 1987., Srikanth, 1989., Madhavilatha, 1997., Bhuvaneswari, 1999., Gopi, 2003.) from different parts of the country. Das Gupta and Bhatt (1946) observed the presence of *A. niger* on the outer surface of the mango fruits in different stages of maturity. BMR on 'Pairi', 'Alphonso' and 'Totapuri' mango fruits in Gujarat was mainly due to *A. niger* (Patel, 1972). *A niger* was considered to be the principal causal organism of internal rot of mango fruits in Lucknow (Raoof and OmPrakash, 1983). Rath and Mohanan (1986) found the involvement of either *A. niger* or *Colletotrichum gloeosporioides* with black irregular lesions at the stalk end of fruits collected from the orchards. However, Garcha and Singh (1980) claimed that mango fruit rot in Punjab was due to *A. niger*, *A. fumigatus* and *A. terreus*. Thus, it was well established that the principal causal pathogen of black mould rot of mango fruit is *A. niger*, particularly in the rainy season in India (OmPrakash and Srivastava, 1987). Spoilage of mango fruits by *A. flavus* was reported by Mujumdar and Modi (1980) who observed increased susceptibility of fruits with the ripening.

Disease Cycle

The fungus persists in the soil as a saprophyte on decaying plant material and its spores are air-borne and are blown up on to the fruits. Infected plant matrial on the trees (twigs, inflorescence stalks) also serve as primary source of inoculum. Srivastava *et al* (1965) isolated *A. niger* from soil as well as dead or decaying twigs or leaves of different mango varieties (Dashehari, Safeda Lucknow, Mekhan, Taimurthy and Langra) from Lucknow, Allahabad and Varanasi. The infected and mummified fruits having black spores also serve as source of primary inoculum for the next season, as the fungus can perennate for a long time saprophytically. These are the potential sources of inoculum in the field. Packing houses and storage chambers also harbour the inoculum causing post-harvest infections (Patel, 1972). *A. niger* also grows parasitically on different plant parts particularly on fruits.

Most infections occur during and after harvest via wounds caused by crawling of insects or by pecking of birds or during harvesting and post-harvest handling of fruits, from where the hyphae transverse into the stalk region (Dodd *et al* 1997). Injury to fruit is necessary for infection and development of the disease besides favourable environmental (temperature and relative humidity) conditions. Injured fruits are more infected than uninjured ones.

Epidemeology

Environmental conditions like, temperature and relative humidity play an important role in the epidimeology of the black mould rot disease in mango. Variation in temperature and relative humidity was attributed for variation of the disease incidence and extent in different parts of the country (Bhargava *et al* 1965).

The heat generated due to accelerated rates of respiration and high temperature in transport containers and godowns help the incidence and spread of the disease. The development of the disease is favoured by warm temperature (30 °C) in transit and storage (Dodd *et al* 1997). Laxminarayana and Reddy (1977) observed severe incidence of BMR disease at a temperature of 22-30 °C and relative humidity of 90-100% under tropical conditions. Infection of the disease, its development and appearance of symptoms were optium at temperature around 30 °C. Fruits stored (incubated) at 30 °C and 100% relative humidity showed highest disease intensity. Rotting was least at 30% relative humidity, but increased with increase in relative humidity.

Neither activity, spore germination, nor symptoms expression of the pathogen was noticed at 10 °C (Prasad and Sinha, 1981., Bagwan and Yeole, 2003). Spore germination did not occur at 10 °C even after 24 hours of inocubation, but there was an increase in spore germination with time and temperature. The highest spore germination was observed at 30 °C and 100% relative humidity when fruits were incubated for 24 hours.

The above observations suggest that the storage of fruits at low temperature (10 °C) and low relative humidity (30%) helps in reducing post-harvest BMR of mango fruits (Bagwan and Yeole, 2003).

Control/Management

Black mould rot of mango may be controlled by either pre-harvest sprays of fungicides and post-harvest treatments with fungicides, plant extracts, hot water treatments, biological control etc. The aim of pre-harvest control/management is to prevent the infections, whereas the objective of post-harvest treatments is both prevention of infection and prevention of the fungus already seated in the injuries/fruits.

Pre-harvest Management

In practice, the diseases are prevented with pre-harvest fungicidal sprays (Subramanyam *et al* 1972). However, cultural management practices may also be followed for prevention of the incidence of diseases.

Harvesting of fruits is a crucial event, which often causes cuts, scratches, bruises and wounds if proper care is not taken during harvesting operation, which leads to disease infection as wounded fruits are more prone to infections. Therefore, greater care is necessary in harvesting and subsequent handling, transport and storage of fruits so as to avoid mechanical injuries and sap burn damage to the fruit.

Post-harvest Management

Since the wounds occur on fruits during and after harvest it becomes imperative to take up post-harvest management measures to prevent fresh infection and also to check further growth and development of pathogen on the fruit. This is achieved by different approaches such as physical, chemical and biological control methods.

Physical Methods

Hot water Treatment

Hot water treatment (HWT) is one of the earliest methods of disease control of fruits and vegetables. HWT, with or without fungicides is employed for the control of post-harvest diseases as well as for regulating ripening. HWT also delays the onset of rots. HWT at 55 °C for 5 minutes delayed the incidence of BMR for 6 days (Pathak and Shekhawat, 1976). Combination of hot water and fungicides (benomyl) was effective to keep the disease under check (Pandey *et al* 1980). Aerated steam treatments at 60 °C for 15 or 20 minutes completely inhibited the development of rot (Vyas and Pathak, 1995).

During the course of HWT, it is very important to maintain suitable temperature and duration of exposure/treatment, since high temperature under long exposures results in injuries to the fruit, which enable the pathogen to enter the fruit through injuries.

Radiation

Efforts were made to reduce the incidence of BMR of mangoes through exposure to radiation. All the radiation treatments resulted in the reduction of disease incidence in treated fruits. The incidence was least when the fruits were exposed to ultra-violet radiation for 25 min. The disease appearance was delayed by 11 days when mango fruits were exposed to ultra violet radiation while fruits without exposure to radiation showed disease incidence with in 5 days. The exposure of fruits to ultra violet and infra red radiation for 20 min. was equally effective and better (Gopi, 2003).

Coatings

Coatings of fruits with wax etc., act as protective barriers against the invation of pathogens. Krishnaiah *et al* (1986), applied 4 different food grade fruit coatings (D2-202, D2-251, D2-261 and FYK-241) against *A. flavus* and found them effective in reducing the incidence.

Chemical Methods

Chemicals are applied in different ways to harvested fruits like dipping, sprinkling and spraying or in hot water, of which dipping treatments are found to be easy and economical.

Dipping of harvested fruits, prior to transport or storage in benomyl (1500 ppm) controlled the disease (Jacob *et al* 1973., Bhargava and Singh, 1975). Delan (1250 ppm) or Triforine (100 ppm) dip-treatments could control the disease (Pandey *et al* 1980). Tecto 40 (a formulation of TBZ) was effective at 4000 ppm in controlling the BMR, but it deteriorated the flavour and taste of the fruit, hence not recommended (Arya, 2004).

In order to find out an effective fungicide for the pre-harvest prevention as well as post-harvest management of the disease, attempts were made under laboratory conditions. Fungicides were used as pre- as well as post-inoculation treatments. Tests were also made with wound, inoculated and unwounded inoculated fruits.

The most effective pre-inoculation treatment of mature green fruit was with TBZ, while post-inoculation treatment was with Difolaton (Pathak and Shekhawat, 1977). Srikanth (1989) also tried 12 fungicides against BMR under laboratory conditions on wounded, inoculated and un-wounded inoculated mango fruits. Alitte (1000 ppm), Bavistin (1000 ppm), Bavistin (1000 ppm) + Waxol (3%), Benomyl (1000 ppm) in hot water at 55 ± 2 °C and Captan (2000 ppm) were effective resulting in 100% inhibition of *A. niger* while Aureofungin – Sol (500 ppm) caused 96.67% inhibition, followed by Diathane-M-45 (87.4%), but Blitox (2000 ppm) was ineffective in case of wounded inoculated fruits. In case of unwounded inoculated fruits, Bavistin (1000 ppm) + Waxol (3%), Benomyl in hot water (1000 ppm at 55° ± 2 °C), Aureofungin-Sol (500 ppm) and Captan (2000 ppm) exhibited disease index (D1) of 1.0 Benomyl (1000 ppm), Blitox (2000 ppm) showed a D1 of 8.0 against 10.0 D1 in control inoculated fruits. The range of D1 was 3.33-10.00 with Alitte (1000 ppm), Aureofungin – Sol (500 ppm), Captan (2000 ppm), Diathane-M-45 (2000 ppm), Waxol (6%) and HWT (55 °C ± 2 °C). These findings of Srikanth (1989) indicated that the mango fruit (unwounded) treated with Bavistin (1000 ppm) + wax (3%) remained free from BMR caused by *A. niger*. BMR *in vitro* was effectively reduced by Benomyl (1000 ppm) followed by carbendazim (1000 ppm) and the appearance of the disease was delayed by 14 and 11 days respectively due to the above fungicides, as against 5 days after storage in untreated control fruits (Gopi, 2003).

Borax (10%) provided good control against rot caused by *Aspergillus* sp. and *Penicillium* sp in mango (Garcha and Singh, 1980). Borax (1000 ppm) was effective in reducing the D1 in case of wounded inoculated fruits (Srikanth, 1989).

Frutox plus Benomyl, fruitox plus Bavistin and Paraffin wax plus Benomyl were the most effective treatments against BMR followed by frutox alone and sodium diethyl dithiocarbomate (OmPrakash and Raoof, 1988).

All the chemical sterilants, except salicyclic acid (0.4%) were effective in reducing the incidence of BMR in mango. The lowest incidence was in calcium chloride (4%) treated fruits followed by potassium permanganate (1%) dip. The disease appeared in treated fruits between 11th and 14th day, whereas the disease made its appearance by 5th day of storage in control (Gopi, 2003). The above sterilants showed least residual toxicity, with the increase in storage period, hence ineffective in reducing post-harvest diseases at later stages of storage. Calcium chloride caused phytotoxicity on treated fruits (Tirmazi and Wills, 1981., Gopi, 2003).

Plant Extracts

To avoid the pesticidal residues in fruits, plant extracts have been tried to control pests and diseases including post-harvest diseases of fruits and vegetables. The efficacy of Ocium (Tulsi) (*Ocimum sanctum*) leaf as well as seed extract at 80% concentration in controlling *Aspergillus* rot was demonstrated. Of the two, seed extract showed least severity of the disease (Vyas and Pathak, 1995). Gopi (2003) also attempted the control of *Aspergillus* rot in mango using different plant extracts and observed significant differences in reduction of BMR. Onion bulb extract (5 and 10%), Ocium leaf extract (5%) and ginger rhizome extract (10%) showed maximum inhibition in the growth of the pathogen. The incidence of rot was delayed by 9 to 11 days by all the plant extracts screened, while untreated control fruits showed disease in 5 days after storage. On 20th day of storage, onion bulb extract (5%) resulted in lowest D1 (5.20) against D1 of 26.04 in untreated control fruits. It was also observed that Ocium leaf at 5% concentration was effective in reducing the incidence of BMR when compared to 80% concentration of Ocium leaf extract (Vyas and Pathak, 1995).

Oils

Smearing the harvested fruits was also demonstrated as effective measure to ward off the post-harvest diseases (Raoof and OmPrakash, 1983., Vyas and Pathak, 1995). Mustard oil (pure) gave maximum control of decay (90.5%) followed by Dalda (88.76%), groundnut oil (87.19%), coconut oil (82.33%) and linseed oil (74.04%), which indicated the possibility of using oils from plants instead of harmful synthetic fungicides for prevention of post-harvest rots of fruits (Raoof and OmPrakash, 1983). Bavistin (1000 ppm) and a mixture of mustard oil plus Bavistin (1000 ppm) at 30 °C proved most effective against *Aspergillus* rot of mango (Vyas and Pathak, 1995).

Biological Control

Of late bioagents (yeast, fungi, bacteria) are being tried against diseases including post-harvest diseases of perishables in the wake of conciousness of health hazards due to the use of chemicals to prevent/control the diseases.

Sharma *et al* (1995) attempted biological control of BMR with yeast (*Debromyces hanseii*) isolated from citrus fruits, which showed strong antagonistic activity against the disease *in vitro*. *Trichoderma viride* was effective against the BMR in mango. A mixture of mycellial extract and culture wash of antagonists with Bavistin (500 ppm) proved best against BMR (Vyas and Pathak, 1995).

All the bioagents (*Trichoderma viride, T. harzianum* and *T. virum*) were equally effective in reducing the incidence of BMR, least reduction of the disease being with

T. harzianum. Bio-agents delayed the incidence of the disease by 11 days in all the treated fruits against 5 days after storage in case of untreated fruits (Gopi, 2003).

Storage

One of the aims of storage is to avoid fruit rots during storage period. This is generally achieved by placing the fruit at low temperature. Storage of mango fruits at 10-15 °C prevents the disease from developing, particularly in case of latent infections (Dodd *et al* 1997).

However, packing materials often serve as sources for disease infections. Therefore, there is every need to sterilize the packing material with protective chemicals to make them free of pathogens. Packing materials in cold storage rooms are sterilized by either 5% aqueous solution of hysol or formalin (2%) solution to get rid of surface contaminants (Chakravarthi, 1982). Unfortunately the extent of protection offered by these fungicides was neither adequate nor consistent. Moreover, the fungicides, themselves are not free from toxin risk. Obviously a non-toxic fungicide is needed in such cases, as the mango fruits are consumed fresh. Mangiferin (1, 3, 6, 7 – tetrahydroxy Xanthone – [2-B-D glucoside] minimizes the disease, when the fruits are soaked for 3-4 hours at 1.5 g/l. The treatment did not show disease symptoms up to 4 days even when a heavy dose of fungal spray was administered on a mechanical injury on the fruit immediately after harvest. In case of longer duration of storage, occasional sprinkling of mangiferin solution after 3-4 hours of soaking would keep the fruits safe and healthy. As mangiferin is a natural constituent of mango and has a very low order of toxicity, the soaking of mangoes in a much diluted solution of mangiferin for 3-4 hours neither affects the quality of the treated fruits nor leaves any residual toxicity in the treated fruits (Chakravarti, 1982). However, no further attempts were made to confirm the above.

The usual practice to minimize the spoilage in bruised fruits is to sprinkle fungicides such as SOPP (1%), Flit 406 (0.2%) or borax on the surface of the fruits prior to storage (Chakravarti, 1982).

PHOMOPSIS ROT

This is another post-harvest rot of mangoes causing severe losses (28-36%) in storage. The pathogen also infects banana, papaya, guava and rambutan of which the last one is most susceptible.

Phomopsis rot on mangoes was reported from Australia, Brazil, China, Cuba, India, Malaysia and United States of America (Johnson, 1994) and of late from Thailand (Ko and Liu, 2009).

Symptoms (Photo 10.11)

Phomopsis rot occurs during storage of fruits. Symptoms differ between the causal pathogens. In case of *Phomopsis mangiferae* the symptoms in storage initially appeared as light to dark brown lesions surrounding the pedicel, slightly irregular in outline, soft, but not watery. At an advanced stage fruit got detached from the stalk and dropped. The infection initiated as a descrete and discoloured areas all over the fruit surface, which gradually turned to dark brown to black finally. Often neighbouring spots, coalesced to form larger spots. The symptoms advanced slowly but eventually penetrated the mesocarp (pulp). After disease advanced, the under laying tissue got roughed and turned to soft juicey mass (Laxminarayana and Reddy, 1975). In advanced stages, the fungus formed abundant picnidia on the infected fruits. Under laboratory conditions *Phomopsis* sp. on inoculation produced large (2-3 cm) brown to dark brown or blakish discoloured patches on the peel at the stalk end of green, unripe or semi-ripe mango fruits (Bhuvaneswari, 1999).

P. multirostrata produced characteristic sunken brown coloured spots, which gradually covered the whole fruit surface. Numerous small pin-head like fruiting bodies (pycnidia) developed on the fruit surface in advanced stages (Arya and Mathew, 1991).

Photo 10.11: Black spot rot due to *Phomposis*

Varietal Susceptibility

The phomopsis rot disease was observed commonly on 'Langra', 'Chausa', 'Bombai', 'Dashehari', 'Himsagar', 'Rataul', 'Safeda' and 'Taimuria' in India (Tandon, 1967a). 'Totapuri' was found to be more susucptible (Laximinarayana and Reddy, 1975). In Thailand, 'Kaew' was highly susceptible, while 'Nam Dok Mai' was least susceptible

(Boonraung *et al* 1993). In Taiwan, 'Keitt' and 'Irwin' were highly susceptible, while' 'Haden' was moderately susceptible (Ko and Liu, 2009).

Etiology

Three species of *Phomopsis* were reported as causal agents of the rot *viz*., *P. mangiferae* Ahmad (Laxminarayana and Reddy, 1975), *P. multirostrata* (Mathur *et al*) Dorenboseh and Boerema (Arya and Mathew, 1991) and *P. amraii* Srivastava *et al* (Tandon *et al* 1965) of which *P. mangiferae* is more common.

Control/Management

Since the disease appears in storage only, the control measures should be directed to prevent the incidence in storage by post-harvest treatments.

Systemic fungicides *viz*., Bavistin, Derosal and Delan were found to be effective of which Derosel at 750 ppm was most effective in controlling the disease up to an extent of 95% (Arya and Mathew, 1991). Use of Bavistin and Benlate at 1000 ppm can control the rot (Arya, 2004).

Dipping of inoculated fruits in carbendazim (750 ppm) or hot water (58 °C) and incubating for 14 hours, checked the rot in storage. Other fungicides (Diathane, Imazalil, Mycobutanil and Prochloraz each at 750 ppm) were slightly less effective. However, hot benomyl (0.05% at 50 °C) was an effective control measure of the rot for two weeks (Boonraung *et al* 1993).

PESTALOTIA ROT

Pestalotia fruit rot was reported on mangoes from Jabalpur (Ratnam and Nema, 1967) and Bangalore (Ullasa and Rawel, 1988) markets and during storage in India. It is a minor post-harvest disease of mango, but causes substantial loss.

Symptoms

The symptoms of the rot as exhibited by different species of pathogen were described by different authors. Apperance of the infection on mature green fruit is characterized by the presence of brown small spots which gradually increase in size, become dark coloured with greyish white centre having pin-head like dots of acervuli. Such a fruit is shed when the disease advances to the stalk-end. Wounding results in quick attack of the fungus and early spreading of lesion (Sarkar, 1960).

Symptoms on a mature fruit include dark brown, circular lesions extending allover the fruit, at advanced stage it assumes dark blackish colouration and sticky fluid comes out commonly on the ripe fruit. Sometimes hyphal growth of the fungus with black pustulues covers the infected fruit.

Pestalotia mangiferae affected fruits show areas becoming olivaceous black and shrinking (Pandey and Mahmmad, 1974-75). Tandon *et al* (1955) reported that the pathogen caused brown spots on mango fruits and was capable of infecting injuried fruits. *P. glandicola* caused water soaked areas around the stemend of the fruits which turned into pale yellow in colour, followed by production of cotton white mycelium, spreading radicals from the pedicel end and subsequent rotting of infected fruit (Ullasa and Rawal, 1988).

On aritifical inoculation of *Pestalotia* sp infected fruit showed small, circular, black lesions on the outer rind of the semi-ripened fruits. These lesions increased in size and coalesced to form large spots measuring 1-2 cm in diameter. Inner pulp turned black and finally rottened (Bhuvaneswari, 1999).

In case of infection of *P. versicola*, decayed portions were olivaceous black, shrank in the central region and fruits became soft and watery (Ratnam and Nema, 1967).

Varietal Susceptibility

'Chausa' is resistant to pestalolia rot. Aritificial inoculation of *P. mangiferae* caused severe rotting of fruits of 'Langra', 'Dashehari', 'Bangalora' and 'Fazli' varieties (Tandon, 1967b).

Etiology (Table 10.5)

Table 10.5: The different species of *Pestalotia* reported as causal agents on mango

Causal pathogen	**Reference**
P. funerea var. *mangiferae*	Uppal *et al* (1935)
P. funerea	Wardlaw and Leonard, 1936
P. dicheta Speg.	Viegas, 1946
P. guepini	Steyaert, 1953
P. sp (*Calonectiria mangiferae*)	Sechet, 1955
P. versicola (Speg.) Stem	Anon, 1974 – 1985
P. glandicola (Cest) Steyent	Ullasa and Rawal, 1988
P. mangiferae (P. Henn) Steyart	Om Prakash and Srivastava, 1987

In addition to *P. mangiferae*, *P. glandicola* is also involved in causing storage rot and leaf disease in India (Ullasa and Rawal, 1988). Their studies indicated that *P. glandicola* is a virulent pathogen of both fruits and leaves and infected them without injury. Ratnam and Nema (1967) reported *P. versicola* causing a rot of mangoes in Jabalpur market, which is a weak parasite causing infection only through injuries. *P. glandicola* was reported earlier from India infecting only living leaves of mango (Sarbhoy *et al* 1968).

Disease cycle and Epidemiology

The infection occurs on leaves throughout the year, which act as the source of primary infection of fruits in the field. The latent infection is carried by mature fruits, which subsequently develops into post-harvest decay during storage. The percentage of decay was highest in August (Ullasa and Rawal, 1988). Moist atmosphere helps in increasing and hastening of infection on fruits. It can cause substantial losses at around 35 °C, temperature (Sarkar, 1960).

Control/Management

Fallen leaves, may be collected and destroyed to avoid as primary source of the disease. Dusting of leaves with zinc sulphate reduced the disease but similar dusting on the fruit failed to control the rot (Tandon *et al* 1955). Spraying of trees with copper oxychloride (0.3%) was effective in controlling disease both on leaves and fruits. Wettable sulphur (0.2%) + Zineb (0.2%) after heavy showers of South West monsoon, followed by wettable sulphur (0.2%) before flowering, carbendazim (0.3%) at pea stage and Zineb (0.2%) before maturation of stone reduced the incidence (Jadeja and Vaishnav, 1984). Of the 5 antibiotics tried, Aureofungin showed greatest inhibitory effect (Sinha and Bilgrami, 1987). Efficacy of homeopathic drugs was tested in controlling the fruit rot of mango by *P. mangiferae*. Lycopodium charatum potency 190 was effective in reducing the post-harvest rot (Khanna and Chandra, 1978). Out of the 5 adjuvants, soap solution was most effective (Khanna and Chandra, 1989). Their results suggested that homeopathic drugs not only inhibit spore germination and growth of fungi *in vitro*, but are effective also in the field. Fruits should be stored at lower temperature to avoid further spread of the disease. Post-harvest dip of fruits, in systemic fungicides prior to placement in storage is also effective in checking the incidence and spread of the disease.

CHARCOAL ROT

This is a minor post-harvest disease of mango, but of late as leaf blight it assumed importance. Avema (1922) from Brazil reported a species of *Macrophoma* causing black rot of mango fruits. Patel *et al* (1949) described a species of *Macrophoma* on branches and fruits of mango from India. The cause of the disease was *M. mangiferae* (Anon, 1953-54). A full account of this disease was provided by Hingorani and Sharma (1956) from Delhi. The disease was subsequently found in Maharashtra (Patel *et al* 1949), Delhi (Patel *et al* 1949, Chakravarty., 1961), Gujarat (Desai and Patel, 1963), Uttar Pradesh (Anon, 1974-1985), Bihar (Chakravarty, 1961), West Bengal (Mandal, 1981), Lucknow (Dhawan, 1982) and Andhra Pradesh (Bhuvaneswari, 1999). From the above account it appears that the disease is slowly spreading throughout the country.

Symptoms

Generally fruits are not found to be infected in the field (Desai and Patel, 1963), but in storage the disease appears as water-soaked, round lesions, which enlarge rapidly to cover the entire fruit surface (Pathak, 1980).

The rot due to *M. phaseolina* appeared from the stalk end and later covered the whole surface of the fruit. Rotting occurs from the pedicel portion, internally shriveled and showed blackning of the internal tissue, which finally bore numerous black oval to irregular small sized pycnidia (Dhawan, 1982). The symptoms of charcoal rot caused by *M. phasiolina 'in vitro'* appeared in the form of large, brown to blakish spots on the outer rind, around the stem end region of the fruit. In advanced stages, the infected fruits developed deep longitudinal cracks exposing the internal pulp (Bhuvaneswari, 1999).

Etiology

Two species of *Macrophoma viz., mangiferae* Hingorani and Sharma (Hingorani and Sharma, 1956)., Chakravarty, 1961., Mandal, 1981) and *phaseolina* were reported as the causal organisms of charcoal rot on mango in India (Dhawan, 1982).

The disease cycle and epidemiology of charcoal rot are not studied and known, except that the infected leaves and twigs in the tree canopy may serve as source of the disease on fruits and the disease extends from the field and appears in storage.

Control/Management

The effect of Burgundy mixture, Perenox, Lime Sulphur and Diathane on spore germination has been tested *in vitro*, indicating the possibility of these fungicides to be utilized for the control of the disease in field (Hingorani *et al* 1960). Since the pathogen survives in the leaves, sanitation by way of removal and destruction of infected parts may be useful in reducing the disease inoculum.

RHIZOCTONIA ROT

Rhizoctonia fruit rot of mangoes is another minor disease reported in India (Wadia and Manoharachari, 1979).

Two species of *Rhizoctonia viz., solani* Kuhn and b*ataticola* were reported as causal pathogens of the rot (Assnami *et al* 1971., Chhatpar *et al* 1975). In Gujarat *R. bataticola* was responsible for about 40 per cent of spoilage in 'Alphonso' in Baroda fruit market (Chhatpar *et al* 1975).

Symptoms

The disease initially appears as an irregular brown spot at the stylar end and spreads towards the stalk end of the fruit. In advanced stages, the rot sets in with abundant sclerotial bodies of the fungus. Infected fruits may appear perfect when seen from above, with a clear line of demarcation between healthy and affected tissues. The tissues become soft as the disease advances. Profuse growth of the fungus with abundant black sclerotial bodies is seen on infected fruits. Black spots appear in partly ripe and ripe mangoes only. Low pH (2.5) of the unripe fruits was considered as one of the causes in the poor development of black spots.

Changes

The development of black spots on the fruit was accompanied by higher levels of starch, ascorbic acid, citric acid and lower levels of reducing sugars and starch. These changes were associated with a considerable decrease in the amylse activity and 2-fold increase in the invertase activity. The fungus may utilize sugars for its growth resulting in a decrease in the sugar content (Assnami *et al* 1971).

Disease Cycle

The infection is probably carried from fruits heaped after harvest with adhered soil over them. The fungus penetrates fruits which happen to be in contact with the ground.

Much is not known about the epidemiology of the rot. The optimum pH and temperature for the growth of *R. bataticola* were 5 and 30 °C (Assnami *et al* 1971).

Control/Management

1. Minimize the contact between fruit and soil by not heaping the harvested fruit on bare soil.
2. Cool the harvested fruits in the orchard, rapidly to check the spread of the disease.
3. Sterilize/disinfect the package boxes, storage rooms and transporting vechicles before putting them to use.
4. Treat the fruits, having field infection with fungicides immediately after harvest as follows :
   (i) A typical method to reduce the infection is to dip the harvested fruits in aureofungin (antifungal and antibiotic) solution (Assnami *et al* 1971).
   (ii) Out of the several chemicals and fungicides tested (aureofungin, copper oxychloride, Zineb, borax, calcium propionate, potassium sorbate and vitamin K). Vitamin-K, borax and aureofungin were most effective in inhibiting the growth of the fungus *in vitro*. When unripe mangoes were dip treated in solutions of aureofungin (10 mg/l for 2 min.) and kept for ripening, the normal colour appearance, taste and flavour were maintained, hence could be employed successfully against the rot (Chhatpar *et al* 1975).

From the above, it is clear that aureofungin is more preferable for the control of rhizoctonia rot.

SCLEROTINIA ROT

This is a storage rot caused by *Sclerotium rolfsii* (Rot) Sacc (*Corticiumnts rolfsii*) or *Sclerotina sclerotiorum* (Lib) De Bary.

The diseae is characterized by light brown patches near the pedicel region of the fruit. These patches are covered with white cottony mycelium, scattered on the entire surface of the fruits. Rotting is seen on the diseased portion which imparts foul smell. Fungus produces black coloured sclerotia scattered on the white mycelial mat on PDA medium (OmPrakash and Srivastava, 1987). Rotted tissue is light brown to pinkish, soft but not watery. The lesions have a sharply defined margin in early stages. Latter rotting becomes intensive, with a substantial amount of internal tissue becoming water soaked, yet firm, hard and stringy. In humid conditions, there is copious development of white mould, which eventually gives rise to small (1-2 mm) spherical resting bodies (sclerotia), white at first and later turning to brown. Skin becomes discoloured and soft (Vir and Sharma, 1965).

The disease appears in fruits harvested at relatively late stages and subsequently heaped on the bare soil. The fungus persists in the soil in the form of sclerotia and mould strands. Infection usually occurs near the soil surface, since the fungus requires good aeration. The disease is most severe under warm moist conditions (Vir and Sharma, 1965).

Disease Management

1. Deep ploughing of field is recommended as a means of burying sclerotia.
2. Avoid contamination in new areas by inadvertent movement of infested soil/fruits.
3. Improve the drainage of the garden.
4. Discard and destroy the infected and fallen fruits promptly.
5. Maintain strict field hygiene.
6. Place the harvested fruits in clean places to avoid contamination.
7. Treat the fruits with a fungicide like thiophenate methyl (0.1%) after harvest.
8. Adopt cold storage, as it is effective in arresting the development of rot.

THIELAVIOPSIS ROT

This is a minor post-havest disease of fruits like banana, guava, mango and pineapple etc. On mango, this rot was first reported by Srikanth (1989) from Anhra Pradesh, who reported 0.25 to 100% incidence due to this rot. Sundara Raman *et al* (1928) could develop this rot on inoculation of isolates of the causal pathogen from arecanut palm on mango etc.

Symptoms and Etiology

The mango fruit artificially infected with soft rot exhibited brown discolouration and rotting around the stalk, which gradually extended over the entrie fruit resulting in soft rot (Srikanth, 1989). Similar symptoms were described earlier (Ellis, 1971., Singh, 1982). Wounded inoculated fruit exhibited the symptoms on second day, whereas unwounded inoculated fruits showed disease symptoms on 4th day (Srikanth, 1989). Symptoms on mango both under market and laboratory conditions were similar to those described on pineapple (Linford and Spiregalberg, 1933) and banana fruits (Stover, 1972).

Srikanth (1989) identified the causal organism causing soft rot of mango fruits collected from fruit markets of Tirupati, Andhra Pradesh as *Thielaviopsis paradexa.*

However, information on disease cycle and epidemiology of the fruit rot is lacking and need to be studied.

Control/Management

Under laboratory conditions, the effect of different fungicides with or without the combination of hot water was studied (Srikanth, 1989). Diathane M – 45 (2000 ppm) inhibited fungal growth of the fungus to a maximum of 91.11% followed by Captan (2000 ppm) with 89.25% inhibition *in vitro.* Bavistin (1000 ppm), Bavistin (1000 ppm) + Wax (3%) and benomyl showd 85.55, 75.27 and 67.78% inhibition of fungal growth in case of wounded inoculated fruits. Aureofungin (500 ppm) was moderately effective with 44.44% inhibition of fungal growth. Aliette was least effective with 7.41% inhibition of fungal growth. Blitox (2000 ppm) and Borax (1000 ppm) were ineffective as they were similar to control treatment in inhibiting the fungal growth.

The disease index (DI) was minimum (1.00) when inoculated unwounded fruits were treated with Aliette (1000 ppm), Bavistin (1000 ppm) + Wax (3%), Bavistin (1000 ppm), Benomyl in hot water (1000 ppm at 55 $\pm$ 2 °C), Benomyl (1000 ppm), Captan (2000 ppm), Diathane-M-45 (2000 ppm) and Wax (6%). Hot water (55° $\pm$ 2 °C) and Borax (1000 ppm) alone were moderately effective showing a DI of 2.33 as against D1 of 6.67 in inoculated control (SriKanth, 1989).

BACTERIAL ROT

This is a bacterial post-harvest disease that appears in transit and storage. It was first reported from erstwhile Bombay State in India (Patel and Padhye, 1948). Later from Coimbatore the rot was reported on 'Neelum' and 'Bangalore' transported from Kanya-Kumari to Coimbatore in bamboo baskets cushioned with paddy straw, showing 4.45% and 16.13% decay in ripened fruits respectively (Palaniswami *et al* 1972). It was also reported from South Africa (Doidge, 1915).

Symptoms

The bacterium infects leaves, petioles, tender stems and fruits. On young fruits, water soaked lesions develop which turn dark brown to black later. Infected fruits may show cracks in the skin and badly affected ones also drop prematurely. Tear stains and exudation of gummy fluids are other symptoms.

Bacterium infected fruits of 'Bangalora' and 'Neelum' developed light brown, water soaked areas initially. Later these areas enlarged and covered the whole fruit and such fruits rotted completely within 2 to 3 days after the appearance of initial symptoms. The infected areas were extremely soft to touch, depressed and light brown in colour. In advanced stages of infection, the infected areas exuded gummy fluid and developed longitudinal cracks. Epidermis of these areas under severe infection was loose and came off as a thin papery covering (Sundararaj *et al* 1972). The diseased fruits decayed during ripening. Bacteriosis of pulp started in the vascular tissue connecting the seed and extended later around the seed, without affecting the embryo (Diaz *et al* 1971).

In 'Neelum', the extent of damage was 73.1% in storage, indicating that it was more prone to bacterial infection than 'Bangalora' (Palaniswamy *et al* 1972). However, information on the relative susceptibility of different varieties of mangoes is lacking.

Etiology

Doidge (1915) from South Africa, isolated and descrbied the pathogen of bacterial rot as *Bacillus mangiferae*. Patel and Padhye (1948) reported that the disease was incited by *Bacterium cartovorus* (Syn : *Erwinia cartovorum*). Later Patel *et al* (1950) considered *Bacillus subtilis* as the causal agent of the rot. Yet the pathogen differs in several characteristics to justify assigining it a species rank (Patel and Padhya, 1948., Patel *et al* 1950). Rangaswami (1962) and Palaniswami *et al* (1972) claimed *Pseudomonas mangifera indicae* sp., as the causal organism of bacterial rot of stored fruits of 'Bangalora' and 'Neelum' varieties shipped from Kanyakumari to Coimbatore. According to Diaz *et al* (1971) bacteriosis of mango pulp was due to *Xanthomonas* sp. Steyn *et al* (1974) concluded that *Erwinia mangiferae* had never been responsible for bacterial rot of mangoes and that *Pseudomonasmangiferae indicae* was the causal agent of all out breaks of this disease in South Africa, India and Pakistan.

Disease Cycle

The disease develops during transit and in storage. It is a wound parasite. The disease spreads through contamination from infected fruits to healthy fruits kept along with diseased ones. Infection takes place through wounds or lenticles and initiates a rot dissolving the middle lamella of the cell walls through the action of enzyme (pectionase) which is secreated by the bacteria (Patel and Padhye, 1948). Development of infection on uninoculated fruits suggests that the primary infection takes place in the field, long before harvest (Palaniswami *et al* 1972).

Epidemiology

Bacterial rot was observed from May to August particularly in wet areas. The optimum temperature for bacterial growth was 25 °C, whereas thermal death point was 32 °C (Patel and Padhye, 1948). 'Bangalora' and 'Neelum' fruits developed severe infection of rot in 10 days of storage at room temperature (84-90 °F) and after 30-32 days in cold storage at 40-45 °F and 45-55 °F (Palaniswami *et al* 1972). Polythene film covers (with or without perforation) accelerated the disease incidence.

Control/Management

1. Orchard sanitation will go long way in preventing the disease
2. Since the disease appears on leaves as well as on fruits control measures need to be directed on the leaves as well as on the early stages of fruit development in the garden.
3. However, Palaniswami *et al* (1972) could not obtain control of the disease in storage with antibiotics such as acidione, aureofungin, tetracycline, agrimycin, chloraphenicol, streptomycine sulphate and streptocycline.

REFERENCES

Ahmad, E.M., Denninson, R.A., and Merkley, M.S. 1969. Effects of low levels of irradiation upon the preservation of food products. Pages 9-12, 68-76. In : University of Florida, Department of Food Science. Annual Report. Division of Isotops Development, U.S. At. Energy Committee.

Akamine, A.K. 1977. Tropical and Subtropical crops in Hawaii. CSIRO Food Research Quarter, 37 : 13-20.

Akamine, E.K. and Goo, T. 1979. Effects of ionizing irradiation on 'Haden' mangoes. Hawaii Agriculture Experiment Station, Research Report, 205 : 11 p.

Alabi, 1986. quoted by A.K. Ghosh, 1998. In : Chapter 4. Post-harvest diseases of Mango, Guava, Banana and Papaya. An overview. Neeta Sharma and M.M. Alam, Post-harvest diseases of Horticultural Perishables. 91 p. International Book Distributing Co.; Lucknow, India.

Alicbusan, R.V. and Schaffer, L.A. 1958. Diplodia rot of mango. Philippine Agriculturist, 42 : 319-22.

Alvarez – Garcia, L.A. and Lopez-Garcia, J. 1971. Gummosis, die-back, and fruit rot disease of mango (*Mangifera indica* L.) caused by *Physalospora rhodina* (B&C)CKe in Puerto Rico. Journal of Agricultural University. Puerto Rico, 55: 435-450.

Amresh, Y.S. 1997. Studies on foliar diseases of sunflower with special reference to *Alternaria* leaf blight caused by *A. helianthi* (Hanst) Tuballi and Nishihara. M.Sc. (Agri.) Thesis, University of Agricultural Science, Dharwad, India.

Anonymous, 1947. Report of Hawaii Agricultural Experiment Station, 1944-46. p. 67.

Anonymous, 1953-54. Report of the Division of Mycology and Plant Pathology. Scientific Report of Agriculture Research Institute, New Delhi. 87-99 p.

Anonymous, 1974-1985. Studies on diseases of mango and their control. Reports of Plant Pathologist, Central Mango Research Station, Lucknow, India.

Anonymous, 1975-85. Studies on post-harvest diseases of mango and their control. Annual Research Report. Central Mango Research Station, Lucknow (India).

Anonymous, 1983. Israel Meteorological Science. Average of temperatures and relative humidity, 1964-79. 41 : 1-84.

Anonymous, 1988. World Meteorological Organization. Agrometerological aspects for operation of crop protection (WMO, NO 687), WMO Technical Note, 192:1-165.

Anonymous, 2005. Production, Post-harvest Management and Marketing of fresh mangoes for export. Indian Institute of Horticultural Research, Bangalore. Technical Bulletin, 20. India.

Aragaki, M., Kimato, W.S. and Uechida, J.Y. 1981. Limitation of hot water treatment in the control of Phytophthora fruit rot of papaya. Plant Disease, 65: 744-745.

Arauz, L.F. 1995. Enfermedades poscoceha del mango y su combate. Escucla de Fitotecnia. Universidad de Costa Rica. San Jose, Costa Rica.

Arauz, L.F. 2000. Mango Anthracnose – Economic Impact and Current Options for Integrated Management. Plant Disease, 84(6) : 600-611.

Arauz, L.F. and Gonzalez – Lobo, M. 1986. Fuentes de inoculo de algunas enfermedades de poscosecha del mango. Agronomy Costa Rica, 10 (1/2) : 217-220.

Arauz, L.F. and Mora, D. 1983. Preliminary evaluation of post-harvest problems in six tropical fruits in Costa Rica. Agronomia Costaricense, 7 : 43-53.

Arya, A. 2004. Tropical Fruits : Diseases and pests. Kalyani Publishers, Ludhiana, New Delhi.

Arya, A. and Mathew, D.S. 1991. Control of Phoma and Botryodiplodia rots of mango. Geobios New Report, 10 : 173-174.

Assnami, M.B., Chhatpar, H.S., Kadri, M.H., Matoo, A.K. and Modi, V.V. 1971. Microbial spoilage of mangoes. Journal of Food Science and Technology, 8 : 208-210.

Astua, G., Arauz, L.F. and Umana, G. 1994. Sensibilidad reducida and triabendazole en *Colletotrichum gloeosporioides* austedo de papaya. Agronomy Costa Rica, 18 : 35-39.

Aulakh, K.S. and Grower, R.K. 1969. Use of oils for controlling ripe fruit rots of tomato caused by *Phoma* and *Curvularia lycopersicii*. Plant Protection Bulletin, F.A.O., 17: 90-91.

Aulakh, K.S., Sokhi, S.S. and Rattan, G.S. 1998. Post-harvest diseases of tropical and subtropical fruits. p. 42-76. In : Post-harvest diseases of Horticultural Perishables (Ed). N.Sharma and M.M. Alam International Book Distributing Co., Lucknow (India).

Avema, L. 1922. Bot. Agric. Sao Paulo. XXIII 82 (cited from report on the export of mango to Europe in 1932 and 1933). Bulletin Dept. Ltd., Rec. Agriculture, Bombay 172 of 1932, 1935, 17 p.

Awasthi, D.P., Sarkar, S. and Kaiser, S.A.K.M. 2004. Reaction of different cultivars of mango fruits against post-harvest rots caused by *Colletotrichum gloeosporioides*. Penz and *Diplodia natalensis*. Pole Evans under market conditions (Abst). Indian Phytopathology, 57(3) : 354.

Bagwan, N.B. 2010. Post-harvest pathogens of mango (*Mangifera indica*) and their effect on fruit quality. Journal of Mycology and Plant Pathology, 40(3) : 352-355.

Bagwan, N.B. and Yeole, R.D. 2003. Effect of temperature and relative humidity on incidence of Post-Harvest Rot of mango. Journal of Mycology and Plant Pathology, 33(1) : 136-137.

Baka, M., Mercier, J., Corcuff, R., Castaigne, F. and Aru, J. 1999. Photochemical treatment to improve storability of fresh strawberries. Journal of Food Science, 64: 1068-1072.

Baker, R.E.D. 1938. Notes on the control of mango anthracnose. Tropical Agriculture, Trinidad, 15 : 12-14.

Baker, R.E.D., Crowdy, S.H. and McKee, R.K. 1940. A review of latent infections caused by *Colletotrichum gloeosporioides* and allied fungi. Tropical Agriculture, Trinidad, 17 : 128.

Baker, R.E.D. and Wardlaw, C.W. 1937. Studies in the pathogenicity of tropical fungi. I. On the type of infection encountered in the storage of certain fruits. Annals of Botany, New Series, 1(1) : 59-66.

Banerjee, B.N., Karmarkar, D.V. and Row, G.R. 1934. Investigation on the storage of mangoes. Agriculture and Livestock in India, 4(1) : 36-53.

Banik, A.K., Kaiser, S.A.K.M. and Dhuva, R.S. 1998. Evaluation of some systemic and non-systemic fungicides against *Colletotrichum gloeosporioides* and *Diplodia natalensis* causing post-harvest fruit rot of mango. Journal of Soils and Crops, 8(2) : 119-122.

Barmore, C.R., McMillam, R.T. and Spalding, D.H. 1973. Post-harvest control of anthracnose on mango fruit as affected by a pre-harvest application of an anti-transparent plus benomyl. Proceedings of Tropical Region of American Society for Horticultural Science, 17 : 74-80.

Barmore, C.R., Purvil, A.C. and Fetters, P.J. 1983. Polyethylene film packing of citrus fruit : Containment of decaying fruit. Journal of Food Science, 48 : 1558-1559.

Barquero, C. and Arauz, L.F. 1996. Variation de La sensibilidad a benomil en poblaciones de *Colletotrichum gloeosporioides* aislado de mango en Costa Rica (Abstr.). Memories IX Congress National Agronomy, Recursos Naturales, III. Cong. Nacional Fitopatol, II. Congress Nacional Suelos. San Jose. Costa Rica, vol. II. P. 133.

Berger, R.D. and Bartz, J.A. 1982. Analysis of monocyclic pathosystem with *Erwinia lycopersicon* as the model. Phytopathology, 72 : 365-369.

Bhargava, S.N., Ghosh, A.K., Srivastava, M.P., Singh, R.H. and Tandon, R.N. 1965. Studies on fungal diseases of some tropical fruits. VIII. Effect of temperature, on the decay of mango, banana and guava caused by some important pathogens. Proceedings of National Academy of Science, India, 35 : 393-398.

Bhargava, S.N. and Singh, A.P. 1975. *Aspergillus* rot of mango – its control. Indian Journal of Horticulture, 32(3-4) : 190.

Bhuvaneswari, V. 1999. Studies on post-harvest pathogens of mango. M.Sc.(Ag.) Thesis, Acharya N.G. Ranga Agricultural University, Hyderabad, India.

Biggs, A.R. 1995. Detection of latent infections in apple fruit with paraquat. Plant Disease, 79 : 1062-1067.

Bitancourt, A.A. 1938. A anthracnose de mangueira (mango anthracnose). Biologico, 4 : 43.

Boonraung, C., Farungsang, U. and Sangchote, S. 1993. Phomopsis Fruit rot of mango and its control. Proceedings of International Conference on Post-harvest handling of Tropical fruits held at Chinag, Mai, Thailand, 19-23, July, 1993:469 p.

Bose, S.K., Sridhar, G.S. and Pandey, B.N. 1973. Studies on the die back disease of mango in the Tarai region of Kumaon. Progressive Horticulture, 5 : 41-53.

Boulet, M., Kane, O. and Castaigne, F. 1982. Effect of chilling-injury on texture and fungal rot of mangoes (*Mangifera indica* L.). Journal of Food Science, 47 : 992.

Brodrick, H.T. and van der Linde, H.J. 1981. Technological feasibility studies on combination treatments for sub-tropical fruits. Proceedings of International Symposium on Combination Processes in Food Irradiation, Columbo, Srilanka.

Brodrick, H.T. and van der Westhizen, G.C.A. 1975. *Hendersonia creberrima*, the cause of soft brown rot of mango in South Africa. Phytophylactica, 8 : 13-16.

Brun, J. 1951. L'. anthracnose due manguier en Guinee. Fruits d' Outré mer, 6 : 475.

Bryan, H.H., McMillan, R.T. Jr. and Mato, S.E. 1972. Effect of benomyl on mango colour and anthracnose. Proceedings of Tropical Region of America Society for Horticultural Science, 16 : 161-170.

Bugante, R. D. Jr. and Lizada, M.C.C. 1997. Disease control in Philippine `Carabao' mango with pre-harvest bagging and post-harvest hot water treatment. Acta Horticulturae, 455 : 797-804.

Burditt, A.K. Jr., Moshoned, M.G., Hatton, T.T., Spalding, D.H., von Windeguth, D.L. and Shaw, P.E. 1981. Low-dose irradiations as a treatment for grapefruit and mangoes infested with caribbean fruit fly larvae. U.S. Dept. Agric. Agric. Res. Serv. Agric. Res. Results, ARS-S-10, 9 pp.

Butler, E.E. 1957. Effectiveness of copper treated paper wrapping in preventing spread of *Rhizopus nigricans* in mature green tomatoes. Plant Disease Reporter, 14:474-477.

Camera, E.S.Da and Da Luz, C.G. 1941. Mycetes aliquot Lustaniae. IV. Agronomy Lusit, 3 : 25.

Campbell, R.J. 1992. A guide to mangoes in Florida. Fairchild Tropical Garden, Miami, Florida.

Cavelier, N., Pineau, C. and Prunier, M. 1994. Characteristics of *Pseudocercosporella herpotrichoides* isolates resistant to prochloraz. Pages 66-70, In : Br. Crop Drot. Cosy, Mangro. 60.

Chacko, E.K. and Ananthanarayanan, T.V. 1985. Accumulation of reserve substances in *Mangifera indica* L. during flower initiation *Z. Pfan* Zen Physiology, 106 : 281-285.

Chakravarthy, D.K. 1961. Survey of fruits and vegetables in Market. M.Sc.(Ag.) Dissertation, Indian Agricultural Research Institute, New Delhi, India.

Chakravarty, D.K. 1982. Control of black mould rot of mango. Indian Horticulture, 27(2) : 15, 28.

Chakravarthy, D.K. and Srivastava, D.N. 1964. Stem-end-rot of mango and orange fruit incited by *Diplodia natalensis.* Pole Evans. Current Science, 33(9) : 285.

Chandra, J. and Pathak, V.N. 1989. Studies on the stem-end rot (*Diplodia natalensis* Pole Evans) disease of post-harvest mango (*Mangifera indica* L.) fruits. Indian Journal of Mycology and Plant Pathology, 19 : 37-43.

Chandra, J. and Pathak, V.N. 1992a. Effect of plastic film wrapping on post-harvest fungal rot of mango fruits. Indian Phytopathology, 45 : 1-4.

Chandra, J. and Pathak, V.N. 1992b. Effect of plastic film wrapping on post-harvest fungal rot of mango fruits. Indian Phytopathology, 45 : 126-127.

Chandra, J. and Pathak, V.N. 1992c. Incidence, infection process and management of Anthracnose of mango fruits. Indian Journal of Mycology and Plant Pathology, 22: 35-38.

Chang, C.C. 1975. Hot water treatment of Irwin mango fruit to reduce anthracnose decay. Taiwan Agriculture Quarter, 11 : 69-78.

Cheema, G.S., Karmakar, D.V. and Joshi, B.M. 1937. The cold storage of Nagpur oranges (*Citrus aurantium*). Indian Journal of Agriculture Science, 7 : 169.

Cheema, G.S., Karmakar, D.V. and Joshi, B.M. 1950. Investigation on cold storage of mangoes. Indian Journal of Agricultural Science, 20(5) : 259-265.

Chhatpar, H.S., Modi, V.V. and Pandit, V.B. 1975. Studies on the control of blackspot formation on mangoes. Current Science, 44(18) : 674.

Clara, E.M. 1927. Anthracnose disease of mango in the Philippines. Philippine Agriculture Review, 20 : 271-273.

Coates, L.M. and Johnson, G.I. 1993. Quoted by R.B. Golan. In: Post-harvest Diseases of Fruits and Vegetables. Development and Control. Elseveir, Amsterdam, Newyork, p. 283.

Coates, L.M., Cooke, A.W. and Dean, J.R. 1997. The response of mango stem-end-rot pathogens to heat. Acta Horticulturae, 455 : 762-772.

Coates, L.M., Johnson, G.I. and Cooke, A.W. 1993. Post-harvest disease control in mangoes using high humidity hot-air and fungicide treatments. Annals of Applied Biology, 123 : 441-448.

Cojocaru, M., Droby S., Glotter, E., Goldman, A., Gottlib, H.E., Jacoby, B. and Prusky, D. 1986. 5 (12-Heptadecenyl) – resorcinol, the major compound of the antifungal agent from the peel of mango fruit. Phytochemistry, 25 : 1093-1095.

Conover, R.A. 1969. Results in recent experiments on control of mango anthracnose. Proceedings of Florida State Horticulture Society, 78 : 364-369.

Conway, W.S., Gross, K.C., Boyer, C.P. and Sam, C.E. 1988. Inhibition of *Penicillium expansum* polygalacturonase activity by increased apple cell wall calcium. Phytopathology, 78 : 1052-55.

Cooke, A.W., Johnson, G.J., Farungsang, U. and Farungsang, N. 1997. Evaluation of phosphonate for control of post-harvest mango pathogens (Abst). Phytoparsitica, 25(1) : 55-56 p.

Cordoba, M.A. 1992. Prueba, de fungicides para el. Combate Quifmico de la antracnosis (*Colletotrichum gloeosporioides*) on mango (*Mangifera indica* L.). Ing. Agric. Thesis, Universidad de Costa Rica, San Jose, Costa Rica.

Cuendis, B. 1988. Tratamientos de immerssion contres fungicides de poscosecha en mango (*Mangifera indica* L.). Ing. Agric. Thesis Universidad de Costa Rica. San Jose. Costa Rica.

Dalvi, M.B., Patil, P.D. and Raut, S.P. 2010. Pre-harvest fungicidal application for control of *Colletotrichum gloeosporioides* in Alphonso mango fruits. Annals of Plant Protection Science, 18(1) : 261-262.

Dangan, J.M., Sanchez, F.F. and Pordesimo, A.N. 1988. Integrated control for mango anthracnose. National Crop Protection Centre Leaflet Series No. 14, University of Philippines, Los Banos.

Darvas, J.M. 1993. *Dothiorella dominicana* an important mango pathogen in South Africa. Acta Horticulturae, 341 : 321-328.

Das Gupta, M.K. and Mandal, M.C. 1989. Control of post-harvest diseases of perishables. In : Post-harvest Pathology of Perishables. Oxford and IBH Publishing Co. Pvt.Ltd., New Delhi, India.

Das Gupta, S.N. and Bhatt, R.S. 1946. Studies in the diseases of *Mangifera indica* L. VIII. Latent infection in the mango fruit. Journal of Indian Botanical Society, 25 : 187-203.

De Jagar, E.S., Hall, A.N., Wehner, F.C. and Korsten, L. 2001. Microbial ecology of the mango phylloplane. Microbial Ecology, 42 : 201-207.

Desai, M.V. and Patel, K.P. 1963. Relative susceptibility of several varieties of mango to blight disease. Indian Phytopathology, 16 : 239-240.

Dharamvir., Rayachaudhuri, S.P., Batra, S.K. and Bajaj, B.S. 1967. Aureofungin as fruit dip and wrap treatment for the control of *Diplodia* rot of mango and *Alternaria* rot of tomato fruits during transit. Indian Phytopathology, 20(4) : 301-303.

Dhawan, S. 1982. Charcoal rot of mango. Current Science, 51 : 140.

Diaz, P.C., Figueroa, M. and Aponte, O. 1971. Bacteriosis of mango pulp caused by *Xanthomonas* sp. Agronomia Tropical, 21 : 17-24.

Dodd, J.C., Bugante, R., Koomen, I., Jeffries, P. and Jager, M.J. 1991a. Pre- and post-harvest control of mango anthracnose in the Philippines. Plant Pathology, 40 : 576-583.

Dodd, J.C., Estrada, A.B., Matcham, J., Jefrries, P. and Jager, M.J. 1991b. The effect of factors on *Colletotrichum gloeosporioides*, causal agent of mango anthracnose in the Philippines. Plant Pathology, 40 : 568-575.

Dodd, J.C., Jeffries, P. and Jeger, M.J. 1989. Management strategies to control latent infection in tropical fruit. Annals of Applied Biology, 20 : 49-56.

Dodd, J.C., Prusky, D. and Jeffries, P. 1997. Fruit diseases. In : R.E. Litz (Ed.). The Mango : Botany, Production and Uses. CAB International, Wallingford, UK, pp. 257-280.

Doidge, E.M. 1915. A bacterial disease of the mango. *Bacillus mangiferae* n.Sp. Annals of Biology, 2 : 1-45.

Doidge, E.M. 1932. Black spot of mangoes. Farming in South Africa, 7 : 89-91.

Domsch, K.H., Gams, W. and Anderson, T.H. 1980. Compendium of soil fungi. Academic Press, Newyork.

Droby, S., Prusky, D., Jacoby, B. and Goldman, A. 1986. Presence of antifungal compound and its relation to the latency of *Alternaria alternata* in unripe peel of mango fruits. Physiological and Molecular Plant Pathology, 29 : 173-183.

Droby, S., Prusky, D., Jacoby, B. and Goldman, A. 1987. Induction of antifungal resorcinols in mesocarp of unripe mango fruits and its relation to latent infection by *Alternaria alternata*. Physiological and Molecular Plant Pathology, 30 : 285-292.

Du Camp – Cokin, M.N., Loiseau, G., N, Guyen, D.D. Le. and Domier, M. 2009. Effect of the Lacto Peroxidase System against causal agents of post-harvest mango diseases. Acta Horticulturae, 820 : 521-528.

EcKert, J.W. 1981. Quoted by R. Gopi, 2003. In : Management of post-harvest diseases of mango. M.Sc.(Ag.) Thesis, Acharya N.G. Ranga Agricultural University, Hyderabad, Andhra Pradesh, india.

Eckert, J.W. and Kolbezon, M.J. 1962. Controlling spread of *Rhizopus* in packed tomatoes by tissue wraps impregnated with active halogen compounds. Phytopathology, 52 : 180-183.

Eckert, J.W., and Ratnayake, M. 1983. Host-pathogen interactions in post-harvest diseases. In : Post-harvest Physiology and Crop preservation. M. Laberman (ed). Plenum Publishing Co., Newyork.

El – Ghaouth, A., Amul, J., Asselin, A. and Benhamou, N. 1992. Antifungal activity of Chitosan on post-harvest pathogens : Induction of morphological and cytological alterations in *Rhizopus stolonifer*. Mycological Research, 96 : 769-779.

Ellis, M.B. 1971. Dematiaceous Hyphomycetes. CMI, Kew Survey, England, p. 608.

English, *et al* 1975. Quoted by Mason, S.J. and Jeffries, P. 1993. The effect of calcium on activation and induction of the pectolytic enzymes produced by *Colletotrichum gloeosporioides*. Acta Horticulturae, 341 : 376-386.

Esguerra, E.B. and Lizada, M.C.C. 1990. The post-harvest behaviour and quality of `Carabao' mangoes subjected to vapor heat treatment. ASEAN Food Journal, 5 : 6-1.

Esguerra, E.B., Valenio- Traya, R.F. and Lizada, M.C.C. 2004. Efficacy of different heat treatment procedures in controlling diseases of mango fruits. Acta Horticulturae, 645 : 551-556.

Eshel, D., Miara, 1., Ailing, T., Dinoor, A. and Prusky, D. 2002. pH regulates endoglucanase expression and Virulence of *Alternaria alternata* in Persimon fruits. Mol. Plant. Mic. Interact, 15 : 776-779.

Estrada, A.B., Dodd, J.C. and Jeffries, P. 1993. Effects of environment on the *in vitro* growth and development of *Colletotrichum gloeosporioides* isolates from the Philippines. Acta Horticulturae, 341 : 360-369.

Estrada, A.B., Jeffries, P. and Dodd, J.C. 1996. Field evaluation of a predictive model to control anthracnose disease of mango in the Philippines. Plant Pathology, 45 : 294-301.

Fallik, E., Grinberg, S., Alkali, S. and Lurie, S. 1996. The effectiveness of post-harvest hot water dips on the control of gray and black moulds in sweet pepper (*Capsicum annuum*). Plant Pathology, 45 : 644-649.

Farungsang, U. and Farungsang, N. 1992. Benomyl resistance of *Colletotrichum* sp. associated with rambutan and mango fruit rot in Thailand. Acta Horticulturae, 321: 891-897.

Fernando, M. 1937. A note on the soft rot of stored mangoes caused by *Botryodiplodia thebromae* Pat. Tropical Agriculturist, 89 : 381-387.

Fernandoz, A.V. and Arauz, L.F. 1992. Reduction of mango anthracnose by *Bacillus* sp. (Abst.) IV International Mango Symposium, University of Florida and International Society for Horticultural Science. Miami Beach, U.S.A.

Fitzell, R.D., 1979. *Colletotrichum acutatum* as a cause of anthracnose of mango in New South Wales. Plant Disease Reporter, 63 : 1067-170.

Fitzell, R.D. 1987. Epidemiology of anthracnose disease of avocadoes. South African Avocado Growers' Association Year Book, 10 : 113-116.

Fitzell, R.D. and Peak, C.M. 1984. The epidemiology of anthracnose disease of mango-Inoculum sources, Spore Production and Dispersal. Annals of Applied Biology, 101 : 53-59.

Fitzell, R.D. and Peak, C.M. 1986. Management strategies for control of anthracnose and bacterial spot in northern New South Wales. Proceedings of First Australian Mango Research Workshop, Melborne, Australia, CSIRO : 253-257.

Fitzell, R.D., Peak, C.M. and Darnell, R.E. 1984. A model for estimating infection levels of anthracnose disease of mango. Annals of Applied Biology, 104 : 457-458.

Flaishman, M.A. and Kolathukudy, P.E. 1994. Timing of fungal invasion using host's ripening hormone as a signal. Proceedings of National Academy of Science, USA, 91 : 6579-6583.

Fonseca, M.J.de O., Salomao, L.C.C., Cecon, P.R. and Pushmann, R. 2004. Fungicides and wax in post-harvest preservation of mango 'Haden'. Acta Horticulturae, 645 : 557-563.

Franco, E. 1946. Doencas e Pragas Constatadas no. Maranho. Bol. Fito, 3-91.

Freeman, S., Katan, T. and Shebi, E. 1998. Characterization of *Colletotrichum* species responsible for anthracnose diseases of various fruits. Plant Diseases, 82 : 596-605.

Fry, W.E. 1982. Principles of Plant Disease Management. Academic Press, USA, Orlando, FL.

Gadgile, D.P. and Chavan, A.M. 2009. Changes in calcium content of mango pulp due to infection from *Aspergillus niger* isolates. Journal of Mycology and Plant Pathology, 39(1) : 166-168.

Gadre, U.A. 1979. Control of mango anthracnose (*Colletotrichum gloeosporioides*). Mango workers' meeting, Panaji, Goa (AICFIP), 2-5 May, p. 197-198.

Gandhi, S.R. 1955. The mango in India. Farm Bulletin No. 6. Indian Council of Agriculture Research, New Delhi.

Ganfotli, B.V. and Davis, M.J. 1993. Pectic Zymogram analysis for characterizing genetic diversity of the mango anthracnose pathogen. Acta Horticulturae, 341: 353-359.

Garcha, H. and Singh, V. 1980. Post-harvest diseases of fruits in Punjab. Indian Phytopathology, 33 : 42-47.

Ghosh, A.K. 1998. Post-harvest diseases of mango, guava, banana and papaya : An overview. In : Post-harvest diseases of Horticultural Perishables (ed.). N. Sharma and M.M. Alam, pp. 77-132. International Book Distributing Co., Lucknow.

Ghosh, A.K., Bhargava, S.N. and Tandon, R.N. 1966. Studies on fungal diseases of some tropical fruits. VI Post infection change in ascorbic acid contents of mango and papaya. Indian Phytopathology, 19 : 262-268.

Golan, R.B. 2001. Post-harvest diseases of fruits and vegetables. Development and Control. Elserier, Amsterdam, London, Newyork, Oxford, Paris, Shanon, Tokyo, p. 283.

Gopi, R. 2003. Management of post-harvest diseases of mango. M.Sc.(Ag.) Thesis, A.N.G.R. Agricultural University, Hyderabad, Andhra Pradesh, India.

Gosbee, M.J., Johnson, G.I. and Joyce, D.C. 1997. Infection pathway of the stem-end-rot fungus *Dothiorella dominicana* in 'Kensington' mango fruit. Acta Horticulturae, 455 : 595-599.

Grower, R.K. and Aulakh, K.S. 1968. Oospora rot of tomato fruits and its control by oils. Indian Phytopathology, 21 : 139-143.

Guest, D.I. and Bompiex, G. 1990. The complex mode of action of phosphonates. Australian Plant Pathology, 19 : 113-115.

Halos, P.M. and Divinagracia, G.G. 1970. Histopathology of mango fruits infected by *Diplodia natalensis*. Philippine Phytopathology, 6(1-2) : 16-28.

Hatton, T.T. Jr., Beraha, H. and Wright, W.H. 1961. Preliminary trials of gamma irradiation on mature green Irwin and Sensation mangoes. Proceedings of Florida Mango Forum, pp. 15-17.

Hatton, T.T.J., and Reeder, W.F. 1964. Hot water as a commercial control of mango anthracnose. Proceedings of Caribbean Region of American Society for Horticultural Science, 8 : 76-84.

Hatton, T.T. Jr., Reeder, W.F. and Campbell, C.W. 1965. Ripening and storage of Florida mangoes. U.S. Department of Agriculture Marketing Research Report, 725-9.

Hayes, W.B. 1953. Fruit growing in India. Kitabisten, Allahabad, India

Hendrix, J.W., Matsura, M. and Kikuta, K. 1947. Fungicide investigations. Report of Hawaii Agriculture Experiment Station, 1944-46; 104-107.

HingoRani, M.K. and Sharma, O.P. 1956. Blight disease of mango. Indian Phytopathology, 9

HingoRani, M.K., Sharma, O.P. and Sohi, H.S. 1960. Studies on blight diseases of mango caused by *Macrophoma mangiferae*. Indian Phytopathology, 13 : 137-143.

Hofman, P.J., Smith, L.G., Joyce, D.C., Johnson, G.I. and Meiburg, G.F. 1997. Bagging of mango (*Mangifera indica* cv. Keitt) fruit influences fruit quality and mineral composition. Post-harvest Biology and Technology, 12 : 83-91.

Hord, M.J., Arauz, L.F. and Rodriguez, W. 1992. Evaluaction de dos metodos, basados en condicions meteorologicas par el combatae qufmico de antracnosis (*Colletotrichum gloeosporioides* Penz.) de mango (*Mangifera indica*) en Costa Rica. Bol. Tec. Estacion Exp. Fabio Baudritm Universidad de Costa Rica. 25(2) : 52-61.

Huang, S. and Liu, X. 1995. Control of post-harvest stem-end-rot of mango. Tropical Science, 35 : 321-326.

Hunter, J.E. and Kummoto, R.K. 1974. Dispersal of *Phytophthora palmivora* sporangia by wind-blown rain. Phytopathology, 64 : 202-206.

Ilag, L. and Co. workers (unpublished) quoted by Prusky, D. Ilag, L., Ackermam, M., Kobiler, I. and Fuchs, Y. 2004. New approaches for the control of post-harvest disease of mango fruits. Acta Horticulturae, 645 : 129-136.

Ingham, J.L. 1973. Disease resistance in higher plants. The concept of pre-infected and post-infectional resistance. Phytopathology Z, 78 : 314-335.

Jacob, C.J., Brodrick, H.T., Swart, H.D. and Mulder, N.J. 1973. Control of post-harvest decay of mango fruit in South Africa. Plant Disease Reporter, 57(2) : 173-176.

Jadeja, K.B. 2000. Ecological association of two post-harvest pathogens of mango fruits. Journal of Mycology and Plant Pathology, 30 : 406-407.

Jadeja, K.B. and Bhatt, H.R. 2008. Biological control of mango fruit rot with fructoplane bacteria. Journal of Mycology and Plant Pathology, 38(2) : 325-327.

Jadeja, K.B. and Vaishnav, M.V. 1984. Effective spray schedule for control of anthracnose and leaf blight of mango. Indian Journal of Plant Botany, 12 : 93-96.

Jadeja, K.B. and Vaishnav, M.V. 1998. Two important post-harvest pathogens on mango Cv. Kesar fruits (Abst.). Paper presented at National Symposium Mango, Lucknow, India, 25-27 June, p. 68.

Jadeja, K.B. and Vaishnav, M.V. 2000. Post-harvest fruit rot diseases in mango cv. Kesar. Indian Phytopathology, 53(4) : 492.

Jagdish, C. and Pathak, V.N. 1989. Studies on the stem-end-rot (*Diplodia natalensis* Pole Evans) disease of post-harvest mango (*Mangifera indica* L) fruits. Indian Journal of Mycology and Plant Pathology, 19(1): 37-43.

Janisicwicz, W.J. and Kartsun, L. 2002. Biological control of post-harvest diseases of fruits. Annual Review of Phytopathology, 40 : 411-441.

Jeffries, P., Dodd, J.C., Jeger, M.J. and Plumbley, R.A. 1990. The biology and control of *Colletotrichum* species on tropical fruit crops. Plant Pathology, 39 : 343-366.

Jeffries, P. and Koomen, I. 1992. Strategies and prospects for biological control of diseases caused by *Colletotrichum.* pages. 337-357. In : *Colletotrichum* : Biology, pathology and control. J.A., Baily and M.J. Jeger (eds). CAB International, Oxon, U.K.

Jeger, M.J., Plumbley, R.A., Prior, C. and Persad, C. 1987. SS-2 post-harvest aspects of crop protection. Manila Philippines, pp. 2.

Jimenez, R., Hemandez, R.L. and Dauz, L.F. 1989. Prueba de fungicides para prevenir enfermedades en el frut del mango (*Mangifera indica* L.). Bol. Tec. Estacian Experiment Fabio Bandrit, 22(4) : 11-20.

Johnson, G.I. 1992. Biology and control of the stem-end-rot pathogens of mango. Ph.D Thesis. University of Quensland, Australia.

Johnson, G.I. 1994. Stem-end-rot. pages. 39-41. In : Compendium of Tropical Fruit diseases (eds). R.C. Ploetz., G.A. Zentmyer., W.T. Nishijima., K.G. Rohrbach. and H.D. Ohr, 1994. American Phytopathological Society, St. Paul. MN.

Johnson, G.I., Boag, T.S., Cooke, A.W., Izand, M., Paritz, M. and Sangchote, S. 1990. Interaction of post-harvest disease control treatments and gamma irradiation on mangoes. Annuals of Applied Biology, 116(2) : 245-257.

Johnson, G.I. and Coates, L.M. 1993. Post-harvest disease of mango. Post-harvest News and Information, 4 : 27N-34N.

Johnson, G.I., Cooke, A.W. and Mead, A.J. 1993. Infestation and Quiescence of mango stem-end-rot pathogens. Acta Horticulturae, 341: 329-336.

Johnson, G.I., and Cooke, A.W., Mead, A.J. and Wells, I.A. 1991a. Stem-end-rot of mango in Australia : Causes and Control. Acta Horticulturae, 291 : 288-295.

Johnson, G.I., Mead, A.J., Cooke, A.W. and Dean, J.R. 1992a. Mango stem end rot pathogen. Fruit infection by endophytic colonization of the inflorescence and pedicel. Annals of Applied Biology, 120 : 225-234.

Johnson, G.I., Mead, A.J., Cooke, A.W. and Wells, I.A. 1992b. Stem end rot diseases of tropical fruits-mode of infection in mango and prospects of control. Acta Horticulturae, 321 : 882-890.

Johnson, G.I., Mead, A.J., Cooke, A.W. and Dean, J.R. 1991b. Mango stem-end-rot pathogens – Infection levels between flowering and harvest. Annals of Applied Biology, 119(3) : 465-473.

Johnson, G.I. and Muirhead, I.F. 1988. Post-harvest disease control of mangoes. Queensland Fruit and Vegetable News, 59 : 16-17.

Johnson, G.I., Muirhead, I.F., Mayers, P. and Cooke, A.W. 1989a. Disease P. 1-9. In : Mango – pests and disorders. Department of Primary Industries, Queensland, Australia.

Johnson, G.I., Muirhead, I. and Rappel, L.M. 1989b. Mango post-harvest disease control. A Review of Research in Australia, Malaysia and Thailand. ASEAN Food Journal, 4 : 139-141.

Johnson, G.I. and Sangchote, S. 1994. Control of post-harvest diseases of tropical fruits : Challenges for the 21st century. In : G.I. B.R., Champ., E. Highley, and G.I Johnson. (Ed), Post-harvest handling of fruit. AICAP Proceedings No. 50 pp. 140-161.

Johnson, G.I., Sangchote, S. and Cooke, A.W. 1990. Control of stem-end-rot (*Dothiorella dominicana*) and other post-harvest diseases of mango (Cv. Kensington Pride) during short and long term storage. Tropical Agriculture, 67 : 183-187.

Johnson, G.I., Sharp, J.L., Milne, D.L. and Oosthuyse, S.A. 1997. Post-harvest technology and quarantine treatments. In : R.E. Litz (Ed.). The mango-Botany, Production and Uses. CAB International, Wallingford, U.K., pp. 447-508.

Joshi, M.S., Gadre, U.A., Kadam, J.J. and Borkar, P.G. 2010. Comparative efficacy of existing and modified fungicide formulations for control of Anthracnose of mango caused by *Colletotrichum gloeosporioides*. Penz. Pestology, 34(10) : 38-39.

Kalyansundarm, P. and Parthasarathy, V.A. 1979. Effect of thiobendazole and ethrel on ripening and stem-end-rot of mango Cv. Pairi. Annamalai University Agriculture Research Annals, 7/8 : 161-164.

Kantilal, N. and Reddy, S.M. 1981. Aeromycoflora of store houses of fruits in relation to post-harvest diseases of apples and mangoes. Proceedings of National Conference on Environment and Biosphere, 169-172.

Kapse, B.M. 1993. An integrated approach to post-harvest handling of mango Cv. Kesar. Ph.D Thesis, Gujarat Agricultural University, Navasari, Gujarat, India.

Kapse, B.M. and Katrodia, J.S. 1997. Studies on hydrocooling in Kesar mango (*Mangifera indica* L.). Acta Horticulturae, 455 : 707-717.

Kapse, B.M., Pawar, V.M. and Sakhale, B.K. 2009. Post-harvest disease management in mango (*Mangifera indica* L.). Cv. Kesar. Acta Horticulturae, 820 : 493-501.

Kapse, B.M., Pawar, V.M., Sakhale, B.K., Yadav, M.V. and Hingole, D.G. 2004. Studies on control of post-harvest diseases of mango. Cv. Kesar (Abst.) Paper presented at First Indian Horticulture Congress, New Delhi, 6-9 November, 32, p.

Khanna, K.K. and Chandra, S. 1978. A homeopathic drug controls mango fruit rot caused by *Pestalotia mangiferae* Henn. Experientia, 34 : 1167-1168.

Khanna, K.K. and Chandra, S. 1989. Further investigations on the control of storage rot of mango, guava and tomato fruits with Homeopathic drugs. Indian Phytopathology, 42 : 436-440.

Khannan, R. and Subbaraja, K.T. 1998. Biassay of fungicides against leaf blight of onion caused by A. *alternata* Fr Keisslur. Pestology, 22 : 44-46.

Ko. Y. and Liu, C.W. 2009. First Report of Stem-end-rot of mango caused by *Phomopsis mangiferae* in Taiwan. Plant Disease, 93 : 764.

Kobiler, I., Reved, R., Artez, L. and Prakash, D. 1998. Antifungal compounds regulating quiescent diseases in mango. Post-harvest handling of Tropical Fruits. Australian Centre for International Agricultural Research Proceedings, 80 : 109-114.

Kobiler, I., Shalom, Y., Roth, I., Akerman, M., Vinokur, Y., Fuchs, Y. and Prusky, D. 2001. Effect of 2, 4-dichlorophenoxy acetic acid on the incidence of side-and stem-end-rot in mango fruits. Post-harvest Biology and Technology, 23(1) : 23-32.

Koomen, I. and Jeffries, P. 1993. Effects of antagonistic microorganisms on the post-harvest development of *Colletotrichum gloeosporioides* on mango. Plant Pathology, 42 : 230-237.

Korsten, L., De-Villiers, E.E. and Lonsdale, J.H. 1993. Biological control of mango post-harvest disease in the pack house. South African Mango Growers' Association Year Book, 13 : 117-121.

Kotze, J.M. 1978. Anthracnose of avocados. South African Avocado Growers' Year Book. 2 : 45-7.

Krishnaiah, J., Satya Prasad and Tirupathaiah,V. 1986. Protection of mangoes from post-harvest fungal decay by using fruit coating. Indian Phytopathology, 40 : 426-428.

Ladger, S.N. 1991. Post-harvest handling of mangoes. Acta Horticulturae, 291 : 479-488.

Laxminarayana, P. and Reddy, S.M. 1975. A new post-harvest disease of mango. Indian Phytopathology, 28(4) : 529-530.

Laxminarayana, P. and Reddy, S.M. 1977. Observation on certain market diseases of mango from Warangal (India). Egyptian Journal of Phytopathology, 9 : 29-33.

Laxminarayana, S., Krishnaprasad, C.A. and Shetty, M.S. 1974. Hot-water-treatment to regulate ripening and reduce spoilage of Alphonso mangoes. Journal of Horticultural Science, 49 : 365-371.

Lim, T.K. 1980. Chemical control of mango anthracnose in Malaysia. *In vitro* fungi toxicity of selected chemicals. Pertanika, 3 : 5-9.

Lim, T.K. and Khoo, K.C. 1985. Diseases and disorders of mango in Malaysia. Tropical Press, Kaula Lumpur, Malaysia.

Linford, M.B. and Spiegelberg, C.H. 1933. Illustrated list of pineapple fruit diseases, blemishes and malformation. Pineapple Quart, 3 : 135-180.

Lingaraj, D.S. 1969. Mango diseases and pests. Journal of Mysore Horticultural Society, 14 : 15-18.

Lingk, W. 1991. Health risk evaluation of pesticides contamination in drinking water. Gesunde Pflangen, 43 : 21-24.

Litz., R.E. (ed.) 1997. The Mango : Botany, Production and Uses. CAB International Willingford, U.K.

Lizada, M.C.C., Agravamte, J.U. and Brown, E.O. 1986. Factors affecting post-harvest disease control in 'Carabao' mango subjected to the hot-water-treatment. Philippine Journal of Crop Science, 11(3) : 153-161.

Lizada, M.C.C., Bautista, M.L., Artes, L.A. and Bacalangco, N.S. 1997. Use of hot-water-treatment to Enhance Natural Resistance of Carabao Mango against anthracnose disease (Abst.). Phytoparasitica, 25(1) : 55.

Lonsdale, J.H. 1992. Pre-harvest fungicide sprays for the control of post-harvest diseases of mango. South African Mango Growers' Association Year Book, 12: 28-31.

Lonsdale, J.H. 1993. Strategies for the control of post-harvest diseases of mango. South African Mango Growers' Association Year Book, 13 : 109-116.

Lonsdale, J.H. and Droamer, L. 1994. The application of short wave infra-red radiation to control post-harvest decay of mangoes. South African Mango Growers' Association, Year Book 14 : 59-61.

Louis, I., Chew, A. and Lim, G. 1988. Influence of spore density and extra cellular conidial matrix on spore germination in *Colletrichum capsici*. Transactions of the British Mycological Society, 91 : 694-697.

Lurie, S. 1998. Post-harvest heat treatments of horticultural crops. Horticulture Review, 22 : 91-121.

Lynch, S.J. and Mustard, M.J. 1950. Mangoes in Florida. Florida Department of Agriculture Bulletin, 20:135p.

Madan, M.S. 1988. Economic assessment of harvest and post-harvest losses in Horticultural Crops. Annual Report. Indian Institute of Horticultural Research, Bangalore, India, pp. 93.

Madhavilath, K. 1997. An integrated approach for the management of post-harvest diseases of mango and citrus in Andhra Pradesh. M.Sc.(Ag.) Thesis, Acharya N.G. Ranga Agricultural University, Hyderabad, Andhra Pradesh, India.

Madrigal, R. 1997. Evaluation de la sensibilidad del hongo *Colletotrichum gloeosporioides* al prochloraz en cultivars de mango Y papaya. Ing. Agric. Thesis, Universidad de Costa Rica. Esuclea de Fitotecnia, San-Jose, Costa Rica.

Majumdar, G. and Modi, V.V. 1980. Spoilage of mango by *A. flavus*. Current Science, 49(21) : 821-822.

Malo, S.E. 1977. The Mango in Florida. HortScience, 12(4) : 286 and 367.

Mandal, N.C. 1981. Post-harvest diseases of fruits and vegetables in West Bengal. Ph.D. Thesis, Viswa Bharati, Santinikatan, West Bengal, India.

Mandal, N.L. and Das Gupta, M.K. 1985. Role of host pathogen interaction in the efficacy of benzimidazole. Indian Phytopathology, 38 : 365-367.

Mathews, H. and Litz. R.E. 1991. Mango. In : Biotechnology of Perennial Fruit Crops (F.A. Hammerschlag, and R.E. Litz, Eds). CAB International, Kew, UK, 433-466.

Mathur, P.B., Singh, K.K. and Kapur, N.S. 1953. Cold storage of mangoes. Indian Journal of Agricultural Sciences, 23 : 65-67.

McGuire, R.G. 1991. Concomitant decay reduction when mangoes are treated with heat to control infestation of Caribbean fruit flies. Plant Disease, 75(9) : 946-949.

McGuire, R.G and Campbell, C.A. 1993. Imazalil for post-harvest control of anthracnose on mango fruits. Acta Horticulturae, 341 : 371-376.

McGuire, R.G. and Kelman, A. 1986. Calcium in potato tuber cell walls in relation to tissue maceration by *Erwinia Carotovora atroseptica*. Phytopathology 76 : 401-406.

McKee, R.K. 1940. Experiments on the control of mango anthracnose by spraying. Tropical Agriculture, Trinidad, 17 : 115-117.

McMillan, R.T.Jr. 1973. Control of anthracnose and powdery mildew of mango with systemic and non-systemic fungicides. Tropical Agriculture, 50 : 245-248.

McMillan, R.T. Jr. 1987. Effectiveness of various post-harvest treatments for mango decay control. Proceedings of Florida State Horticultural Society, 100 : 7-9.

McRae, W. 1924. Economic Botany. Part III. Mycology, Annual Report, Board of Scientific Advice, India, 1922-23 : 31-35.

Meah, M.B., Plumbley, R.A. and Jeger, M.J. 1991. Growth and intensity of *Botryodiplodia theobromae* causing stem-end-rot of mango. Mycological Research, 95 : 405-408.

Meijiao, H., Gao Zhao Yin, Li., Min Yang., Zheng, F.Z. and Fu Cong., 2005. Effect of hot water and 1-MCP treatments on storage of mango fruit. Journal of Fruit Science, 22(3) : 243-246.

Miller, W.R., Spalding, D.H. and Hale, P.W. 1986. Film wrapping of mangoes at advancing stages of post-harvest ripening. Tropical Science, 26 : 9-17.

Minz, G. and Yohebeo, B.M. 1952. Comparative pathogenicity of potential causal agents of stem-end-rot in Shamouti oranges. Ktavim, 2-3 : 141-145.

Morino, Q.L., Micheri H, S.J., Silva, J.B., Pedrosa, R.A., Silveria, N.S. and Tavaves, L.A. 1997. Post-harvest bio control of *Lasio diplodia rot* of mango fruits by sapro phytic yeasts. Arquivos de Biology-e-Technologia, 40(1) 29-37.

Moreno, R.L. and Paningbatan 1995. Quoted by Golan 2001. In : Post-harvest diseases of fruits and vegetables. Development and Control. Mango p. 281-283. Elseveir, Amsterdam, London, Newyork, Oxford, Paris, Tokyo.

Moss, M.O. 2002. Mycotoxin, Review I *Aspergillus* and *Pencillium*. Mycologist. 16 : 116-119.

Muirhead, I.F. 1974. Chemotherapy of latent infection disease caused by *Colletotrichum* sp. in tropical fruits. Australian Plant Pathology Society News Letter, 3 : 36.

Muirhead, I.F. 1976. Post-harvest control of mango anthracnose with benomyl and hot water. Australian Journal of Experimental Agriculture and Animal Husbandry, 16: 600-603.

Muirhead, I.F. and Gatirdge, R. 1984. Post-harvest diseases of mango – the Queensland experience. Proceedings of First Australian Mango Research Workshop. Cairns, Queensland, 23-30 November, 1984. Mulbourne, CSIRO, 248-252.

Mukherjee, S.K and Bhattacharya, S.K. 1965. A new disease of *Mangifera indica* L. in India caused by *A. tennussima*. Plant Disease Reporter, 49(5) : 405-407.

Muller, A.T. and Bart, J.R. 1989. Post-harvest storage control of mango stem-end-rot with fungicidal dips. Australian Journal of Experimental Agriculture, 29 : 125-127.

Muller, A.T. and Wood, P. 1991. Post-harvest disease of mango, Carnarvon Western Australia. Acta Horticulturae, 291 : 358-361.

Munishamanna, K.B., Ranganna, B. and Chidanand, D.V. 2008. Impact of post-harvest injuries on Disease Incidence and Severity in mango and Sapota fruits. Mysore Journal of Agricultural Science, 42(4) : 625-628.

Munishamanna, K.M., Ranganna, B., Subramanya, S. and Chidananda, D.V. 2009. Influence of source of infection on spoilage of mango fruits at orchards and markets. Mysore Journal of Agricultural Science, 43(3) : 477-481.

Murthy, S.K. and Rao, K.P.G. 1983. Post-harvest control of spoilage in mango (*Mangifera indica* L.) with hot water and fungicides. Journal of Food Science and Technology, 20 : 74-77.

Narsi Reddy, T., Ranga Reddy, T., Shankaiah, V. and Babu, J.D. 2004. Effect of pre-harvest fungicidal sprays in controlling important post-harvest diseases of Baneshan Mango in Andhra Pradesh. Andhra Agricultural Journal, 51(3 & 4) : 431-433.

Neergaard, P. 1945. Danish species of *Alternaria* and *Stemphylium*. Taxonomy, Parasiticism, Economical significance. Einar Munkgoard, Copenhagen.

Nel, A., Krans, M. and Khelwanlall, N. 2003. A guide for control of plant diseases. Directorate of Agricultural Production Inputs, National Development of Agriculture, RSA.

Nigam, M. and Sharma, N. 2000. Yeast-a potential antagonist of stem-end-rot of mango. International Journal of Tropical Plant Disease, 18(1-2) : 131-140.

Nunez – Elisea, R., Davenport, T.L. and Caldeira, M.L. 1993. Bud initiation and morphogenesis in 'Tommy Atkins' mango as affected by temperature and triazole growth retardants. Acta Horticulturae, 341 : 193-198.

Nyamjage, M.O., Wainwright, H. and Bishop, C.F.H. 1998. The effects of hot water treatments in combination with cooling and/or storage on the physiology and disease of mango fruits (*Mangifera indica* Linn.). Journal of Horticultural Science and Biotechnology, 73(5) : 589-597.

Ocfemia, G.O. 1931. Notes on some economic plant diseases new in the Philippine Islands. Philippines Agriculturist, 19 : 581.

Okigbo, R.N. and Osuinde, M.E. 2003. Fungal leaf spot disease of mango (*Mangifera indica* L.) in South Eastern Nigeria and Biological control with *Bacillus subtilis*. Plant Protection Sciences, 39(2) : 70-77.

OmPrakash, 1990. Report of the Plant Pathologist. Annual Report of CIHNP (CMRS), Lucknow for the year 1989-90.

OmPrakash. 1993. Post-harvest diseases of Indian mango and their control (Abst.) Golden Jubilee Symposium, Horticulture Research, Changing Scenario, Bangalore, May 24-28, 1993 : 287.

OmPrakash. and Misra, A.K. 1988. Evaluation of different fungicides for the control of anthracnose of mango. International Journal of Tropical Plant Diseases, 6 : 263-264.

OmPrakash. and Pandey, B.K. 2000. Control of mango anthracnose by hot water and fungicide treatment. Indian Phytopathology, 53 : 92-94.

OmPrakash. and Raoof, M.A. 1985. Blossom blight disease of mango. Indian Journal of Plant Pathology, 3 : 271-272.

OmPrakash. and Raoof, M.A. 1988. Control of mango fruit decay with post-harvest application of various chemicals against Black rot, Stem-end-rot and Anthracnose disease. International Journal of Tropical Plant Diseases, 6 : 99-105.

OmPrakash. and Srivastava, K.C. 1987. Mango diseases and their control – A World Review. Today and Tomarrow's Printers and Publishers – New Delhi, India.

Orian, G. 1960. Mango melanosis. Review. Agric. Suer Maurice, 39(3) : 128-130.

Padwal Desai, S.R., Ghanekar, A.S., Thomas, P. and Sreenivasan, A. 1973. Heat Radiation combination for control of mold infection in harvested fruits and processed used foods Acta Alimentaris, 2 : 189-207

Palaniswamy, A., Muthuswamy, S. and Krishnamurthy, C.S. 1972. Bacterial rot of stored mangoes. South Indian Horticulture, 20 : 102-103.

Pandey, K.K., Viswakarma, S.V. and Chaube, R.S. 2000. Fungicidal control of brinjal leaf spots caused by *Alternata alternata*. Pestology, 24 : 42-47.

Pandey, R.S., Bhargava, S.N. and Shukla, D.N. 1980. Two new fungicides for the control *Aspergillus* rot of mango. National Academy of Science Letters (India), 114 (3) : 243-248.

Pandey, R.S., Bhargava, S.N., Dwivedi, D.K. and Shukla, D.N. 1983. Studies on Botryodiplodia rot of mango. Indian Journal of Horticulture, 40(3-4) : 271-273.

Pandey, R.S., Bhargava, S.N. and Shukla, P.N. 1983. Control of Botryodiplodia rot of mango by oil-soap emulsion. International Journal of Plant Pathology, 2 : 1983.

Pandey, S.C. and Mohammad, A. 1974-75. Studies on fruit rot of mango III Brown spot caused by *Pestalotia mangiferae* Butl. Proceedings of Bihar Academy of Science, 22-23, 89-95.

Pandey, S.C. and Mohammed, A. 1976. Varietal behaviour of mango against fruit rotting fungus. *A niger*. Proceedings of Bihar Academy of Agricultural Science, 24 : 106-107.

Papavizas, G.C. and Lumsden, R.D. 1980. Biological control of soil borne fungal propagules. Annual Review of Phytopathology, 18 : 389-413.

Passam, H.C. 1982. Storage of some local and introduced mango cultivars grown in Trinidad. Scientia Horticulturae, 16 : 171-177.

Patel, J.G., Patel, S.T. and Patel, S.K. 1985. Occurrence of stem-end-rot of mango fruit in Gujarat. Gujarat Agricultural University Research Journal, 11(1) : 59-60.

Patel, M.K. and Padhye, Y.A. 1948. Bacterial spot rot of mango in Bombay. Indian Phytopathology, 1 : 127-128.

Patel, M.K., Kamat, M.N. and Bhidey, V.P. 1949. Fungi of Bombay. Supplement 1. Indian Phytopathology, 2. 42-145.

Patel, M.K., Kamat, M.N. and Padhye, Y.A. 1950. *Bacillus subtilis* causing soft rot of mango. Indian Phytopathology, 3 : 153-154.

Patel, R.B. 1972. *Aspergillus* rot of mango fruits in Gujarat, India. Current Science, 41 : 194-195.

Patel, R.B. and Goswami, N.P. 1984. Post-harvest diseases of mango fruits (*Mangifera indica*) and their control. Indian Journal of Microbiology, 24 : 140-141.

Patel, S.M. and Patel, B.K. 1998. Inhibitory effect of different fungal bioagents on *Fusarium oxysporum* f.sp. *cumini* causing of cumin wilt. Annals of Plant Protection Sciences, 6 : 25-27.

Pathak, V.N. 1966. Studies on diplodia stem-end-rot of mango fruit with emphasis on the epidemiology and prevention of the disease. Ph.D. Thesis, Indian Agricultural Research Institute, New Delhi.

Pathak, V.N. 1970. Diseases of mango fruit. Indian Farmer's Digest, 3(12) : 43-45.

Pathak, V.N. 1973. Diplodia stem-end-rot of mango and its control. All India Workshop on Fruit Research, Solan.

Pathak, V.N. 1980. Diseases of fruit crops. Oxford and IBH Publishing Company, New Delhi.

Pathak, V.N. 1997. Post harvest fruit pathology – Present status, Future possibilities. Indian Phytopathology, 50 : 161-185.

Pathak, V.N. 1998. Post-harvest fruit diseases – Pathosis and Management. Journal of Mycology and Plant Pathology, 28(2) : 87-113.

Pathak, V.N., Goyal, J.P. and Sharma, H.C. 1971. Screening trials for control of Diplodia and Rhizopus rots of mango fruits. Plant Disease Reporter, 55 : 752.

Pathak, V.N. and Khandalwal, G.S. 1969. Radiation and chemicals in the control of Diplodia stem-end-rot of mango fruits. Tropical Agriculturist, 125(3&4) : 99-100.

Pathak, V.N., Majumdar, M.L., Kumar, P. and Godara, S.L. 1988. Biocontrol of post-harvest rots of fruits. Proceedings of International Congress of Research in Plant Sciences and Its Relevance to Future, UGC Centre of Advanced Study in Botany. University of Delhi, India, March 7-11, 1988, pp. 159.

Pathak, V.N. and Shekhawat, P.S. 1977. Efficacy of some fungicides and hot water in control of anthracnose and *Aspergillus* rot of mango fruits. Indian Phytopathology, 29(3) : 315-317.

Pathak, V.N. and Srivastava, D.N. 1967a. Mode of infection and prevention of Diplodia stem-end-rot of mango fruits (*Mangifera indica* L.). Plant Disease Reporter, 57 : 744-746.

Pathak, V.N. and Srivastava, D.N. 1967b. Mango losses due to Diplodia stem-end-rot. Tropical Agriculturist, 123 : 75-77.

Pathak, V.N. and Srivastava, D.N. 1968. Controlling stem-end-rot of ripe mango. Indian Farming, 18(5) : 43-45.

Pathak, V.N. and Srivastava, D.N. 1969a. Pectionolytic enzymes of *Diplodia netalensis* Pole Evans. Phytopathology Zhurnal, 65(3) : 263-266.

Pathak, V.N. and Srivastava, D.N. 1969b. Epidemiology and prevention of diplodia stem-end-rot of mango fruits. Phytopathology Zhurnal, 65(27) : 165-175.

Patibanda, A.K. and Prasad, R.D. 2004. Screening of *Trichoderma* isolates against wilt pathogens of safflower, *Carthamus tinctorius* L. Journal of Biological Control, 18: 103-106.

Patil, R.K. 1992. Studies on Rhizopus rot and Botrydiplodia rot of mango fruits. Ph.D. Thesis, Rajasthan Agricultural University, Bikaner, India.

Patil, R.K. and Pathak, V.N. 1993. Influence of temperature and relative humidity on incidence of *Botrydiplodia* and *Rhizopus* rot of mango fruits. Indian Journal of Mycology and Plant Pathology, 23 : 296-301.

Pavichitrasakul, S. 1980. Stem-end-rot caused by *Diplodia natalensis* Special problem. Kasetsart University, Bangkok, 28 p.

Peak, C.M., Fitzell, R.D., Hannah, R.S. and Batten, D.J. 1986. Development of micro-processor based data recording system for predicting plant disease based on studies on mango anthracnose. Comput Electron Agriculture, 1 : 251-262.

Pelser, P. Du., T. 1985. Control of post-harvest decay in mangoes. South African Mango Growers Association Year Book, 5 : 47-54.

Pelsar, P. Du., T. 1987. Control of soft brown rot and anthracnose in mangoes during cold storage by propiconazole and prochloraz. South African Mango Growers' Association Year Book, 7 : 25-29.

Pelser, P. Du., T. and Lesar, K.H. 1989. Decay control in South African mangoes by Flusilazol, Pemconazole and prochloraz during simulated shipment to Europe. Aspects of Applied Biology, 20 : 41-47.

Pelser, P.Du.,T and Lesar, K.H. 1990. Post-harvest decay control in mangoes under South African conditions. South Africa Mango Growers' Association Year Book, 10 : 11-17.

Pennock, W. and Maldonaldo, G.L. 1962. Hot-water-treatment of mango fruit to reduce anthracnose decay. Journal of Agriculture of the University of Puerto Rico, 46 : 272-283.

Person, R.C. and Hace, D.H. 1975. Factors affecting the occurrence and severity of black mold of ripe tomato fruits caused by *A. alternata*. Phytopathology, 65 : 1352-1359.

Peterson, R.A. 1986. Mango disease. In : Proceedings of the CISRO, 1st Australia Mango Research Workshop, 1984, CSIRO, Melbourne, pp. 233-247.

Phillips, D.J. 1984. Mycotoxin as a post-harvest problem. In : Moline, H.E. (ed.) Post-harvest pathology of Fruits and Vegetables : Post-harvest losses in perishable crops. Agricultural Experimental Station, University of California, Berkley Publication, NE, pp. 50-54.

Pinto, A.C. de Q., Andrade, S.R.M., Amaro, A.A. and Gomes, U. 2004. Mango industry in Brazil. Acta Horticulturae, 645 : 36-50.

Ploetz, R.C. 1994. Mango. In : Compendium of Tropical Fruit Diseases (Ed.), R.C. Zentmyer, G.A., Nishijima, W.T., Rohrbach, K.G. and Ohr. H.O. pp. 33-44, A.P.S. Press, St. Paul, USA.

Ploetz, R.C. 2003. Diseases of mango. In : Diseases of Tropical Fruits (Ed.) R.C. Ploetz. CAB International Publishing. Wallingford, UK, pp. 327-367.

Ploetz, R.C. 2004. The Major Diseases of Mango : Strategies and Potential for Sustainable Management. Acta Horticulturae, 455 : 137-150.

Ploetz, R.C., Zentinmeyer, G.A., Nishijima, W.T., Rohrbach, K.G. and Ohr, H.D. (eds). 1994. Compendium of Tropical Fruit Diseases, APS Press, St. Paul, USA

Pordesimo, A.N. and Barredo, F.C. 1976. Effect of fungicide and insecticide sprays on mango fruit setting and post-harvest rots. Philippine Phytopathology, 12 : 51-59.

Prasad, S.S. and Sinha, A.K. 1981. Effect of temperature and humidity on fruit rot of mango. National Academy of Science Letters (India), 4 : 345-347.

Prasanna Kumar, M.K., Nargund, V.B. and Khan, A.N.A. 2006. Laboratory evaluation of fungicides and Botanicals against *Alternaria alternata* causing post-harvest disease in mango. Mysore Journal of Agricultural Science, 40(1) : 21-26.

Prasanna Kumar, M.K., Nargund, V.B. and Reddy, C.N.L. 2007. Severity of post-harvest diseases in Mango. Journal of Plant Disease Science, 2(1) : 73-76.

Prior, C., Elango, F. and Whitwell, A. 1992. p. 326-336, Chemical control of *Colletotrichum* infection in mangoes. In : J.A. Bailey and M.Jegar (eds), *Colletotrichum* : Biology, Pathology and Control. CAB, International Wallingford, UK.

Prusky, D. 1994. Part III. Mango : Mango Diseases caused by Fungi. In : Compendium of Tropical Fruit Diseases (Ed.) R.C. Ploetz, G.A. Zentlmyer, G.A., W.T. Nishijima, K.G. Rohrbach, and H.D Ohr. American Phytopathological Society, St Paul, Minnesota, pp. 34-35.

Prusky, D. 1996. Pathogen quiescence in post-harvest diseases. Annual Review of Phytopathology, 34 : 413-434.

Prusky, D. 2009. Mechanisms of Resistance and Susceptibility of Mango Fruits to Fungal attack and Their modulation by post-harvest handling and storage. Acta Horticulturae, 820 : 487-492.

Pruskey, D., Falick, E., Kobiler, I., Fuchs, Y., Zauberman, G., Pesis, E., Roth, I., Weksler, A., Akerman, M., Ykutiely, O., Waisblum, A., Keinan, A. and Ofek, G. 1997. Hot water brush : A new method for control of post-harvest disease caused by *Alternaria* rot in mango fruits. Acta Horticulturae, 455 : 780-785.

Prusky, D., Fuchs, Y., Kobiler, I., Roth, I., Weksler, S.Y., Fallik, E., Zauberman, G., Pesis, E., Akerman, M., O., Waisblum, R., Regev, O. and Artes, L. 1999. Effect of hot water brushing, prochloraz treatment and waxing on the incidence of blackspot decay caused by *Alternaria alternata* in mango fruits. Post-harvest Biology and Technology, 15 : 165-174.

Prusky, D., Fuchs, Y. and Yanko, U. 1983. Assessment of latent infections as a Basis for control of post-harvest disease of mango. Plant Disease, 67 : 816-818.

Prusky, D., Fuchs, Y. and Zauberman, G. 1981. A method of pre-harvest assessment of latent infections on fruits. Annals of Applied Biology, 98 : 79-85.

Prusky, D., Gat, Z. and Burd, P. 1992. Effect of relative humidity during mango growth on the incidence of quiescent infection of *Alternaria alternata*. Plant Disease, 77 : 249-252.

Prusky, D., Ilag, L., Ackerman, M., Kobiler, L. and Fuchs, Y. 2004. New Approaches for the control of post-harvest diseases of mango fruits. Acta Horticulturae, 645 : 129-136.

Prusky, D., Karni, L., Kobiler, I. and Plumbly, R.A. 1990. Induction of the antifungal diene in unripe avocado fruits : effect of inoculation with *Colletotrichum gloeosporioides*. Physiological and Molecular Plant Pathology, 37 : 425-435.

Prusky, D. and Keen, N.T. 1993. Involvement of pre-formed antifungal compounds in the resistance of sub-tropical fruits to fungal decay. Plant Disease, 77 : 114-119.

Prusky, D. and Keen, N.T. 1995. Inducible pre-formed compounds and their involvement in the resistance of plant pathogens, p. 139-152. In : R. Reuveni, (ed.) Novel Approaches to Integrated Pest Management. Lewis Publishers, Boca Raton.

Prusky, D., Kobiler., Akerman, M. and Miyara, I. 2006. Effect of acidic solution and acidic prochloraz on the control of post-harvest decay caused by *Alternaria alternata* in Mango and Persimmon fruit. Post-harvest Biology and Technology, 42 : 134-141.

Prusky, D., Kobiler, I and Jacoby, B. 1988. Involvement of epicatechin in cultivar susceptibility of avocado fruits to *Colletotrichum gloeosporioides* after harvest. Journal of Phytopathology, 123 : 140-146.

Prusky, D., Kobiler, I., Zauberman, G. and Fuchs, Y. 1993. Post-harvest conditions and post-harvest treatments affecting the incidence of decay in mango fruits during storage. Acta Horticulturae, 341.

Prusky, D., McEvoy, J.L., Leverentz, B. and Conway, L.S. 2001. Local modulation of host pH by *Colletotrichum* species as a mechanism to increase virulence. Mol. Plant Mic. Interact, 14 : 1105-1113.

Prusky, D. and Yakoby, N. 2003. Pathogenic fungi : leading or led by ambient pH. Molecular Plant Pathology, 4 : 509-516.

Quimio, A.J. and Quimio, T.H. 1974a. Post-harvest control of Philippine mango by hot water treatment. Philippine Agriculturist, 58(3&4) : 138-146.

Quimio, A.J. and Quimio, T.H. 1974b. Post-harvest control of Philippine mango anthracnose by benomyl. Philippine Agriculturist, 58 : 147-155.

Quimio, T.H. and Quimio, A.J. 1975. Pathogenicity of mango anthracnose. Philippine Agriculturist, 58 : 322-329.

Quroshi, S.U. and Meah, M.B. 1991. Post-harvest loss in mango owing to stem-end-rot. International Journal of Tropical Agriculture, 9 : 98-105.

Rada, G.G. 1939. La enfermedad de La antracnosis del mango. Circ. Estac. Exp. Agric. La, Molina, 50 : 6.

Rameshwar, A. 1988. Mango flowering – Stress induced. Acta Horticulturae, 231 : 433-439.

Rangaswami, G. 1962. Leaf spot of mango (*Mangifera indica* L.) 38, Chapter, 3. p. 127.8 in Bacterial Plant Diseases in India. Asia Publishing House, Bombay, India.

Rao, V.G. 1968. Studies on market and storage diseases of fruits and vegetables in Maharashtra. Journal of University of Poona, No. 34 : 21-50.

Raoof, M.A. and OmPrakash, 1983. Evaluation of some fixed oils for the control of black rot of mango fruits. Indian Journal of Mycology and Plant Pathology, 10: 348-349.

Raoof, M.A. and OmPrakash, 1986. Effectiveness of pre-harvest field sprays of various chemicals against post-harvest diseases of mango. Indian Journal of Plant Pathology, 4(2) : 189-190.

Rappel, L.M., Cooke, A.W., Jacobi, K.K. and Wells, I.A. 1991. Heat treatment for post-harvest disease control in mangoes. Acta Horticulturae, 291 : 362-371.

Rath, G.C. and Mohanan, M.K. 1986. Fungi inciting fruit rot of mango. Indian Phytopathology, 39(4) : 612-613.

Ratnam, C.V. and Nema, K.G. 1967. Studies on market diseases of fruits and vegetables. Andhra Agricultural Journal, 4 : 60-66.

Rawal, R.D. 1997. Management of fungal diseases in tropical fruits. In : Tropical Fruits in Asia. Proceedings of IPGRI-ICAR-UTFANET Regional Training Course, IIHR, Bangalore, India, p. 217-218.

Rawal, R.D and Ullasa, B.A. 1988. Efficacy of pre-harvest field sprays of fungicides against stem-end-rot of mango. Acta Horticulturae 231: 537-539.

Rios, R. and Arauz, L.F. 1997. Evaluaction de dos antioxidants Y un antitranspirances of el combate de antracnosis (*Colletotrichum gloeosporioides* Penz.) en mango (*Mangifera indica*) var. Tommy Atkins. En poscosecha (Abstr.). Memories XXXVII Congr. Assoc. Am. Phytopathology Dicision Centre – San Jose, Costa Rica.

Rios, R., Arauz, L.F. and Hord, M.J. 1999. Relation de variables elimatices con el desarolla de la antracnosis en fruitos de mango en Costa Rica. Page. 41. In. Memories XI Congr. Nacional Agron. Recursos Naturales, V. Congress Nacional Entomol. IV Congr. Nacional Fitopatol, III. Congr. Nacional Suclos I. Congress, Nacional Ext. Agric. Forestal, San Jose. Costa Rica, Vol. II.

Rotem, I. 1964. The effect of weather on dispersal of *Alternaria* species in a semi-arid regions of Israel. Phytopathology, 54 : 628-629.

Rotem, I. 1965. Sand and dust storms as factors leading to Alternaria blight epidemics on potato and tomatoes. Agricultural Meterology, 2:281-288.

Rotem, I. 1978. Climate and weather influence on epidemics. Pages 317-337 In. Plant Disease. Academic Press, Newyork.

Rotem, I. 1981. Fungal diseases of potato and tomatoes in the Neger Desert. Plant Disease, 65 : 315-318.

Roy, B.N., Biswas, S., Saha, S.K. and Pramanik, S. 1980. Effects of hot water with Captan on the storage and quality of mango fruits (*Mangifera indica*) Cv. Langra and Safder Pasand. South Indian Horticulture, 28 : 5-7.

Roy, S.K. and Pal, S. 1991. Multilocational studies to reduce post-harvest losses during harvesting, handling, packing, transport and marketing of mango in India. Acta Horticulturae, 291 : 499-507.

Ruehle, G.D. 1953. Organic fungicides for control of anthracnose of mango. Proceedings of Florida Mango Forum, 12-14.

Ruehle, G.D. and Ledin, R.B. 1955. Mango growing in Florida. Florida Agriculture Experiment Station Bulletin, 574 : 90 pp.

Ruehle, G.D. and Wolfen Barger, D.O. 1949. Diseases and pests of mango in Florida. Memiries Report of Florida Subtropical Experiment Station, 13: 6.

Saaiman, W.C. 1996. Evaluation of pre-harvest fungicides for the control of post-harvest diseases in mangoes. South African Mango Growers' Association Year Book 16 : 65-69.

Saaiman, W.C. 1997a. Orchard Sanitation as a means of controlling post-harvest diseases. South African Mango Growers' Association Year Book, 17 : 73-74.

Saaiman, W.C. 1997b. Shortwave infrared radiation as an alternative to the hot water bath to control post-harvest decay in mangoes. Acta Horticulturae, 455 : 773-776.

Saaiman, W.C. and Lonsdale, J.H. 1994. The effect of prochloraz in combination with guazatine on post-harvest diseases of mangoes. South African Mango Growers' Association Year Book, 14 : 62-64.

Saaiman, W.C. and Smith, Z. 1997. Selective medium for isolating *Nattrassia mangiferae*. Acta Horticulturae, 455 : 587-593.

Saini, P.K. and Pathak, V.N. 1991. Economic importance of *Aspergillus* rot of mango fruits in Udaipur. Annual Conference of Society of Mycology and Plant Pathology, Haryana Agricultural University, Hissar, Haryana. India, April 19-21, 1991. C-236.

Salunke, D.K. and Desai, B.B. 2001. Post-harvest biology of fruit. Vol. II CRC Press, Boca Raton, Florida, 168.

Sampaio, V.R., Borbin, D. and Dametrio, C.G.B. 1981. Post-harvest control of mango decay. Revista de Agricultura, Piracicabo, 56(3) : 173-182.

Sangchote, S. 1987. Post-harvest diseases of mango fruits and their losses. Kasatsart Journal (Natural Science), 21 : 81-85.

Sangchote, S. 1989. Effects of post-harvest treatments on anthracnose, *Colletotrichum gloeosporioides* Penz and stem-end-rot, *Dothiorella dominicana*, Pet. etcif. of mangoes stored in air or modified atmosphere. ASEAN Food Journal, 4 : 142-144.

Sangchote, S. 1991. Botryodiplodia stem-end-rot of mango and its control. Acta Horticulturae, 291 : 296-303.

Santha, K. and Gopal Rao, K.P. 1983. Post-harvest control of spoilage in mango (*Mangifera indica* L.) with hot water and fungicides. Journal of Food Science and Technology, 20 : 74-77.

Sarbhoy, A.K., Agarwal, D.K. and Varshney, J.L. 1968. The new host records of *Pestalotiopsis* from India. Indian Phytopathology, 31 : 535-536.

Sarkar, A. 1960. Leaf spot disease of *Mangifera indica* L. caused by *Pestalotia mangiferae*. Butl. Lloydia, 23 : 1-7.

Sattar, A. and Malik, S.A. 1939. Some studies on anthracnose of mango caused by *Glomerella cingulata* (Stonem) Spauld Sch (*Colletotrichum gloeosporioides* Penz.). Indian Journal of Agriculture Science, 9 : 511-521.

Saxena, A.K. and Rawal, R.D. 1998. Efficacy of safer and less toxic chemicals for the management of post-harvest rots in mango fruits (Abst.). Paper presented at National Symposium on Mango. Lucknow, 25-27, June, 1998, p. 64.

Schiffmann – Nadel, M., Waks, J., Gutter, Y. and Chalutz, 1983. Alternaria rot of citrus fruit. Proceedings of International Society of Citriculture, 2 : 791-793.

Schirra, M., D'hallewin, G., Ben-yehoshua, B. and Fallick, E. 2000. Host – pathogen interaction modulated by heat treatment. Post-harvest Biology and Technology, 21: 71-85.

Sechet, M. 1955. Note on the existence of perfect forms in cycle of *Pestalolia* sp. on fruit trees. Fruits de Outremer, 10(7) : 276-277.

Sepaiah, M. 1986. Effectiveness of hot water, hot benomyl and cooling on post-harvest diseases of mango Cv. Harumanies. ASEAN Food Journal, 2 : 117-120.

Sharma, A. and Verma, K.S. 2007. *In vitro*, cross pathogenicity and management of *Colletotrichum gloeosporioides* causing anthracnose of mango. Annals of Plant Protection Science, 15 : 186-188.

Sharma, I.M. and Badiyala, S.D. 1994. Effect of pre-harvest fungicidal sprays against stem-end-rot of mango fruits in storage caused by *Botryodiplodia theobromae.* Indian Journal of Mycology and Plant Pathology, 24 : 141-142.

Sharma, I.M., Raj, H. and Kaul, J.L. 1994. Studies on post-harvest diseases of mango and chemical control of stem-end-rot and anthracnose. Indian Phytopathology, 47: 197-200.

Sharma, N., Nigam, M. and Ghosh (Nayek), R. 1995. Biocontrol of black mould rot of mango with *Debaryomyces hansenii*. Indian Journal of Plant Pathology, 13(1&2) : 45-61.

Sharma, S.K. 2002. Post-harvest diseases of mango. In : V.K. Gupta and S.K. Sharma (Ed.). Diseases of Fruit Crops. Chapter 17 : 213-222. Kalyani Publishers, Ludhiana.

Sharp, J.L. and Spalding, D.H. 1984. Hot water as a quarantine treatment for Florida mangoes infested with Caribbean fruit fly. Proceedings of the Florida State Horticultural Society, 97 : 355-7.

Shellic, K.C. and Mangan, R.L. 1994. In : Post-harvest Handling of Tropical Fruits (Champ, B.R. Ed.). AICAR Proceedings, pp. 304-310.

Shiekh, I.A., Ali, S.S., Ehteshamuddin, A.F.M. and Haque, M.Y. 1977. Preservation of mangoes with fungicide wax emulsion. Pakistan Journal of Scientific and Industrial Research, 20 : 198-200.

Shu, Z.H. and Sheen, T.F. 1987. Floral induction of axillary buds of mango as affected by temperature. Scientia Horticulturae, 31 : 81-87.

Simmonds, J.H. 1963. Studies on the latent phases of *Colletotrichum* species causing ripe rots of tropical fruits. Queensland Journal of Agriculture, and Animal Science, 20 : 373-424.

Simmonds, J.H. 1965. A study on the species of *Colletotrichum* causing ripe fruit rots in Quensland. Queensland Journal of Agricultural and Animal Science, 22 : 437-439.

Singh, B.P. and Tandon, R.N. 1967. Pathological studies on some isolates of *Alternaria tenusis* causing leaf spot of *Musa pardisiala* L., *Psidium guajava* L., *Mangifera indica* L, and *Carica papaya* L.. Proceedings of National Academy of Science, India, 37 : 109-116.

Singh, L.B. 1960. The Mango. Leonard Hill (Book) Ltd., London.

Singh, R.H. 1968. Cultural and pathological studies of certain fungi. D. Phil. Thesis, Allahabad University, Allahabad, India.

Singh, R.S. 1968. Plant diseases. Oxford and IBH Publishing Co., New Delhi.

Singh, R.S. 1982. Plant pathogens. The fungi. Indian Book House, New Delhi, pp. 443.

Singh, R.S. 2000. Mango diseases. In : Diseases of Fruit Crops. Oxford and IBH Publishing Co., Pvt. Ltd., New Delhi, Calcutta, 153-182.

Sinha, I.B.P. and Bilgrami, R.S. 1987. Effects of antibiotics on growth, sporulation, spore germination and germ tube elongation of *Alternaria tenuis* Necs and *Pestalotia mangiferae*. P. Henn. Perspectives in Mycological Research Vol. I. Today and Tomorrow Printers and Publications, 131-138.

Sinha, S. 1945. Studies in the diseases of *Mangifera indica* L. V. The structure and development of lenticels in the mango fruits. Journal of Indian Botanical Society, 24(3) : 199-127.

Sinha, S. 1948. The pathogenicity of certain fungi attacking mango fruit in storage. Journal of Indian Botanical Society, 27 : 105-110.

Sive, A. and Resinzky, D. 1985. Experiments on CA storage of a number of mango cultivars in 1984. Alon Hantea, 39 : 845-855.

Slippers, B., Johnson, G.I., Cooke, A.W., Crous, P.W., Coutinho, T.A., Wingfield, B.D. and Wingfield, M.J. 2001. Taxonomy of *Botryosphaeria* sp. causing stem-end-rot of mango. Proceedings of 13th Biennial Australian Plant Pathology Conference, 24-27, September, 2001, Cairns, Australia.

Small, W. 1924. Annual Report of the Govt. Mycologist, Department of Agriculture, Uganda for the year ending 31st December, 1922, p.29.

Smoot, J.J. and Segall, R.H. 1963. Hot water as a control for decay of orange anthracnose. Plant Disease Reporter, 47 : 739-742.

Sohi, H.S. 1975. Anthracnose in Tropical Fruits. In : Advances in Mycology and Plant Pathology, (ed. Dr. Rayachaudhri), 193-204.

Sohi, H.S., Sokhi, S.S. and Teware, R.P. 1973. Storage rot of mango (*Colletotrichum gloeosporioides*) and its control. Phytopathologia Mediteranea, 12 : 114-116.

Sornsrivichaai, J., Anusandorn, P., Oogaki, C. and Gemma, H. 1989. Storage life and quality of mango (*Mangifera indica* L.). Cv. Keaw Sawoey fruit stored in seal packaging by plastic films and under low pressure at different temperatures. Japanese Journal of Tropical Agriculture, 33 : 6-17.

Spalding, D.H. 1982 Resistance of mango pathogens to fungicides used to control post-harvest diseases. Plant Disease, 66 : 1185-86.

Spalding, D.H., King, J.R. and Sharp, J.L. 1988. Quality and decay of mangoes treated with hot water for quarantine control of fruit fly. Tropical Science, 28 : 95-101.

Spalding, D.H. and Reeder, W.F. 1972. Post-harvest disorders of mangoes as affected by fungicides and heat treatments. Plant Disease Reporter, 56(9) : 751-753.

Spalding, D.H. and Reeder, W.E. 1986. Decay and acceptability of mangoes treated with combination of hot water, imazalil and radiation. Plant Disease, 70 : 1149-51.

Spalding, D.H., Reeder, W.F., Windeguth, D.L. Von. and Burdett, A.K. Jr. 1976. Effect of ethylene dibromide on appearance of decay and ripening time of mangoes. Proceedings of Florida State Horticulture Society, 88 : 471-476.

Sreenivasan, A.P., Thomas, P. and Dharkar, S.D. 1971. Physiological effects of gamma radiation on some tropical fruits. In : Disinfestation of Fruit by Irradiation. I.A.E.A., Vienna, pp. 65-91.

Srikanth, T. 1989. Studies on market diseases of mangoes (*Mangifera indica* L.). M.Sc. (Ag.) Thesis. Andhra Pradesh Agricultural University, Hyderabad, Andhra Pradesh, India.

Srivastava, D.N. 1972. Epidemiology and prevention of diplodia stem-end-rot of ripe mango fruits. Acta Horticulturae, 24 : 235-236.

Srivastava, D.N. and Durgapal, J.C. 1965. Mode of infection and control of Diplodia stem-end-rot of mango (*Mangifera indica* L.). Indian Journal of Horticulture, 22 : 77-79.

Srivastava, M.P., Chandra, S. and Tandon, R.N. 1964. Post-harvest diseases of some fruits and vegetables. Proceedings of National Academy of Science. India, 34(4) : 339-342.

Srivastava, M.P. and Tandon, R.N. 1966. Effect of *Botryodiplodia* infection on the Vitamin-C content of mango fruit. Current Science, 35 : 419.

Srivastava, M.P., Tandon, R.N., Bhargava, S.N. and Ghosh, A.K. 1965. Studies on fungal diseases of some tropical fruits III. Some post-harvest diseases of mango. Proceedings of National Academy of Science, India. 35 B : 69-75.

Srivastava, M.P., Tandon, R.N., Bilgrami, K.S. and Ghosh, A.K. 1964. Studies on fungal diseases of some tropical fruit I. A list of fungi isolated from fruits and fruit trees. Phytopathology Zhurnal, 50 : 250-261.

Stevens, F.L. and Pierce, A.S. 1933. Fungi from Bombay. Indian Journal of Agricultural Science, 3 : 912-916.

Stevens, H.E. 1936. Control of mango blossom blight and anthracnose. Proceedings of Horticulture Society, 49 : 125-130.

Steyeart, R.J. 1953. *Pestalotiopsis* from the Gold Coast and Togaland. Transactions of British Mycological Society, 36 : 235-242

Steyn, P.L., Viljoen, N.M. and Kotze, J.M. 1974. The causal organism of bacterial spot of mangoes. Phytopathology, 64 : 1400-1404.

Stover, R.R. 1972. Banana, Plantain and Abaca Diseases. Common Wealth Mycological Institute, England, p. 315.

Su, M.T. 1933-34. Report of the Mycologist, Burma, Mandalay for the year ending with 31st March 1934. Report of Department of Agriculture, Burma, 1934 : pp. 25-33.

Subramanyam, B. 1985. Studies on anthracnose and twig blight of cashew. M.Sc.(Ag.) Thesis, Andhra Pradesh Agricultural University, Hyderabad, India.

Subramanyam, H. and Narayana Murthy, N.V. 1973. Control of spoilage and ripening in mango fruit by zineb and sodium diethyl dithio carbanate. Pesticide Science, 4: 25-31

Subramanyam, H., Narayana Murthy, N.V., Lakshminarayana, S. and Dalal, V.B. 1972. Control of fungal spoilage in Alphonso mangoes by pre-harvest application of fungicides. Acta Horticulturae, 24 : 224-226.

Sumbali, D. and Badyal, K. 1991. Relationship between fungal air-spora of fruit shops and incidence of fruit rots. Indian Phytopathology, 44 : 214-218.

Sumbali, G. and Mehrotra, R.S. 1980. Evaluation of some fixed oils for the control of certain temperate fruit rot fungi. Indian Phytopathology, 33 : 517.

Sundaraj, J.S., Muthuswamy, S. and Palaniswami, A. 1972. Bacterial rot of stored mangoes. Acta Horticultuae, 24 : 213-219.

SundaraRaman, S., Nayar, C.K. and Ramakrishnan, T.S. 1928. The stem bleeding disease of Arecanut (*Areca catechu*) caused by *Thielaviopsis paradexa* Vantoohn. Agriculture Research Institute, Pusa, Bulletin, 169, p. 12.

Suvarna, J., Basha, S.T., Hemalatha, T.M. and Eswar Reddy, N.P. 2009. Efficacy of fungicides against *Colletotrichum gloeosporioides* causing mango anthracnose. Indian Journal of Plant Protection, 37(1&2) : 201-203.

Swart, S.H., Serfontein, J.J. and Kalimowski, J. 2002. Chemical control of post-harvest diseases of mango – the effect of Prochloraz, Thiabendazole and Fludioxonil on Soft brown rot, stem-end-rot and anthracnose. South African Mango Growers' Association Year Book, 22 : 55-62.

Swart, S.H., Serfontein, J.J., Swart, G. and Labuschange, C. 2009a. Chemical control of post-harvest diseases of mango : the effect of Fludioxonil and Prochloraz on Soft brown rot, Stem-end-rot and Anthracnose. Acta Horticulturae, 820 : 503-509.

Swart, S.H., Swart, G. and Labuschange, C. 2009b. The effect of strategically timed pre-harvest fungicide application on post-harvest decay of mango. Acta Horticultuae, 820 : 511-519.

Swart, S.H. and van Brockhuizen, W. 2004. Die effect Van Strategies plasing Van Vooroesswamdo derbes puilings op. na-oessiektes by mangoes. South African Mango Growers' Association Year Book, 24 : 56-60.

Swinburne, T.K. 1983. Quiescent infections in post-harvest diseases. In : Post-harvest pathology of fruits and vegetables. (Ed). Dennins, C. Academic Press, London, pp. 1-21.

Tandon, I.N. and Singh, B.B. 1968a. Control of mango anthracnose by fungicides. Indian Phytopathology, 21 : 212-216.

Tandon, I.N. and Singh, B.B. 1968b. Control of mango anthracnose by hot water treatment. Indian Phytopathology, 21 : 331-336.

Tandon, R.N. 1967a. Observations on storage diseases of certain fruits. Indian Phytopathology, 20 : 1-12.

Tandon, R.N. 1967b. Post-harvest diseases of tropical and subtropical fruits. Final Technical Report, PL 480 Scheme FG-in-133, Allahabad University, India.

Tandon, R.N. 1970. Certain problems of post-harvest diseases of fruits and vegetables. Indian Phytopathology, 23 : 1-15.

Tandon, R.N., Sisodia, U.S. and Bilgrami, K.S. 1955. Pathological studies of *Pestalotia mangiferae*. Proceedings of Indian Academy of Science, 42 : 219-255.

Tandon, R.N., Srivastava, M.P., Bhargava, S.N. and Ghosh, A.K. 1965. Studies on fungal diseases of some tropical fruits (Abs). 34th Annual Session of National Academy of Science, India, 58-59 p.

Thakur, D.P. 1972a. A new host for *Rhizopus stolonifer*. Indian Phytopathology, 25 : 158-160.

Thakur, D.P. 1972b. Factors influencing storage rots of certain fruits and vegetables caused by three species of *Rhizopus*. Indian Phytopathology, 25 : 354-358.

Thakur, D.P. and Chenulu, V.V. 1966. Survey for storage and market diseases of certain fruits and vegetables in Delhi (Abs). International Symposium on Plant Pathology, Dec. 27, New Delhi, 35-36, p.

Thakur, D.P. and Chenulu, V.V. 1967. Survey of Delhi market for fungal disease of fruits and vegetables. Plant Disease Problems. Proceedings First International Symposium of Plant Pathology, India, New Delhi, 1970, pp. 308-315.

Thakur, D.P. and Chenulu, V.V. 1970a. Chemical control of soft rot of apple and mango fruits caused by *Rhizopus arrhizus*. Indian Phytopathology, 23 : 58-61.

Thakur, D.P. and Chenulu, V.V. 1970b. Survey of Delhi market for fungal diseases of fruits and vegetables. In: Plant Disease Problems (Ed. S.P. Royachaudhuri *et al*), pp. 308-315.

Thomas, A.C. 1975. The application of ionizing radiation to the shelf life extension of mangoes in South Africa. At. Energy Board. Rep. South Africa, PEL, 244, 15 pp.

Thomas, A.C. and Brodrick, H.T. 1977. Current status of the South African research program on the radiation preservation of sub-tropical fruits. At Energy Board Rep. South Africa. Pelindaba, PER-9, 19 pp.

Thomas, P. and Dalal, V. 1968. A note on the effect of some chemicals on control of stem-end-rot in mangoes after harvest. Science and Culture, 34(7) : 311.

Thomas, P. and Desai, S.R.P. 1976. Improvement of shelf life and quality of mangoes by gamma irradiation. Indian Food Packer, 30(5) : 83-86.

Thompson, A.K. 1987. The development and adoptation of methods for control of anthracnose. In : R.T. Prinsley, and G.Tucker, (Eds.) Mangoes : A Review, Common Wealth Science Council. London, pp. 29-38.

Tirmazi, S.I.H. and Wills, R.B.H. 1981. Retardation of ripening of mangoes by post-harvest application of calcium. Tropical Agriculture, 58 : 137-141.

Tomkins, R.G. 1936. Wraps for the prevention of rotting of fruits. Great Britan, Department of Science, Industry and Research. Food Industry Board Report, 129-151.

Traub, H.T. and Robinson, T.R. 1938. Improvement of subtropical fruits other than citrus. Year Book of U.S. Department of Agriculture,:1589.

Udhaykumar, R. and Usha Rani, S. 2010. *In vitro* efficacy of bacterial antagonists on the incidence of mango anthracnose and development of its pathogen. Agriculture Science Digest, 30(4) : 293-295.

Ullasa, B.A. 1989. Plant Disease Research, 4 : 196-197.

Ullasa, B.A. 1991. Fitopatol Brasseleria, 16 : 286-288.

Ullasa, B.A. and Rawal, R.D. 1982. Annual Report, I.I.H.R., Bangalore, India, p. 207.

Ullasa, B.A. and Rawal, R.D. 1988. Occurrence of a new post-harvest disease of mango due to *Pestalotiopsis glandicola*. Acta Horticlturae, 231 : 540-543.

Uppal, B.N., Patel, M.K. and Kamat, M.N. 1935. The fungi of Bombay. Department of Agriculture, Bombay Bull. No. 176 : 203-207.

Van Etten, H.D., Mansfield, J.W., Bailey, J.A. and Farmer, E.E. 1994. Two classes of plant antibiotics : Phytoalexins versus Phytoanticipins. Plant Cell, 9 : 1191-1192.

Verhoeff, K. 1974. Latent infections by fungi. Annual Review of Phytopathology, 12 : 99-110.

Verma, G.S. and Kamal, M. 1951. Rot of *Mangifera indica* L. caused by *Aspergillus*. Current Science 20 : 68-69.

Viegas, A.O. 1946. Some fungai of Brazil (St.Paulo). Bragantia, 6 : 1-37.

Villiers, E.E. de and Korsten, L. 1994. Biological treatments for the control of mango post-harvest diseases. South African Mango Growers' Association Year Book, 14: 48-51.

Vir, D. and Sharma, D. 1965. A note on mango fruit rot in storage. Indian Phytopathology, 18 : 390.

Vyas, R.K. and Pathak, V.N. 1995. Management of *Aspergillus* rot of mango fruits. Journal of Mycology and Plant Pathology, 25 : 133.

Wadia, K.D.R. and Manoharachary, E. 1979. Studies on fruit rot disease from Andhra Pradesh, India. Indian Journal of Botany, 2(1) : 121-125.

Wang, B., Chaozhen, L.I.V., Pakun, Q.I., Wang, B.S.G.V.C.Z. and Qipk, 1994. Identification of mango stem-end-rot pathogens and its control in post-harvest. Journal of South China Agricultural University, 15 : 355-360.

Wardlaw, C.W., Baker, R.E.D. and Crowdey, S.H. 1939. Latent infection in tropical fruits. Tropical Agriculture, Trinidad, 16 : 275-276.

Wardlaw, C.W. and Leonard, E.R. 1936. The storage of West Indies mangoes. Memoirs of Low Temperature Research Station, 2 : 47-48.

Waskar, D.P. and Gaikwad, R.S. 2005. Post-harvest hot-water-treatment for disease control in Kesar mango fruits. Indian Journal of Agriculture Research, 39(3) : 186-191.

Wells, I.A., Peacock, B.C., Brown, B.I. and Littlemore, J. 1989. Post-harvest mango sap activity of the Australian Cultivar 'Kensington Pride'. Acta Horticulturae, 291.

West, E. 1934. Anthracnose of mango. Bulletin of Florida Agriculture Experiment Station, 463.

Wisniewski, M.E. and Wilson, C.L. 1992. Biological control of post-harvest diseases of fruits and vegetables : Recent Advances. HortScience, 27 : 94-98.

Yadava, A.S. and Udainarayana 1970. Effect of micro-succession on the fallen stem of *Mangifera indica* L. II. The ability of microfungi to cause actual decay of the woody stem *in vitro* and the factors which influenced. Journal of Indian Botanical Society, 47(3-4) : 274-281.

Zauberman, G., Schiffmann – Nodel, M. And Yanko, U. 1976. Control of fungi causing decay of mango fruit during storage. Special Publication of Agricultural Research Organisation, No. 65 : 105-106.

CHAPTER 11

Post-Harvest Physiological Disorders of Mango Fruits

K.M. Yuvaraj

Exposure of fruits to various environmental, cultural and nutritional stresses and imbalances often leads to tissue abnormalities within the fruit which are quite different from those caused by pathological and entomological agents. Such abnormalities are generally termed as 'physiological disorders' which are essentially the result of imbalances in metabolism induced by some factors in the pre-harvest or post-harvest environment that leads to cell collapse and the appearance of water-soaked or brown areas in some parts of the fruit, externally or internally (Lizada *et al* 1984). Such physiological disorders are also observed in mango in most of the major growing areas of the world. Knowledge on the metabolic changes causing physiological disorders in mango is limited. Briefly, pre-harvest factors that have been found to pre-dispose mango fruits to physiological disorders including growing location, orchard conditions, tree nutrition and conditions at harvest, while post-harvest factors (temperature, oxygen, and CO_2) are contributing to the physiological disorders. Identification of the causes has not been possible as there are poor correlations between the outbreak of physiological disorders and specific factors.

The most extensive description of physiological disorders in mango were given by Malo and Campbell (1978) and discussed and reviewed by Wainwright and Burbage (1989), Lizada *et al* (1984) and Lim and Koo (1985) from different mango production areas. The physiological disorders of mango fruits are detailed in the following pages.

SPONGY TISSUE

'Alphonso' is one of the premier cultivars of Indian mangoes. It is the choicest variety with multiple good characteristics. It is the most suitable commercial mango for export from India. It has excellent eating and keeping qualities and attractiveness as compared to other cultivars. Inspite of all these good characteristics, it suffers from two problems viz; alternate bearing and spongy tissue.

"Spongy tissue" is the development of a non-edible sour patch with or without air pockets in the mesocarp (pulp) of fruits, which renders the fruit unfit for human consumption and processing. It is a serious problem in domestic trade, overseas trade and processing industry. The disorder, also called "soft centre", "white corky tissue" and "internal breakdown" has become a bottle neck in the expansion of its cultivation in Maharashtra, Karnataka, Goa and Andhra Pradesh (Rane *et al* 1976), as well as in export of this fruit to global markets. Spongy tissue of mango is considered to be a type of internal flesh breakdown (Wainwright and Burbage, 1989).

Geographical Occurrence

The occurrence of 'spongy tisue' in 'Alphonso' mango fruits was first noticed in 1932 by Cheema and Dani (1934), when the export of 'Alphonso' mango fruits from erstwhile Bombay province to Europe was detained and examined. A survey conducted in southern states of India indicated that 'Alphonso' grown in Maharashtra, Gujarat, Karnataka, Goa and Andhra Pradesh is also affected by this disorder. The disorder has been reported from Malaysia (Kwee and Chong, 1985), Nepal and Philippines (Gautam and Lizada, 1984., Nuevo *et al* 1984) and Australia (Chaplin, 1986).

Symptoms (Photo 11.1)

The peculiar nature of the disorder is that it does not exhibit any external symptoms either at picking time or when ripe. Hence, it is difficult to sort out the affected fruits from the unaffected fruits. The symptoms of spongy tissue are internal and are seen only when the fruit is cut open, as pale yellow flesh, soft or spongy with or without fissures (air pockets) and with an off-flavour. X-ray images of affected fruits show dark brown (grey) patches corresponding to internal cavities in contrast to light areas of healthy flesh (Thomas, *et al* 1993). The intensity of spongy tissue in a fruit varies from a spot to the whole fruit pulp. However, different workers described variable symptoms of the disorder as follows:

Cheema and Dani (1934) described three kinds of abnormalities in affected fruits viz.,

(a) The blemishes occurring on all parts of the fruit when cut open. The flesh of the badly blemished fruits was abnormal, in certain regions being hard and rubbery in texture, pale in colour, lacking in juice and very acid in taste. The abnormal flesh is separable from surrounding normal flesh with ease. Blemished fruit could be separated from normal fruits.

(b) The second type of abnormality was browning of the tissues around the seed (stone), which commences in the vascular tissue to the stone and gradually spreads outwards. In some cases, the brown colour of the vascular tissue stood out very distinctly against the yellow colour of the flesh, while in acute cases, 50 per cent of the total flesh was affected.

(c) The third abnormality was breaking down of tissue around the stone with abnormal flavour. This type of abnormality was observed mainly in the last consignment which was kept in cold storage, before export.

Plate 11.1: Spongy tissue in ''Alphonso'' mango fruit

Amin (1967) described spongy tissue as white sponge-like, corky tissue, slightly desiccated in the pulp between the skin and stone of the ripe fruit. According to Chhatpar, *et al* (1968-69), the main feature of this abnormality is a slight desiccation in the centre of the mesocarp tissue surrounded by a soft hallow between skin and stone of the fruit. Subramanyam *et al* (1971) described the disorder as an internal breakdown and that the external symptoms were not apparent either at the time of picking or at ripening stage; the affected tissue is differentiated from healthy one, by its pale yellow colour, soft, leathery or spongy tissues with or without air-pockets accompanied by an off-flavour.

The location of the spongy tissue in the pulp varies, maximum being with lower part of the fruit (Tare, 1977., Katrodia, 1979., Katrodia and Rane, 1989). The higher incidence at the lower part (apex) of the fruit was attributed to the heat coming from soil surface (due to solar radiation in the months of April and May) as convective flux, first touching the apex of the fruit, facing towards the soil and entering the fruit pulp and causing the damage.

Fruits affected by this disorder possess a few black or brown veins originating from the exterior of the endocarp. The mesocarp surrounding these discoloured strands often turns soft and watery and is white to yellowish spongy-like at maturity, with embedded air pockets (Murthi and Upreti, 2000).

Post-harvest exposure of fruits to sunlight results in internal breakdown in which the flesh appears pale yellow in colour, with more air pockets, harder, leathery than the normally noticed spongy tissue that develops prior to picking (Gunjate *et al* 1982).

Causes

Though spongy tissue was reported in 1932 (Cheema and Dani, 1934) and exports were jeopardized, very little work was done until 1970's regarding its causes and control when the disorder approached alarming proportions threatening the cultivation of 'Alphonso' mango and its export. Much of the light on its causes, symptoms, factors and control was thrown through students research (Joshi, 1975., Rangwala, 1975., Tare, 1977., Patkar, 1978., Katrodia, 1979., Sheth, 1981). However, there is much controversy surrounding the cause(s) of spongy tissue in mango fruits, necessitating further research. The causes so far put forth are discussed under the following heads:

Mineral Nutrition

Spongy tissue was initially thought to be due to deficiency of nutrients like N, P, K and Ca, particularly the last one. However, studies on these assumptions proved otherwise. Katrodia (1979) observed that the soil of mango orchards with affected fruits, was well supplied with both macro-and micro-nutrients at fruitset as well as at fruit harvesting stages, indicating that probably the nutrients have no role in causing spongy tissue in mango. Rangwala (1975) observed that N, P and K contents in the leaves of affected fruits (of Alphonso) were higher and 'Ca' and 'Mg' were lower than in the leaves of healthy ones. Tare (1977) observed that 'Ca' content in "Alphonso" leaf increased from 1.63 per cent at fruit set stage to 1.86 per cent at fruit harvest stage. Katrodia (1979) reported that leaf and stalk (petiole) from spongy tissue affected "Alphonso" fruits and those from healthy fruits did not differ significantly in respect of macro and micro-nutrients indicating that probably nutritional status of leaf and stalk had no relation with spongy tissue development. The lack of reduction of the incidence of spongy tissue, following the application of N,P and K nutrients to soil as well as to the foliage, support that N, P and K contents either in the soil or in the leaves have no relationship with spongy tissue (Amin, 1967., Joshi and Limaye, 1984). On the contrary, their application increased the incidence of spongy tissue (Subramanyam *et al* 1971., Limaye *et al* 1975).

Many fruit disorders are related to deficiency of calcium. Comparative study of healthy and affected fruit pulp tissues suggested that calcium deficiency may be involved in the disorder (Subramanyam *et al* 1971., Gunjate, *et al* 1979a). Reports on the contents of 'Ca' and 'P' in affected and healthy tissues of fruits were not uniform (Subramanyam *et al* 1971., Gunjate, *et al* 1979a., Santha, 1981., Katrodia and Sheth, 1988). Results with pre- and post-harvest dips or sprays of calcium were also inconsistent indicating that calcium may not be involved in the spongy tissue development (Gunjate *et al* 1979 b., Santha, 1982., Katrodia and Sheth, 1988). However, it appears that the calcium deficiency in certain parts of the pulp of mango fruit might be leading to metabolic changes and skeletal weakness of tissue leading to this disorder.

Raymond *et al* (1998b) also found that there was no relationship between calcium concentration and internal breakdown in 'Tommy Atkins' mango fruits. However, their results indicated that a nutritional imbalance during fruit ontogeny might be responsible for internal breakdown, since at the early stages of disorder N, P Ca and B concentration were higher in disordered fruit than in healthy fruits. They also opined that an unknown factor triggered the biochemical processes that resulted in these elevated concentration of N and P in the disordered fruit, resulting in the early ripening of the mesocarp. On the contrary, studies on "Tommy Atkins" mango fruit harvested at phenological maturity and stored at 12 °C for up to 28 days showed internal break-down and a relation with calcium content of the fruit pulp (Lima *et al* 2000). According to Burdon *et al* (1991) internal breakdown may be the result of redistributed or compartmentization of nutrients within the fruit.

Santha (1981) suggested two possible reasons for spongy tissue in mango fruits viz.,

(i) physiological causes involving improper distribution of minerals, hormones etc., during growth and development of fruit, since a similar type of corking disorder existing in apple and pear was overcome by pre-harvest sprays of minerals ('Ca' and 'B') or a post-harvest dip treatment of fruits and

(ii) pathological causes in which bacteria might be involved.

Physiological Causes

Katrodia (1979) described the spongy tissue as a physiological disorder in which the fruit remains unripe because of the unhydrolyzed starch due to physiological and biochemical disturbances caused by heat in immature fruit at pre- and post-harvest stages. This ripening disorder occurs because of inhibition of ripening in certain portions of the pulp, caused by inactivation of ripening enzymes due to high temperature convective heat, post-harvest exposure to sunlight etc., (Chadha, 1989).

Transpiration is one of the reasons for some fruit disorders in other crops when grown under high humidities (Bangerth, 1979., Holder and Cockshell, 1990). In Maharashtra, 'Alphonso' mangoes are grown under high humid conditions. This was evident by the high incidence of spongy tissue observed in fruits harvested after a spell of heavy rains. In this case, high humidity associated with rains could have reduced the transpiration rate of fruits, thereby increasing the incidence of spongy tissue in 'Alphonso' mangoes. This was mainly due to the lower transpiration from the leaves and fruits, leading to the lower transfer of mineral nutrients to the fruits (Clarkson, 1984., Menzal and Kirkby, 1987).

Based on the above, the relationship of leaf and fruit transpiration rates with the incidence of spongy tissue was investigated. The significant and negative relationship observed between the fruit transpiration rates suggest that the lower fruit transpiration rates in 'Alphonso' cultivar were varietal specific traits which result in slowest movement of water and minerals to the fruit from soil leading to the development of spongy tissue.

A significant reduction in spongy tissue when the fruit transpiration was increased by using perforated polythene covers as well as the slight increase in the incidence, when fruit transpiration was checked substantiate the theory that fruit transpiration is involved in the development of spongy tissue in mango (Shivashankar and Mathai, 1999).

Biotic Causes

Chhatpar *et al* (1968-69) isolated *Bacillus sp* of bacteria from spongy tissue and reproduced the disorder on inoculation of healthy mango fruits and concluded that the bacteria, *Bacillus sp* is associated with the disorder. The presence of *Bacillus sp* was also reported by Subramanyam *et al* (1971) and Assani *et al* (1971) in the spongy tissue of 'Alphonso' mangoes. The probable source of infection (entry) of bacteria may be at flowering stage (Modi, 1975). However, Desai (1966), Subramanyam *et al* (1971) and Katrodia (1979) ruled out the possibility of such association of bacteria in the absence of establishment of pathogenicity and owing to unfarouralde conditions (low pH, high acidity) of ripe fruit, for microbial growth and suggested that the physiological abnormality of the tissue pre-disposed the fruit to secondary infection. The association of other pathogenic organisms (fungus, yeasts or virus) was not proposed with the incidence of spongy tissue. Thus, biotic causes for spongy tissue were ruled out.

Environmental Causes

Among the different environmental parameters heat and temperature appear to be playing a major role in the development of spongy tissue. Desai (1966) observed that, when atmospheric temperature was high, fruits exposed to direct sunlight were normally affected by this disorder. He further, opined that exposure of fruit on western side was more detrimental and allowed the development of spongy tissue in such exposed fruits.

Contrary to this, Katrodia (1979) reported that direction of sunlight had no effect on the incidence and intensity of spongy tissue, since shaded fruits were also affected, while those exposed to direct sunlight were hardly affected. Thus, sunlight had an insignificant or no role in the development of spongy tissue.

Hence, the possibility of heat from soil surface inducing spongy tissue was investigated (Katrodia, 1979). He reported that convective heat arising from the soil was the main cause of spongy tissue in 'Alphonso' mango fruits. Because of this convective heat the lower part (apex/distal) of fruit was more affected than upper and middle part and entire fruit. The reduction of incidence of spongy tissue by mulching the tree basins or practicing sod culture in the garden (Katrodia and Rane, 1979., Katrodia and Sheth 1988., Burondkear *et al* 1994) supports the claim that convective heat from the soil is the principal cause of spongy tissue in mango fruits.

The development of spongy tissue in mango fruits, due to convective heat from the soil was explained by Katrodia and Rane [1981]. According to them the soil surface below the tree canopy gets heated up by solar radiation to the extent of 58 ^{o}C at noon during day time due to prolonged exposure in April, May and June months in the garden. Under such situation, the mode of action, in transference heat jointly by radiation, conduction and convection causes continuous currents to flow as cenvective heat from soil towards the fruits on the tree. This convective heat penetrates fruit pulp through the skin and reaches the stone more where it gets accumulated raising the temperature of the fruit pulp to the extent of causing the development of spongy tissue. This high temperature in fruit pulp near stone disturbs the normal ripening process of fruit pulp by destroying the functions of ripening related enzymes (Treshow, 1970). Secondly, high temperature in combination with high relative humidity under the tree, lowers the rate of evoparation from the soil, consequently resulting into a high sensible heat (Critchfield, 1975) which causes more damage to the fruits. This explains as to why the fruit from vigorous trees showed more incidence as compared to those from weak trees. Similarly blowing wind also, might have helped to some extent the weak tree in keeping the micro-climate milder around the fruits as against the vigorous tree where winds usually blow touching the crown are entrapped in tree canopy and get heated up. Then the convection flux of heat in combination with high humidity and temperature caused more damage in the fruits harvested from the vigorous trees than those harvested from weak trees.

The reason of high incidence of spongy tissue in fruits under shade as compared to direct sunlight was also explained by Katrodia (1979). According to him, exposed fruits are resistant to heat, whereas the fruits under shade are more susceptible to connective heat as these are not acclematized to heat.

Artificial production of spongy tissue by applying proper amount of heat by exposing mature fruits to sun, infrared rays, hot-plate and incubation at pre- and post-harvest steps also indicated that convective heat is responsible for the development of spongy tissue in mango fruits (Katrodia, 1979., Sheth, 1981).

Thus it was demonstrated that the development of spongy tissue in mango fruits was mainly due to convective heat coming from the soil surface of tree basins during the final stage of fruit maturity. The temperature at the surface of the stone and at the centre of the pulp (1.5 cm. below the skin) was as high as 52.5 °C and it was above 47 °C where spongy tissue has developed. This clearly indicates that high temperature caused denaturation of protein and destroyed the catalytic function of enzymes and led to spongy tissue (Katrodia, 1988).

Factors in Relation to Spongy Tissues

Several factors related to the plant, fruit, nutritional and environmental ones were attributed to aggregate the disorder. Irrigation at maturity stage, late harvest and clean cultivation favour the incidence of spongy tissue in mango (Katrodia, 1979).

Varietal Susceptibility

"Spongy tissue" was found in all the grafted and country mango varieties in Gujarat (Desai, 1966). Originally, "spongy tissue' was reported in "Alphonso" mango (Cheema and Doni 1934). But later studies indicated that spongy tissue was not restricted to 'Alphonso' cultivar alone, but also occurs in other cultivars (Limaye *et al* 1975). They reported spongy tissues in 'Vanraj' (36.7%), 'Olour' (26.6%), 'Goa Manker' (13.33%), 'Vellai-kollamban' (10%), 'Suvarnarekha' (6.69%) and 'Fernandin' (3.33%) and the cultivars 'Pairi', 'Kesar', 'Doodh Peda', 'Neelum' and 'Dasheri' were free from spongy tissue at Vengurla (Maharastra, India). Earlier Desai (1966) found 'Kesar' as less susceptible to spongy tissue. Amin (1967) found that 'Jamadar' cultivar was susceptible. Katrodia (1979) studied the reaction of 'Rajpuri', 'Vanraj', 'Afus', 'Kesar', a hybrid of 'Alphonso' and 'Baneshan', 'Jamadar' and country (local) varieties to artificial production of spongy tissue, among which 'Rajpuri' cultivar showed resistance even at 4 hours exposure to sun between 11.30 and 15.30 hours at pre-harvest stage. Rest of the cultivars showed different degrees of susceptibilities. This variable reaction of mango cultivars was attributed to different physico-chemical properties of the fruit skin which might have acted as an insulating material against heat.

Though not reported specially, spongy tissue is also often encountered in "Neelum", 'Banglora', 'Baneshan', 'Mulgoa', 'Janardan Pasand' etc., in South India.

Spongy tissue according to Schaffer (1994) and Young (1957) prevails in Indian cultivars or cultivars with an Indian pedigree. Spongy tissue was reported in cultivars – 'Alphonso', 'Fan Siamese', 'Harumanies' in Malaysia (Kwee and Chong 1985). 'Carabao' in Nepal and Philippines (Gautam and Lizada 1984., Nuevo *et al* 1984) and Kensington in Australia (Chaplin 1986).

Root Stock

It is well established that root stock influences several aspects of scion including the incidence of physiological disorders, either aggregate or inhibit. Joshi (1975) observed that 'Alphonso' fruit and 'Alphonso' seedling root stocks showed maximum percentage of damage (57.39%) and those on 'Vellaikolamban' root stock showed the least (35.43%)., The remaining root stocks viz., 'Shahabuddin' (55.93%), Pairi (48.59%), 'Neelum' (46.16%), control mixed seedling (46.09%), Peswha (43.05%) and 'Totapuri' (43.06%) also showed slightly varying effect on the occurrence of spongy tissue in 'Alphonso'.

Tree Age

Tree age has direct relationship with spongy tissue (Joshi 1975). He observed 31.11, 34.46 and 52.22 per cent spongy tissue affected fruits in 20, 40 and 70 years old "Alphonso" mango trees respectively and described the relationship : 'linear'. On the other hand, Subramanyam *et al* (1971) reported no relationship of tree age with that of spongy tissue.

Tree Vigour

Higher incidence (50%) of spongy tissue disorder in fruits from vigorous trees than those from less vigorous trees (9%) was attributed to the dense foliage of the vigorous trees, which might have retained more sensible heat in longer periods in their umbrella like canopy holding the fruits much like an incubator, while the less vigorous trees could not retain the sensible heat due to their sparse foliage and more free space in between. Further, high temperature coupled with higher relative humidity under vigorous trees with more resultant sensible heat might have contributed to the development of the disorder (Katrodia and Rane, 1981).

Time of Harvest

Cheema and Dani (1934) and Katrodia (1979) noted that early pickings escaped the incidence of spongy tissue to some extent. Cheema and Dani (1934) further noticed that the tendency for the internal breakdown became more evident at the end of the season. Subramanyan *et al* (1971) found higher incidence of spongy tissue (43%) in fruits harvested in mid-season as compared to late (37%) and early (17%) harvested fruits. However, Joshi (1975) recorded opposite observation, that the late harvesting renders the fruit more susceptible to the disorder. Similar observations were made by Katrodia (1979), that late harvested fruits showed higher incidence of disorder as compared to mid and early-season harvested fruits, obviously early harvest of fruits helped to escape the incidence but the quality of such fruits on ripening would be inferior (Rangwala 1975., Limaye *et al* 1978., Patkar, 1978). In a study at Bangalore (India) more than 70 per cent incidence was observed in fruits harvested late, after 15th June at mature-ripe stage as compared to almost negligible incidence (2%) in early-harvested fruits i.e., in the fifth week of May at immature stage (Anon 2003-2004).

Stage of Maturity

Earlier studies indicated spongy tissue as a ripening disorder, occurring only in ripe 'Alphonso' fruits (Subramanyam, *et al* 1971., Santha 1980). However, in later studies it was not found true. Thomas *et al* (1993) observed spongy tissue in fully matured unripe fruits also indicating that the manifestation of this disorder occurs towards the final stage of fruit growth and maturation. The studies at Bangalore (India) confirm that spongy tissue incidence had correlation with fruit maturity stages rather than age of the fruit [Anon 2003-04].

With the advancement of the maturity the fruits become increasingly susceptible to spongy tissue development (Subramanyan *et al* 1971., Joshi 1975., Santha 1980., Joshi and Roy 1989). Joshi (1975) found that in first and second year of his studies, the percentage of spongy tissue affected fruits were 5 to 10 per cent at ¾th maturity, 20.0 and 23.3 per cent at 7/8th maturity and 55.0 and 53.33 per cent at full maturity respectively. Tree ripe fruits showed 73.5 and 85.0 per cent affected fruits in the first and second year of studies respectively.

Thus it was confirmed that with the advancement of the maturity of fruits, percentage of spongy tissue incidence increases. The disorder increases when ripe fruits are harvested, instead of mature but unripe fruits. The occurrence and intensity of spongy tissue was highest in tree ripe fruits, whereas it was least in fruits harvested at ¾th stage of maturity.

Fruit Size

In general, the incidence of damage due to internal breakdown increases with increase in fruit weight (Subramanyam *et al* 1971., Joshi 1975., Joshi and Limaye 1986). 'Alphonse', mango fruits weighing 200 g or less showed, less damage (19%) as compared to those weighing more than 300 g (47%) especially after rains when the fruits are in the growth terminating stage (Subramanyam *et al* 1971., Joshi 1975., Joshi and Limaye, 1986). Raymond *et al* (1998b) on the other hand, did not observe definite pattern in the relationship between fruit weight and internal breakdown of 'Tommy Atkins' fruits and attributed this discrepancy to the fruit size variation between 'Alphonso' and 'Tommy Atkins' cultivars.

Specific Gravity of the Fruit

The fruits with specific gravity more than 1.02 were more susceptible (46%) than those with 1.00 to 1.02 specific gravity. Fruits with specific gravity less than 1.00 (immature fruits) were completely free from the disorder (Santha, 1980., Joshi and Roy 1985). Thus the incidence of spongy tissue is positively correlated with specific gravity of fruits.

Location of Cultivation

Variation in the extent of spongy tissue in 'Alphonso' fruits grown at different locations was also reported and attributed to variation in agro-climatic condition prevailing at different locations (Subramanyam *et al* 1971., Deshapande, 1975., Joshi and Limaye 1986). Studies conducted in different states on the issue indicated that the locality (place of cultivation) played an important role in the incidence of spongy tissue and that the incidence was higher in fruits grown in coastal regions as compared to that of inland areas. Similarly, the incidence decreased with increase with elevation at which mango garden was grown. The trees located at 5 meters elevation showed maximum percentage (56.67) followed by those at 63 meters elevation (46.67%) and at 120 meters elevation (35.56%) (Joshi and Limaye 1986).

Environmental Factors

Among the different environmental factors, exposure to sunlight, temperature and rain were considered and recongnized as influencing the incidence and intensity of spongy tissue in mango.

Sunlight

Incidence of spongy tissue is known to be influenced by the exposure of fruits either on the tree or after the harvest to sunlight. As the period of exposure of harvested fruits to sunlight increased, there was an increase in the incidence. Fruits exposed to sunlight after harvest for more than 120 minutes showed maximum (100%) incidence. Thus it was concluded that post-harvest exposure of fruits to sunlight was responsible for internal break down in "Alphonso" fruits (Gunjate *et al* 1982).

Temperature

High temperature during the later stages of fruit development was considered to be the main environmental factor responsible for the development of spongy tissue (Gunjate *et al* 1982., Katrodia and Sheth 1988., Chadha 1989). When the temperature rises above 40.5 °C consistently for 2 or 3 days, the proportion of fruits with spongy tissue increased to an extent of 75 per cent (Joshi, 1975). However, the occurrence of spongy tissue was also noticed in fruits borne inside the tree canopy and in cooler geographical areas, indicating that other environmental factors such as light, relative humidity etc., may be playing important roles in the development of the disorder. According to Dalal (1975) spongy tissue may also occur in those mangoes, which are subjected to very chill weather during their critical stage of cell expansion. Mango cultivators of Ratnagiri district of Maharastra state (India) believe that spongy tissue occurs only during ripening process, because of heat developed within containers during transit to distant places (Deshapande, 1975.).

Rains

Rains, one or two weeks prior to harvest render the 'Alphonso' fruits more susceptible to spongy tissue development (Subrymanyam *et al* 1971). Rain at the time of fruit development increases the occurance of internal breakdown in mango (Katrodia *et al* 1988). Under dry and warm climatic conditions, the incidence of spongy tissue was observed to be lower (Shiva Shankar and Mathai, 1999).

Soil Moisture Conditions

Relatively high soil moisture content increased susceptibility of fruits to the disorder. It was suggested that the reason for more incidence of spongy tissue in irrigated trees was due to higher sensible heat owing to the higher relative humidity under the tree canopy. Increased irrigation at the time of fruit development has been considered to increase the incidence of internal breakdown in mango. (Malo and Campbell, 1978). However, over a six-year period, no differences in the percentage of disordered 'Sensation' fruit were observed among irrigation treatments (Farre and Hermoso 1993). The fruits grown on well fertilized soils and irrigation with more moisture content at the time of harvest were more susceptible to spongy tissue development than those growing on less fertile and lighter soils with lesser moisture content (Rane *et al* 1978).

Mineral Status of Healthy and Affected Fruits

In an attempt to understand the various reasons of spongy tissue, healthy and affected pulp was analysed and compared for mineral content. Preliminary studies on comparison of spongy tissue affected pulp showed low Ca (0.074%) and high 'P' (0.12%) in the breakdown tissue as compared to 0.085 per cent Ca, and 0.096 per cent 'P' in the healthy pulp (Subramanyam *et al* 1971). In 'Carahao' mango fruits, on the other hand, 'Ca' levels were higher and 'K' levels were lower in affected pulp opposed to unaffected pulp (Gautam and Lizada (1984)., Gunjate *et al* (1979a) found a 'Ca' gradient from the skin to stone and from skin to beak end. The maximum amount of spongy tissue was in the areas of low Ca i.e. near the stone and beak. Therefore, it was concluded that the occurrence of spongy tissue was associated with lower calcium content in the fruit and there is a possibility of reducing the incidence by treating the fruits with calcium.

Rangwala (1975) reported that N,P,K and Mg contents were higher in spongy tissue pulp than in healthy one. Mineral analysis of tissue surrounding spongy and healthy tissue of 'Alphonso' fruits, indicated that spongy tissue had low levels of K, Ca, Na and higher N, P, Mg, Zn and Fe contents as compared to healthy tissues (Selveraj 1998).

Katrodia (1979) observed that the spongy tissue and healthy pulp around the spongy tissue did not differ in their macro and micro elemental composition. Santha (1981) did not observe significant difference between healthy and breakdown pulp in respective of Ca, Na, and Mg contents, but 'K' level was lower and the 'P' level was higher in the affected pulp. However, the available information on the subject shows that the mineral composition of mango fruits suffering from spongy tissue is extremely variable and do not follow a pattern.

Activity of Enzymes

Spongy tissue occurs due to inactivation of enzymes. Functional activity of different types of enzymes was attributed to the development of spongy tissue. Chhatpar *et al* (1968-69) observed high invertase and ascorbic acid oxidase and decreased amylase activity in the extract of spongy tissue, which resulted in high breakdown and more ascorbic acid oxidase and low starch degradation. The spongy tissue exhibited much lower activity of amylase, glutamate dehydrogenase, glutamate oxalo acetate transaminase and catalase and much higher activity of invertase and ascorbic acid oxidase compared with healthy tissue of 'Alphonso' fruits (Gupta, *et al* 1985., Gunjate 1989). Healthy mango fruits (cv. Tommy Atkins) showed lower activity of paraxidase, polyphenol oxidase and phenyl alanine ammonialyase throughout storage period of 28 days at 12 °C and 90 per cent relative humidity as compared to affected fruits. The activities of polygalacturonase and pectine methyl esterase were lower in affected than in healthy pulp (Lima *et al* 1999). The fruit affected with spongy tissue exhibited a much lower amylase activity and reducing and non-reducing sugars, but exhibited a much higher starch content during storage at 12 °C and 90 per cent relative humidity as compared to healthy tissue of fruits of "Tommy Atkins (Lima *et al* 2000).

Katrodia (1979) observed amylase activity as higher (3.27) in healthy tissue while the same was very much decreased (0.39) in spongy tissue and claimed that the accumulation of heat near the stone in the fruit pulp at pre- and post-harvest stages might have retarded the enzyme activity and this action resulted in retention of more unhydrolysed starch in both the stages of affected pulp. Invertase activity has been lower in both sun-desiccated tissue and spongy tissue as compared to that of healthy fruit which was due to heat as in case of amylase. Amylase activity in spongy tissue was about three times lower than in the healthy tissue (Katrodia *et al* 1988).

According to Gunjate *et al* (1982) internal breakdown of mango fruits after exposure to sun light was due to inactivation of ripening system as a result of heat or accumulation of phenols.

Biochemical Changes

Incidence of spongy tissue induced biochemical changes in the affected fruits as follows.

Acidity and pH

Pulp of the spongy tissue affected fruits was found to contain higher acid content than the pulp of the unaffected healthy fruits (Rangwala 1975., Patkar 1978., Katrodia 1979., Katrodia *et al* 1988). On the contrary, Chhatpar *et al* (1968-69) reported reduction in acidity in healthy of unaffected tissue and spongy tissue of affected fruits as compared to healthy tissue of unaffected fruits, which was ascribed to the utilization of citric acid by *Bacillus sp.* Spongy tissue had lower pH and high acidity as compared to normal tissue (Subramanyam *et al* 1971., Rangwala 1975., Patkar 1978., Katrodia 1979).

Ascorbic Acid

Several workers observed a greater reduction in ascorbic acid content in the affected tissue than in healthy tissue of mango fruits (Amin, 1967., Chhatpar *et al* 1968-69., Rangwala 1975., Katrodia 1979). The reduction in ascorbic acid content in disordered fruit was attributed to its utllisation by bacteria (Chhatpar *et al* 1968-69) and assigned to the losses by oxidation owing to higher temperature in the affected pulp (Katrodia 1979) or due to activation of enzymes like ascorbic acid oxidase which perhaps have upset the ascorbic dihydro ascorbic acid balance by oxidation (Sharma 1991).

Total Soluble Solids

Lower content of total soluble solids (TSS) was observed in spongy tissue affected portion than unaffected portion of the fruits (Patkar *et al* 1984., Subramanyam *et al* 1971), which appears to be due to the reduced hydrolysis of starch into sugar (Patkar *et al* 1984).

Sugars

Appreciable reduction of non-reducing sugars was observed in the affected fruit and unaffected part of spongy tissue in affected ripe fruit as compared to normal fruit (Amin, 1967., Chhatpar *et al* 1968-69., Subramanyam *et al* 1971., Rangwala 1975., Patkar, 1978., Katrodia, 1979., Katrodia *et al* 1988). It was opined that the lower content of sugars in the spongy tissue pulp was due to the presence of higher content of starch in the affected pulp, which remained unhydrolysed under the low activity of starch converting enzyme (amylase) during ripening. Thus the reduction in reducing-and non-reducing sugars in spongy tissue was assigned to lower activities of invertase and amylase.

Starch

Until the start of ripening starch accumulates in the pulp and gets converted into sugar afterwards under normal conditions. It appears that starch does not get converted in the affected portion of the pulp due to inactivation of ripening enzymes (invertase and amylase) at high temperature (Katrodia 1979). Rangwala (1975) observed 1.47 per cent starch in affected tissue and no starch in heallhy ripe pulp. Patkar (1978) on the other hand, recorded lower starch content (8.25%) in spongy tissue and higher starch content (13.80%) in the mature unripe normal fruit. Much of the starch in affected tissues remained unhydrolyzed and it was higher in spongy tissue (7.98%) against that in healthy pulp (0.25%) and in the mature fruit (12.87%) (Katrodia 1979., Joshi and Roy 1989). Decreased amylase activity in spongy tissue might have resulted in the reduced hydrolysis of starch into sugar (Chhatpar *et al* 1968-69).

Carotenoid Pigments

Significant decrease in carotenoid pigments in spongy tissue as compared to healthy tissue in ripe fruits was observed. The reduction of beta-carotene in spongy tissue was claimed to be due to its lesser production or its degradation probably due to higher temperature in the fruit pulp because of convective heat received by the fruit on the tree from the soil. (Chhatpar *et al* 1968-69., Subramanyam *et al* 1971., Rangwala 1975., Katrodia 1979).

In addition to the above, other bio-chemical constituents were also reported as affected by spongy tissue. Higher amounts of pyruvic acid, diacetoglutaric acid (Santha 1981) polyphenol (Gupta *et al* 1985), and proteins (Chhatpar *et al* 1968-69) and lower contents of soluble pectins (Gupta *et al* 1985) sucrose and hexose (Chhatpar *et al* 1968-69) and TSS; acid ratio (Selvaraj, 1998) were reported in the spongy tissue of ripe mango fruits than in healthy fruits. The spongy tissue contained three times higher phenols than the healthy fruits. Accumulation of phenols may be responsible for inactivation of ripening enzymes in case of spongy tissue of 'Alphonso' mango fruits (Patkar *et al* 1984).

Losses

Spongy tissue in mango causes greater economic losses to the grower, because it occurs, mainly during ripening, when economic returns are expected hopefully. Further, loss due to spongy tissue is not only to the individual grower but also to the nation, since reputation of the country and internal and external trade are seriously impaired and affected.

According to a report, the loss of 'Alphonso' mango fruits due to spongy tissue development alone was to the extent of 35-55 per cent in different years (Anon 1968-70). The overall losses caused by spongy tissues ranged between 1 and 88 per cent, averaged to about 30 per cent in different plantations (Subramanyam *et al* 1971). The losses vary from place to place and season to season depending upon the orchard and agro-climatic conditions (Katrodia, 1979). Further the damage (loss) varies with the weight of fruit, time of picking, location of garden and orchard conditions.

Detection of Affected Fruit

The export of mangoes from India has been increasing in recent years. The lion share of mango exports from India, constitutes 'Alphonso' fruits. However, in recent years, the export 'Alphonso' mangoes from the country has dwindled because of spongy tissue. The need for quality assurance is of utmost importance to retain existing markets and to expand trade to other world markets. Quality control for internal trade is also warranted in view of the high price being paid by the consumer for this cultivar. Therefore, the foremost thing to be considered is separation of healthy fruits from spongy tissue affected fruits. This however, becomes difficult in view of the absence of external symptoms of the disorder on fruits and symptoms can be seen only after being cut open. Therefore, a non-destructive method of detection of affected fruits is needed, which would be of immense advantage to the mango industry and such methods could be used as quality control measures.

Preliminary studies at Bangalore (India) indicated the fruits with higher specific gravity (1.02 and above) had a higher percentage of incidence of spongy tissue (Santha, 1979). Light transmission studies may throw some light on the aspect. At Bangalore X-ray and salt water dippings were found to be satisfactory as non-destructive methods for detecting the spongy tissue affected fruits (Anon 2003-04).

Thomas *et al* (1993) reported X-ray imaging for the detection and rapid separation of 'Alphonso' mangoes affected by spongy tissue, having no external symptoms. X-ray photography and X-ray images of affected fruits show clear indication of mesocarp damage, as dark grey patches, corresponding to the internal air-cavities in the flesh in contrast to light grey areas of healthy flesh. There is a close agreement with the internal examination of the cut fruits and the detection based on X-ray photography indicating the reliability of the technique. A correlation exists between the number and extent of dark grey patches on the X-ray photograph with the number and size of the air cavities observed in the flesh of the affected fruits on cutting open of the fruits. It was envisaged that this technique could be used for quality control for on line detection and separation of spongy tissue affected fruits prior to packaging.

Control or Remedial Measures

Unlike pests and diseases, spongy tissue hits directly the finances of the grower and export trade and mango industry. Hence, this disorder should be avoided by adopting remedial measures. Keeping this in mind, attempts were made to control or at least minimize the incidence particularly, in 'Alphonso' mangoes, since its discovery in 1932, through application of minerals, through soil, through dipping of fruits in calcium solutions and cultural measures.

Mineral Application

This measure is based on the assumption that spongy tissue was due to deficiency of nutrient(s). However, application of N, P and K either to the foliage or through the soil did not workout to reduce the incidence (Amin 1967., Katrodia 1979., Joshi and Limaye 1984). On the contrary their application increased the incidence (Subramanyam *et al* 1971., Limaya *et al* 1975).

The application of different levels of N, P, K and micronutrient (B) individually and in combination to soil as well as to foliage did not help to overcome the incidence of spongy tissue in 'Alphonso' (Amin 1967., Katrodia, 1979).

Spraying of shoots with chemicals like zinc sulphate, Borax, and GA or with micronurients Cu, Fe and Mn (in sulphate form) alone or in combination with zinc did not show significant effect. Spray with zinc alone or in combination with Fe or Cu caused blackening and rusetting on the fruit skin, indicating that the disorder is not due to nutritional deficiency or imbalances (Katrodia, 1979).

Soil application of magnesium in the form of dolomite and a foliar formulation (developed at Indian Institute of Horticultural Research, Bangalore, India) in December, January and February after pruning reduced the incidence of spongy tissue significantly in addition to inducing early flowering and better fruit set in 'Alphonso' mango at Bangalore (Anon, 2003-04).

Calcium Treatment

Deficiency of calcium was supposed to be a cause of spongy tissue in mango (Subramanyam *et al* 1971., Gunjate *et al* 1979b) and therefore, correction of calcium deficiency was expected to check the development of this disorder in the fruits. As a measure of control, pre- and post-harvest dipping of fruits were tested to rise the content of "Ca" in fruits and overcome the disorder. Gunjate *et al* (1979b) tried both pre- and post-harvest dips in "Ca" solution. Single and double pre-harvest dips of fruits in 5 or 20 g per litre of calcium chloride (or) calcium nitrate solutions increased the 'Ca' content in fruits and reduced the incidence of spongy tissue in 'Alphonso' mangoes.

Calcium chloride was more effective than calcium nitrate. By contrast, the post-harvest dips of fruits in calcium solutions did not significantly reduce the incidence (Gunjate *et al* 1979b). Katrodia (1979), however, observed that calcium chloride dips or sprays had no effect on spongy tissue development or its control, but its higher concentrations (2 and 3%) caused blackening and roughening of the fruit skin. Results with pre- and post-harvest dips or sprays of calcium were, however, inconsistent. The contradictory results with respect to 'Ca' sprays were mainly due to the immobility of 'Ca' through phloem and may also be due to the effect of environmental factors. The post- harvest dips as compared to the pre-harvest treatment are more cheaper and easier. Therefore, it is necessary to find out the possibility of increasing the efficiency of post-harvest treatments by manipulation of certain factors like the temperature of treatment solution, use of wetting/sticking agents, time of dip etc. (Gunjate *et al* 1979b).

Ethephon Treatments

The ripening of the mango fruit is inhibited by inactivation by ripening enzymes (Catalase, peroxidase and amylase) by the endogenouly produced ethylene in the fruit during ripening process (Mattoo *et al* 1968., Mattoo and Modi, 1969b). When ethephon (Ethrel) is applied to fruits or leaves it releases ethylene inside the tissue. It is possible that elhylene released by ethephon inside the fruits might be interacting with the ripening inhibitors inside the fruit, which are responsible for the inactivation of above ripening enzymes during the process of ripening (Mattoo *et al* 1968). Therefore, Lad *et al* (1985) tested the effect of post-harvest ethephon treatment on the occurrence of spongy tissue in 'Alphonso' mango at Dapoli (Maharashtra, India). Dipping of fully matured fruits immediately after harvest in 500, 750 or 1000 ppm ethephon solution separately for 15 min. and ripening in wooden crates between layers of paddy straw and newspaper, reduced the incidence and also advanced the ripening of fruits by 2 to 3 days. The keeping quality of fruit treated with 750 and 1000 ppm ethephon solution was however, slightly reduced. Whereas treatment with 500 ppm ethephon solution considerably reduced the incidence of spongy tissue and developed good colour with desirable keeping quality.

Cultural Practices

Amongst the various factors studied pre- and post-harvest exposure of fruits to higher temperature has strongly been associated with this disorder (Gunjate *et al* 1982., Katrodia and Rane, 1979). Katrodia (1979) concluded that convective heat from the soil is mainly responsible for spongy tissue in mango fruits. Therefore, it was hoped that by regulating soil and atmospheric temperatures the incidence of the disorder could be reduced. The provision of wind breaks for protecting the orchard from warm wind during fruit maturation period (May) was considered to be helpful in mitigating the problem. It was also hoped that spongy tissue could be alleviated by mulching or sod-culture, which reduce and protect the orchard soil from excessive heating by solar radiation (Katrodia and Rane, 1979).

Katrodia and Sheth (1988) tested the effect of cover crops (mung bean and cowpea) and mulches (paddy straw, and mango dry leaves) against clean cultivation (control) in Gujarat (India), of which mulches were found more effective in protecting the orchard soil from excessive heating by solar radiation, through keeping the soil surface temperature down (12.1 to 15.9 °C). As a result of this, there were small amount of convective heat currents from the soil, which reduced the occurrence of spongy tissue. Due to mulching heat was not absorbed by the soil and this resulted in negligible incidence. Cover crops raised with three irrigations did not prove beneficial and economical, although they reduced the soil surface temperature considerably, but irrigations favoured the disorder because of raise in humidity (86.75%) and high atmospheric temperature (34.6 °C). Sod-culture having natural vegetation of *darbha* (*Engrosties cynosuorvides* L) could control the occurrence of spongy tissue in fruits on tree to the extent of 100 per cent by keeping the temperature of soil surface down to 20.28 °C as compared to that of clean cultivation. In view of the above it was concluded that the occurrence of spongy tissue could be checked by dry mulches at pre-harvest stage and by protecting the fruits from convective heat at post-harvest stage.

Incidence of spongy tissue was significantly lowered due to mulch (20 cm thick) during March-April, done at half-grown stage of fruits and sod-culture (retaining existing vegetation) during cropping season, as a result of reduction in heating of orchard soil by solar radiation and due to reduced convective heat currents from the soil (Katrodia and Rane 1979., Burondkar *et al* 1994). From the results it was concluded that the incidence of spongy tissue in ripe 'Alphonso' mango fruits could be reduced by providing grass mulch under the tree canopy or following sod culture provided adequate precautionary measures are taken against possible fire.

The beneficial effect of mulches and sod-culture was mainly through the reduction of soil heating from solar radiation resulting in less currents of convective heat reaching the fruit, which does not allow the temperature rise in the fruit to the extent of inactivating the ripening enzymes and development of spongy tissue in the fruit pulp.

Thus mulching or green vegetation (sod-culture) seems to be the proper answer to save mango fruits from this serious disorder of spongy tissue at pre-harvest stage in the field.

Low Temperature Ripening

Mango growers of Maharashtra (India) believed that spongy tissue developed because of heat development in the container during transit, when the ripening was triggered (Deshpande, 1975). Attempts made to get reduction in the incidence of spongy tissue by ripening the fruits at low temperature (20 °C) with high relative humidity (85%) and higher concentration of Oxygen (30%) at Bangalore were successful (Anon 2003-04).

Breeding Resistant Varieties

Inspite of the above successful results, satisfactory control of spongy tissue has yet to be developed and standardized. The above remedial measures are temporary only and depend upon several intricate factors for giving success. Therefore, a permanent solution is needed and this can be achieved by breeding resistant varieties.

A genetical approach was made at Bangalore to develop resistant varieties (by hybridization), similar to "Alphonso", but free from spongy tissue (Iyer and Subramanyam, 1992). As a result, two hybrids viz; Hybrid 10 (Banganapalli × Alphonso) and Hybrid 13 (Alphonso × Banganapalli) were identified as free from spongy tissue. Hybrid 13 was more akin to "Alphonso" in terms of colour, size, flavour and quality and hence, could be a substitute for "Alphonso" in areas where spongy tissue is a problem.

A hybrid named as "Ratna" was developed by crossing 'Alphonso' with 'Neelum' at Dapoli (Maharastra, India) which was reported to be resistant to the development of spongy tissue in addition to having all good characters of 'Alphonso' (Limaye *et al* 1984).

Recommendations

Based on the results obtained so far, the following recommendations to over come / avoid the problem of spongy tissue in mango, can be made :

1. Mulch the tree basins at least during fruiting season (Mar-May) to reduce the flow of convective heat from the soil to the fruit and there by reduce the incidence.
2. Harvest the fruits at $3/4^{th}$ or $7/8^{th}$ stage of maturity having a specific gravity of 1.00 to 1.02 i.e.; harvest the fruits early.
3. Do not expose the harvested fruits to direct sunlight and higher temperatures but move then to shade after harvest.
4. Dip the harvested fruits in 500 ppm solution of ethephon (ethrel) for 5 minutes before moving to ripening or storage chambers or before packing.
5. Ripen the fruits at low temperature (20 °C), higher RH (85%) and higher oxygen concentration (30%).
6. Use non-destructive methods (x-rays etc) to separate the unaffected healthy fruits from affected fruits before packing to send them to markets.
7. In future plantations grow resistant varieties akin to "Alphonso" wherever spongy tissue is problematic, particularly when the fruit is meant for world markets.

SOFT NOSE

'Soft Nose' is a fruit physiological disorder of mango in which breakdown of flesh (pulp) occurs towards apex of the fruit before ripening on the tree itself. Verma (1950) first described this as a physiological disorder of fruit in mango and called it "tip pulp" from

India. A similar disorder termed as "Soft nose" was later described by Young (1957) in nearly mature, "Tommy Atkins" fruits in Florida, in which a premature softening of the mesocarp (pulp) develops at the distal end of the fruit.

'Soft nose' was reported from Canary islands (Sauco *et al* 1984). From Malaysia, a physiological disorder of mango fruits under the name "yeasty fruit" or "insidious fruit rot" which appear to be identical with "soft nose" of Florida was reported (Kwee and Chong, 1985). The incidence of `soft nose' in 'Neelum' mango has been observed in Andhra Pradesh (India) and is on the increase causing great concern to the grower.

Symptoms (Photo 11.2)

Typically, fruits nearing maturity show characteristic yellowing of the skin at the tip of the fruit, followed by softening of the pulp inside at this site. External symptoms are not always obvious, but yellowing of the green skin between the apex and the stigma point with the loss of firmness often occurs. Internally, affected flesh appears overripe and brown with a bitter taste; which in several cases becomes spongy greenish black mass (Wainwright and Burbage, 1989). Burden *et al* (1992) made a preliminary examination of the 'soft nose' in 'Kent' mango fruits. The mesocarp of the affected fruit showed marked cell separation and cell wall degeneration and dissolution of vascular connection between the endocarp at the fruit tip and then spread out into the rest of the apex and from there to

Plate 11.2: Mango fruit with soft-nose disorder (right) and healthy fruit (left)

the endocarp (stone/seed) and mesocarp. The first sign of the disorder occurred next to the endocarp at the fruit tip and then spread out into the rest of the mesocarp. The border between the disordered and healthy tissues exhibits a translucent appearance. When the affected fruits are cut open the flesh on the ventral side towards the apex in some fruits, nearly appeared overripe, while that around the shoulders and on the dorsal side is unripe. In most advanced cases, some of the tissues may become spongy, greyish – blackness, which in extreme cases may extend through most of the fruit.

Causes

No pathogenic organism was found in the disordered material of the fruit, which has led to the suggestion that the disorder "soft nose" should be regarded as a physiological breakdown of the fruit pulp (Wainwright and Burbage, 1989). No increase in soft tissue was observed in 'soft nose' affected growth in storage for 10 days at room temperature and they ripened normally. This also supports the view that 'soft nose' is not due to any pathogen.

Fertilizer trails by Young and Miner (1961) suggested that the disorder was the result of calcium deficiency, although there was no firm evidence for this suggestion (Wainwright and Burbage, 1989). However, high calcium levels in leaves (2.5% or greater) retarded soft nose incidence in Florida (Young and Miner, 1961). In a nutritional study high levels of nitrogen in the leaves were linked with increased incidence of soft nose in 'Kent' mango fruits (Young, 1957., Young and Miner 1961). Though conclusive evidence was lacking there were sound indications that high calcium levels in the trees of higher nitrogen status retarded the development of soft nose. There was no consistent relationship between the incidence and P, K or Mg content in leaves (Young and Miner, 1961). A later report confirmed that soft nose due to high nitrogen status of the tree tended to decrease with increased 'K' fertilization and a heavy application of 'Ca' in the form of lime stone or gypsum. Even so manipulation of fertilizer application failed to give control and seasonal effects were suggested as an indirect factor influencing the frequency of soft nose (Young *et al* 1965). However, further research on seasonal influence on the incidence of soft nose is lacking. Increased incidence of soft nose was correlated with high levels of nitrogen and potassium fertilization on acid soils as opposed in calcareous soils. (Young and Miner, 1961., Young *et al* 1962., Malo and Campbell, 1982). Both Young *et al* (1962) and Katrodia and Rane (1981) did not observe relationship between soft nose in 'Tommy Atkins' and 'Alphonso' respectively and K, Mg and P concentration in leaves. Burdon *et al* (1991) observed low levels of Ca and Mg in disordered 'Kent' fruits as compared to healthy fruits. Katrodia and Rane (1981) further suggested that the flesh breakdown in 'Alphonso' cultivar was associated with poor seed development.

Thus the available evidence points out that minerals other than N, P, K and Mg are involved in the incidence of soft nose. Therefore, soft nose has been cited as a "Ca" deficiency disorder. Soils of affected fruits are generally low to deficit in Ca and Mg contents. Soft nose has been claimed due to the low content of Ca at the affected area of the fruit. The reduction of incidence following the sprays of Ca containing chemicals also supports that the Ca may be involved in the soft nose disorder.

However, according to Galan *et al* (1984) and Winston (1986) the low fruit "Ca" concentration at the site prone to soft nose is a possible contributing factor for the development of the disorder although cultivar susceptibility clearly plays a significant role.

The influence of orchard location and environment were suggested as the causes of soft nose. However, the incidence in Florida grown mango appears to have little correlation with environmental factors, such as wet or dry seasons, young or old trees. Soft nose was less prevalent on the rocky alkaline soils than on acidic sandy soils of Florida (Young, 1957). 'Soft nose' of 'Neelum' is largely confined to Chittoor district of Andhra Pradesh and not reported from any other mango growing districts in the state/country.

Fruits harvested at the green-ripe stage were affected significantly less than that have been harvested when fully ripe (Wainwright and Burbage, 1989., Sauco *et al* 1984). Delay in harvesting the fruits appears to be one of the reasons for 'soft nose' in 'Neelum' mango fruits in Andhra Pradesh (India).

Varietal Susceptibility

Mango cultivars show great variation in their susceptibility to soft nose. Verma (1950) observed the incidence of soft nose in four mango cultivars in Uttar Pradesh in India. 'Neelum' and 'Bangalora' cultivars showed susceptibility in Andhra Pradesh (India). Young (1957) observed soft nose in cultivars. 'Brooks', 'Davis', 'Haden', 'Irwin', 'Keitt', 'Kent', 'Mulgoa', 'Stensation', 'Springtels' and 'Zill' in Florida (which owe their origin to India) of which 'Kent' is more susceptible. In Canary islands cultivars- 'Ahpingh', 'Ameeri', 'Anderson', 'Fascell', 'Gouvein', 'Kent', 'Lippins', 'Sensation', 'Smith', 'Tolbert', 'Tommy Atkins' were affected, of which 'Kent' and 'Ameeri' were more sensitive (Sauco *et al* 1984).

Losses

Verma (1950) observed that a fourth of the total crop is lost due to 'tip pulp' in India. In Florida the leading cultivar 'Tommy Atkins' has suffered losses as much as 50 per cent due to soft nose (Malo and Camphell, 1982). Burdon *et al* (1991) reported that 90 per cent 'Kent' and 30 per cent 'Beverly' fruit were lost due to soft nose. Though the losses due to 'soft nose' have not been estimated in India, it is clear that huge losses are incurred causing great economic losses to the grower.

Control

Alleviating 'soft nose' in 'Tommy Atkins' fruits was attempted by the application of calcium in the form of calcium nitrate, calcium bicarbonate and gypsum to the soil, which reduced 45, 35 and 27 per cent of incidence respectively in Florida (Young *et al* 1962). Hydrated calcium nitrate (0.5%) was suggested against soft nose in Andhra Pradesh. The sprays should be given soon after the appearance of symptoms. These sprays check the further development of the soft nose and thus minimize the losses. Incidentally these sprays, increase the size and weight of the fruit, and also allow the tree storage of fruits. The post-harvest life of the fruits was also improved (Singh *et al* 1987). Timely harvesting of fruits also avoids soft nose.

INTERNAL FRUIT NECROSIS

'Internal fruit necrosis' is a physiological (nutritional) disorder of mango fruits. It was first noticed in 1972 in mango orchards of Lucknow district of Uttar Pradesh (India) and subsequently in different districts of the state (Ram *et al* 1978). However, the disorder was reported neither in any other part of India nor world so far.

Symptoms

Ram (1988) and Ram *et al* (1988) have described the symptoms, causes and control of this disorder in detail. The disorder is found in immature fruits while on the tree and the affected fruit drops before attaining maturity. The first visible symptom in 'Dashehari' was the appearance of dark green colour in lower half of the fruit on the epicarp skin. Soon it is accompanied by browning of seed and mesocarpic (pulp) tissue in a small isolated area of indefinite out line before hardening of the endocarp (seed). The brown tissues then turn into brownish black necrotic lesions which later on extended towards the epicarp. At this stage water soaked isolated necrotic areas appear on the surface of the fruit, exuding gummy substances below the green tip. These brown areas develop into dark brown gummy cavities in the mesocarp. At advanced stages the affected tissues of the entire lower half of the fruit turn necrotic and result into longitudinal cracking of the fruit through the necrotic region exposing the seed. Some fruits develop seed necrosis without any external symptoms. Moderately affected fruits may develop taper-tip and are deformed. The affected fruits are hard and dry. The fruit growth gets restricted even at the earliest stages of the disorder. Affected fruits fail to grow further and drop and fail to ripen during storage if they are carried to maturity. Similar symptoms have been exhibited by other cultivars viz., 'Chausa', 'Bombay Green', 'Safeda Lucknow' and 'Fajri' (Ram, 1988).

Causes

Ram *et al* (1978, 1988) found that the disorder was not a pathological one, since the attempts made to reproduce the symptoms of disorder on healthy fruits by inoculating the extract from the affected fruits were unfruitful. Further, investigations indicated that internal necrosis of mango fruits was due to deficiency of boron in fruits of susceptible cultivar. Affected fruit and leaves of 'Dashehari' and 'Bombay Green' had lower levels of boron than healthy and black tip affected leaves and fruits of the above cultivars. A linear relationship between internal necrosis and boron levels in leaf and fruit was observed. The decrease in per cent fruit necrosis by increase in the levels of boron in the fruit also supports the view that internal necrosis of fruits was due to deficiency of boron (Ram *et al* 1988).

Ram (1988) discussed the different factors that cause and aggravate the disorder. Mango varieties studied showed variable susceptibility to the disorder. 'Dashehari' fruits were most susceptible and were severally affected by internal necrosis followed by 'Safeda Lucknow'. 'Fajri', 'Bombay Green' and 'Chausa' while, 'Langra' was free from the malady. Soil factors like boron level in the soil, soil pH, exchangeable calcium, calcium carbonate etc., were investigated. Low soil boron at increasing soil depth, neutral pH (6.6-7.3), high exchangeable calcium (3.2 to 10.7 me/100 g soil) and high calcium carbonate (1.2 -2.6%) were found to affect the disorder. Low soil boron coupled with high level of calcium, free carbonates and high pH favour the low uptake of boron in leaves and fruit leading to deficiency of boron and in turn to internal fruit necrosis. Soils with pH 6.5 to 8.5 scale in general, restrict the supply of boron to plants (Olson and Berger, 1947). Presence of high exchangeable calcium coupled with high soil pH, further reduce the boron uptake (Fox, 1968). Further, when mango trees (Dashehari) are fertilized with urea, boron content is decreased in leaves and fruits and incidence of percent necrosis increased in fruits with higher doses of nitrogen (1000 g/tree and above). Soil in mango orchards at a depth of 75 cm, where most of the feeding roots are located (Bojappa and Singh, 1973) does not provide sufficient boron to the trees, owing to low boron levels in soil, particularly when coupled with high calcium, free carbonates and pH which restrict the uptake of born (Ram *et al* 1988).Treatments of nitrogen, potassium and calcium imposed, showed higher incidence of internal fruit necrosis in 'Dashehari'. Thus effects of the interaction of nitrogen, potassium, calcium and boron were found to be additive on the disorder (Ram and Sirohi, 1979). The disorder is more common in sandy soils (Negi, 2000).

Control

Internal fruit necrosis causes heavy economic losses to the mango grower. Up to 37.5 per cent loss was estimated in different cultivars (Ram *et al* 1988). Therefore, the disorder warrants control.

Spray of blitox, copper sulphate and sodium hydroxide did not provide relief from the disorder (Ram *et al* 1988). On the other hand, foliar as well as soil application of borax decreased the percentage of necrotic fruits by increasing the boron levels in leaves and fruits of sprayed trees. Boron levels of leaf and fruit showed inverse relationship with the percentage of necrotic fruits (Ram *et al* 1988). Thus, internal fruit necrosis can be effectively controlled by soil application of 10-25 g borax per tree and by foliar spray of borax (0.05 -0. 1%) after fruit set twice at monthly intervals (Negi, 2000).

STEM-END CAVITY

'Stem-end-cavity' is another physiological disorder of mango fruits. It was reported mainly in case of 'Kensington Pride' cultivar of mango in all the major mango producing areas of Australia and Queensland (Mead and Winston, 1991).

Symptoms

Externally there may be or may not be evidence of the disorder. In cases, where the disorder has progressed to a severe rate and the cavity has extended to the surface, a sunken grey/brown area appears opposite the suture near the point of skin attachment. This is the first sign of the disorder. In cases with no external symptoms, the disorder may not be detected until the fruit is cut open. Affected fruits fail to exude sap as in the case of normal fruit. Stem button of affected fruit feels spongy under pressure (Chaplin, 1986).

Internally, the disordered fruit shows a distinct cavity in the flesh at the stem-end between the peduncle and the endocarp (seed), so that the seed becomes physically and physiologically separated from the tree. Flesh surrounding the cavity may show a great discolouration and be of a different texture (similar to wound tissue) than the rest of the fruit. Torn tissue strands may be visible. Damaged tissue is confined to the stem end of the fruit. Vascular fibre that connects the mango fruit stem to the flesh and top of the seed snap or tear thereby forming a cavity with in the flesh under the skin. This results in a breaking of sap pressure in the fruit and pre-mature ripening of the flesh around the tear and seed. There is no sour smell, only the aroma of ripening mango fruit. Pre-mature ripening is followed by fruit shedding prior to normal harvest.

Causes and Factors

There is much confusion between stem-end cavity, Jelly seed, soft nose, stem-end-rot etc. Hence pin-pointing evidence for stem-end cavity is not available. However, as in case of other physiological disorders, mineral deficiencies have been attributed as cause of stem-end-cavity. But as yet there is no direct evidence. Trees with high leaf calcium showed low incidence of the disorder, while trees with high calcium and high nitrogen exhibited normal incidence in North Queensland.

The extent of incidence could be influenced by factors such as soil moisture holding capacity, soil aeration, soil structure etc., that may cause leaching of nutrients or nutrient stress. However, adequate data is not available to support conditions favourable to the disorder including cultivars, high nitrogen and low calcium levels, wet, windy conditions and harvesting of tree ripe fruits.

The cultivars 'Akbar', 'Early Gold', 'Harunmanis', 'Jakarta', 'Kensington Pride', 'Kensington white', 'Kent', 'Sensation', 'Tommy Atkins' and 'Van Dyke' showed higher level of stem-end-cavity than others. 'Kensington Pride' was highly susceptible.

Stem-end-cavity is more in fruits harvested later than at the green mature stage. Major financial losses occurred when harvest was delayed beyond normal harvest time by environmental factors or when growers delayed harvest for financial gain or when growers were endeavouring to harvest tree-ripe fruits.

Mean stem-end cavity was negatively correlated with mean fruit size and shoot growth of the tree (Hermose *et al* 1997). The disorder affects the fruits during early antogeny. The presence of calcium oxalate crystals was observed in affected fruits (Raymond *et al* 1998a).

Control

The presumption that deficiency of calcium in leaves and fruits is responsible for stem-end-cavity prompted the scientists to try the correction of the disorder through the application of calcium in different forms with or without other chemicals. Leaf application of calcium chloride, Ca-EDTA, potassium chloride or magnesium chloride, however did not affect mesocarp calcium content in 'Sensation' mangoes grown in Spain. Single application of Ca-EDTA also did not influence leaf calcium or stem-end-cavity. Ca-EDTA in vaseline applied at the base of the fruit peduncle significantly decreased the incidence of stem-end-cavity and also potassium content of bark and wood of the peduncle. Zinc, boron and copper injected into the trunk in spring season, slightly increased the incidence (Hermosa, *et al* 1997).

However, the best methods to reduce the incidence of stem-end-cavity include, early harvest and discarding of all fruits that fail to exude sap in the desapping operation (Mead and Winston, 1991).

JELLY SEED

'Jelly seed' is a descriptive name assigned to a physiological disorder of mango fruits that initially appear as watery, translucent tissue around the seed at the proximal end of the fruit, but spreads later to the distal section. It is associated with ripening of the fruit on the tree and unless already present at harvest, symptoms do not develop after picking. In jelly seed disorder, flesh (pulp) around the seed completely disintegrates into a jelly like mass (Malo and Campbell, 1978., Winston, 1986). However, Malo and Campbell (1978) considered 'jelly seed' to be the first stage of internal breakdown and concluded that this disorder is an inherent weakness in certain cultivars of mango, whereas Winston (1986) recognized it as potentially separate and distinct disorder. Similar view was expressed by Raymond *et al* (1998a).

The incidence of 'Jelly seed' was reported from Australia and Queensland (Winston, 1986) and South Africa (van Lely Velled and Smith, 1979., Broadrick and Thord Grey, 1982).

Symptoms (Photo 11.3)

The disorder is associated with fruit ripening on the tree. It is initiated at earlier or later stages in different cultivars. The symptoms of jelly seed are not visible without cutting open the fruit i.e. external symptoms are difficult to detect, except a possible softening around the beak area. However, this is not readily obvious except in more severe cases. At first it appears as watery, translucent tissue at the proximal end of the fruit around the seed, spreading to the distal part of the fruit in later stages, giving on overripe appearance.

Plate 11.3: Mango fruit with jelly seed disorder

Causes

The causes of the incidence of "Jelly seed" are obscure and not known/understood fully, though deficiencies of minerals were suggested in general. South African Mango research workers tried unsuccessfully to relate the ripening enzyme-pectin esterase with the content of reducing sugars in the fruit pulp within the 'Jelly seed" disorder.

Varietal Susceptibility

In Australia, 'Akbar', 'Fascell', 'Kamerunga White', 'Monik', 'Ono', 'Sabre' 'Santa Alexandra', Sensation. 'Tommy Atkins' and 'Van Dyke' cultivars are mostly affected by jelly seed, of which 'Tommy Atkins' and 'Sensation' are particularly prone to the disorder (Winston, 1986). In South Africa Keitt and 'Sensation' are affected (van Lelyveld and Smith 1979., Bradrick and Thord Grey 1982).

Miscellaneous Fruit Physiological Disorders

Wainwright and Burbage (1989) gave an account of details of new miscellaneous physiological disorders affecting mango fruits in certain mango growing countries of the east, while Khader (1988) described disorders of bearing in mango from India and Zhang and Haung (1998) detailed a fruit physiological disorder of mango from China.

In Thailand and in Philippines a disorder termed “Lumpy tissue” in mango fruits was reported (Lizada *et al* 1984). Externally, on the peel indication as seen at the time of ripening, whilst the mesocarp contains white lumps and cells with high density of starch grains (Phukadikun-Samphan and Thong-Amphai 1970., Lizada *et al* 1984). In Thailand, Cvs. ‘Namtan’, ‘Pong-Graborg’ and ‘Pinsen Dang’ and in the Philippines Cv. ‘Pico’ were affected with ‘Lumpy tissue” (Lizada *et al* 1984).

Another disorder of unknown origin common in ‘Carabao’ mango is ‘ricy disorder’, which does not exhibit external indication but small lesions about the size of rice grain surrounded in cotton-like tissue were found in the mesocarp of even green unripe fruit (Lizada, *et al* 1984).

‘Pre-mature ripening’ as a physiological disorder of some mango cultivars was reported from Australia (Winston, 1986). In dissordered fruits, in pulp ripens prematurely and flesh adjacent to endocarp becomes soft. It has been observed in cultivars, ‘Akbar’, ‘Carabao’, ‘Early Gold’, ‘Florigen’, ‘Golden Tropic’, ‘Haden’, ‘Harumanis’, ‘Hongsa’, ‘Kamerunga White’, ‘Nam Domai, Okrong’, ‘Pairi’, ‘Pico’, ‘Sabre’, ‘Santa Alexandrinal’ ‘Sensation’, ‘Springlets, ‘Tong Dun’, ‘Van Dyke’, and ‘Zill’ in Australia (Winston, 1986).

SAP BURN

Mango exporting countries, notably Mexico and Australia face a major problem in exporting mangoes viz., ‘Sap burn’ on the fruits. Sap burn causes browning, blackening and necrosis of the skin in a sequence. The damage is superficial, but reduces their marketability and reduces the chances of export. The sap burn is considered to be a quality constraint in the export of mangoes to global markets. The damaged areas of the skin can subsequently become a site for secondary infections, causing breakdown of flesh, which further depreciates the value and market life of the fruit (Lim and Kuppelweiser, 1993). Fruits showing `Sap burn’ may be invaded by *Aspergillus* sp. especially in hot weather (Johnson and Coates, 1993).

Sap burn is a major problem in the mango fruits in Australia, Thailand and Mexico.

Causes

While harvesting the fruit, the stalk is severed from the fruit, which results in the exudation of large quantities of latex (sap) with low pH and high oil content and flows over the surface of the fruit (Bagashew and Brown, 1989). The primary cause of sap burn is the entry of volatile sap components (terpinolene etc.,) through the lenticels of the fruit skin and cuticular cracks, resulting in tissue damage and enzymatic browning of the skin (Loveys *et al* 1992., Holmes *et al* 1993., Robinson *et al* 1993).

The sap is a viscous colloidal suspension clear or slightly milky blue liquid and contains several chemical compounds.

Damage (Photo 11.4)

The sap exuded from the cut or broken stalk (pedicel) during harvesting and/or subsequent handling on contact causes damage to the fruit skin. Sap in the fruit ducts, is under pressure and when the pedicel is cut or broken spurts out and falls on the fruit skin. Damage/injury occurs when the sap flows on the fruit skin. Browning of fruit skin induced by the sap is predominantly catalysed by polyphenol oxidase in the skin (Robinson *et al* 1993).

Plate 11.4: Sap damage on Baneshan mango fruit

Sap injury depends upon the cultivar, season of harvest, as well as growing region, age of the tree, maturity of fruits etc., (Lim and Kuppelweiser, 1993., Robinson *et al* 1993). The extent of sap injury is influenced by other several factors including, total sap flow, sap oil content and lenticel distribution on the fruit. Fruits having greater sap flow and oil content are damaged more and fruits with high lenticel densities are more prone to sapburn (O' Hare and Prasad, 1992).

Varietal Differences

Varietal differences occur in response to sapburn. Most susceptible cultivars in Mexico, are: 'Stantto' and 'Haden' (Bosquez *et al* 2000). Sap burn is particularly severe with Kensington Pride in Australia (O' Hare, 1994) and less severe in Florida cultivars. 'Haden' is less susceptible to sapburn than 'Tommy Atkins' (Bosquez *et al* 2000). Thailand mango cultivars viz. 'Nam Dok Mai', 'Nangklang Wum', 'Tong Dum' and 'Keow Savocy' were less susceptible to sapburn injury than 'Kensington Pride' (O' Hare and Prasad, 1992).

The susceptibility of cultivars to sap burn was attributed to the amount of sap flow and amount of oil in the sap. Cultivars which produced greater amounts of sap and oil were more susceptible to sap injury than those which produced lesser amounts of sap flow and oil. 'Haden' and 'Tommy Atkins' are the examples of former type while Thailand cultivars are the examples of latter type.

The susceptibility of mango cultivars to sapburn is also dependent on cultivars and harvesting season. 'Tommy Atkins' fruits are more susceptible to sap burn during the early season than in the late season, while 'Haden' fruits showed less susceptibility during early season than in the late season. During the mid-season of harvest both showed similar level of susceptibility (Diaz *et al* 2000).

The susceptibility to sap burn appears to depend on the levels of terpinolene and 3-carenene in the sap. The concentration of both chemicals tended to increase in 'Tommy Atkins' while with 'Haden' opposite tendency was observed with season. The concentration of 3-carenene was always higher than that of terpinolene in all the three harvesting seasons in both cultivars. Inspite of having similar terpinolene concentrations in the early season, the 'Tommy Atkins' was more susceptible than the 'Haden'.

In the last harvest season the levels of terpinolene in 'Tommy Atkins' increased but its susceptibility diminished. The 'Haden' cultivar showed a high susceptibility in the last harvest season while its terpenes concentrations were significantly lower. From the above results it was inferred that there was no correlation between the changes in the concentration of terpinolene and 3-carnene and the susceptibility in the three different harvesting seasons (Diaz *et al* 2000).

Control

As pointed earlier, sap burn, which causes browning and blackening of the skin, is a bottle neck in exporting mangoes to the world market, as the market demands a cleaner fruit without blemishes, blotches, spots, browning etc of skin. To avoid sapburn and get cleaner fruit it becomes inevitable to carry certain pre- and post-harvest handling techniques. The problem of sapburn can be some what alleviated if the time of picking is scheduled between 9 AM and 3 PM when the fruit has lost some turger pressure and latex flow is minimal (Kosiyachinda and Mendoza, Jr. 1984). Harvesting the fruit with longer stalk (10-15 cm) will also be useful, as longer stalks do not bleed much at the time of harvest. Post-harvest sap burn is avoided by a process called "desapping" or "bleeding" (see desapping). Another method is to wash the fruit prior to desapping (Mendoza, 1978). Other ways of avoiding sap burn are:

(i) Dipping of fruits in detergent solution prior to removal of the pedicel.
(ii) Desapping in Calcium hydroxide solution (1%)
(iii) Packing of fruits with short stalk.
(iv) Washing fruits in aluminum potassium sulphate solution (1%).
(v) Applying surface coating to fruit prior to desapping
(vi) Using harvesting aids having a sloping canvas catching device soaked in a detergent and dipping or spraying with detergent immediately.

Of the above, the harvesting aids proved to be the most effective in reducing sapburn incidence to about 16 per cent and using only half the labour of other methods (Holmes *et al* 1993). Although harvest aids have reduced the incidence of sap burn, other problem such as skin browning and bruishing have come in the way of their use (Holmes *et al* 1990).

De-stemming fruits in 1 per cent 'cold power' (a detergent) solution reduced sap injury, but not as effective as calcium hydroxide solution (O' Hare and Prasad, 1992). 'Cold power' increases fruit wettability allowing penetration of sap into cuticle cracks and spread the injury. Dipping fruits in 'Cold power' immediately prior to de-stemming them was partially effective in reducing sap injury and may have acted as a physical barrier to sap entry (O' Hare and Prasad, 1992).

Lim and Kuppellweiser (1993) obtained excellent protection against sap burn with DC-Tron oil. Dipping fruits in DC-Tron oil at 100 ml/litre of water concentration for 30- 60 seconds gave better results than similar rates of other highly refined parafinic oils. DC-Tron treated fruits ripened normally with no off-flavours and reduction in quality, but fruits appeared dirty.

Cultivars that are prone to sapburn may be harvested with 10-20 cm long stem attached and re-trimmed at the pack house for desapping. Latex does not usually exuade from the longer stalks because there is no continuity between the fruits and stem resin ducts (Joel, 1980) and the lactifiers are not severed. However, a portion of the stalk may break during transit, resulting in leakage of latex and skin burn. Thus, long stalk of fruit reduced the risk of sapburn (Johnson *et al* 1993).

High nitrogen levels in the fruit have been linked with more severe sap burn. Nitrogen should therefore, only be applied immediately after harvesting the fruit, so that levels are relatively low during fruit set and development. Additionally, this results in better fruit colour and reduced sap burn.

LENTICEL DISCOLOURATION

Post-harvest discolouration of mango fruit lenticels is a major problem and detracts from the aesthetic value of fruits, since the resultant black markings on the fruit surface are unacceptable to the consumer (Tamijinda *et al* 1992) and consequently depreciates the economic value of the fruits. Thus 'lenticel discolouration' has serious implications on the marketability of mangoes especially on export market.

Symptoms

On susceptible cultivars, the lenticels become dark and thin, surrounded by a very dark green border. As the background colour changed to yellow this border is faded. The virtual damage would be greatest on green fruits. Initially, lenticel discolouration occurs in the form of light purple spot surrounding the lenticel and there will be vascular accumulation of pigmentation. In black discoloured lenticels, there is no vascular pigmentation, but the cell walls are discoloured by natural pigmentation (Bejuidenhout *et al* 2005).

Causes

Several causes have been advanced for lenticel discolouration. O'Hare and Prasad (1992) showed that one of the most important causes of lenticels discolouration was leakage of sap from resin ducts in to the lenticels. Bejuidenhout *et al* (2005) observed a correlation between lenticel discolouration and resin ducts. Mango sap consists of two distinct components: an oil fraction and a protein-polysacchoride fraction (Joel and Fahn, 1980). On contact with the fruit skin, the components of sap get separated and it was the oil component that was responsible for lenticel discolouration (O'Hare and Prasad 1992). Loveys *et al* (1992) Tamijinda *et al* (1992) reported that discolouration of the surrounding tissue of lenticels was due to the entry of air and water through lenticels. According to Robinson *et al* (1993) discolouration of fruit lenticles was largely the result of enzymatic oxidation of phenolic compounds in the vacuoles in the presence of polyphenol oxidase in cell walls and gaseous oxygen. Robinson *et al* (1993) maintained that lenticel discolouration could be induced by terpenoids from mango resin which causes the vacuoles to lyse and the phenolic compounds to come in contact with polyphenol oxidase. As a result of either of these discolouration processes, lenticels change from purple to dark brown, giving typical discolouration of lenticels. The absence of a cork cambium and cork cells in the mango fruit lenticels could be one of the most important reason for lenticel discolouration. An interaction between naturally occurring pigments and sap from the resin ducts in the exocarp (peel) was attributed as another reason for lenticel discolouration (Bejuidenhout *et al* 2005).

It has been reported that some post-harvest treatments also cause and increase lenticel discolouration in mango. Post-harvest treatments shown to cause and increase lenticel damage include : dipping of fruits in hot water (45 °C) for 30 minutes (Jacobi *et al* 2001) a combination of hot water and hot air (Jacobi *et al* 1996), washing fruit in one of several disinfectants or soaps (Bally *et al* 1997), washing fruit in ambient water (O'Hare *et al* 1999), application of partial pressure of –66kpa to 99 kpa (Shorter and Joyce, 1998) or post-harvest infiltration with calcium (Joyce *et al* 2001).

Factors

The incidence of lenticel damage is thought to be affected by many of the conditions to which mango fruits are exposed prior to, during and after harvest. The effects of air temperature, relative humidity, duration of exposure to solar radiation, wind and amount of rain fall on the day of harvest, on post-harvest incidence of lenticel discolouration were accessed (Osthuyse, 1998). Strong positive correlation with maximum humidity was observed. The results clearly indicated that cool, humid or wet conditions on the day of harvest strongly favoured the lenticel discolouration. Conversely, dry hot conditions discouraged its occurrence. A delay prior to cold storage led to increase the incidence of lenticel discolouration in South Africa (Oosttuyse, 2000).

Exposure to radiation also results in lenticel discolouration, because it is not possible to control spoilage fungi on mangoes without injuring the fruit. Irradiation with doses higher than 75 Gy, particularly from 300 or 600 Gy caused lenticel discolouration severely (McLauchlon, *et al* 1990). Irradiation of 'Kensington Pride' mangoes in Australia at 0.6 and 1.2 KGy showed unacceptable levels of lenticels spotting and skin damage. Fruits which were partially ripe and in their pre-climacteric rise were largely unaffected by irradiation (Boag *et al* 1989). However, there is still much uncertainity concerning the mechanism of lenticel discolouration in mango.

Control

Since lenticel discolouration is resulting in economical losses to the grower, it is imperative to avoid such incidence or at least reduce it, to permit marketability. The treatments likely to reduce irradiation injury in mangoes include lowering temperature, blanketing in a nitrogen atmosphere and lowering the dose rates (Dharkar *et al* 1966). They showed that damage to "Alphonso" mangoes (skin spotting or blackening) by doses of 0.5 to 2 KGy, could be regulated by equilibriating the fruits in nitrogen or carbondioxide for 3 hours prior to treatment followed by irradiation in the same atmosphere. Fruits irradiated under nitrogen atmpsphere could withstand doses as high as 2 KGy without apparent injury while fruits exposed to these doses in air turned completely black. Contrary to the above, Mc Launchlan *et al* (1990) could not get relief from lenticel discolouration caused by irradiation at 300 Gy or higher doses by altering the condition of irradiation (lower temperature, nitrogen atmosphere, lower dose rate).

CHILLING INJURY

Attempts made to use low temperature for extension of storage of mango were not successful because of its susceptibility to 'chilling injury', a storage disorder when mangoes are stored for longer periods below critical temperature of 10 °C (Hatton *et al* 1965.,) or 13 °C (Chaplin *et al* 1991). It has become a major problem for storage of mangoes. The chilling injury (CI) of the fruits is responsible for a great loss during the storage as the chilling damaged fruits are unmarketable due to discolouration of the skin and poor quality. The CI restricts the prolonged refrigerated shipments to the international markets by sea or road, leaving air transport as the only alternative means of transport. CI also reduces the duration of storage. It is the most important obstacle to the expansion of trade in tropical fruits including mango in the world markets. In the presence of CI, disease development may be earlier and more extensive (Johnson *et al* 1997). For instance susceptibility of mango fruit to soft brown rot increased with chilling injury (Oosthuyse, 1990). Exposure to extremely cold temperature can also lead to delayed ripening, poor flavour and water soaking in pulp tissue after ripening (Nair *et al* 2003b).

Symptoms

The symptoms of CI usually develop on transfer from low temperature to high temperature for ripening. The most common visual symptoms of CI in mango fruits are: dark scald like discolouration in the peel and pitting or sunken lesions (Wardlaw and Leonard 1936). Dark colouration of the skin, prominence of lenticels, uneven ripening, development of off-flavour, softening of tissues and deterioration of fruit quality (Chhatpar *et al* 1971., Nair, *et al* 2001) are other effects of CI.

However, symptoms vary with different storage temperatures and cultivar. Storage of mangoes below 45 °F leads to CI, and symptoms include manifestation of pitting, failure to colour normally, inability to ripen satisfactorily and lowered resistance to decay organism (Singh 1974). Besides, mangoes stored below 10 °C show reduction in the level of carotenoides, aroma and flavour (Abou Aziz *et al* 1976., Thomas and Oke, 1983).

CI appeared as pitting in "Raspuri" (followed by discolouration at the end), but in the case of 'Badami' only discolouration was noticed without pitting (Mathur *et al* 1953). The CI symptoms in 'Kensigton Pride' mango are: skin darkening, prominence of lenticels, development of poor colour and flavour and uneven ripening and increase in titrable acidity (Lederman *et al* 1997). In 'Tommy Atkins' CI showed red spots on the peel and pitting and black spots. (Pesis *et al* 1997). In Thailand, CI manifested initially as of red to brown discoloration of the peel, followed by colour change in the pulp and endocarp. The seed tissue became brittle during storage at 4 °C and 8 °C and seeds failed to germinate. External symptoms on the peel appear earlier than the internal symptoms. CI symptoms were observed in the fibrous parts of the flesh directly around the seed coat and both showed a discoloration varying from greyish to brownish tinge (Phakawatmongkol *et al* 2004).

CI symptoms develop during ripening and progress after the transfer of fruit from chilling to non-chilling temperature. The severity of symptoms increases with storage time and ripening period (Chaplin *et al* 1991).

Factors Influencing Chilling Injury

Different factors like cultivar, fruit maturity, degree of ripeness, temperature, duration of exposure to chilling temperature, duration of storage etc., influence the incidence and severity of CI in mango.

Cultivar: Compared to other cultivars "Alphonso" was more sensitive to low temperature injury (Thomas and Oke, 1983) Chaplin *et al* (1986a) found 'Kensington Pride' more sensitive than 'Zill'. Among the cultivars. 'Okrong' mangoes were the least and 'Nam Dok Mai' were the most sensitive to CI. (Phakawatmongkol *et al* 2004).

Maturity of fruits: Sensitivity of mango fruits to CI is influenced by the stage of maturity of fruits at harvest. Green fruit of 'B' stage maturity (shoulders out grow the stem end and the skin colour is dark green) was susceptible to CI (Cheema *et al* 1950). Fruits harvested when nearly ripe are less prone to CI (Medlicott, 1985), while fruits harvested at an earlier stage of maturation are more prone to CI and other storage related disorders. (Oosthuyse, 1993). Early season fruits stored at 10 °C and 75-85% relative humidity showed more CI, lesser TSS and total carotenoids and more acidity as compared to late season fruits (Narayana and Singh, 1998).

Degree of Fruit Ripeness

The sensitivity of mango fruits to CI also depends on the degree of ripeness at the time of storage. Half-ripe fruits (cv. Totapuri) at 13 °C showed less severe symptoms of CI and better ripening qualities than fruits in advanced ripening. Ripe mangoes are much more sensitive to CI, than pre-climactric ones. (Thomas and Oke, 1983). On the other hand, fruits at more ripened stages show more tolerances to CI than unripened ones (Medlicott *et al* 1990). The work of Chaplin *et al* (1991) indicated that fruits, in which ripening had commenced were less sensitive to low temperature storage and CI. Green mangoes (stage I) when stored for more than 15 days and transferred to higher temperature (25 °C) showed, the first symptom of CI while no incidence of CI was observed in fruits of stage II (more green than yellow). These observations indicated that more ripened mangoes could be kept at much lower temperature without CI where as full green and immature mangoes cannot be stored at low temperature as they are susceptible to CI. However, fruits of all stages at 45th day of shelf showed same CI symptoms (Wangchai *et al* 2000).

Temperature

Low temperature is the major factor that influences the incidence of CI in mango. Harvested mangoes are susceptible to CI when exposed to temperatures below 13 °C. However, fruits that were held and ripened at tropical ambient temperature (27-34 °C) developed CI (skin staining or browning) when they were subsequently stored at 5,10 or 15 °C but fruits held and stored at a lower temperature (20 ± 1 °C) showed little evidence of CI, when subsequently stored at 5 or 10 °C for up to 14 days (Thomas and Joshi, 1988). Several workers reported CI development in mangoes at temperature regimes ranging from 3° to 12 °C (Sadasivam *et al* 1971., Thomas and Oke, 1983; Chaplin *et al* 1986 a, b., Thomas and Joshi 1988., Medlicott, *et al* 1990). However, Farooki *et al* (1985) could store fruits of 'Sensection' and 'Samara Bahist' for 4 days at 6-12 °C without CI in Pakistan. Working with 'Kensington' mangoes, Chaplin *et al* (1991) found that when fruit was stored at 1° or 5 °C, CI increased with time and the symptoms increased even further, when the fruit was transferred to 30 °C for ripening. However, in a later study Chaplin *et al* (1991) observed no visual symptoms of CI in fruits stored at 10° or 15 °C in contrast to the findings of other workers who have shown that certain other varieties developed symptoms, when stored at the above temperature (Subramanyam *et al* 1972., Veloz *et al* 1977., Thomas and Oke 1983). This indicates varietal difference in the incidence of CI in mango. 'Kent' mangoes showed typical CI injury at temperatures below 13 °C (Veloz *et al* 1977), while 'Manila' mangoes showed symptoms even at 16 °C (Veloz and Lakshminarayana 1977). Totapuri mango fruits could be stored up to 8 weeks at 5-7 °C and 'Alphonso' mangoes up to 4-7 weeks at 7°-10 °C without CI (Srivastava, 1967). Contrary to this "Alphonso" mangoes in Sudan showed CI symptoms in storage at 7°-10 °C (Musa and Mirghani, 1974).

In China, Ji *et al* (1998) observed CI symptoms in fruits of cultivar 'Zinna' after 15 days at 2 °C and after 34 days at 5 °C indicating that the development of CI is faster and needs less time at lower temperature than at higher temperature. 'Manila' fruits maintained at 6° or 12 °C for 12 days in storage in Mexico, showed symptoms of CI by 4th day of ripening and the symptoms were more severe in fruits stored at 6 °C. It was concluded that 'Manila' fruits could be stored at 6° or 12 °C for 6 days without CI (Hidalgo *et al* 1997). CI symptoms developed in 'Kensington Pride' mango fruits stored at 0, 5 and 10 °C but not at 15 or 20 °C in Australia. CI increased as the storage temperature decreased from 10 to 5 °C and the storage period was prolonged from 3 to 28 days (Nair *et al* 2004a). Storage at lower temperature induced more CI and therefore, reduced the duration of storage (PhakawatmongKol *et al* 2004). 'Haden', 'Irwin', 'Edward' and 'Kent' mangoes were stored for 10 days at 5 °C and ripened at 24 °C without cold damage (Laborem *et al* 1992). Similarly, storage of 'Tommy Atkins', 'Spring fields' and `Ford' at 12 °C were not affected by CI in Venzuaela (Manzamo-Mundez, 1995). Thus, the temperature at which CI symptoms develop appear varing with the cultivar and place.

Just how the lower temperatures induces CI in fruits is not clear. A few hypotheses were advanced. Under low temperature conditions, the normal metabolism of mango tissue is prevented and biochemical reactions associated with respiration proceed in an alternate pathway (Farooki *et al* 1985). Mukherjee and Srivastava (1979) proposed that the critical chilling temperature of the mango fruit is governed by TSS of its juice and concluded that the chilling temperature for 'Langra' and 'Dashehari' mangoes was lower with a high TSS. Ramana *et al* (1984) also suggested a relationship between TSS and the susceptibility to CI in mango and the occurrence of CI during storage at low temperature was directly proportional to the TSS content of fruits. This caroborates with the finding that the fruits showing CI are less sweet than those unaffected fruits (Roy, 2001). This is in contrast to the findings of Thomas and Oke, (1983) that ripe fruits were much more sensitive than pre-climacteric ones because ripe fruits contain higher content of TSS. In a study, Nair *et al* (2004a) showed that the suppression of ethylene bio-synthesis during ripening of fruits stored at chilling temperature was due to reduced amounts of ACC. (l-amino cyclo propane-1-carboxylic acid) and reduced activities of ACC oxidase and ACC synthesis in skin and pulp during storage and ripening period and these may be associated with the development of CI symptoms in 'Kensington Pride' mangoes in Australia. It may also be possible that by increasing or decreasing the activity of certain enzymes needed for ripening process at low temperature is causing CI in mango fruits. The lower concentration of water-soluble pectin in low-temperature-stressed mango fruits correlated with delayed softening. This suggests that softening of ripening mangoes results from an increase in the degradation and polymerization of cell wall pectin (Huber, 1983). In low-temperature-stressed fruits polygalacturonase (PG) and β-galactosidase activities increased during ripening to a lesser extent than in non-stressed mango fruits. Thus, in the chilled fruits, there may be reduced depolymerization and solubilization of pectic substances (Taylor *et al* 1994).

Changes Due to Chilling Injury

Various physiological and biochemical changes have been reported in fruits in response to chilling stress. Under low temperature conditions the normal metabolism of mango tissue is prevented and biochemical reactions associated with respiration proceed in an alternative pathway (Farooki *et al* 1985).

Changes in cell wall constituents: Chilling injury in a wide range of fruits and vegetables has been reported and related to the transition of cell membranes from a liquid crystalline to a gel state (Lyons, 1973). Much evidence accumulated to support the concept that this phase transition of membrane lipids leads to modification in the activity of enzymes and alters the permeability of membranes (Lyons, 1973., Lyons *et al* 1979). Kane *et al* (1978) concluded that chilling induced fatty acid changes in mango mitochandria resulted in reduced respiration and tissue damage.

The cell wall contents of chilled fruit contained less water soluble pectin, more ammonium oxalate soluble pectin and less alkali soluble pectin than in non-chilled fruits. The increase in ammonium-oxalate soluble pectin esterase was at the expense of the alkali soluble pectin and correlated with high pectin in the chilled fruits. The ammonium-oxalate-soluble pectin in chilled fruits appears to have a major structural role in maintaining the firmness of chilled fruit (Ketsa *et al* 2000). The increase in ammonium-oxalate-soluble pectin in chilled fruit may be a consequence of the high pectin esterase activity found during low temperature stress. This high level of ammonium oxalate soluble pectin may be responsible for the slower softening of chilled fruits (Ketsa *et al* 2000).

Enzymes and their activities: Solubilization of insoluble pectin due to increased pectin esterase (PE) activity in chilling injured fruits seems to be one of the reason for the development of softness in CI. During this stage invertase activity increased and that of amylase decreased (Mattoo and Modi 1969a). It has been reported that the activity of invertase, which is involved in carbohydrates metabolism increased during development of CI in mango peel (Chhatpar *et al* 1971). They also reported a decrease in amylase activity with a corresponding decrease in sugar content and lowered starch breakdown in 'Alphonso' mangoes due to CI. There was two-fold decrease in amylase activity whereas invertase activity was two times more indicating that low temperature injury results in more sucrose breakdown than starch (Chhatpar *et al* 1972). Kane *et al* (1978) claimed that there was a disturbance in the succinate oxidation capacity of mitochondria, isolated from "Amelie" mango fruit stored at four temperature regimes between 4 and 20 °C. Peroxidase and cellulase activities in the peel of chilled fruits (at 5 °C) rose during the development of CI to much higher levels than in non-chilled fruits. The activity of these two enzymes started to increase before any changes in TSS and acid contents of the pulp could be observed. Based on the results, it was suggested that the increase in the activity of above two enzymes is part of the CI syndrome that develops during storage of mango fruits at chilling temperature (Zauberman *et al* 1988). Fruits of 'Zhihna' stored at 2 °C for 15 days developed CI and the fruits showed increased super oxidase-dismutase activity and a decline in catalase indicating that membrane lipid peroxidation was enhanced (Wang *et al* 1995). PE is generally considered not to be involved in fruit softening since its activity decreases during ripening (Abu – Sarra and Abu-Gowkh 1992., Awad and Young 1980., Roe and Bruemmer, 1981). The increased PE activity in response to low temperatures may cause the increased ammonium-oxalate pectin content, but does not appear to be retarded to the rate of fruit softening. Increased PE activity in mangoes after exposure to low-temperature stress may be related to CI (Ketsa *et al* 2000). ACC oxidase, ACC synthesis and ACC contents in the skin and pulp were reduced in fruit stored at 5 °C compared with those stored at 15 °C prior to ripening (Nair *et al* 2003b).

Aroma volatile compounds: Suppression of ethylene biosynthesis in mango fruits stored at low temperature (5 °C) was attributed to reduction of aroma volatile compounds (Lakshminarayana and Subramanyam 1970., Nair 2002). This indicates that ethylene biosynthesis may be involved in the biosynthesis of aroma volatiles and the synthesis of some volatile esters requires continuous ethylene action and a high rate of ethylene production. Suppression of ethylene biosynthesis with increased CI and the reduction of CI with the exogeneous application of ethrel suggested that the ethylene is involved in CI development. (Nair et al 2004b) Chill-injured fruits exhibited significant reduction in the production of total aroma volatile compounds during ripening as compared to normal fruit showing an inverse relationship between mean content of various aroma volatile compounds and mean CI induced with different storage temperatures and durations (Nair *et al* 2003c, 2004b., Singh *et al* 2004).

The reduction of production of aroma volatile compounds due to CI may be attributed to the inhibition of the enzymes involved in several metabolic pathways. Lipids have been considered to be the main precursors of volatiles in mangoes and in fruits in general via their breakdown to fatty acids. An increase in the concentration of fatty acids during ripening was observed in 'Alphonso' mangoes (Bandopadhya and Gholap, 1973). Therefore, the reduction of volatiles in chill-injured fruits may be attributed to the inhibition of biosynthesis of fatty acids from which the aroma volatiles are biosynthesised.

Other changes: Chilling injury of mango fruits at 0°-5 °C caused a decrease in the soluble sugar content (mainly sucrose) in the peel. There was no significant change in hexose content, but there was less starch breakdown and accumulation of ascorbic acid (Mattoo and Modi, 1969a). Chill-injured tissues contained more starch and ascorbic acid and less TSS, with practically no change in the free sugar and total nitrogen as compared to non-injured healthy tissue (Chhatpar *et al* 1972). However, Farooki *et al* (1985) reported that CI had no significant effect on the biochemical constituents of the fruit pulp and the pulp colour and flavour were normal in the chilled fruits. The CI causes leakage of metabilities such as amino acids, sugars and minerals from the cells. The succinate oxidation capacity of mitochondria also decreases in chill-injured fruits.

Accumulation of polyamines (putrescine) due to CI was observed in mango fruits (Wangchai *et al* 2000). In chill-injured fruits of 'Kensington Pride' the accumulation of putrescine and reduction in endogenous levels of spermidine and spermine in the peel and pulp were observed (Nair *et al* 2003 a).

The chill affected fruits also showed increased accumulation of minerals (Chhatpar *et al* 1971). (Alcohols and aldehydes are formed as breakdown products in the fermentative decarboxylation in chill-injured mangoes, particularly in high CO_2 atmospheres (Lakshminarayana and Subramanyam, 1970).

Prevention or Alleviation of Chilling Injury

Prevention or alleviation of chilling injury in fruits and vegetables is essential to avoid economic losses in storage and also prolong their storage life. It also assumes significance in view of the ever increasing exports and the internal trade of mango fruits. The preventive methods should at least delay the development of CI symptoms, so that it would be feasible to store perishable products at lower temperature as long as possible to reduce the rate of deterioration. During past 60 years or so, various methods have been attempted and adopted to prevent/reduce CI in different fruits and vegetables. These methods can be categorised as temperature conditioning, intermittent warming, controlled atmosphere, application of plant growth regulators, other chemical treatments, waxing and other coatings and the packaging. The first three approaches manipulate and modify the storage environment while the others involve direct treatment to the commodities (Wang 1993). A perusal of available information on the subject indicates that all the above techniques were not tired with mango. The various methods attempted in case of mango CI can be dealt under (i) physical and (ii) chemical methods broadly.

Physical Methods

Heat treatments: Warming of harvested fruits prior to cold storage was found to be effective in reducing CI as the tolerance to chilling increases after heat treatments. Pre-storage high-temperature conditioning to reduce CI was reported as early as in 1936 by Brooks and Mc Colloch. They found that pre-storage heating at 38 °C for 17-22 hours, significantly decreased pitting in grapefruit [*Citrus paradisica*] stored at 2° or 4-5 °C. High temperature pre-storage conditioning also favours lignin formation and the healing of wounds at harvesting and transportation (Brown *et al* 1978). Maintaining a high humidity level is important to obtain desired results in heat treatments. Protection against CI by pre-storage heating may relate to the accumulation of heat-shock proteins (Lurie and Klein, 1991).

Heat treatment as a method of reducing CI in horticultural crops to extend the storage period has gained much interest. High temperature treatments have been shown to increase the chilling tolerance of chilling sensitive tisssues. (Klein and Lurie 1991., Lurie and Klein, 1991). Hot water dipping allows a more homogenous heating of the product and it has been used successfully to prevent decay in a variety of fruit species.

Mature fruits of 'Langra' and 'Dashehari' subjected to hot water treatment at 50 °C for 15 min. cooled and stored at 9°-10.5 °C for up to 4 weeks for a further period of 14 weeks and next held at room temperature for 2 days, did not show CI (Mukherjee and Srivastav, 1979). Dipping of pre-climacteric fruits of Cv. 'Palmer' in hot water at 38 °C significantly reduced CI., but dipping at 46 °C had no effect on CI, but at 54 °C significantly increased CI symptoms (Zambrano and Materanao, 1998, 1999). Based on their studies they concluded that temperature manipulation before storage can prevent the development of CI on mature green mangoes, however, temperatures above 46 °C may cause heat damage. A post-harvest heat treatment applied prior to low temperature storage can reduce the incidence of CI in cold sensitive fruits like avocado (Woolf *et al* 1995). Heating the fruits of 'Nam Dok Mai" before cold storage decreased decay and CI and prevented the development of off flavours (Ketsa *et al* 2000). Heating 'Zinhna' mangoes in hot air at 38 °C for 12 hours significantly reduced the development of CI following storage up to 12 days at 2 °C. Prolonging the heat treatment up to 72 hours gave similar results and this was linked with increased chilling tolerance due to high level of absicic acid in treated fruits (Zhu and Ji, 2002). However, heat treatments (fruit conditioning followed by hot water treatment or hot air treatment) did not show reduction in CI development after 21 and 35 days storage at 5 °C in "Kensington Pride" mangoes in Australia (Nair *et al* 2001).

Cold adoption: Cool temperature conditioning or hardening is also called 'cold adoption' (Thomas and Oke, 1983). Temperatures that fruits and vegetables are exposed in just prior to chilling treatment affect significantly the chilling tolerance of these commodities. [Salt Veit and Cabrera, 1987). Conditioning at temperatures slightly above the critical chilling range increases the tolerance of commodities to chilling during subsequent low temperature storage and delays the development of CI symptoms [Wang, 1993].

Cold adoption technique is recommended for an improvement in the quality and storage life of the mango by preventing CI. 'Totapuri' fruits which were gradually cooled to 13 °C showed less severe symptoms of CI in Sudan than fruits stored directly at 13 °C. Similarly half-ripened fruits and gradually cooled fruits had less sever CI symptoms (Musa, 1974). Pre-climacteric hard green 'Alphonso" mangoes stored directly at 10 °C for 30 days developed CI as manifested by pitting and a reduction in the level of carotenoids, aroma and flavour during ripening. On the other hand, pre-climacteric fruits adopted to 10 °C by step-wise exposure to 20 and 15 °C for one and two days respectively could withstand storage at 10 °C for longer duration and developed better flesh colour and organoliptic qualities on ripening. Pre-climacteric fruits previously stored at 10 °C and subsequently ripened at 27° – 32 °C did not show CI, when placed in temperatures as low as 4° -7° for up to 3 weeks thus extending the shelf life (Thomas and Oke, 1983). Pre-climacteric fruits of 'Harumanis' in Malaysia with low-temperature adoption did not suffer CI after 45days storage at 7 °C. However, pre–climacteric fruits without cold adoption suffered CI (Lam and Ng, 1986). CI in 'Alphonso' mango fruits was prevented by holding pre-climacteric fruits for a minimum period of 30 days at 10 °C before ripening at 27°-34 °C with acceptable quality (Thomas and Joshi, 1988). Temperature conditioning by gradually decreasing the temperature from 20° to 17 °C or 14 °C during the two days after harvest and before storage at 9 °C reduced the number of red spots by 50 per cent in 'Tommy Atkins' mango fruits compared with the controls (Pesis *et al* 1997).

'Cold adoption' technique is now recommended for an improvement in the quality and storage life of the mango fruits by preventing CI (Roy, 2001). However, the mechanism by which cold adoption alleviates the CI is not clear. Apparently, the temperature conditioning treatment induces an adoptive response in fruits and vegetables to chilling stress. This adoption to lower temperatures is the result of various physiological modifications induced by the conditioning treatment. Many biochemical and physiological alternations in chilling sensitive plants have been associated with cold adoption treatments. Conditioning or adoption of plant tissues have been shown to induce changes in the composition of membrane lipids (Wang and Baker, 1979., Pike and Berry (1980). It is possible that a similar mechanism may be operating in cold-adopted mangoes which however, needs confirmation (Thomas and Oke, 1983).

Intermittent warming: Since, continuous storage of mangoes at lower temperatures (13 °C) is inducing CI, the possibility of using intermittent heating for prevention of CI was explored successfully. It was found that interrupting low-temperature increases the storage life of some chilling-sensitive commodities. The brief warm treatments must take place before injury become irreversible. If the critical time at chilling temperature has been exceeded and CI progresses beyond recovery, then raising the temperature would only accelerate the degradation processes and hasten the development of injury symptoms. Conversely if the warming treatment is applied too early or too frequently tissues will become excessively soft and valneurable to bruising or invasion by decay organisms. Therefore, correct timing and duration are essential for successful intermittent warming treatments. Optimum treatments vary with different species and cultivars.

Intermittent warming has increased the ability of fruits to withstand CI (Zhang *et al* (1997). Dai *et al* (1999) investigated the effect of intermittent heating on the incidence of CI in mangoes stored at 2 ± 5 °C and found that CI was reduced by intermittent warming of fruits.

Intermittent warming is hypothesized to remove toxic or inhibiting substances that accumulate during chilling (Pentzer and Heinze, 1954). Temperature increases, apparently induce higher metabolic activity that removes excess intermediates and replenishes deficiencies that developed during chilling. Warming of chilled tissues for short periods helps to repair damage to membranes, organelles or metabolic path ways (Lyons and Breidenbach, 1987).

Modified controlled atmosphere: Responses of fruits and vegetables to chilling temperatures can be altered by modifying the storage environment, particularly the atmosphere surrounding the commodities (Forney and Lipton, 1990). So some attempts were made to eliminate or minimize CI symptoms through atmospheric modification. Chaplin *et al* (1991) was partially successful in reducing skin browning in mango cultivars by storing fruit at 1 °C in polyethylene bags. Under these conditions CO_2 was enhanced and O_2 was lowered without the development of off-flavours although it was uncertain as to which gas was active in symptom alleviation. From their studies 'O' Hari and Prasad (1993) concluded that elevated CO_2 was responsible for alleviation of chilling symptoms in "Kensington Pride", mango and reduced CO_2 concentration had no significant effect. However, higher concentrations of CO_2 (>10%) are injurious to tissue resulting in skin discolouration and high levels of ethanol in the pulp. The delay in the development of CI in PVC film wrapped mango fruits may be due to a modified atmosphere generated, within the packages (Chhatpar *et al* 1986a., Ketsa and Raksritong, 1992). A modified atmosphere was created (5%. O_2 and 10% CO_2) in 4 kg film-lined cartons by using micro-perforated polyethylene bags or Klend film (XF). After 3 weeks of storage at 12 °C plus one week at 20 °C both modified and cantrolled treatments were effective in reducing CI. The most effective reduction was in fruits packed in the XF bags. A second advantage of using XF bags was the reduction in the level of sap inside the package due to the lower relative humidity in these bags (<90%) compared with polyethylene packaging (99%) (Pesis *et al* 2000). Marcellia and Kane, (1978) however, found no advantage of modified atmospheric storage (5% O_2 & 95% CO_2) in 'Amelie' mangoes.

The reduction of CI by film packaging is attributed largely to the modification in the micro-environment with in the package. Modified atmospheres in enclosed packages also extend ripening time, improve firmness and maintain quality. Shrink wrapping which appears to serve as an artificial exocarp greatly modifies the internal gaseous composition and reduces CI in some commodities.

Chemical Methods

Ethylene treatment: Since ethylene is a ripening hormone it was envisaged that fruits that increase their resistance to CI with ripening would benefit from pre-storage ethylene treatment. The suppression of endogenous ethylene production during storage and ripening was exhibited by high incidence of CI (Nair *et al* 2003b., 2004 a). In other words, lower level of endogenous ethylene (lower than threshold level) during cold storage may be resulting in CI. Hence, by increasing the level of ethylene in fruits through exogenous application of ethylene, it may be possible to prevent CI. Efforts have been made in this direction in mango (Satveit. 1999., Mohammad and Brecht. 2000., Nair and Singh, 2003., Nair *et al* 2004a).

Mohammad and Brecht (2000) investigated the effects of exogenous application of ethylene on alleviation of CI in storage. Half-mature fruits of mango were placed in air at 20 °C for two days or exposed to 100 ppm ethylene for two days at 20 °C. After 15 days at 5 °C, fruits were transferred for one and three days at 20 °C to examine the manifestation of CI in treated fruits. Exogenous application of ethylene at 100 ppm for one day at 20 °C plus one day in air prevented CI in fruits during 15 days storage at 5 °C plus one or three days at 20 °C. A two day ethylene treatment though effective in reducing CI, it resulted in fruit decay (10%). Meanwhile fruits exposed to air at 20 °C for two days before storing for 15 days at 5 °C plus one or three days at 20 °C developed moderate symptoms of CI. Based on these results, they concluded that post harvest loses in mango due to CI can be effectively reduced by gassing mature fruit for one day at 20 °C prior to storage at 5 °C. However, it was cautioned that handling of more mature fruit or ripe fruit would require steps for decay control. Pre-storage ethrel (500 ppm) dip-treatment for 5 min reduced CI and improved fruit quality and taste of "Kensington Pride" mango fruits (Nair and Singh. 2003).

How ripening or ethylene treatment can alter fruit physiology and indirectly change sensitivity of fruits to CI is still unclear. Published reports have indicated that ethylene treatment can reduce, increase or have no effect at all on CI development (Wang. 1989). It is well established that low temperature stimulates auto-catalytic ethylene production, a common feature of climacteric fruits (Burg and Brug 1962., Nair *et al* 2004 a). Chilling induced ethylene production also follows general path way of ethylene biosynthesis. The amieloration of CI in mango fruit with the exogenous application of ethylene prior to low temperature storage clearly indicates that CI is associated with suppression of ethylene biosynthesis (Nair *et al* 2004a). And exogenous application of ethylene to mango and other climacteric fruits prior to storage at chilling temperatures has significantly reduced

CI by encouraging endogenous ethylene biosynthesis (SaltVeit 1999; Nair and Singh 2003., Nair *et al* 2004a).

Polyamines

The involvement of polyamnes in the reduction of CI has been suggested based on the successful results of exogenous application of polyamines in preventing CI (Zhang *et al* 2000., Nair and Singh, 2004). Accumulation of putrescine during exposure to chilling stress is attributed for reduction of CI. Previous researches suggested that endogenous polyamines to alleviate chilling stress by increasing their level in citrus (Mc Donald and Kushed 1986) and cherimoya (Escribano and Meradio 1994). Indirect evidences also point to the involvement of polyamines in alleviating CI. Post-harvest treatments that elevate polyamines levels in tissues have been shown to lead to the reduction of CI. Temperature conditioning inhibits the development of CI with concomitant increase in polyamines in different fruits (Gonzalez *et al* 1997., Mc Donald 1989., Rodov *et al* 1995). Temperature conditioning or low O_2 storage significantly increases spremidine and spermine levels and reduces the development of CI in squash (Mc collum and Mc Donald 1992). Post-harvest treatments with exogenous polyamines before cold storage increases internal polyamies levels and reduce chilling injury (Kramer and Wang 1989). The reduction of CI with polyamines may relate to their antioxidant activity and stabilizing effect on membranes (Ballass *et al* 1983., Drolet *et al* 1986).

Nair *et al* (2003a) investigated the relationship between CI and endogenous polyamine levels in the peel and pulp of "Kensington Pride" mangoes and exogenous application of polyamines. Pre-storage application of 3 polyamines (Putrescine, spermidine and spermine) reduced CI symptoms and among the 3 polyamines, spermine was most effective in reducing CI. Putrascine has accumulated, whilist spermidine and spermine levels declined in skin and pulp due to CI. The depletion in endogenous levels of spearmidine and spermine with CI and the reduction of CI with pre-storage exogenous application of spermidine and spermine indicates that polyamines are associated with CI development in "Kensington Pride" mango fruit. In another study, Nair and Singh (2004) found that total free polyamines in the skin and pulp were higher in the chill-injured fruits than in the non-chilled fruits during storage and ripening. Total free polyamines were initially lower and then increased as the storage period was prolonged in chill-injured fruits as compared to non-chilled fruits which exhibited a reverse trend. Total free-polyamines were higher during the initial days of ripening in the chill-injured fruits. Based on their results, Nair and Singh (2004), concluded that the accumulation of endogeneous polyamine-putrescine did not inhibit CI in mango fruits, rather chilling stress promoted its accumulation in the early stages of ripening. Depletion of other two polyamines viz., spermidine and spermine in the skin and pulp of chill-injured fruits and a substanial reduction in CI symptoms with pre-storage application of polyamines indicate that CI development in "Kensington Pride" mango fruits seems to be associated with the biosynthesis of polyamines.

Wangchai *et al* (2000) investigated the chilling sensitivity of mango fruits at three ripening stages viz., Stage I (Green fruits), Stage II (Greener than yellow fruits) and Stage III (Yellower than green fruits) under two different temperatures (5° and 13 °C) to establish the possible relationship between polyamine accumulation at different stages and degree of CI in mango fruits. Stage II fruits were treated with acetylene gas two days after harvest, while stage III fruits were treated with acetylene gas four days after harvest. The study has shown possible relationship between polyamine accumulation and degree of CI suggesting that CI may affect the formation of polyamines by increasing at initial time of storage and catabolism of activation by a decline of polyamine levels among the 3 stages of fruits. For the stage I fruit putrescine level increased reaching to the maximum values during 15 days storage at 5 °C and then it decreased followed by other two stages with a similar trend. In contrast, stage II and III fruits held at 13 °C had higher putrescine levels than stage I. Spermidine levels in stage II and III fruits tended to decrease with storage times, while in Stage I fruits it increased after 15 days storage when fruits were transferred to 25 °C for ripening. Putrescine and spermidine levels decreased in all ripening stages of fruits.

Treatment with exogenous putrescine postponed the occurrence of CI in mango fruits by increasing endogenous polyamine levels, inhibiting ethylene production and maintaining superoxidase dismutage activity in the pericarp (peel) at a higher level, thus retarding the increase of melondialdehyde content and membrane permeability (Zhaug *et al* 2000). On the other hand, Nair and Singh (2004) achieved subsubstantial reduction of CI symptoms on ripe fruit of "Kensington Pride" which were stored for four weeks at 5 °C by pre-storage exogenous application of polyamines.

Methyl Jasmonates

Recently methyl jasmonates (MJ) were tested for reduction of CI in mango (Gonzalez *et al* 2000). Exposure of "Tommy Atkins" fruits to MJ vapours (10 μm) for 24 hours at 25 °C reduced CI during subsequent storage for 21 days at 7 °C and afterwards 5 days of shelf life at 20 °C. The CI tolerance induced by MJ was correlated with the reduction in percent ion-leakage of mango tissue. MJ treatment is assumed to prevent CI symptoms of mango fruits without altering the ripening process (Gongclez *et al* 2000). Treatment of 'Kent' mangoes with MJ vapour for 20 hours at 20 °C not only reduced the symptoms of CI, but also enhanced the skin colour. MJ at 10^{-4M} was the most effective concentration for reducing CI and decay in fruits stored at low temperatures. Based on the above results, it was concluded that MJ treatment could be used with advantage to reduce CI and decay symptoms and also to improve colour development of mango fruits without adverse effects on fruit quality (Gonzalez *et al* 2001).

Conclusion

Most of the research on CI emphasizes the prevention of CI through various methods. The latest approach developed for pineapple (Graham *et al* 1998) uses a molecular approach aimed at manipulating the activity of polyphenol oxidase which is believed to be responsible for CI after removal of pineapple fruits to warm temperatures. However, this molecular approach can also be considered to merely prevent the development of the symptoms but not the injury.

A better approach should aim at preventing the injury itself. Unfortunately, the mechanism of the CI especially the primary event when temperature drops into the chilling range is not well understood. The hypothesis on lipid phase transition proposed by Lyons (1973) and later modified by Raison and Lyson (1980) by redefining terms has not yet been proved or disapproved, but a newer hypothesis was advanced in 1991 (Shewfalt and del Rosario, 2000). According to this hypothesis, membranes are also involved in the development of the injury but was not regarded as a primary event but as a secondary event. It was proposed that chilling stress induced reactive oxygen species, which damage membranes and other molecules at a sub-cellular level. As a result membranes breakdown causing further damage to cells and finally, leading to the appearance of symptoms. It was also suggested that antioxidants could not act as a defence mechanism to prevent injury by destroying the reactive oxygen species. These antioxidants could be non enzymic such as β-carotene, lycopene, α-tochophenol, ascorbic acid and glutathione or enzymic like catalase, peroxidase and superoxidase dismutage.

REFERENCES

Abou-Aziz, A.B., Nabaway, S.M.E., Abdel – Wahabs, F.K. and Kader, A.S. 1976. The effect of storage temperature on quality and decay percentage of 'Pairi' and 'Taimour' mango fruit. Scientia Horticulturae, 5:65-72.

Abu-Sarra, A.F. and Abu-Gowkh, A.A. 1992. Changes in pectinesterase, polygalcturonase and cullulase activity during fruit ripening. Journal of Horticultural Science, 67:561-568.

Amin, H.D. 1967. Development of white corky tissue in a mango fruit (*Mangifera indica* L.). Navsari Agriculture College Magzine, 6 : 14-17.

Anonymous, 1968-70. Annual Report, Central Food Technological Research Institute, Mysore, India.

Anonymous, 2003-04. Research Achievements – Mango, pp. 18. Annual Report of Indian Institute of Horticultural Research. Bangalore for the year 2003-2004.

Assani, A.B., Chhatpar, H.S., Kadri, M.H., Mattoo, A.K. and Modi, V.V. 1971. Microbial spoilage of mangoes. Journal of Food Science and Technology, 8: 208-09.

Awad, M and Young, R.E. 1980. Avacado pectin methyl esterase activity in relation to temperature, ethylene and ripening. Journal of the American Society for Horticulture Science, 105:638-641.

Bagashaw, B. and Brown, B. 1989. Disorders. In : R. Ridgeway (ed.). Mango pests and disorders. Queensland Department of Primary Industries. Brisbane, pp. 22-28.

Ballas, S.K., Mohandas, N., Marton, L.J., and Shohet, S.B. 1983. Stabilization of erythrocyte membranes by polyamines. Proceedings of National Academy of Science (USA), 80:1942-46.

Bandopadhyay, C. and Gholap, A.S. 1973. Changes in fatty acids in ripening mango pulp (Variety 'Alphonso'). Journal of Agriculture, Food and Chemistry, 21:496-97.

Bangerth, F. 1979. Calcium related physiological disorders of plants. Annual Review of Phytopathology, 17 : 97-122.

Boag, T.S., Johnson, G.I., Izand, M., Murrary, C. and Fitzsimmom, K.C. 1989. Effects of gamma irradiation on mangoes Cv. Kensington Pride I. Physiological responses to irradiation treatment as affected by fruit maturity and ripeness. Annals of Applied Biology, 116 : 177-187.

Bojappa, K.M. and Singh, R.N. 1973. Studies on the root activity and soil feeding zones of mango (*Mangifera indica* L.) using 32 P. Newsletter, Indian Society of Nuclear Technology Agriculture and Biology, 2 : 112-113.

Bosquez, E., Figueroa, S.C., Dominguez, J., Perez, L., Kerbel, C. and Diaz de leon, S.F. 2000. Sap burn control by the application of different chemical compounds in Mexican mango fruit with export quality. Acta Horticulturae, 509 : 687-696.

Broadrick, H.T. and Thordgrey, R. 1982. Irradiation in perspective - the significance for the mango industry. South African Mango Growers' Association Research Report, 2 : 23-36.

Brooks, C. and Mc Colloch, L.P. 1936. Some storage diseases of grapefruit. Journal of Agriculture Research, 52:319-351.

Brown, G.E., Ismail, M.A. and Barmore, C.R. 1978. Lignification of injuries to citrus fruit and susceptibility to green mold. Proceedings of Florida State Horticulture Society, 91:124-126.

Burdon, J.N., Morre, K.G. and Wainwright, H. 1991. Mineral distribution in mango fruit susceptible to the physiological disorder softnose. Scientia Horticulturae. 48 : 329-336.

Burdon, J.N., Morre, K.G. and Wainwright, H. 1992. A preliminary examination of the physiological disorder 'Soft nose' in mango fruit. Acta Horticulturae, 298 : 15-22.

Burg, S.P. and Burg, E.A. 1962. Role of Ethylene in fruit ripening. Physiologia Plantarium, 72:572-78.

Burondakar, M.M., Gunjate, R.T. and Govekar, M.A. 1994. Effects of mulching and mulch regulated soil and atmospheric temperature on ocurrence of Spongy tissue in Alphonso mango. Journal of Maharashtra Agricultural Universities, 19(3) : 408-409.

Chadha, K.L. 1989. Mango research in India – new developments. Indian Journal of Horticulture, 46(3) : 279-294.

Chaplin, G.R. 1986. Post-harvest physiology of mango fruit. In : First Australian Mango Research workshop. November, 1984. Proceedings, CSIRO, Melbourne, Australia, 264-270.

Chaplin, G.R., Cole, S.P., Landrigan, M., Nuevo. P.A., Lam, P.F. and Graham, D. 1991. Chilling injury and storage for mango (*Mangifera indica* L.) fruit under low temperature. Acta Horticulturae, 291:461-471.

Chaplin, G.R., Graham, D. and Cole, S.P. 1986a. Reduction of Chilling injury in mango fruit by storage in polyethylene bags. ASEAN Food Journal, 2:139-142.

Chaplin, G.R., Nuevo, P.A., Graham, D and Cole, S.P. 1986b. Chilling responses of 'Kensington' mango fruit stored under variable low temperature regimes. ASEAN Food Journal, 2:133-138.

Cheema, G.S. and Dani, P.G. 1934. Report on the export of mangoes to Europe in 1932 and 1933. Department of Agriculture, Bombay, Bulletin No. 170, pp. 1-31.

Cheema, G.S., Karmakar, D.V and Joshi, B.M.1950. Investigation on the cold storage of mangoes. Indian Journal of Agricultural Science, 20:259-325.

Chhatpar, H.S., Gai, G., Mattoo, A.K. and Modi, V.V. 1972. Some problems pertaining to storage and ripening in mango fruit. Acta Horticulturae, 24 : 243-250.

Chhatpar, H.S., Mattoo A.K. and Modi, V.V 1971. Biochemical studies on chilling injuries in mango. Phytochemistry, 10 : 1007-1009.

Chhatpar, H.S., Modi, V.V. and Vasavada, P.C. 1968-69. Bacterial production of spongy tissue in mango. J.M.S. University of Baroda, 12 and 13 : 49-53.

Clarkson, D.T. 1984. Calcium transport between tissues and its distribution in the plant. Plant Cell and Environment, 7 : 449-450.

Critechfield, H.J. 1975. General climatology. 3rd Edition. Prentice Hall of India Pvt. Ltd., New Delhi.

Dalal, V.S. 1975. Personal communication. Scientist, CFTRI, Mysore. Quoted by Rane et al. (1976). Problem of spongy tissue development in mango (*Mangifera indica* L.) Cv. Alphonso A review. Journal of Mharashtra Agricultural Universities, 1(2-6) : 89-94.

Dai, H.F., Ji, Z.L. and Zhaug, Z.Q. 1999. The effect of intermittent warming on chilling injury and metabolism of glutathione, ascorbic acid of mango fruit. Journal of South China Agricultural University, 20(2) : 51-54.

Desai, M.C. 1966. Sponginess in mango fruits. Junagadh Agriculturae College Magazine, 5 : 4-6.

Deshpande, V.A. 1975. Personal communication, Head, Dept. of Horticulture, P.K.V. Akola. Quoted by Rane et al. (1976) Problems of Spongy Tissue Development in Mango (*Mangifera indica* L.) Cv. Alphonso. A review. Journal of Maharashtra Agricultural Universities, 2(26) : 89-94.

Dharkar, S.D., Saragaon, K.A., Srirangarajan, A.N. and Srinivasan, A. 1966. Irradiation on mangoes. I. Radiation-induced delay in ripening of Alphonso mangoes. Journal of Food Science, 31 : 863-869.

Diaz de Leon, S.F., Lifshitz, C.K., Perez, F.L.G., Feguerua, T.S.C., Sanchiz Diaz, L.D.M. and Bosquez, M.E. 2000. Relationship between sap chemical composition and incidence of sap burn in Mexican mango. Acta Horticulturae, 509 : 321-328.

Drolet, G., Dumbroff, E.B., Legge, R.L and Thompson, J.E. 1986. Radical scavenging properties of polyamines. Phytochemistry, 25:367-371.

Escribano, M.I. and Meradio, C. 1994. The relevance of polyamine levels in Cherimoya (*Annona cherimoya* Mill) fruit ripening. Journal of Plant Physiology, 69:701-706.

Farooki, W.A., Sattar, A. Jr., Dand, K. and Hussain, M.1985. Studies on the post-harvest chilling sensitivity of mango fruit (*Mangifera indica* L). Proceedings of the Florida State Horticultural Society, 98 : 220-221.

Farre, I.M. and Hermoso, J.M. 1993. Mulching and irrigation effects on growth, cropping and fruit quality of the mango Cv. Sensation. Acta Horticulturae, 341 : 295-301.

Forney, C.F. and Lipton, W.J. 1990. Influence of controlled atmospheres and packaging on chilling sensitivity. P. 257-267. In: C.Y. Wang.(ed). Chilling Injury of Horticultural Crops, CRC Press, Boca Raton, Florida. USA.

Fox, R.H. 1968. The effect of calcium and pH on boron uptake from high concentrations of boron by cotton and alfalfa. Soil Science, 106 : 435-439.

Galan, S.V., Galvan, D.F. and Galvo, R. 1984. Incidence of 'Soft nose' in mangoes in the Canary islands. Proceedings of the Florida State Horticultural Society, 97:358-360.

Gautam, D.M. and Lizada, M.C.C. 1984. Internal breakdown in 'Carabao' mango subjected to modified atmospheres. IV. Ca and K levels in the spongy tissue. Post-harvest Research Notes, 1 : 90-2.

Gonzalez, A.G.A., Buta, J.G. and Wang, C.Y. 2001. Methyl Jasmonate reduces chilling injury symptoms and enhances colour development of 'Kent' mangoes. Journal of the Science of Food and Agriculture, 81(13):1244-1249.

Gonzalez, A.G.A., Fortiz, J., Cruz, R., Raez., R. and Wang, C.Y. 2000. Methyl Jasmonate reduces chilling injury and maintains post-harvest quality of mango fruit. Journal of Agricultural and Food Chemistry, 48(2): 515-519.

Gonalez, A.G.A., Zacarias, L., Mulas, M. and Lafunte, M.T. 1997. Temperature and duration of water dips influence chilling injury decay and polyamine content in 'Fortune' mandarins. Post-harvest Biology and Technology, 12:61-9.

Graham, M., Ko, L., Hardy,V., Robinson, S., Sawyer, B., O' Hare, T., Jobin, M., Dahler, J., Underhill, S. and Smith. M. 1998. The development of black heart resistant pineapples through genetic engineering. Pp. 133-136. In: S. Subhadrabandhu and P. Chaimichai. (eds) proceedings of the third International Pineapple symposium, Acta Horticulturae, 529.

Gunjate, R.T. 1989. Personal communication. Quoted by Chadha, K.L. 1989. Mango research in India – New Developments. Indian Journal of Horticulture, 46(3) : 279-294.

Gunjate, R.T., Tare, S.J., Rangwala, A.D. and Limaye, V.P. 1979a. Calcium content in Alphonso mango fruits in relation to occurrence of spongy tissue. Journal of Maharashtra Agricultural Universities, 4 : 159-161.

Gunjate, R.T., Tare, S.J., Rangwala, A.D. and Limaye, V.P. 1979b. Effect of pre-harvest and post-harvest calcium treatments on calcium content and occurrence of spongy tissue in Alphonso mango. Indian Journal of Horticulture, 36 : 140-144.

Gunjate, R.T., Walimbe, B.P., Lad, B.L. and Limaye, V.P. 1982. Development of internal breakdown in Alphonso mango by post-harvest exposure of fruit to sunlight. Science and Culture, 48(5): 186-190.

Gupta, D.N., Lad, B.L., Chavan, A.S. and Salvi, M.J. 1985. Enzyme studies on Spongy Tissue : A physiological ripening disorder in Alphonso mango. Journal of Maharashtra Agricultural Universities, 10(3) : 280-282.

Hatton, T.T., Reeder, W.F. and Campbell, C.W. 1965. Ripening and storage of Florida mangoes. USA Marketing Research Report No:275 Agriculture Research Service. Washington D.C.

Hermose, J.M., Guirado, E., Gomez, R., Castilla, A., Velasco, R. and Farre, J.M. 1997. Effects of nutrients and growth substances on internal breakdown of Sensation mango fruits. Acta Horticulturae, 455 : 92-99.

Hidalgo, M., Cruz, J de la., Parkin, K.L. and Garcia, H.S. 1997. Refrigerated storage and chilling injury development of Manila mangoes (*Mangifera indica* L). Acta Horticulturae, 455:718-725.

Holder, R. and Cockshell, K.E. 1990. Effects of humidity on the growth and yield of tomato. Journal of Horticultural Sciences, 65 : 31-39.

Holmes, R.J., Campbell, T. and Ledger, S.N. 1990. Mango picking guide-'Kenington Pride' Queensland Department of Primary Industries, Brisbane.

Holmes, R.J., Ledger, S.N. and Maclood, W.N.B. 1993. Handling system to reduce sap burn. Acta Horticulturae, 341 : 528-532.

Huber, D.J. 1983. The role of cell wall hydrolases in fruit softening. Horticultural Reviews, 5 : 169-219.

Iyer, C.P.A. and Subramanyam, M.D. 1992. Possibilities of overcoming physiological disorders in mango by breeding. Acta Horticulturae, 317 : 241-243.

Jacobi, K.K., Macrae, E.A. and Hetherington, S.E. 2001. Effect of fruit maturity on the response of 'Kenington' mango fruit to heat treatment. Australian Journal of Experimental Agriculture, 41 : 793-803.

Jacobi, K.K., Wong, L.S. and Giles, J.E. 1996. Effect of hot air disinfestation treatment in combination with simulated air-fright conditons on quality of 'Kensington' mango. Australian Journal of Experimental Agriculture, 36 : 739-745.

Ji, Z.L., Dai, H.F. and Zhang, Z.Q. 1998. Changes of ascorbic acid and glutathione contents during chilling injury of mango fruit. Acta Horticulturae. Sinica, 25(4):324-328.

Joel, D.M. 1980. Resin ducts in the mango fruit. A defense system. Journal of Experimental Botany, 125 : 1707-1798.

Joel, D.M. and Fahn. A. 1980. Ultra structure of the resin ducts of *Mangifera indica* L. (Anacardiaceae) 2. Resin secretion in the primary stem ducts. Annals of Botany, 46: 779-783.

Johnson, G.I. and Coates, L.M. 1993. Quality assurance in tropical fruits – future disease control prospects. In : Sustainable Management of pests, diseases and weeds. Australian clean and green in the 90s. Proceedings 1st National Conference. Australian Society of Horticultural Science. Macquire University, Sydney, pp. 71-76.

Johnson, G.I., Cooke, A.W. and Mead, A.J. 1993. Infection and quiesence of mango stem end rot pathogens. Acta Horticulturae, 341 : 329-336.

Johnson, G.I., Sharp, J.L, Milne, D.L. and Oosthuyse, S.A. 1997. Post harvest Technology and Quarantine Treatments. In:R.E.Litz(ed) The Mango, Botany, Production and Uses. CAB International, Willingford. 447-507.

Joshi, G.D. 1975. Studies on spongy tissue of mango (*Mangifera indica* L.) fruits. M.Sc. (Ag.) thesis, Konkan Krishi Vidya Peeth, Dapoli (India).

Joshi, G.D. and Limaye, V.P. 1984. Effect of manuring and harvest time on spongy tissue occurrence in Alphonso mango. Journal of Maharashtra Agricultural Universities, 11 : 104.

Joshi, G.D. and Limaye, V.P. 1986. Effects of tree location and fruit weight on spongy tissue ocurence in Alphonso mango. Journal of Maharashtra Agricultural Universities, 11(1) : 104-105.

Joshi, G.D. and Roy, S.K. 1985. Spongy tissue in mango, a physiological disorder. Indian Horticulture, 29(4) : 21-22.

Joshi, G.D. and Roy, S.K. 1989. Studies on spongy tissue in Alphonso mango. Acta Horticulturae, 231 : 649-661.

Joyce, D.C., Shorter, A.J. and Hockings, P.D. 2001. Mango fruit calcium levels and the effect of post-harvest calcium infiltration at different maturities. Scientia Horticulturae, 91 : 81-99.

Kane,O., Morcellin, P., and Mazaliak, P. 1978. Incidence of ripening and chilling injury on the oxidative activities and fatty acid compositions of the mitochondria from mango fruits. Plant Physiology, 61:634-638.

Katrodia, J.S. 1979. The study into the causes of the development of spongy tissue in mango (*Mangifera indica* L.) fruit of cultivar Alphonso. Ph.D. thesis (un published) submitted to Marathwada Agricultural University, Parbhani, India.

Katrodia, J.S. 1988. Spongy tissue in mango – causes and control measures. Acta Horticulurae, 231 : 814-826.

Katrodia, J.S. and Rane, D.A. 1979. Study into the causes of development of spongy tissue in mango (*Mangifera indica* L.) fruit Cv. Alphonso. Research Bulletin, Marathwada Agricultural University, 3(7) : 88-92.

Kartodia, J.S. and Rane, D.A. 1981. Effect of tree vigour in relation to spongy tissue development in Alphonso mango fruit. Haryana Journal of Horticultural Science, 10(3-4) : 151-154.

Katrodia, J.S. and Rane, D.A. 1989. Pattern of distribution of spongy tissue in the affected Alphonso fruits at different locations. Acta Horticulturae, 231 : 873-877.

Katrodia, J.S. and Rane, D.A. and Salunkhe, D.K. 1988. Biochemical nature of spongy tissue in Alphonso fruits. Acta Horticulturae, 231 : 835-839.

Katrodia, J.S. and Sheth, I.K. 1988. Spongy tissue development in mango fruit of Cv. Alphonso in relation to temperature and its control. Acta Horticulturae, 231 : 827-834.

Ketsa, S., Chitragool, S. and Lurie, S. 2000. Effect of pre-storage heat treatment on post – storage quality of mango fruits. HortScience, 35(2) : 247-249.

Ketsa, S. and Raksritong, T. 1992. Effect of PVC Film wrapping and Temperature on storage life and quality of 'Nam Dok Mai' mango fruits on ripening. Acta Horticulturae, 321:756-763.

Khader, S.E.S.A. 1988. Physiological changes in healthy and aborted mango at various stages of growth. Indian Journal of Plant Physiology, 31 : 316-319.

Klein, J.D. and Lurie, S. 1991. Prestorage heat treatment as a means of improving post-storage quality of apples. Journal of American Society for Horticultural Science, 115:265-269.

Kosiyachinda, S. and Mendoza, Jr. D.B. 1984. Harvesting of mango. In : Mango. Fruit Development. Post harvest – Physiology and Marketing in ASEAN (Ed). D.B. Mendoza Jr. and R.B.H. Wills, ASEAN Food Handling Bureau, Bangsar, Kaulalumpur, Malaysia, 80-83.

Kramer, G.F. and Wang. C.Y. 1989. Correlation of reduced chilling injury with increased spermine and spermidine levels in Zucchini squash. Physiology Plantorium, 76:479-484.

Kwee, L.T. and Chong, K.K. 1985. Diseases and disorders of mango in Malaysia. Tropical Press. Sdn. Bhd, Kaulalumpur.

Laborem, G., Reyes, F.J. and Rangel, L. 1992. Maduracion del mango previo almacenamiento a baja temperature. Fruits, 47:419-423.

Lad, B.L., Gunjate, R.T. and Salvi, M.J. 1985. Effect of post harvest Ethephon dipping of fruits on occurrence of spongy tissue disorder in Alphonso mango. Indian Journal of Plant Physiology, 28(1) : 85-87.

Lakshminarayana, S. and Subramanyam. H. 1970. Carbondioxide injury and fermentative decarboxylation in mango fruit at low temperature. Journal of Food Science and Technology, 7:148-152.

Lam, P.F. and Ng, K.H. 1986. Influence of temperature adaptation and physiological stage on the storage of 'Harumanis' mangoes. First Australian Mango Research Work shop. Proceedings : 274-277.

Lederman, I.E., Zauberman, G., Weskler, A., Rot. I. and Fuchs, Y. 1997. Ethylene forming capacity during cold storage and chilling development in 'Keitt' mango fruit. Post-harvest Biology and Technology, 10:107-112.

Lim, T.K. and Koo, K.C. 1985. Diseases and disorders of mango in Malaysia. Tropical Press, Kaulalumpur, Malaysia.

Lim, T.K. and Kuppelweisor, W. 1993. Mango sap burn ameliaration in the Northern Territory. Acta Horticulturae, 341 : 518-527.

Lima, L.C. de O., Chitarra, A.B. and Chitarra, M.I.F. 2000. Chemical studies on spongy tissue. A physiological ripening disorder in 'Tommy Atkins' mango. Revista Brasileira de Fruit cultura, 22 (special issue) : 96-99.

Lima, L.C. de O, Chitarra, A.B., Chitarra, M.I.F., Silva, E.E. and Michalezuk, L. 1999. Enzymatic activity changes in spongy tissue : a physiological ripening disoroder of 'Tommy Atkins' mango. Acta Horticulturae, 485 : 255-258.

Limaye, V.P., Gunjate, R.T. and Joshi, J.D. 1975. Studies on the occurrence of spongy tissue in some varieties of mango. Agriculture College Magzine, Dapoli 10: 36-37.

Limaye, V.P., Joshi, J.D. and Gunjate, R.T. 1978. Occurrence of spongy tissue in relation to stage of harvest of Alphonso fruits. Journal of Maharashtra Agricultural Universities, 1 (Addl.) : 292-293.

Limaye, V.P., Patil, M.M., Lad, B.L., Magdum, B.M., Gunjate, R.T. and Salvi, P.V. 1984. Ratna – a new mango hybrid. Journal of Maharashtra Agricultural Universities, 2 : 60-1.

Lizada, M.C.C., Kosiyachinda, S. and Mendoza, D.B. 1984. Physiological disorders of mango. In : Mango, (Eds) D.B. Mendoza and R.B.H. Wilis. ASEAN Food Handling Bureau, Kaulalumpur, 68-74.

Loveys, B.R., Robinson, S.P., Brophy, J.J. and Chacko, E.K. 1992. Mango sap burn : components of fruit sap and their role in causing skin damage. Australian Journal of Plant Physiology, 19 : 449-457.

Lurie, S. and Klein, J.D. 1991. Acquisition of low-temperature tolerance in tomatoes by exposure to high-temperature stress. Journal of American Society for Horticultural Science. 116:1007-1012.

Lyons, J.M. 1973. Chilling injury in plants. Annual Review of Plant Physiology. 24 : 445-466.

Lyons, J.M. and Breidenbach, R.W. 1987. Chilling injury. p.305-326. In:J. Weichmann (ed). Post-Harvest Physiology of Vegetables. Marcel Dekker, New York.

Lyons, J.M. Raison, J.K., and Steponkus, P,L. 1979. The plant membranes in response to low temperature: An overview. In: J.M. Lyons, D. Graham. and J.K. Raison (ed). Low temperature stress in crop plants. The role of the membrane. Academic Press, New York. pp 1-24.

Malo, S.E. and Campbell, C.W. 1978. Studies on mango fruit breakdown in Florida. Proceedings of American Society for Horticltural Science : Tropical Region, 22 : 1-15.

Manzano-Mendez. J.E. 1995. Response of several mango cultivars to gibberellic acid. Harvest and post-harvest technologies for fresh fruits and Vegetables : 477-487.

Marcellia, P. and Kane, O. 1978. Chilling injury of Amelie mangoes: Physiology and Prophylaxis. Abstract No:1214. XXII International Horticultural Congress, Sydney, Australia.

Mathur, P.B., Singh. K.K., and Kapur. N.S. 1953. Cold storage of mangoes. Indian Journal of Agricultural Sciences, 23:65-77.

Mattoo, A.K. and Modi, V.V. 1969a. Biochemical aspects of ripening and chilling injury in mango fruit. Proceedings of Conference on Tropical and Subtropical Fruits, Tropical Production Institute, pp. 111-115.

Mattoo, A.K. and Modi, V.V. 1969b. Ethylene and ripening of mangoes. Plant Physiology, 44 : 308-310.

Mattoo, A.K., Modi, V.V. and Reddy, V.V.R. 1969. Oxidation and cartenogenesis regulating factors in mangoes. Indian Journal of Biochemistry, 5 : 111-112.

McCollum, T.G. and Mc Donald, R.E. 1992. Temperature Conditioning inhibits chilling injury in summer squash (*Cucurbita pepo*) fruit. Proceedings of Florida State Horticulture Society, 100:105.

McDonald, R.E. 1989. Temperature conditioning affects polyamines of lemon fruit stored at chilling temperatures. HortScience, 24 : 475-477.

McDonald, R.E. and Kushad, M.M. 1986. Accumulation of putriscine during injury of fruits. Plant Physiology, 82 : 324-326.

McLauchlon, R.L., Mitchell, G.E., Johnson, G.I. and Wills, P.A. 1990. Irradiation of Kenington Pride mangoes. Acta Horticulturae, 269 : 469-476.

Mead, A.J. and Winstar, E.C. 1991. Description of the disorder 'Stem-end-cavity' possible causes and suggestions for reducing incidence in packing sheds. Acta Horticulturae, 291 : 265-271.

Medlicott, A.P. 1985. Mango fruit ripening, and the effects of maturity, temperature and gases. Ph.D thesis, Wolver hampton Polytechnic, Florida, USA.

Medlicott, A.P., Sigrist, J.M.M. and Sy. O. 1990. Ripening of mangoes following low temperature storage. Journal of American Society for Horticultural sciences, 115:430-434.

Mendoza, Jr. D.B. 1978. Post harvest handling of mango fruits in the Philippines. In : Aspects of post-horvest in ASEAN (Ed.), D. Graham, R.B.H Wills., Th Lee and W.B McGlasson: 23-32. Australian Development and Assistance Bureaue, Australia.

Menzel, K. and Kirkby, E.A. 1987. Principles of Plant Nutrition. International Potash Institute, Beran, pp. 687.

Modi, V.V. 1975. Personal Communication, Microbiologist. M.S. University of Baroda. Quoted by Rane et al. (1976) Problem of Spongy Tissue Development in Mango (*Mangifera indica* L.), Cv. Alphonso. A Review. Journal of Maharashtra Agricultural Univerities, 1 (2-6) : 89-94.

Mohammad, M and Brecht, J.K. 2000. Influence of ethylene treatments on ripening and chilling injury alleviation in mango (CV. Palmer) Fruit. Acta Horticulturae, 509:437-446.

Mukherjee, P.K. and Srivastava, R.B. 1979. Increasing storage life of mango (*Mangifera indica* L) by lowering the critical temperature. Progressive Horticulture, 10(4) : 63-69.

Murthi, G.S.R. and Upreti, K.K. 2000. Physiology of mango. In : Plant Physiology, Biochemistry and Plant Molecular Biology (Ed) A. Hemantaranjan, Scientific Publishers (India), Jodhapur, pp. 397-419.

Musa, S.K. 1974. Preliminary investigations on the storage and ripening of 'Totapuri' mangoes in the Sudan. Tropical science, 16:65-73.

Musa, S.K, and Mirghani, M.A. 1974. Preliminary investigation of grading, storage and ripening of Alphonso mangoes. Sudan Journal of Food Science and Technology, 6:18-23.

Nair, S. 2002. Biochemical aspects and control of chilling injury in mango fruit. Ph.D. Thesis, Curtain University of Technology, Australia.

Nair, S. and Singh, Z. 2003. Pre-storage ethrel dip reduces chilling injury, enhances respiration rate, ethylene production and improves fruit quality of 'Kenington' mango. Journal of Food, Agriculture and Environment, 1(2) : 93-97.

Nair, S and Singh, Z.2004. Chilling injury in mango fruit in relation to bio-synthesis of free polyamines. Journal of Horticultural Science and Bio-technology, 79(4): 515-522.

Nair, S., Singh, Z., and Tan,S.C. 2001. Heat treatments effect development of chilling injury, respiration and ethylene production and fruit quality of mango. Acta Horticulturae, 553 : 549-50.

Nair, S., Singh, Z., and Tan, S.C. 2003a. Chilling injury development in 'Kensigton Pride' mango fruit in relation to free polyamines, Australian Post-harvest Horticulture Conference, Brisbane, Australia, 278-279 .

Nair, S ., Singh, Z., and Tan, S.C. 2003b. Chilling storage temperature affects ethylene biosynthesis in mango fruit during ripening. Acta Horticulturae, 628 (1) : 285-93.

Nair, S., Singh, Z. and Tan, S.C. 2003c. Aroma valatile emission in relation to chilling injury in 'Kensington Pride' mango fruit. Journal of Horticultural Science and Biotechnology, 78(6) : 866-873.

Nair, S., Singh, Z. and Tan, S.C. 2004a. Chilling injury in relation to ethylene biosynthesis in 'Kensington Pride' mango fruit. Journal of Horticultural Science and Biotechnology, 79(1) : 82-90.

Nair, S., Singh, Z. and Tan, S.C. 2004b. Chilling injury adversely affects Aroma Volatile Production in Mango during Fruit ripening. Acta Horticulturae, 645 : 529-536.

Narayana, C.K. and Singh, B.P. 1998. Effect of maturity on chilling injury development during cold storage of Dashehari mangoes. Paper presented at National Symposium on Mango production and Export. Lucknow, June 25-27, 1998(Abstract), p. 80.

Negi, S.S. 2000. Mango Production in India. Acta Horticulturae, 509 : 69-77.

Nuevo, P.A., Cua, A.U. and Lizada, M.C.C. 1984. Internal breakdown in 'Carabao' mango subjected to modified atmospheres. III Starch in spongy tissue. Post harvest Research Notes, 1 : 34-5.

O' Hare, T.J. 1994. The susceptibility of Thai and Australian mango cultivars to sap injury and possible means of control. In: G.I. Johnson, and E. Highley, (Eds) Development of Post harvest Technology for Tropical Tree Fruits. ACIAR Proceedings 58 : Camberra, pp. 21-24.

O' Hare, T.J. and Prasad, A. 1992. The alleviation of sap induced mango skin injury by calcium hydroxide. Acta Horticulturae, 321 : 372-381.

O' Hare, T.J. and Prasad, A. 1993. The effect of temperature and carbondioxide on chilling symptoms in mango. Acta Horticulturae, 343 : 244-250.

Olson, R.V. and Bergar, K.C. 1947. Boron fixation as influenced by pH, organic matter content and other factors. Soil Science Society of American Proceedings, 11 : 216-220.

Oosthuyse, S.A. 1990. Effect of lowering the temperature of the cold storage regime on the quality of export mangoes after four weeks of storage. Year book of South African Mango Growers' Association, 10 : 38-56.

Oosthuyse, S.A. 1993. Disorders of fibreless mangoes grown in South Africa for export. Yearbook of South African Mango Growers' Association, 13 : 80-88.

Oosthuyse, S.A. 1998. Effect of environmental conditions at harvest on the incidence of lentical damage in mango. Yearbook of South African Mango Growers' Association, 18 : 15-17.

Oosthuyse, S.A. 2000. Effect of cool storage delays after harvest on the extent of ripening during and fruit quality after cool storage. Acta Horticulturae, 509 : 395-400.

Patkar, V.R. 1978. Studies on chemical composition during growth and development and effect of growth regulators on ripening of mango fruits with particular reference to spongy tissue M.Sc(Ag) Thesis. Konkan Krishi Vidyapeeth, Dapoli, India.

Patkar, V.R., Gunjate, R.T. and Lad, B.L. 1984. Effect of spongy tissue on chemical composition of ripe Alphonso mango fruit, South Indian Horticulture, 32(2): 83-85.

Pentzer, W.T. and Heinze, P.H. 1954. Post harvest physiology of fruits and vegetables. Annual Review of Plant Physiology, 5 : 205-224.

Pesis, E., Aharoni, D., Aharon, Z., Ben-Arie, R., Aharoni, N. and Fuchs, Y.2000. Modified atmosphere and modified humidity packaging alleivates chilling injury symptoms in mango fruit. Post-Harvest Biology and Technology, 19(1) : 93-101.

Pesis, E., Faure, M. and Marinansky Ben- Arie, R 1997. Induction of chilling tolerance in mango by temperature conditoining, heat, low O_2 and ethanol vapours. Acta Horticulturae, 455 : 629-634.

Phakawatmongkol, W., Kesta, S. and Van Doorn, W.G. 2004. Variation in fruit chilling among mango cultivars. Post-harvest Biology and Technology, 32 : 115-118.

Phukdikun-Samphan, R. and Thong-Amphai, P. 1970. Preliminary study on the lumpy tissue in mango (*Mangifera indica* L.) Pimsaendang. Warasan Phutsum (Horticulture Journal), 14 : 49-52.

Pike, C.S. and Berry, J.A. 1980. Membrane phospholipid phase separations in plants adapted to or acclimated to different thermal regimes. Plant Physiology, 66 : 238-241.

Raison, J.K. and Lyons, J.M. 1986. Chilling injury: a plea for uniform terminology. Plant Cell Environment, 9:685-686.

Ram, S. 1988. Factors associated with black tip and internal necrosis in mango and their control. Acta Horticulturae, 231 : 797-804.

Ram, S., Bist, L.D. and Dwivedi, T.S. 1978. Internal fruit necrosis. A new physiological disorder in mango (*Mangifera indica* L.), Pantnagar Journal of Research, 3 : 196-209.

Ram, S., Bist, L.D. and Sirohi, S.C. 1988. Internal Fruit Necrosis of mango and its control. Acta Horticulturae, 231 : 805-813.

Ram, S. and Sirohi, S.C. 1979. Effects of N, K, Ca and B on the internal fruit necrosis in mango CV. Dashehari. Paper presented at Mango Workers Meeting, Panaji, Goa, 2-5th May, 1979, pp. 435.

Ramana, K.V.R., Aravinda Prasad, B., Krishna Prasad, C.A. and Patwardan M.V. 1984. Temperature of storage and the ripening behaviour of 'Alphonso'. Indian Food Packer, 38:58-61.

Rane, D.A., Katrodia, J.S. and Kulkarni, D.N. 1976. Problem of spongy tissue development in mango (*Mangifera indica* L.). Cv. Alphonso. A Review. Journal of Maharashtra Agricultural Universities, 1(26) : 89-94.

Rangwala, A.D. 1975. Changes in chemical composition of Alphonso fruits during ripening with particular reference to spongy tissue. M.Sc. (Ag) thesis. Konkan Krishi Vidyapeeth, Dapoli, India.

Raymond, L., Scheffer, R., Brecht, J.K. and Crane, J.H. 1998a. Internal breakdown in mango fruit. Symptomology and histology of Jelly Seed. Post-harvest Biology and Technology, 13(1) : 59-70.

Raymond, L., Scheffer, R., Brecht, J.K. and Crane, J.H. 1998b. Internal breakdown, Mineral Element Concentration and weight of mango fruit. Journal of Plant Nutrition, 21(5) : 871-889.

Robinson, S.P., Loveys, B.R. and Chacko, E.K. 1993. Polyphenol oxidase enzymes in the sap and skin of mango fruit. Australian Journal of Plant Physiology, 20(1) : 99-107.

Rodov, V., Ben-Yehoshua, S., Albagli, R. and Fang, D.O.1995. Reducing chilling injury and decay of stored citron fruits by hot water dips. Post-Harvest Biology and Technology, 5:119-127.

Roe, B. and Bruemmer, J.H. 1981. Changes in pectic substances and enzymes during ripening and storage of 'Keitt' mangoes. Journal of Food Science, 46:186-189.

Roy, S.K. 2001. Storage of tropical and subtropical fruits. Indian Journal of Horticulture, 58(1-2): 164-169.

Sadasivam, R., Muthuswamy, S., Sundaraj, J.S. and Vasudevan,V.1971. Note on Chilling injury in mango (*Mangifera indica* L) fruits in refrigerated storage. Indian Journal of Agricultural Science, 41:715-716.

Santha, K. 1979. Studies on internal breakdown in Alphonso mango (*Mangifera indica* L.). Paper presented at Mango Workers' Meeting. Panaji, Goa, 2-5th May 1979. p. 436-439.

Santha, K. 1980. Internal breakdown during ripening of Alphonso mango (*Mangifera indica* L.) in relation to specific gravity of the fruit. Journal of Food Science and Technology, 17 : 198-199.

Santha, K. 1981. Chemical studies on internal breakdown in Alphonso mango. Journal of Horticultural Science, 56 : 247-250.

Santha, K. 1982. Effect of calcium and boron on the incidence of itnernal breakdown in Alphonso mango. Journal of Food Science and Technology, 19 : 80-81.

Saltveit, M.E. Jr. 1999. Effect of ethylene on quality of fresh fruits and vegetables. Post-Harvest Biology and Technology, 15:279-292.

Saltveit, M.E. Jr. and Cabrera, R.M. 1987. Tomato fruit temperature before chilling influences ripening after chilling. HortScience, 7:394-395.

Sauco, V.G., Galvam, D.F. and Calvo, R. 1984. Incidence of 'Soft nose' on mangoes in the Canary islands. Proceedings of the Florida State Horticultural Society, 97 : 358-360.

Schaffer, B. 1994. Mango disorders caused by abiotic factors. Pp. 43-44. In : R.C. Plotz, G.A. Zontmyer, W.T. Nishijima; K.G. Rohrbach and H.D. Ohr (eds). Compendium of Tropical Fruit Diseases. APS Press, St. Paul, M.N.

Selveraj, Y. 1998. Investigations on the biochemistry of Internal Breakdown in Alphonso mango (*Mangifera indica* L.) fruit. Paper presented at National Symposium on Mango Production and Export. June 25-27, 1998 at Lucknow, India, p. 72 (Abstract).

Sharma, M.R. 1991. Towards the understanding of spongy tissue development in mango cv. Alphonso. Orissa Journal of Horticulture, 19(1&2) : 50-56.

Sheth, I.K. 1981. The study of temperature in relation to spongy tissue development in mango fruit of cultivar Alphonso and its control. M.Sc. (Ag.) thesis (unpublished), Gujarat Agricultural University, Navsari, India.

Shewfelt, R.L. and del Rosario, B.A, 2000. The role of lipid peroxidation in storage disorders of fresh fruits and vegetables. HortScience, 35(4):575-579.

Shivashankar, K.S. and Mathai, C.K. 1999. Relationship of leaf and fruit transpiration rates to the incidence of spongy tissue disorder in two mango (*Mangifera indica* L.) cultivars. Scientia Horticulturae, 82 : 317-323.

Shorter, A.J. and Joyce, O.C. 1998. Effect of partial pressure infiltration of calcium into 'Kensington' mango fruit. Australian Journal of Experimental Agriculture, 38 : 287-294.

Singh, K.K. 1974. Refrigerated storage of mango in India. Climate Control, 7(4): 33, 35-39. (Abstract).

Singh, R.N., Singh, G., Mishra, J.S. and Rao, O.P. 1987. Studies on the effect of pre- and post-harvest treatment of calcium nitrate and calcium chloride on the storage life of 'Amrapali' mango. Progressive Horticulture, 19(1-2) : 1-9.

Singh, Z, Herianus, J.D.L and Nair, S. 2004. A Review of Mango Fruit Aroma Volatile compounds. State of the Art Research, Acta Horticulturae, 645 : 519-527.

Srivastava, H.C. 1967. Grading, Storage and Marketing. In: The Mango : A Hand Book. Chapter 5. Indian Council of Agricultural Research, New Delhi.

Subramanyam, H., Krishnamoorthy, S., Subhadra, N.V., Dalal, V.B., Randhawa, G.S. and Chacko, E.K. 1971. Studies on breakdown, a physiological ripening disorder in Alphonso mango (*Mangifera indica* L.). Tropical Science, 13(3) : 203-210.

Subramanyam, H, Moorthy, N.V.N. Lakshminarayana, S. and Krishnamurthy, S. 1972. Studies on harvesting, transport and storage of mango. International Society of Horticultural Science, 24:260-264.

Tamijinda, B., Siriphanich, J. and Nobuchi, T. 1992. Anatomy of lenticels and the occurrence of their discolouration in mangoes (*Mangifera indica* cv. Nam Dok Mai). Kasetsort Journal, 26 : 52-64.

Tare, S.J. 1977. Studies on calcium in relation to occurrence of spongy tissue in Alphonso mango fruits. M.Sc. (Ag.) Thesis (unpublished) Konkan Krishi Vidyapeeth. Dapoli (Ratnagiri), India.

Taylor, M.A., Rabe, E., Dodd, M.C. and Jacobs, G. 1994. Effect of storage regimes on pectolytic enzymes, pectic substances, internal conductivity and gel breakdown in cold stored 'Songold' plums. Journal of Horticultural Science, 69:527-534.

Thomas, P. and Joshi, M.R. 1988. Reduction of chilling injury in ripe Alphonso mango fruit in cold storage by temperature conditioning. International Journal of Food Science and Technology, 23(5) : 447-455.

Thomas, P. and Oke.M.S. 1983. Improvement in quality and storage of 'Alphonso' mango by cold adaption. Scientia Horticulturae, 19 : 257-262.

Thomas, P., Saxena, S.C., Chandra, R., Rao, R. and Bhatia, C.R. 1993. X-ray imaging for detecting spongy tissue, an internal disorder in fruits of 'Alphonso' mango (*Mangifera indica* L.). Journal of Horticultural Science, 68(5) : 803-806.

Treshow. M. 1970. Environment and Plant response. McGraw Hill Publishing. Co., New York, pp. 51-76.

Van Lely Velled, L.J. and Smith. J.H.E 1979. Physiological factors in the maturation and ripening of mango (*Mangifera indica* L) fruit in relation to the jelly-seed, physiological disorder. Journal of Horticultural Science, 54:283-287.

Veloz, C.S. and Lakshminarayana, S. 1977. Effects of different temperatures de almacenamiento en la damuracio mango (*Magifera indica.* L) Var. Manilachapingo. Nueva Experos, 3:27-36.

Veloz, C.S., Torres, F.E. and Lakshminarayana, S. 1977. Effect of refrigerated temperatures on the incidence of chilling injury and ripening quality of mango fruit. Proceedings of Florida state Horticultural Society, 60:205-210.

Verma, G.S. 1950. Tip pulp of the mango fruit. Current Science, 19 : 246.

Wainwright, H and Burbage, M.B. 1989. Physiological disorders in mango (*Mangifera indica* L) fruit. Journal of Horticultural Science, 64(2): 125-135.

Wang, C.Y. 1993. Approaches to reduce chilling injury of Fruits and Vegetables. Horticulture Reviews, 15: 63-95.

Wang, C.Y. and Baker, J.E. 1979. Effect of two free radicals scavengers and intermittent warming on chilling injury and lipid composition of cucumber and sweet pepper fruits. Plant Cell Physiology, 20:243-251.

Wang, S.F. 1989. Biosynthesis and action of ethylene. HortScience., 20 : 41-45.

Wang, Y., Li, X.P and Ji, Z.L., 1995. Effect of chilling injury on two scavengers of free radicals of mango fruit. Acta Horticulturae, Sinica, 22(3): 235-239.

WangChai, K., Uthaibutra, J. and Iwohari, S. 2000. Endogenous polyamines in 'Nam Dok Mai' mangoes with different ripening stages and its relation to chilling injury during storage. Acta Horticulturae, 509: 429-436.

Wardlaw, C.W. and Leonard, E.R.1936. The storage of West Indian Mangoes. Imperial College of Tropical Agricultural, Trinidad, Low Temperature Research Station Memorandum No.3. 47pp

Winston, E.C. 1986. Observation of internal mango flesh breakdown : need for standardization of terminology. Proceedings of First Australian Mango Research Workshop. November, 1984. Melbourne. Australia, 77 : 81.

Woolf, A.B., Watkins, C.B., Bowar, J.H., Layyee, M., Maindonalad, J.H. and Ferguson., I.B., 1995. Reducing external chilling injury in stored 'Hass' avocados with dry heat treatment. Journal of American Society for Horticultural Science, 120: 1050-1056.

Young, T.W. 1957. 'Soft nose', a physiological disorder in mango fruit. Proceedings of Florida State Horticultural Society, 70:280-283.

Young, T.W., Koo,.R.C.J and Miver, T.J. 1962. Effect of nitrogen, potassium and calcium fertilization on 'Kent' mangoes in deep, acid and sandy soil. Proceedings of the Florida State Horticultural Society, 75: 324-328.

Young, T.W., Koo, R.C.J and Miver, J.T. 1965. Fertilizer trials with 'Kent' mangoes. Proceedings of the Florida State Horticultural Society, 78:369-375.

Young, T.W. and Miver, J.T. 1961. Relationship of nitrogen and calcium to 'soft nose' disorder in mango fruit. Journal of American Society for Horticultural Science, 78 : 201-208.

Zambrano, J. and Materano, W. 1998. Effect of heat treatment on post-harvest quality of mango fruits. Trocpical Agriculture, (Trinidad), 75 (4):484 -487.

Zambrano, J. and Materano, W. 1999. Effect of hot water immersion treatment on the development of chilling injury in mango fruits. Agronomica Tropical, Maracay, 49(1): 81-92.

Zauberman, G., Fuchs, Y., Rot, I. and Wexler, A. 1988. Chilling injury, Peroxidase and Cellulase Activities in the peel of Mango Fruit at low temperature. HortScience, 23(4): 732-733.

Zhang, C.L. and Huang, H.B. 1998. The stunted fruit disorder – A physiological anomoly in mango (*Mangifera indica* L.) caused by a fluorine pollutant. Journal of Horticultural Science and Biotechnology, 73(4) : 513-516.

Zhaug, Z.Q., Hong, H.J., Li, H.P., and Ji .,Z.L.1997. Effects of intermitant warming on chilling injury and physiological and bio-chemical responses of mango fruits. Acta Horticulturae, Sinica, 24(4): 329-332.

Zhaug, Z.Q., Li, X.P.; Hong, H.J. and Ji, Z.L. 2000. Alleviation of chilling injury of mango fruit in cold storage by using exogenous putrescine. Fujian Journal of Agricultural Science, 15(2): 32-36.

Zhu, S.J. and Ji, Z.L. 2002. The relationship between heat-induced chilling tolerance and endogenous ABA levels in mangoes. Agricultural Sciences in China, 1(5):568-571.

CHAPTER 12

Post-Harvest Losses

G. Jyothi

'A post-harvest loss' is any change in the quantity and quality of a product (here mango fruit) after harvest that prevents or alters its intended use or decrease in its nutritive and economic values (Sharma and Alam, 1998). Post-harvest begins at the moment of separation of fruit or vegetable from the plant and ends when it is delivered to the consumer or utilized. In between, there are greater chances for the loss in quality and quantity of harvested produce. This is because, the separated produce continues to function metabolically even after harvest. However, their metabolism is not identical to that of the parent tree and therefore, they are subjected to physical, physiological and pathological deterioration and losses. Physical and physiological losses are sequel of growth and development of fruits and vegetables travelling towards senescence. Physiological losses may sometimes be due to occurrence of disorders as a result of unfavourable temperature or other environmental factors. Harvested fruits and vegetables, being pack houses for nutritive foods, are attacked by a number of microorganisms, leading to decays/rots, resulting in greater post-harvest losses. Thus, the post-harvest deterioration of fruits occurs as a result of endogenous biochemical and physiological changes, microbial and insect spoilage, over ripening, dehydration and mechanical injury.

The problem of spoilage is often most severe in the tropical and subtropical regions, where high ambient temperature and relative humidity prevail during the harvesting and post-harvesting seasons. Mangoes ripen during transport, storage and marketing, resulting in losses. Both quantitative and qualitative losses of extremely variable magnitude occur at all stages in the post-harvest system from harvesting, through handling, storage, processing and marketing to final delivery to the consumer.

Impact of Post-harvest Losses

The post-harvest losses have an impact both at macro- and micro-levels of the country. All that is being produced is not available 'in toto' to the consumer. Due to post-harvest losses, per capita availability of produce (e.g. mangoes) is reduced. In an assessment, Roy and Pal (1991) found that the actual availability of marketable ripe 'Dashehari' mangoes as 76.2% of the total fruit harvested. The physical post-harvest losses result in economic loss to the producer and distributors at various levels. Careless and improper handling reduce the market value and keeping quality causing huge losses and deprives right benefits to both grower and consumer. Post-harvest losses during transnit may increase the transport cost sizeably. The losses in transit have an impact on the ultimate consumer's price as the cost of transport per unit of produce (fruits etc.,) is directly proportional to the quantity lost in transit. A loss of 1% results in increasing the transport by about 1.24% (Subramanyam, 1986). Besides these quantum losses, the loss in quality before actual consumption is enormous, which can hardly be estimated. Further, these huge losses and resultant financial losses discourage the farmers to take up the cultivation of fruits and to abandon their gardens by the current growers and go for more remunerative crops. This results in a huge gap between supply and demand.

Extent of Post-harvest Losses

The magnitude of post-harvest losses are often substantial, but vary greatly with cultivars, production areas, season of production, cause and different levels (field, harvesting, handling, transport, market level etc.). They are difficult to assess, quantify and the prevailing status of supply and demand have a decisive influence on the extent of physical and monetary losses. The magnitude of post-harvest losses in perishables is influenced by a wide range of factors like genetical, physiological aspects of harvest, storage, transport, handling operations, storage temperatures, packing systems and environment. Storage temperatures affect both the physiology and pathology of produce, while some packing systems may alter the rate of metabolism and diseases spectrum (Sharma and Alam, 1998). The extent of post-harvest losses in mango in developing countries like India, is substantially high due to lack of improved post-harvest technology.

Surveys were conducted to assess the extent of causes and estimate the post-harvest losses in mango from field level to processing units to consumer level in different varieties, different states etc. (Madan and Subramanyam, 1987., Madan and Ullasa, 1991., Roy and Pal, 1991., Srinivas *et al* 1997 etc), which revealed variable causes, alone or in combination contributing post-harvest losses, variation in the extent of losses at different levels, with different cultivars, places etc.

Provisional estimate revealed that 25-30% of the produce goes waste (lost) during picking, handling, transport and storage (Subramanyam and Moorthy, 1972., Salunke and Dasai, 1984) in India. The post-harvest loss was around 14.30% of the total produce harvested, consisting of injured and unmarketable fruits and the losses went up to 42.08% when losses in assembling centres were also included. These losses were entirely due to pre-harvest factors like faulty cultural practices, absence of plant protection measures and defective harvesting practices in Kolar district of Karnataka in India (Madan and Subramanyam, 1987). In addition, ripening loss of 7.28%, weight loss of

8.59% and half damaged fruits due to diseases (anthracnose and/or stem-end-rot) 11.95% were also recorded (Madan and Ullasa, 1991). It was estimated that 8.5% of harvested 'Dashehari' produce was lost with an economic loss of Rs. 141 per tonne of harvested fruit. Roy and Pal,(1991) assessed the post-harvest loss in 'Dashehari' mangoes as follows (Table 12.1).

Table 12.1: Assessment of post-harvest losses in 'Dashehari' mangoes

S.No.	Particulars	Extent of loss (%)
1.	Discarded in the field (Cracked/damaged/diseased)	1.30
2.	Cull fruit (undersized/deformed)	15.00
3.	Transportation losses (mainly loss in weight)	3.65
4.	Losses during ripening in boxes (after transportation)	
	(a) Weight loss	2.57
	(b) Rotting	1.32
	Total	**23.84**

Srinivas *et al* (1997) conducted a survey of mangoes at different levels of post-harvest handling in case of 'Bangalora' and 'Alphonso' cultivars in Kolar district of Karnataka and reported a total loss of 17.9%, which consisted 3.5% loss at field level, 4.9% in transportation, 4.1% in storage and 5.4% at retail level in case of 'Bangalora' and a total post-harvest loss of 14.4% in case of 'Alphonso' which comprised of 1.9% at field level, 3.7% in transportation, 3.5% in storage and 5.3% at retail level. The low extent of post-harvest losses in case of 'Alphonso' was attributed to the extra care taken at the time of harvesting and subsequent handling of the fruits as it is an export oriented and high value mango crop, as compared to "Bangalora" cultivar. In a latest survey, the following post-harvest losses were observed at different levels in mango in India (Table 12.2).

Table 12.2: Post-harvest losses in mango at different levels (Anon, 2005)

S.No.	Level	Extent of loss (%) in	
		Banganapalli	Bangalora
1.	Field level	15.60	22.00
2.	Market level	8.89	5.30
3.	Retail level	5.25	0.00

Post-harvest losses vary from cultivar to cultivar and also vary for the same cultivar at the market and processing units and the causes of post-harvest losses also vary with the same cultivar (Table 12.3).

Table 12.3: Post-harvest losses in different mango cultivars due to different diseases

S.No.	Cultivar	Cause	Extent of loss (%)	Reference
1.	Alphonso	Fungal infection	40.00	Chhatper *et al* (1972)
2.	Alphonso	Fungal infection	25.30	Subramanyam and Moorthy, 1972
3.	Alphonso	Fungal infection	14.40	Srinivas *et al* (1997)
4.	Bangalora	Fungal infection	17.90	Srinivas *et al* (1997)
5.	Banganapalli	Fungal infection	20.18	Madan, 1988
6.	Banganapalli	Fungal infection	29.79	Anonymous, 2005
7.	Carabao	Pests	20.00	Leilua, 1979
8.	Dashehari	-	23.84	Roy and Pal, 1991
9.	Langra	Stem-end-rot	6.20	Pathak and Srivastava, 1968

Post-harvest losses may vary from market to market, state to state, perhaps due to differences in weather conditions, storage conditions, conditions prevailing at the markets and time taken for disposal of the produce, besides the inherent capacity of cultivars to withstand various detrimental hazards (Table 12.4).

Table 12.4: Post-harvest losses in mango in different states of India

S.No.	State	Cultivar	Extent of loss (%)	Reference
1.	Andhra Pradesh	Banganapalli	29.79	Anonymous, 2005
2.	Andhra Pradesh	Bangalora	18.50	Anonymous, 2005
3.	Gujarat	Kesar	30.00	Patel and Vyas, 2003
4.	Karnataka	Bangalora & Neelum	17.36	Madan and Ullasa, 1993
5.	West Bengal	-	22.00	Mitra, 2008

Post-harvest diseases are considered as the main causes of post-harvest losses in mango (Cheema *et al* 1959., Eckert and Ogaraw, 1988., Madan and Ullasa, 1991., Arauz *et al* 1994). Chenulu and Thakur (1968) reported that 17.7% mangoes were decayed due to fungal diseases in transit, storage and ripening. The estimated average loss due to multiple infections in the markets of Delhi (Thakur and Chenulu, 1970), Andhra Pradesh (Lakshminarayana and Reddy, 1977) and Karnataka (Sohi *et al* 1973) were 17.70, 16.90 and 10.30% respectively (Table 12.5).

Apart from the above, post-harvest losses may also occur due to fruit flies, sap burn etc. (Arauz *et al* 1994).

Madan and Ullasa (1991) in a survey of markets in Karnataka (India) state, estimated post-harvest losses up to 4% at the processing end of marketing system. During transportation of 'Dashehari' mangoes from Lucknow to Delhi, the ripening loss was around 7.54% (Anon, 1990a).

Table 12.5: Post-harvest losses in mango due to certain post-harvest diseases

S.No.	Disease	Extent of loss (%)	Reference
1.	Powdery mildew	15-20	Cheema *et al* (1959)
2.	Anthracnose, Stem-end-rot, Aspergillus rot and Rhizopus rot	20-30	Madan and Ullasa, 1991
3.	Anthracnose	64.60	Arauz *et al* (1994)
4.	Stem-end-rot	7.20	Arauz *et al* (1994)
5.	Aspergillus rot	5.20	Arauz *et al* (1994)
6.	Post-harvest diseases	3.00	Eckert and Ogawa, 1988
7.	Stem-end-rot	6.20	Pathak and Srivastava, 1968
8.	Stem-end-rot	12.50	Quroshi and Meah, 1991

The post-harvest losses may occur in quality and quantity of produce. The quality loss may be due to physiological and biochemical changes and spoilage, while quantitative loss may be due to physiological loss of weight as a result of loss of moisture from the fruit.

Causes

Post-harvest losses of horticultural produce (fruits, vegetables, flowers etc) occur due to a variety of causes both in the field, at and after harvesting and till it reaches the consumer. Post-harvest losses in mango are due to physical, physiological or pathological, sometimes entomological causes, singly or in various combinations. Probably the highest and most important single factor is the attack of fungal or bacterial pathogens after harvest. The causes also differ from cultivar to cultivar, production area to area. Thus the causes of post-harvest losses could be many and be grouped in two broad categories viz., pre-harvest and post-harvest causes.

A. Pre-harvest causes

A substantial amount of post-harvest losses have their origin in the pre-harvest stage of the crop, e.g. genetical, microbial infections, pest infestation, disease infections, environmental causes and cultural practices adopted during the production stage. The post-harvest losses are generally ascribed to adverse pre-harvest conditions and abnormal weather conditions at the time of fruit development, maturity and harvest of fruits. One of the most serious causes of spoilage of mango fruits is incipient fungal infection at different stages of fruit development. The fruits which appear normally at the time of harvest, carry latent infection from the field and suffer spoilage during the ripening process, resulting in huge losses. The spoilage is essentially due to stem-end-rot, lateral rot and anthracnose and stone weevil in South India and sooty mould infection is common in coastal areas (Subramanyam and Moorthy, 1972). The practice of harvesting the entire orchard at a time irrespective of maturity of fruits in different orchards is considered as the prime cause resulting in post-harvest losses. The pre-harvest causes get aggrevated when certain factors prevail/appear.

B. Post-harvest causes

Post-harvest losses are huge due to post-harvest reasons, such as improper harvesting, improper (rough) handling (like exposure to sun after harvest for longer periods), improper packaging and mode of transportation, unhygienic conditions in pack houses and storage chambers, disease incidence, physiological disorders (spongy tissue, internal breakdown etc), lack of proper infrastructure facilities like pack house for sorting, grading and packing of harvested produce, non availability of commercial cold storage facilities, maintained at varied (low) temperatures, lack of cool chain during transport and storage and marketing, lack of suitable shipping containers (Munishamanna *et al* 2002., Dhemre and Waskar, 2004., Santha and Rao, 2001).

Lack of awareness, knowledge and skills of produce handlers and markets aggrevate the losses. Dalal (1989) listed the following causes of post-harvest losses in fruits;

(i) improper picking,
(ii) rough handling,
(iii) improper container/package,
(iv) heaping of fruits in the vehicle and riding over them,
(v) bruising during sorting, grading, washing, brushing;
(vi) long lapse of time between harvesting and storage at low temperature,
(vii) inadequate treatment of produce for long-term storage,
(viii) improper stacking of produce and
(ix) improper density of loading in the storage chamber and vehicles/wagons etc.

The major causes of post-harvest losses in mango in Kolar district of Karnataka (India) state in the order of their occurrence were, mechanical injury (in harvesting), spoilage, either over mature/shrivelling or immature/unmarketable sizes, pilferges and damage by birds/bats/hail storm (Srinivas *et al* 1997). In Bangladesh, post-harvest handling, especially the transport system and facilities for storage are poor, resulting in post-harvest losses in mango and the post-harvest spoilage was mainly due to stem-end-rot (Quroshi and Meah, 1991). Thus in developing countries like India, Bangladesh etc. the causes of post-harvest losses in mango are lack of infrastructure, needed for post-harvest management/handling and lower utilization of fruits for processing. In Brazil post-harvest losses of mangoes were mainly due to improper harvesting and post-harvest handling, poor transport facilities, lack of proper storage structures and processing and post-harvest diseases and disorders (Dondia, 1996).

The causes of post-harvest losses in mango can also be grouped as follows, which may individually or in various combination contribute to losses.

(i) ***Mechanical:*** Bruising, cutting etc., are the mechanical causes that usually occur at harvest due to improper/faulty harvesting. These injuries disrupt the internal metabolism leading to an increased respiration and transpiration and consequent loss of moisture and to weight loss. Damaged tissues become infected by microorganisms through the injuries in the peel and stalk-end and lead to rotting of the fruits. High moisture content and soft texture of fruits make them vulnerable to mechanical injuries. The other causes would be, inferior quality of container, crates, cartons etc which may have splintered wood, sharp edges, poor nailing, stapling and rivetting, over packing or under packing and careless handling during grading, sorting, transport, storage, packing or marketing.

(ii) ***Physical:*** Excess or insufficient heat or cold can spoil fruits and cause losses both in quality and quantity. Loss of moisture causes wilting/shrivelling. As little as 5% loss in water causes the fruits to wilt and shrive and eventually make the fruits unmarketable. Qualitatively there will be loss of stored energy (carbohydrates) and other nutritive constituents like vitamins (Verma and Joshi, 2000). Improper atmosphere in closely confined storage, at times causes losses. Low temperature below 12 °C, will result in chilling injury, which increases water loss, susceptibility to decay and detrimental changes in flavour and quality.

(iii) ***Chemical:*** Many of the naturally occurring chemical constituents present in fruits change/react, causing loss of colour, flavour, texture, taste and nutritive value.

(iv) ***Bio-chemical:*** A number of enzyme-activated reactions occur in fruit giving rise to off-flavours, discolouration and softening.

(v) ***Physiological:*** Fruits even after separation from parent plant continue their metabolic activities like respiration. This respiration accounts to a significant level of weight loss and generates heat. Physiological changes which occur during ripening, senescence including wilting may increase the susceptibility of the fruit to mechanical damage or infection of pathogens. A reduction in nutritional level (quality change) and consumer acceptance may also arise with these changes. Some of these changes are beneficial and others may be detrimental. Physiological disorders (spongy tissues, jelly seed, internal breakdown etc.) also result in post-harvest losses.

(vi) ***Biological:*** Both insect pests and diseases cause post-harvest losses in mango. Insects attacking the mango fruits like fruit borers, fruit flies, seed weevil etc. cause post-harvest losses by burrowing into the fruit, by consumption of the fruit, leaving excreta inside and causing unpleasant odours and secondary infection by pathogens. Though pest infestation occurs before harvest in the field itself, their larvae and pupae remain in the fruit and cause losses.

Post-harvest diseases like anthracnose, stem-end-rot and other fruit rots also result in post-harvest losses. They invade and breakdown the fruit tissue systems followed by subsequent invasion of weak pathogens. Infections after harvest can occur at any time between the field and the final consumption. The post-harvest diseases can be spread by field boxes/crates contaminated by soil or decaying produce or both. Contaminated water used to wash fruit and when both fresh healthy and contaminated fruits are packed in the same packages the disease incidence is aggraved and lead to post-harvest losses.

The various causes of post-harvest losses in mango tend to increase cumulatively as the produce moves from the producer to the consumer; anywhere from the point of harvest of produce up to the point of consumption. The various points/stages/levels at which losses are likely to occur are detailed below with probable causes.

(i) Field level (Photo 12.1): The losses at the field level are mainly during harvesting, operations due to the improper/faulty harvesting methods followed. Drop of the fruits to the ground during harvesting causes bruising and/or cracking. Faulty harvesting and handling operations were the main reasons for the damage at this stage which account for 2.37% of the total harvested fruit. Losses at field level were also due to delay in moving the harvested produce to market, some post-harvest factors, diseases and physiological disorders. Small fruits and over ripe fruits also contribute to the losses (Madan and Subramanyam, 1987).

Plate 12.1: Post-harvest loss in the field

(ii) Marketing system: Generally the grower/farmer does not himself dispose of his produce in the consumer market but approaches and sells to the middle men like pre-harvest contractors/whole salers through commission agents. At every middle man level, the post-harvest losses occur due to various causes, mentioned above. In a survey of mango markets in Karnataka state (India) post-harvest losses were observed at two stages in the distribution channel:

(i) at the assembling markets where the produce was brought from the field and sorted out for sale or transport,

(ii) at the time of storage for ripening at the whole sale/retailer/processing factories level (Madan and Subramanyam, 1987).

(a) ***Commission agent:*** No losses were observed at commission agent level as he does not hold the fruit and only facilitates the transfer of produce from farmer/pre-harvest contractor to the wholesaler.

(b) ***Wholesaler:*** The losses of wholesale market level depend on the length and number of participants in the marketing channel followed. Usually the losses at this level are meagre as in the case of commission agent, as the wholesaler also acts as facilitator for the transfer of produce to retailers and small traders.

(c) ***Retailer:*** Retailer is the last participant in the marketing system and incurs considerable post-harvest losses at his level. The causes for the losses at this level are compression injuries, disease infection and injuries due to mechanical and physical factors. The main damage at this level was due to overripening of the fruits, which accounted for 6.79% of the fruits procured from the wholesaler (Table 12.2 and 12.9). The damage due to rotting in ripening was estimated at 12.91% of the total fruits kept for ripening (Table 12.7 and 9). The other reason for damage at this level was pressing of fruits during transportation, which accounted for 2.57% of the total fruits, handled by the retailer. Thus, a total of 14.9% fruits were damaged at the retailer level. The losses at the retailer level are higher because the functionary was located at the end of the market channel and the damaged fruits at earlier stages also get accumulated at this stage. Moreover, the retailer also has to hold the fruit till such time, they are sold or became unfit for marketing.

Plate 12.2: Post-harvest loss at the market

Market level (Photo 12.2)

Post-harvest losses in the marketing process were high particularly when damaged or diseased fruit is kept along with healthy fruits. The market level losses are mainly due to handling and transport practices. The physical damage caused during harvesting, which are not visible at sorting of fruits, contributes for the loss at this level. Mechanical damage and diseases are the major causes of losses at this stage. The physical loss at the market level is around 5.13% of which the immature and malformed fruits account for about 4% (Anon, 2005).

At processing level (Photo 12.3): The losses at this level are mainly due to damage during transport and also secondary infection during ripening. In a survey of mango processing units in Karnataka state, post-harvest losses up to 4% were recorded and the major causes of losses were identified as diseases like stem-end-rot, anthracnose, aspergillus and rhizopus rots (Madan and Ullasa, 1991).

When the losses experienced at different levels were pooled, it indicates that on an average only 63.79% fruits could reach the consumer in sound, marketable and consumable condition and rest were lost (Tables 12.6 to 12.11).

Plate 12.3: Post-harvest loss at the processing unit.

Table 12.6: Post-harvest losses in mango at field level

S.No.	Causes	Loss (%)
1.	Mechanical injury	2.37
2.	Bird/bat damage	1.31
3.	Rotting	0.23
	Total	**3.91**

Table 12.7: Post-harvest losses in mango at ripening stage

S.No.	Causes	Loss (%)
1.	Physiological loss in weight	9.07
2.	Ripening	12.91
	Total	**21.98**

Table 12.8: Post-harvest losses in mango at retailer level

S.No.	Causes	Loss (%)
1.	Rotting	5.37
2.	Over-ripened	6.97
3.	Press marks	2.57
	Total	**14.91**

Table 12.9: Post-harvest losses in mango (Pooled)

S.No.	Levels	Loss (%)
1.	Field	3.91
2.	Ripening	21.98
3.	Retailer	14.91
	Total	**40.80**

Table 12. 10: Post-harvest losses in different cultivars of mango (kg) under different items

S.No.	Item	Totapuri	Neelum	Banganapalli	Dashehari
1.	Fruit harvested in orchard	100.0	100.0	100.0	100.0
2.	Fruits sorted out	14.0	6.0	5.0	3.5
3.	Fruits kept for ripening	86.0	94.0	95.0	96.5
4.	Ripened fruits sorted out	9.0	7.0	8.2	3.0
5.	Ripened fruits discarded	6.3	4.0	6.8	2.6
6.	Good fruits marketed	63.3	78.0	80.0	90.9
7.	Post-harvest loss (physical)*	36.7	22.0	20.0	17.1
8.	Total loss (Rs. / tonne harvested)	620.0	758.0	500.0	141.0

* including weight loss

Table 12.11: Post-harvest losses in mango in different regions at different stages of handling

S.No.	Cultivar	Physical loss (%)	Stage of handling	Regions (s)	Reference
1.	Totapuri	36.7*	Overall	Karnataka Tamilnadu Andhra Pradesh	Madan (1988)
2.	Neelum	22.0*	Overall	Tamil Nadu	Anon (1989)
3.	Banganapalli	20.0*	Overall	Andhra Pradesh	Anon (1989)
4.	Mulgoa	20.0*	Ripening	Bangalore	Anon (1987)
5.	Raspuri	14.3*	Ripening	Bangalore	Anon (1987)
6.	Alphonso	16.4*	Ripening	Bangalore	Anon (1987)
7.	Dashehari	4.7	Retail	Delhi	Anon (1990a)
8.	Langra	5.6	Retail	Delhi	Anon (1990a)
9.	Chausa	5.3	Retail	Delhi	Anon (1990a)
10.	Sindhuri	6.5	Retail	Delhi	Anon (1990a)

* including physiological weight loss

Factors Affecting Post-harvest Losses

Dalal (1989) listed the following pre-harvest factors for post-harvest losses of fruits :

(i) poor orchard sanitation,

(ii) non-judicious application of plant growth regulators,

(iii) untimely irrigation and spraying of urea to get better size,

(iv) scant attention to the environmental factors during growth and development of fruits,

(v) irregular cultural practices and

(vi) poor management of orchard.

The important factors which influence the quantum of loss of fresh mangoes from the time of harvest till it reaches the consumer, according to Pruthi (1992) are :

(i) prevailing climate and cultural practices of the region of production,

(ii) variety, stage of maturity and incidence of disease infection at harvest,

(iii) methods of packaging, handling and extent of physical damage,

(iv) time lapse between harvesting and storage,

(v) number of handlings before packing,

(vi) chemical and physical treatments,

(vii) method of transportation,

(viii) type and quality of containers, method of packaging and suitability of packing materials used,

(ix) suitability of rail wagon/truck/stacking of packages and

(x) method of storage.

Recently, Shikhamany (2004) listed and detailed the following contributing factors for post-harvest losses in fruits in addition to seasonality and perishability of fruits.

Biochemical Changes in Fruits

Fruits contain 50-90% moisture, with high rate of respiration and high generation of heat. They are soft textured (succulent) with a short shelf life. A number of physiological and biochemical processes occur in the harvested, non-processed fruits that influence the post-harvest losses and appearance of the materials (Sharma and Alam, 1998) (Table 12.12).

Table 12.12: Biochemical changes in harvested mango fruits

S.No.	Change	Process	Significance
1.	Water loss	Transpiration	Unattractive appearance, textural changes, weight loss, shrivelling
2.	Carbohydrate conversion	Enzymatic	Starch to sugar, beneficial
3.	Flavour	Enzymatic	Beneficial
4.	Softening	Pectic enzymes	Detrimental
5.	Colour	Pigment synthesis	Beneficial
6.	Vitamins	Enzymatic	Gain in vitamin A
7.	Decay and rots	Physiological and pathological	Detrimental

Lack of Inherent Keeping Quality

Basically keeping quality is the inherent character of a genotype. It can, however, be slightly altered by cultural practices including nutrition, water management and use of growth regulators. Keeping quality depends on the fruit firmness and the stage of harvest. Increasing fruit firmness through providing optimum leaf area per fruit and adequate calcium nutrition can help in increasing the keeping quality. The stage of maturity at harvest also determines the keeping quality of the commodity. Pre-harvest sprays of growth regulators such as cytokinin and auxin can also improve the keeping quality.

Improper Harvesting and Handling Processes

In general the fruits are harvested mostly by hand (at least in India), except the fruits at a higher level, which are harvested with a long bamboo pole having a cloth bag at the top-end along with a knife. In some instances, the fruits are harvested by shacking the fruits or by beating the branches with sticks. These methods result in the fall of the fruits to the ground, leading to bruises and cracks and subsequent incidence of decays and rots. Bruises are the main source of increased spoilage and reduction of storage life in mango.

Rough handling by labourers provides more chances of damage to the fruit. Most of the farmers follow traditional methods of harvesting, using the minimal mechanical aids. Growers are also unaware of the right stage of maturity at which the fruits are to be harvested to get good post-harvest quality. The labourers employed for harvesting are unskilled.

Spoilage

Spoilage due to microbial pathogens contributes to post-harvest losses to a large extent in fruits. Spoilage leads to qualitative losses (Khurdiya and Roy, 1986). The predominant spoilage diseases identified in the surveys of mango markets in India are :

Black mould rot (*Aspergillus niger*), anthracnose (*Colletotrichum gloeosporoides*), soft rot (*Rhizopus oryzae*) and stem-end-rot (*Botrodiplodia theobromae*) (Srinivas *et al* 1997), *Pencillium cyclopium* (Palijwala *et al* 1988) and yeast (*Saccharomyces cerevisiae*) (Manjiri *and* Modi, 1988).

Environmental Factors

Low as well as high temperature injury causes direct loss and renders the produce more susceptible to microbial decay. High relative humidity reduces dehydration and avoids condensation on their surface. However, if free water is allowed to condense on produce or under very high humidity, the incidence of diseases is greatly increased (Sharma and Alam, 1998).

Operational Factors

In India, handling of fruits is mainly done in bulk. The present post-harvest process besides causing damage, bruising and cracking, makes the fruits more prone to microbial spoilage.

Market Glut

Major cause for market glut is the seasonality of fruit crops. Although there are early-,mid-and late-maturity cultivars, their harvesting period cannot be extended beyond two months in a given agro-climatic region. Market glut is also due to lack of clear cut early, mid and late varieties as they overlap and the same variety comes to harvest at different times under different agro-climatic regions and even in the same agro-climatic region depending upon the cultural practices adopted. Lack of purchasing power and meagre quantities of fruits diverted for processing are other contributing factors for market glut. Market glut is difficult to handle and therefore, sometimes lead to post-harvest losses, in addition to unremunerative prices to the grower.

Improper Market Infrastructure

Market is the place, where major losses occur due to lack/improper infrastructure for handling the produce. Lack of cold storage facilities at markets, particularly in case of market glut, is the major problem. Similarly lack of net-work of local markets is another reason for unprecedented and un-regulated arrivals in the markets, leading to distress sales and post-harvest losses.

Inappropriate Package and Transport Conditions

Usually, growers transport their produce in a bullock cart, truck or in a tempo, loose, without packing (Srinivas *et al* 1997). Large quantities of mangoes are transported to local markets in open conveyances. While the produce is transported in open conveyances without covering the produce at top and cushioning at the bottom and sides (Photo 2.5). In either case, the weight of the fruits of the upper (top) layers causes bruises on the bottom layer of the fruits, which are sometimes crushed by the top layers. The humidity build up due to the transpiration of fruits causes decays. Major production of mango is consumed within the country, while a meagre quantity is exported to overseas markets. While packaging and storage of exportable fruits are as per the standards, the fruits meant for domestic markets are not packed systematically and transported appropriately. Release of ethylene, condensation, transportation of fruits, mutual pressure in a package, the shocks that the fruits experience during transport and storage are the basic causes for spoilage and losses of fruits sent to long distance markets. Longer routes and hours of transportation means higher quantum of loss, particularly when the produce is transported in bulk and loose without packing. Rough roads, non-availability of padded and refrigerated and closed trucks are additional factors for spoilage and post-harvest losses. The possible damage and most frequent losses incurred during non-refrigerated transport are generally mechanical damage and overheating. The principal hazards in road transport are vibration and bouncing of packages due to poor road conditions. Spoilage and deterioration of mango fruits in transport and storage are attributed to harvesting at immature stage, physical damage during transport and consequent fungal infection and improper methods of storage resulting in loses (Ramakrishna and Rao, 1989).

Pathological Factors

Infection by microorganisms is generally considered as the most serious cause of post-harvest losses in mango. Latent infections of anthracnose, stem-end-rot and spoilage due to species of *Aspergillus* and *Rhizopus* are some of the important post-harvest diseases observed during the ripening process (Madan and Ullasa, 1991). Physical loss due to these diseases were estimated around 20 to 30% of the total quantity kept for ripening.

Meagre Quantity Processed

One of the factors aggravating post-harvest losses of mango is the utilisation of very meagre quantities for processing (<2%). The main reasons for this situation are :

(i) lack of suitable varieties for processing,

(ii) lack of internal demand for processed fruit products, and

(iii) high overhead charges involved in the processing and marketing of fruit products.

Lack of awareness and knowledge and skill on the part of the producer, handling and markets aided by high ambient temperatures and non-availability of cool-chain are other factors influencing the post-harvest losses in mango.

PREVENTION OF POST-HARVEST LOSSES

Post-harvest losses assume significance from both economic and nutritional points. But the problem of prevention of the losses is technologically and scientifically complex. Nevertheless, greater success in increasing the net production of finally utilisable produce may be achieved by reducing such losses than by increasing production. Prevention of post-harvest losses of fruits also assumes greater significance in view of the huge losses and negative impact on the economics, financial loss, availability of produce, enhanced market costs etc. Prevention of post-harvest losses has been recognized as an effective means of increasing the availability of mango for domestic as well as export markets. Assuming proper scientific distribution of fruits from areas of production to areas of consumption curtailing losses in transit, will help in augmenting availability of fruits. Besides, increasing the net availability of produce, prevention of post-harvest loses decreases the area needed for production and conserves nature's resources. Minimizing the post-harvest losses of already produced fruits is more sustainable than increasing production to compensate their losses. Reduction in losses is cheaper than equivalent increases in fruit production, when the cost is expressed in terms of economic energies or adverse environmental impact. Even if the losses are reduced by 2%, there will be a net gain of at least Rs. 100 to 150 crores annually (Narayana *et al* 1989).

Sethi and Maini (1989) suggested technology for reducing post-harvest losses in fruits and vegetables. According to them post-harvest losses can be reduced to some extent by breeding of fruits and vegetables for longer shelf life; improvement of pre-harvest factors and harvesting techniques, proper methods of handling, marketing, packaging and transport, appropriate storage techniques, development of appropriate processing technology and utilization of waste from fruits and vegetables.

Madan and Ullasa (1991) opined that post-harvest losses of mango could be reduced by following the recommended technologies of production. But they observed that the farmers as well as other participants in the marketing system are not adopting the recommended technology to reduce the post-harvest losses and that less than 1% of the total income from an orchard was spent on plant protection measures by farmers.

Post-harvest losses can be minimized by selection of a cultivar having longer shelf life, adopting proper cultural practices in the orchard and use of proper harvesting technology, handling, packaging, storage and transportation methods. Control of fruit environment, use of chemicals for pre- and post-harvest treatments and proper marketing procedures etc., also help to reduce losses. Reduction in qualitative and quantitative losses could be achieved by harvesting fruits at optimum stage of maturity, ripening, sorting, grading and suitable packaging of the graded fruits, pre-cooling of the fruits prior to storage, adoption of refrigerated transport and storage under controlled conditions and pre- and post-harvest treatments. The principal post-harvest technology now being adopted the world over to minimize the post-harvest losses include: pre-harvest treatments, washing, cleaning and grading, packaging, pre-cooling, cold storage, processing, transport and marketing (Santha and Rao, 2001). Reduction in the period from harvest to consumption or processing by efficient marketing system can considerably lower the losses. Alternate distribution system such as direct sale to the consumer or processors by the producer or producer cooperatives can considerably reduce the losses.

Technology for Minimizing Post-harvest Losses

The technology suggested for minimizing the post-harvest losses in mango can be dealt under two broad sections viz.,

(i) pre-harvest technology,

(ii) post-harvest technology.

Pre-harvest Technology

Pre-harvest crop management of mango is of greater significance, since considerable amount of produce (12-18%) is rejected in the field as culled fruits (Roy and Pal, 1991).

Maturity: It is often suggested to harvest the fruits at proper/optimum maturity to get a good quality produce for consumption/processing. Optimum stage of the maturity in mangoes has been worked out based on physico-chemical parameters. Mangoes with a specific gravity of 1.0 to 1.02 are considered to be at an optimum stage of maturity for best storage and subsequent marketing. However, many of the parameters are not still adopted and the growers determine the optimum maturity of fruit by its colour, shape, size etc. Besides, proper maturity often depends on the intended use of the produce, market distance and mode of transport. It is therefore, advisable to determine the harvest indices for local, long distance and overseas markets, variety wise and harvest accordingly. Stage of maturity affects keeping quality of produce. Immature fruits lack optimum colour, flavour and taste and have less nutritive value. On the other hand, over-mature fruits have shorter storage life; their transportation is also difficult because they become very soft.

Harvesting: Generally manual harvesting is followed. Harvesters have been developed by different organisations in the country as well as elsewhere in the world. Harvesting by right technologies of using harvesting devices increases the efficiency of labourers, reduces mechanical injuries and enhances post-harvest life. The spoilage of manually harvested fruit due to anthracnose and stem-end-rot was higher (21.8%) compared with fruits harvested using harvesting devices (0.01%) (Waskar *et al* 1997). Therefore, it is advisable and better to use harvest devices to reduce post-harvest losses due to diseases etc.

Time of harvesting is also important in minimising the losses due to desiccation etc. The loss of moisture by transpiration (through stomata and lenticels) results in shrivelling of fruits. In case of afternoon harvests, when the temperature is around 36 °C, the fruits should be pre-cooled to 12 °C within 6 hours of harvest rapidly. To prevent desiccation and achieve rapid cooling the fruits should be harvested during the cooler part of the day, before the field temperatures reach 30 °C i.e. before 10 AM or after 4 PM in the evening. Any delay in harvest shall be at the reduction of their shelf life.

Harvested fruits should be protected from direct sun to avoid loss of moisture. They must first be moved to shade preferably in plastic crates/put in cool water immediately after picking. This helps to reduce their field heat and post-harvest losses and increase their storage life.

Protective treatments: Pathological infections are the chief sources of post-harvest losses in mango, some of which come from the field during pre-harvest stage. These can be reduced by good orchard management, pre-harvest cultural practices and orchard sanitation. Bagging of individual fruits, one month prior to harvest with news paper/ brown paper bags has been effective in minimizing the microbial load on fruit surface, latent infection and getting better quality fruit (Pandey and Tandon, 2007).

Pre-harvest sprays using chemicals have reduced the disease incidence and thereby reduced post-harvest losses. Pre-harvest sprays of thio-phenete methyl (0.05%) effectively controlled post-harvest diseases and post-harvest losses in 'Dashehari' mangoes. Pre-harvest sprays of fungicides (Topsin, Bavistin or Benlate) at periodical intervals reduce the post-harvest loses by checking the post-harvest spoilage due to fungal diseases in mango. According to Rawal and Ullasa (1985), the percentage of post-harvest loses in mango due to stem-end-rot can be reduced from 30.77% to 3.84% by a pre-harvest application of Bavistin (0.1%) as a pre-harvest spray. Decay loss, particularly due to stem-end-rot could be minimized by harvesting the fruits with stalk.

Post-harvest technology (Treatments)

The main purposes of the post-harvest treatments are:

(a) prevention of infection through harvest injuries,

(b) eradication or alteration of established infection and

(c) maintenance of host resistance by retarding senescence.

Control of diseases: The incidence of post-harvest diseases and losses due to rotting and weight loss can be controlled by the use of certain senescence retardants-ethylene absorbants, growth regulators, fungicides, fungal wax emulsion and anti-transpirants as post-harvest treatments or along with pre-cooling, hot water dips, irradiation etc. Latent infection can be eradicated by the application of fungicides or by subjecting the produce to hot water treatment. Bhatnagar and Subramanyam (1973) listed several chemical fungicides and antibiotics used as post-harvest treatments to control the decay of mangoes. From time to time new chemicals are added to the list. These fungicides are either used in conjunction with cold or hot water and both on pre-harvest and post-harvest diseases of mango. Among the post-harvest treatments, Bavistin (1000 ppm) plus hot water dip (52 °C) for 5 min. was the most effective for minimising the post-harvest losses and enhancing the shelf-life (Roy and Joshi, 1985). Efficacy of post-harvest treatments with different fungicides (Benlate, Thiabendazole, Captan, Rovral etc., at 500 ppm) showed that Benlate was the best to control the fungal spoilage from 30 to 10% during ripening of 'Alphonso' and 'Pairi' mangoes.

Infection by microorganisms resulting in decay can be reduced to minimum by maintaining sanitation during post-harvest handling, packaging, storage and transportation. All equipments, tools, machines, containers, packaging boxes/cartons, transport vehicles and storage rooms must be kept clean. Human labour should be hygienic and keep clean their cloths, hands etc. while handling the produce. Spoilage of mangoes due to yeast can be reduced by post-harvest treatment of fruits with anti-fungal compounds (Bavistin/Aureo fungin) or gamma irradiation at 50 Krad (Singh, 1983).

It is gainful to adopt post-harvest chemical treatments to reduce post-harvest diseases and consequent losses, before they are dispatched for long distance transportation. This got double benefit of not only increasing the returns to the traders, but also benefits the consumer by minimising the losses during transit and reducing the transport cost per unit of produce and making the availability of produce at reduced/less expensive prices (Subramanyam, 1986).

In post-harvest handling system fruits are washed in the pack house. Washing the fruits in static water tanks will increase the disease incidence due to the increase in inoculum from infected fruits from repeated batches. Therefore, water has to be changed frequently and the water should contain sodium hypochlorite (100 ppm) and/or fungicide to minimize the microbial load and post-harvest losses due to the diseases.

Heat has fungicidal as well as insecticidal action. Hot water dips have generally been used for fungal pathogen control. Fungal spores and latent infections are located either on the surface of the fruit or within first few cell layers under the peel. Therefore, post-harvest hot water dips to control latent infections are applied for only few minutes (5-10 min.) with or without suitable fungicides, using temperatures higher than those for hot air or vapour heat, because only the surface of the fruit needs to be heated.

The natural waxy cover on the fruits is partly removed during washing, necessitating the artificial waxing. The wax should be sufficiently thick and consistency to stick to the fruit surface on drying and prevent water loss and entry of decay organism into fruit. The minute particles of wax emulsion clog the pores (stomata and lenticels) of the fruit skin. Waxing of fruits serves three purposes: (i) reduces the water loss, (ii) prevents the entry of pathogen and (iii) provides glossy appearance to the fruits.

Reduction of Post-harvest Losses

Pre-cooling

Pre-cooling of harvested fruits also aids in reducing post-harvest losses by slowing down the metabolism, respiration, transpiration and ripening processes, thereby minimizing the susceptibility of fruits to invasion of microorganisms, besides reducing water loss. It also eases the load in the cooling system of transport and storage chamber. However, to achieve the above goals, the produce should be pre-cooled as quickly as within 4-6 hours of harvest.

Market glut: Market glut sometimes leads to post-harvest losses. To tide over this situation, methods such as delayed picking of fruits, staggering of pickings or withholding the stocks for a few days at the market are adopted. But these are not very effective measures as they give rise to some problems in storage. Thus, once the fruits start ripening on the tree, the picking cannot be delayed and when the fruit is over mature, the stock cannot be held for long. However, by proper methods of storage, the produce can be disposed off over longer periods.

Packaging: Packaging of produce is a fundamental tool for the post-harvest management of highly perishable produce like mango. Packaging is necessary to retain freshness, quality of fruits and their protection. Packaging makes it easy to market the produce and fetch better price. Packaging material should facilitate easy transportation, help to retain the shape of the produce and keep the produce cool. The use of CFB boxes, woven cases, moulded plastic trays, re-usable containers reduce the post-harvest losses. MAP, where O_2, CO_2, humidity and temperature are maintained at recommended levels retain freshness of the fruits for longer time and reduces the post-harvest losses. Use of pallets can also reduce the post-harvest losses.

Grading: Systematic grading coupled with the scientific packaging and storage reduces the post-harvest losses and marketing costs substantially, which enables the producer to fetch a comparatively better price.

Transport: Post-harvest losses during transit are mainly due to harvesting of fruits at improper maturity, vibration, compressions, mechanical injuries. Fast and economical transport with minimal damage during transit is the most important factor in the success to reduce post-harvest losses of perishable horticulture produce and for successful marketing and processing. To prevent transport losses, the moisture released by the produce should be contained in the packaging or the carrier should be lined with proper plastic film. The packages should allow proper ventilation of contents with minimum waste of space in the vehicle. Loading and unloading of vehicle/wagon should be properly supervised to prevent careless handling of packages to avoid losses. Insulated and/or refrigerated vehicles/wagons with elaborate air intake capacity are ideal for long distance transport. Road transport should be preferred to rail transport for perishable commodities.

Storage: Proper storage of fresh mango fruits also minimizes the post-harvest losses. In tropical regions like India, prevailing high temperature and high humid weather during monsoon reduce very much the storability of fruits and cause heavy losses. By adopting low temperature storage after harvest the metabolic activities (respiration, transpiration, ethylene production), build up of respiration heat, thermal decomposition and microbial spoilage could be reduced with the retention of quality and freshness for a longer period and reduction of post-harvest losses.

However, certain precautions should be observed in the storage to prevent losses. Fruits from diseased trees should not be stored along with fruits from healthy trees. Similarly bruised/cut fruits should not be stored along with sound fruits. Cold storage rooms should be sterilized either with 5% aqueous solution of lye solution or sprayed with 2% formalin solution and left for a couple of days. Packaging materials also should be treated in a similar way so that the spore load of the fungi could be minimized to bare minimum. To minimize the spoilage during transport and cold storage, individual fruits may be wrapped with diphenyl impregnated wrappers/tissue paper. With these precautions spoilage and resultant losses will be minimized (Pruthi, 1992).

Processing: The conversion of perishable fruits into value added products also reduces the post-harvest losses; therefore, top-priority to processing of fruits both in public and private sectors should be given.

Conclusion and Future

Strategies to minimise the post-harvest losses should include the pre-harvest, harvest and post-harvesting practices. In addition, development of packaging methods, pre-cooling techniques and post-harvest treatments to increase the storage life and minimizing the losses, infrastructure development including the establishment of cold chain, mobile pre-cooling units, pre-cooling and cold storage facilities in the viscinity of production sites is also essential. Staggering of harvest, cold storage and diversion of produce to processing helps in the reduction of market glut and eventually the post-harvest losses. Developing a net work of local markets, efficient dissemination of market information, organisation of producers groups for collective marketing shall go a long way in avoiding the market glut. Utilisation of large proportion of produce for processing can also help minimising the post-harvest losses, besides adding value to the produce. It is required to identify or evolve suitable varieties for processing and to feed the processing factories round the year/for extended period than at present in order to bring down the over-head cost in processing (Shikhamany, 2004).

Post-harvest losses in mango contribute an important problem that needs increased and continuing attention at national and regional level and requires application of appropriate technology to maintain the quality and quantity of production to prevent the losses. The physical post-harvest loss apart from creating an economic loss to the producer and distributor also leads to the less availability of the produce resulting in an increase in consumer's price. A substantial amount of post-harvest losses have their origin in the pre-harvest stage ex., genetic factors, microbial infections, pest infestations, environmental factors and cultural practices during the production stage. Poor sanitation in all post-harvest operations is the key factor for the source of infestation and high level of contamination. Thus the high level of post-harvest loss of mango is largely due to neglect of certain recommended practices for harvest, handling, transport, storage and protective treatments in the field and post-harvest treatments in the markets or processing units.

REFERENCES

Anonymous, 2005. Post-harvest practices for Mango, Banana, Tomato and Capsicum. Indian Institute of Horticultural Research, Bangalore (India). Tech. Bulletin, No. 27.

Arauz, L.F., Wang, A., Duran, J.A. and Monterry, M. 1994. Causes of post-harvest losses of mango at the wholesale market level in Costa Rica. Agronomia Costar ricense. 18(1) : 47-51.

Anonymous, 1987. Proceedings of the workshop-cum-field tests on Assessment of Post-harvest losses, held at IARI, New Delhi and CIHNP, Lucknow (India), 3-27th June 1987.

Anonymous, 1989. Annual Report on Post-Harvest Technology, Indian Institute of Horticultures Research, Bangalore, India.

Anonymous, 1990. Mango in India. Production. Presentatin and Processing. CFTRI Mono graph 21.

Bhatnagar, H.C. and Subramanyam, H. 1971. Some aspects of preservation, processing and export of mango and its products. Exports group meeting. United Nations Industrial Development Organisation, 88/15.

Cheema, G.S., Bhat, S. and Naik, K.C. 1959. Commercial Fruits of India, Macmillan, New York.

Chenulu, V.V. and Thakur, D.P. 1968. Survey for Storage and Market diseases of certain fruits and vegetables in Delhi and control of soft rot due to *Rhizopus* sp. Indian Phytopathology, 21 : 140-144.

Dalal, V.B. 1989. Losses in Fresh Fruits and Vegetables and Their Remedial Measures. Indian Food Industry, 7(3) : 1-13.

Dhemre, J.K. and Waskar, D.P. 2004. Effects of post-harvest treatments on shelf-life and quality of Kesar mango fruits, during storage. Journal of Maharashtra Agricultural Universities, 29(2) : 159-163.

Dondio, L.C. 1996. Quoted by Santos *et al* 2002. Storage of Mango fruits Cv. Rosa treated with calcium chloride after harvest at different maturity stages. Acta Horticulturae, 645 : 663-670.

Dalel V.B. 1989. Losses in Fresh Fruits and Vegetables and their Remedial Measures. Indian Food Industry, 7(3): 1-13.

Eckurt, J.W. and Ogawa, J.M. 1988, Annual Review of Phyto-pathology, 26: 433-469.

Khurdiya, D.S. and Roy, S.K. 1986, Studies on Ripening and Canning of mangoes, Indian Food Paker, 40 (1): 45-48.

Lakshiminarayana, P. And Reddy, S.M. 1977. Observation on certain market diseases of mango from Warangal (India) Egyption Journal of phyto pathology, 9:29:33.

Leilua, I. 1979, Econonmic losses of Philippines mango cv. "Carabao" due to fruit rots. M.S. Thesis, University of the Philippines, Los Banos, College, Laguna, Philippines.

Madan, M.S. and Subramanyam, K.V. 1987. A commodity loss profile for mango in Karnataka. Indian Journal of Agricultural Marketing, 1(2) : 80-86.

Madan, M.S. and Ullasa, B.A. 1991. Post-harvest losses in mango : Causes and Control Measures. Mysore Journal of Agricultural Science, 25 : 458-463.

Madan, M.S. and Ullasa, B.A. 1993. Post-harvest losses in fruits, In : Advances in Horticulture, vol.4 : pp. 1795-1810. K.L. Chadha and O.S. Pareek (eds). Malhotra Publishing House, New Delhi.

Manjiri, V.T. and Modi, V.V. 1988. Studies on spoilage of Alphonso mangoes by *Saccharomyces cerevisiae*. Acta Horticulturae, 231 : 697-708.

Mitra, S.K. 2008. Harvesting, Packaging, Transportation and Marketing of Tropical fruits in West Bengal, India : Present Practices and Future needs. Acta Horticulturae, 768 : 301-306.

Munishamanna, K.B., Ranganna, B. and Kumar, C. 2002. Post-harvest Microbial spoilage of Mango during storage. Current Research, 31: 185-186.

Mandan, M.S. 1988. Economic assessment of harvest and pre-harvest losses in horticultural crops. Annual report, Indian Institute of Horticultural Research Bangalore (India) pp: 93-36.

Narayana, C.K., Roy, S.K. and Pal, B. K. 1989. Use of high molecular weight high density polyethylene film (HM Film) in transnit of mango (cv. Baneshan). IIIrd International Symposium on Mango held at Darwin (Australia), (Abstract), Nov. 1989, p. 89.

Palijwala, V.A., Sattar, A.P. and Modi, V.V. 1988. Strategies for the control of post-harvest spoilage of mangoes by *Penicillium cyclopium*. Acta Horticulturae, 231 : 688-696.

Pandey, B.K. and Tandon, D.K. 2007. Protocol for post-harvest management of mangoes. Indian Horticulture, May-June, 18-19.

Pruthi, J.S. 1992. The Mango Book – I. Agricultural and Processed Food Products Export Development Authority Monograph, New Delhi, p. 144.

Pathak, V.N. and Srivastava, D.N. 1968, Mango losses due to dipludia-stem-end-rot, Tropical Agriculturist, 123: 75-77.

Quroshi, S.U. and Meah, M.B. 1991. Post-harvest loss in mango owing to stem-end-rot. International Journal of Tropical Agriculture, 9(2) : 98-105.

Ramakrishna, M. and Rao, M.R. 1989. Post-harvest changes in Sapota under different storage treatments. Andhra Agricultural Journal, 36 : 166-171.

Rawal, R.D. and Ullas, B.A. 1985. Efficacy of pre-harvest field sprays of fungicides against stem-rot of mango due to *Botryodiplodia theobromae*. IInd International Symposium on Mango, Bangalore (India), 20-24, May, 1985.

Roy, S.K. and Joshi, G.D. 1985. An approach to integrated post-harvest handling of mangoes. IInd International Symposium on Mango, Bangalore (India), 20-24, May, 1985.

Roy, S.K. and Pal, R.K. 1991.Multilocational studies to reduce post-harvest losses during Harvesting, Handling, Packaging, Transportation and Marketing of Mango in India. Acta Horticulturae, 291 : 499-503.

Salunke, D.K. and Desai, D.B. 1984. Mango, In : Post-harvest Biotechnology of Fruits. CRS Press, Boca Raton, Florida, USA.

Santha, K. and Rao, D.V.S. 2001. Status of post-harvest management of fruits. Indian Journal of Horticulture, 58(1&2) : 152-163.

Sethi, V. and Maini, S.B. 1989. Appropriate Technology for Reducing Post-harvest losses in Fruits and Vegetables. Indian Food Packer, 43(2) : 42-56.

Sharma, N. and Alam, M.M. 1998. Post-harvest diseases of Horticultural perishables. International Book Distributing Co., Lucknow (India), pp. 1-22.

Shikhamany, S.D. 2004. Strategies to Improve Productivity, Quality and Post-harvest management in Fruit crops. Souvenir. 1st Indian Horticulture Congress, New Delhi.

Singh, H. 1983. "Horticultural Industry and substitution of packaging". Paper presented at National Workshop on "Substituting Forest Timber for Packing cases for Fruits".

Srinivas, R.N., Reddy, T.V., Ravi, P.C., Achoth, L. and Reddy, B.V.C. 1997. Post-harvest Loss Assessment of 'Totapuri' and 'Alphonso' mangoes. Journal of Food Science and Technology, 34(1) : 70-72.

Subramanyam, H. and Moorthy, N.V.N. 1972. Control of fungal spoilage in Alphonso mangoes by pre-harvest application of fungicides. Acta Horticulturae, 24 : 224-226.

Subramanyam, K.V. 1986. Post-harvest losses in horticultural crops : An appraisal. Agricultural Situation in India, XLI : 339-343.

Sohi, H.S. Sokhi, S.S. and Tewari, R.P. 1973. Storage rot of mango (*Colletotrichum* gloeosporioides) and its controls. Phytopathologia mediteranea, 12:144-146.

Thakur, D.D and Chenulu, V.V. 1970. Survey of Delhi market for fungal diseases of fruits and vegetables. In: Plant disease problems. (S.P. Rayachouhri chel (Ed) pp. 307-308, Indian Phythologics Society; New Delhi.

Verma, L.R. and Joshi, V.K. 2000. Post-harvest technology of Fruits and Vegetables : An overview. In : L.R. Verma and V.K. Joshi (ed) Post-harvest Technology of Fruits and Vegetables. P. 2, Indus Publishing Co., New Delhi.

Waskar, D.P., Masalkar, S.D., Gaikwad, R.S. and Damame, S.V. 1997. Effect of method of harvesting on post-harvest loss reduction and extension of shelf life of mango. Acta Horticulturae, 455 : 662-668.

CHAPTER 13

Utilization of Mango Waste Material

M. Hanuman Naik

Huge quantities of mango wastes occur both at pre- and post-harvest stages varying between 35 and 55 percent of the total production. In the course of processing alone, large quantities of waste accounting to 40 to 50 percent of the fruit processed come out in the form of seed stones, peel etc. Mango waste can occur at various stages right from fruit set to consumption.

From fields: Wastes from field include dropped, damaged, injured, bruised fruits due to pests and diseases, faulty picking, unutilized or underutilized ones, fruit pulp and parts there of.

From Market, Pack houses, Storage etc: Unsold, wilted, spoiled or damaged fruits.

From kitchen and other consumption centres: Stones and peels leftover after consumption.

From Processing factories: During the processing of mango, stones, peels, pulp, pulper waste, spilled processed products are generated as waste (40 to 50 percent of total fruit weight), They are rich in nutrients, fibre etc., and quite ameneble for obtaining many value added products.

Maurya and Chaudhury (1977), Tandon and Kalra (1989), Kalra *et al* (1995), Sri Hari Babu(1996) and Garg, (2004) have reviewed the work done on mango waste utilization in India.

Economic Importance: The amount of raw fruits falling from the trees, spoiled whole fruits, inedible wastes (stones, peel etc.) from processing factories, kitchen etc., when combined together may exceed 5-6 million tonnes annually. From ten million tonnes of annual mango produce about four million tonnes of mango stones,and one million tonnes of kernels are obtained, which yield 40,000 tonnes of fat valued at Rs. 500 million and 16,000 tonnes of starch which are lost. Beerh *et al* (1976a) estimated that out of 10,000 tonnes of mangoes processed, 1500 tonnes of peel or about 500 tonnes of dried peel are available annually in the country. According to another estimate 0.3 million tonnes of mango kernels are available annually in the country from which 30,000 tonnes of fat, valued at Rs. 200 millions can be obtained (Narasimhachari *et al* 1977).

Efficient disposal/utilization of these waste material will not only help to minimize pollution, but also provides valuable raw material for extraction of useful products, gas reactors and organic manures (Table 13.1), Fruit waste can be utilized by producing either upgraded material of higher value (animal feeds, ensilages etc.,) or quite distinct new products, since these wastes are rich sources of vital constituents like proteins, starch, fats, fibre, pectin etc., (Sri Hari Babu,1996).

Table 13.1: Possible bye-products from solid waste in mango from processing factories (Sethi and Maini, 2001)

S.No.	Type of Waste	Quantity (%)	Nature of waste	Possible bye-products
1.	Peels	12-15	Peel and pulp	Proteins,cattle feed, alcohol
2.	Pulper waste	5-10	Fibre	Wine, vinegar, juice
3.	Stones	15-20	Hull and kernel	Fats, tannins, starch.

Hazards from the waste material: Any waste material from fruits and vegetables causes great health hazards to humans, animals, and even plants and create pollution problems. Mango waste from the above sources, particularly from the market yards and processing factories is generally dumped on the roadside, in road side ditches in the viscinity of factories or waste lands, a little away from the living habitates. This dumped waste on decay or fermentation emits foul smell up to a distance of 3-4 Km in all directions for about 5 to 10 months, making it extremely difficult for the nearby population to live. If this waste is not removed or left unattended, it could result into a dreadful source of air, water, and soil pollution. The dumped waste becomes a breeding place for insects, worms etc. Stray animals (pigs, dogs, cattle etc.) feed on the waste, fall sick and may eventually die (Photo 13.1). During rains, the waste is washed off into nearby tanks and canals causing contamination of water leading to the death of fish etc. The liquid waste also percolates into the ground water and pollutes it.

However, in case of some large scale processing units these wastes are taken away by the farmers for their use as cattle feed and/or manure.

Government have to incur huge expenditure in the treatment and disposal of wastes generated by the processing factories. Of late, Government enacted “The Environmental Protection Act 1986” under which dumping solid and liquid wastes out side the factory premises and in the municipal sewage system directly, respectively attracts a fine and/or imprisionment.

Plate 13.1: Mango pulp and seed waste consumption by cattle

Utilization of mango wastes: Utilization of fruit wastes is both a necessity and a challenge. Utilization not only economizes the cost of finished product and reduces the pollution levels, but also leads to complete utilization of the raw material (Sri Hari Babu, 1996., Garg, 2004). Processing units located in congested area with limited space and inadequate water supply find it difficult to law-fully manage. The processing wastes have high biological oxygen demand. Therefore, the disposal of huge quantities of the waste material is a serious problem to the processing factories as well as the civic bodies of Government from the sanitary and the economic point of view. Utilization of solid wastes from certain fruit crops (apple, citrus, pineapple etc,) as cattle feed directly or after partial treatment is well known in developed countries. But waste utilization for human consumption and in the industry itself will be a more profitable proposition (Sri Hari Babu, 1996).

The waste can be put to use in two ways, it may be used as such directly or after drying for feeding animals, it may be converted into a higher valued product either by chemical treatments or by fermentation (Garg, 2004). Studies have shown the possibility of utilization of mango waste for recovering useful products and simultaneously avoid the serious disposal of waste. The possible ways of utilization of mango wastes from different places are discussed below.

From Field: Fruits get dropped at different stages of their development due to various reasons. Large quantities of dropped mango fruits will be available in fields after dust/hail storms, high winds and rains together with bird/bat/insect damaged fruits. These can be used for pickling, dried mango slices and powder (amchur). Collected raw mango wastes are washed, peeled, sliced and dried in the sun to make 'amchur' which is widely used in North India in curries etc, 'Amchur' is also exported. In 1995-96, India exported about 350 tonnes of 'amchur' with a value of Rs. 623/- million (Maini, 2001). Other best uses of fallen, discarded and sorted out mango fruits from the field are: making of candied pulp, jam, jelly, paste, cheese, squash, chutney and pickles depending on the soundness and size of the fruits (Maurya and Chaudhury, 1977).

From Kitchen: Mango waste from kitchen mostly consists of peel and stones. Since less than 2 percent of total produce is processed, bulk of the waste comes from the table consumption in the houses. However, collection of this kitchen waste is difficult as most of it is thrown in garbage and dust bin and hence do not find place in utilization, However, concerted efforts must be made for collection of kitchen waste as it forms the largest volume of waste, for its profitable utilization.

From Processing Factories: Processing industry generates huge quantities of waste, next to waste from kitchen. Each factory processing 30 tonnes of mango pulp a day leaves a waste of about 10 tonnes stones, 8 tonnes of peels and 6 tonnes of residual sludge. In other words a factory processing 1000 tonnes of mango fruit generates 30 tonnes sugar solids, 4 tonnes of tannins and 3-5 tonnes of fat besides considerable quantity of carotinoid colouring matter from mango waste (Sri Hari Babu, 1996). However, processing factories have no organized method of either waste disposal or utilization.

Two types of wastes viz., solid and liquid wastes come out of processing factories. Solid waste consists of mango peel stones, stalks, trimmings and fibrous material obtained during the preparation of raw materials. This constitutes about 40-50 percent of total fruit waste, of which 5-10 percent is pulper waste, 12-15 percent is peels and 15-20 percent is kernel. Liquid waste comes out of factory after washing of fruits, blanching, cooling, and machinery clean up etc (Garg, 2004).

Utilization of solid waste

Seeds (stones) and peels are the two important wastes of mango, which constitute 35-55 percent of fruit (Bhatnagar and Subramanyam, 1973). The chemical composition of mango pulp and kernel is furnished in Table 13.2.

The different potential approaches for utilization of mango solid waste are as follows (Garg, 2004) (Table 13.3).

Table 13.2: Chemical composition of peel and kernel of Mango (on dry weight basis)

S.No.	Constituent	Pulp (%)	Kernel (%)
1.	Mositure	68.5 a	55.0 b
2.	Total sugar	48.1 a	8.0 b
3.	Reducing sugars	40.8 a	2.9 b
4.	Starch	2.9 a	57.8 b
5.	Pectin	12.9 a	0.8 b
6.	Protein	3.9 a	7.1 c
7.	Crude fibre	8.4 a	3.0 c
8.	Tannins	2.3 a	10.6 b
9.	Ash	2-9 a	1.8 c
10.	Fat	-	13.7 c

Note: a -Beerh, 1976; b – Pruthi and Susheela, 1963; c- Mittal and Singh 1972.

Table 13.3: Approaches of Mango solid waste utilization

Pulp		Kernel	
Chemical	Microbial	Chemical	Microbial
Extraction	Fermentation	Extraction	Fermentation
Fibre	Vinegar	Fat	Enzyme (Amylase)
Pectin	Citric acid	Starch	Alcohol
Dyes	Single cell protein	Flour	
Pigments	Enzymes (Cellulose, Pectinases etc)		
	Composting, Biogas		
	Mushroom cultivation		

Drying of mango kernel and peel makes a high class energy food for animals (Cooke,1949; Bose, *et al* 1952). Solid wastes are also useful as organic manures.

Mango Stones (Seeds) and Kernels: Mango seed stones are available both from house-hold consumption and processing factories (Photo 13.2). However, collection of seeds from house-hold consumption is difficult and is a major challenge. There are no organized efforts from any quarter to collect such thrown away seed stones from kitchen etc., except by mango nurseries. Mango nurseries engage beggars and nomadic people etc., to collect the seed stones from market places, dust bins, garbages etc., and purchase them for raising seedling root stocks. Where there are no nurseries, there will be no collection and the garbage collection made by civic bodies including mango waste material goes unutilized and wasted. On the other hand, a negligible quantity of mango stones from processing factories is disposed off to nurseries to raise seedling root stocks. One of the problem associated with mango stones from processing indurity for raising seedling root stocks is the leftover pulp adhering the stones, which sometimes affects the germination. The adhering pulp (6-7 percent) attracts the microorganisms, especially fungi and yeast as the pulp contains sugar. These fungi and yeast cause diseases and thus affect the germination. Therefore, unless this adhering pulp is removed from the seed stones from processing industries, they cannot be used in nurseries. Mechanical or manual cleaning of stones with water will be costly as well as generates liquid waste which in turn increase the BOD of water streams. So alternate measures to remove the adhering pulp on the stones were sought. Fruit pulp, because of its high TSS(22 °B) and favourable acidity(pH 5.0-5.5) is one of the best substrate for yeast growth. Based on this Garg (2006) developed a process to remove the pulp from seed stones as well as to produce yeast. Dry backer's yeast @ 25 g is dissolved in 100 ml water and mixed well with mango stones in a stainless steel vessel, which is loosely covered with a lid to prevent entry of dust, flies, mosquitoes etc. Inoculum mixed with curd (Yoghurt) is added and mixed uniformly so that it reaches every stone and left at room temperature. By this simple process, the pulp adhering to stones is completely removed in such a way that the veins on the stone are clearly visible after the process. This is a simple and easy to follow technique which can be used in the processing factories. It serves duel purpose firstly, it helps in cleaning the stones from adhering pulp and provides value added Backer's yeast. Further, this technique does not affect the seed germination. It is a simple and cheap technique which can be adopted as a cottage industry without requirement of much infrastructure. The cleaned seed stones are not only useful for nurseries but also for other purposes.

Lakshminarayana *et al* (1983) carried out systematic studies on varietal variations in content, characteristics and composition of seeds and gave a wealth of information. The seed content varies from variety to variety and ranges from 9 to 23 percent of the fruit weight with an average of 15 percent and consist of shell (hull) and kernel. (Palaniswamy *et al* 1974).

Kernel: The main product of mango seed stone is the 'kernel' (cotyledons) which consituties between 45.7 and 72.8 percent of the seed depending on variety and size (Hemavathy et al 1988). Kernels are comparable to most of the cereals in respect of carbohydrates, proteins, fats and minerals (Ca and P). Kernel is obtained by breaking the hard seed coat of the stone (decortication). Decortication of mango stones is a major problem, coming in the way of its utilization and extraction of fat etc. Because of their asserted, odd sizes, shapes and toughness of shell wall, mango stones are most difficult to decorticate. However, decorticators, exclusively for mango stones were developed by Oil Technological Research Institute, Ananthapur, Andhra Pradesh (India) (Narasimhachari, *et al* 1975).

Mango seed kernel (MSK) has several uses. Paramar and Sharma (1984) reviewed the chemical composition and edible uses of MSK. It is rich in proteins, fat, starch, fibre, vitamin-A, tannin, sterols and terpentine alcohol (Table 13.2). According to Maurya and Chaudhary (1977) MSK is a rich source of protein (8.5 percent), fat (8.85 percent), carbohydrates (79.2 percent), minerals (2.6 percent), fibre (2 percent), calcium (0.2 percent), and phosphorus (0.21 percent), and its fat and calcium contents are higher than that of cereals, but the protein content is less than that of wheat, maize and barley. However, Mark and Stewart (1976), opioned that MSK is comparable to most of the cereals particularly in respect of carbohydrates, proteins, fat, vitamins, calcium, and phosphorus. MSK is a very poor source of vitamin-C (Sastry and Krishnamurthy, 1974).

Kernels can not be used directly for consumption. They are roasted, powered and cooked as a gruell by tribal people. It can, however, be used directly as a manure (Vimal and Adsule, 1976).

Plate 13.2: Discarded mango seed stones from processing unit

Protein: The MSK is a rich source of proteins of high biological value with different kinds of useful amino acids. The total amino acid content in MSK proteins was found to be 88-97 percent of which 31-35 percent are glutamic, asportic, and leucine, while sulpher containing amino acids are present in small amounts (Laszlity *et al* 1988). The kernel proteins thus can be used to produce food mixtures of higher nutritive value.

There is shortage of proteins the world over. The problem is acute among population which consumes cereals as their staple diet. Mango processing waste was suggested as a good raw material for the production of simple cell proteins (SCP). The production of proteins through animal sources is very much expensive than through the microorganisms (Gray, 1962). Sethi *et al* (1977) used mango peel as carbon source for fungal protein, where as Sood *et al* (1980) reported the possibility of using MSK as a good substrate for SCP production with the help of fungi. The protein of MSK has high biological value and therefore, can be utilized for making nutritious food (Mark and Stewert, 1976).

Fat: The fat content of MSK ranges from 8.2 to 14.3 percent on dry weight basis. (Hemavathy *et al* 1985). The fat that MSK contains on an average about 10 percent, has raised many hopes in the vegetable oil world. The fat from MSK has physical characteristics similar to those of cocoa butter. Both are rich in stearic acid but differ in palmitic acid and glyceride acid content. The main components of fatty acids of MSK of Malaysian mango cultivars were palmitic (6.9-7.3%), stearic acid (44.3-44.4%), oleic (39.9-42.1%) and linolic (4.5-7.4%) acids (Augustin and Ling, 1987) where as the lipid material of kernels is composed of variety of natural liquids, phospho lipids and unsaponijiable matter. The quality of lipids from MSK is suitable for human consumption and also as a suitable substitute to cocoa butter in the preparation of confectionary products (Garg, 2006). Due to its blandness and absence of toxic substances, the kernel fat enjoys a potential use as edible fat in sweet meats (Narasimhachari *et al* 1977). It can also be used as a substitute for tallow in the preparation of quality soaps and other cosmotic products and cocoa butter (Baliga and Shitole, 1981., Hussain, *et al* 1983., Krishnamurthy, 1996). In India where vegetable oils are scarce/in shortage, this fat can be used in place of vanaspathi . Addition of 1 percent kernel fat to buffalo milk ghee acts as an antioxidant (Paramar and Sharma 1986).

Fat from MSK is extracted by solvent extraction method. Decorticated kernels are coarsely powdered after drying, cooked and passed through an expeller using edible grade hexane as solvent (Narasimhachari *et al* 1977).

Dry MSK contains 9-13 percent oil (Narasimhachari and Reddy 1977). It is extracted by solvent extraction by expellers and is semi-solid with cream colour. The oil is a potential source of vegetable oil and is used as a cocoa substitute in chocolate manufacturing. Yahia and Moustafa (1982) did not find differences in the texture, taste and flavour of toffees prepared from MSK oil and cocoa butter and suggested the use of MSK oil as a cocoa substitute or partial replacer in the preparation of confectinery products. Kernel oil can also be used as a medicine against diarrhoea and dysentry etc. (Sri Hari Babu, 1996). It can also be used in paints and chemical industry. The seed meal (cake) left over after the extraction of oil is used as organic manure (Hayes, 1970) or as cattle feed (Awasthi and Singh, 1998).

Starch: Mango seed kernel (MSK) is a rich source of good starch and carbohydrates. It contains 47-63 percent starch of which 19-22 percent is amylase (Roy and Mitra, 1970). Wilkins (1942) reported 69.22 percent of carbohydrates in MSK, while Pruthi and Susheela (1963) estimated it to be 57 percent.

The MSK are kept in water 5-10 times their volume with a small quantity of SO_2 for 24 hours. After draining off water, the kernels are crushed in stone grinders and strained through a fine cloth after one more washing with water. The strained water containing starch is treated with SO_2 and is allowed to settle for 6 hours. The sediment is separated by syphoning off the supernatent liquid. The pH of the sediment is adjusted to between10 and 11, and is dried and prepared in the usual way (Das and Benerjee, 1952)

The pure MSK starch is similar in its characteristics to tapioca starch and hence, recommended for food use. (Bhatnagar and Subramanyam, 1973). Starch from MSK can also be used for industrial purposes such as textiles, paper, and jute industries, in the manufacture of adhesives, syrups, glucose, ceramics, printing etc (Mittal and Singh 1972). The starch after mixing with wheat or maize flour is used in *chapaties.*

Tannins: The tannin content in MSK varies from 10.6 to 18.0 percent, Tannins are extracted by soaking and thermal treatment is employed during production of flour (Das and Benerjee, 1952., Pruthi and Susheela 1963). The kernel fat is further extracted with denatured fat to remove tannins, the tannins, then are recovered by evaporating solvent (Krishnamurthy, 1996). Poly-phenols such as ethyl gallate calactium and leucocyanidium can also be isolated from the seed coat of kernels. However, some of these gallo-tannins were found to be toxic (Kalra *et al* 1995).

Flour: MSK can be made into flour for food use. However, the tannins in the MSK give astringency (bitter taste), hence the tannins are to be removed before the flour is put to food use, The kernels are crushed and washed in running water for a few hours and soaked in 2 to 3 times water for 20 min. at 80 °C. The material is then dried and made in to flour. This removes the tannins and astringency (Mittal and Singh,1973). As compared to the cereals, mango kernel flour contains more fat and calcium (Anon, 1962). Tannin-free kernel flour can be utilized in making traditional Indian food items such as *chepathi, parota* etc, by replacing wheat flour to an extent of 10-40 percent without effect on the quality and acceptability (Dhingra, and Kapoor 1985, a,b). The flour could also be used in animal feeds, as manure, in textile, jute and paper industries (Narasimhachari *et al* 1988).

The flour can be hydrolyzed by enzyme treatment or by acid hydrolysis and the dilute sugar syrup obtained is decolourized with activated carbon and then concentrated to 70 °B. A light coloured syrup, thus obtained can be used in confectionary. The uses of Kernel and MSK are as follows:

Human Consumption: In some parts of India, kernels are eaten after boiling, and backing and relished most (Wilkins 1942., Musahib-ud-din and Bhutani, 1946). They and Hayes (1970) suggested the uses of mango kernels for making pickles, fried products and flour respectively. A delicious pickle was also prepared from kernels by mixing necessary spices and condiments and boiling them in mustard oil and was found to be like any other pickle made from green mangoes and relished well (Wester, 1920., Hayes, 1970).

The MSK flour is equal to rice in food value and hence, finds use in food items. It is potentially a rich source of good quality starch and proteins of high biological value. It is eaten in a variety of ways: made into cakes, put between green leaves and baked in the fire or may be mixed and cooked with rice. Another way is to cover with green leaves and cook in earthen pot. The defatted kernel flour is also used in the preparation of *chapaties* etc. by replacing wheat flour in North Western parts of India (Mark and Stewart, 1976). Preliminarly results showed that 20 percent kernel flour, after the removal of tannins can be incorporated in the food items like *pakori*, *parotas* and *chapaties* without loss of quality (Dhingra and Kapoor, 1985 b). In Java, the flour is used in the form of a parridge with coconut milk (Hayes, 1970). The cake made from MSK can be used as an alternate to wheat and maize flour, however, if the MSK has to be used for human consumption, the kernels have to be hygienically preserved.

Animal Rations: The mango seed kernel (MSK) has been used in animal ration since long (Dhingra *et al* 1948). Studies of Dhingra and Kapoor (1985 b) and Das *et al* (1988) have elucidated the potential of MSK in animal nutrition in place of maize, particularly in case of chicken feed. Poultry feeds contain 40-50 percent cereals and hence compete with humans for cereals (Chand and Shukla 1973). Maize is generally used as chicken feed to an extent of 50-70 percent of the total feed in Nigeria (PAN 1995). However, there is shortage of maize because much of it finds use in the production of ethanol and to some extent for domestic purposes. In view of the above, a search for alternative energy feed resources was made (Aina,1990). The resources should have a lesser agronomic demand, high energy, limited processing method and have a low competeting demand for human consumption. MSK has satisfied all the requirements, hence it was considered as a suitable substitute for maize in chicken feed. MSK contains 5.59 percent crude protein, 2.3 percent ether extract, 78.62 percent nitrogen free extract and 3.17 K ca/g ME. The above composition showed that MSK was lower in protein, ether extract and metabolized energy (ME), but slightly higher in ash, fibre and carbohydrate fraction as compared to maize, which demonstrated that optimum replacement of maize by relatively underutilized MSK in broiler chicken feed should not exceed 20 percent (Odunsi and Farinu, 1997). Shakawat, *et al* (1992) found that use of mango stone kernel did not affect the quality of broiler meat. On the other hand, Ravindran and Shiva Keaneshan (1996) found that raw MSK was unsuitable as a feed ingredient in chicken diets, unless they are soaked and boiled to overcome the negative effect of raw kernels.

The dried kernels, with less than 10 percent moisture showed promise as a good fodder supplement in cattle rations without any further treatment, because of balanced proportion of amino acids and can replace wheat or maize to an extent of 60 percent (Kehar and Chanda, 1946). Cake made out of MSK can substitute for wheat and maize flour in animal feed.

MSK are greatly relished by pigs, so processing factories can have piggeries as subsidiary occupation or supply the feeds to the farmers having piggeries. Similarly owners of piggeries can obtain mango seed stones from garbage, consumer centers and market yards as well as from the mango processing units.

Defatted MSK powder with 64.5 percent starch, 4.1 percent protein and 4.9 percent fat could constitute 40 percent rat diet together with wheat or rice flour (Geetha and Devi, 1986).

Fish flour or fish protein, though a rich source of protein (85%) is often blended with bengal gram (chick pea) flour, ground nut (Pea nut) flour, maize (corn) starch and tapioca (cassava) starch which are costly. As an alternate, mango kernel as a source of protein (4.4-4.8%) was suggested for blending with fish flour for feeding fish (Chand and Saroj, 1970).

Industrial Utilization: A light coloured syrup obtained from MSK free from tannins and fat can be incorporated into squash, jams, confectionary items etc (Sri Hari Babu 1996). It can be hydrolysed to glucose by enzymatic method or acid hydrolysis. The dilute syrup thus obtained can be concentrated to 70 ° Brix, after treating with activated carbon and filtered (Krishnamurthy, 1996).

MSK can be used for 'ethanol' production by co-culture fermentation using *Aspergillus niger* (amylolysic fungus) along with yeast (*Saccharomyces cerevisiae*). About 10 percent alcohol could be obtained by this method (Garg, 2006).

MSK can also be used for the production of 'amylase' using amylolysic microorganism. Pre-treatments like steeping or defatting of MSK results in increased production of amylase. *Aspergillus oryzea*, and *Syncephalastrum raumosum* are the efficient 'amylase' producers with MSK as substrate. MSK can also be used for the production of citric acid and lactic acid using *Aspergillius niger* (Garg, 2006).

MANGO PEEL

Mango peels, generally termed as 'Total waste' is the second most important waste generated in the processing factories as well as in the kitchen. Mango peel consists of 12 to 15 percent of the total fruit. From a factory processing 50 tonnes of mangoes (cv.Totapuri) per hour (8 hrs/day) 6 tonnes of peel as waste is available (Photo 13.3). However, bulk of the peel waste comes out from kitchen (since peels are not consumed), which is difficult to collect as the total quantity is thrown into the garbase (see above). Disposal of peel is a problem as it is not easily bio-degradable and thus adds to environmental problem. Although soluble sugars (Table 13.4) present in the peel could initiate bio-degradation, complete degradation is difficult due to ligno-cellulosic nature of mango peel fibre and it may take very long time for complete degradation. Since, lignins are known to be resistant to anaerobic digestion, oxidative degradation such as composting has been suggested as more suitable for these substrate (Alfani and Cantaralla, 1987).

Plate 13.3: Mango peel waste from processing unit

Pre-treatment of such substances with acid/alkali was suggested for breaking down the complex structures into simple components to make them more susceptable for bio-degradation (Sounder and Chandra, 1987). However, acid/alkali pre-treatment did not bring about siginificant degredation of mango pulp fibre (Garg *et al* 1998). Co-composting with other easily degradable wastes has been suggested to enhance bio-degradation (Anderson and Smith, 1987), Co-composting with cowdung (75% peel + 25% cowdung, w/w) brought about 33 percent reduction in the fibre content of mango peel after 40 days of composting under ambient condition (30 $\pm$ 5 °C). Natural micro-flora in the peel are not capable of hydrolysing the fibre in the peel. Cowdung a rich source of ligno-celluloytic microorganism when added (wet dung at 333 g/kg of peel) promoted degradation of mango peel. Thus, co-composting of mango peel waste with cowduing resulted in complete bio-degradation of fibre in the peel, making it possible to produce a good quality compost in 40 days. This finding will be of great use to mango processing industries for utilization of mango peel whose disposal has been a problem hither-to (Garg *et al* 1998).

The chemical composition of mango peel waste is furnished in the Table 13.4.

Table 13.4: Chemical composition of mango peel waste

S.No	Consilment	Content (%)
1.	Total sugars	48.0
2.	Reducing sugars	40.8
3.	Starch	29.0
4.	Protein	0.8
5.	Pectin	3.5
6.	Fibre	8.4
7.	Cellulose	14.2
8.	Hemi cellulose	8.01
9.	Lignins	13.82
10.	Tannins	2.3
11.	Ether Extract	6.4
12.	Ash	2.9

Thus, mango peel is a complex organic substrate comprising of pectin, fibre, sugars, pectin etc. Peel containes more vitamin 'C' than flesh and kernel (Sastry and Krishnamurthy, 1974).

The utilization of mango peel has much economic potentiality as it can be used for the preparation of vinegar. The water extract from mango peel on concentration and fermentation forms vinegar (Vimal and Adsule, 1976). Mango peel can be utilized as a carbon source for fungal protein, carboxymethyl, cellulose and polyglacturonase production by fungi (Sethi *et al* 1977).

Peel fibre: Mango peel is a good source of edible fibre (5.4%) which is a complex organic substance of ligno-cellulose nature associated with semi-cellulose, pectin, proteins etc (Alfani and Cantaralla, 1987). Mango peel fibre has been found to have better *in vitro* properties than lemon peel fibre due to its higher values for antioxidative activity and glucose retardation index (Garg, 2004).

Fibre from mango peel can be extracted by different methods. In one method, fresh mango peel is wet milled with water and dried to get fibre. A peel particle size of 15 mm in wet milled and a washing time of 5 min. removes soluble sugars and increases total polyphenals and soluble dietary fibre content (Larrauri *et al* 1996). Fibre can also be extracted from residue leftover after pectin extraction without any difference in the amount of fibre than that extracted directly from the peel. The cost of production of pectin and fibre can further be lowered if the processing factory itself develops the facility of their extraction in the same premises, as an ancillary industry.

Peel Pectin: Pectin is an essential and versatile ingradient in the manufacture of fruit products like jam, jellies, marmalades etc. Besides, it has various uses in pharmaceutical products also. As yet the production and availability of pectin is not up to the requirements of our country and large quantities of pectin have to be imported to meet the country demand. India imports 160-200 tonnes of pectin valued at Rs.10-15 crores annually. Pectin is recovered from a number of fruits like jackfruit, wood apple etc (Jain and Girdhari 1955), lime (Srinivas *et al* 1963 a), Assam oranges (Chaliha *et al* 1963b), lemon (Chaliha *et al* 1963a), papaya (Bhatia *et al* 1959), orange waste (Agrawal and Pruthi 1972), apple pomace (Johar *et al* 1960) and guava (Pruthi *et al* 1960). Besides, use of mango peel as a source of pectin has been suggested by number of workers to meet the country's demand as well as to avoid the imports and at the same time to break the monopoly of Denmark (Sethi and Maini, 2001). Pectin obtained from mango peel has good jelly grade (6.1%) and the yield is also high, hence can be commercially exploited to extract the pectin. Work done at the Baba Atomic Research Centre, Mumbai (India) has also indicated that mango peel is a good source of pectin, a polysaccharide of high commercial and industrial importance (Narasimhachari *et al* 1988).

It has been estimated that mango peel can yield about 13 percent pectin on an average. The quantity and quality of pectin varies with the cultivar. The pectin from the peel of 'Chausa' and 'Saheb Pasand' had the best jelly grade (180) while that from the peel of 'Langra Ghorakpuri' had the poorest (140) (Tandon and Garg, 1998). The pectin with 8.9 percent methaloxyl and with jelly grade of 200-220 is considered to be of good quality. (Beerh *et al* 1976a., Srirangarajan and Shrikhande 1977). Pectin from mango peel had 4.8 percent moisture, 0.94 percent ash, 8.3 percent methaloxyl contents and over 200 jelly grade and is comparable with pectin from other well known sources of pectin (Jain and Girdhari, 1955., Bhatia *et al* 1959., Johar *et al* 1960., Pruthi *et al* 1960., Chaliha *et al* 1963a,b., Srinivas, *et al* 1963., Agrawal and Pruthi, 1972).

Beerh *et al* (1976b) evaluated two methods of extraction of pectin from mango (Totapuri) peel viz (I) Aluminium precipitation and (II) Alcohol precipitation,which yielded good quality (grade 200-240) of 10.3 percent and 15.7 percent pectin respectively. The Aluminium precipitated pectin was dull gray in colour as compared to attractive white colour of alcohol precipitated pectin. Tandon and Kalra (1991) extracted pectin from dry peel and kernel of immature and ripe 'Dashehari' mangoes using 0.05N NaoH; 0.05N HCl and 0.3 percent sodium hexametaphosphate (SHMP) and obtained a maximum yield of 12.8 percent pectin from ripe mango peels through SHMP extraction, The kernel yielded very little pectin. Quality estimated in terms of moisture equivalent weight, metholoxyl content, relative viscocity and jelly grade indicated it as a good pectin. Alcohol in-soluble pectins are maximum in kernels from ripe fruits. The water soluble pectins are more in the peel of ripe fruit, while alkali-soluble fraction was maximum in the peel of immature fruits.

Peel wax: Mango peel is a good source of wax. Narasimhachari *et al* (1988) extracted wax from mango (cv.Neelum) peel with a n-hexone in cold and obtained a yield of wax ranging from 2.5 to 3.0 percent. The extracted wax was dark cream coloured (yellowish) solid and had a pleasant odour, reminiscent of ripe mango and the following characteristics (Table 13.5).

Table 13.5: Characteristics of Mango peel wax.

S.No.	Parameter	Value
1.	Melting point	62 °C
2.	Acid value	39
3.	Saponification value	197
4.	Ester value	158
5.	Iodine value **(wijs)**	44
6.	Unsaponiable matter	17.7%

Utilization of Mango Peel

Jam: Peel could be made in to jam as follows:

Inactivation of polyphenol oxidase by immersion of peel in boiling water, extraction of peel soluble compounds at 80 °C, addition of sugar, citric acid and pectin as required, concentrating, cooling and packging are the steps. The jam thus prepared was rated as good in sensory tests and sweets made from it did not differ from those made traditionally (Larrauri *et al* 1996).

Syrup: Mango peels after pectin enzyme treatment and pressing yield 50-55 percent juice having 15-16° Brix, which on refining with hydrated lime or gelatin treatment gives good quality sugar syrup. The dilute syrup on further concentration to 70° Brix yields a light brown golden yellow coloured syrup which finds use in confectionary products. The dilute syrup can also be directly incorporated into mango squash, juice or nector without going for costly processes of concentration (Krishnamurthy, 1996).

Carotinoids: Mango peel is rich in carotinoids, particularly after ripening. The pressed out mango peels after PE treatment (for juice recovering) were found to be rich in carotinoid pigments (14-18 mg/100g). These on extraction with a suitable solvent and subsequent removal of the solvent, yield a concentrate of carotinoids pigments, The coloured paste of carotinoids can be made into a soluble solution with the help of suitable emulsifing agents, the emulsion on addition to a coloured mango juice and nectar gives attractive orange yellow colour. Since it is a natural colouring matter and is quite safe, it can be used in bottling, bakery and confectionary industries in place of synthetic coal tar dyes (Krishnamurthy, 1996). Hameed (2000) studied the utilization of carotenoids extracted from mango (cv. Balady) peel waste in Ezypt and indicated the possibility and suitability of utilizing the separated pigments (carotinoids) in colouring fat based foods such as butter and water based foods such as ice-creams instead of artificial colourants which are harmful to human health.

Enzymes: In fruit processing industries, various enzymes such as cellulases, pectinases and amylases are invariably used for pulp lignification, juice clarification etc. Of late, there is increased interest in enzyme production from food processing wastes. Cellulases can be produced from mango peel waste using *Colletotrichum gleoesporioides*, a pathogen causing anthracnose disease in mango. Pectinase can be produced from mango peel using *Aspergillus niger*. Similarly polygalacturonase (PG) can be obtained from peel at 30 °C and pH 4.5. Peel was a better substrate than pure pectin for PG production (Garg and Tandon, 1998), Amylase from mango peel could be produced by microbial fermentation (Anon, 2003).

Wine and vinegar: The mango peel waste can also be utilized for production of wine and vinegar after partial precipitation of tannins to reduce the astringency (Beerh *et al* 1976b). Vinegar was produced from mango peel using immobilized cells of *Acetobactor aceli*. Ethiraj and Suresh (1992) explored the possibilities of utilizing mango peel and seed stones for production of vinegar.

Direct fermentation yielded slightly more alcohol as compared to fermentation of cold or hot water extracts and a good quality vinegar with 4.5 to 5.0 percent acetic acid and characterized by mild flavor, was obtained. Alcoholic fermentation of these waste products was carried out using two methods. Peels or stones are mixed with cold or hot water (50% w/w) and saccharine. The extracts of peel or stones (obtained by repeated washing in hot or cold water) were treated with 50 ppm sulpher dioxide, inoculated with yeast (*Saccharomyces cerevisiae* var, *ellipsoideus*) and fermented at 24 $\pm$ 4 °C for 72 hours. Direct fermentation of peels and stones yielded more alcohol than the saccharine extract (Ethiraj and Suresh, 1992).

A good quality mango wine can be obtained from the press out juice by fermentation with yeast and on further acetic fermentation of mango wine, a quite palatable mango vinegar can be obtained (Krishnamurthy, 1996). Acceptable quality vinegar (5.2% acetic acid) and citric acid (20 g/kg peel) could be obtained through microbial fermentation (Anon, 2003).

Industrial Utilization: A few acids can also be obtained from mango peel waste using microorganisms.

Citric acid: In processing and pharmaceutical industries, citric acid is used widely and can be produced from mango peel waste using *Aspergillues niger*. Addition of methanol at 3 percent level enhances the production of citric acid from mango peels.

Lactic Acid: Mango peel and its molasses are sources of lactic acid, obtained by solid state fermentation using *Rhizopus oryzae*. Addition of phosphrous (0.01%) nitrogen (0.1%) to molasses increases the yield of lactic acid from 2.2 to 2.7 percent.

Animal Nutrition: Mango peel waste also finds use in animal nutrition. But, mango peel as animal feed has poor nutritive value because of its low protein content. However, its protein value can be enhanced four times by solid state fermentation, using *Aspergilles niger* (Garg, 2006). Mango peel can also be used as supplement in fish diet (Mahadevaswamy and Venkataramana,1990). Dried mango peels can be fed to cattle. Mango peel feeding gives more milk but the peel should be used on the day of peeling, other wise it gets infected with fungi and be-comes unfit as cattle feed.

Vermicompost: Mango peels can also find place in the preparation of vermicompost. It does not cause any type of pollution to air, water or soil. A mixture of half rotton farm yard manure and peel gives good quality vermicompost. In one year 4 cycles of vermicompost can be obtained. The ultimate waste resulting from the extraction of oil from mango peel waste can also be used for making vermicompost. Now-a-days vermicompost is increasingly used in crop production, but there is shortage for its extensive use. Processing factories themselves can have vermicompost units to make the compost from the generated peel waste and supply the same to the needy farmers and gain additional income. Peel does not ferment or decompose readily, hence it is advised to boil before composting.

Biogas: Mango peel can also be utilized for biogas production with urea at 20-30:1 ratio (Garg, 2004). Preliminery studies revealed that microbial pre-treatment of mango peel offers possibility of improving the overall economics of bio-gas production (Devi and Nand, 1989). Biogas effluent of mango peels, when used at 34 kg/100 m area in ponds as the sole source of fish feed yielded 8.35 kg fish per 100 m area, which had acceptable colour, flavour and taste (Mahadeva swamy and Venkatramana, 1990).

Miscellaneous uses

1. Pickles from unripe mango peels can be prepared.
2. Mango peel can be used as a substrate for mushroom cultivation after supplementing with paddy straw (Garg, 2004).
3. Fire bricks can be made from peels by pressing technique.
4. Mango peel waste can be converted into yeast (Rashid *et al* 1990).
5. Raw mango peels can be used for preparation of chutney and pickle, after soaking in 15 percent salt solution and 3 percent acid for a month, dehydrated and stored for year and reconstitution (Mehta and Tomer, 1982).

PULPER WASTE

The coarse and fibrous pulp coming out from the pulpers constitutes pulper waste up to 5-10 percent on dry weight basis. The chemical composition of pulper waste is shown in Table 13.6.

Table 13.6: Chemical composition of Mango pulper waste (Beerh *et al* 1976 b)

S.No.	Constituent	Content (%)
1.	Moisture	83.20 3.2*a
2.	Total soluble solids	12.50 3.2*b
3.	Total sugar	51.30
4.	Acidity	2.30
5.	Pectin	10.60
6.	Starch	5.20
7.	Crude fibre	6.00
8.	Tannins	Traces
9.	Ash	2.40
10.	p^H	3.2*c

* a,b,c, on fresh weight basis, rest on dry weight basis.

Pulper waste could be utilized to make juice, nectar, vinegar etc. It can also be used as colouring agent, as it contains carotenoids and as flavouring agent (Beerh *et al* 1976b., Ethiraj and Suresh, 1992).

The recovery of the juice from the pulper was 75-80 percent and 51percent of the peel. The possibility of treating this waste with pectic enzymes to express the juice, which can be used in the preparation of nectar and vinegar has been shown (Beerh *et al* 1976b). The quality of the nectar prepared from the pulper waste supplemented with juice recovered from mango waste was acceptable (Garg, 2004). Other value added products like syrup, wine and colouring matter can be obtained from the pulper waste. The pressed out pulper waste, after PE treatments is rich in carotinoids (6-9 mg/100g) which on extraction with solvent yields concentrated carotinoids pigments and this concentrate can be used as natural colouring matter (Krishnamurthy, 1996). On treatment with PE and pressing as described under mango peel waste (see above) the pulper waste yields about 50 percent juice with 15 ° Brix. This juice being sweet with mango flavour and completely free from bitterness, can be directly incorporated into squash, juice or nectar. Neither lime nor gelatin treatment is required for this. The dilute syrup can also be concentrated to 70 ° Brix and bottled. Further, wine and vinegar of delicious nature can be prepared from the dilute syrup by yeast and acetone fermentation respectively (Krishnamurthy, 1996).

Fruit vinegar is produced mostly from grapes (*Vitis* spp) throughout the world, while in India, it is generally produced from sugar cane (Tewari *et al* 1991) and jamun (*Syzygium cumunii* L.) (Anon, 1976). Vinegar can also be produced from mangoes. However, despite being a rich source of sugars (Kalra and Tandon, 1983) mango has not been used for vinegar production. Garg *et al* (1995) produced vinegar having 5.5 acidity as acetic acid from mango pulp waste by twin processes of fermentation and oxidation using yeast, *Saccharomyces ceravisiae* and *Acetobactor aceti.* Such vinegar had acceptability parameters in terms of colour, odour and taste. The recovery of vinegar was 15 percent, but the cost of preparing mango vinegar was found to be high. It was suggested that this high cost of production can be compensated by the quality of vinegar in terms of colour and taste. The cost of production can further be lowered by using the residue left over after ethanol production as a source of single cell protein in animal feeds (Goyal *et al* 1984).

The pulper waste along with peels can also be used for bio-gas production and carotinoid extraction.

UTILIZATION OF LIQUID WASTE

Liquid waste of mango coming out of processing units contains both soluble and suspended organic matter prinicipally, carbohydrates, lipids, cellulose and hemi-cellulose materials etc. as its constituents. In addition it also contains pesticide residues, detergents (used for washing fruits) and sanitizing agents etc. Hence, it has high BOD and COD values. These effluents sometimes become the carriers of pathogenic microorganisms. In order to recycle or re-use the waste water and to comply with the pollution control standards, treatment of waste water is necessary. Land application system is the simplest and cheapest way of liquid waste disposal.

The other ways are to subject the waste water biologically to aerobic or anaerobic treatments. Aerobic treatments of liquid waste include ponds, aerobic filter, activated sludge, bio-filters and aerobic lagooning etc. However, anaerobic treatment has been found to be more effective, in which the anaerobic digestion is brought about in two steps, by two distinctive types of microorganisms. In the first stage, complex organic compounds are degraded, hydrolysed and converted into organic acids, primarily acetic acid by acid forming bacteria. Methanogenic bacteria shall then convert it into methane and CO_2. The resulting sludge is well stabilized and can be disposed of readily as a fertilizer. The advantages of anaerobic treatments are many:

(i) less use of electrical energy to achieve 70-90 percent BOD reduction,

(ii) lower nutrient requirement,

(iii) minimum amount of sludge generation with constant effluent quality and

(iv) production of methane.

However, it has some limitations as anaerobic bacteria are more sensitive. Their growth is slow and sensitive to changes in pH, temperature and oxygen level.

An integrated approach of using processing industry waste water for getting value added products and materialsing pollution control operations shall optimize the production cost as well as protect the environment.

Constraints in the Utilization of Mango Waste Material

The following constraints in the utilization of mango waste material are enumerated:

1. Collection of waste material at one place, particularly in the case of table consumption in the house holds.
2. Presence of large number of small scale processing factories/units scattered allover the region/country giving too small quantities of waste for commercial utilization.
3. Highly seasonal nature of processing and variation in the quantum of waste available for cost-effective utilization.
4. Perishability of waste material.
5. Health hazards involved in the collection of waste material.
6. Unawareness of the ways of utilization of wastes by the public and factory owners.
7. Non-availability of technologies for utilization of waste material.
8. Financial constraints and difficulities in creating necessary machinery.
9. Sentiments and apprehensions from the people coming in the way of utilization of products from the waste materials.
10. The processing industry has neither the resources nor technology nor manpower to convert the waste into organic manure or biogas or non-edible oil or cattle feed etc.

Impact of Utilization of Waste Materials: The vermicompost and organic manures from the waste material will be a viable alternation to chemical fertilizers in saving the soil and its fertility and valuable foreign exchange can be saved by reducing the import of chemical fertilizers and pectin. Soil and water pollution can be minimized/avoided. Health of inhabitants near the processing factories can be saved. Additional income from the sale of mango stones and vermicompost to the processing factories and more employment are certain additional benefits of utilization of waste material. By establishing bio-gas plants using the mango peel waste and pulper waste, electricity consumption can be reduced with attendant benefits like saving the power and reducing the electricity bill amounts. By including the waste materials and their derivatives in animal ration/feeds valuable maize can be saved or diverted for other purposes. Utilization of waste will certainly help stabilising prices as it would provide some marginal revenues out of bye product recovery. Waste utilization would certainly reduce the cost of production and increase competetiveness of products. Utilization neutralizes the high cost of production of processed mango products. Efficient disposal and recycling of wastes bring down the cost of production of processed products besides minimizing the pollution hazards.

REFERENCES

Agrawal, P.C. and Pruthi, J.S.1972. Effect of different methods of dehydration of mandarin orange waste (peel and pomace) on the quality and recovery of pectin therefrom. Indian Food Packer, 26(2): 9-16.

Aina, A.B.J. 1990. Replacing maize, cassava peels in finisher rations for cockerel: effects on cut-up pieces of eviscerated carcass.Nigerian Journal of Animal Production, 17:17-22.

Alfani, F. and Cantaralla, M. 1987. Lignin and Cellulose Biodegradation. In: J.M. Sidwick and R.S. Holdom (Eds.). Biotechnology of waste treatment and Exploitation. Ellis Harwood, Ltd., England, pp.256-274.

Anderson, J.E., and Smith, J.E, 1987. Composition. In: J.M. Sidwick and R.S. Holdom (Eds.). Biotechnology of waste Treatment and Exploitation. Ellis Harwood Ltd., England. pp. 301-324.

Anonymous, 1962. The Wealth of India – Raw Materials. Vol-VI (L-M) Council of Scientific and Industrial Research. New Delhi. P.248 and 282.

Anonymous, 1976. *Syzygium curmini* L, In: Y.R., Chopra (Ed.). The Wealth of India – Raw Materials Vol-X, SP-W, Publication and Information Directorate, C.S.I.R. New Delhi, P-100-104.

Anonymous, 2003. Mangoes: Packaging and Processing. Agriculture and Industry Survey, 13(9):23-24.

Augustin, M.A. and Ling, E.T.1987. Composition of mango seed kernel. Pertanika, 10(1):53-59.

Awasthi, O.P. and Singh, R.K. 1998. Mango Kernel: Potential Source of Vegetable Oil, National Symposium on Mango Production and Export, held at Lucknow, June 25-27 (Abstract), P-87.

Baliga, B.P. and Shitole, A.D. 1981. Cocoa butter substitutes from mango fat. Journal of American Oil Chemistry Society, 58:110.

Beerh. O.P., Raghuramaiah, B. and Murthy, G.V.K. 1976a. Utilization of mango waste: peels as a source of Pectin. Journal of Food Science and Technology, 13:96-97.

Beerh. O.P., Raghuramaiah, B. and Murthy, G.V.K. 1976b. Utilization of mango waste: Recovery of juice from waste pulp and peel. Journal of Food Science and Technology, 13:138-141.

Bhatia, B.S., Krishnamurthy, G.V. and Girdhari, L. 1959. Preparation of pectin from raw papaya (*Carica papaya*) by Aluminium Chloride Precipitation. Food Science, 8:314.

Bhatnagar, H.C. and Subramanyam, H. 1973. Some aspects of preservation, processing and export of mango and its products. Indian Food Packer, 27(4): 33-52.

Bose, S., Thukrai, B.M. and Narayana, S.1952. Utilization of mango seed kernel and jamun seed meal in a simplified poultry ration for growing chicken. India Journal of Veterinary Science, 22(4): 247.

Chaliha, B.D., Barua, A.D. Gohain, A. and Siddappa, G.S. 1963a. Assam lemon as a source of pectin. Part-I. Effect of method of extraction, drying and storage of peels and pomace on the recovery and quality of pectin. Indian Food Parker, 17:4.

Chaliha, B.D., Barua, A.D., Mahanta , D. and Siddappa, G.S. 1963b. Food science, 12:243.

Chand, G. and Shukla, P.C. 1973. Use of unconventional feeds in broiler ration. Indian Journal of Animal Science, 43:1013-1017.

Chand, S. and Saroj, K.K. 1970. Chemical studies on mango kernel and jack fruit-Part-I. Indian Food Parker, 24(2): 40-42.

Cooke, F.C. 1949. Pineapple industry of the Hawaiin Island. Department of Agriculture, Federation of Malaya.

Das, D.C., Sahu, B.L., Panda N.C. Dehuri, P.K. and Mahapatra, H.C. 1988. Mango seed-kernel as a feed ingredient in chick ration. Indian Journal of Animal Production and Management, 4(2):92-95.

Das, N.B. and Benerjee, R.M. 1952. Utilization of mango seed kernel as a source of starch. Science and Culture, 17(8):339.

Devi, S.S. and Nand, K. 1989. Microbiological pre-treatment of mango peel for biogas production, Journal of Microbial Biotechnology, 4(2): 110-115.

Dhingra, D.R., Kapoor, S.N. and Chandra, 1948. Proceedings of Annual convention of oil Technology Association, India, 3:1-9.

Dhingra, S. and Kapoor, A.C. 1985a. Acceptability of mango seed kernel flour in conventional food items. Indian Journal of Agriculture Science, 55(8): 550-552.

Dhingra, S, and Kapoor, A.C. 1985b. Nutritive value of mango seed kernel. Journal of Science of Food and Agriculture, 36(8):752-756.

Ethiraj, S. and Suresh, E.R. 1992. Studies on the utilization of mango processing wastes for production of Vinegar. Journal of Food Science and Technology, 29:48-50.

Garg, N. 2004. Value Addition of Mango Processing Waste. Processed Food Industry, 8-11.

Garg, N. 2006. Production of Backer's yeast from mango stones. Indian Food Parker, 60(1): 61-62

Garg, N. and Tandon, D.K. 1998. Cellulose production from mango peel using *Colletotrichum gloeosporoides* National Symposium on Mango Production and Export, held at Lucknow, June-25-27. (Abstract).P-88.

Garg, N., Tandon, D.K. and Kalra, S.K. 1995. Production of mango vinegar by immobilized cells of *Acetobactor acati.* Journal of Food Science and Technology, 32(3): 216-218.

Garg, N., Tandon, D.K. and Kalra, S.K. 1998. Biodegradation of mango peel fibre. Indian Food Packer, 52(5): 32-34.

Geetha, A. and Devi, C.S.S. 1986. Utilization of mango kernel powder (defatted) by rats, Indian Veterinary Journal, 63(9): 767-770.

Goyal, N., Tauvo, P. and Satya, V. 1984. An integrated process for fuel and feed production from sugar beet. Urja, 10: 272-273.

Gray, W.D.1962. Microbial protein for space age. Journal of Laboratory and Clinical Midicine, 27:955.

Hameed, H.S. 2000. Studies on the utilization of natural corotenoids extracted from Balady Mango wastes. Egyptian Journal of Agriculture Research, 78(3):1265-1278.

Hayes, W.B.1970. Fruit Growing in India. Kitabistan Allahabad, India.

Hemavathy, J., Prabhakar, J.V. and Sen, D.P. 1988. Drying and storage behaviour of mango *(Mangfera indica)* seeds and composition of kernel fat. ASEAN Food Journal, 4(2):59-63.

Hussain, M.G., Haque, M.E., Gafur,M.A., Ali, M.H. and Ali, M.M. 1983. Studies on the kernel fat of mango of Rajashahi region. Bangladesh Journal of Scientific and Industrial Research, 18(1-4):146-149.

Jain, N.L. and Girdhari, L. 1955. Bulletin of Central Food Technology Research Institute, Mysore 4: 287.

Johar, D.S., Krishnamurthy, G.V. and Bhatia, B.S. 1960. Utilization of apple pomace, Food Science, 9: 82.

Kalra, S.K. and Tandon, D.K. 1983. Ripenning behaviour of 'Dashehari' mango in relation to harvest period. Scientia Horticulturae, 19:263-269.

Kalra, S.K. and Tandon, D.K. and Singh, B.P. 1995. Mango. In: DK. Salunkhe and S.S, Kadam(eds) Hand Book of Fruit Science and Technology, Marcel Dekkar Inc. Newyork, Hongkong, 156-158.

Kehar, N.D. and Chand, R. 1946. Mango seed kernel – a new source of fat. Current Science, 15: 46-48.

Krishnamurthy, G.V. 1996. Utilization of Fruit and Vegetable Processing Waste. Paper presented at one day seminar on "Fruit Processing Waste Convertion into profitable business" at Chittoor, July 10th 1996.

Laxminarayana, G., Rao, T.C. and RamalingaSwamy, P.A. 1983. Journal of American Oil Chemists Society, 60:88.

Larrauri, J.A., Barroto, B., Hombre, R.De. and Cruz. H.dola. 1996. Manufacture of Jam from mango peel. Alimentaria, No: 277: 53-56.

Lasztity, R., El-Shagel, M.A., Abdul-Somai, M.B., Hatour, F.S. and Labib, M. 1988. Biochemical studies of some more conventional sources of protein. IV. Proteins of mango waste stone kernels, Nahrung, 32(9);867.

Madhavaswamy, M. and Venkataramana, L.V. 1990. Integrated utilization of fruits. processing waste for biogas and fish Production. Biological Wastes, 32(4): 243-251.

Maini, S.B. 2001. Utilization of the processing waste from tropical and subtropical fruits. Indian Journal of Horticulture, 58(2): 183-190.

Mark, E.M. and Stewart, G.F. 1976. Advances in Food Research. Academic Press, Newyork, 21:293.

Maurya, K.R. and Chaudhury, D.N. 1977. Utilization of Mango waste. Indian Food Packer, 31(3): 83-84.

Mehta, G.L. and Tomar, M.C. 1982. Dehydration and Utilization of raw mangoes. Indian Food Packer, 36(6): 75-79.

Mittal, J.P. and Singh, B. 1972. Mango seed decorticator. Journal of Agricultural Engineering, 9(4):52.

Mittal, J.P. and Singh, B. 1973. Utilization of Mango kernel. Research and Industry, 8(1), P.17.

Musahib-ud-Din and Bhutani, R.C. 1946. Mango products. Punjab Fruit Journal, 10:42.

Narasimhachari, B.L., Azeemoddin,G., Ramayya, D.A. and Tirumala, Rao. S.D. 1988. Extraction and Analysis of mango peel wax. Acta Horticulturae, 231: 749-750.

Narasimhachari, B.L. and Reddy, B.R. 1977. Processing mango stones for fat. Oil Technology Research Institute, Anathapur (AP), India.

Narasimhachari, B.L. Reddy, B.R. and Tirumala Rao, S.D. 1975. Research India, New Delhi, 20(2): 79.

Narasimhachari, B.L., Tirumala Rao, S.D. and Reddy,B.R. 1977. Processing mango stones for fat, Journal of American Oil Chemistry Society, 54:494.

Odunsi, A.A. and Farinu, G.O. 1997. Assessment of Nigerian mango (*Mangifera indica*) seed kernel as a substitute for maize in finishing broiler diets. Indian Journal of Animal Sciences, 67(7): 605-607.

Palaniswamy, K.P., Muthukrishnan, C.R. and Shanumugavelu, KG. 1974. Physico-chemical chacteristics of some varieties of mango. Indian Food Parker, 28(5): 12-19.

PAN, 1995. Annual Report 1995. Poultry Asociation of Nigeria, Lagos, Negeria.

Paramar, S.S. and Sharma, R.S. 1984. Mango seed kernels: A review in chemical composition and edible uses. Indian Food Packer, 38: 40-47.

Paramar, S.S. and Sharma, R.S. 1986. Use of mango seed kernel in enhancing the oxidation stability of ghee. Indian Journal of Dairy Research, 5(2):91.

Pruthi, J.S., Mookherji, K.K. and Girdhari, L. 1960. A study of the factors affecting the recovery and quality of pectin from guava. Indian Food Packer, 14(7):7-13.

Pruthi, J.S. and Susheela, R. 1963. Studies on the utilization of mango waste kernels: some chemical technological aspects. Punjab Horticulture Journal, 3(2-4): 272.

Rashad, M.M., Moharib, S.A. and Jwanny,E.W, 1990. Yeast conversion of mango waste or methanol to single cell protein and other materials. Biological Wastes, 32(4): 277-284.

Ravindran, V. and Sivakanesan, R. 1996. The nutritive value of mango seed kernels for starting chicks. Journal of the Science of Food and Agriculture, 71(2): 245-256.

Roy, N. and Mitra, A.K. 1970. Amylase content of starch from unfamiliar source of food. Journal of Food Science and Technology, 7:164.

Sastry, M.V. and Krishnamurthy, N. 1974. Studies on Indian Pickles-I, Preparation and storage of pickles from mangoes. Indian Food Packer, 28(1): 32-34.

Sethi, R.P. and Sood, S.M. 1977. Orange peel as carbon cource for Cx and fungal protein production. Indian Journal of Ecology, 4;236.

Sethi, R.P., Tewari, H.K. and Sood, S.M, 1977. Note on the mango peel as a carbon source for fungal protein, Carboxy methyl cellulose (Cx) and poly galactro (PG), Production. Indian Journal of Agriculture Research, 11(1): 45-47.

Sethi. V. and Maini, S.B. 2001. Utilization of Fruit and Vegetable wastes for value addition. Indian Food Industry, 4(12): 19-21.

Shekawat, A., Zia-ur-Rehman., Khan, A.D. and Shah, F.H. 1992. Utilization of mango waste in poultry feed. Pakistan Journal of Scientific and Industrial Research, 35(9): 360-361

Sood, S.M., Ghai, S.K. and Sharma, B.N. 1980. Fungal protein from canning industry waste II mango stone kernel as substitute, Indian Journal of Agriculture Research, 14(4):203-208

Sounder, S. and Chandra, J.S. 1987. Cellulose degradation by a mixed bacterial culture. Journal of Indian Microbiology, 2:257-265.

Sri Hari Babu, R. 1996. Utilization of Mango Waste Material. Financeing Agriculture, 28(3): 3-5.

Srinivas, S.R., Pruthi, J.S. and Siddappa, G.S. 1963. Food Science, 12:340.

Srirangarajan, A.N. and Shrikhande, A.J. 1977. Characterization of mango peel pectin. Journal of Food Science, 42(1): 279.

Tandon, D.K. and Garg, N. 1998. Chemical Extraction of value added products from mango waste, National Symposium on Mango Production and Export, held at Lucknow, June, 25-27, Abstract, P. 87.

Tandon , D.K. and Kalra, S.K. 1989. Utilization of mango waste. Beverage Food World, 16(1): 21.

Tandon, D.K. and Kalra, S.K. 1991. Characterization of pectin from mango fruit waste. Indian Food Packer, 45(4): 9-12.

Tewari, H.K., Marwarla, S,S., Gupta, A. and Khanna, P.K. 1991. Quality vinegar production from juice of sugar cane var. COJ.64, Journal of Research, Punjab Agriculture University, 28:77-84.

Vimal, O.P. and Adsule, P.G. 1976. Utilization of Fruit and Vegetable wastes, Research and Industry, 21(1).

Wester, P.J. 1920. Bulletin of Bureau of Agriculture 15.18, cited in 'The Mango' byL.B. Singh(1960), Leonard Hill (Books) Ltd, Inter science Publisher Newyork, P-393.

Wilkins, E.G. 1942. Mango kernels as Food. Indian Farming, 2:63-6.

Yahia, G.M. and Mousletta, A.M. 1982. Utilization of mango seed kernel (*Mangifera indica*) as a source of Oil. Food Chemistry, 8; 269-276.

CHAPTER 14

Marketing of Mango Fruits and its Products

B. Neeraja Prabhakar

The production of agricultural produce or industrial products is incomplete until it reaches the consumer who is the ultimate user. Production efficiency depends not only on the basis of how best one produces but also on how well one sells (markets) his produce/product. Marketing therefore, forms an integral part of the successful production. In a way production is incomplete without marketing, particularly in case of perishables like horticultural crops such as fruits, vegetables and flowers etc (Thorat *et al* 1990). The fruit and vegetable marketing in particular is characterized by high perishability, seasonality and bulkiness of the produce, which make them distinctive from other agricultural produce like cereals, pulses etc. Hence, marketing of fruits and vegetables is more complex and needs special care and immediate disposal after particularly harvesting. Marketing, therefore, is as important as production particularly perishable fruits like mango.

Though considerable efforts are made and are being made to improve the production of mango, very little attention is being paid to its marketing and improvement of its marketing. Though some studies were made on its marketing and valuable suggestions were made, unfortunately increase in production of mango instead of being a boon, is ending up as a curse to the grower (Subramanyam, 1999). Marketing of mango as well as other fruits is beset with several problems. Unless these special problems of these crops, which are highly perishable and seasonal and grown mostly by small and marginal growers, The benefits of the new technology of production and Government development programmes will be of little help to the growers.

Of late, a commercial outlook is being developing among the mango growers in the country, due to increase in its demand and increase in the purchasing power. In addition, the export of fresh fruits as well as processed mango products offer a solid source of foreign exchange to the country. Now-a-days mango is being looked upon as an asset and commercial crop, rather than a subsistence crop; its cultivation is not considered as a hobby, but a commitment on the part of the grower. Therefore, crop continues to be a higher value, prestigious and prosperous crop today and hopefully in future also.

MARKETING CHANNELS

Two extremes of the marketing process in general include a 'producer' and a 'consumer'. Producer supplies the produce/product in the market and consumer creates a demand by purchasing it. Between these two extremes, the produce/product passes through many hands from the 'producer' to the 'consumer' end. The number of hands (intermediaries/ middle men) *i.e.* channels vary from 2 to 7. Raju and Voneppan (1987) described the 'Marketing Channels' as the chain of intermediaries through which the commodities pass between the producers and the consumers.

Srivastava (1967) in the 'Mango: A Hand Book" by Indian Council of Agriculture Research, New Delhi has depicted in Chart 14.1 the different marketing channels of mango, being followed in India.

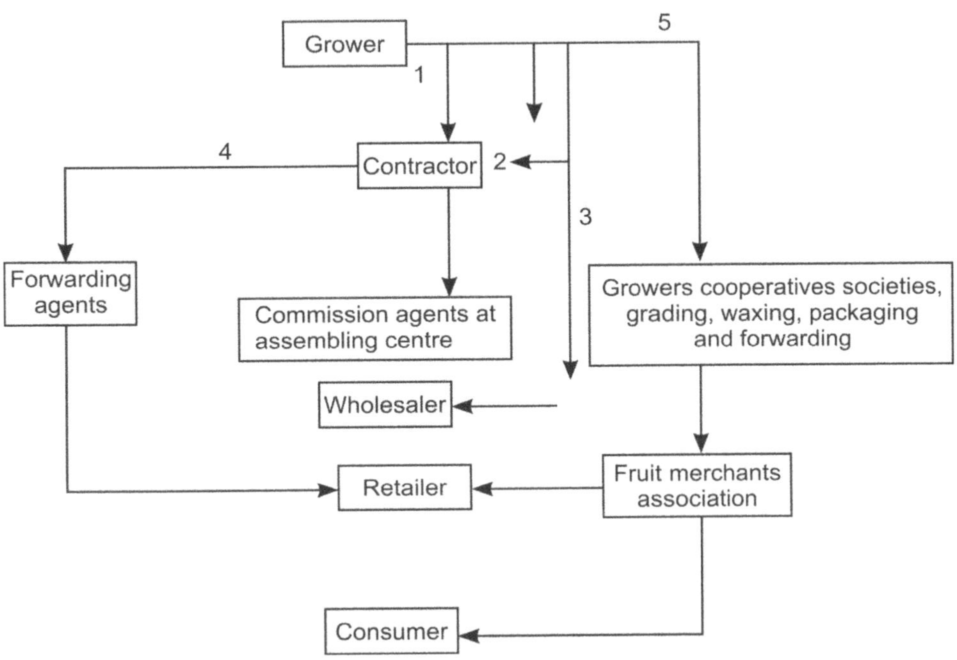

Chart 14.1: Marketing channels for mango in India

Note: The marketing channels NOs 1 and 2 are generally followed in India. Whereas NOs 3 and 4 are in vogue only in some places, channel 5 is an improved method of marketing. Which eliminates a number of middle agencies. Thereby it gives better return to the grower and the fruits would be available at a cheaper price to the consumer.

However, the marketing channels followed for the disposal of mangoes differ from state to state as well as variety to variety. According to Wadkar *et al* (1994) and Patil (1995), there are four channels for disposal of "Alphonso" mangoes in Ratnagiri and Sindurgh districts of Maharashtra (India) as follows:

Channel I: Producer → Consumer.

Channel II: Producer → Wholesaler/Commission Agent → Retailer → Consumer.

Channel III: Producer → Pre- harvest contractor → Wholesaler/Commission Agent → Retailer → Consumer.

Channel IV: Producer → Cooperatives → Consumer.

The most important market channel for mango in Karnataka State (India) is (Rame Gouda *et al* 1993) :

Producer → Pre-harvest contractor → Wholesaler → Retailer → Consumer.

Sailaja (1994) noted the following two channels for 'Banganapalli' mangoes at Hyderabad market (Andhra Pradesh):

Channel I: Farmer → Commission Agent → Retailer → Consumer.

Channel II: Farmer → Pre-harvest contractor → Commission Agent → Retailer → Consumer.

She found that channel I was most effective from the point of price spread, marketing efficiency, index and producer's share in consumer's rupee.

The market channels for 'Kesar' mangoes in vogue in Gujarat are (Singh, 1996).

Channel I: Grower → Pre-harvest Contractor → Commission Agent → Retailer → Consumer.

Channel II: Grower → Commission Agent → Retailer → Consumer.

Channel III: Grower → Commission Agent → Wholesaler → Retailer → Consumer.

Channel IV: Grower → Cooperative Society → Wholesaler → Consumer.

However, of late the following market channel is being followed for 'Kesar' mangoes in Gujarat State (Deodhar and Pandit, 2002).

Grower → Mediator → Trader/Agent → Wholesale Agent → Retailer → Hawker → Consumer.

Following two channels were reported to be in vogue for distribution of mangoes in Rampur district of Uttar Pradesh State (India) (Singhal and Agarwal, 1997):

Channel I: Producer → Pre-harvest Contractor → Commission Agent /Wholesaler cum Commission Agent → Retailer → Consumer.

Channel II: Producer → Wholesaler/Commission Agent → Retailer → Consumer.

In case of sales to processing plants, following three different channels are followed:

Channel I: Grower → Processing plant.

Channel II: Grower → Mediator → Trader /Agent → Processing plant.

Channel III: Grower → Mediator → Processing plant.

Extent of use of Different Channels

The different studies/surveys conducted indicated variation in the extent of use of different market channels for sale of mangoes in India. Patil *et al* (1969) observed that out of the total produce, pre-harvest contractors handled 40 per cent and the rest is sold directly by the growers. Mango growers of Ratnagiri district of Maharashtra State, sold their mango produce by four methods with variation in quantity as presented in Table 14.1 (Temphare and Borude, 1976).

Table 14.1: Extent of use of different channels for mango marketing

S.No.	Method	Extent (%)	Quantity of product (%)
1.	To pre-harvest contractors on long term basis	37.70	12.66
2.	To pre-harvest contractors on annual basis	23.20	25.88
3.	To local merchants after harvest	1.80	0.14
4.	To whole sale merchants after harvest	39.30	61.32

About 20 per cent of 'Alphonso' mango growers sold their orchards to pre-harvest contractors in Maharashtra (Wadkar *et al* 1994). Around 56 per cent of mango growers have sold their produce to pre-harvest contractors and the rest (44%) sold directly in Madhurai district of Tamilnadu (Subramanyam, 1986). In Karnataka state 80 per cent of the total harvest of mango is marketed by pre-harvest contractors. Whereas major portion of mango in Lucknow area of Uttar Pradesh (India) is disposed as follows (Raj Kishore *et al* 1999).

1. To pre-harvest contractors -------------------------- 68.32%
2. To local merchants --------------------------------- 15.90%
3. To distant markets (like Delhi, Kanpur etc.) ------13.48%

In Kolar district of Karnataka state, 78 per cent of mango growers sold their crop to pre-harvest contractors (Rame Gowda *et al* 1993) while 87 per cent of mango orchardists of Rampur district of Uttar Pradesh disposed their gardens through pre-harvest contractors (Singhal and Agarwal, 1997).

In Andhra Pradesh about 61 per cent of mango growers leased their gardens to pre-harvest contractors, 23 per cent sold directly through commission agents and 3 per cent through societies and balance to retailers (Raju, 2002).

The above presentation on extent of use of different agencies/channels indicates the pre-dominance of pre-harvest contractors in marketing of mango in India, as a large percentage of growers depended on the pre-harvest contractors for sale of their gardens.

Choice of the Market Channel

The choice of the marketing channel for disposal of mango depends on certain considerations by the grower. Among the different marketing channels, pre-harvest contractor channel is most preferred, because it shifts the burden of production and marketing risk from the farmer to the pre-harvest contractor and provides the farmer advance payment and time to spare for other business and/or cultivation of other crops (Malleswari, 1998). The sale to pre-harvest contractor reduces the risk of maintenance of garden during fruiting, loss of crop from biotic (pests and diseases, fruit drops) and abiotic (environmental hazards-untimely rains, gales, hail storms, pilferage etc.,) besides providing advance payments and payments without risk. Hence, this channel is mostly preferred by the farmer.

In the case of the use of next channel, *i.e.* sale at the marketing place through commission agents/wholesalers, the farmer faced the problems like cheating, exploitation/mal-practices even in regulated markets. Hence it is not considered as better choice since the producer gets minimum share in consumer's rupee.

In case of mango, regulated market is most preferred channel after pre-harvest contractors. Large sized farmers preferred regulated markets, while small and marginal farmers opted for corporative societies followed by pre-harvest contractors in the marketing of 'Kesar' mangoes in Gujarat State. The main reason for the selection of cooperative societies as marketing channel, is the quality based, timely and higher prices and payments.

The sale of the produce directly to the processor will reduce the number of intermediaries involved in the above channels, provides higher returns and brings down the percentage of post-harvest losses, highest percentage of consumer's rupee. This channel is preferred for processing (e.g. Bangalora, Alphonso, Dasheheri) and also by big farmers who produce large quantities of the above varieties (Malleswari, 1998). However, this channel is not much preferred, since processors generally do not buy all the varieties grown by farmers (Rame Gowda *et al* 1993).

Leasing Orchards to Pre-harvest Contractor

Leasing out the fruit orchards to pre-harvest contractors is an age-old practice, which is still continued in spite its demerits and disadvantages to the growers due to various following reasons:

Reasons for selling fruit gardens to pre-harvest contractors:

The fruit growers leased out their gardens to pre-harvest contractors for one reason or the other. The reasons varied in different situations and states.

1. **Financial reasons:**

   (i) Most of the fruit growers are small and marginal farmers who lack finances for various operations of productions and marketing. The most important reasons in case of above small and marginal farmers are: (i) lack of credit facilities from government institutions (Balakrishnan, 1979., Subramanyam, 1986., Rajkishore *et al* 1999) to meet their monitory needs for various field operations etc.,

(ii) Willingness of contractors to advance money even before the crop is ready. The advanced money is generally used either for purchasing inputs (fertilizers, pesticides etc.,) required for the mango crop or other crops or for meeting certain necessary domestic requirements,

(iii) Indebtedness of farmers that forces the sale of gardens, sadly at unfavourable prices and terms (Balakrishnan, 1979).

2. **Reasons of Risk:** Besides the above financial reasons,

   (i) the risk of loss of crop due to biotic and abiotic reasons seems to be a major reason for leasing out the orchards (Subramanyam, 1986). Nearly 85 per cent of the mango growers leased out their gardens to avoid the risk of crop failures due to uncertainty of weather, pests and diseases.

   (ii) Production risks like thefts also compelled the growers to sell their gardens to contractors (Subramanyam, 1998).

3. **Non-availability of labour:** In more than 90 per cent cases, shortage of labour for watch and ward, for harvesting the crop etc., is the main reason that forces the growers to lease out their gardens to pre-harvest contractors. Over lapping of mango and sapota harvesting is the main reason in Gujarat (Singh 1996., Rajkishore *et al* 1999)

4. **Market Problems:**

   (i) Complicacy in marketing (Rajkishore *et al* 1999),

   (ii) cheating/exploitation by intermediaries like commission agents/ wholesalers/traders (Subramanyam, 1998),

   (iii) malpractices at the market (Subramanyam,1998),

   (iv) price fluctuations (Subramanyam, 1998), were observed to be some of the reasons for leasing out the gardens to contractors.

5. **Size of the garden and quantum of produce:** Small and marginal farmers producing small quantity of crop, prefer to sell their gardens to contractors due to reasons presented in the Table 14.2.

Table 14.2: Farmer's reasons for selling gardens to pre-harvest contractors (Subramanyam, 1986)

S.No	Reasons	Pre-harvest contractor group size (%)		
		Up to 1.00 ha	1.00 to 2.00 ha	Above 2.00 ha
1.	Need for money	45.45	37.50	-
2.	Risk	27.28	12.50	25.00
3.	Small Quantity	9.09	12.50	-
4.	Difficult to market	9.09	12.50	-
5.	No time	9.09	25.00	25.00
6.	Not interested	-	-	50.00

6. Other reasons:

(i) Farmer is engaged in other business or in cultivation of other crops,

(ii) Farmers having large holdings due to lack of time and interest in marketing (Subramanyam, 1986),

(iii) Farmers, particularly big farmers and absentee-land lords having their gardens for off to markets and have no time to bestow personal attention (Thorat *et al* 1990).

Leasing Process

The orchards are leased/sold to pre-harvest contractors at different stages of flowering and fruit development, right from pre-bloom stage up to maturity stage, but mostly at flowering or at fruit set stage the gardens are leased out. In Kerala (India) the mango orchards are leased after harvest of the crop, during June-July itself (Radha and Nair, 2000). In other states, the farmers prefer to dispose off their orchards either at the time of fruit set or earlier to avoid loss of crop due to biotic and abiotic reasons and prices and other reasons mentioned above. When the contract was made at the time of flowering stage, contractors pay less and pay more when the fruits have developed and there is less chance of risk of loss. The contract amount is either payed in lump sum at the time of contract or in instalments. Much depends on the mutual trust of the farmer and contractor.

Fruit orchards including mango are generally leased out either on annual basis or long term basis. Contracts are fixed by negotiations on the basis of duration of contract, bearing intensity of the orchard, history, stage of fruit development, condition, location of the orchard, variety and its demand in the market. In annual contract the orchards may be leased out at advanced stage of fruit development to fetch higher contract amount (Temphare and Boruda, 1976). Many growers prefer to lease out their gardens on long term basis (2 or more years) for obvious reasons. Mango in general being an alternate bearer, the grower benefits from long-term lease, which helps him cover the uncertainty of returns in alternate years. The contractor similarly prefers long-term lease (at least 2 years) for the same above reason, but offers low price because of alternate bearing habit of the crop. Long-term leasing is done due to pressing need for the money by small and marginal farmers and by absentee-land owners with large holdings, as their gardens are generally far away from markets and or engaged otherwise. In long-term lease, the production and protection of the crop and its maintenance are undertaken by the contractor. While in case of annual lease, the grower undertakes the above. The long term contracts are generally in writing, while the short-term contracts are generally oral agreements. Annual contracting allows the fixation of price according to the prevailing market prices, unlike in long-term contracts, in which exploitation of farmer is possible (Temphare and Boruda, 1976).

Comparison between self-marketing and Contracting

The leasing out of fruit gardens (mango, citrus, guava etc.,) to contractors for a fixed period and amount without bothering for the yield or the price that may be expected is called "contracting." The process of selling the produce by taking it to the market or to the processers is referred to "self marketing". Both these types of marketing have their own advantages and disadvantages in respect of costs, returns, share etc.

Costs associated with Self-marketing and Contracting

Self-marketing demands some extra costs, which are absent when gardens are sold on contract. Besides the marketing costs like transport, loading and unloading, commission charges etc., the self- marketing growers have to bear additional expenses towards items like watch and ward, harvesting, manuring, irrigation, plant protection measures etc., required for the mango crop, which are generally borne by the leasee in the contracting system. The total expenses incurred by the self-marketing growers for maintenance of garden and marketing are more than those in contracting system. This may be one of the reasons for majority of small and marginal farmers/growers preferring contracting system.

Extra returns through self-marketing

The gross returns through self-marketing were more than two and half times than the returns realized through contracting. The net returns realized after accounting for the extra costs (as mentioned above) involved in self–marketing were also considerably more (>50%) than the returns realized from contracting, which shows the desirability of taking up of self-marketing by the growers (Thorat *et al* 1990). Further, self-marketing gave on average a benefit: cost ratio of 2.37 higher than that of contracting (Subramanyam, 1986). Therefore, from the growers point of view, it was opined that it was not profitable to lease out the orchards to contractors as they made huge profits at the cost of growers and all mango growers should take up self-marketing of their produce to reap the benefit of better prices and resultant higher returns.

Share: Elimination of pre-harvest contractors, will enable the grower to enhance his share from 36 per cent of the sale proceeds to 70-80 per cent. Contracting of the gardens is viewed with disfavour by many as it gives the lowest share in the consumers rupee to the grower (Malleswari. 1998).

Because of the higher returns and higher share, self-marketing should be encouraged as follows (Subramanyam, 1986).

(i) Awareness among the mango growers for self-marketing may be created by the extension mechanism.

(ii) Providing market credit facilities to meet the extra costs both at orchard and market levels.

(iii) Making arrangements for advancing loans through institutional agencies against hypothecation to encourage the growers, especially small and marginal growers, to adopt self-marketing.

(iv) Provision of crop insurances to the mango crop in case alternate bearers (Alphonso, Dashehari, Langra etc.,) and in states like Andhra Pradesh, Tamilnadu, Orissa, West Bengal, which are frequently affected by cyclones to protect the grower from the vagaries of yield uncertainty and natural hazards respectively. The investment needed and the associated risks involved in the independent marketing (self-marketing) are so high, which place the small and marginal growers of the fruit crops, at a particular disadvantage in the absence of crop insurance and marketing credit facilities (Malleswari, 1998).

However, self-marketing itself is not free from defects, such as unauthorized deductions, charging higher commission rates than the official rates and other mal-practices at the market yard.

INTERMEDIARIES/MARKET FUNCTIONARIES

A chain of intermediaries (middlemen) and market functionaries in the marketing system exists for every commodity. They provide some service or perform some functions and help the movement of produce/product from the producer to the consumer. The traditional intermediaries in fruit marketing system are: pre-harvest contractors, commission agents, wholesalers, traders, retailer/hawker (venders). In some instances growers also act as market functionary.

Growers

Some producers take active part in assembling the produce and selling them through commission agents. Some growers may transport, assemble and pack the produce (fruits) themselves and sell their produce through commission agents either in the local markets or in the distant markets. However, majority of the growers sell their orchards to pre-harvest contractors, who are responsible for assembling and every other process involved in the marketing of mangoes.

Pre-harvest Contractors

In India, pre-harvest contractors play a pre-dominant role in marketing of fruits. They serve a useful purpose as the growers are not in a position to handle orchards themselves, not willing to take the risk of marketing and expenses of hiring reliable watchmen to protect the crop when large. Contractors purchase the gardens on lease from the growers after estimating the crop either at pre-flowering, flowering or fruit set stage.

The yields of the orchards are estimated taking into account:

(i) the number of bearing trees,

(ii) their age,

(iii) varieties grown and

(iv) the previous year's crop.

The allowances for losses that may occur during the season an account of natural calamities, such as pests and diseases, gales, hail storms, untimely rains *etc.,* are given importance while arriving at a price for the garden. Having formed an idea of the yield of a particular orchard, an estimate of the value of the produce is made, taking into account the prices obtained by them in the previous season and the condition of the crop in other mango producing areas about which the contractors receive information from their commission agents. The time of the contract also influences the contract prices to a great extent, when the contract is made at the time of flowering or when the fruit is very small, the price offered is comparatively low since the fruit takes 4-5 months to mature and during this period it may be attacked by pests and diseases or other natural calamities. In case where contracts are made after the fruit has attained full size, higher prices are offered, as at this stage the crop is less susceptible to pests etc. After the bargain has been struck between the grower and the pre-harvest contractors, the contractor pays a nominal amount of bargain as advance to the grower and thereafter takes up all activities needed for the maintenance of orchard till marketing of produce at his own cost. Pre-harvest contractor sells the produce to wholesaler through commission agents. Sometimes he disposes the produce directly to retailers or to the processors (Rana, 1984). Usually these contractors are financed by commission agents/wholesalers. Thus the pre-harvest contractors are obliged to sell the produce to them. Though the pre-harvest contractor is an important link in the marketing system he can be eliminated with advantage if the grower desires.

Commission Agents

Commission agents are another most important functionaries in the marketing of fruits and two-thirds of the total market is controlled by these agents. They are useful when the fruit is dispatched to terminal markets, that are far away from the producing areas. These agents generally finance the pre-harvest contractors. They perform the function of dispatching fruits from the terminal markets to the ultimate retailer outlets. They bring together buyer and seller and settle the price through negotiations, amicable to both parties. For this function/service they charge commission. Commission agent looks after only distribution function but do not engage himself in assembling the produce. Occasionally, commission agent acts as a wholesaler in other transaction, using an additional commission agent or selling directly to retailers (Rana, 1984). Commission agents also include forwarding agents, who own the responsibilities of packaging and transport. Commission agents are located in both the assembling (producing) and consumption centres. At some places they not only help to sell fruits on commission basis but also transact wholesale business on their own account (Abid Hussain, 2009).

Wholesalers

Wholesale merchants operate in the terminal markets. They purchase the produce in lots from the grower/pre-harvest contractor through commission agents. They are generally not in personal touch with the growers or the agent in the far dispatching points. The grower or the merchant who consigns produce to the terminal markets, depends solely on the word of the receiver, regarding the market price or the condition of the produce on arrival at the destination. Wholesalers operate through their own establishment through the agents in the assembling as well as terminal markets and sell the produce to retailers. Wholesalers sometimes finance the pre-harvest contractors and also retailers.

Retailers (Photos 14.1 to 14.3)

Retailers include growers, contractors, commission agents, wholesalers, stall-holders, shop keepers and hawkers/vendors/cart *wallahs*. The growers generally do not sell mangoes in retail, except in some cases, where the orchards are adjacent to the main roads. The pre-harvest contractors also generally do not enter into retail trade, except on road side near orchards or by hawking as in Uttar Pradesh, Maharashtra, Orissa etc. The commission agents and wholesalers also do not take up retailer role, unless they own stalls or shops in the same city. The bulk of the fruits, including mango is sold in retail by stall-holders, shop-keepers and vendors on carts. They sort out their produce into 3 or 4 classes according to the size and quality and display them in neatly arranged heaps. The street vendors also act as important retailers who purchase on a day-to-day basis from commission agents, wholesalers, or stall-holders and carry the fruits as head loads or in hand-push carts, either from house to house or settle at some busy centres and do their business. The above different types of retailers obtain fruits from wholesalers/commission agents either on loan or financed by them (Srivastava, 1967).

Others

Other market functionaries include, fruit growers association, cooperative sale societies and regulated markets.

Plate 14.1: Mango fruit stall

Plate 14.2: Retailing of mango fruits on a push cart

Plate 14.3: Retailing of mango fruits in baskets

Fruit Growers Association

These associations function to facilitate proper assembling, packaging and applying various treatments and also arrange transport. They help the reduction of the cost of marketing of fruits and yield better returns to the grower, while reducing the price for consumers. For this service, the association charges a fixed amount to the members of the association, to meet the transport etc. The rate charged varies year to year depending on the railway freight, labour wages etc. The savings, after apportioning some amount for its reserve fund are distributed among the members, according to the number of packages/quantity of produce sent by each (Srivastava, 1967).

Agriculture Produce Market Committees

Factually the AP market committee can also be viewed as an additional middle man in the regular agricultural produce markets, since they supervise the transaction and charge market fees.

Method of Disposal/sale of fruits in the market

In India, mangoes are sold by auction in the market and there are three ways of conduction of auction *viz.,* open auction, under-cover auction and private negotiations (Srivastava, 1967).

Open auction

The fruit baskets/ heaps are exposed for sale in the open yards and the buyers bid for the quantity and the highest bidder gets the produce.

Under-cover auction

In this system, the prices are settled by secret signs under the cover of a piece of cloth. The commission agent covers his own hand and that of the buyer with a piece of cloth and the latter indicates the price, he is prepared to pay, by pressing the fingers of the commission agent. In this way the buyer conveys the price secretly to the agent who sells the produce to the highest bidder. This system has the advantage that it does not prejudice or influence the other buyers, each of who assesses the value of the produce independently and participates in the auction. This system of auction for disposal of mango and other fruits is practiced in Maharashtra.

Private negotiations

Under this system, the intending buyer approaches the commission agent to settle the price with him by private negotiation. The buyer before approaching the commission agent, acquaints himself regarding the price trend of the market. This method is followed when the quantity of produce is not very large.

Unit of sale

Mangoes are either sold on weight basis or on number basis (dozen or hundred). Mangoes are also sold by other than weight and number basis like in baskets containing variable number of fruits (Anon, 1985) or vehicle loads (Subramanyam, 1988). The main advantages of this type of sales are that neither the buyer nor the seller will know the exact quantity and the exact price and there is no way to compare the price when they are sold on the basis of number as the prices generally depend on size/weight of the fruits. So there is a need to make it compulsory to sell the produce on the basis of weight only (Subramanyam, 1988). From the point of fairness and consumer, selling fruits in baskets should be discouraged because smaller, poor quality and culled fruits are placed at the bottom of the basket up to a height over which larger, eye appealing and quality fruits are placed (topping) and this amounts to cheating and loss to the consumer.

PRICING

Price is an incentive to the producer and is considered as an elixir. High prices encourage the producer to increase the production as well as supply more quantum to the market. On the contrary low prices disappoint him to increase the production and less supplies to the market.

Fixation of price of the produce is neither in the hands of grower nor in the hands of pre-harvest contractor nor in the hands of retailer and consumer. In reality, the price is fixed by either commission agents and/or wholesalers depending upon several factors. They play a predominant role in price fixation both for the grower and retailer. Retailer fixes the price after the produce comes to his hand.

There is a wide disparity in the prices of standing crop from place to place, variety to variety and even year to year and from one orchard to another in the same area. This is mainly due to alternate bearing habit of mango. Other reasons include bearing capacity of the gardens, depending upon the orchard management and exposure to biotic and abiotic factors. Seasonality and perishability of the produce have greater bearing on the determination of prices at the market yard. There are not only wide variations in mango prices during the season, but also they varied during the same week/same day depending the arrivals and demand. Ordinarily the prices would be highest at the start of the season due to arrival of less quantity, declining gradually as the arrivals increase. Later when the arrivals decrease at the end of the season, the prices tend to recover and reach a peak in the last days of season, irrespective of the variety and it's quality. At the end of the season even medium to poor quality varieties (Neelum, Totapuri, Rumani etc.,) fetch very high prices. Late arrivals of 'Kesar' mango as compared to varieties like 'Totapari'. 'Alphonso', 'Rajapuri', 'Pairi' etc., also influence the prices of 'Kesar' mango (Deodhar and Pandit, 2002).

Price is also influenced by the variety irrespective of time of arrival. 'Alphonso' mangoes always fetch higher prices than any other variety. Similarly, 'Banganapalli' variety commands higher price than other varieties like 'Suvarnarekha', 'Totapuri', 'Neelum', 'Rumani' etc., in Andhra Pradesh. However, 'Himayat' (syn: Himayuddin) being of higher quality fruit, and supplies being less (as it is a shy-bearer) gets higher price than 'Banganapalli'.

Mango retail price varies great deal from year to year depending upon each year's total production and various other factors like prevailing wholesale prices, supply-demand, transport and marketing facilities. Wholesale prices of mangoes also vary considerably depending upon supply- demand of particular varieties, periods of availability, health condition, transport facilities, time of arrivals (early to late), quality *etc.* Daily arrivals also have a direct bearing on the prices. The price is also influenced by the quality and size of the fruit and its grade at the market.

Producer's share in the Consumer's Rupee/Price

Similar to price, producer's share in the consumer's rupee/price is an incentive to the grower. The producer's share is influenced by the type and size of channels, kind and number of intermediaries, location of the market, variety etc.

Producer's share in consumer's rupee varies from 25 to 60 per cent in various marketing channels. It is least (25% or less) in marketing channels where more number of intermediaries are involved *i.e.* more the number of intermediaries less will be the producer's share (Patil, 1995). The producer's share was the lowest (12.61%) in Group I growers (leasing for longer periods) than in Group II growers (leasing on annual basis)

(23.49%) (Temphare and Boruda, 1976). Wadkar *et al* (1944) did not consider selling of mangoes through wholesalers/commission agents (channel II) as the best practice since producer gets only 40 per cent share. The producer's share was highest in channel I (producer → consumer) and lowest in channel II (producer → pre-harvest contractor commission agent/wholesaler → retailer → consumer). The share of the producer/pre-harvest contractor was 37.44 per cent and 39.61 percent in channel I and in channel II respectively at the market of Rampur district in the Uttar Pradesh (Singhal and Agarwal, 1997).

In direct marketing, producers received maximum share of 30.63 per cent in Maharashtra State (Temphare and Boruda, 1976). The share of the producer was highest in direct sale to consumer (83.74%) followed by cooporative sale (73.01%) and sale through pre-harvest contractors (21.86%). The producer's share in respect of direct sale was in highest due to total elimination of marketing functionaries and their margins and cost of marketing. Next to this, the sale through co-operatives was more profitable, because cooperatives directly dealt with the consumer, who were benefitted from the purchase through cooperatives as the prices were lower (Patil, 1995). The net share received by the grower from sales through cooperative channel was 52.3 percent of the price paid by the consumer. On the other hand, the net share/price received by the producer is quite small in the non-cooperative channel (Singh, 1996). In view of the above promotion of direct and cooperative sales should be encouraged in order to gain better share by the grower in the final price of fruit crops (Raj Gopal, 1987).

The price paid by the consumer is shared among the different agencies involved in marketing. Consequently the share of the price paid by the consumer gets reduced substantially when it reaches the producer (Parida and Naik, 1988., Singh and Singh, 1996). In a country wide survey, it was observed that in case of 'Alphonso' mango in Ratnagiri District of Maharashtra State producer's share was 42.3 per cent, that of wholesalers was 2.8 per cent and that of retailers was 25.6 percent (Anon, 1962). In a later study Patil *et al* (1983), while analyzing the mango marketing, observed that the pre-harvest contractors and retailers got more share of the consumer's rupee as compared to the producer in Ratnagiri district (Rame Gowda *et al* 1993). The retailer sells mangoes at 150 per cent higher price than the wholesaler's price. It means, out of consumer's price, growers get only 40 per cent and the rest (60%) goes to the intermediaries.

Location (distance from the production centre) of markets also influences the producer's share. It is likely to vary grately with the distance of the orchards from the markets due to transport costs *etc.* In mango marketing in case of producer → consumer channel, the producer's share came to 21.4 percent and 26.75 percent at Delhi and Lucknow markets respectively. In case of producer- contractor, the producer's share was 33.18 per cent and 47.10 percent at Delhi and Lucknow respectively (Garg and Singh, 1971).

The low share received by the producers and high wastage of mango fruits were observed to be due to the following factors:

(i) the producers are unorganized and scattered, owned small holdings which produce low marketable surplus, lack of financial support and poor or no access to market information.

(ii) The investment in market structure is still far shortage of providing economical service to farmers for packaging, grading, refrigerated storage and transport,

(iii) inadequate facility of processing, makes the demand for this product further elastic (Singh and Singh, 1996).

Profit Margins

The receipt of profit margins in marketing of mangoes follows more or less similar pattern as that of producer's share from consumer's rupee. Highest margin was worked at retailing stage followed by wholesaler stage. The highest input : output ratio was in case of wholesaler (1: 146) and lowest in case of producer (1: 1.05). Thus in trading of mango producers are getting lower margin and the intermediary is earning more than what the producer is earning (Swarup *et al* 1985).

Retailers obtained highest margin of profit followed by pre-harvest contractors and wholesalers in mango markets at Samastipur market of Bihar (Singh *et al* 1999). The middlemen's margin accounted for 48.78 and 48.22 per cent at Delhi and Lucknow markets respectively (Garg and Singh, 1971). The profit reaped by wholesalers and retailers was around 81 and 45 per cent on their investment respectively after taking into account the transit losses (Abid Hussain, 2009).

MARKETING EXPENSES

Marketing of any produce/product either agricultural or horticulture or industrial, involves expenses to dispose off the same to a consumer. The different categories of marketing costs (expenses) differ with the commodity, market channel, intermediary type of service, type of produce/product etc.

The marketing costs differ under different market channels. The total marketing expenses in channel I was greater than that in channel-II at Hyderabad (Sailaja, 1994). In self -marketing channel the producer will be incurring expenses on watch and ward of the garden, picking, handling, packing, grading of fruits, transportation, check-post charges and unloading charges at the market yard, before getting into contact with commission agent/ wholesaler. Therefore, market expenditure includes commission to the commission agent and other miscellaneous expenses till the proceeds of sales are received. Producer also incurs unauthorized expenditure like charity, electric, weighment, *hamali* charges, telephone and, cleaning charges at the market yard (Raju, 2002). Among the major marketing costs borne by the producer, the cost accounted for losses in weight, was the highest one followed by transportation, rent of cold storage godowns and spoilage of produce (Raju and Soria, 1989).

Similar to the producer following self-marketing, the pre-harvest contractor incurs expenditure both at orchards and the markets. At orchards, the expenses would be contract amount, wages for watch and ward of the garden, picking, sorting and grading and loading charges. At the market yard the expenses include unloading charges, transportation and handling charges, cost of packaging material and packaging expenses, commission to the commission agents, storage charges, market cess, weighment charges, market license fee, professional tax *etc.* The share of contract money in the total market expenditure was the highest (47.7%) followed by marketing charges (19.4%), packaging (17.92%), transportation (8.77%) watch and ward (4.47%), harvesting charges (1.46%) and grading charges (0.73%) (Thorat *et al*1990).

Commission agents do not incur any market expenditure. (Khunt *et al* 2001). On the other hand, the wholesale traders will be incurring expenses on transportation for carrying the produce from market yard to his premises/distant markets, loading and unloading (Raju, 2002). The major costs incurred by the wholesaler in Gujarat were losses in weight, spoilage and the rent of storage godowns (Raju and Soria, 1989).

Retailers would be incurring transport charges for carrying the produce from wholesale market to his premises, loading and unloading charges, ripening expenses etc. (Raju, 2002).

Results of a survey showed that marketing costs incured by the grower were the highest followed by wholesalers and retailers. The huge marketing expenses borne by the growers reflect on the producer's share in the consumer's rupee.

In general, marketing of mangoes particularly in regulated markets includes the following items of marketing expenses:

License fee, rent, professional tax, commission charges, postage, interest on capital, labor wages for loading and unloading, transportation charges, weighment charges, costs of packaging materials, cost of packaging, fines/penalties, market cess, losses and wastage of produce etc., (Wadkar *et al* 1994).

There is lot of scope of reducing the marketing costs by adopting appropriate handling, packing technology, provision of storage facilities and following suitable regulatory measures (Khunt *et al* 2001).

MARKETING PROBLEMS

High perishability, seasonality and bulkiness of fruits and vegetables make their marketing quite complex, risky and challenging. Their marketing poses many problems in their marketing, which again differ from producer to consumer. Each functionary involved in the marketing expressed different problems in the disposal of the produce. However, before going into the marketing problems of mango let us learn first about the nature and characteristics of mango market in our country, which makes it easy to understand the problems.

Characteristics of Mango Market

Fruits like mango, banana, sapota etc., are highly perishable, seasonal and bulky. Irregular supply of mangoes, temporary market structures (at assembling centres), congestion in market yards, wide spread terminal markets, indispensible role of intermediaries, lack of cheap, quick and efficient transport and vulnerable position of producer- seller are salient features of mango markets in the country (Anon, 1985). Irregular supply of produce to market is due to its alternate bearing habit, seasonality and short season (2-4 months) besides exposure to damaging biotic and abiotic factors.

The marketing system of mango in the country is very unorganized because of the large number of intermediaries, mainly private individuals, lack of integration of growers, and lack of strong and sincere intervention and implementation by Government. Deodhar and Pandit (2002) reported that the present marketing system of mango is not "growers oriented" since the growers get little profit out of marketing of their produce and the market environment is not quite suitable for the bargaining process. Mango markets are monopolised by the intermediaries. Because of inadequate packaging, transportation and processing facilities and presence of many intermediaries, growers receive lower price/profit for their produce. The main feature/short coming of the present fresh produce market system is the absence of effective participation of most producers and retailers in the market process and their indifference attitude (Arora, 2005).

Marketing problems faced by producers, intermediaries, and Consumers

Several studies found that the fruit growers are marketing their produce through pre-harvest contractors and commission agents, wholesalers, which is the evil of all problems of marketing of fruits including mango. The marketing problems or constraints are evaluated to supply of produce, pre-cooling and storage, transportation, grading and packaging, credit facilities, market place, malpractices, labour, price, market intelligence, market information etc. The market problems vary with different functionaries as follows:

Marketing Problems faced by Producer

1. Location of markets, being at a long distance from the production centres, increases transportation costs and spoilage chances.
2. The dominant role of middlemen (commission agents, traders etc.,) *i.e.* monopoly of the intermediaries.
3. Malpractices in the market

4. Majority of the growers reported that six aspects related to transportation of produce, such as, existence of metal or tar roads, cost and timely transportation, spoilage of produce in transit, availability of specially designed vehicles and location of market centres as constraints.
5. Lack of good facilities at the market places like, pre-cooling, cold storage, grading, packaging, handling etc.
6. Higher marketing costs.
7. Lack or inadequate credit facilities and crop insurance.
8. Shortage of labour, particularly skilled labour for grading, packaging etc., their availability in time at desired place and at reasonable wages.
9. Receipt of low price, share and margin due to non-transparent functioning of the market, lack of grading and storage facilities at the markets, lack of processing facilities in the production areas, exploitation by the middlemen and lack of proper supervision by the market committee officials during transaction and their collusion with the intermediaries.
10. Collection of higher rates of commission charges, than what is specified in the market regulations.
11. Lack of market information, regarding prices and supplies to the market, resulting in quoting lower prices than the actual prices.
12. Not following closed tender system.
13. Part or delay in payment of sale proceeds.
14. Long waiting (>12 hours) for disposal of produce particularly in regulated markets, resulting in loss of its weight and freshness and increase in spoilage.

To overcome the above problems measures like establishment of intermediate product processing units in production areas, advancing credit by institutional agencies for bearing orchards as in the case of seasonal crops, linking marketing through cooperative societies and market committees, increasing over-seas exports through "Fruit Growers Associations" and insurance of the crop against natural calamities were suggested (Subramanyam , 1998). Kahlon and Randhawa (1957) suggested that the cultivation of early and late varieties, processing of the raw produce and development of storage and freezing industries could play a significant role in the orderly marketing, better distribution of the produce and better returns to the producers and elimination of intermediaries and retailers and to form grower cooperative societies to avoid the market problems faced by the growers.

Problems faced by Commission Agents/Traders

1. The major problem is the lack of finances for meeting the market expenses as well as purchase of the produce.

2. Lack of market facilities, like facilities for grading, storage, particularly during peak periods of arrivals to reduce the spoilage losses of produce, packaging etc. It is expressed that due to lack of grading facilities and in the absence of grading by the producer the trader has to do himself grading incurring extra expenditure and the quality of the produce becomes questionable.
3. Lack of proper supervision by market officials during market transactions.
4. Spoilage and wastage disposal, hygienic conditions of market place.

Problems faced by Retailers

1. Losses during transit and high transportation costs.
2. Cheating by middle men forcing them to pay higher price on display of good sample and in actual delivery giving them inferior quality produce.
3. Lack of regular customers like wholesalers.
4. Traffic rules and regulations of police, disturbing them move from pillar to pillar.

Solutions to the Market Problems/Constraints

In nut shell, the above presentation of different problems or constraints or short comings in marketing of perishables like mango, faced by the producers, intermediaries reveal a lot of scope to improve the marketing system and to increase the profitability of mango orchard by reducing the spoilage losses, by providing suitable infrastructural facilities for transport, proper packaging, grading, handling technology as well as cold- chain facilities and reducing marketing costs by decreasing the role of intermediaries. This needs proper and planned intervention by government to safeguard the interest of fruit growers. The other alternate is creation of a marketing organization, an indirect step to control the activities of the middlemen by offering more competition. Well organised and established multi-purpose cooperatives will play a pivot role not only at the stage of wholesaling, but also at retailing stage. Contract farming, particularly in case of processing units is another solution for improving the revenue of growers.

Provision of credits by banks etc., supervision of market transactions by the market officials as per the rules of the market committee; provision of adequate storage facilities, especially during peak arrivals to reduce the spoilage losses; starting of processing units by market committee itself for utilizing the culled and damaged fruits to prevent the spoilage losses; prevention of transport losses, through a better means of transportation and provision of packing material by the market committee by charging nominal rates (Sailaja, 1994), improvement of roads, more number of procurement centres, provision of concessional transportation charges, fixing minimum wages for labour, providing lodging facilities at market place, standardization of grades and grading system, provision of garbage disposal facilities, display of prices at the market place, access to market information, fixing minimum/support prices for the produce, policy for support price, provision of cold chain/cold storage, provision of low cost packaging materials are some of the suggestions made by producers and market intermediaries for consideration by government authorities and policy makers to improve the efficiency of marketing system of perishables like mango (Shivamurthy and Girija, 2004).

Preparation of Fruits for Marketing

Preparation of mango fruits for marketing in the domestic markets of India consists of operations like desapping, sorting, grading, packaging and ripening. These operations are carried out either by the growers at the field or more often by market intermediaries at the market.

Desapping

Cutting or severing of stalk of mango fruit causes exudation of sap during harvesting or subsequent handling. Spilling of sap on the fruit surface causes stains (sap injury,) giving bad appearance to the fruits besides increasing the incidence of diseases and reducing the quality and post-harvest life (Photo 11.4). To avoid all the above, desapping (removal of sap) is practiced. In India, desapping is done as follows (Photo 2.1).

Fruits are harvested with a long stalk (8-10 cm) and transported immediately to pack house. At the pack house desapping benches are made available. The stalks are clipped to about 1 cm length with sharp scissors holding the fruit in inverted position with stem–end down and placed on desapping benches, made of bamboo, with gaps between poles. Desapping is being done under shade. Desapping requires 2-3 hours for completion. After every desapping of a lot, these bamboo racks are cleaned with water, to ensure that they are free of spilled latex. Care is taken that while cutting the stalk or during sapping that latex does not get sprayed, spilled or dripped on to the fruits being desapped or other adjacent fruits.

Sorting and Grading

Some of the fruits borne even on the same tree and/or in the same orchards show different shapes and sizes, which demands sorting for uniformity in shape and sizes, The extent of sorting and grading depends upon the type of markets. In a regulated market after the sale of the produce, the wholesaler gets the fruits sorted out based on the size and appearance (Photo 14.2). The best quality and good sized fruits are exported to other states/countries. While the lower grade or low quality fruits are sold in the domestic/local markets to retailers. In case of un-regulated markets, wholesaler/traders will purchase the fruits in bulk, which are sorted out later on the basis of quality and size, the high grade fruits are sent to other state markets in the country and lower grade fruits are left over to the grower for sale. The basic distinguishing feature between the two markets in the context of grading of fruits is that, in case of 'regulated markets', the purchasers have to take the entire stock and grade himself according to his choice and market condition and need, while in case of 'unregulated markets,' the trader will purchase from the lot, only the specified quality and grades of fruit.

However, in general practice only when the produce reaches the retailer or hawker or vender, the fruits are graded according to their size mainly. Many times the fruits are roughly graded and sold without any grading even today, particularly at the wholesale markets.

Packing

In most cases, harvested mangoes are transported to markets, loose without any packaging in bullock carts, tractors, trucks etc., (Photo 2. 9 and 2.10). Sometimes, however, they are packed in different types of containers. The most popular containers used for packing of mangoes are baskets of various makes and sizes, mostly made of bamboo. However, wooden crates and boxes are used for certain specified high value mango varieties like' Alphonso' at certain places. Of late, CFB boxes are being used for the above variety within the country and for export of 'Alphonso' and other varieties. The sizes of the packing containers differ with varieties and place and purpose. In packing of mangoes, grass or straw are used as cushioning material in the containers. Ventilated wooden boxes with paper covering in alternate layers as cushioning material are used for transportation of mangoes to distant markets. For export to overseas markets, CFB boxes with holes on sides are used. Inside the box, it is lined with cross-arranged card board strips to hold one fruit each. Fruits wrapped in tissue paper are generally packed in the boxes (See Chapter 2 also).

Ripening

Generally, hard and semi-mature or mature green fruits are harvested depending upon the market requirement. In this condition they are transported to the godowns or *mandies* by growers or pre-harvest contractors where they are artificially ripened. Fruits are kept in a room, where paddy straw is spread to a thickness of 7-10 cm on the floor and fruits are spread on this paddy straw. Paddy straw is preferred because it gives more heat and helps in uniform ripening. When the fruits have to be transported over a long distance the ripening process is carried during transit or after reaching the destination. In this case calcium carbride is used for ripening. The pellets of calcium carbide are placed between the layers of the fruits. However, calcium carbide is prohibited for artificial ripening in India. Instead, ethrel (ethylene) is recommended now for artificial ripening of mangoes. This chemical gives uniformly ripened and coloured fruits (See Chapter 4 also).

Mango Marketing Practices in India

Field sale is the most common practice in the disposal of fruits like mango, citrus, banana, guava, sapota *etc* in the country. The cultivators sell their crop to pre-harvest contractors (Srivastava, 1967). More than 90 per cent of the growers sell their crop to these pre-harvest contractors, who visit the mango gardens at the time of pre-flowering

/flowering stage. The garden is sold/leased out for varying periods (1-5 years) to the same contractor. The contractor will pay advance amount (10-50% of the agreed amount) at the time of agreement and the balance amount in installments or after the sale of the produce. Sometimes growers allow harvesting of produce, after the payment of balance amount only. After the agreement and payment contractor takes over the care of the orchard (under long-term contract) and arranges watch and ward protection of the crop; irrigation, manuring of crop and finally harvesting. The produce is brought to the assembling centres or markets by grower or pre-harvest contractor. Generally each grower or contractor will go to a particular commission agent or wholesaler, because of the previous contacts or loan taken from them etc. The produce is sold through commission agent to wholesale merchants at the market place, who in turn either send a major portion of the produce to distant markets within the country and the remaining portion of the produce is distributed to the consumers through retailers or hawkers. In some places, Fruit Grower's Association enters the scene of mango marketing. In some places, regulated markets also handle the produce for further distribution (Bhatnagar and Subramanyam, 1973).

The above traditional procedure of marketing mango has not changed much basically even today, except, the large corporates entered the market. These corporate organizations directly purchase the produce from the growers and sell the produce through their wide net work of retailing outlets in towns and cities.

However, slight variations in the above general pattern of marketing of mangoes in different states were reported as follows: Majority of the mango growers (73.44%) in Punjab sold their orchards to pre-harvest contractors on contract basis for 1 to 3 years. About 18 percent growers sold their orchards to village merchants. About 10 percent growers sold their mango fruits themselves without leasing or selling their orchards to pre-harvest contractors or traders (Singh and Sidhu, 1978).

In Karnataka, farmers sell their mango crop to pre-harvest contractors for a period of 1 to 4 years. The contractors enter into agreement with the growers soon after flowering season and pay half of the agreed amount on the day of agreement and the remaining amount just before the harvest of the fruits. The contractor arranges his own watch and labour for harvesting the fruits. The contractor makes his own transport arrangements and sells the produce to the wholesalers from terminal markets like Bangalore, Hyderabad, Pune, Nagpur, Baroda *etc* (Rame Gowda *et al* 1993).

Mango growers in Chittoor district of Andhra Pradesh have the option of either to lease out their gardens to pre-harvest contractors or to processing units or do self-marketing. In the later case, fruits are sold to country traders at the farm gate itself through commission agents/traders. For 'Totapuri' mango the processing channel is the best choice as the producer gets highest returns compared to other channels. However, large sized growers have greater access to this channel than the smaller-sized group growers, because of the larger volume of the produce (Malleswari, 1996). In North-coastal area of Andhra Pradesh, particularly in Vishakapatnam and surrounding areas, growers lease out their gardens to contractors when the fruits have half-matured. Taking into account the expenses in engaging watch and ward, loss due to fruit drop and for plant protection measures, contractors offer about half of the actual price of the crop. Then the contractor approaches commission agent available in his own village or in surrounding area or representative of the commission agent and sells the produce. Commission agents erect temporary sheds during the season and carry the business. There, the fruits are sorted out, packed suitably and fruits are dispatched to Cuttack, Kolkatta, Nagpur, Punjab etc. Agents of traders often visit the production areas during the season from the above markets. They offer advances to the local commission agents, a part of which is also advanced to contractors. Commission agent collects his advance from both the contractor and wholesaler or trader. At the destination market, through another agent the produce reaches the traders, who will distribute the same to retailers. Thus there will be 5-6 middle men between the grower and consumer (Jogi Raju and Padmanabha Rao, 1973).

Marketing system of mango in Kerela is interesting, in which contractors and agents play a significant role between the growers and consumers, almost all the growers dispose their produce on contract basis. The contractors approach the growers and enter into a contract with them for the next two consecutive years during the months of June-July, well in advance of flowering season. The two–year contract is expected to compensate the losses if any due to the alternate habit of mango trees. This type of contract business results in high profits to the contractors and low returns to the growers. However, inspite of this disadvantage, the growers prefer this type of contract and are happy since they receive payment well in advance and the crop failures will not affect them. Once the contract is agreed upon, the maintenance of the garden becomes the responsibility of the contractor (Radha and Nair, 2000).

Marketing of ‘Kesar’ mangoes in Saurashtra region of Gujarat state is also an interesting one. The fruits after harvesting are packed in wooden boxes. These boxes are brought by the grower to the market and stacked against the shop of trader or agent before 3 PM of the day. These are the open markets, where the growers can sell their produce to any one and at any rate. No amount is paid by the growers to the market for the dealing, but the agent has to pay commission (6%) and charges for loading and unloading of boxes according to the prevailing government rates. After 3 PM auction starts. The wholesaler, trader, agent and grower get together along with market authority person. The market authority person records the auction details; the auctioneer pulls out the mango fruit from any one box at random and exhibits to the crowd. After fixing the price in auction, the boxes are loaded into vehicles and sent to wholesalers place or processing plants. Wholesaler inturn will sell the produce to retailers, through whom the consumers get the fruits (Deodhar and Pandit, 2002).

In West Bengal, mango growers generally do not undertake wholesale distribution. It is a common practice to lease out the orchards to pre-harvest contractors, who take care of the crop till maturity and then dispose the produce as it suits them. Only a small number of growers have direct dealings with consumer or sell their produce through commission agents. Usually contractor are financed by commission agents or wholesalers. Thus, the contractor is obliged to sell the produce through commission agents or to the wholesalers. Commission agents also include the forwarding agents (in the marketing system) who look after packing and transport of the produce as their responsibility. They are the most important link in marketing chain and 75 per cent of total market is controlled by these agents. At some places they not only sell fruit on a commission basis, but also transact whole business on their own account. Retail distribution is done by growers, contractors, commission agents, wholesalers, stall-holders, shop-keepers and hawkers in varying degrees (Mitra, 2008).

Main Distributors and Marketing Centres of Mango in India

Mangoes grown in different parts of India are sent to different markets for marketing. For instance, mangoes produced in Andhra Pradesh, Tamil Nadu and Karnataka find markets in Nagpur, Mumbai, Delhi, Kolkatta, Kanpur, Ahmedabad etc. The important wholesale markets in India are: Kolkatta, Delhi, Mumbai, Chennai, Ahmadabad, Pune, Kanpur, Nagpur, Varanasi, Lucknow, Hyderabad, Bangalore etc. Mangoes for these big markets are usually assembled at the central places in all mango growing areas as follows:

1. Andhra Pradesh: Hyderabad, Vijayawada, Chittoor, Cuddapah, Vijayanagaram, Khammam, Rajahmundry, Mylavaram etc.
2. Assam: Gawhati
3. Bihar: Patna
4. Gujarat: Bulsar district, Ahmadabad, Junagadh, Narova, Valsad, Vopi, Unai etc
5. Himachal Pradesh: Simla, Sambha, Mandi, Dhyamsar, Bari
6. Jammu and Kashmir: Srinagar, Kethwa, Rajori
7. Karnataka: Bangalore, Hubli, Banganpet etc
8. Kerala: Trivendrum
9. Madhya Pradesh: Bhopal
10. Maharashtra: Mumbai, Ratnagiri, Vengurla, Pune
11. Odissa: Bhubaneswar
12. Punjab: Ludhiana, Chandigarh, Jallundhar
13. Rajasthan: Jaipur
14. Tamil Nadu: Dharmapuri, Vellore, Salem, Chennai
15. Uttar Pradesh: Lucknow, Varanasi, Sahranpur
16. West Bengal: Kolkatta
17. Jarkhand: Jamshedpur
18. Delhi: Azadpur, Moghulpura, New Delhi

Marketing of Mango Processed Products

Mango is a versatile fruit and allows its use right from immature (raw) stage to fully ripe stage. Several products are prepared from mangoes. Mango leather, mango juice, jam, pickles chutneys etc. Though India is the largest producer of mango in the world, very less (1%) percentage of its production is processed. Till recent years fruits including mango in India are grown mainly for fresh market, but of late with change in life style, resultant changes in food taking habits and establishment of processing plants, some quantity of production is diverted for processing. And substantial quantity of processed products, mostly mango pulp are being marketed domestically and exported to foreign countries. The chart 14.2 illustrates the distribution channel being adopted by most manufacturers of processed products in the country (Anand and Naik, 1981).

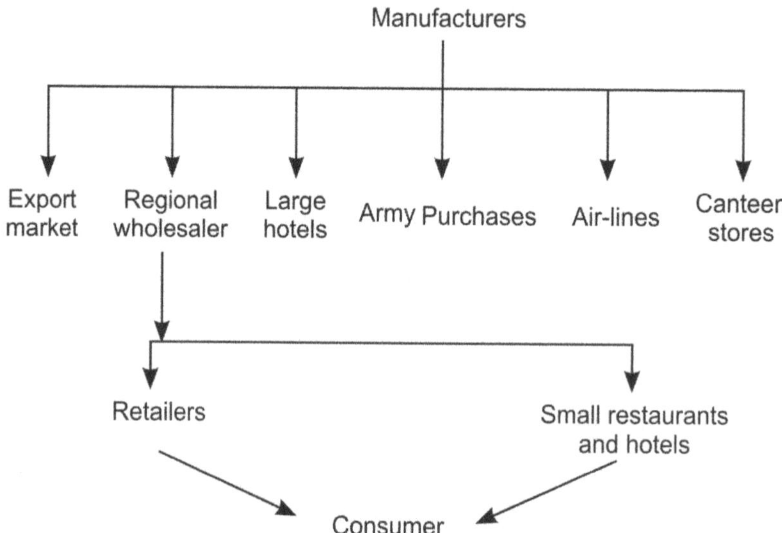

Chart 14.2: Chart showing the distribution of mango processed products

The market outlets for processed fruit and vegetables can be grouped into 3 distinct categories (Anand and Naik, 1981) as follows:

(i) Domestic (local) consumer markets

(ii) Domestic Institutional markets

(iii) Over-seas export markets

Domestic (local) Consumer Markets

The purchase of most processed products of fruits and vegetables is largely confined to urban areas and is concentrated in towns and cities. Selling of mango processed products at factory level and in various local markets is also a common feature now-a-days

(Veerker and Boruda, 1993). However, because of the following reasons mango processing does not appear to be a brighter and expanding one:

(i) Strong preference for fresh fruits
(ii) A very high price elasticity of processed products
(iii) Strong regional variation in taste and preference
(iv) General apprehensions regarding the use of synthetic ingredients and side affects

Domestic Institutional Markets

(i) Army purchase organization
(ii) Canteen stores of different departments
(iii) Hotels and Restaurants
(iv) Airlines and Railways

Overseas Export Markets

These are the main and the largest outlets of processed fruit and vegetables. In fact, our processing industry of fruits and vegetables is surviving on over-seas export markets, without which there exists very little scope for processing of fruits and vegetables in the country.

Out of the several processed products of fruits and vegetables, mango processed products occupy the largest space and rank first. However, this market is not steady as the major items of export *viz.*, mango pulp, mango juice, have erratic demand. The sales to the export market are largely to wholesaler/government and very little retailing is done for these products. However, India's share in these markets is very little of the total potential because of the following (Anand and Naik, 1981):

(i) Packaging – absence of flip top cans for juice etc
(ii) High unit costs, with the absence of incentives the product cannot be exported economically
(iii) Lack of awareness of type of products in demand

The following problems are encountered in marketing processed fruits and vegetables (Anand and Naik, 1981).

(i) High cost of production of processed products
(ii) Need to carry stocks over a longer period of time
(iii) High marketing costs within the country
(iv) High direct taxes
(v) Low average sales leading to lack of budget for marketing functions of advertising, sales promotion and attractive packaging and
(vi) Low purchasing power of Indian consumer

FUTURE

Agricultural marketing system, particularly for horticultural crops (fruits, vegetables and flowers) in India is gloomy. It is neither producer-oriented nor consumer-oriented. There are several lacunae and demerits in the present marketing system of fruits including mango. Several measures to improve marketing system in order to help both the grower and consumer in a fair way and in the process to reduce the number and role of intermediaries at least, if not entirely abolish intermediary system were, suggested and steps were taken during the last couple of decades. Some of the suggestions made for improvement of fruit and vegetables marketing are as follows (Arora, 2005):

1. Modernization of markets by establishing modern grading, pre-cooling (forced), packing houses for effective post-harvest management.
2. Establishment of modern processing units for pulp and juice manufacturing to overcome glut.
3. Development of techniques for bulk preservation of pulp.
4. Reduction of post-harvest losses during picking, storage, handling and transportation.
5. Elimination or reduction of role of intermediaries.
6. Prohibition of frequent transport strikes and labour strikes in the market yards and *bunds* by different organizations affecting the transportation and functioning of the markets.
7. Amending the APMC produce and Market Act to help the farmers to sell their produce to any one, to encourage direct marketing (self-marketing) and contract marketing.
8. Following Codex standards for export of mangoes and products and making them certified to ensure safety of food products.
9. Popularization of regular use of fresh and processed products to tap non-traditional markets.
10. Minimizing wastages, thereby augmenting the income of farmers.
11. Development and improvement of backward and forward linkages.
12. Development of alternate markets, and
13. Establishment of 'Mango Marketing Board'.

Subramanyam (1999), Veera Kumaran and Satheesh (2002) and Randev (2008) suggested the following strategies to overcome marketing problems of mango in future.

(i) **Establishment of Mango Marketing Board:** In order to improve the existing marketing system and address the problems therein, a separate organization of an institutional agency, a public corporation like 'Mango Marketing Board' is absolutely necessary. This institutional body should cover all the marketing functions, starting from harvesting, assembling of produce, grading, packaging, transport, storage, market intelligence etc., and arranging marketing finances in order to avoid leasing of orchards to pre-harvest contractors *etc*. Further more, this body may open its own retail shops in important cities and towns and may undertake export of mangoes and mango products to realize more benefits to the grower. It should be grower- and consumer-oriented and may be a wing of much needed and much awaited 'Mango Board' or a separate body itself (Randev, 2008).

(ii) Provision of Infrastructural facilities: Most of the farmers and even other intermediaries expressed that infrastructural facilities at the market place like pre-cooling equipment, storage, grading, packaging, basic amenities (dormitories/rest houses, canteen) etc., are either lacking or inadequate and need improvement. In some cases, the infrastructural facilities available at present such as credits etc., are not matching with the production potential of mango in the country. Modern marketing system as well as efficient marketing system of agricultural/horticultural produce requires well and properly developed marketing infrastructure facilities. The existence of an adequate infrastructure is an essential pre-requisite for marketing. Infrastructure development paves way to overall development of agro-economy of the country. The perishability of produce (fruits and vegetables) coupled with seasonality in production and the distance between production site and consumption centres, warrant an effective linkage of producers and consumers through infrastructural facilities such as motorable roads, cold- storage, refrigerated transport, packaging, processing facilities, credit support, market information, research and development.

The cooperative sector can play a moderate role in building up the infrastructural linkages for collection, processing and marketing of perishables. The cooperative sector has a net work of marketing facilities throughout the country. However, the fruit and vegetable cooperative marketing societies do not have adequate infrastructural support to match their production (Singh, 1996). Further, such infrastructural facilities are concentrated in certain developed states (like in Gujarat) in India and not in all states. This calls for strengthening of fruits and vegetable marketing cooperatives in the interests of the growers and consumers.

Government have developed infrastructural facilities for extending the institutional credit to markets, which, however, are not matching the production of perishables (Malleswari, 1996). National Horticulture Board (NHB) is one such Government body, which developed some marketing infrastructural facilities for horticultural crops. Although these new marketing infrastructural facilities are engaged to provide incentives to growers to produce more and enhance their incomes, they faced constraints while using these facilities. In view of the above, the following suggestions were made to alleviate the constraints and for development and revitalisation of present horticultural marketing infrastructure in the country (Kshir sagar, 2006).

(i) Establishment/rejuvenation of pre-cooling and cold storage capacity.

(ii) Adoption of mechanization to overcome the shortage of labour.

(iii) Establishment of units in the production areas/market places to ensure supply of packaging material at a reasonable price.

(iv) Adoption of appropriate policy and priority measures for easy and timely availability of credit for marketing.

(v) Evolving a suitable transport system that ensures effective movement of horticulture produce for marketing and distribution on one hand and to reduce costs and losses on the other hand.

(vi) Provision of timely and accurate market information to farmers in regular basis by NHB

(vii) Establishment of large number of processing units in the country for more value addition.

(viii) Promotion of transparent functioning of markets to alleviate selling constraints faced by the farmers.

(ix) Rejuvenation and revitalization of extension and training services to ensure enhanced awareness and orientation among farmers regarding the effective use of marketing infrastructural facilities to increase their profits.

(iii) Transportation: The growers as well as the pre-harvest contractors expressed the need of better roads, easy availability of vehicles/railway wagons at lower freight rates and removal of irregularities in the movement of the produce from the production sites to the ultimate markets. These needs are to be investigated and suggestions and remedial measures are to be made so that the growers and contractors carry their produce in their quantity and quality intact to their ultimate consumer, through improving and strengthening of existing system of transportation (Randev, 2008).

(iv) Grading and Packaging: Grading needs to be improved so as to bring uniformity in the produce and higher returns. Present level and way of packaging also need to be improved with respect to desired quality, so as to maintain quality. Similarly packaging materials need to be improved still further at international standards not only for export market, but also for domestic markets (Veera Kumaran and Satheesh, 2002).

(v) Storage: Storage facilities, especially for fresh and perishable produce (fruits, vegetables, flowers) in urban areas have been created by central/state governments as well as by cooperative marketing organization etc., in different states. However, the existing facilities are inadequate when viewed from the point of production. Depending upon the production capacity and future production potential, scientific storage structures like refrigerated structures, control atmospheric storage (CAS) and modified atmosphere storage (MAS) structures need to be promoted in future, when the fruits in 'bulk' can be stored in a cost effective manner and supplied whenever needed by the market. Storage facilities are also needed to avoid wastage of the produce and distress sales (Randev, 2008).

(vi) **Market Intelligence:** The functions of market intelligence include: collection of information on the quantum of arrivals in each market and prices at different intervals of time, number of markets available, their location and facilities at the markets and conduct of survey to learn the trends of markets and to pass on the same to the growers and other interested ones.

Most of the existing market intelligence covers field crops only. Banana and papaya among fruits; potato, sweet potato, cassava, onion and garlic among the vegetables are included for collection of regular data on prices, arrivals, procurement, distribution, stocks, marketing margins etc. Other commercially important horticulture crops like mango have not received much attention so far and not included. The need for collection of market information on the above aspects on regular basis has been emphasized time and again by different committees and commissions and based on their recommendation a market intelligence organization for agricultural produce was set up. The organization has regional offices and market intelligence units "ECOSTAT" at different locations in the country. National Horticulture Board has been entrusted with market intelligence on horticulture crops. At the state level this work is mostly assigned to Agriculture and Marketing departments and Bureau of Economics and Statistics in different states. And this organisation needs to be improved and strengthened in future for efficient functioning and serving the growers for enhancing their returns.

(vii) **Market information:** Farmers/growers are not getting adequate and accurate market information regarding the arrivals (supplies) and demand of their crop, and the prices prevailing across the markets and times, resulting in the estimation, often either less or disproportion and losses to the grower. Complete exploitation of market, to the best advantage of growers is not possible due to non availability of information on market prices/trends from various markets. Lack of information adds to the miseries of the growers, but the traders, who are in constant touch with other markets take full advantage of the situation. Therefore, information on the above items should be made available to the growers. It should be quick, trust worthy, transparent and effective.

To disseminate the market information among the farmers and buyers about price trends, supplies (arrivals) and demands in domestic as well as over-seas markets, a separate organization should be created. It will help the farmers in taking price decision, choice of time and place for the sale of their produce.

In view of the fact that for horticultural crops (like mango) timely information on regular basis about prices, market trends and behavior is very necessary. Government of India assigned this task to the NHB with the following objectives:

(i) To generate useful information on price trends in various markets of India, for at least six commercially important fruits and vegetables.

(ii) To help the growers and consumers to orient marketing plans to optimise their returns and

(iii) To over view the fruit and vegetable marketing situation which will help in analyzing the past trends of arrivals, prices and requirements.

NHB publishes monthly bulletins on prices, market trend/behavior, crop forecast for circulation among different user organization.

Besides the above, a state-owned website for information should be created to provide market information to the concerned. Electronic trading can be encouraged for direct-marketing between sellers and buyers (Subramanyam, 1988).

(viii) Reducing the risk of low price: One of the main problems faced in the marketing of perishables is "Price risk" i.e., wide price fluctuation due to seasonal glut. To overcome this following measures were suggested (Subramanyam, 1999):

(i) Linking marketing with processing by establishing intermediate processing units in production areas

(ii) To prevent post-harvest losses as well as wide fluctuation in price, the best way is to direct produce as much as possible for processing purposes.

(ix) Price Support by Government: Because of the highly seasonal nature, horticultural commodities like mango face seasonal glut, resulting in very low price/distress sale. Hence, there is need to extend price support to these crops, as in the case of certain agronomy crops. At present NAFED is extending this facility to a few crops like potato, onion and ginger, which need to be extended to other fruits and vegetables. The market intervention through "price support" should be taken up by State Government also in case of perishable crops through either cooperative societies or government run agro-industries. The Government can also think and consider 'fixing prices' as in the case of sugar cane for raw material for fruit and vegetable processing units, treating them as raw materials (Subramanyam, 1999).

(x) Provision of credit to fruit crops: Lack of financial resources and urgent need of money, contractors' readiness to advance money even before harvesting the crop, are the important reasons for continuance of age-old practice of selling/leasing out the orchards to pre-harvest contractor even today. In the absence of advance payment, growers approach private financiers for cultivation of crop. So unless some alternate arrangement is made for providing this advance to the growers, this practice of selling/leasing to pre-harvest contractors will not stop. At present Institutional Agencies are advancing credit in case of fruit crops for establishing the orchards up to economic bearing stage, but not for maintenance of existing bearing orchards, nor for meeting the marketing expenses (in case of self marketing/pre-harvest contractors). Hence, there is a need to advance short-term credit to the growers, based on the future crop, which can be repaid after harvesting and sale of the crop. The quantum of advance can be decided based on the area, variety and average yield of the crop. For elimination of pre-harvest contractors in fruit marketing, steps like advancing production and marketing credit against hypothecating the future crop, is needed (Subramanyam, 1988).

(xi) **Issues related with Government agencies in marketing:** To protect the interests of the various segments/sections of society, Government intervention in the marketing mechanism is necessary. Although Government has taken many programmes and measures to improve the marketing system, yet fruit growers complain, lack of adequate infrastructural facilities, transport, market information etc. Since majority of the growers belong to small and marginal category, and have small holdings and low trend of production and lack awareness in marketing, they are not able to dispose their produce profitably. So Government agencies should undertake market related jobs (infrastructure; storage etc.,) where heavy investments are needed and cannot be afforded by individual fruit grower (Randev, 2008).

(xii) **Development of alternate market channels:** The existing marketing system for perishables is neither producer-oriented nor consumer-oriented. It has several short comings and disadvantages to both producer and consumer, who are exploited to the highest level by market intermediaries. The markets for perishables are mostly in the hands of private persons, who monopolize in every aspect of marketing to the disadvantage of both farmer and consumer. Hence, there is every need to develop alternate market channel(s) to overcome the above, in the interest of both producer and consumer.

Efforts were made in this direction in the past, but were unsuccessful. A brief account of such efforts is presented below:

Establishment of Horticulture Produce Marketing Development Corporation (HPMDC) is one such attempt made to provide an alternate market support. This also attempted to establish processing units in the selected areas to provide remunerative price and generate employment opportunities in rural areas. However, its functioning was not satisfactory and was constrained with several problems in implementation (Parida and Naik, 1988).

Attempts were also made to encourage self-marketing, and Cooperative Growers' Association/Society and to establish regulated markets as alternate market channels for perishables like mango etc.

(a) **Encouraging self-marketing:** The existing marketing system is an indirect one involving intermediaries. The age-old practice of selling/leasing out the fruit gardens to pre-harvest contractors still prevails, which is considered to be the evil of all problems and disadvantageous to producers and consumers in marketing. Therefore, urgent steps are needed to stop this practice by encouraging self-marketing (Subramanyam, 1988). For encouraging the growers to take self-marketing, which provides highest share in the consumers' rupee, steps like:

(i) Providing marketing credit,

(ii) improving the marketing efficiency,

(iii) changing the attitude of the growers towards fruit cultivation as the primary avocation etc., are to be taken up.

Besides the above, knowledge about the marketing costs borne by the growers is also necessary to reduce them (costs) and drive the growers towards self-marketing.

There is a need to control the activities of market intermediaries for encouraging self-marketing, for which some market regulations were extracted. However, mere regulation of marketing, by a simple legislation does not solve the problems and what is needed is strict supervision by the concerned, which is lacking at present; for this purpose there is a need to provide, auctioning platform, fixing up timings for auctioning, grading of the produce, selling by weight in the market yard, besides making market committee fully responsible for any violation of the market regulations (Subramanyam, 1988). The measures taken so far like regulation of trade in fruits and vegetables by notifying them to overcome the marketing problems, however, did not bring the desired results, mainly due to the fact that the measures suggested could not tackle the real problem like selling produce to pre-harvest contractors for want of urgent money or the fear of exploitation of growers by the intermediaries in the market. Hence, there is a need to re-orient the marketing strategies in the interest of producer and consumer (Subramanyam, 1999).

(b) **Promotion and Development of Cooperatives:** Fruits (60-95%) are generally marketed through pre-harvest contractors and other market intermediaries, who carry away exorbitant profits at the cost of growers, in some cases as high as 60 percent on their investment, besides exploiting the growers through malpractices. This can be avoided to a large extent if marketing is undertaken by the grower's organizations/cooperatives and production is taken up under cooperative farming system or contract farming system (which is popular, particularly for export and processing). Among these, to ensure better returns to growers and produce (fruits) at fair rates to the consumer, formation of growers' cooperative sale societies deserves encouragement (Kaul, 1997).

Sale of the produce through cooperatives offers certain advantages, as there are no malpractices and exploitation of intermediaries. Majority of the fruit growers in the country belong to smaller and marginal category and their production levels are also low. Hence, the best suggestion is cooperative marketing. Cooperatives should be developed as real alternate channels of trade by formulating of polices which will attract the farmers to its fold. For this, steps like procuring the produce without prior indent, accepting all the grades of the produce, locating purchasing centres at the main wholesale market places, and wide publicity will be helpful.

In order to improve perishable marketing, farmers need to be encouraged to form "Producer's Marketing Societies" and there by promoting group marketing which would not only reduce the marketing cost, but also increase the producer's share and also avoid inconveniences faced by the growers in marketing the produce.

Grower's organizations play a significant role in establishing a good trend of production and marketing. In addition to ensuring remunerative prices to the growers, Cooperatives also provide know-how relating to production of the crop and post-harvest handling techniques (Patil, 1995). The study of Veera Kumaran and Satheesh (2002) clearly indicated the prominent role played by cooperative marketing societies in guarding the interests of producer and consumer. Cooperative marketing societies alone could provide an effective and remunerative alternative channel of marketing. It was found to be most effective one followed by direct selling and selling through commission agents in regulated markets (Patnaik, 1985). The marketing of mango in Bulsar district of Gujarat is mostly done by cooperative societies of which all the growers and commission agents are members. These societies advance about 50 per cent of the cost of the produce to the growers which is cleared after the sale of the produce.

Even though, government is encouraging intensively the formation of growers' cooperatives, a limited response has been observed in India over the years although the number of cooperative societies has grown over the years. Although the number of cooperative societies has grown over the time in the country, yet they need to be extended further with sound footing so that not only quality product marketing is spread, but also even inputs supply to the member-farmers, the desired quantity and quality in time are ensured. There is need to make the existing cooperative structure more competent and to regulate the market operations to curb market malpractices (Singh, 1996).

The in-depth studies on functioning of both successful and not-so-successful cooperative marketing societies of fruit and vegetables. (Subramanyam *et al* 1994., Gajanana and Subramanyam, 1993) revealed that not only over-head costs need to be minimized, but also the trading should also be improved by making majority of the cultivators to participate in the activities of the societies. This would happen only when the society could make large quantities of produce as at present most of the societies depend only on local sales through their retail outlets at the Head Quarters of the society. Hence there is a need to establish a chain of cooperative marketing side outlets on the lines of public distribution system of cereals, sugar etc., by inter linking the activities of cooperative societies both within the state as well as across the states (Randev, 2008).

With the support of National Horticulture Mission, National Horticulture Board and other relevant public and private agencies, small mango growers' cooperative societies with common facilities may be established for small and marginal growers of mango (Mitra, 2008).

(c) **Starting of Regulated Markets for Horticulture crops:** Organized marketing of fruits and vegetables is almost absent in the country. Consequently fluctuation in daily prices and large margin between growers and intermediaries is a common feature. In organized markets (like HOPCOM, SAFAL etc.,) the farmers are least exploited and distribution of produce is quite wide spread despite high degree of perishability of the fruits (Kaul, 1997). As stated earlier in this chapter the present marketing system of fruits and vegetables is neither grower-oriented nor consumer-oriented. The marketing of perishable fruits, including mango in the country as a whole is on the mercy of private middle men, due to the absence of well organized cooperatives and regulated markets. Further, the producers share in consumer's rupee is as low as 25 per cent in mango in unregulated markets, which hopefully increases in regulated markets. All these pointed to the urgent necessity of regulated markets, Growers are forced to sell their produce either to contractors in local markets or lease to the contractors due to perishability of fruits. Hence efforts should be made to establish regulated markets for perishables like mango.

The concept of regulation of markets as part of reforms has been accepted in India since long. The advent of regulated markets has helped in mitigating the market handicaps of producer-seller by releasing them from the monopolistic power of traders in the villages. In addition to this protective function, the regulatory apparatus has also helped in making the marketing system more efficient in functioning. Market regulation is a necessary reform not only for the benefits of the producer-seller, but also of various others participating in the marketing process (Ramaswamy, 1980). In a regulated market the growers can do away with the usual market problems, due to the presence of impartial third body *i.e* officials of market committee, which alone could check on the declining average and production trend of mango (Chakravarthy and Mandal, 1988).

The very objective of establishment of regulated markets is to regulate marketing practices, procedures, and operation of marketing to ensure free competition and fair play in the market transaction to protect the interests of grower, trader and consumer. The main objectives of regulated markets are to help the farmers to escape various malpractices and undue deductions which have been prevalent in the trade since long (Swarup and Singh, 1987).

The move for establishment of regulated markets for fruits and vegetables has always been resisted principally by traders, commission agents etc., engaged in the trade of these commodities, under the apprehension of undue interferences in their trading activities by the regulatory authority and also by present owners of the existing markets, who now enjoy monopoly over locations and who fear of the loss of income from the rent for the market stalls with the regulation of the trade. Both these apprehensions are ill-founded and required to be dispelled. What regulation seeks to do is only to ensure fair trading and therefore, the trading community has no genuine cause to oppose this market reform (Ramaswamy, 1980).

India has about 4000 regulated markets, for fruits and vegetables, which function under APMC's of individual states (Kaul 1997). However, regulation of markets has not solved any of the problems of fruits and vegetables which marketing anticipated at the start of the reforms. Similarly various corporation boards or similar other bodies established for the benefits of growers have not been able to come up to the expectations (Tiwari, 1987). This indicates further improvement in functioning of these reformed markets. Any attempt to introduce organizer marketing using modernised system at the wholesale levels will have to be done under the perciew of APMC Act in the stiff opposition of vested interests (Kaul, 1997).

With the expansion of area under fruits like mango, following the implementation of the National Horticulture Mission and also its rejuvenation programme of old and senile orchards, production is expected to increase many folds in the coming decades. To tackle this huge production, improved marketing systems with new reforms should be developed simultaneously keeping in mind the interests of both grower- producer and consumer.

REFERENCES

Abid Hussain, 2009. Mango's marketing ordeal, Agriculture Today, June 2009: 50-51.

Anand, D.R. and Naik, K.G. 1981. Marketing of Processed Fruits and Vegetables in India, Indian Food Packer, 35: 112-119.

Anonymous, 1962. Marketing of Mango in India, Marketing Series No.153. Ministry of Food and Agriculture .Govt. of India. pp-55.

Anonymous, 1985. Production and Marketing of Mangoes in Gujarat. Agriculture Situation in India, March 1985: 949-952.

Arora, V.P.S. 2005. Marketing and Export of Horticultural Produce of Uttaranchal: Status Potential and Strategies. Indian Journal of Agricultural Marketing (Conference Special), 19(2):194-216.

Balakrishnan, V.N. 1979. A study of markets and marketing of major farm products in Puthukottai division of Tanjavur district. Tamil Nadu Agricultural University, Coimbatore.

Bhatnagar, H.C. and Subramanyam, H. 1973. Some Aspects of Preservation, Processing and Export of Mango and its products, Indian Food Packer, 27: 33-52.

Chakravarthy, S.K. and Mandal, R.C. 1988. Regulated Market to Arrest the Decline of Mango cultivation in Malda, West Bengal, Indian Journal of Agricultural Marketing, (Abstract) 2(1): 89.

Deodhar, S.Y. and Pandit, P.S. 2002. Quality Issues in Supply chain: A case of Kesar mango at Saurashtra Region. Agriculture Marketing, October-December, 2002: 27-34.

Gajanana, T.M. and Subramanyam, K.V. 1993. Cooperative Marketing of Fruits and Vegetables – Present Status and Future requirements. Indian Cooperative Review, 31(1): 34-48.

Garg, J.S. and Singh, G.N. 1971. Economics of Production and Marketing of Mango. Indian Horticulture, 29: 47-51.

Jogiraju, G. and Padmanabha Rao, K. 1973. Mango family In: Fruits II part (in Telugu). Andhra Grandhalaya Trust, Vijayawada, Andhra Pradesh, India: 341:344.

Kahlon, A.S. and Randhawa, G.S. 1957. Price spread in the marketing of citrus fruits in India, Indian Journal of Horticulture, 14: 106-118.

Kaul, G.L. 1997. Horticulture in India – Production, Marketing and Processing, Indian Journal of Agricultural Economics, 52(3):561-573.

Khunt, K.A., Gajipara, H.M. and Kakadiya, B.H. 2001. Marketing of Kesar mango in Saurashtra region of Gujarat State. Indian Journal of Agricultural Marketing, 15(1): 47-54.

Kshirsagar, K.G. 2006. Availing Marketing Infrastructure for Horticultural crops: Emerging Prospects and Future polices. Indian Journal of Agricultural Marketing, 20(2): 69-80.

Malleswari, M.N. 1996. Mango Processing in Andhra Pradesh. Potential, Infrastructure and Constraints. Indian Journal of Agricultural Marketing, 10(2): 18-27.

Malleswari, M.N, 1998. Marketing Risks : Justification for Pre-harvest Contracting of Mangoes. Indian Journal of Agricultural Economics, 53(1-4): 399 (Abstract).

Mitra, S.K 2008. Harvesting, Packaging, Transportation and Marketing of Tropical Fruits in West Bengal. Acta Horticulturae ,768: 301-306.

Parida, G. and Naik; D. 1988. An Economic Analysis of Marketing of Mango in Bhuvaneswar City Market. Indian Journal of Agricultural Marketing, 2(1): 133 (Abstract).

Patil, H.N. 1995. Temporal Variation in price spread of Alphonso mangoes in Maharashtra. Indian Journal of Agricultural Marketing (Conference Special), 66-67 (Abstract).

Patil, H.N., Kumar, P. and Muralidharan, M.A. 1983. Marketing margin and price spread in the marketing of Alphonso mangoes in Ratnagiri District of Maharashtra, Indian Journal of Marketing, 14: 21-24.

Patil, T.Y., Hinge, B.J. and Rajmane, K.D. 1969. Economic problems of horticultural plantations in Maharashtra. Indian Journal of Agricultural Economics, 24(4): 253-254.

Patnaik, K.I.S, 1985. Economic Performance of Groundnut marketing channels – A case study of Rayalaseema region of Andhra Pradesh. Indian Journal of Agricultural Economics, LX: 26-33.

Radha, T. and Nair, S.R. 2000. Status of Mango cultivation in Kerala. Acta Horticulturae, 509: 117-121.

Rajgopal. 1987. Economics of fruit production and marketing in India : A study of selected crops. Indian Journal of Agricultural Economics, 42(3): 405.

Rajkishore, K.P. Singh T.R. and Singh, D.S. 1999. Pattern of utilization of disposal and marketing of mango in Lucknow district. Indian Journal of Agricultural Marketing, (Conference Special) 13(2): 39 (Abstract).

Raju, V.T, 2002 . *Fruit Marketing in India.* Daya Publishing Home, Delhi, India.

Raju, V.T. and Soria, C.N. 1989. Marketing of mangoes in Junagadh district of Gujarat State, Indian Journal of Agricultural Marketing, 3(1): 34 (Abstract).

Raju, V.T. and Voneppen, M. 1987 Market channels of SAT crops and Cooperative market efficiency in India. A Case study proposal. Economics Unit, ICRISAT, Hyderabad pp, 1-20.

RamaSwamy, P. 1980. Market Regulation as applied to Terminal markets for Fruit and Vegetables – Some issues. Agricultural Marketing, 22(4): 19-23.

Rame Gowda, P.V., Reddy, P.N.V. and Chengappa, P.G. 1993. Production and Marketing Practices of mango in Karnataka. Mysore Journal of Agricultural Science, 27: 189-194.

Rana, R.S. 1984. Fruit Production and Marketing in India. Agricultural Marketing, 27(3): 21-31.

Randev, A.K. 2008. Researchable Issues for improving marketing efficiency in case of Fruits and Vegetables in Himachal Pradesh. Agricultural Situation in India, LXV (4): 211-216.

Sailaja, N. 1994. Marketing of Fruits in Hyderabad city. A case study of Gaddiannaram Fruit Market M.Sc (Ag.) Thesis, A.P. Agricultural University, Hyderabad, Andhra Pradesh, India.

Shivamurthy, M. and Girija, P.R 2004. Constraints in Marketing of Fruits and Vegetables as experienced by farmers and Market intermediaries. Land Bank Journal, 43(2): 35-39.

Singh, B. and Sidhu, D.S. 1978. Market Integration among the Mango Trading Firms. Agricultural Marketing,21(1): 21-24.

Singh, B. and Singh, K. 1996. Future Requirements of Cold Storage Facility in Punjab. Indian Journal of Agricultural Marketing, 10(3): 42-45.

Singh, B.B., Yadav, R.N., Yadav, N.K. and Israfil, Md. 1999. A price spread analysis of Mango Marketing in samastipur district of Bihar. Indian Journal of Agricultural Marketing (Conference Special), 13(2): 57 (Abstract).

Singh, S. 1996. Marketing infrastructure and Agro-processing Development. A case study of Gujarat. Indian Journal of Agricultural Marketing, 10(2): 1-9.

Singhal, A.K. and Agarwal, S.K. 1997. Economics of Production and Marketing of Mango in Rampur district of Uttar Pradesh. Indian Journal of Agricultural Economics, 52(3): 662-663 (Abstract).

Srivastava, H.C. 1967. Grading, Storage and Marketing. In:Mango - A Hand book. Indian Council of Agricultural Research, New Delhi: 124-129.

Subramanyam, K.V. 1986. Marketing of Fruits. Self Marketing Vs Contract sale-Case study of Mango marketing in Madhurai district (TamilNadu), Agricultural Marketing, 29(3): 21-22.

Subramanyam, K.V. 1988. Marketing of Horticultural Crops in Karnataka. Present status and Future requirements. Indian Journal of Agricultural Marketing, 2(1): 15-22.

Subramanyam, K.V. 1998. Mango Marketing: Present status and Future requirements. Paper presented at National Symposium on Mango Production and Export, held at Lucknow, June 25-27 1998: 89 p (Abstract).

Subramanyam, K.V. 1999. Risk reducing and efficient marketing strategies for Perishables, Fruits and Vegetables. Indian Journal of Agricultural Marketing (Conference Special), 13(2): 28-29 (Abstract).

Subramanyam , K.V., Gajanana, T.M and Sudha, M. 1994. Cooperative Marketing of Fruit and Vegetables. Polices and Performance-A case study of HOPCOMS, Bangalore. Indian Journal of Agricultural Marketing , 8(2): 186-193.

Swarup, R. and Singh, R. 1987. Impact of Regulation on Fruit Marketing in Northern India. Indian Journal of Agricultural Marketing, 1(1): 31-36.

Swarup *et al* 1985. Price-spread and Marketing Margin for Himachal apples. Indian Journal of Agricultural Economics, 45(3): 376-381.

Temghare, P.B. and Boruda, S.G. 1976. Marketing of Mango in Ratnagiri District of Maharashtra State. Journal of Maharashtra Agricultural Universities, 1 (Additional): 206-210.

Tewari, S.C. 1987. Rapporteur's report on Horticultural Development. Indian Journal of Agricultural Economics, 42: 494-501.

Thorat, V.A., Patil H.N. and Boruda, S.G. 1990. Efficiency of Self-marketing Vs Contracting for Alphonso mangoes in Maharashtra. Agricultural Marketing, 33(2): 46-48.

Veera Kumaran, G. and Satheesh, C.K. 2002. Marketing of Fruit and Vegetables through Cooperatives – An analysis of Consumer Behavior. Indian Journal of Agricultural Marketing, 16(2): 53-59.

Veerker, P.D. and Boruda, S.G. 1993. Comparative Economics of Scale of Processing Alphonso Mango into Pulp in Ratnagiri District (MS). Indian Journal of Agricultural Marketing, 4(1): 60-65.

Wadkar, S.S., Talathi, J.M. and Thakur, G.G. 1994. Price spread and Market channels of Alphonso mango in Ratnagiri and Sindurgh district of Konkan region in Maharashtra State. Indian Journal of Agricultural Marketing, 8(2): 250-257.

www.ingramcontent.com/pod-product-compliance
Ingram Content Group UK Ltd.
Pitfield, Milton Keynes, MK11 3LW, UK
UKHW060019200726
13845UKWH00046B/86

9 789385 433832